THE COMPLETE
HANDBOOK
OF MAGNETIC
RECORDING

THE COMPLETE HANDBOOK OF MAGNETIC RECORDING

Fourth Edition

Finn Jorgensen

TAB Books
an imprint of McGraw-Hill

New York San Francisco Washington, D.C. Auckland Bogotá
Caracas Lisbon London Madrid Mexico City Milan
Montreal New Delhi San Juan Singapore
Sydney Tokyo Toronto

McGraw-Hill

A Division of The McGraw·Hill Companies

Library of Congress Cataloging-in-Publication Data

Jorgensen, Finn.
   The complete handbook of magnetic recording / by Finn Jorgensen. -
- 4th ed.
      p.   cm.
   Includes index.
   ISBN 0-07-033045-X
   1. Magnetic recorders and recording.   I. Title.
TK7881.6.J66   1995
621.382'34—dc20   94-36652
                                                               CIP

hc 1 2 3 4 5 6 7 8 9   DOC/DOC   9 0 0 9 8 7 6 5

The acquisitions editor for this book was Roland S. Phelps. The executive editor was Robert E. Ostrander, the book editor was Aaron G. Bittner, and the director of production was Katherine G. Brown. It was produced in Blue Ridge Summit, Pennsylvania. Printed and bound by R.R. Donnelley & Sons Company.

Product or brand names used in this book may be trade names or trademarks. Where we believe that there may be proprietary claims to such trade names or trademarks, the name has been used with an initial capital or it has been capitalized in the style used by the name claimant. Regardless of the capitalization used, all such names have been used in an editorial manner without any intent to convey endorsement of or other affiliation with the name claimant. Neither the author nor the publisher intends to express any judgment as to the validity or legal status of any such proprietary claims.

ISBN 0-07-033045-X

McGraw-Hill books are available at special quantity discounts to use a premiums and sales promotions, or for use in corporate training programs. For more information, please write to the Director of Special Sales, McGraw-Hill, 11 West 19th Street, New York, NY 10011. Or contact your local bookstore.

HB1B
033045X

In order to receive additional information on these or any other McGraw-Hill titles, in the United States please call 1-800-822-8158.
In other countries, contact your local McGraw-Hill representative.

CONTENTS

Chapter 5. Fundamental Recording Theory 111

Chapter 6. Playback (Read) Theory 145

Chapter 7. Materials for Magnetic Heads 163

Chapter 8. Fields from Magnetic Heads 197

The Erase Process and Overwrite
Stray Flux Computations
Flux Linkage
Flux Leakage
Shielding

Chapter 9. Manufacture of Magnetic Head Assemblies   225

Heads with Laminated Metal Cores and Solid Ferrite Cores
Heads with Ferrite Cores
Thin-Film Heads
Specifications

Chapter 10. Design and Performance of Magnetic Heads   263

Design Procedure
Selecting Upper Frequency, Gap Length, and Depth
Electromagnetic Design
The Head Reluctance Model—A Tutorial Head Analysis
Metal-in-Gap (MIG) Heads
Saturation in Thin-Film Heads
Back Gap Benefits
Effects of Core Geometry
Side-Reading Effects (Crosstalk)
Effects of Coating Permeability
Crosstalk in Multitrack Heads and Transformer Coupling between Cores
Side-Reading Effects
Erase Heads
Core Losses
Radio Frequency Interference from Write Heads
Susceptibility to External Fields
Variations in Permeability (Production Tolerances)
Changes in Head Parameters with Wear
Toward Higher Data Rates and Storage Densities

Chapter 11. Tape and Disk Materials   317

Particulate Media Components
Magnetization of a Powder Magnet
Magnetic Particles
Thin Magnetic Films
Flexible Media Substrates
Rigid Disk Substrates

Chapter 12. Magnetic Properties of Tapes and Disks   341

Measurements of the MH-Loop
Sample Size
Correction for Demagnetization in Sample
Coercivity, Remanent Flux, and SFD
Factors Influencing the Magnetic Properties
Measurements on Single Particles
Behavior of an Assembly of Particles
ac-Bias Sensitivity
Effects of Time

Effects of Temperature
Mixture of Two or More Particle Types
Spatial Variations in Magnetic Properties
Noise from Particulate Media
Noise from Thin-Film Media

Fabrication of Particulate Media, Tapes, and Disks
Fabrication of Thin-Film Media, Tapes, and Disks
Quality Control Methods

Mechanical Properties of Substrates
Mechanical Properties of Coatings
Electromagnetic Properties
Eraseability and Overwrite
Dropouts
Storage Stability
Audio and Instrumentation Response Using ac-Bias
Print-Through
Contact Duplication of Tapes

Layout of a Tape Transport
Direct Drive Capstan Configurations
Capstan Motors
Servo-Controlled Speed
Capstans and Pinch Rollers
The QIC Data Cartridge
Tape Guidance and Skew
Reeling and Winding of Tapes
Head Mounting
Design for Low Flutter
Helical Scan Tape Transports
Scanners
Tracking

Flexible Disk Drives and Removable Rigid Disks
Head Positioning in Flexible Drives
Rigid Disk Drives
General Spindle Motor Considerations
Head Positioning in Rigid Disk Drives
Head Suspensions
Track Accessing in Rigid Disk Drives
Track-Following in Rigid Disk Drives
Position Error Signal (PES)
Servo Writer
Servo Design
Testing Disk Drives

Chapter 17. Vibration in Tapes, Disks, and Suspensions

Chapter 18. Tribology and Head/Media Interface

Chapter 19. Advanced Topics in Recording

Chapter 20. Recording with ac Bias

The MO-Disk and Media Magnetics
The Minidisc (MD)

Chapter 31. Care and Maintenance   777

Causes of Failure in Media
Handling of Media
General Storage of Media
Archival Storage of Media
Maintenance of Recording Equipment
Troubleshooting Tape Recording Equipment

ACKNOWLEDGMENTS

This fourth edition brings the handbook up to date in magnetic tape and disk technologies. The resource material has swelled to more than 10,000 technical books and papers, and the writing task was overwhelming.

I am fortunate in having many friends that have encouraged me to carry on, and in many cases, discuss with me (and teach me) new topics. I specifically wish to thank the following: Sid Damron (Datatape), Fred Jeffers (Datatape), Jack Judy (Univ. of Minn.), Klaas Klaassen (IBM), Tim Perkins (AMC), Ed Packard (HMT), Ed Williams (Read-Rite), and Roger Wood (IBM). For help with earlier editions, I thank Eric D. Daniel (Memorex) and Steve S. Jauregui (USN), Joseph Judge (Aerospace Corp.), and John McKnight (Mag. Ref. Lab.). And I wish to remember Len Johnson and my son Morten for their help in their time.

A thank-you also goes to the many companies in the storage industry who supported this work by sponsoring in-house courses by the author. Thanks also go to students who came to classes wherever in the world I presented a course.

Thanks go to my daughter Tina for great help with the manuscript's syntax, to my wife Bodil for her unfailing support, and to Martin Vos of the 3M Company for many suggestions and his painstaking proofreading of the book's treatment of the theory of magnetism.

McGraw-Hill's editorial staff, headed by Roland Phelps, did a great job in transmitting all the words, formulae and drawings to the final book; and thanks to Aaron Bittner for editing the entire book.

PREFACE TO THE FOURTH EDITION

This new edition reflects recent years' dramatic progress in magnetic storage technologies, for tape as well as disk applications. Sections on recorders with rotating head assemblies (DAT, 8 mm) or using 3M Company's QIC cartridge have been expanded, and Chapter 16 on disk drives is essentially new. These updates carry on in corresponding sections on heads, media and related topics.

The book is, like its predecessors, written to be easily understood by the technically inclined person working in the field, and for the equipment user who wants to acquire an in-depth knowledge of magnetic recording and storage.

The book is a textbook as well as a reference. It is divided into logical sections:

2 chapters with introduction	(Ch. 1–2)
4 chapters on magnetism, writing, and reading	(Ch. 3–6)
4 chapters on heads	(Ch. 7–10)
4 chapters on media	(Ch. 11–14)
4 chapters on tape/disk drives	(Ch. 15–18)
3 chapters on advanced write/read	(Ch. 19–21)
4 chapters on signals and codes	(Ch. 22–25)
4 chapters on applications	(Ch. 26–29)

An additional chapter on magneto optical storage (Chapter 30) has been included to complete the treatment of magnetic storage. Chapters 26 through 30 are necessarily kept brief, and the reader is referred to volume two of McGraw-Hill's *Handbook of Magnetic Recording*, edited by E.D. Daniel and C.D. Mee, published 1995. The chapters in their two-volume set are written by specialists in their fields, and are highly recommended for further studies. A good overview of the book's content is found in the second half of Chapter 1, Prospectus.

The technical units in the book are SI (System International, meter-kilogram-second); cgs units are often found listed in parenthesis. The symbols used match the ones in Daniel and Mee's handbook whenever possible.

Numerous technical papers are referenced and listed in short reference and bibliography sections. They are culled as the best from a database of nearly 10,000 papers.

CHAPTER 1
HISTORY AND OVERVIEW

The possibility of using magnetism in a recording device was first discussed in a paper by the American Oberlin Smith in 1888. He envisioned a cotton thread impregnated with steel dust, and being magnetized by passing through a current-carrying coil, with the current controlled by a microphone. But he discarded the idea of using a steel wire, because "the magnetic influence would probably distribute along the wire in a most totally depraved way." There were no experiments.

A working recording machine was invented by a Danish engineer, Valdemar Poulsen (1869–1942). His invention of the Telegraphone in August 1898 has indeed today spread into every home in the world, but the progress has been by leaps and bounds and with numerous setbacks. During certain periods technology was lacking (wire, tape, ac-bias) and at other times manipulating business interests or national interest seemed to hold back universal progress. Poulsen's invention was the outcome of a simple experiment where he stroked out a line along an iron plate and found that iron filings would gather along the line. The next experiment involved a strung-out piano wire and a primitive electromagnet connected to a microphone. Poulsen moved the electromagnet along the wire as he spoke into the microphone, and by later connecting the wires from the magnet to the telephone receiver he heard his voice reproduced.

THE TELEGRAPHONE

One version of his recording device is shown in Fig. 1.2. The magnetic wire is wound from one end to the other of a brass drum, which in turn is rotated through a belt coupling to a motor. The head assembly has two small pole pieces that protrude from its surface. These parts grab each side of the wire, and guide it along the drum. When it reaches the end it is lifted from the drum surface, and then quickly moved back to the start position. Poulsen was received with great enthusiasm at the World's Fair in Paris in 1900, and was awarded a Grand Prix for the Telegraphone.

The electromagnetic head assembly could be either single-pole or double-pole, and Poulsen later advised how the two poles should be offset to produce a longitudinal magnetization rather than a perpendicular one. When he conceived of a disk recorder he selected a diameter of 5.25 inches, and used a needle to write and read—on both sides. So he predicted the correct dimension, opted for perpendicular recording, and made it double-sided. Finally, Poulsen is also credited with applying a dc-cur-

FIGURE 1.1 Danish engineer and inventor Valdemar Poulsen (1869–1942).

FIGURE 1.2 Poulsen's Telegraphone from 1900. The large drum has a single layer of magnetic wire wound from end to end. The magnetic head follows the wire from one end of drum to the other when the drum rotates.

rent bias for improvement of the recording (see Chapter 5, dc bias). This came about by first erasing the steel wire with a permanent magnet, and then adjusting a superimposed field upon the record signal field to provide optimum quality.

In 1902 Poulsen made another invention, described in his paper "System for Producing Continuous Electric Oscillations". This was a high-powered, continuous oscillator operating at a frequency of many kHz (kiloHertz, or thousands of cycles per second), up to one MHz (MegaHertz, or millions of cycles per second). The oscillations were modulated by shunting the series-tuned LC-circuit with an electric arc, burning in an atmosphere of hydrogen. This system replaced many of Marconi's On-Off transmitters, which spread signals over the entire broadcast spectrum. This work soon absorbed all of Poulsen's time, and his company sold transmitters in many parts of the world. Lee de Forest's vacuum tube transmitters replaced these transmitters after 1920.

Poulsen sold his patent rights to the Telegraphone in 1905. It is, in retrospect, a shame that just a little fraction of the signal from his arc-generator did not stray over to the Telegraphone: Poulsen could then have added ac bias to his patents.

The first decades of the twentieth century were tumultuous for the Telegraphone. Several manufacturing companies were formed, and many varieties of the recorder promoted. There were dictating machines, message repeaters, telephone answering machines, small disks were tried for tape letters, and so on. But there was no technology to improve the quality or playing time. The invention and the associated ideas were clearly too early. So the companies struggled and either lost money or changed hands.

Accusations were voiced against the American Telegraphone Company. People wondered if the president of the company was paid to suppress production that the phonograph and the telephone companies feared. Or worse, was he or others in a part with the Germans, who successfully used telegraphones onboard their submarines in World War 1? The Germans had made message recordings at normal speed and then transmitted them backwards, at higher speed.

The MAGNETOPHONE

Whatever the cause, leadership in magnetic recording went to the Germans. In the twenties they manufactured and sold steel-tape recorders, designed by Stille. A similar machine was made in England and called the Blattnerphone.

In 1928 Pfleumer filed a patent for coating iron particles on a strip of paper as a recording medium, and a machine using such a tape, the German Magnetophone, was exhibited in Berlin in 1935. And then no more was heard of it until 1943.

At that time the U.S. Army Signal Corps, stationed in England, was puzzled by sometimes hearing radio broadcasts of operas and music during the middle of the night (noting no record scratches and other such deficiencies) and then hearing Hitler speaking from different parts of Germany almost within the hour.

The answer was found in 1945 in Frankfurt by one of their officers, John Mullin. In a radio station he found several A.E.G. Magnetophones, all equipped with ¼" plastic tape, some of which later would serve for the Bing Crosby radio shows.

The sound quality was far better than any other machine in those days, and a close examination revealed that the Germans used high-frequency ac bias. Chapter 5 will explain how it works, but today it is universally used in recorders. History will have it that W.L. Carlson and G.W. Carpenter of the U.S. Navy discovered and patented ac bias in 1927, but it was obviously not used to any extent. The Magnetophone ac bias patent was in 1946 granted in Germany (retroactive to 1940) to H.J. Von Braunmuhl and Dr. W. Weber, and in the United States to Marvin Camras of Armour Research.

AFTER WORLD WAR II

The next three decades bring us through a rapid growth period with innovations, products, people, and companies too numerous to list in a few pages. The Minnesota Mining Manufacturing Co. (3M

FIGURE 1.3   A German magnetophone brought to the USA by John Mullin.

Co.) finished their first oxide tapes in 1947, under Dr. W. Wetzel. Ampex, founded by Alexander M. Poniatoff, started delivering audio recorders in 1948. Mincom, a division of 3M Co., pushed the state-of-the-art instrumentation recorders, and demonstrated television recording in 1951. RCA followed their lead in 1953.

Other early pioneers were Dr. Marvin Camras of Illinois Institute of Research (then Armour Research), Otto Kornei of Brush/Clevite, and S.J. Begun, who wrote the first book on magnetic recording. The result was an industry that flourished with a large selection of sound tape and sound film recorders.

The breakthrough in television came from Ampex, where in 1955 Charles Ginsburg and Ray Dolby (father of today's Dolby System) unveiled the rotating-head video recorder (the readers of this book are certainly aware of the perfection this technique has achieved today, as the basis of all commercial video recorders). Instrumentation recording jumped ahead in 1961 when Wayne Johnson at Mincom conceived of a tape drive virtually free of timing errors (low-TDE).

Industry standards were always needed to provide interchangeability of recorded tapes. Audio tapes experienced rapid developments from full-track recorded tapes to 2, 4, and 8 tracks on ¼" tape, and now 4 tracks on 0.150" wide tape in cassettes. Such transitions could not have happened in an orderly fashion if it were not for the cooperative work of manufacturers and standard committees. There are today numerous standards groups, and of these the following played a key role in the past 40 years of development:

- ANSI (American National Standards Institute)
- CCIR (International Radio Consultative Committee)
- IRIG (Interrange Instrumentation Group)
- NAB (National Association of Broadcasters)
- SMPTE (Society of Motion Picture and Television Engineers)

The most difficult standards to reach agreement on were those for the equalization of the recorder response. Thus, there are still several standards today for essentially the same thing (see Chapters 26–29).

INTO THE FUTURE

New generations of recorders are always on their way. These inevitably follow improved technologies and new methods, such as using digital encoding of data, music, or video. Home video recorders are now available that reproduce music without noise and flutter. An estimated 400 million VCRs are in use for home entertainment.

They employ analog-to-digital converters, and the signal is recorded in its digital form. And today digital audio recorders using the Philips DCC cassettes are available for personal use. Magneto-optical recording is applied in both data and music disk storage.

Magnetic recording is enjoying an accelerated growth in the field of computer storage. Several graphs in Chapter 2 illustrate this. A 2-inch-diameter diskette drive can replace photographic film in cameras. Home video on 8 mm tapes in cassettes has replaced 8mm film equipment. What next?

An entire book could be written about the evolution of the magnetic recording industry, its people and the chain of developments. It would be most interesting reading. This overview is well-concluded with a saying an old friend of mine had about us in the industry:

"It is not really a business—it's a way of life."

PROSPECTUS

This chapter, *History and Overview*, provides a brief history of magnetic recording, from Poulsen's Telegraphone to today's disk drives, video recorders, and 4 and 8 mm tape systems. The following prospectus provides the reader with a quick summary of the books material.

Fundamentals and Evolution (in Chapter 2) provides a survey of data density growth on various devices with a forecast of the future, and discussion of the technologies that make this evolution possible.

Chapters 3 through 6 are devoted to the basics of magnetism and the write/read (record/playback) processes. We start with Chapter 3 on *Magnetism*. General to advanced magnetic topics are explained using electric currents as the fundamental source of fields and magnetism in solids. Magnetic properties of head and media (tape, disk) are explained in a way that allows for elegant modelling of the fundamental record and playback functions.

Necessary details on *Magnetization in Recorded Bits and Heads* is the topic of Chapter 4. The remanent magnetization in recorded bits is always less than the maximum saturation value. This loss is due to demagnetization associated with the bit geometry. Perpendicular recording is introduced as a means to reduce demagnetization at very short bit lengths, i.e., when the bit length becomes smaller than the recorded thickness. Introduction of magnetic resistance makes analysis of magnetic heads easy, resulting in expressions for a head's efficiency and impedance, optimizing design.

Fundamental *Recording Theory* in Chapter 5 explains the nonlinear magnetic recording process. Simple and elegant modelling shows resulting media magnetization after recording. This allows the engineer/technician to optimize record conditions for several different applications. Examples are included.

Chapter 6, *Playback Theory*, examines and explains the scanning losses that decrease the available read voltage from a playback head. These are quantified into useful terms for the equipment designer/user. Examples are included.

The next four chapters cover magnetic heads, starting with *Materials for Magnetic Heads*, Chapter 7. Drawing on the experience from Chapter 3, the reader learns how a nickel-iron alloy is suitable for heads. Its limits at high frequencies are analyzed (limiting factors being eddy currents and changes in magnetization) and core fabrication from laminated NiFe or the use of ferrites is discussed. Thin film heads use thin deposited NiFe layers, and next-generation read heads will use a magnetoresistive element. These MR elements are analyzed with examples.

Chapter 8 covers *Magnetic Fields from Heads*. The magnetic field pattern in front of the gap in a head controls the record as well as the playback processes. A fundamental introduction to fields allows the reader to analyze field patterns. A clear distinction is made between long pole (conventional) and short pole (thin film) heads. Side writing and reading are analyzed, because this will be the ultimate limitation of reduced track widths. Overwrite of old data while new are being recorded is not perfect, and the reader is shown why.

Chapter 9 describes the *Fabrication of Heads*. Three fabrication processes are described: 1. The making of a laminated multitrack head, 2. A solid ferrite core head assembly (slider) for a disk drive, and 3. Thin film heads for disk or tape systems, fabricated by thin film chemical deposition methods (wet) or sputtering. Sliders are fabricated from ceramics by slicing, grinding, or ion milling.

Design and Performance of Heads are discussed in Chapter 10. Examples of several head types are analyzed, measured and commented. This will aid the engineer or technician in optimizing write/read circuitry. Some unexpected behaviors of heads are explained (popcorn noise, remanent magnetization noise, and other noise).

Storage media, magnetic tapes and disks, are the topics of Chapters 11 through 14. *Materials for Media Magnetics* are covered in Chapter 11. Knowledge from Chapters 3 and 4 (Magnetism) and Chapter 5 (Recording) is used to specify the magnetic properties that will make an optimum media for a given application. A range of traditional and new materials is listed and described.

Chapter 12, *Magnetic Properties of a Coating*, shows methods of magnetic measurements, illustrated for samples of recording materials (tapes, disks.) Variations in magnetic properties occur with time and with changes in temperature, and will affect the tape or disk performance. These subtle changes and/or variations are playing a major role in the ultimate advancements in packing densities over the next couple of decades.

Chapter 13 covers *Manufacturing of Media*. There are two manufacturing methods for flexible media: coating of particulate media or evaporation of metal films onto flexible substrate is commonly used for tape, and thin film disks are made by sputtering a magnetic material onto small disk substrates. This latter process is still evolving as demands for higher output (higher coercivity) and less noise persist. Equipment and some process details are covered.

The achieved *Characteristics of Magnetic Tapes and Disks* are discussed in Chapter 14. Mechanical properties are examined, including surface smoothness, which affects signal response and noise. Electromagnetic characteristics of data storage media are examined. Dropouts, print-through, and stress demagnetization for tapes are discussed, along with optimum storage conditions. Rigid disk lubrication and overcoats are examined.

Tape transports and disk drives are important mechanical devices because mechanical vibrations and tolerances tend to affect the write/read signal quality. Chapter 15 covers *Tape Transports*. Two fundamentally different transport mechanisms are described and analyzed. First the longitudinal transports used in reel-to-reel recorders, in the Philips cassette format (better known as the Walkman) and the data cartridge known as QIC (Quarter Inch Cartridge). Next, the helical scan format that is used in VCR equipment (Studio, D-1, D-2, D-3, VHS, 8 mm, and R-DAT) and their applications for data storage are considered. The mechanical functions of these transports are analyzed, and the reader is introduced to the powerful technique of electromechanical analogies.

Disk Drives are the topics of Chapter 16. Three fundamental disk drives are described. First the traditional 5¼" and 3½" flexible disk drives are explored. Then I discuss the large 8" to 14" rigid disk drives with linear head positioning motors, and finally the smaller 3½" to 2" rigid disk drives with rotary head positioning systems. Servo designs are outlined and drive testing examined.

A very important area in both hard and floppy drives is the head/media interface, where vibrations and direct contact create problems for reliability and signal fidelity. We'll examine both in two chapters.

Chapter 17, *Vibrations in Tapes and Disks*, examines the minute high-frequency vibrations in tapes and disks that are sources of flutter plus servo and data errors. Vibrations are also catalysts for head and/or media wear. The reader is introduced to methods of analysis and vibration reduction.

The contact problems, head wear in tape drives, and CSS (Contact-Start-Stop) in disk drives are covered in Chapter 18, *Head/Media Interface (Tribology)*. Two media/head contact functions are described and analyzed. First, the flexible tape/head interface (as found in longitudinal and in helical scan recorders) is discussed. Head and tape wear are described, analyzed, and measured. Next, the rigid-disk-to-head-slider air-bearing design is described and analyzed with different flying attitudes. The air bearing will soon become an incontact bearing, although there are difficult wear and damage control issues to be solved. Possible solutions are discussed.

The write/read processes are reviewed in the next three chapters. Chapter 19, *Advanced Topics in Recording*, introduces the reader to the inclusion of demagnetizing fields during the recording (write) process. Advanced modelling with self-consistent iterations are introduced. Also described are recent findings in recording research done at major companies and universities.

ac Bias Recording is analyzed in Chapter 20 using the Preisach model, which demonstrates the linearity of this recording process. ac bias is used in all instrumentation and studio recorders, and at times found in the recording of digital signals.

Chapter 21, *Playback (Read) Models and Observations* expands on the computational method (introduced in Chapter 6) for finding read waveforms. It permits prediction of read waveforms based on knowledge of anisotropies in the media. Perpendicular write/read methods are discussed. Dibit and Tribit pulse interference are analyzed.

Four chapters are devoted to signal recovery and codes, starting with Chapter 22, *Analog Signal Response and Equalization*. A summary of the theory of the write and read processes illustrates the signal path "through" a magnetic recorder channel. The chapter recaps findings from earlier chapters and provides the reader a good survey of write/read parameters. Methods of equalization (signal restoration) are described for analog and for digital signals. Amplitude variations and data accuracy are analyzed.

Next, *Digital Signal Recovery and Detection* are covered, in Chapter 23. Digital (PCM) recording principles and A-D conversion are introduced. The reader is introduced to amplitude detection and peak detection. Pulse interference based on the write/read theory, plus fundamental linear system concepts allow for signal analysis. Pulse interferences are shown to contain data information. This evolves into partial response signalling. Bit error rates and the effects of timing errors are described. This leads to a discussion of window margin, its measurement, and time interval analysis. Several sources of errors are discussed. System measurement techniques and instruments are presented. Methods of improvement of SNR are described.

Chapter 24 covers the *Modulation Codes* that are used to change the NRZI data stream. The ordinary magnetic recording channel does not reproduce dc levels, and signal processing is needed to make the digital signal dc-free. This is done by rearranging the sequence of transitions in the digital signal and simultaneously assuring regularly occurring clock signals. This has led to the evolution of new codes from FM to MFM to RLL(1,3), and recently RLL(1,7) and RLL(2,7) codes. Group codes (4,5), (8,9) are also applicable.

The last "textbook" chapter, Chapter 25 (*Error Detection and Correction Codes*) introduces the reader to error detection and correction. This was early accomplished by adding parity check bits to blocks of data. Today both block codes and convolution codes are used. Excellent error protection is further achieved by implementing interleaving of data. The fundamental theory of ECC gives background for appreciating new ECC code designs.

Finally, four areas of applications are outlined, starting with Chapter 26, *Digital Data Storage*. Data organization for SAS (Sequential Access Storage on tape) and DAS (Direct Access Storage on disk) are explained, along with tape and disk drives. Very high data rates use HDDR recorders, either multiplexed or many-channel longitudinal and helical recorders.

Chapter 27, *Analog Data Storage*, explores how instrumentation recorders for storage of analog data suffer from amplitude and tape speed stability. Their impact on data is analyzed and quantized. FM recordings are a means of improvement, and will simultaneously extend the bandwidth to dc. Servo-controlled speed corrects for time-based errors.

Audio Recording is the topic of Chapter 28. The energy spectrum of audio signals require special emphasis of low and high frequencies. This is accomplished by equalization, for which standards are explained and illustrated. The proper recording level must be observed and general hints for recording are given here. The new digital DAT and DCC recorders are introduced.

Video Recording is expanding its applications like never before. Chapter 29 explains how video signal spectra require special pre- and de-emphasis for best record/play signal-to-noise ratio. The luminance and chroma video signal components are examined, and the best recording methods are outlined. Standards for VHS, S-VHS, Beta, and 8 mm home formats, as well as broadcast D-1, D-2, and D-3 formats, are illustrated.

Then comes a brief chapter on *Magneto-optical Storage*, Chapter 30. The magneto-optical principle based on Kerr readout is introduced, and densities analyzed. Future improvements (such as the use of a blue laser) are described. Comparisons with magnetic and optical disk recordings are made.

Finally we come to Chapter 31: *Care and Maintenance*. General preventive maintenance for recording equipment, along with some hints of proper equipment alignment, are given. Storage of tapes, disks and MO (magneto-optical disks) is discussed in terms of environment and archival longevity.

To help the reader access this vast array of information the book is provided with a table of contents, front, and in the back with two lists:

List of Symbols
Subject and Name Index

BIBLIOGRAPHY

Bate, G. Sept. 1978. Bits and Genes: A Comparison of the Natural Storage of Information in DNA and Digital Magnetic Recording. *IEEE Trans. Magn.* MAG-14 (5): 964–965.

Begun, S.J. July 1936. The New Lorenz Steel Tone Tape Machine. *Electronic Communication* Vol. 15, pp. 62–69.

Camras, M. April 1985. Origins of Magnetic Recording Concepts. *Jour. ASA.* 77 (4): 1314–1319.

Camras, M. Nov. 1943. A New Magnetic Wire Recorder. *Radio News*, Radionics Section Vol. 30, pp. 3–5, 38, 39.

Drenner, D.V.R. Oct. 1947. The Magnetophone. *Audio Engineering* Vol. 31, pp. 7–11.

Dunlop, D.J. July 1977. Rocks as High-Fidelity Tape Recorders. *IEEE Trans. Magn.* MAG-13 (5): 1267–1272.

Fankhauser, C.K. Jan. 1909. The Telegraphone. *Jour. Franklin Inst.* Vol. 16, pp. 37–46.

Hamilton, H.E. Dec. 1935. The Blattnerphone—Its Operation and Use. *Electronic Digest* pp. 347.

Hammar, P., and D. Ososke. March 1982. The Birth of the German Magnetophon Tape Recorder 1928–1935. *db—the Sound Engineer's Magazine* pp. 47–52.

Hickman, C.N. July 1937. Sound Recording on Magnetic Tape. *Bell System Technical Journal* Vol. 16, pp. 165–177.

Hickman, C.N. Sept. 1937. Magnetic Recording and Reproducing. *Bell Laboratory Record* Vol. 16, pp. 2–7.

Hoagland, A.S. Jan. 1980. Trends and Projections in Magnetic Recording Storage on Particulate Media. *IEEE Trans. Magn.* MAG-16 (1): 26–29.

Holst, H. 1906. *Elektriciteten*, Copenhagen, Gyldendal, pp. 553–561.

Larsen, A. 1950. The Telegraphone (in Danish), *Ingenioer Videnskabelige Skrifter* No. 2, 305 pages.

Mallinson, J.C. 1987. Recording Limitations. Chapter 5 in *Magnetic Recording*, Ed. by C.D. Mee and E.D. Daniel, McGraw-Hill, pp.337–375.

Mallinson, J.C. May 1985. The Next Decade in Magnetic Recording. *IEEE Trans. Magn.* MAG-21 (3): 1217–1220.

Mooney, Jr., M. 1957. The History of Magnetic Recording. *Hi-Fi Tape Recording.*

Mullin, J.T. Nov. 1972. The Birth of the Recording Industry. *Billboard* 6 pages.

Poulsen, V. 1904. System for Producing Continuous Electric Oscillations. *Trans. el. Congress St. Louis.* Vol. II, pp. 963–971.

Poulsen, V. Nov. 1900. The Telegraphone. *Ann. d. Phys.* Vol. 3, pp. 754–760.

Roizen, J. 1956. Project Videotape. *Ampex Bulletin AI.*

Rust, N.M. Jan. 1934. Marconi-Stille Recording and Reproducing Equipment. *Marconi Review* Vol. 46, pp. 1–11.

Selsted, W.T., and R.H. Snyder. Sept. 1954. Magnetic Recording—A Report on the State of the Art. *IRE Convention Transactions* 8 pages.

Shea, J.R. 1967. The Untold Story of Tape Recording. *Tape Recording*, Nos. 3, 4, 5, and 6.

Smith, O. Sept. 1888. Some Possible Forms of Phonograph. *The Electrical World*, 3 pages.

Snyder, R.H. April 1953. History and Development of Stereophonic Sound Recording. *Jour. AES* Vol. 1, pp. 176–179.

Stevens, L.D. Sept. 1981. The Evolution of Magnetic Storage. *IBM Journal of Research and Development*. 25 (5): 663–675.

Stille, K. March 1930. Electromagnetic Sound Recording. *Elektrotech. Zeitschrift* Vol. 51, pp. 449–451.

Thiele, H. July 1983. On the Origin of High-Frequency Biasing for Magnetic Audio Recording. *Jour. SMPTE* pp. 752–754.

Volk, T. Sept. 1935. A.E.G. Magnetophone. *AEG Mitteilung* pp. 299–301.

CHAPTER 2
FUNDAMENTALS AND EVOLUTION

Magnetic recording has come to play a major role in society. Speech, music, measurement data, bookkeeping, computations, and live pictures are currently recorded, stored, and replayed on magnetic tapes or disks.

The success of magnetic recording is due to its convenience of use, low cost, and reuseability of the media. Paradoxically, however, many tapes are recorded only once. When the voice of a family member or a distant friend is heard, or a favorite musical number is finally recorded, one is quite hesitant to erase it. This also applies in the recording of scientific data, where the tapes and disks normally are stored in libraries.

The principles of magnetic recording are based on the physics of magnetism, a phenomenon that relates to certain materials; magnetization of these occurs when they are placed in a magnetic field. If the material is in the group of so-called "hard" magnetic materials, it will hold its magnetization after it has been moved away from the field.

Figure 2.1 is a simplified diagram of a sound recorder. An incoming sound wave is picked up by a microphone (1) and amplified (2) into a recording current I_r (3) that flows through the winding in the record head. The ring-shaped record head has a "soft" magnetic core (so that magnetization is not retained) with an air gap in front. The current I_r produces magnetic field lines that diverge from the air gap (4) and penetrate the tape, moving past the record head from the supply reel (5). The tape itself is a plastic ribbon coated with a "hard" magnetic material that retains its magnetization after it has passed through the field from the record head gap.

The tape passes over the playback head, which also is a ring core with a front gap. The magnetic field lines (flux) from the recorded tape permeate the core and produce an induced voltage e (6) across the winding. The voltage modulates with variations in the flux on the tape. This modulated voltage, after suitable amplification (7), reproduces the original sound through a speaker (8).

This elementary record and playback process is limited by a poor fidelity in music and data recording. It is used extensively, however, in computer applications, where the criterion for performance is the presence or absence of a signal. In high-fidelity music recordings and in instrumentation recordings, an additional bias current I_b (9) is added to the record current flowing through the record head winding. The bias is a high-frequency current that provides a great improvement in recording fidelity and a simultaneous reduction of background noise.

Figure 2.1 also illustrates the components that are found in all recorders:

- Magnetic heads for recording (write) and playback (read); these functions may both be served by a combination record/play (write/read) head.
- A magnetic head for erasure of any signal previously recorded on the tape. This erase head is an optional feature that is mainly used in audio and television recorders. Old information in computer disk drives is generally overwritten by the recording field. Instrumentation tapes are bulk degaussed.
- A transport mechanism for moving the media (tape or disk) past the magnetic heads at a uniform speed. Reeling mechanisms or motors are required for providing smooth supply and takeup of tape.
- Record and playback amplifiers for processing of signals to and from the magnetic heads.
- Control logic circuitry for start, stop, and fast winding of tape.
- Servo control circuits for precise tape or disk motion.
- Power supply assemblies for the transport motors, solenoids, and relays, and for the amplifiers and control logic.

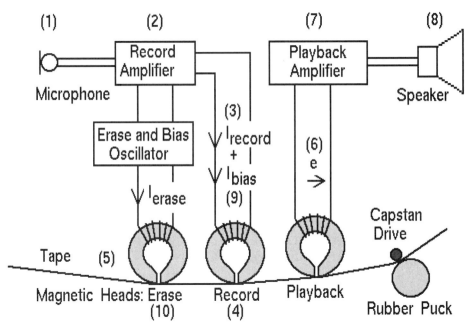

FIGURE 2.1 Fundamental elements of a sound recorder.

The media is currently not only the version found in audiocassettes, or the half-inch wide VCR cartridge tapes, but comes in various grades and widths, as magnetic stripes on film and cards, and coatings on rigid disks and flexible disks ("floppies").

Each recorder application and its media configuration dictates how that recorder is configured. Anyone familiar with recording equipment is aware of the wide variety of equipment designs.

The matter of selecting, specifying, using, or designing a piece of recording equipment will, therefore, involve tradeoff decisions. These decisions will be better, and easier to make, if the person involved has a knowledge of magnetic recording.

It is not possible or practical to provide a survey of various recorders and data storage files within the scope of this book. So there is no detailed treatment of conventional amplifier circuits, control logics, power supplies, overall packaging, and the many possible operational features.

The much-talked-about optical and magneto-optical disks have slowly entered the digital storage market and will find applications in 10 to 15 percent of the total market. Disk drive applications and tape backup systems will keep on growing. Floppies are, in many instances, being replaced with CD-ROMs. Videotapes are made and sold in an enormous volume, but with a flattened growth. The audiocassette market may get a boost from the new digital sound recording technique (DCC) introduced by Phillips in 1993.

Projections for the future developments and growth in the computer data storage, video, and sound recording fields are displayed in graphs as shown in Figs. 2.2 through 2.5. The evolution of magnetic recording has clearly gone beyond the wildest projections made in the 1960s and 70s, and the cost of magnetic data storage is well below one dollar per megabyte.

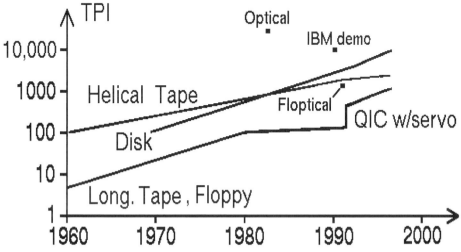

FIGURE 2.2   Track densities for all storage methods are still increasing.

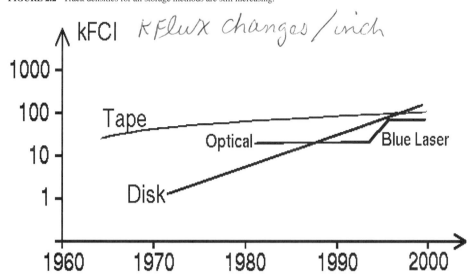

FIGURE 2.3   Flux densities along a track are increasing dramatically for disk drives. The main cause is the decreased distance between the head and disk (flying height).

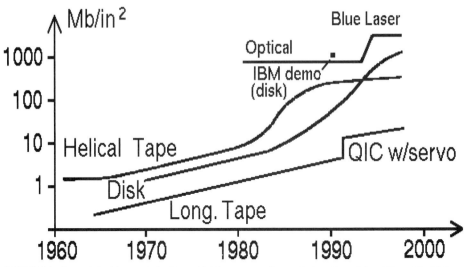

FIGURE 2.4 The areal densities are obtained by multiplying corresponding graphs from Figure 2.2 and Figure 2.3. All show a continued rise.

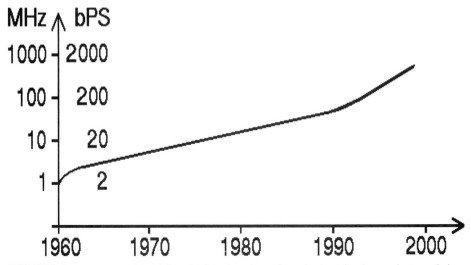

FIGURE 2.5 Larger programs and more information throughput place an increasing demand on data rates in storage devices.

The main theme of this book is the nature of magnetism, recording (write), playback (read), and the performance of magnetic heads and media, in a variety of applications. Of prime concern is that the playback signal be a faithful reproduction of the input signal.

The second concern is the achievement of the highest possible packing density on the media which provides the minimum storage volume (length of tape or size/number of disks).

The original (and basic) audio recorder is a good example to illustrate the signal record/reproduce process in further detail. Later chapters will deal with instrumentation, video, computer storage, coding, and modulation techniques.

FREQUENCY RESPONSE AND NOISE

When the magnetic tape leaves the record head, it has a permanent magnetic record of the sound or data signal, with flux lines extending from the surface (see Fig. 2.6). These flux lines pass in sequence through the reproduce-head core and induce a voltage, e, which after amplification reproduces the original signal. The number of flux lines is proportional to the recorded signal strength and their duration is inversely proportional to the recorded frequency. Their duration represents a certain wavelength λ on the tape and can be expressed as:

$$\text{Wavelength } \lambda = \text{Tape speed } v \text{ / Frequency } f$$

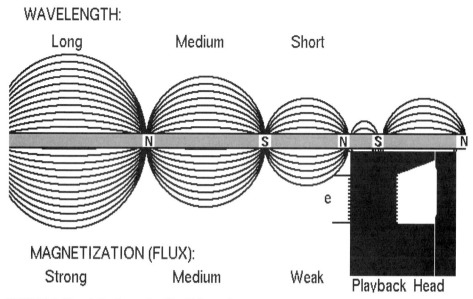

FIGURE 2.6 Magnetic flux lines go from bit to bit in recordings.

Unless stated otherwise, wavelength means recorded wavelength, not the electromagnetic wavelength (where v then is the speed of light). It is a common practice to run tape at slow speeds to reduce the quantity of tape needed, and to design magnetic heads and tapes for good short-wavelength performance. For example, if a high-frequency response of 15 kHz is desired, as in the reproduction of high-fidelity music, the wavelength should be as short as possible to save tape. If you want to record and reproduce 15 kHz at a tape speed of 9 cm/sec (3¾ IPS), the wavelength is

$$\lambda = 90 \text{ mm}/15{,}000$$
$$= 0.006 \text{ mm}$$
$$= 6 \text{ μm } (240 \text{ μin})$$

That is an extremely short length—only one-quarter of the thickness of a long-play magnetic tape. And the reproduce (playback) head gap must be shorter than half of that wavelength; otherwise, the high-frequency response will suffer, and the fidelity of the recording will be lost.

The induced voltage (e) in the reproduce head winding is very low, and requires amplification of 10,000 to 100,000 times to provide a useful output. This is particularly true in modern recorders with micro-gaps, narrow track widths (four tracks are frequently recorded on a 3.84 mm {.15 inch} wide tape in a cassette), and low tape speeds. These factors complicate the noise problem, a design consideration that is aggravated by the fact that the induced voltage is lowest at low and high frequencies, where the noise is highest. If all signal frequencies were recorded at the same level, we would expect that the corresponding wavelengths on the tape would be of equal strength. But in reality, we find that the flux from the tape falls off at short wavelengths (high frequencies), as shown in Fig. 2.7.

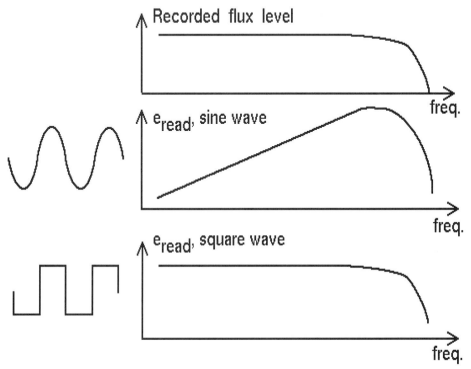

FIGURE 2.7 The recorded flux is constant at low frequencies (long bit lengths). The read voltage is proportional to the magnitude and rate of change of flux. It will therefore double when the signal frequency is doubled, at lower frequencies. The read voltage in reading square waves is generated at the transitions, and is therefore independent of the frequency of occurance of transitions.

Thus the voltage (e) induced in the playback head is not only proportional to the tape flux but also to the frequency. In order to achieve a constant output voltage over the entire frequency range, more amplification must be provided for the low and high frequencies. This is what is technically referred to as equalization. An otherwise flat noise level will therefore increase at the low and high frequencies, which is unfortunate because it emphasizes any amplifier hum and tape hiss.

In an unequalized digital recording, the signal is a sequence of alternating transitions where the magnetization changes from one polarity to the opposite. The levels are typically at saturation. Because all transitions are alike and the read voltages are the differential of these, there is no change in read voltage level. At high packing densities (high frequencies) the individual pulses overlap each other, i.e., interfere, and the voltage falls off.

It is standard practice in audio recording to boost the record current in the high and low frequency regions so that less playback equalization is required. Too much boost may cause an overload condition with subsequent distortion, so the audio industry has established standards for equalization (see Chapter 28). These standards rely on the knowledge of sound level vs. frequency in music and speech. When this knowledge does not exist, as may happen in instrumentation recording, the boost of the record current cannot be tolerated and instrumentation recorders, therefore, show an apparently worse signal-to-noise ratio (i.e., the separation between the maximum record level and quiescent noise level). However, the signal-to-noise spectrum should always be evaluated for the more realistic details, such as in the "weighted" signal-to-noise ratio for audio recorders and noise spectrum level for instrumentation recorders. (A weighted signal-to-noise ratio is measured with an instrument that compensates for the ear's sensitivity to low and high frequencies.)

As a guideline in evaluating audio recorder performance, the following signal-to-noise ratios are typical:

35–40 dB	78 rpm phonograph records
50–65 dB	Modern LP phonograph records
35–45 dB	Inexpensive home tape recorders
45–55 dB	Good home tape recorders
50–65 dB	Professional tape recorders
65–75 dB	Studio tape recorders
85–95 dB	Digital studio tape recorders

Signal-to-noise ratios are reduced at least 3 dB each time the track width is cut in half (in going, for example, from 2 tracks to 4 tracks). In high-quality audio recorders, the frequency response should cover from 20 to 16,000 Hz. This corresponds to the limits of the human ear, but consider that few musical instruments can produce frequencies above 10,000 Hz, and the hearing of high frequencies generally decreases after the age of 20.

Values of signal-to-noise ratios for instrumentation recorders vary widely according to tape speed and the range of the frequency response. However, after equalization of the reproduce head voltage, the frequency response is essentially flat. The frequency response of a tape recorder is never perfectly flat, and may be specified as 50–16,000 Hz, plus or minus 3 dB. The term dB is an abbreviation of decibel (decibel is the logarithmic ratio between two signal levels: $20\log(Level1/Level2)$ and one dB corresponds fairly well to a change in sound level that can barely be noticed. As a rule of thumb, we can use the following indications of sensitivity to level changes:

1 dB	Barely noticed by an expert
3 dB	Noticed under normal listening
6 dB	A definite change in level

A specification of plus or minus 3 dB corresponds quite well to what the ear can tolerate. Frequency response applies to the record and reproduce electronics, using a good grade magnetic tape; it does not include microphones or speakers. And the requirements for the frequency response vary with the applications and program source:

Live recordings (Live performance, CDs, FM broadcast)	20–20,000 Hz at 38 cm/s (15 IPS))
Recordings of FM programs	30–15,000 Hz at 19 cm/s (7½ IPS)
Recordings of AM programs	50–6,000 Hz at 4.8 cm/s (1⅞ IPS)

The most frequently used tape speeds are indicated. Very few instruments have sound signals (or even harmonics) above 10,000 Hz, and a speed of 4.8 cm/s (1⅞ IPS) is in many cases fully adequate for home recordings of FM programs. As the above table indicates, the frequency response does improve with increasing tape speed and it is in fact more logical to speak of how many wavelengths can be recorded per cm of tape. One thousand to four thousand wavelengths per cm of tape length is common practice for audio recorders, and the present limitation is 20,000 wavelengths per cm. This requires high-quality tapes and good magnetic heads, but it is an improvement of ten times over the past twenty years. (The frequency-response requirements of instrumentation recorders and video (TV) are discussed in Chapters 27 and 29.)

NARROW TRACKS

Thirty-five years ago, all magnetic recordings were made on 6.25 mm (¼") wide tape with the full width of the tape being used for one channel (monaural). Improvements in tape quality came rapidly during the sixties, and the amateurs and high-fidelity enthusiasts took advantage of this improvement and began recording on only half the width of the tape (called half or two-track recording). After playing one side, the reel was simply turned over for the other half (Fig. 2.8). By cutting the track width in half, only half the flux is available, and the signal-to-noise ratio should suffer, but the improved tapes made up for the difference. (Otherwise, the signal-to-noise ratio would have decreased by a factor of 3 to 6 dB.)

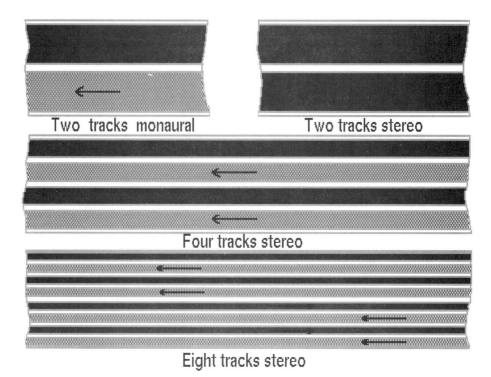

Two tracks monaural   Two tracks stereo

Four tracks stereo

Eight tracks stereo

FIGURE 2.8   Track arrangements on ¼-inch magnetic audio tape.

With the advent of stereo (or binaural) recordings, the two tracks were used for the two channels. With further improvements in tape quality, the standards were once again changed to what now is called four (or quarter) track tapes, two tracks being used for stereo recording in one direction and the remaining two tracks for the reverse direction. Latest developments have made possible the recording and playback of four tracks on 3.8 mm (0.15") wide tape, the method used with prerecorded stereo music cassettes. (Eight-track recordings were primarily used in car installations, where the noise level is somewhat high and the worse signal-to-noise ratio of the narrow track width (21 mils) is not too critical.)

The improvements in tapes that made this evolution possible are dramatically illustrated in Figs. 2.9 through 2.12. They show magnetic particles from then and now, and the refinements achieved in providing a smooth tape surface.

Today we are entering an era where magnetic recording is made in digital format on DAT recorders and DCC units (from Phillips). These techniques will be described in Chapter 28. The reader is undoubtedly already enjoying the quality and convenience of CD recordings. Ninety percent of these recordings were nevertheless made on ANALOG studio tape recorders, edited and THEN digitized for CD production. Notice the tiny label on a CD that should say DDD, but usually reads ADD, meaning Analog recording, Digital editing, and Digital CD signal.

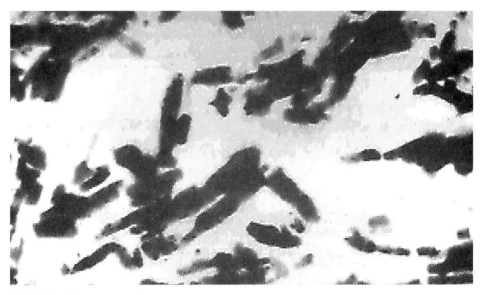

FIGURE 2.9   Early, coarse magnetic particles.

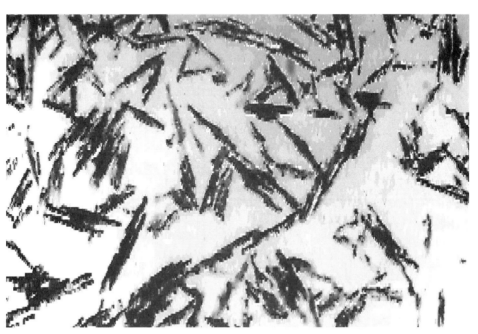

FIGURE 2.10   Modern small, uniform magnetic particles.

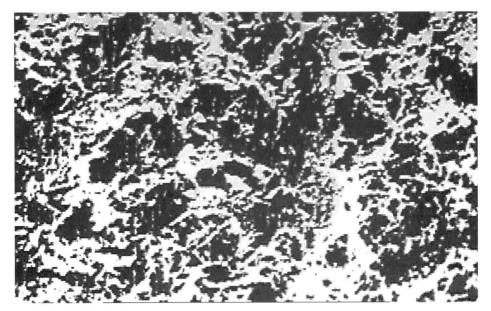

FIGURE 2.11   The coarse finish of a typical early magnetic tape surface.

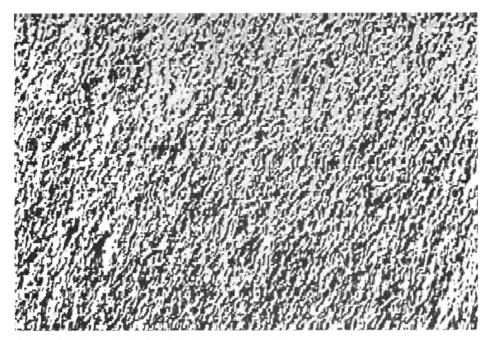

FIGURE 2.12   The smooth finish of a typical modern magnetic tape surface.

TAPE TRANSPORT and FLUTTER

Among the variety of recorders manufactured today, there are probably not two with transports that are exactly alike. But the differences in general are reflected only in cost, quality, and style, while the basic functions are common to them all. The art of designing and mass producing recorders has come a long way over the past decade, with ever-increasing performance. There have been reductions in price and many features have been added.

It is basic to all transports that they move the tape at a constant speed past the magnetic heads while recording or playing back. (Small speed variations, called flutter and wow, do exist, as will be discussed later.) All tape transports have provisions for fast winding of the tape, either for reaching a certain portion of the tape in a short time or for complete rewinding.

The tape transport mechanism in its simplest and most reliable form consists of a metal plate with three motors mounted on it, one motor for driving the tape at a constant speed and the other two for reeling the tape. Savings in the manufacturing costs of a tape transport are sometimes sought by using only one motor with friction or clutch drives for the reeling mechanism. The layout in general is as shown in Fig. 15.4, where the tape passes from the supply reel through the guide and over the heads; the tape speed is controlled by the capstan, against which the tape is held by a rubber pinch roller. The capstan shaft rotates at a constant speed and is a critical item in the transport mechanism. It must rotate in good bearings and should be perfectly concentric.

A constant tape speed is generally best obtained by using a hysteresis-synchronous motor for the capstan drive. A hysteresis-synchronous motor follows the power supply frequency (50 or 60 Hz), which is quite accurately controlled. These motors exhibit hunting characteristics, which are usually smoothed out by a flywheel. This is obtained in the better-grade recorders using the so-called inside-out hysteresis-synchronous motors. Such motors usually have two speeds, which are electrically switchable, and this, in a simple way, provides for a choice in tape economy (and/or playing time). As mentioned earlier, other tape transports use a single motor for the capstan drive and reel drive systems. In this arrangement the speed change is accomplished by an arrangement of pulleys or belts.

Later years have seen many recorders emerge with servo-controlled capstan speed, a technique that will be described in Chapter 27.

The selection of tape speed depends entirely on program material bandwidth to be recorded. "Program material" bandwidth ranges from direct current to a few Hz for underwater oceanographic investigations, to a few thousand Hz for taped letters and voice communications, to 15,000 Hz for high-fidelity music reproduction, to a few MHz for wideband recordings, and up to 10 MHz for color video recording. In data recording, floppies use a data frequency of 125 kHz, while rigid disk drives approach 10 MHz.

Because most users of magnetic tape recorders have requirements for a variety of bandwidth ranges, recorders generally have from two to six speed ranges. In home entertainment recorders, this is achieved by a two or three-speed hysteresis-synchronous motor or by an appropriate pulley arrangement, and in instrumentation recorders it is achieved by the use of servo-controlled tape drive systems.

The tape transport mechanism, in addition to providing a constant speed, must ensure that good contact is maintained between the magnetic tape and the heads. This contact requires a certain amount of pressure, which can be obtained through hold-back tension on the supply reel (a felt pad against the ingoing tape guide, or felt pads that press the tape against the heads).

The felt pad arrangement is inexpensive, but can easily cause excessive wear on the heads, which increases overall operating costs through the necessity of replacing worn-out heads. The felt pad is best used against the tape at a guide post and will, if properly adjusted, give a more constant speed throughout a reel of tape than the hold-back torque applied at the feed reel. In a single-motor tape deck torque is provided by a slip-clutch arrangement, and in a three-motor deck by reverse current to the feed-reel motor.

No capstan drive system is perfect and, as mentioned earlier, this will give rise to tape speed variations that affect the playback quality of music. (Its effect on instrumentation data is discussed in Chapter 27.) Even minute speed variations are noticeable, although it does depend on the type of program material. The human ear is particularly sensitive to speed variations during playback of pure tones, like those of bells, the flute, and the piano. Such speed changes are called wow and flutter. Speed variations up to 10 Hz are called "wow," and speed variations above 10 Hz are considered "flutter."

Wow is primarily caused by capstan shaft eccentricity or dirt buildup on the capstan, or possibly a poor reel drive system with a varying hold-back. Motor cogging and layer-to-layer adhesion in the tape pack can also cause wow. In cases of wow, tones are frequency-modulated, which gives rise to a singing sound. Flutter, on the other hand, is generated by the tape itself. Magnetic tape has elastic properties and when it moves over guides and heads, a scraping, jerky motion can take place, which causes longitudinal oscillations in the tape. This destroys otherwise pure tones, and gives them a raw and harsh sound that is often mistaken for modulation noise.

Typical amounts of wow and flutter (up to 250 Hz) that can be tolerated are:

Speech max.	0.6 percent peak
Popular music max.	0.3 percent peak
Classical music max.	0.15 perc. peak

Wow is just barely noticeable at max. 0.1 percent peak (value). In this area the digital recorders are superior. They do have about the same flutter in the transports, but the digitized signal stream can be digitally meshed into perfect synchronism with a clock, and flutter is removed.

Recording equipment from various manufacturers is described in terms of many specified items so a comparison between similar units can be made. A two-page listing of recorder specifications is quite common, and the media, tape or disk, may have a one-page listing. This book will throughout its chapters make clear to the reader what such specifications cover, and what they imply for the recorder's overall performance.

BIBLIOGRAPHY

Altrichter, E., G. Boden, H. Lehmann,. and H. Volz. 1972. *Grundlagen der Magnetischen Signalspeicherung III: Anwendung fur Fernsehen, Film, Messtechnik und Akustik sowie eine Geschichliche Entwicklung*. Berlin. Akademie-Verlag. 124 pages.

Arnoldussen and L.L. Nunnelley (Editors). 1992. *Noise in Digital Magnetic Recording*. Singapore. World Scientific. 280 pages.

Begun, S.J. 1949. *Magnetic Recording*. Rinehart & Co. New York.

Borwick, J. (Editor). 1980. *Sound Recording Practice*. London. Oxford University Press. 503 pages.

Camras, M. (Editor). 1985. *Magnetic Tape Recording*. New York. Van Nostrand Reinhold. 443 pages.

Camras, Marvin. 1988. *Magnetic Recording Handbook*. New York. Van Nostrand Reinhold. 718 pages.

Davies, G.L. 1961. *Magnetic Tape Instrumentation*. New York. McGraw-Hill. 263 pages.

Geller, S. B. Oct. 1974. Archival Data Storage. *Datamation* 20, pp. 72–80.

Gregory, S. 1988. *Introduction to the 4:2:2 Digital Video Tape Recorder*. London. Pentech Press. 200 pages.

Hoagland, A.S. 1963. *Digital Magnetic Recording*. New York. Wiley. 154 pages.

Hoagland, A.S., and J.E. Monson. 1991. *Digital Magnetic Recording*. 2nd Edition. New York. John Wiley & Sons. 230 pages.

Iwasaki, S., and J. Hokkyo. 1991. *Perpendicular Magnetic Recording*. Tokyo. Ohmsha. 211 pages.

Jorgensen, Finn. 1970. *Handbook of Magnetic Recording*. 1st Edition. Blue Ridge Summit, PA. 192 pages.

Jorgensen, Finn. 1980. *Handbook of Magnetic Recording*. 2nd Edition. Blue Ridge Summit, PA. 448 pages.

Jorgensen, Finn. 1988. *Handbook of Magnetic Recording*. 3rd Edition. Blue Ridge Summit, PA. 740 pages.

Krones, F. 1952. *Die Magnetische Schallaufzeichnung*. Technischer Zeitschriftenverlag. Wien. 216 pages.

Lindsay, H.W. Oct. 1950. Precision Magnetic Tape Recorder for High-Fidelity Professional Use. *Elec. Mfg.* 46 (?):135–139.

Lowman, C.E. 1972. *Magnetic Recording*. New York. McGraw-Hill.

Mallinson, J.C. 1987. *The Foundations of Magnetic Recording*. 1st Edition. San Diego. Academic Press. 175 pages.

Mallinson, J.C. 1993. *The Foundations of Magnetic Recording*. 2nd Edition. San Diego. Academic Press. 217 pages.

Mallinson, J.C., and V. E. Ragosine. Sept. 1971. Bulk Storage Technology: Magnetic Recording. *IEEE Trans. Magn.* MAG-7 (1): 98–600.

Marchant, A.B. 1990. *Optical Recording*. Reading, MA. Addison-Wesley. 408 pages.

Mee, C.D., and E.D. Daniel. 1987. *Magnetic Recording.* 1st Edition. Vol. 1–3. New York. McGraw-Hill.

Mee, C.D., and E.D. Daniel. 1990. *Magnetic Recording.* 2nd Edition. New York. McGraw-Hill.

Mee. C.D. 1964. *The Physics of Magnetic Recording*. New York. Elsevier. 271 pages.

Moon, J., and L.R. Carley. 1992. *Sequence Detection for High-Density Storage Channels.* Boston. Kluwer Academic Publishers. 153 pages.

Moulin, P. 1975. *L'Enregistrement Magnetique d'Instrumentation.* Paris. Editions Radio. 416 pages.

Patent Digest. 1984. *IBM Class 360, Dynamic Magnetic Information Storage or Retrieval*, The Boston Patent Co. 490 pages.

Phillips, W.B, and H.P. McDonough. April 1974. Maximizing the Areal Density of Magnetic Recording. *IEEE COMPCON Proc.* pp. 101–103.

Ruigrok, J.J.M. 1990. *Short-Wavelength Magnetic Recording.* Oxford. Elsevier Advanced Technology. 564 pages.

Schneider, C. and H. Volz. 1970. *Grundlagen der Magnetischen Signalspeicherung II: Magnetbander und grundlagen der Transportwerke.* Berlin. Akademie-Verlag. 133 pages.

Scholz, C. 1968. *Magnetbandspeichertechnik.* Berlin. VEB Verlag Technik. 284 pages.

Scholz, C. 1980. *Handbuch der Magnetband speichertechnik*. Muenchen-Wien. Hanser Verlag. 392 pages.

Schouhammer Immink, K.A. 1991. *Coding Techniques for Digital Recorders.* Hertfordshire, UK. Prentice Hall. 297 pages.

Siakkou, M. 1985. *Digitale Bild-und Tonspeicherung.* Berlin. VEB Verlag Technik Berlin. 296 pages.

Sierra, M. 1990. *An Introduction to Direct Access Storage Devices.* San Diego. Academic Press. 260 pages.

Stewart, W. E. 1958. *Magnetic Recording Techniques.* McGraw-Hill. 272 pages.

Watkinson, J. 1991. *RDAT.* Oxford. Focal Press. 244 pages.

Westmijze, W.K. April 1953. Studies in Magnetic Recording, I: Introduction. *Philips Res. Rep.* 8 (3): 148–157.

Westmijze, W.K. Sept. 1953. Principles of Magnetic Recording and Reproduction of Sound. *Philips Tech. Rev.* 15 (?): 84–96.

White, R.M. 1985. *Introduction to Magnetic Recording.* New York. IEEE Press. 307 pages.

Winckel, F. 1960. *Technik der Magnetspeicher.* Berlin. Springer-Verlag. 614 pages.

Winckel, F. 1977. *Technik der Magnetspeicher*, 2nd Edition. Springer-Verlag. 402 pages.

CHAPTER 3
FUNDAMENTAL MAGNETISM

In this chapter, magnetism is introduced in a manner that will give the reader an understanding of the physics underlying magnetic recording and playback (or write and read). As the reader goes through the sections, he or she will learn how electricity and magnetic materials relate to magnetic write/read heads and tapes/disks. This chapter focuses on magnetism from electric currents and magnetism in materials, while Chapter 4 will expand on the technical aspects of magnetism when magnetic materials are applied in heads and media coatings (media, plural of medium, is the common term for tapes or disks).

Magnetism comes in two forms: as field lines from permanent magnets (tape and disk recordings), and from electric currents (in record, or write, fields). Chapters 3 and 4 will provide the reader with the necessary tools for understanding magnetism in recording and playback. The particular magnetic materials used in heads are useful because they can conduct (or guide) and concentrate magnetic fields.

We start the chapter by learning about the fields from a magnetic pole, a current-carrying straight wire, a loop, and a coil. Next we form an electromagnet by inserting a piece of iron in the coil, and we find that this greatly increases the field strength. To understand why, we must next learn about magnetic materials.

From physics we know that all materials may be divided into diamagnetic, paramagnetic, and ferromagnetic classes by their magnetic properties. Details on these properties are postponed for Chapters 7 and 11, which cover magnetic materials for heads and for tapes or disks. The ferromagnetic materials are evaluated by tracing their hysteresis loop, which leads to an introduction of the domain hypothesis and the magnetization process of ferromagnetic materials. These can roughly be divided into soft and hard materials, corresponding to their use in heads and media.

Media coatings are finally analyzed in this chapter, based on models for the individual particles they are made from. Statistical tools are well-suited for describing the overall properties of these assemblies of particles, and the important switching field distribution is introduced. This chapter concludes with a brief introduction to solid metallic coatings.

EARTH MAGNETISM AND PERMANENT MAGNETS

Recording technology is founded on magnetism and on electromagnetic induction. The earliest description of magnetism is obscure, but a mineral called magnetite was known centuries before the birth of Christ. It would attract iron, and would also magnetize a piece of iron if it was rubbed against it.

The sailor's compass could be made from a properly shaped piece of magnetite, free to turn about a pivot. It would turn to point in the north-south direction and was named lodestone, which means "waystone" or "leading stone," because its property made it useful for navigation. The first to use this principle were apparently the Chinese, although they only used it to maneuver about in China; it was European sailors who first used the lodestone as a compass for oceanic navigation (Fig. 3.1). Another legend concerning lodestone depicts its use in defense (Fig. 3.2).

FIGURE 3.1 The early sailor's compass, the lodestone.

The first scientific study of magnetism was made by the Englishman William Gilbert (1540–1603), who published a classic book, *De Magnete*. All his experiments had to be carried out using iron or steel samples that were rubbed with a lodestone. He rationalized that the earth itself was a magnet, and his experimental samples were therefore shaped as spheres. And he did not recognize a North and a South pole, but rather a centered source of magnetic force.

A certain amount of mysticism was associated with magnetism. In one experiment a piece of iron was carefully weighed and placed in a box. The empty space was next filled with iron filings and their weight determined. The box was closed and stored for a few years. Then the box was opened and the weight measurements repeated: the amount of filings now weighed slightly less than before, and the iron piece more. The conclusion: lodestones eat iron powders to sustain themselves.

Gilbert believed that the lodestone possessed a soul, but discounted the old superstition that "Onyons and Garlick are at odds with the lodestone" (steersmen were forbidden to eat the vegetables when on duty!). The medical healing power of magnets was, apparently, discarded by Gilbert, as it was later on by Edison, who found "no effect from strong fields, on himself nor on his dog"; and yet, in recent years it has been discovered that surgical wounds heal much more nicely when a strong magnet is placed in the bandage near the wound.

René Descartes (1596–1650) exorcised the soul out of the lodestone and established a constructive philosophy. He postulated that magnetism had "threaded parts" that were channeled through the earth through pores in the poles. These parts would further seize on the opportunity to cross any lodestone in the way.

FIGURE 3.2 The legend about defense using a large lodestone to capture an ironclad warrior.

One of Descartes' students, the Swedish scientist and theologian Emanuel Swedenborg (1688–1772), envisioned an ordering within a magnetized versus a non-magnetized piece of iron (see Fig. 3.5). Today's domain theory is in remarkable agreement with this illustration (see later).

The lodestone was found to point one end toward north, and this was hence named the north seeking pole or, in brief, north pole. The compass needle does the same, and we are, with the early definition of polarities, stuck with the fact that the earth magnet has its south pole located near the geographic north pole, and its north pole near the south pole (see Fig. 3.3).

The earth poles are not stationary; they have moved about quite a bit, and still do so. They are about 750 miles from the geographic poles, and move approximately one mile per year. There is also evidence that they have alternated: when molten rock containing iron cools and solidifies, it becomes magnetic. The direction of its magnetization will follow any external field that may be present, such as the earth field. It will also be of polarity opposite the external field (in accordance with Le Chatelier's Law, when a system in equilibrium is disturbed, it always reacts in such a manner as to oppose the forces that upset the balance).

During the mining of iron ore magnetizations have been found, varying in strength and direction. Later examination of these data revealed the changes in pole positions and polarities. We can correctly say that the earth was the first magnetic recorder.

The origin of the earth's magnetic field is obscure. All magnetization ceases to exist above a certain temperature called the Curie temperature, being in the order of a few hundred degrees Celsius. Most of the earth's mass is in the molten state, and hence nonmagnetic. One hypothesis explains the magnetization as arising from circulating currents that somehow are related to the earth's rotation.

A snapshot of the earth's magnetic field is provided by the northern lights (Aurora Borealis). They are caused by very high-energy particles, chiefly electrons, plunging from space into the atmosphere along the outermost closed field lines of the earth's magnetic field. When these electrons hit atoms of the high atmosphere, the atoms glow in colors indicating their composition and the electron energies (see Figs. 3.6 and 3.7).

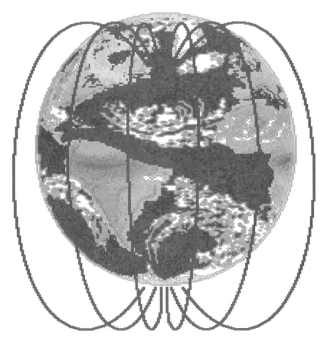

FIGURE 3.3 The earth's magnetic field.

FIGURE 3.4 British physician and physicist William Gilbert.

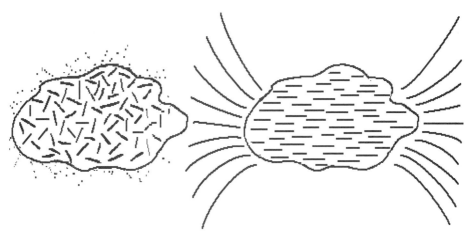

FIGURE 3.5 Swedenborg's vision of a nonmagnetic and a magnetic lodestone.

FIGURE 3.6 The northern lights (aurora borealis) may produce dancing, colored veils on the night sky over places above 60 degrees of latitude.

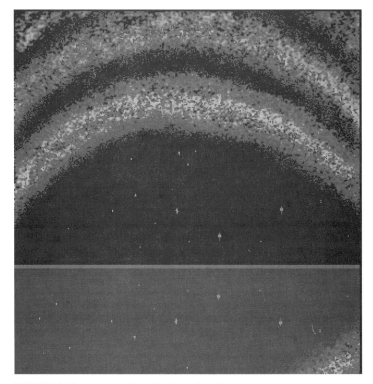

FIGURE 3.7 Fluorescent arcs formed by the northern lights.

Field lines from magnets may be displayed by means of iron powder patterns. This can easily be done by the reader by first obtaining some iron powder from a hobby store. Sprinkle a small portion thereof on a 5-×-8-inch index card, as evenly as possible. Next place this card on top of a small permanent magnet (or construct a battery-powered electromagnet like that shown in Fig. 3.8), and tap the card lightly with a finger. This will free the powder to move about, and the iron particles will behave like small compass needles: soon a field pattern develops. (Some magnets are so strong that they literally saturate the powder with them, and destroy the pattern. Try placing a spacing between the magnet and the index card, e.g., a piece of corrugated board, or wood, if this happens.) An example is shown in Figure 3.9.

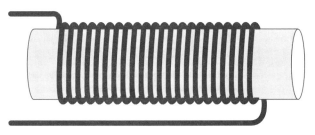

FIGURE 3.8 An iron bar wound with 200 turns of electric wire (magnet wire).

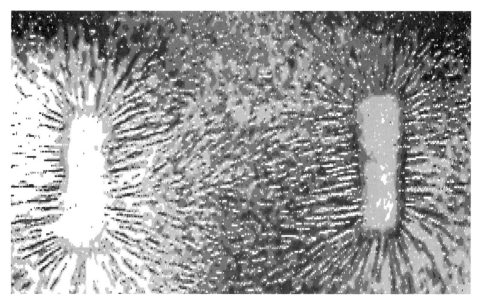

FIGURE 3.9 A field line picture produced by sprinkling iron powder on a card placed above an energized electromagnet from Figure 3.8.

NOTE: For observation of recorded tracks on tapes or disks, a liquid containing fine magnetic particles can be sprayed onto the surface. One such liquid, Part no. PF-SM1, is available from Spraque, Anaheim, CA, USA. (818-994-6602.)

MAGNETIC FIELD STRENGTH FROM A POLE

During the latter half of the eighteenth century Charles Augustin Coulomb carried out experimental work in order to determine the field strength from magnetic poles. He had proven that the electrostatic force between two electric charges varied in strength in inverse proportion to the distance squared. He proved that the same law holds for magnetic poles and used long bar magnets in order to isolate the forces from the poles at the ends of the magnets (see Fig. 3.10). A distance of one unit between neighboring poles and ten between the end poles would introduce an error of less than two percent, assuming an "inverse squared" force law.

Magnetic field strength is defined as the force the field exerts on a unit pole. This leads to the expression for the field strength from a pole of strength p_1:

$$H_1 = \frac{p_1}{4\pi r^2} \text{ A/m} \tag{3.1}$$

where:
    H = field strength in A/m
    r = distance in m
    p = pole strength in Am

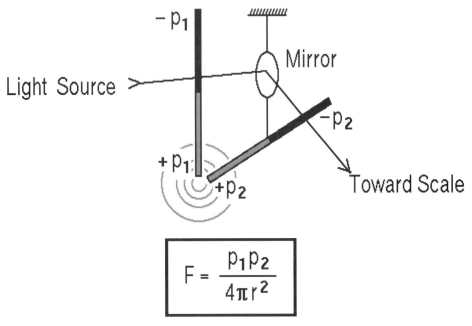

FIGURE 3.10 Coulomb's experiment.

Coulomb's expression for the force F on a pole p in the field H then becomes $F = \mu_o H p$, where μ_o is a constant ($4\pi 10^{-7}$ Hy/m). We may replace the field strength H with another expression for the magnetic field, namely the equivalent number of field lines per unit area, B, by using

$$B = \mu_o H \text{ Wb/m}^2 \tag{3.2}$$

We can now determine the torque that a uniform field exerts on a compass needle of length L, making an angle of α degrees with the field. The force on each pole is $B \times p$, and the torque arm is $(L/2) \times \sin\alpha$; the total torque is therefore

$$T = B \times p \times L \times \sin\alpha \tag{3.3}$$

The product $p \times L$ is defined as the magnetic moment m.

$$m = pL \text{ Am}^2 \tag{3.4}$$

We will find that this quantity corresponds to the permanent magnetization in a recorded track. We will be able to calculate this magnetization, and hence find the field associated with it.

In digital recording, bits of alternating polarity represent ones and zeros. The boundary between the magnetized bits is named the transition zone and is actually a short and wide pole, stretching from one side of the track to the other. We will return to this in Chapter 4.

A modern example of utilization of pole forces is found in the magnetic force scanning tunneling microscope (Gomez, Adly, Mayergoyz, and Burke). Here a small magnetic probe is scanned along a magnetized surface and there will be minute deflections of the probe tip due to forces between it and the magnetization of the surface. A suitable servo mechanism can be designed to keep the force constant by moving the probe down and up.

The probe deflects as it interacts with the surface magnetic fields and consequently changes the probe tip to sample separation. The feedback compensates for this change, and this signal now contains information about the surface topology as well as magnetics. It is possible by design of the probe tip to enhance the magnetic contribution over the topology. Examples are shown in Figs. 3.11 and 3.12.

The field strength from an infinitely long line source decreases in inverse proportion to the distance (not squared, as for a point source), and the formula for the field strength that we will find useful is:

$$H = \frac{p'}{4\pi r} \text{ A/m} \tag{3.5}$$

where:
  H = field strength in A/m
  p' = pole strength per unit length in Am/m = A
  r = distance in meters

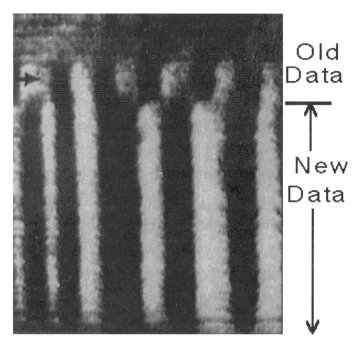

FIGURE 3.11 Observation of insufficient overwrite of a track due to the misalignment of the head (*courtesy of Dr. E.R. Burke*).

The visualization of field lines by iron powder can now be explained: These patterns show up where the field is changing. A small iron whisker placed in a field will have a pole induced in both ends. If the field strength is the same everywhere, then the pulling forces in these poles are identical in strength, but in opposite directions. This will cause the dipole to turn like a compass, but without lateral motion. When the field strength varies, then the pole in the strongest portion of the field will be pulled more than the pole in the weaker portion, and now the particle will move toward the stronger field.

Interaction among particles in a particulate media (tape) or the magnetic grains in a thin metal film is another example of the action of magnetic forces in nature (Fig. 3.13.)

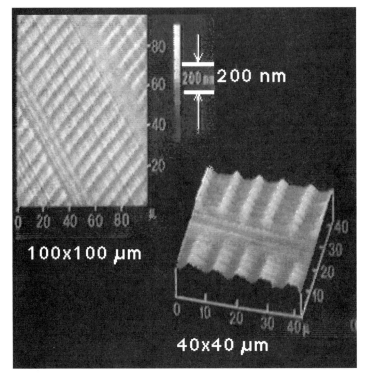

FIGURE 3.12   Observation of tracks recorded on a disk (*courtesy of Dr. Gomez*).

FIGURE 3.13   Computed interaction field lines among particles or grains in a magnetic recording medium.

FIGURE 3.14 Danish scientist and philosopher Hans Christian Oersted.

MAGNETIC FIELD STRENGTH FROM A CURRENT

A very important discovery occured in 1820 when the Danish pharmacist and philosopher Hans Christian Oersted (1777–1851) found that an electric current produces a magnetic field (Fig. 3.15). He had (like other scientists in those days) pondered over the nature of electricity and magnetism, and a possible connection between the two. A compass was not influenced by placing it near a battery's terminal, like a charged piece of amber was.

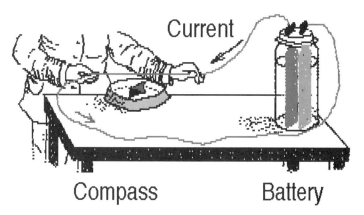

FIGURE 3.15 Oersted's experimental setup to show the influence of a current upon a compass needle.

FIGURE 3.16 French mathematician and physicist Andre-Marie Ampère.

One of the many experiments gave a clue: Oersted discovered that a compass needle would jitter ever so slightly when it was placed near a metal wire that, in turn, was connected to the terminals on a battery. More experiments revealed that a thicker wire caused a measurable deflection, and Oersted now revealed his discovery to the world's scientific community. He also discovered that the forces acting on a compass had circular patterns, centered at the wire.

French mathematician Andre-Marie Ampere (1775–1836) found that the wire connecting the battery terminals conducted a current, and he also found that two currents have a mutual magnetic effect, and hence a force between them (Fig. 3.17). One could now manufacture electromagnets and motors. Ampere also postulated that all magnetism has circulating currents as their origin; we will return to this topic later in the book.

The field lines from the current I in a straight wire are shown in Fig. 3.18. Ampere showed that the field is generated by the current I, and its strength can be determined by taking the line integral along a closed path surrounding the wire.

The simplest path around the wire is a circle, centered at the wire, and the value of the line integral can be found to equal $H \times 2\pi r$, where r is the circle's radius. This results in an important mathematical expression that tells us that the field strength is inversely proportional to the distance from the wire.

$$H = \frac{I}{2\pi r} \quad \text{A/m} \tag{3.6}$$

where:
  I = current in amperes
  r = distance in meters

In short, the field strength decreases proportionally to the distance.

It appears that the field strength becomes infinite at the center of the wire ($r \to 0$), but bear in mind that the line integral now is taken along a path with radius $\to 0$, and therefore encircles a current of

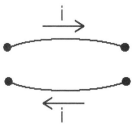

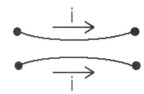

Force f = B·l·i (N).

(B is flux density near wire, here produced by current i in adjacent wire. l is length of wire exposed to field).

Opposite currents repel.

Currents in same direction attract.

FIGURE 3.17   The mutual influence of two currents.

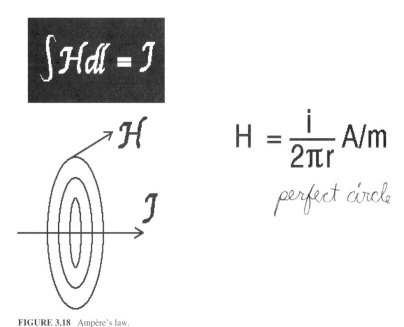

$$\int H \, dl = I$$

$$H = \frac{i}{2\pi r} \text{ A/m}$$

perfect circle

FIGURE 3.18   Ampère's law.

value $\to 0$. As a matter of fact, when the circular path has a radius that equals or is less than the wire radius, then we find a lesser field strength:

$$H = \frac{I \times r}{2\pi R^2} \tag{3.7}$$

where:

R = radius of the wire

r = distance in meters

Figure 3.19 illustrates the variation in field strength with distance from the wire. The same curve applies for the field strength from a line pole source of length L (or row of poles p), and the formula for that field strength is:

$$H = \frac{p/L}{2\pi r} \tag{3.8}$$

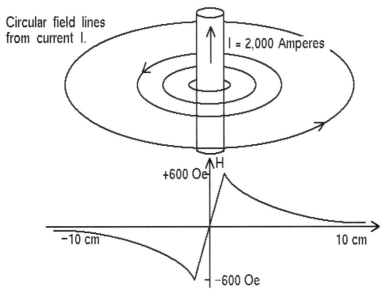

FIGURE 3.19 A magnetic field from a current-carrying wire.

The early unit for magnetic field strength was the oersted, which still is widely used. In this book all technical formulas and computations are in SI units, while materials will be described using cgs units; in many places the complementary unit will be shown in parentheses.

It is not difficult to translate from one unit to another. For magnetic field strength the numbers are:

$$1 \text{ A/m} = 4\pi/1000 \text{ Oe} \tag{3.9}$$

$$1 \text{ Oe} = 1000/4\pi \text{ A/m}$$

The A/m is a rather small unit, being approximately ⅟₈₀ of one oersted. And the oersted itself is a small unit: the earth field strength varies from 0.6 Oe near the poles to 0.3 Oe near the equator. And the field strength required to erase an ordinary audiocassette must be greater than 300 Oe = 24,000 A/m.

Example 3.1: A person inside an elevator carries a cassette tape in his pocket, and is leaning against the wall. Just outside are cables carrying heavy currents for the lift motor. Will the tape be erased if the distance between the tape and one cable is 10 cm, and the in-rush motor current is 2000 A?

Answer: The field strength is, using (3.6): H = 2000/2π × 0.1 = 10,000/π A/m = (4π/1000) × 10,000/π Oe = 40 Oe, well below the tapes coercivity of a few hundred Oe. The tape is not erased. Figure 3.19 illustrates the circular field lines indicating the field strength variation. Figure 3.20 shows a topographic map of the field magnitude around the wire. At a distance of 1 cm the strength is 400 Oe, and that would erase most tapes of standard coercivity (around 300 Oe).

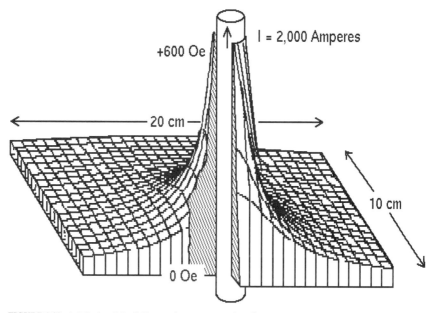

FIGURE 3.20 A 3-D plot of the field around a current-carrying wire.

FORCES IN MAGNETIC FIELDS

Oersted made another significant discovery: a wire in a magnetic field will seek to move when a current flows through it.

Experiments showed that the force was proportional to the product of the field strength, the current, and the length of the wire, assuming the wire is placed perpendicular to the field lines. In addition we must multiply by the sine of the angle α between wire and field directions.

$$F = B \times l \times i \times \sin\alpha \text{ N} \tag{3.10}$$

where:
  B = flux density in Wb/m^2
  l = length of wire in m
  i = current in A

The direction of force is given by a simple rule by Fleming, shown in Fig. 3.21.

Example 3.2: A linear motor has a circular winding with 50 turns and a diameter of 6 cm. What is the axial force if the winding is placed in a uniform field of 5000 Oe, and the coil current is 2 A?

Answer: Converting the field strength gives $H = 5000 \times 1000/4\pi = 1.25 \times 10^6/\pi$ A/m. Next $B = 4\pi 10^{-7} \times H = 0.5$ Wb/m^2. Finally $F = 0.5 \times 50 \times \pi \times 0.06 \times 2 = 9.42$ N.

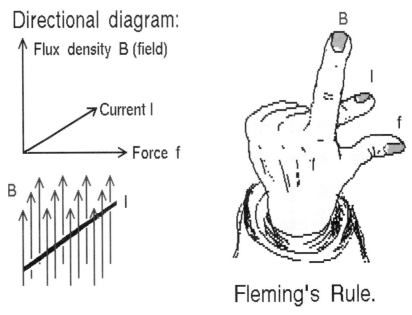

FIGURE 3.21 Fleming's rule for finding the direction of the magnetomotive force when the current and field directions are known.

Head positioner motors in disk drives are examples of applied magnetic forces. They are of the linear (Fig. 3.22) or rotary construction (Figs. 3.23 and 3.24.)

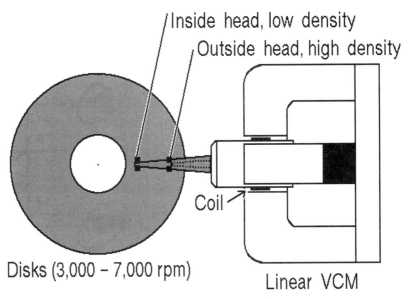

FIGURE 3.22 Linear voice-coil motor for head assemblies in a disk drive.

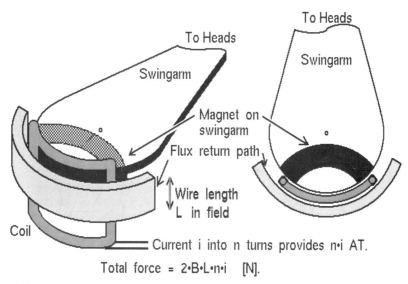

To Heads

Swingarm

Magnet on swingarm

Flux return path

Wire length L in field

Coil

Current i into n turns provides n·i AT.

Total force = 2·B·L·n·i   [N].

To Heads

Swingarm

FIGURE 3.23   Rotary actuator (or head arm) with the magnet movable and the coil stationary.

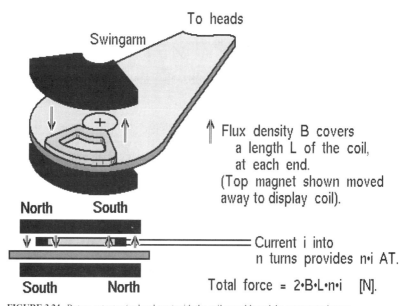

To heads

Swingarm

Flux density B covers a length L of the coil, at each end. (Top magnet shown moved away to display coil).

North   South

South   North

Current i into n turns provides n·i AT.

Total force = 2·B·L·n·i   [N].

FIGURE 3.24   Rotary actuator (or head arm) with the coil movable and the magnet stationary.

FIELD FROM A COIL

There are several instances where we wish to generate and control a magnetic field: in the gap of a recording head, or in a test instrument to measure magnetic properties of materials. These are just a few applications of what are known as electromagnets, i.e., where magnetism is created by currents in electric wires.

Each of the many turns of wire in a coil generates a flux that all add up to a final, larger flux. We will find that the flux lines run in patterns through the length of the coil and close themselves (Fig. 3.25). It is characteristic for the magnetic flux lines from currents to form closed patterns around these currents, in contrast to electrical field lines that begin and end on electric charges. It is also in contrast to field lines from permanent magnets, which run from pole to pole on open paths as shown in Fig. 3.25.

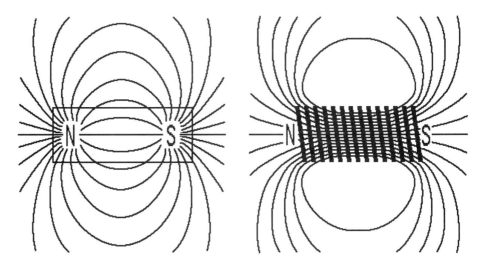

FIGURE 3.25   Magnetic field lines inside and around a magnet and a solenoid.

We cannot use formula (3.6) to calculate the field strength in and about the coil because it only holds for an infinitely long wire. We must start with the field from a single, closed turn, which we derive by summing up the contributions from many small sections dl of this loop. Each section dl produces a field dH that is found by Biot-Savart's Law.

$$dH = \frac{I \times dl \times \sin\alpha}{4\pi r^2} \ \text{A/m} \tag{3.11}$$

where:
  I = current in A
  dl = length of wire segment in m
  α = angle measured clockwise from direction of current along dl to direction of radius vector r, from dl to point P
  r = distance from dl to P, in m

For a long, straight wire we find formula (3.12), by summing (integrating) the contributions from minus to plus infinity. For a single turn loop we find for the field strength on the axis of the loop.

$$H = \frac{I \times a^2}{2 \times (a^2 + x^2)^{3/2}} \ \text{A/m} \tag{3.12}$$

where:

I = current in A

a = wire loop radius in m

x = distance from loop center, in m

A coil with n turns can be thought of as comprising n single loops. The field intensity well inside the coil is proportional to the current as well as the number of turns n, and is inversely proportional to the coil length l. The computed field pattern is illustrated in Fig. 3.25. The field strength in the center of the coil can be calculated. For a coil with many evenly wound turns (n) and length (L) much greater than its diameter, it is:

$$H = \frac{ni}{L} \text{ A/m} \tag{3.13}$$

Other coil configurations are listed at the end of this section.

Figure 3.25 illustrates another interesting point: it appears that we must deal with two distinct sources of magnetic fields. From the earth, the lodestone, and other permanent magnets we have field lines that are "open," and directed from a south pole toward a north pole. And from a current in a straight wire or a coil, we have "closed" field lines.

TABLE 3.1 Magnet Wire Data

AWG size	Diameter mm	Turns/cm.	Ohms/meter	I_{max} amps
14	1.628	5.9	0.0084	5.87
16	1.291	7.4	0.0134	3.69
18	1.024	9.3	0.0213	2.32
20	0.821	11.6	0.0339	1.46
22	0.644	14.6	0.0540	0.918
24	0.511	18.3	0.0858	0.577
26	0.405	22.8	0.1365	0.363
28	0.321	28.6	0.2170	0.228
30	0.255	35.6	0.3449	0.144
32	0.202	44.5	0.5485	0.890
34	0.160	56.3	0.8721	0.057

When we move away from the permanent magnet and the coil, as shown in Fig. 3.26, we find that the fields look very much alike. Further away, in Fig. 3.27, they are identical, and are both named the magnetic dipole field. Near the dipole (magnet or current-carrying coil) this field decreases in inverse proportion to the distance cubed.

Similar to the way the magnet would turn in a magnetic field, so will the current carrying coil. And the torque is found from:

$$T = B \times i \times A \times n \times \sin\alpha \text{ Nm} \tag{3.14}$$

where:

B = flux density in Wb/m2

i = current in A

A = area of the winding "window"

n = number of turns

α = angle between field direction and coil axis

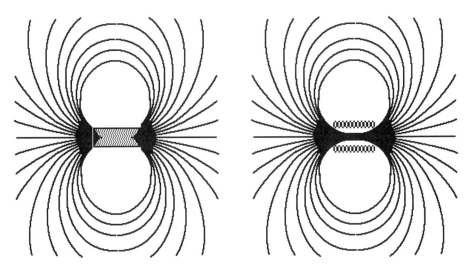

FIGURE 3.26 Field lines from a permanent magnet and from a solenoid, near the source.

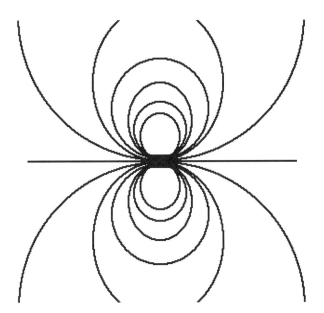

FIGURE 3.27 Magnetic field lines from a dipole. The field from a permanent-magnet dipole is identical to the field from a solenoid, provided they have the same magnetic moment. The field strength from a dipole decreases with the distance to the third power (see Figure. 3.33).

This formula has the same form as (3.3). Because the behavior of the magnet and the coil are the same, and the torque equations are alike, then we say they are models of each other. The dipole moment of the current carrying coil is:

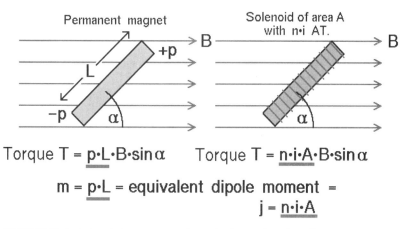

$$\text{Torque } T = \underline{p} \cdot L \cdot B \cdot \sin \alpha \qquad \text{Torque } T = \underline{n \cdot i \cdot A} \cdot B \cdot \sin \alpha$$

$$m = \underline{p} \cdot L = \text{equivalent dipole moment} = j = \underline{n \cdot i \cdot A}$$

FIGURE 3.28 Equivalence between a permanent magnet and a current-carrying coil.

$$j = iAn \; \text{Am}^2 \tag{3.15}$$

We will use the identity between the dipole moments when we seek to find the read voltage from a magnetized media coating. Finding the flux lines that link through the coil winding on the read head is exceedingly difficult if we try to follow the field lines from the magnetization in the coating. By substituting the permanent magnetization dipoles with the equivalent coil dipoles we find the problem readily solved by using the analogy to coupled coils. We shall return to this topic in Chapter 6.

Example 3.3: A uniformly wound coil has n = 100 turns, a length l of 10 cm, and a current I = 2A running through the wire. Find the flux density at the coil center.

Answer: $H = nI/l = 100 \times 2/0.1 = 2000$ A/m. $B = \mu_o H = 4\pi 10^{-7} \times 2000 = 8\pi \times 10^{-4}$ Wb/m^2.

NOTE: μ_o equals one in the cgs system; remember, however: Never use cgs units in MKS formulas: $H = ni/l = 100 \times 2/10 = 20$ Oe is wrong! The true number is $(4\pi/1000) \times 2000 = 8\pi = 25.13$ Oe, where we have used the conversion factor from (3.9) to convert A/m to Oe.

ELECTROMAGNETS

The flux density near a coil is increased when the turns are wound onto a magnetic core of, for instance, iron. We are actually generating poles at the end of the iron, with polarities as shown in Fig. 3.29. The coil-plus-core now forms what is called an electromagnet.

The field patterns in and around this electromagnet are a combination of the left and right side of Fig. 3.25, and we will discuss the resulting field pattern in Chapter 4. The contribution to the field from the magnetic core (poles) adds outside the coil, but subtracts inside where they run counter to the coil field lines. They actually tend to demagnetize the core magnetization.

Practical electromagnets applied in magnetic recording are shown in Figs. 3.30 and 3.31, which illustrate a conventional core-type head and a thin film head (TFH).

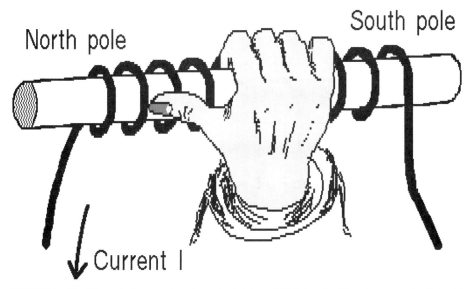

FIGURE 3.29 The right-hand rule shows where the north pole is located: Grab the solenoid (or electromagnet) with the right hand, and wrapping the fingers in the direction of the current. The thumb will then point toward the north pole.

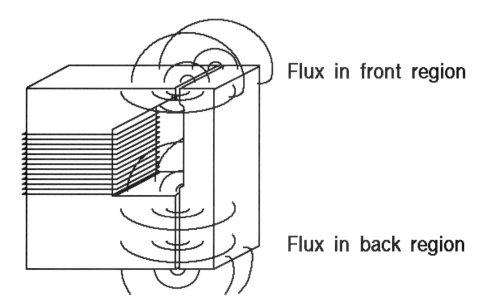

FIGURE 3.30 Magnetic flux in a conventional-core magnetic head.

wider coils at Top To dissipate
heat by lowering resistance

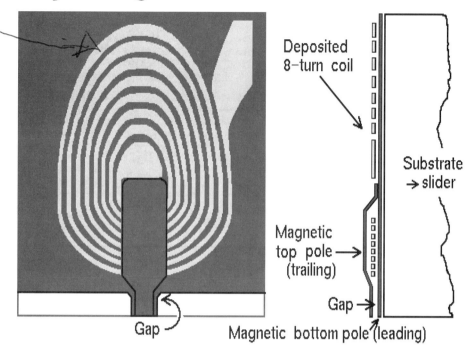

FIGURE 3.31 Construction of an eight-turn thin-film head.

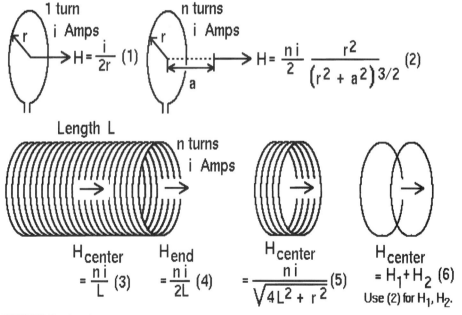

FIGURE 3.32 Fields from several coil arrangements.

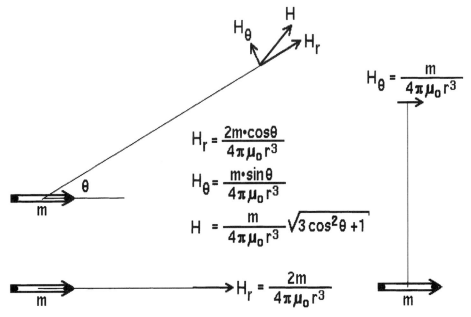

FIGURE 3.33 The near field from a dipole.

INDUCTION OF VOLTAGE: FARADAY

A current can generate a magnetic field that, in turn, can magnetize a piece of iron permanently. This permanent magnet will generate an electric voltage when it is dropped through a coil, bringing about a change in the number of flux lines. Figure 3.35 shows how Faraday carried out the experiment by changing the flux using a magnetized coil.

The magnitude of the generated voltage is

$$e = -n \frac{d\phi}{dt} \text{ volts} \tag{3.16}$$

where:

n = number of turns
ϕ = number of flux lines in Wb

The unit Wb (Weber) equals volts × seconds (verify: differentiate volts × seconds with respect to time, and voltage results).

We can appreciate this voltage generation by recalling Le Chatelier's law, which states that when any system in equilibrium is disturbed, it always reacts in such a manner as to oppose the forces that upset the balance. This situation arises when the field through the coil is changed, and the electrons will start moving to generate a current that, in turn, creates a field that opposes the change. The minus sign in the formula expresses the reaction.

Example 3.4: The flux level through the core of a composite read head is 0.01 nWb, at a data rate that corresponds to a frequency f = 1 MHz. The coil has 25 turns. What is the induced voltage (disregarding losses)?

FIGURE 3.34 British experimental genius Michael Faraday.

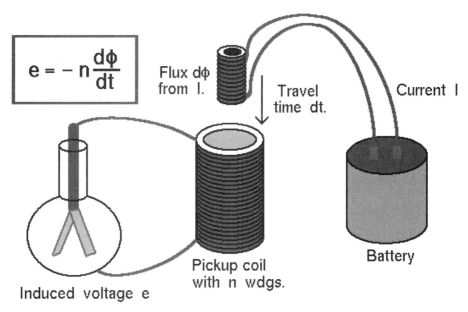

FIGURE 3.35 Faraday's discovery of the electromagnetic induction of voltage.

Answer: Assume a sinusoidal variation in the flux, so it can be expressed as $\phi = \phi_{max} \times \sin \omega t$, where $\omega = 2\pi f$. Then we find, after differentiation: $e = -n \times \phi_{max} \times \omega \times \cos \omega t$. The peak voltage is then $E = n \times \phi_{max} \times \omega = 25 \times 0.01 \times 10^{-9} \times 2 \times \pi \times 1 \times 10^6 = 1.57$ mV.

The discovery of electromagnetic induction was made by Michael Faraday (1791–1867) in 1831—eleven years after Oersted's discovery of electromagnetism. The two discoveries together laid part of the foundation for the Industrial Revolution of the 19th century: electric generators and motors. The circumstances and methods of their discoveries may seem trivial in this day and age, but both scientists were diligently pursuing a possible connection between magnetism and electricity, as testified by Oersted's lectures and Faraday's notebook entries.

Magnetism and electricity are also the principles behind magnetic recording and playback, as shown in Fig. 3.36. The field from the write coil is concentrated inside, and in front of, the gap in the write head. This is achieved by the use of a magnetic material, and the head is therefore an electromagnet. Similarly, the flux lines from the media are collected by a magnetic read head core and conducted through the read coil.

We have now covered the fundamentals we need to know regarding the relations between electricity and magnetics, and their role in the write/read processes. Next we turn our attention to magnetic materials as they are used in write and read heads and media coatings.

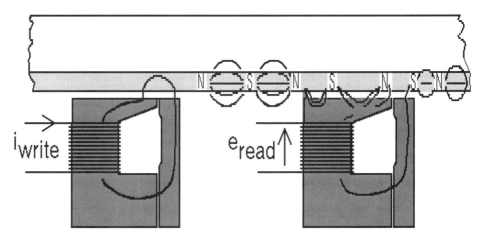

FIGURE 3.36 Principles in magnetic write/read processes (record/playback): The write current I_{write} produces a field that magnetizes the magnetic coating; field lines from transitions (N,S) induce a voltage e_{read} in the read head.

ORIGIN OF MAGNETISM

The knowledge of magnetism was at the time of Valdemar Poulsen's invention very shallow, and Poulsen's invention was therefore quite significant. He received a gold medal in recognition thereof at the World's Fair in Paris in 1900.

Ampere had earlier speculated that magnetism in lodestones and other magnetic rocks had its origin in "some" molecular circulating currents inside the material. Modern quantum physics explains magnetism as originating in electron-spins that produce interacting fields among themselves, giving rise to magnetization in certain materials.

Ampere was quite right in his hypothesis, as we see when we consider a simple model of an atom (see Fig. 3.37). Electrons are moving in circular orbits around the positively charged nucleus. Such a moving charge represents a current: $i = dq/dt$, and we know that currents generate a magnetic field. The magnetic moment generated by an orbiting electron is shown in Fig. 3.38. The electrons are also spinning about their axes, and with their charge distributed over their surfaces these again represent currents, giving rise to an electron-spin magnetic moment.

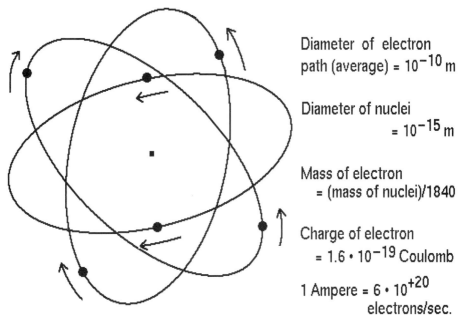

Diameter of electron
path (average) = 10^{-10} m

Diameter of nuclei
$= 10^{-15}$ m

Mass of electron
= (mass of nuclei)/1840

Charge of electron
= 1.6 • 10^{-19} Coulomb

1 Ampere = 6 • 10^{+20}
electrons/sec.

FIGURE 3.37   A simplified model of an atom with 6 orbiting electrons.

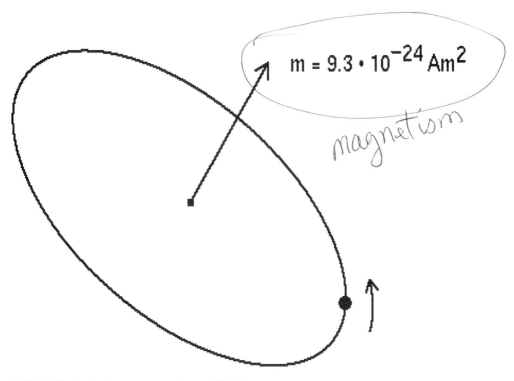

$m = 9.3 \cdot 10^{-24} \, Am^2$

magnetism

FIGURE 3.38   The dipole moment generated by an orbiting electron.

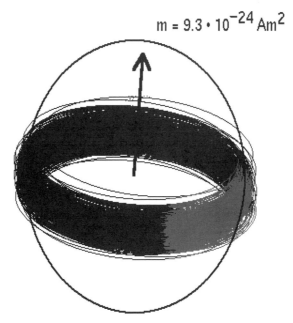

$$m = 9.3 \cdot 10^{-24}\, Am^2$$

FIGURE 3.39   The dipole moment generated by a spinning electron.

A magnetic field, therefore, always originates from currents. When the current runs in a loop we have a magnetic dipole (see Fig. 3.25), and it always has TWO poles: cutting the magnet in half results in two smaller magnets, each with half the original moment. Similarly, slicing a single turn loop will result in two single turn loops, each with half the original ampere-turns, and therefore also half the original moment—and each with a pair of poles.

DIAMAGNETISM

A circular conductor with a current i running in it will generate a magnetic vector with a dipole moment $j = iA$ Am^2. Each orbiting electron will therefore generate a magnetic moment, and this would be a plausible explanation for the origin of magnetization in all materials. But it so happens that in diamagnetic materials two electrons travel in each orbit, in opposite directions, so no net magnetic moment is produced, and the number of electrons is always even.

When a magnetic field is applied, and links through the electron's orbit, then one electron will resist this field by speeding up, and the other by slowing down. Now the balance is destroyed and the material is said to be diamagnetic, a property of ALL materials.

This magnetic property is very weak, and manifests itself by the fact that a diamagnetic material will be pushed away by a magnetic field. Some diamagnetic materials are copper, lead, and water.

PARAMAGNETISM

We now consider the other motion: the electron spin. We will find that the charge of q (= 1.6022 × 10^{-19} coulomb) is distributed only on the surface of the electron because like charges repel themselves. The charge distribution that moves in a circular pattern around the electron's equator does again represent a current, and is associated with a magnetic moment, also called the electron spin. Its magnitude equals one Bohr magneton, after the atomic physicist Niels Bohr.

$$m = 9.3 \times 10^{-24} \, Am^2 \tag{3.17}$$

The electrons' orbits within shells, as indicated in Fig. 3.37, are for most materials in perfect balance. Because of thermal vibrations, the axes of the spins are distributed over all possible directions, and the net magnetic moment is equal to zero.

The application of a magnetic field will unbalance the pairing of the spins, the material will become slightly magnetic, and it will be attracted into the field. Air and aluminum are paramagnetic materials.

FERRO- AND FERRIMAGNETIC MATERIALS

It now happens that at room temperature there are three materials where there are unpaired electron spins: four electrons in iron, three in cobalt, and two in nickel. These materials are the main ingredients in the magnetic materials used in modern recording technology.

Owing to the presence of such uncompensated spins, some of the electrons in one atom are located so closely to the nucleus of another atom that there is an exchange of electrons between the two atoms. The resulting forces are called *quantum mechanical forces of exchange*, and are electrostatic in origin. They are associated with the exchange energy and exchange field, the latter being either positive or negative. Ferromagnetism is displayed when the exchange field is positive and has a certain definite value.

These exchange forces cause a parallel alignment of neighboring spins, and very strong magnetization levels (10^7 Oe) are present inside these so-called ferromagnetic materials. The spins will tend to align themselves within small volumes, called domains, in such a way that the domain magnetizations are pointing in different directions. The material may, therefore, appear nonmagnetic until influenced by an external field. We will shortly proceed by investigating what happens when such a field is applied (and discuss no further quantum mechanics).

The spin alignment is perfect at the very lowest temperature, –273 degrees Celsius, or 0 degrees Kelvin (Fig. 3.40). At higher temperatures, thermal agitation causes excursions of the angle of mutual orientation (Fig. 3.41), and at high enough temperatures the exchange forces lose control. The spins are now in total disarray, pointing in all directions, and the material is no longer ferromagnetic. It has become diamagnetic.

The temperature where this disorder happens is called the Curie temperature T_c. The process is reversible: when the material cools, it again becomes ferromagnetic—but it has no memory of its prior magnetization. We will encounter this phenomenon in coatings made from chromium dioxide particles (T_c = 110 Celsius), and in certain high permeability ferrites in heads (T_c near room temperature).

For a perfect ferromagnetic material, one should therefore expect that all spins line up in one direction, resulting in a very strong magnet as shown in Fig. 3.42.

The energy stored in the field outside the magnet is quite large, and it can only be lowered if the magnetization splits up into domains. The formation of two domains, as shown at left in Fig. 3.43, reduces the outside field energy while adding magnetostatic energy between the two domains. The overall energy is nevertheless smaller. There is a large number of atomic moments, typically 10^{12} to 10^{15} in each domain. With a lattice spacing of 1–3 Å the smallest domains are $(10^{12})^{1/3} \times (1 \text{ to } 3) =$ 10,000 to 30,000 Å = 1 to 3 μm cubes. In thin-film disks, the domains may be as small as 300 Å across. This may with time (and heat or impact excitation) lead to further division into small domains as shown on the right in Fig. 3.43. The magnetization is still there, but the block appears nonmagnetic to the outside because there are no external field lines. This verifies both Swedenborg's and Ampere's hypothesis.

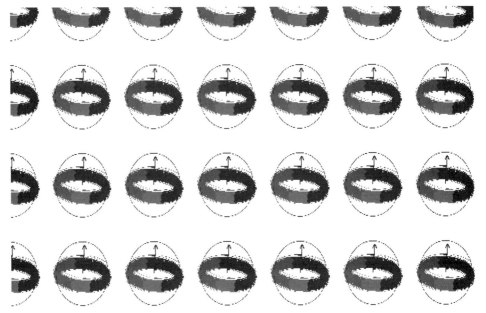

FIGURE 3.40 Alignment of spins is perfect at 0 K (–273 C).

FIGURE 3.41 Brownian movements and other disturbances cause imperfect alignment of spins, and a resulting reduction in magnetization level.

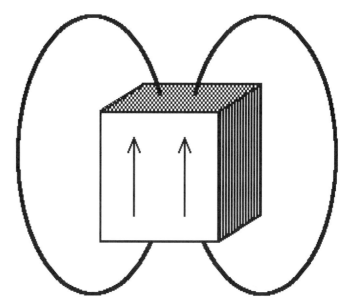

FIGURE 3.42 Perfect alignment of all spins results in a strong magnet.

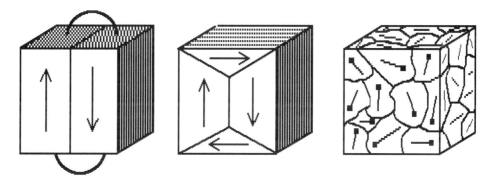

FIGURE 3.43 A lowering of the total energy occurs when the magnet from Figure 3.42 divides its magnetization into smaller domains.

The saturation magnetization M_s is proportional to the number n_B of Bohr magnetons m per molecule, multiplied by the number of molecules per volume:

$$M_s = m \times n_B \times (number\ molecules)$$

For iron $n_B = 2.2$, the density is 7.87 g/cm^3, and the atomic weight is 53.85 g/mole:

$$\begin{aligned}
No.\ molecules &= (\text{Avogadro's number} \times \text{density})/(\text{weight per molecule}) \\
&= 0.602 \times 10^{24} \times 7.87/55.85 \\
&= 0.0848 \times 10^{24} \text{ atoms/cm}^3 \\
&= 0.0848 \times 10^{30} \text{ atoms/m}^3.
\end{aligned}$$

The saturation magnetization of iron is $M_s = 9.3 \times 10^{-24} \times 2.2 \times 0.0848 \times 1030 = 1710$ kA/m ($=$ emu/cm^3).

Values for several recording materials' M_s are found in Chapter 11.

MAGNETIZATION CURVE AND THE HYSTERESIS LOOP

The field lines associated with a magnetic field H are called flux lines, given the symbol ϕ. A stronger field has more flux lines running through a given area than a weak field, and it is convenient to measure the intensity in terms of flux density B (or incorrectly: induction B). Its unit of measure is weber/m^2 or tesla (gauss is used in the cgs system—abbreviated G). The conversion formula is

$$1 \text{ Wb/m}^2 = 10^{+4} \text{ G} \tag{3.18}$$

From quantities in Wb/m^2 (or gauss) we can deduce that the number of flux lines for a given cross section is measured in webers (or maxwells in cgs; 1 Wb = 10^{-4} Mx).

When a magnetic material is placed in a magnetic field H then the number of flux lines inside the material is multiplied by its relative permeability μ_r, defined as.

$$\mu_r = \frac{\text{flux density in material}}{\text{flux density in air}} \tag{3.19}$$

If the field strength is strong enough to saturate the material and the field is turned off, then a certain amount of magnetism remains, called the retention or remanence B_r, in Wb/m^2. A field of the opposite direction is required for reduction of this remanence to zero, and its magnitude is named the coercive force or coercivity H_c, in A/m.

These characteristics are easier to remember if we point them out as we follow a complete magnetization cycle of a magnet.

We will consider a ring core sample as shown in Fig. 3.44. The magnetic field from the current i through the n turns in the coil generates a field $H_{core} = ni/l$, and we will use this field H_{core} as the abscissa in a graph of the flux density B. We will, in other words, plot the flux density B versus field H_{core} (B is measured with a small search coil with n turns placed on the ring core).

The ring core is initially without remanence (i.e., it is in a demagnetized state) and the flux density B will increase with the field H_{core} from the origin (0) (see Fig. 3.44). Its increase becomes more rapid (1–2) as H_{core} increases but for large values of H_{core} it levels off (2–3); we say the material saturates (which is what happens when the record level is too high during an audio recording). This curve is called the initial magnetization curve. The permeability μ is defined as the ratio between the flux density B inside the material and the flux density B we would have when the material is nonmagnetic (e.g., wood, air, etc.).

μ is a function of both material and field strength H_{core}. The starting value at $H_{core} = 0$ is called the initial permeability μ_{init}, and equals the slope of the initial magnetization curve at (0,0). The permeability attains its maximum value in the range 1–2, then it decreases and eventually reaches the value of one for very high fields H.

If we had returned the current to zero before we reached saturation, then the B-value would not go back to the origin, but to a point B_r between 0 and B_{rsat}. Its value would depend on the maximum field value that the material was brought to, although the relationship is nonlinear. It illustrates the principle used by Valdemar Poulsen in his first recordings; the removal of the field was accomplished by moving the steel wire out of the recording field, and the field magnitude was controlled by a carbon microphone that in accordance with the sound pressure field adjusted the current from a battery through the record head winding.

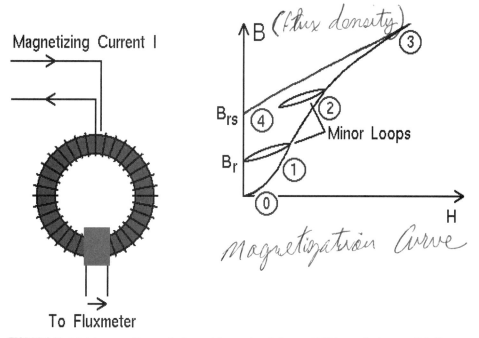

FIGURE 3.44 A test ring core with magnetization- and flux-sensing windings, and initial magnetization curve (1, 2, 3), remanence after saturation (4) and two minor hysteresis loops.

Removal of these remanent magnetizations can be done by reversing the current, and exposing the material to a negative field (Fig. 3.45). Removal of B_{rsat} would require the largest field, which is a property named the coercivity H_c. It depends on the material composition and shape of the sample.

From this point the material can be saturated into the negative magnitude by further increasing the negative field. Again, removing the field brings the magnetization level to $-B_{rsat}$. A positive field of magnitude H_c is required to bring the magnetization back to zero, and further field increases will bring us back to $+B_{rsat}$. The curve thus traced is called the hysteresis loop.

The full theory for the initial magnetization curve, the hysteresis loop, and the minor hysteresis loops is very complicated. Yet, the phenomena that take place in ferromagnetic materials can be conveniently discussed on the basis of the domain hypothesis.

DOMAIN THEORY

In 1907 Pierre Weiss put forward the hypothesis that ferromagnetism might be the result of an unusually strong interaction between the individual atomic magnets, which in some way made them all point in the same direction. Such regions with aligned magnetic moments are called domains, and they may have a variety of shapes and sizes ranging from one micrometer to several centimeters, depending on the size, shape, material and the magnetic history of the sample. In most materials the preferred direction is parallel to one or the other of the major crystallographic axes.

Each domain has a very strong magnetization called spontaneous magnetization M_s (A/m). It is the sum of electron-spins per volume. The direction of these domain magnetizations vary from domain to domain, and the net overall magnetization is zero in a virgin ferromagnetic material. There are various techniques for visible observation of these domains, one by applying a colloidal iron powder on the polished surface of a sample and obtain the so-called Bitter patterns.

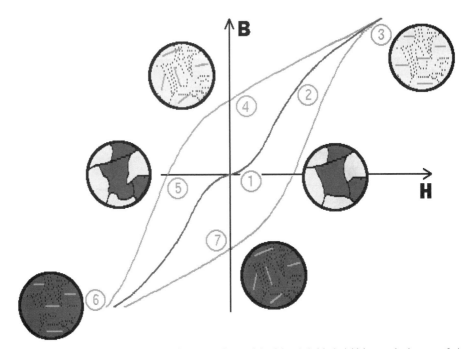

FIGURE 3.45   A complete hysteresis loop for a magnetic material: origin—1, 2, 3 is the initial magnetization curve; 3, 4, 5, 6, 7, 8, 3 is the hysteresis loop.

A series of domain patterns is shown in Fig. 3.46, and how they change under the influence of an external field. The patterns that one normally observes by the Bitter method are two-dimensional; the domains are, like crystal grains, three-dimensional.

The first column shows the simplest case, where the domain orientations are parallel, the center has four preferred directions, and the right column is a more typical random pattern.

A small external field H (corresponding to the range origin–1 in Fig. 3.45) will cause a shift in the areas of preferred and nonpreferred directions of M_s; the areas where M_s follows H will expand at the cost of those with opposing M_s. This causes a shift in the boundary between domains.

These boundaries are called walls, and the electron spin orientation changes gradually across the width of a wall. Wall thickness is in the order of 100–1000 Å (1 Å = 1 Angstrom = 10^{-10} meters), and their location and pattern in a material may follow crystal grain boundaries as well as other preferred patterns dictated by the material's metallurgy, imperfections, and impurities.

The wall movements in the low field region are reversible and will move back if the field is removed. But, as the field is increased (range 1–2, Fig. 3.45) some wall movements are impeded by impurities and imperfections in the crystal structure; the boundary will move discontinuously to a new equilibrium position, and the magnetization process is irreversible. These movements often occur in jumps.

If now the field is reduced to zero, the flux density does not go to zero again, but will have a certain remanence B_r (Fig. 3.44). If next the field strength is increased, then the flux density goes back along a lower curve; this pattern is called a minor hysteresis loop. The number of minor loops within the one and only major hysteresis loop is infinite.

The irreversible wall movements were discovered early in the history of magnetism because in signal transformers and certain magnetic microphones they would generate noise spikes by induction when they suddenly shifted to a new position. They were discovered by Barkhausen and this type of noise is named after him.

Further increase in the field strength H will eventually reduce all opposing domains to zero, and when H approaches saturation values we find that the last increase in B is caused by rotation of the magnetizations M_s into the field direction.

This rotation is reversible, and when the field H is removed two things can happen: the magnetization will rotate back and the walls will recur and move back toward their original positions. But the original domain pattern may not show up again due to anisotropy (preferred orientation of magnetization in the material, caused by crystal structure, impurities, slip planes, and stresses).

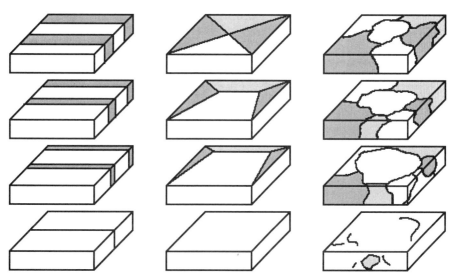

FIGURE 3.46   Domain patterns in three material samples. An applied field favors the clear domain volumes that expand at the cost of the other volumes. The field strength is zero in the top row, maximum in the bottom.

Cobalt

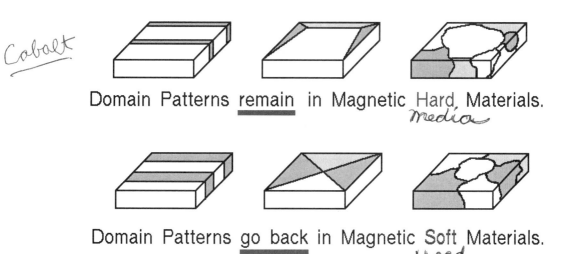

Domain Patterns remain in Magnetic Hard Materials.
media

Domain Patterns go back in Magnetic Soft Materials.
Head

FIGURE 3.47   Remanent magnetization in the three samples from Figure 4.46. When the applied field is turned off, the domain patterns will change little if no demagnetizing field is present (no air gaps, i.e., no free poles).

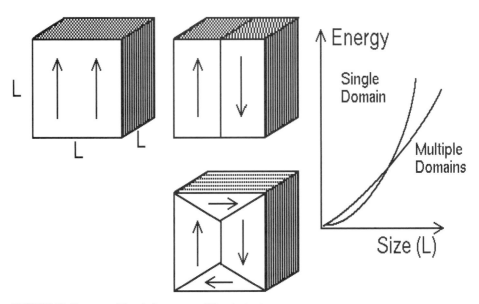

FIGURE 3.48 Energy stored in a single versus a multidomain structure.

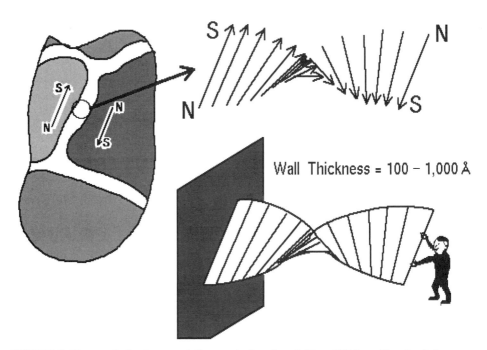

FIGURE 3.49 The magnetization changes gradually across the dimension called the wall thickness. The pattern is the same as in a twisted stake fence.

Figures 3.50 and 3.51 show examples of domain behaviors. Figure 3.50 is a collage of illustrations that show how different domain structures correspond to the steps in magnetization, corresponding to Fig. 3.45.

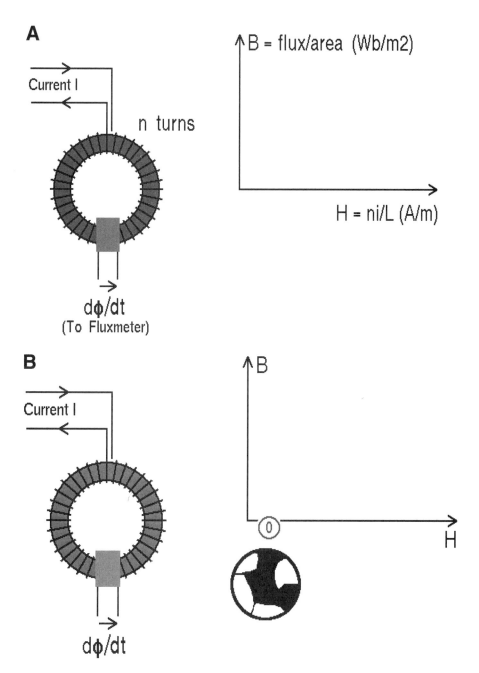

FIGURE 3.50 Domain structures associated with the various stages of magnetization of a ring core. A. The definition of setup parameters. B. The domain structure is originally neutral, i.e., equal dark and light domains are shown. C. Initial magnetization only shifts walls, and is reversible. D. Large values of H cause irreversible wall movements, leading to a remanence after H has been removed. E. An increase in H to value at (2) has greatly decreased the volume of opposing magnetization. F. At saturation (3), there is no further opposing magnetization, and the magnetization direction is further in the direction of the field. G. When H returns to zero, there may still be a large amount of remanent magnetization in direction of the prior field. H. The complete loop shows the changes in domain patterns.

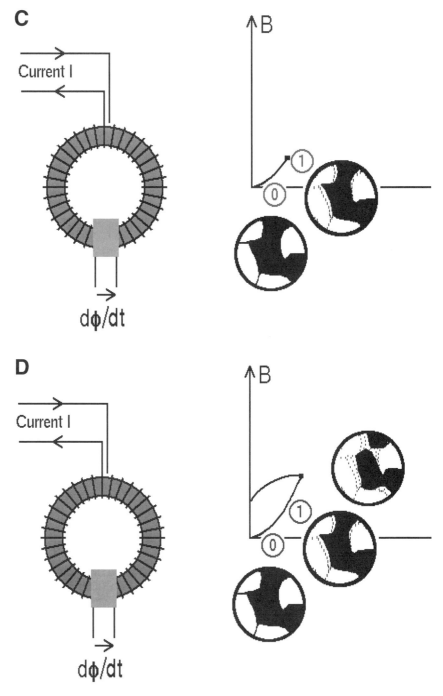

FIGURE 3.50 (*Continued*).

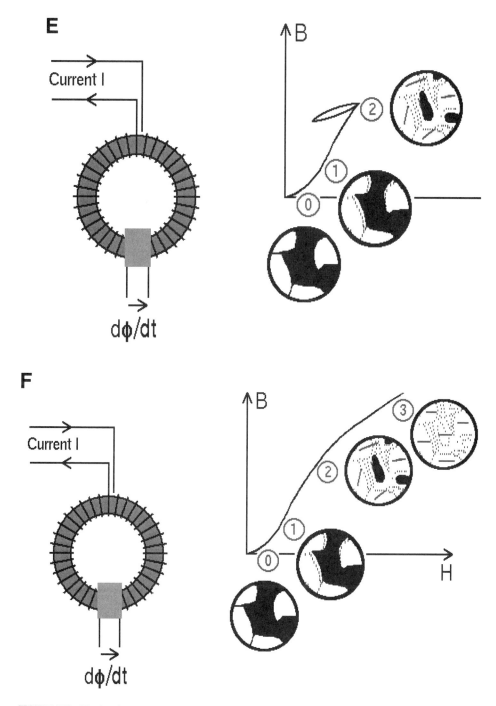

FIGURE 3.50 (*Continued*).

G

H

① is non-magnetic origin

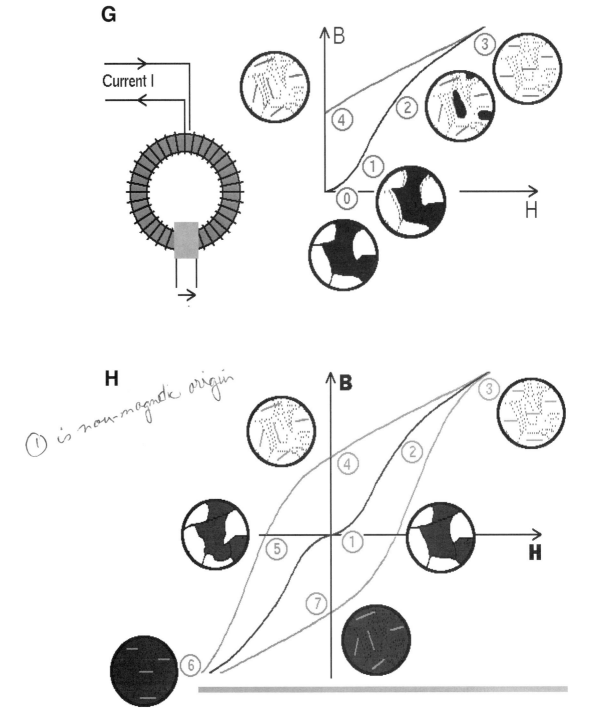

FIGURE 3.50 (*Continued*).

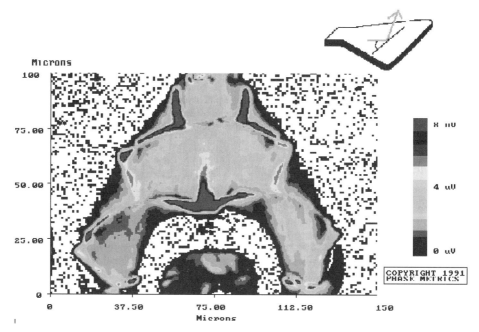

FIGURE 3.51 Observation of domains in TFH top pole structure *(courtesy of Phase Metrics).*

Figure 3.51 shows domain patterns in the top pole piece in a thin film head. They are observed by applying a beam of polarized light onto the film surface. This is typically a laser beam. The reflected light will undergo a slight phase change during the reflection of the magnetic surface by Kerr effect, the same mechanism that is applied in the magneto-optical readout of MO disks. By detection and proper processing of the reflected image, a picture of the domains results.

DOMAINS AND WALLS

It may have puzzled the reader to learn that the spin moments line up to form strong internal magnetization, and then to also split up in groups called domains. A simple experiment will convince the reader that this formation of domains results in a lower overall energy in the magnetization system: take two bar magnets and place them next to each other. If like poles are next to each other, then the bars will move away from each other. Turning one bar over will result in strong attraction; we even say that they serve as "keepers" for each other.

Figure 3-48 shows a similar situation: at top left, a single domain cube is shown, corresponding to the two bars parallel with poles in the same directions. Dividing itself up into two domains corresponds to the bars facing each other with opposite poles. If possible, the cube would further divide itself up into four domains, the two smaller ones called closure domains; that last configuration has the lowest energy of the three, the magnetostatic energy being virtually zero.

A magnet would keep on dividing itself into smaller and smaller domains, until the exchange energy starts rising again because few spins are parallel. Hence, there should be an optimum domain size for a given material, where the total energy is at a minimum.

The division into domains is not favored by the exchange energy, and this will also affect the dimension over which the magnetization direction changes, the so-called wall. Figure 3.49 illustrates how magnetization changes direction across the wall (180 degrees total). The exchange energy is inversely proportional to the wall thickness, and thus favors a thick wall. The magnetostatic energy will, however, be small for a thin wall, and an optimum wall thickness exists when the sum of two energies is minimum. This occurs for the following typical values:

Fe	300 Å (120 atoms)
Ni	720 Å (290 atoms)
Fe_2O_3	1000 Å (approximately)

An analog model to the magnetic pattern in the wall is shown in the stake fence, with two magnets as end pieces. Twisting the fence (exchange energy) is more difficult for a short fence, while the pull between the end magnets then is less.

SOFT MAGNETIC MATERIALS

Materials used in magnetic heads are: Mu-metal, Permalloy, Alfesil, ferrites, hot-pressed ferrites, and deposited thin films of nickel/iron compositions. These are what we call "soft" magnetic materials, and they are easily remagnetized with low magnetic fields. Hysteresis loops for some commonly used head materials are shown in Fig. 3.52.

For small values of H_{core}, near the origin, one can use the formula:

$$B = \mu H_{core} \tag{3.20}$$

$$B = \mu_o \mu_r H_{core}$$

where:
$\mu_o = 4\pi 10^{-7}$ Hy/m
$\mu_r = \mu_{init}$

The relative permeability μ_r is a material property. μ_o equals the absolute permeability of air, and $\mu_r = 1$ for air. Magnetic heads are made from materials that have values of μ_r ranging from several

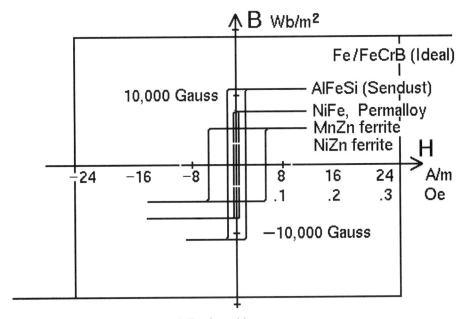

FIGURE 3.52 *BH-loops for common magnetically soft materials.*

hundreds to several thousands. μ_r depends on the flux level in the material, as shown in Fig. 3.53 and 3.54; the field levels are very low in most heads, and the formula listed above can be applied to computations of head performance (see Chapters 7 and 10).

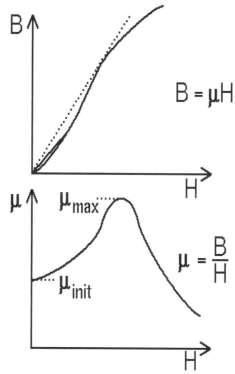

FIGURE 3.53 Permeability is nonlinear.

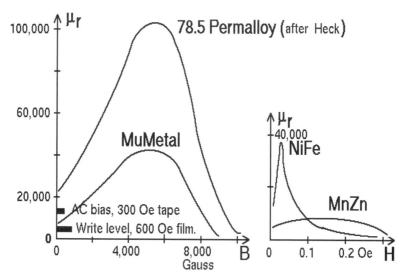

FIGURE 3.54 Measured, nonlinear permeabilities in Permalloy, mu-metal, and ferrites.

NOTE: The flux density levels in short-gap heads for writing on high-coercivity media may approach saturation levels. Then formula (3.20) is no longer valid. See pole-tip saturation and record head distortion.

The cgs-system of units has one convenience to offer: $\mu_o = 1$. That simplifies the formula above to $B = \mu_r H$, or $B = H$ when in air. This means that a given number X Oersteds corresponds to the same number X Gauss. The value of gap flux density is therefore the same for the field strength.

Example 3.5:   A toroidal core (see Fig. 3.44, left) has a mean radius of $R = 10$ cm, and a cross-sectional area of $S = 2$ cm^2. The core is wound with $n = 200$ turns, and carries a current of 1 A. The magnetic flux ϕ is measured to 100 μWb. What is the relative permeability of the core material?

Answer: $H = ni/l = 200 \times 1/2\pi \times 0.1 = 318$ A/m; $B = \phi/area = 100 \times 10^{-6}/2 \times 10^{-4} = 0.5$ Wb/m^2. Also $B = \mu_o \mu_r H = 4\pi 10^{-7} \times \mu_r \times 318 = 0.5$; $\mu_r = 0.5/0.0004 = 1250$.

NOTE: There are materials with a relative permeability of 1,000,000. But in usual engineering applications, the material is handled in different ways, and unavoidable coldworking reduces the permeability. Ni-Fe alloys (Permalloy, Mu-metal, etc.) in finished heads seldom have relative permeabilities over 2000–3000 (see Chapter 7).

HARD MAGNETIC MATERIALS

The other group of materials, the magnetically "hard" ones, are found in the powders used for tape and disk coatings: gamma ferric oxide, cobalt-iron oxide, chromium dioxide, iron ferrites, and barium ferrites. Also used are deposited films composed of nickel, cobalt, phosphorous, etc.

A formula for the entire BH-loop cannot be found since the B values are many, for any single value of H_{core} (with the exception of B_{rs}, the remanence after saturation). The value of B depends on the history of the prior magnetization process.

The BH-loop for air is a straight line with slope μ_o. The higher B-values for a magnetic material show that they have a susceptibility χ to become magnetized:

$$M = \chi \, H_{core} \ \ (\text{A/m}) \tag{3.21}$$

The total flux density in the ring core in Fig. 3.44 is therefore:

$$B = \mu_o H + \mu_o \times \chi \times H_{core}$$
$$= \mu_o(1+\chi)H_{core}$$
$$= \mu_o \mu_r H_{core}, \text{ with } \mu_r = \chi + 1.$$

We can also write:

$$B = \mu_o(H_{core} + M) \tag{3.22}$$

where:
B = magnetic flux density in Wb/m^2
H_{core} = magnetic field in A/m
M = magnetization in A/m

$\mu_o H_{core}$ is the contribution from the field H, while M is from the material's spontaneous magnetization. This is illustrated in Fig. 3.55.

A true magnetization loop for the material alone is obtained by subtracting $B = \mu_o H$ from the BH-loop. The resulting MH-loop is shown in Fig. 3.56.

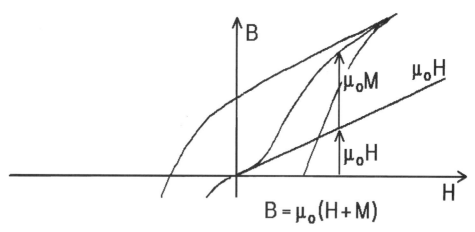

FIGURE 3.55 Measured values of B versus H equal the sum of the field ($\mu_o H$) and the spontaneous magnetization $\mu_o M$.

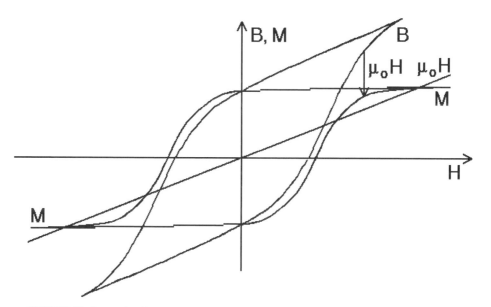

FIGURE 3.56 Construction of *MH*-loop from measured *BH*-loop.

Typical recording materials are shown in Fig. 3.57, showing their *MH*-loops. The values are typical and will be used in the rest of this book. (For exact engineering applications the designer should use the data sheet from the manufacturer of the material, or his or her own measured values. This applies to head materials as well.)

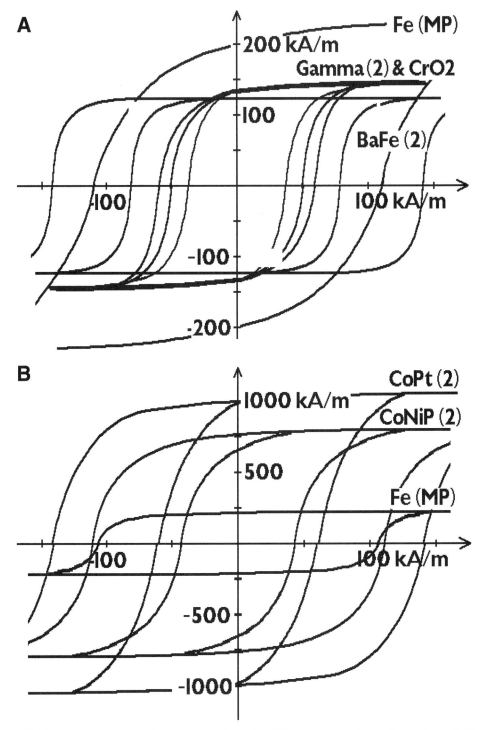

FIGURE 3.57 Magnetization curves for particulate recording media (**a.**) Magnetization curves for thin-film recording media (**b.**)

MAGNETIZATION IN SMALL PARTICLES

We will now apply our knowledge about magnetism to explain what happens when a magnetic tape is recorded. The most commonly used magnetic material in the tape coating is gamma ferric-oxide (γFe_2O_3), almost equivalent to iron rust. It is manufactured in a fine powder form and dispersed in a plastic binder, which is finally applied to a PET film (Polyethylene-terephthalate). The particles are cigar-shaped and are, during the coating process, aligned with a strong magnetic field in the lengthwise direction of the final tape. Another particle is pure iron (Fe); both are shown in Figs. 3.58 and 3.59.

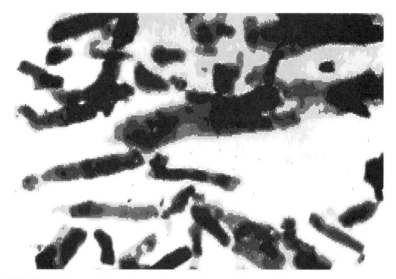

FIGURE 3.58   A transmission electron microscope picture of γFe_2O_3 particles.

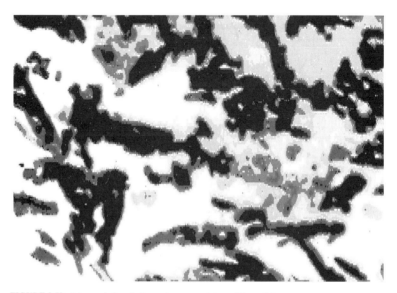

FIGURE 3.59   A transmission electron microscope picture of Fe particles.

They are so small that most of them have a length that is equal to, or a few times larger than, the wall thicknesses. Magnetic theory provides us with a guideline to their domain structure, shown in Fig. 3.48; small particles have lowest energy when they are single domains. Larger particles are multidomains.

A single-domain particle is magnetized to saturation in one or the opposite direction and remagnetization takes place through rotation of the magnetization vector M_s. Let us assume that each particle is a single domain, with shape anisotropy: its preferred direction of magnetization is along the major axis; for a spherical particle there is no preferred direction, and its resistance to changing magnetization directions zero (coercivity $H_c = 0$); only materials with crystalline anisotropy, such as cobalt-treated oxides, can be spherical and have a high coercivity (see Chapter 11).

STONER-WOHLFARTH SINGLE-DOMAIN MODEL

A single domain particle model was proposed and discussed by Stoner and Wohlfarth in 1955. An ellipsoidal particle is assumed, and the shape anisotropy is the determining force in particle's magnetization, overriding exchange, crystalline, and magnetostatic anisotropies. The magnetization will lie along the major axis.

An external field will turn the magnetization away from the major axis. An example is shown in Fig. 3.60, where a particle is placed at a 45-degree angle with respect to the x-axis, while a field H is applied in the negative x-axis direction. The magnetization will begin to turn into the direction of the field, but stop at a certain equilibrium position where the sum energy of the field torque and anisotropy energies is minimum (2). The magnetization will turn toward the field direction until a critical value h_{ci} is reached where it jumps to a new direction, now close to the direction of H ($3 \rightarrow 4$).

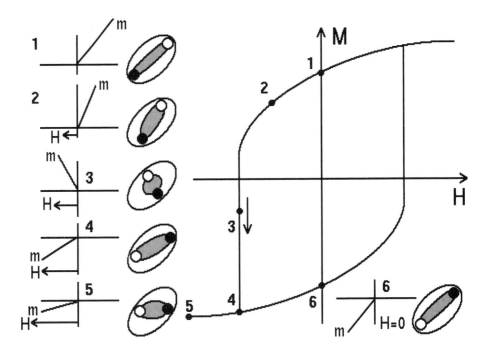

FIGURE 3.60   The direction of particles (and the mechanical model) 45° to the field H.

When H is made very large, the magnetization M lies almost in the H direction (5). If the field is reduced to zero, then the direction changes to that of lowest anisotropy energy, which now is opposite to the original direction (6). We have remagnetized the particle into the direction of the field.

Just to the right of each vector diagram are shown eggshell-like containers with two small balls inside. Imagine the two balls are kept apart by a compression spring, and at rest they will lie on a line equal to the major axis. An external force is now applied so the balls will try to change positions. Again an equilibrium is reached (2), and only a certain force can overcome the spring and flip the balls' positions (3 → 4).

This mechanical model serves well as a tool to remember how the magnetization in a single domain particle changes under the influence of an external field.

If we make a plot of the x-component of the magnetization vector versus H we obtain the MH-loop for a single-domain particle with shape anisotropy. It is shown to the right in Fig. 3.60. Similarly, if the x-components of the displacements and forces for the mechanical model were plotted, an identical loop would be traced. This justifies the analogy.

Another situation exists when the particle's major axis follows the x-direction. This situation is encouraged when the media coating is made: a magnetic field is applied to the still-wet coating, and it will turn the particles like compass needles into the field direction. This process is named orientation.

Figure 3.61 shows how a particle with magnetization in the positive x-direction is affected by a negative H field. The magnetization vector will be "stressed" by the field, and eventually will flip when the field strength exceeds the particle coercivity h_{ci}. This is also explained by the mechanical model: when the external force has compressed the (invisible) spring between the two balls to the point where their distance equals the small axis dimension in the eggshell, then they will flip positions.

The other extreme is a particle placed ninety degrees to the x-axis. It is left to the reader as an exercise to plot its MH-loop; its B_{rsat} value equals zero.

The MH-loop for the particle previously discussed is named a square loop. The MH-loop for media coatings will approach this shape at high levels of orientation.

If we prepare a coating mix and thoroughly stir it, we can assume that the particles will be pointing in all directions. The value of B_{rsat} will then equal one half of maximum magnetization M_{sat}. Orientation will increase the number of particles that are in the x-direction (normal direction of media travel), and hence increase B_{rsat}. Typical values in modern media range from 0.7 to 0.9 times M_{sat} .

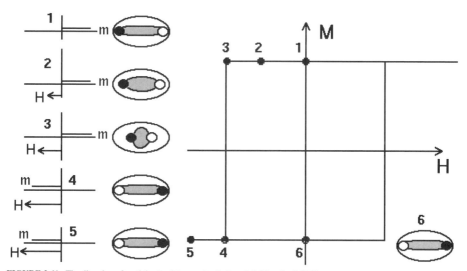

FIGURE 3.61 The direction of particles (and the mechanical model) 0° to the field H.

Now we can explain the initial magnetization curve and the hysteresis loop for a magnetic tape (see Fig. 3.45). The initial magnetization from the erased state (origin) is reversible because the increase in B over H ($B = \mu H$) is caused by small reversible domain rotations.

At higher field levels (1–2) we have irreversible wall movements. The wall movements are completed in small "jumps" (each domain sort of falling into place). These "jumps" show up as a fuzzy line on an oscilloscope, or as crackling noise in a speaker/amplifier connected to an induction coil around a tape-wound toroid.

As we approach saturation (2–3) the additional increase in induction B takes place by reversible rotations, and at 3 all particle magnetizations point along the direction of H.

When H now is reduced to zero it follows that the magnetization rotates back to the state of 2, and the material is permanently magnetized. The rest of the hysteresis loop follows the path 4–5–6–7–8 as described earlier.

JACOBS-BEAN CHAIN-OF-SPHERES MODEL

The Stoner-Wohlfarth model explains very nicely the magnetization behavior of particulate media. So far in the text, however, we neglected one important fact—that of the individual fields among the particles themselves. These are known as interacting fields, and their magnitude varies from zero to h_{ci}. To this field is now added any external field, and the latter need not exceed h_{ci} to flip particle magnetizations.

It is always nice to be able to model various engineering problems, write algorithms, and let the computer calculate a solution. But in the case of interacting fields we must account for the MH-loop of each particle, and also the overall resulting interacting field from the particles poles. The resulting model is difficult if not impossible to program, as the computed field pattern of the interacting field shown in Fig. 3.13 illustrates. We will shortly introduce statistical concepts as a tool for solving the write process problem.

One difficulty exists with the single-domain particle model: consider Fig. 3.62, which illustrates the magnetization in a particulate medium. The bit length is 0.4 μm, which corresponds to a packing density of 25,000 bits/cm (63,500 bits/inch). This sort of recording is found in high-density digital recorders (instrumentation machines), and audio is approaching this density (20,000 Hz at $^{15}/_{16}$ IPS = 21,333 wavelength/inch = 42,667 bits/inch).

The middle of the illustration shows the outline of the particles, and their single domain magnetization pattern. Clearly, a couple of the larger particles cannot resolve the short bits.

A more appealing picture is obtained if we consider a multidomain model introduced by Jacobs and Bean, in the mid 1950s. They proposed that the particle magnetization is represented by small magnetization vectors, like compasses, and that each particle has a number thereof that equals their length divided by their width. The shape anisotropy is accounted for by the fact that the magnetizations will try to pattern themselves in a chainlike fashion, with north poles seeking south poles. They may lie in a straight pattern, or in a zig-zag pattern called fanning. And the particles' relationship with their neighbors are again through the interaction field.

This model allows for a much-improved resolution, as shown in Fig. 3.62. And coercivity measurements made on particle assemblies show close agreement with the theoretical data for the Jacobs-Bean model. A long particle has higher coercivity than a short particle.

This shape anisotropy is easily understood by the mechanical eggshell model: a long distance between the balls requires a higher force to revert them. Calculated values for an assembly of particles are shown in Fig. 3.63. Measured values fell on top of the Jacobs-Bean curve.

Directional variations of the coercivity can be expected in a media coating. Stoner-Wohlfarth's model would prescribe a high value for the coercivity measured along the oriented direction, approaching zero in the direction perpendicular thereto. Jacobs-Bean's model should give almost no variation, if any, because it magnetically is an assembly of individual compasses. Measurements (ref. E.D. Daniel, personal communication) show a 10 to 15 percent increase at angles between 30 to 60 degrees, then decreasing by 10 percent at a perpendicular field (Fig. 3.64). This again corresponds closest to the Jacobs-Bean model.

Recorded bit pattern:

Magnetization, single domain model:

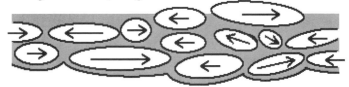

Magnetization, Chain-of-Spheres model:

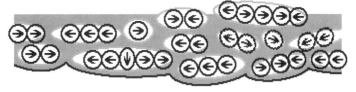

FIGURE 3.62 Media coatings with multiple domains in particles have higher resolution than coating with single-domain particles. Bit length = 0.4 μm (16 μin.).

FIGURE 3.63 Computed values of coercivities for an assembly of particles assuming either the Stoner-Wohlfarth or Jacobs-Bean models. Square points are measured.

A particle configuration that closely resembles Jacobs-Bean's model has recently been found in bacteria (see references; Blakemore, et al.). During a study of a live group of bacteria under a microscope they were found to gather along one side of the droplet containing them (see Fig. 3.65). At first the light from a nearby window was thought to attract them, but they behaved the same way at night.

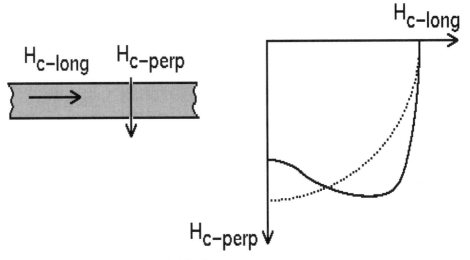

FIGURE 3.64 Variation of coercivity with field direction.

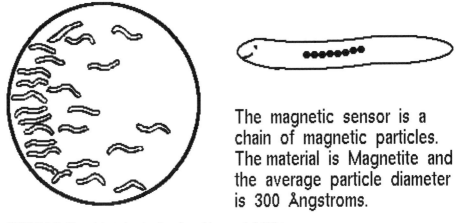

The magnetic sensor is a chain of magnetic particles. The material is Magnetite and the average particle diameter is 300 Ångstroms.

FIGURE 3.65 Magnetic bacteria swim along the earth's magnetic field lines.

The investigator then placed a magnet near the droplet, and now the bacteria swam toward or away from the magnet, depending on its direction. After exposing the particles to a slowly decaying ac field, half swam in one direction, the other half the opposite way! A closer examination in a transmission electron microscope revealed a chain of 20 or so beads, cubic or octahedral in form, each having a diameter of approximately 500 angstrom. And the material was magnetite, very much like the material in lodestones.

This chain of particles serves to guide the bacteria along the earth's magnetic field lines. For the location where these bacteria were found the earth field has an inclination of 60 degrees, and it would therefore guide the bacteria down into the ground. They were found about two feet below the surface, where the temperature is nice and warm during the winter above.

The same mechanism guides homing pigeons and is possibly involved in other animals' navigation. This is currently under investigation. It brings a thought to mind: if your neighbor cultivates homing pigeons, and they are a nuisance to you, just think what you can do with a degausser during the dark of the night?

SWITCHING FIELD DISTRIBUTION

When a group of particles is placed in a magnetic field all opposing domain magnetizations will be acted on by the field and brought into different positions. First of all, they will not rotate in unison because they will exhibit various degrees of resistance to domain rotation (anisotropy). Secondly, they are also influenced by the magnetic fields from their neighbors, called interaction fields. If this was not the case, then they would all, more or less, rotate together, and the tape would become fully magnetized in one direction or the other. No intermediate levels would be available for dynamic range, and the tape would be good for digital pulse recording only.

It is exactly the particle interactions and their different coercivities that make these particle materials useful as a linear recording medium, using ac bias (see Chapter 20). Their influence shows up strongly in the BH-loop for a coating.

We need to introduce yet another property of small magnetic particles, which is a size effect: their coercivity dependence on volume. We have seen that the coercivity is proportional to their length/width ratio. Its dependence on volume is shown in Fig. 3.66. This curve is general for all particles we will encounter. There is a maximum coercivity, which occurs at a diameter near the same values as earlier listed for the wall thicknesses.

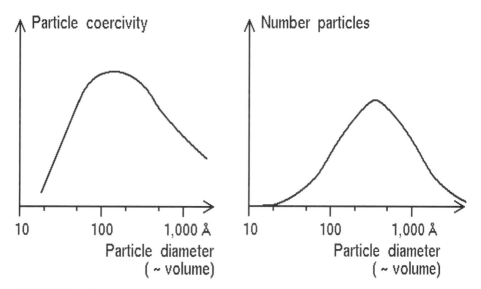

FIGURE 3.66 Particle coercivity h_{ci} as a function of particle size (diameter), and a distribution histogram of particle sizes in a batch of oxide powder.

Below that critical size the coercivity falls off rapidly, and eventually becomes zero. And in this range—from maximum to zero coercivity—a disturbing behavior is found. The particles are unstable, or superparamagnetic. They each contain an inadequate number of atoms to support the exchange force phenomenon, which therefore at times loses its grip on the spin alignments. When this happens, the assembly of particles reduces magnetization.

Some time later the exchange forces gain, and the particle is again magnetic. Needless to say, this behavior is enhanced by thermal fluctuation. It is possible to calculate a time constant for the changing of the magnetism, and it is dependent on not only the particle volume, but also its temperature. It has a range from a few seconds to a few hundred years for a 30 percent change in particle volume only. This naturally plays a role in archival storage of tapes, and also in print-through signal in audio recordings (see Chapters 12 and 14).

We will here restrict the particle size to belong to the right side of the maximum, and will now also find that particles in a batch of material have different sizes. When a histogram thereof is plotted, it is not surprising to find a Gaussian distribution curve that occasionally shows a log-normal tendency, a characteristic of some powders.

When we superimpose the coercivity-versus-size curve on the histogram, we obtain another Gaussian distribution (see Fig. 3.67, top). Assuming an assembly that to the outside is nonmagnetic, then half the particles are negatively magnetized, the other half positively.

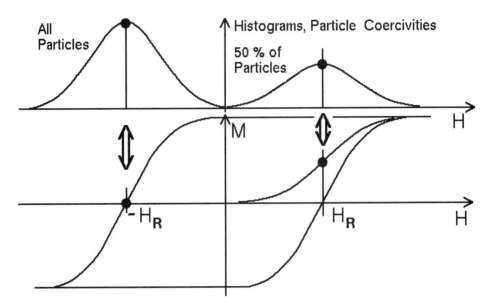

FIGURE 3.67 Distribution of h_{ci}, a bell-shaped curve. The largest value (left) assumes saturation magnetization, and integration results in the side of the ideal magnetization curve. The half-value distribution to the right represents fifty percent of the particles being magnetized positive. An equal amount of particles is magnetized negative and the net magnetization is zero. Integration of the bell-shaped curve results in the ideal magnetization curve.

Now let us repeat the experiment shown in Fig. 3.44 with a coil form made from wound magnetic tape. Assume that it has been completely degaussed, so half the particles have positive remanence, the other half negative. As the positive going field increases in strength, it will switch more and more of the negatively magnetized particles. This will lead to a rise in positive magnetization, eventually reaching M_{sat}. This level will ideally remain when the field is reduced to zero.

Next, apply an increasing negative field, and all particles will eventually be switched, resulting in $-M_{sat}$. Reduce the field to zero, and $-M_{sat}$ remains. Another positive half cycle with a magnetic field will flip all particles to $+M_{sat}$.

We have in essence traced the ideal MH-loop for a particulate medium, assuming perfect orientation of the particles and no interaction. It is a simple matter to arrive at the BH-loop by adding the values $\mu_o H$ to the M values.

The effect of interaction and imperfect orientation is evident in the difference between the MH-loops in Figs. 3.70 and 3.56. The latter has a remanence value that is less than the saturation value.

When we reduce a measured BH-loop to the MH-loop and finally differentiate it, we obtain a double-peaked curve that represents a mix of the particle size effect and the interaction fields. We can add

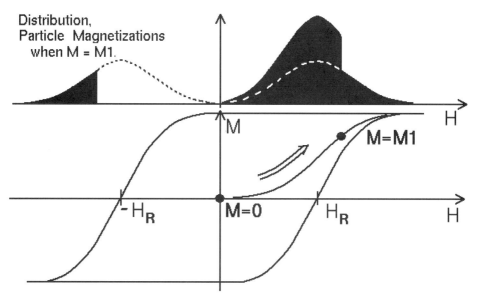

FIGURE 3.68 The change in magnetization distribution during magnetization from 0 to M_1.

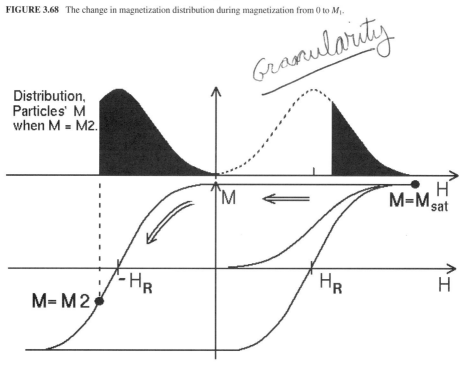

FIGURE 3.69 The change in magnetization distribution during magnetization from M_{sat} to M_2.

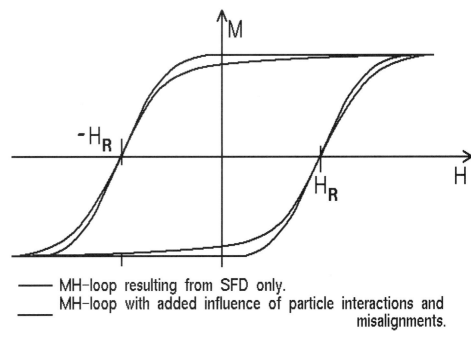

FIGURE 3.70 The addition of interaction fields and imperfect particle alignment to the ideal curve produces the real *MH*-loop. This is difficult to model.

a useful number to the magnetic values such as coercivity, remanence, and saturation magnetization: the switching field distribution. This is the width of either peak in the differentiated *MH*-loop, at the half saturation levels ($+M_s/2$ and $-M_s/2$), as shown in Fig. 3-67, divided by the coercivity H_R, and designated Δh_r. Its usefulness will become apparent in Chapter 5, on recording.

SWITCHING SPEEDS

We have presented various mechanisms for the change, or switching, of magnetization in materials. It seems natural to conclude this chapter with an answer to the question "How fast can we switch the magnetization in (for example) a coating with small particles, or in the core of a head?" Both will eventually limit the data rate we can use in writing and reading signals.

Thornley and Williams (see reference) conducted experiments to record pulses on several particulate tape samples, using a very-high-frequency technique. They found that the recording "efficiency" decreased when pulses used were shorter than tabulated below:

$$t = 4.1 \text{ ns for } CrO_2$$
$$t = 2.6 \text{ ns for } \gamma Fe_2O_3$$
$$t = 1.4 \text{ ns for } Co\text{-}\gamma Fe_2O_3$$

These numbers correspond to upper frequencies $f_u = t/2$:

$$f_u = 120 \text{ MHz for } CrO_2$$
$$f_u = 200 \text{ MHz for } \gamma Fe_2O_3$$
$$f_u = 350 \text{ MHz for } Co\text{-}\gamma Fe_2O_3$$

The change of magnetization direction in bulk materials (mumetal laminations in cores, solid ferrite cores) has a much lower cut-off frequency, which will be discussed in Chapter 7. In general it can be said that the higher the permeability the lower the cut-off frequency; values of μ_r less than 1000 are typical in heads reaching out into the MHz range.

MAGNETIZATION IN METALLIC FILM COATINGS

The recording surface in rigid disk drives is a deposited metal film. It is quite thin, only a few tenths of a μm (5–20 μin), as compared to a few μm (70–200 μin). It appears from what we now know that it is ideal, because each transition (often designated Δx) laid down by the write head represents a wall between two domains (bits).

Given a metallic coating for which the optimum wall thickness is, say, 400 angstroms, then the ultimate maximum packing density is $10^8/400 = 250,000$ transitions per cm (= 635,000 flux reversals per inch) assuming no space for the bit itself, so transitions are touching each other. That is okay, because the recorded information lies in the transitions.

THE ZIGZAG TRANSITION

In reality fewer transitions per cm are achieved. This again has its explanation in the magnetics striving for a state of minimum energy (see Fig. 3.71). The magnetics bordering the transition will, if given the slightest chance, form Néel spikes, which reduce the wall energy. These are similar to closure domains (Fig. 3.48). The Néel spikes are formed so north and south poles move closer to each other, thus reducing the energy stored. They may occur where there are crystalline faults or foreign particles in the metal film. In Néel walls the magnetic moments rotate within the plane of the material (they do not occur in bulk specimens).

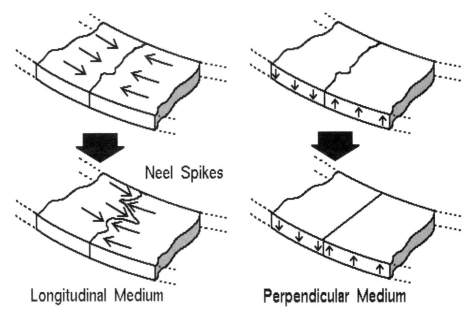

Neel Spikes

Longitudinal Medium   **Perpendicular Medium**

FIGURE 3.71 Transitions between bits in longitudinal and perpendicular media.

Reduce spikes w/higher coercivity

Let us consider a metal film that has perpendicular anisotropy. This encourages the magnetization to form perpendicularly to the surface. This transition is interesting for two reasons: for one, the transition length Δx tends to be short, because the bits' opposite magnetizations attract each other. Also, transitions with wiggles are discouraged: the energy stored in the transition (wall) is proportional to its length, which nature therefore will try to shorten. This results in a straight transition (see references in Chapter 19, Soohoo).

Studies of the zigzag transition continues, because the nature of the micromagnetics around the transitions remains very difficult to get a handle on. The forces at play are: magnetic interaction between the film crystallites (like between particles; see Fig. 3.13), magnetic exchange forces, and crystalline and magnetostrictive anisotropies.

Typical films are Co-Cr, often on a chromium underlayer, which gives high coercive forces by causing growth of cobalt in the hexagonal phase with a relatively narrow distribution of grain sizes (Daval and Randet). The mean grain size ranges from 200 to 500 angstroms; it appears that certain deposition methods onto an underlayer provide control over the granular structure so a film with properties resembling particulate media can be grown (Ref. Jack Judy.)

Zigzag transitions have been studied (Curland and Speliotis), and a large number of experiments has resulted in an expression for the length of the zigzag, or sawtooth, transition (Dressler and Judy):

$$a = \mathrm{K} \times \delta \times H_c{}^{\mathrm{B}} \tag{3.25}$$

where:
a = amplitude
K = constant for the particular film
δ = film thickness
$B = -1.5$
H_c = film coercivity

A short transition is obtained by using a thin film, which incidentally also provides the desired higher coercivity. Quite long transitions in low coercivity permalloy films were observed by Hsieh, Soohoo, and Kelly. They observed the transitions by Lorentz microscopy, an example of which is shown in Fig. 3.72.

Another study estimated the minimum transition length in a series of Co-Re films to be ≈ 1 μm (Chen), but more recent work (Yoshida, et al.) reports transitions as short as 0.2 μm (Yoshida, et al.) and 0.4 μm (Tong, et al.), the latter for a flying head system. When the write head is in contact with the film, a dramatic decrease in the transition length has been observed (Kullmann, Koester and Dorsch.)

The transitions will occasionally touch each other at high packing densities, and this bridging may develop into separate small islands (Arnoldussen and Tong).

A model comparable to Williams-Comstocks has recently been proposed, and the length of the transition found to be determined by the combined action of the head and dipolar field against the coercivity of the film (Muller and Murdock).

MICROMAGNETICS

Numerous factors influence the magnetic behavior of an assembly of grains: Interaction fields, exchange forces, crystalline and magnetostrictive anisotropies, methods of fabricating the film, annealing, imperfections, substrate conditions, temperature, etc.

An analysis may start by simulating the fields among grains and then adding other influences. This becomes very difficult due to, for instance, nonlinearities, and only recently has progress been made in the modeling of micromagnetics to explain the formation of islands and zigzag patterns in recorded transitions. Supercomputing power is always required.

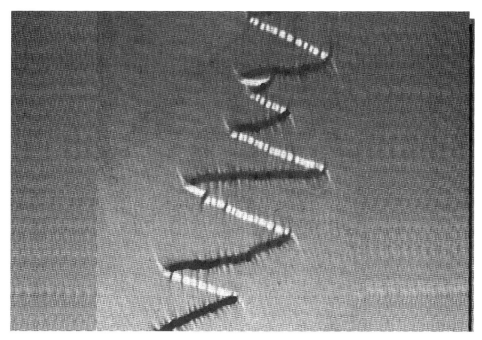

FIGURE 3.72 Zigzag transition (*Hsieh et al.*).

Interaction fields alone are dramatic: field strengths are proportional to the dipole moment, but varies in inverse proportion to the distance cubed, because the near field condition applies. A simple experiment will demonstrate that the transitions shown in Fig. 3.71 will occur due to magnetic field interactions only. A matrix of 6-×-10 dipoles was constructed, using small dime-sized ferrite disks with a center hole and diagonal magnetization (see Fig. 3.73).

The dipoles on the perimeter were first locked into positions to represent a longitudinal magnetization (Fig. 3.74, left). This is the head-on transition as seen from the track edge, and the dipoles in the interior adjusted themselves to the positions shown. They alter directions row by row going from top to bottom (track edge to edge). The zigzag amplitude is maximum, from the left to the right column of fixed dipoles. Because no other anisotropies are present, the coercive force of the assembly is low, and the zigzag amplitude is large (formula 3.25).

Notice the simultaneous formation of many small closure patterns; they remind us of the ripple and vortex patterns described by Chen. Might fractal patterns have a place in micromagnetics?

The edge dipoles were next arranged to represent a perpendicular transition or side view (Fig. 3-74, right). It was then very easy to arrange the dipoles in the pattern shown, which remained stable. This simulates the short transition in the perpendicular magnetization mode.

NOISE FROM ZIGZAG TRANSITIONS

The irregularities found in zigzag transitions are an obvious source for noise. Attempts at correcting these matters began in the mid-1980s (Arnoldussen and Tong). There is an increase in noise at the higher packing densities, probably due to the formation of islands and not the zigzag jitter (Madrid and Wood). Measurements of a number of transitions and determination of the zigzag distribution can be used to determine the source of noise (Tang and Osse). Correlation between domain observations and noise has also been investigated (Aoi, Tsuchiya, and Shiroishi).

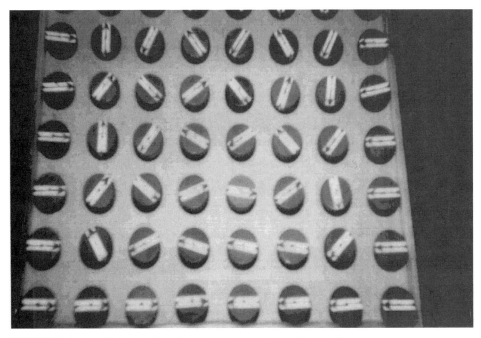

FIGURE 3.73 A board with dime-sized, round ferrite magnets, simulating particle interaction.

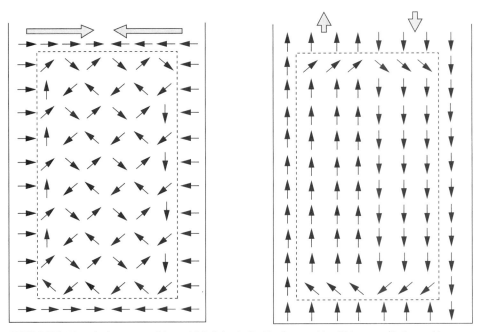

FIGURE 3.74 Magnetization patterns of the model. Left: longitudinal head-on transition. Right: perpendicular transition.

REFERENCES

Blakemore, R.P., and Frankel, R.B. Dec. 1981. Magnetic Navigation in Bacteria. *Scientific American*. 245 (6): 58–65.

Dunlop, D.J. July 1977. Rocks as High-Fidelity Tape Recorders. *IEEE Trans. Magn.* MAG-13 (5): 1267–1272.

Ewing, J.A. 1922. New Model of Ferromagnetic Induction. *Phil. Mag.*(6) Vol. 43, pp. 493–503.

Franksen, O.I. 1981. *H. C. Oersted—a man of the two cultures*, Bang & Olufsen, Denmark, 49 pages.

Friedlander, G.D. Aug. 1975. Ampere: Father of Electrodynamics. *IEEE Spectrum*, pp.75–78.

Gilbert, William. 1600. *De Magnete—On the Lodestone and Magnetic Bodies and on the Great Magnet the Earth*, England. Reprinted by Dover Publications (New York), 1957, and by Encyclopedia Britannica, "Great Books of the Western World", Vol. 28, 1952, pp. 1–121.

Jacobs, I.S., and Bean, C.P. Nov. 1955. An Approach to Elongated Fine-Particle Magnets. *The Phys. Rev.*, 100 (4): 1060–67.

Mapps, D.J. March 1978. Magnetic Domain Demonstrations Using the Kerr Magneto-optic Effect. *Contemp. Phys.* 19 (3): 269–281.

Soohoo, R.F. 1965, *Magnetic Thin Films*, Harper & Row, 316 pages.

Stoner, E.C., and Wohlfarth, E.P. May 1948. A Mechanism of Magnetic Hysteresis in Heterogenous Alloys. *Phil. Trans. Royal Soc.* Vol. 240, Ser. A, pp. 559–642.

Weisburd, S. Nov. 1985. The Earth's Magnetic Hiccup. *Science News*, Vol. 128, pp. 218–220.

THE ZIGZAG TRANSITION

Aoi, H., R. Tsuchiya, Y. Shiroishi, and H. Matsuyama. Nov. 1988. A Study on the Thin-Film Media Noise by Magnetic Domain Observation. *IEEE Trans. Magn.* MAG-24 (6) 2715–17.

Arnoldussen, T.C. and H.C. Tong. Sept. 1986. Zigzag Transition Profiles, Noise, and Correlation Statistics in Highly Oriented Longitudinal Film Media. *IEEE Trans. Magn.* MAG (5): 889–91.

Brown, W.F. Jan. 1984. Tutorial Paper on Dimensions and Units. *IEEE Trans. Magn*, MAG-20 (1): 112–118.

Chen, Tu. March 1981. The Micromagnetic Properties of High-Coercivity Metallic Films and Their Effect on the Limit of Packing Density in Digital Recording. *IEEE Trans. Magn.* MAG-17 (2): 1181–1191.

Curland, N., and D.E. Speliotis. March 1970. Transition region in recorded magnetization pattern. *Jour. Appl. Phys.*, Vol. 41, pp. 1099–1011.

Daval, J., and D. Randet. Dec. 1970. Electron Microscopy on High-Coercive-Force Co-Cr Composite Films. *IEEE Trans. Magn.*, MAG-6 (4): 768–773.

Dressler, D.D. and J.H. Judy. Sept. 1974. A Study of Digitally Recorded Transitions in Thin Magnetic Films. *IEEE Trans. Magn.*, MAG-10 (3): 674–77.

Giacoletto, L.J. Dec. 1974. Standardized Use of SI Magnetic Units. *IEEE Trans. Magn.* MAG-10 (4): 1134–1136.

Hsieh, E.J., R.F. Soohoo, and M.F. Kelly. June 1974. A Lorentz Microscopic Study of Head-On Domain Walls. *IEEE Trans. Magn.* MAG-10 (2): 304–308.

Kullmann, U., E. Koester, and C. Dorsch. March 1984. Amorphous CoSm Thin Films: A New Material for High Density Longitudinal Recording. *IEEE Trans. Magn.* MAG-20 (2): 420–24.

Madrid, M. and R. Wood. Sept. 1986. Transition Noise in Thin-Film Media", *IEEE Trans. Magn.* MAG-22 (5): 892–94.

Muller, M.W. and E.S. Murdock. Sept. 1987. Williams-Comstock Type Model for Sawtooth Transitions in Thin Film Media. *IEEE Trans. Magn.* MAG-23 (5): 2368–70.

Okuwaki, T., K. Yoshida, N. Osakabe, H. Tanabe, Y., Horiuchi, T., K. Shinagawa, A. Tonomura, and H. Fujiwara. Sept. 1983. Observation of Recorded Magnetization Patterns by Electron Holography. *IEEE Trans. Magn.* MAG-19 (5): 1600–04.

Soohoo, R.F. June 1984. Micromagnetics of Domain Walls in Vertical Recording. *Jour. Appl. Phys.* 55 (6): 2211–13.

Tang, Y.S. and L. Osse. Sept. 1987. Zig-Zag Domains and Metal Film Disk Noise. *IEEE Trans. Magn.* MAG-23 (5): 2371–73.

Thornley, F.R.M., and Williams, J.A. Nov. 1974. Switching Speeds in Magnetic Tapes. *IBM Jour. Res. Devel.* pp. 576–78.

Tong, H.C., R. Ferrier, P. Chang, J. Tzeng, and K.L. Parker. Sept. 1984. The Micromagnetics of Thin-Film Disk Recording Tracks. *IEEE Trans. Magn.* MAG-20 (5): 1831–33.

Weismehl, K.R., J.A. Brug, and E. S. Murdock. Nov. 1988. Transition Profile Measurement. *IEEE Trans. Magn.* MAG-24 (6): 2497–99.

Wolf, I., and T. Neuman. July 1989. Recording at High Volumetric Packing Densities. *SMPTE Journal* 98 (7): 515–519.

BIBLIOGRAPHY

Armstrong, R.L., and J.D. King. 1973. *The Electromagnetic Interaction*, Prentice-Hall, 493 pages.

Bickford, L.R. Jan. 1985. Magnetism During the IEEE's First One Hundred Years (1884–1984). *IEEE Trans. Magn.* MAG-21 (1): 2–9.

Bolton, B. 1980. *Electromagnetism and its Applications*, Van Nostrand Reinhold, 157 pages.

Bozorth, R.M. 1951. *Ferromagnetism*, D. van Nostrand, 968 pages. Reprinted 1993 by IEEE Press ISBN 0-7803-1032-2.

Chikazumi, S. 1966. *Physics of Magnetism*, John Wiley & Sons, 554 pages.

Craik, D.J., and R.S. Tebble. 1965. *Ferromagnetism and Ferromagnetic Domains*, North-Holland (Wiley), New York, 319 pages.

Cullity, B.D. 1972. *Introduction to Magnetic Materials*, Addison-Wesley, 666 pages.

Durand, E. 1968. *Magnetostatique,* Masson et Cie, Paris, 673 pages.

Fuller Brown, W. Jr. 1963. *Micromagnetics*, Interscience Publ. (Wiley), New York, 143 pages.

Hammond, P. 1978. *Electromagnetism for Engineers*, Pergamon Press, 290 pages.

Hayt Jr., W.H. 1974. *Engineering Electro-Magnetics*, McGraw- Hill, 496 pages.

Jiles, D. 1990. Introduction to Magnetism and Magnetic Materials. Chapman & Hall, 440 pages.

Kneller, E. 1962. *Ferromagnetismus*, Springer Verlag, 792 pages.

Koch, A.J., and J.J. Becker. Feb. 1968. Permanent Magnets and Fine Particles. *Jour. Appl. Phys.* 39 (2): 1261–1264.

Kraus, John D. 1984. *Electromagnetics*, McGraw-Hill, 775 pages.

Lee, E.W. 1970. *Magnetism, an Introductory Survey*, Dover, 281 pages.

Lerner, E.J. May 1984. Biological Effects of Electromagnetic Fields. *IEEE Spectrum*, pp. 57–69.

Lowther, D.A., and P.P. Silvester. 1986. *Computer-Aided Design in Magnetics*, Springer-Verlag, 324 pages.

Mattis, D.C. 1981. *The Theory of Magnetism I*, Statics and Dynamics, Springer-Verlag, New York (Berlin-Heidelberg), 300 pages.

O'Reilly, W. 1984. *Rock and Mineral Magnetism*, Chapman and Hall, New York (Blackie, London), 220 pages.

Parton, J.E., S.J.T. Owen, and M.S. Raven. 1986. *Applied Electromagnetics*, Springer-Verlag, 288 pages. Plonus, M.A. 1978. *Applied Electro-Magnetics*, McGraw-Hill, 615 pages.

Watson, J.K. 1980. *Applications of Magnetism*, John Wiley & Sons, 468 pages.

Zijlstra, H. 1967. *Experimental Methods in Magnetism*, North-Holland Publ. (Wiley), 295 pages.

CHAPTER 4
MAGNETIZATION IN MEDIA AND HEADS

In Chapter 3 we examined the magnetization M, which is an intrinsic material property equal to the magnetization level within each domain. The remanence M_{rs} depends upon the material composition, while the coercivity H_c depends upon external factors such as shape and stresses. In this chapter I will discuss the technical aspects of the processes whereby a material's magnetization is changed by a shift in domain sizes and magnetization directions. Such changes are caused by air gaps, shaping of the material, application of external fields, and interactions among small magnetic particles.

The BH-loops and MH-loops change in a dramatic way when an air gap is introduced, as shown in Fig. 4.1. When we repeat our experiment from Figs. 3.44 and 3.50 on a core sample from the same material, but now with an air gap, then the BH-loop in Fig. 4.1 results (W/Air Gap). Notice the lowered remanence but unchanged coercivity. In this chapter we will examine this magnetization process called demagnetization.

We will examine how demagnetization reduces the recording level in a coating. This problem has in recent years stirred up many activities and arguments in the recording industry. It is argued that a perpendicular magnetization mode is preferable at short bit lengths, because its demagnetization is less than in the longitudinal mode. This matter will be explained by developing the theory of demagnetization, illustrated with several examples. Magnetic heads are basically simple to analyze, as will be shown by employing an electrical circuit model for the head core. This allows for calculations of head efficiency, inductance, and the write gap field.

THE DEMAGNETIZING FIELD

The introduction of an air gap in the flux path causes the formation of poles, and the field H_d from these poles opposes the applied field H_a from the magnetizing current I. The net effect is an overall decrease in the field acting upon the material by H_d to $H_{core} = H_a - H_d$, and therefore, also a decrease of the corresponding flux densities. We call this effect demagnetization, and the field H_d the demagnetizing field.

Air gaps occur in all applications of magnetics, with the exception of some toroids and the switching cores in magnetic memories. Air gaps exist in moving coil instruments, loudspeaker magnets, motors, generators, relays, and so on.

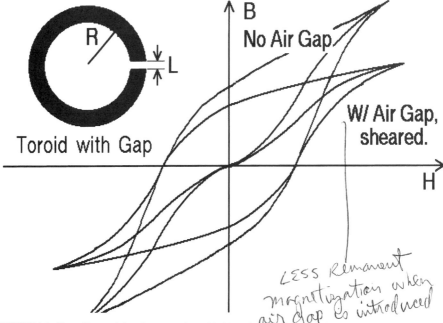

FIGURE 4.1 Sheared hysteresis loop for a magnetic toroid with an air gap. The core material has low permeability and high coercivity—typical for recording media.

Figure 4.2 shows a *MH*-loop for a head material with an air gap. The severe shearing of the curve means a lower effective permeability, and at the same time a linearizing effect on the effective permeability. The effective permeability is still defined as the ratio between B and $\mu_0 H$, but is no longer a material property alone; it includes the effect of demagnetization.

Another advantage for magnetic heads is the reduction of the remanence M_{rs} to M_{rs}', where the added superscript ' stands for the demagnetized value. It depends upon M_{rs}, and is a function of the air gaps and the shape of the magnet. The remanence in a soft magnetic material is not at all small (dotted curve, Fig. 4.2), and would, if not reduced, cause serious dc noise and distortion problems in recordings. A very small air gap reduces the remanence to a tolerable level, and even then, it is good housekeeping to demagnetize a recorder's heads at regular intervals to remove any remanence.

MAGNETIZATION OF MAGNETS

It is instructive to examine the magnetization process of a permanent magnet, such as the bar shown in Fig. 4.3a, top. It can become permanently magnetized by an external field H_a from a coil carrying a current or from another magnet, such as the shown horseshoe magnet. In Fig. 4.3b poles are induced, and their polarities are easy to remember because we know that the bar magnet will be attracted to the horseshoe magnet, and that poles of opposite polarity attract each other. Hence a negative south pole is formed opposite the horse shoe's north pole, and vice versa at the other end.

These poles remain when the external field is removed (Fig. 4.3c). The internal field from these poles is opposite to the external magnetizing field. The net field inside the bar magnet was therefore:

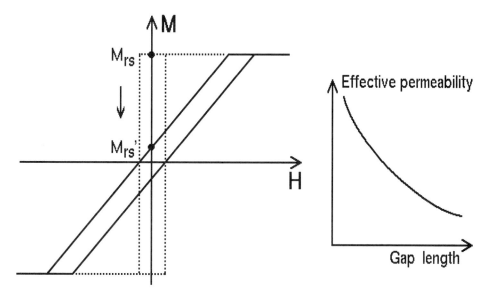

FIGURE 4.2 Sheared hysteresis loop for a magnetic toroid with an air gap. The core material has high permeability and low coercivity—typical for magnetic heads.

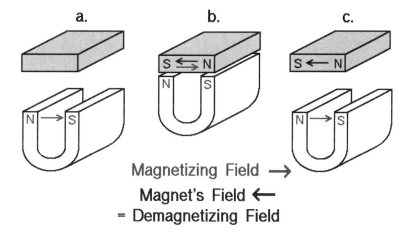

FIGURE 4.3 Steps in the magnetization of a bar magnet (including poles).

$$H_{\text{core}} = H_{\text{a}} - H_{\text{d}}$$

during the magnetization process, and

$$H_{\text{core}} = -H_{\text{d}}$$

after removal of the external field. We want, at this point, to become aware of the fact that the field from the generated magnet is opposite to the field that magnetized the magnet. We have again encountered Le Chatelier's rule.

M AND H FIELDS IN A PERMANENT MAGNET

A recorded track in a magnetic recording consists of alternating magnetism along the track. This is in particular true in digital recordings where the magnetization alternates between positive and negative values. (Figure 4.10 shows the bit pattern.) We wish to find the remanent magnetization M that exists in one such bit, or bar magnet, because this represents the flux level that is available for readout.

The relationship between B and H is:

$$B = \mu H \tag{4.1}$$

where:

$$\mu = \mu_0 \mu_r \tag{4.2}$$

We have no problem envisioning the B and H lines from a current-carrying wire or coil; they are alike, because $\mu_r = 1$ in air. Their magnitudes differ by μ_0, and here we all wish we were still using the cgs system of units because here $\mu_0 = 1$, and gauss equal oersteds in air (but NOT inside ferro- or ferrimagnetic materials).

In a toroid-shaped core the B and H fields are also easy to envision; there are μ_r more B lines than H lines. But cutting an air gap in the flux path immediately complicates matters: We measure a decrease in the flux density B.

Let us first consider the fields inside the bar we magnetized in Fig. 4.3. There is initially no field (Fig. 4.3a). The horseshoe magnet will send a field H_a through it, directed left to right. We will designate this polarity positive, comparable to the x-axis direction in a standard coordinate system.

The field H_a induces the poles as shown. They will, in turn, set up a field H_d going from the north to the south pole. This field is opposite to H_a, and therefore named the demagnetizing field.

The total, true field that acts on the magnetic material is the sum of the two fields:

$$
\begin{aligned}
H_{core} &= H_a + (-H_d) \\
&= H_a - H_d
\end{aligned}
\tag{4.3}
$$

We can now use the MH-loop measured on a closed toroid by using H_{core}, not H_a. Figure 4.4, right, shows the results.

What happens is that the field $-H_d$ remains in the bar when $H_a \rightarrow 0$. This is because H_d originates from the pole. It will move the point for the remanence M_r down in the second quadrant to M_r'.

The point where M_r' ends up is the point where the descending loop intersects a line known as the load line. This comes as no surprise to anyone that has worked with permanent magnet design. Only the second quadrant of the BH- or MH-loop is needed, and the portion of the BH-loop that is located therein is named the demagnetization curve. Our task ahead is to find the load line, but let us first clear up the concepts of B and H fields, and associated magnetization M, in an electromagnet.

B AND H FIELDS IN AN ELECTROMAGNET

Figure 3.25 (right) shows an air coil; here the B field pattern is identical to that of the H field. There are no poles in the air coil, and the field lines are closed patterns.

When a ferromagnetic material is inserted into the coil, two poles are induced. They will in turn produce the demagnetizing field that will reduce the field acting upon the material; see equation (4.3).

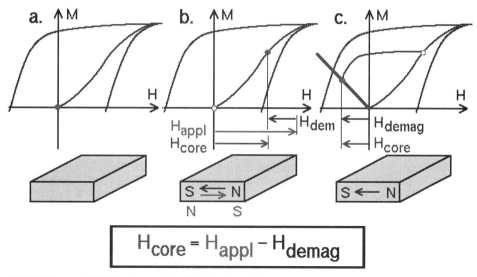

FIGURE 4.4 Applied field H_a, demagnetizing field H_d and core field H_{core} during the magnetization process. The true remanence $B_{rs} <= B_{rs}'$.

The field H_{core} will shift the domains into its direction; this will cause a magnetization M, that increases the flux density term by $\mu_o M$:

$$B = \mu_o(H_{core} + M) \tag{4.4}$$

The result is a net increase in the number of flux lines, inside as well as outside the material. This situation corresponds to Fig. 4.4b, center.

Finally, turn off the magnetizing current. H now changes to H_d, the demagnetizing field from the poles. Inside the material we now have

$$B = \mu_o(-H_d + M) \tag{4.5}$$

Now the H lines go from pole to pole, while the B lines still are closed lines. We can summarize (and should memorize) three rules:

- H field lines are always closed when due to currents.
- H field lines are always open when due to poles.
- B field lines are always closed.

The reader can verify the identity between B and H outside the magnet, as shown in Fig. 4.5. Inside, the picture does not look right; that is because B, H, and M really are vectors, quantities that are magnitudes with directions. Properly written, equation 4.5 is:

$$\mathbf{B} = \mu_o(\mathbf{H_d} + \mathbf{M}) \tag{4.6}$$

Note also that $\mathbf{B}$ and $\mathbf{H_d}$ have opposite signs. That is how we ended up in the second quadrant, and that is where most of our future discussion of magnetism will concentrate.

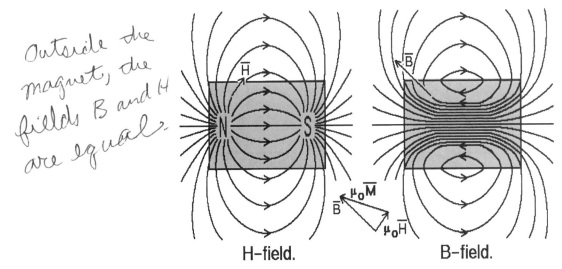

Outside the magnet, the fields B and H are equal.

FIGURE 4.5   The H and B field lines in and around a permanent magnet.

THE DEMAGNETIZING FACTOR N

The magnet's own field H_d needs to be specified better. We can accomplish that by first realizing that its magnitude will be proportional to the magnetization M that is being produced by the outside magnetizing field H_a. Let us introduce a multiplier N so that

$$H_d = -NM, \text{ or}$$

$$M = (-1/N)H$$

This is a line in the MH diagram with slope $-1/N$, or:

$$\tan \alpha_{MH} = -1/N \tag{4.7}$$

A similar slope can be found for use with the BH-loop. We have

$$\begin{aligned} H_{core} &= H_a + H_d \\ &= H_a - NM \\ &= H_a - N(B - \mu_0 H_{core})/\mu_0 \\ &= H_a - N(B/\mu_0 - H_{core}). \end{aligned}$$

When $H_a \to 0$,

$$H_{core} = -NB/\mu_0 + NH_{core}$$

which reduces to

$$B/H_{core} = -\mu_0(1/N - 1), \text{ or} \tag{4.8}$$

$$\tan \alpha_{BH} = -\mu_0(1/N - 1)$$

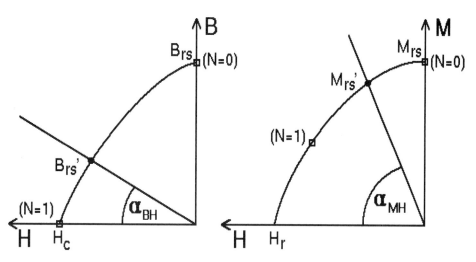

FIGURE 4.6 The remanence assumes a value that is located where the descending *BH*- (or *MH*-) loop branch intersects with the demagnetization line. Note that complete demagnetization (*N*=1) leaves $B_{rs}' = 0$, but $M_{rs}' > 0$.

The demagnetizing field H_d is a fraction *N* of the magnetization *M*. The latter is mainly a material property, so if we can determine H_d then we immediately have the value of *N*.

N is called the demagnetization factor and will be a number that represents the shape anisotropy of the magnet. Because all magnets are three-dimensional, we need to determine *N* for the three rectangular coordinates: *x*, *y*, and *z*.

From a textbook on magnetism we learn that these components always add up to unity (or 4π in the cgs system):

$$N_x + N_y + N_z = 1 \tag{4.9}$$

This formula also reflects that the net sum of magnetizations in the three directions always adds up to the specific magnetization of a material, no more, no less.

By plain, logical reasoning we can now arrive at *N* for some simple shapes. A sphere is simple; the magnetization will not know which direction to settle down to if there are no anisotropies in the material itself. The analogy from Fig. 3.60 applies if we make the eggshell a perfect sphere. The result is that $N = \frac{1}{3}$, in any direction. This will also apply to a cube.

Figure 4.7 shows a sheet magnetized in two different directions, and the values of *N* are not surprising. A sheet is nearly impossible to magnetize from side to side, but easy from one edge to the opposite.

For a closed-ring core, $N = 0$; there are no poles and therefore no demagnetizing field. The slightest air gap will change that, and it can be shown that

$$N \approx L_g/L \tag{4.10}$$

Where L_g = length of the gap and *L* = overall length of the flux path ($L \gg L_g$).

Example 4.1: A magnetic write head is made from NiZn with M_s = 400 kA/m (= 5027 gauss) and H_r = 80 A/m (= 1 Oe). The gap length L_g = 1 μm (40 μin). The gap length L_g = 1 μm (40 μin), and the overall flux path is 2 mm (80 mils). What is the remanent magnetization if the head inadvertently was saturated?

Answer: N $\approx L_g/L$ = 1/2000. arctan(1/N) = arctan (2000) = 89.97°. A graphic determination of M_{rs}' is quite impossible. You would have to draw a graph where each division of one Oe corresponds

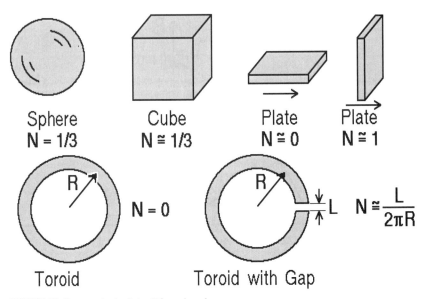

FIGURE 4.7 Demagnetization factor N for various shapes.

to one gauss: if we set 1 Oe = 1 cm then the resulting loop will be 2×0.12 cm wide, and 100 meters tall! Next, find the intersection between the left side of the loop with the demagnetization line.

Figure 4.8 shows the general problem. Recall that the tangent to an angle in a rectangular triangle equals the ratio between the opposite side a divided by the adjacent side b (see Fig. 4.13). By equating the two expressions for $\tan \alpha$ the formula for M_{rs}' is readily found.

Using the formula we find $M_{rs}' = 80 \times 2000 = 160$ kA/m $= 160 \times 4\pi = 2011$ gauss, which results in a gap field $H_g = 2011$ Oe, an excessive value; it will erase all information on most media.

There are two solutions: Select a material with much lower coercivity (like one A/m) or increase the head's back gap length. The latter has been done in some studio recorders' record head. Higher record currents are needed because the head's efficiency is reduced when the back gap length is increased (see the end of this chapter).

Example 4.2: A permanent magnet motor with an Alnico magnet is taken apart to replace a bearing. When reassembled and tested, its torque is far below the specified value. What happened? The length of the flux path is 8 cm and the air gap 5 mm.

Answer: The BH-loop is shown in Fig. 4.9, where the load (or demagnetization) line is also drawn. $\alpha_I = \arctan (1/N - 1) = \arctan (L/L_g - 1) = \arctan 15 = 86°$.

When the motor is taken apart, the gap length increases dramatically, say to 40 cm. Now N increases to 40/8 according to the formula. That exceeds the domain for N—and remember that $N \approx L_g/L$ only holds for $L_g \ll L$. Let us set $N \to 1$, then $\alpha_{II} \to 0°$, and the load line is shown as line II.

Now reassemble the motor. The load line moves back to I, with $B_{rs}' = 7000$ gauss. But the remanence does not move up to its original value; there is no field present to help the domains in shifting. Rather, it moves along a minor loop up to point III, where B_{rs}'' only equals 900 gauss. (Note the use of superscript " for this case.) Hence the motor's torque is only $900/7000 \approx \frac{1}{8}$ of its rated value.

To bring it back, a magnetization process is required. A few turns of a heavy wire are wound around the pole piece(s), and a large field is induced by sending many amperes through the wire, for instance by discharging a large capacitor. This brings the B value up to saturation at IV, which settles to I.

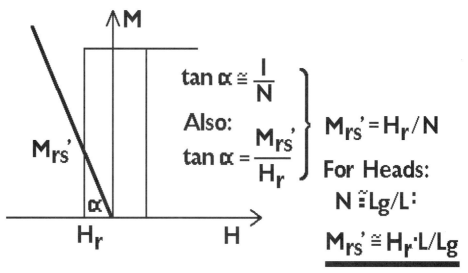

FIGURE 4.8 Remanent magnetization M_{rs}' in a magnetic head.

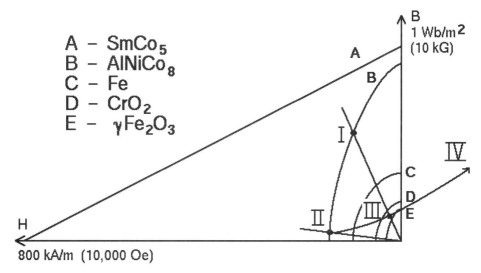

FIGURE 4.9 Comparative demagnetization curves (from trade literature); also, changes in the load line for a permanent-magnet motor (Example 4.2).

Figure 4.9 also provides comparisons between some magnet materials: three particulate media materials (down in the right-hand corner), and samarium cobalt. The latter material does not demagnetize; the working point merely moves up and down the *BH*-loop, which now is a straight line. This is advantageous, but the cost is also 100 times that of Alnico.

DEMAGNETIZATION IN A RECORDING

A realistic picture of the tape magnetization results when we divide the coating up into bars, each representing a bit, or half a wavelength. This is shown in Fig. 4.10, where the arrows show the magnetization directions for longitudinal and perpendicular magnetizations.

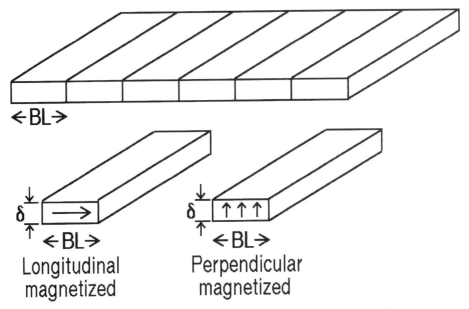

FIGURE 4.10   Geometry of six bits recorded along a track, magnetized longitudinally and perpendicularly.

Our discussions will be in the *x-y* plane only; all conditions shown on this cross section (at $z = 0$) are the same for various other cross sections at different values of z.

Further, each bar element represents a packet of a large number of magnetic particles, normally longitudinal oriented. We will treat the magnetization of each bar as an individual magnet, leaving a discussion of interacting fields for Chapter 11. Each bar represents one bit; two bars represents one wavelength.

It is difficult to find H_d, and then N, for shapes other than the sphere and perfectly symmetrical ellipsoids. The demagnetization factors N_x and N_y are determined as approximations to the same factors for a prolate ellipsoid. Figure 4.11 illustrates how we can stretch the ellipsoid in the *z*-direction and thus make N_z approach 0, and preserve the ratio between N_x and N_y, while keeping the sum of the three equal to one. Good approximations are:

$$N_x = \delta_{rec}/(BL + \delta_{rec}) \tag{4.12}$$

$$N_y = BL/(BL + \delta_{rec}) \tag{4.13}$$

where δ_{rec} = recorded depth ≤ coating thickness δ.

Note the introduction of the recorded depth δ_{rec}, rather than the physical thickness δ of the coating. The case where $\delta_{rec} < \delta$ corresponds to the conditions *partial penetration* or *under-bias* record-

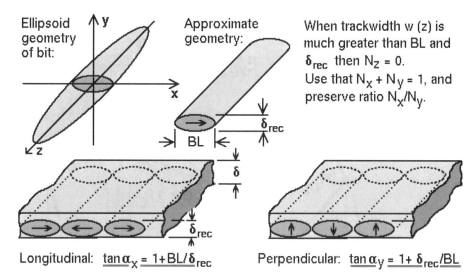

FIGURE 4.11   Load line slope $\tan\alpha$ for longitudinal and perpendicular magnetizations.

ing, which both result in higher packing densities (better high-frequency response). $\delta_{rec} = \delta$ is normal write, or normal bias. (NOTE: Overwrite, or overbias, cannot produce a $\delta_{rec} > \delta$.)

When the terms from (4.12) and (4.13) are substituted in the expressions for tan (4.78), we find, (using cgs-values for the BH-loop, gauss and Oe):

$$\tan \alpha_{xM} = 1 + BL/\delta_{rec} \tag{4.14-x}$$

$$\tan \alpha_{yM} = 1 + \delta_{rec}/BL \tag{4.14-y}$$

$$\tan \alpha_{xB} = BL/\delta_{rec} \tag{4.15-x}$$

$$\tan \alpha_{yB} = \delta_{rec}/BL \tag{4.15-y}$$

The demagnetized level (B_r', or M_r') may play back as a higher level (B_r'' or M_r'') due to recoil: When the track comes in contact with the read head it sees a high-permeability material, which shunts the flux through it. Or, in terms of demagnetization, the air gap from pole to pole in the bit has been replaced with a keeper, and the demagnetization field reduces in the inside of the coating.

When $H_d \to 0$ then the point for B_r' moves along a recoil line with slope $\mu_o \times \mu_{rt}$ in the BH-loop and ($\mu_o \times [\mu_{rt} - 1]$ in the MH-loop) to the point where it intersects the vertical B- or M-axis. μ_{rt} is the relative permeability of the tape, typically between one and two in the direction of the oriented particles, but three to four when measured perpendicular thereto. This magnetization level produces the short circuit flux (shown as M_{rs}'' in Fig. 4.12), that is equal to ($\mu_{rt} = 1$) or greater than ($\mu_{rt} > 1$) the open circuit flux.

The demagnetization losses are now quite simple to determine, either by graphic methods or computations. The latter were treated earlier in the literature (Daniel, Smaller, Mallinson).

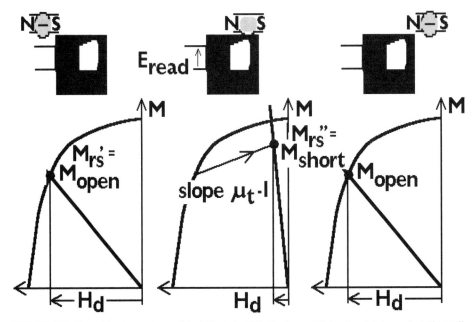

FIGURE 4.12   The read head presents a small load (short circuit) for the flux, and H_d is reduced during readout. The recoil line has a slope that equals that of the tangent to the *MH* curve at $H = 0$.

APPLYING THE DEMAGNETIZATION THEORY

The demagnetization theory has the following practical applications.

DIGITAL RECORDING

A recording of very short bit lengths will be examined. The transition density is 100,000 fpi (flux reversals per inch, along the track.) This results in a bit length of 10 μin = .25 μm. Other densities are easily evaluated by following the steps in the following procedure.

We will initially analyze the demagnetization when using a 2 μm thick tape coating such as used in current backup devices (QIC, 8 and 4 mm systems). The remanent coercivity is initially 300 Oe.

The demagnetization line has a slope of $\tan \alpha = 1 + BL/\delta_{\text{rec}} = 1 + .25/2 = 150 \times 2.25/(150 \times 2) = 337.5/300$. We have multiplied the numerator and the denominator with the same number and have chosen to make the denominator equal to the value of the tape coercivity. This facilitates a simple way of drawing the line without the steps of finding the angle α and then drawing the line on an *MH* plot that has been redrawn to have identical increments for the *M* and *H* axes. The definition of the tangent function is shown in Fig. 4.13. The tangent equals the ratio between sides *a* and *b* in the triangle. By setting side *b* equal to $H_r = 300$ Oe, we merely have to find the length of side *a* (= 337.5). This line is drawn in Fig. 4.14a, showing severe demagnetization.

Reducing the recorded depth δ_{rec} to a fraction of the physical coating thickness δ will reduce the demagnetization. Selecting $\delta_{\text{rec}} = \delta/10 = .2$ mm results in a slope $\tan \alpha = 1 + .25/.2 = .45/.2 = 675/300$. This line is drawn in Fig. 4.14b, showing reduction in the demagnetization loss.

It is reasonable to object to this approach with a question: "Is not the available flux reduced by a factor of ten while the smaller demagnetization loss is reduced by a factor of almost two, bringing a net loss of five?". This would be true if all magnetization of the coating contributes to the read flux.

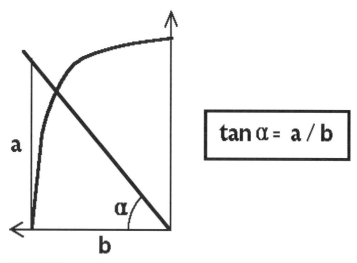

FIGURE 4.13 The definition of the function tangent permits an easy way to draw the demagnetization line.

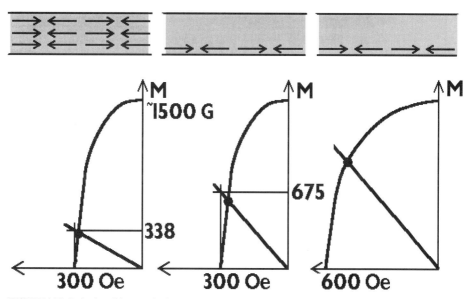

FIGURE 4.14 Reduction of demagnetization losses by partial penetration recording and/or higher coercivity.

But only a portion of this flux is utilized, as will be shown in Chapter 6 on Readout (Playback). There is a deficiency in the read process that results in a coating thickness loss. It means the effective coating that produces 75% of the read flux comes from a surface layer of thickness $\delta_{eff} = 0.22 \times wavelength = 0.44 \times BL$. This level is only 2.5 dB below the total read flux.

The effective coating in our example is $0.44 \times 0.25 = 0.11$ µm, and a thick recording does therefore not produce more flux. The penalty in flux reduction from reduced recording depth is minor, because 75% of the total flux comes from the surface of the recording. In high data density recordings, only 10–20% of the coating is recorded. New tapes are being developed that have very thin coatings, in order to match the required shallow-depth recording required to reduce the demagnetization loss of M.

Note that the effective thickness $\delta_{eff} = 0.11$ μm $= 1100$ Å is comparable to the thickness of the thin-film coating on a rigid disk ($\delta = 500 - 1000$ Å.)

The final step in improving the output is done by increasing the media coercivity to for example 600 Oe, shown in Fig. 4.14c.

Another way of overcoming the demagnetization loss was suggested in 1972 by Prof. Iwasaki. A medium with perpendicular anisotropy should be used since then:

$$\tan \alpha = 1 + BL/\delta_{rec} \tag{4.14-y}$$

For $\delta_{rec} = 2$ μm we find $\tan \alpha = 1 + 2/0.25 = 2.25/0.25 = 2700/300 = 900/100$. This demagnetization line is drawn in Fig. 4.15. Perpendicular recording is clearly an attractive solution that digital tape systems may benefit from.

Demagnetization losses in rigid disk drives are currently very small (if any), as shown in Fig. 4.16 for a thin CoNiPt disk. Rigid disk coatings are very thin, and the perpendicular mode may not be implemented until very high densities are encountered. For further discussion see Chapter 20, Advanced Recording Theory.

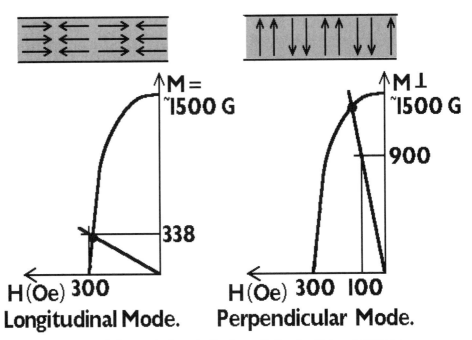

FIGURE 4.15   The perpendicular magnetization mode offers demagnetization when BL < recorded thickness.

ANALOG RECORDING ON TAPE

A worksheet for graphical estimates of the demagnetization losses at several wavelengths is shown in Fig. 4.17, using the MH-loop. Figure 4.18 illustrates the procedure to determine the demagnetization losses in a longitudinally oriented tape, as function of frequency at a speed of 1⅞

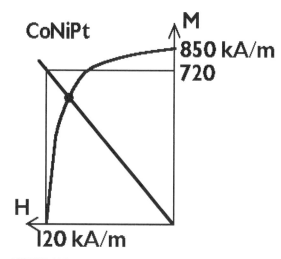

FIGURE 4.16 Demagnetization in thin-film disks is small due to thin media and high coercivity.

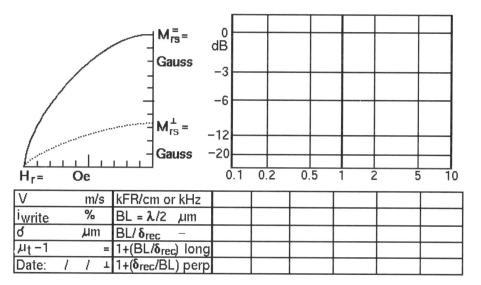

FIGURE 4.17 Template for analysis of demagnetization in analog tape recordings.

IPS. The recorded depth δ_{rec} is equal to the coating thickness δ. The losses will operate on differ-ent levels of remanent maximum magnetization in a tape, in accordance with the squareness ra-tios S_x and S_y. A longitudinally oriented medium will have a predominantly longitudinal remanence (except at short wavelengths). A perpendicularly oriented medium will have little or no demagnetization at short wavelengths, but very little output at long wavelengths. An isotropic tape will offer the best of both worlds.

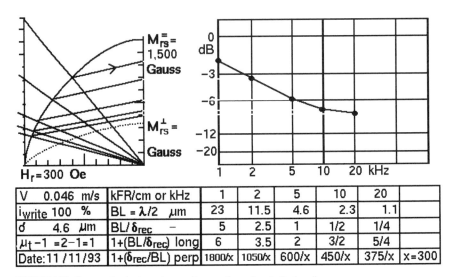

V 0.046 m/s	kFR/cm or kHz	1	2	5	10	20	
i_{write} 100 %	BL = λ/2 μm	23	11.5	4.6	2.3	1.1	
δ 4.6 μm	BL/δ_{rec} –	5	2.5	1	1/2	1/4	
μ_t−1 =2−1=1	1+(BL/δ_{rec}) long	6	3.5	2	3/2	5/4	
Date:11 /11/93	1+(δ_{rec}/BL) perp	1800/x	1050/x	600/x	450/x	375/x	x=300

FIGURE 4.18 Demagnetization losses in an audio recording on longitudinal media.

Figure 4.19 shows the demagnetization losses for a perpendicular medium. A third example is shown for a 3½-inch mini-floppy in Fig. 4.20. The figure captions plus data in the illustrations contain all pertinent data. We will return to a discussion of longitudinal versus perpendicular magnetization modes in Chapters 19 and 20.

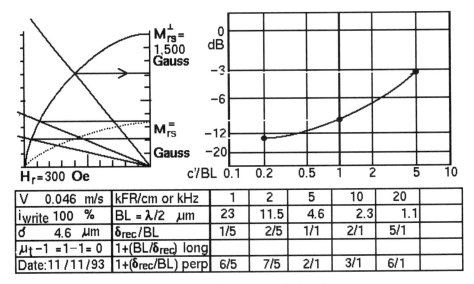

V 0.046 m/s	kFR/cm or kHz	1	2	5	10	20	
i_{write} 100 %	BL = λ/2 μm	23	11.5	4.6	2.3	1.1	
δ 4.6 μm	δ_{rec}/BL	1/5	2/5	1/1	2/1	5/1	
μ_t−1 =1−1= 0	1+(BL/δ_{rec}) long						
Date:11 /11/93	1+(δ_{rec}/BL) perp	6/5	7/5	2/1	3/1	6/1	

FIGURE 4.19 Demagnetization losses in an audio recording on a perpendicular media.

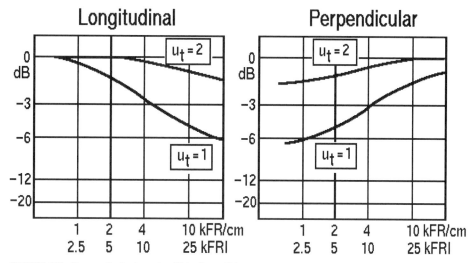

FIGURE 4.20 Demagnetization losses in a 3½-in. floppy disk.

CHARACTERIZATION OF MAGNETIC HEADS

The recording of a signal onto a tape or disk is done by feeding a current representing the signal through the winding in the write head; this will generate the field in front of the record head gap that lays down a permanent magnetization in the coating. The flux from this magnetization will at a later time induce the read voltage in the winding of the playback head.

We will need to design the head so its current requirements, and input impedance, are compatible with our drive electronics. It seems logical to put a large number of turns onto the core, which will result in the lowest current requirement and the greatest induced voltage during read.

We will learn that the head impedance is proportional to the number of turns squared, and that the winding's self inductance will resonate with the combined capacitance of amplifier, cable, and self capacitance. This frequency must, of course, be above the highest signal frequency we plan to use.

There are, finally, certain losses occurring in the head core material at high frequencies. These losses will be discussed in Chapter 7, but we clearly need some method to characterize a head's impedance, current requirements, and voltage output during the read mode.

An elegant tool is found in the analogy between the current flowing in an electric circuit to the magnetic flux flowing in what we now will name a magnetic circuit. The key element is the definition of the magnetic resistance of an element (like an air gap or a core-half) through which the flux flows, identical to the definition of the electric resistance of a wire element.

We will arrive at an equivalent electrical circuit for the head, from which we can compute efficiency, write current, read voltage, and self-inductance. The model will be extended to include high-frequency losses in Chapter 7.

MAGNETIC CIRCUITS

We can treat magnetic circuits as being made up of magnetic flux paths in the same way we consider electric circuits to be made up of resistors. Figure 4.21 shows a battery E connected to a resistor R. The current I is easily found, using Ohm's law:

$$I = E/R \tag{4.16}$$

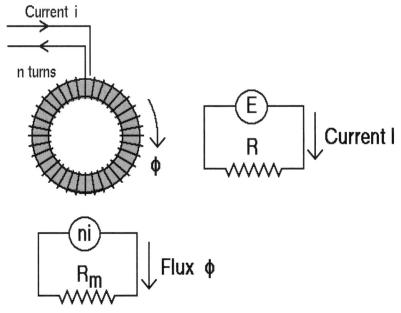

FIGURE 4.21   An electromagnet and its equivalent circuit.

A ring core with relative permeability μ_r, average length L, and cross sectional area A is also shown. With a current I flowing through the winding we find the field strength:

$$H = nI/L \tag{4.17}$$

where nI is the magnetomotive force in ampere-turns. The flux ϕ is:

$$\begin{aligned}
\phi &= BA \\
&= \mu HA \\
&= \mu A \times nI/L \\
&= nI/(L/\mu A)
\end{aligned} \tag{4.18}$$

This compares with an electric current I through a wire of conductivity ρ, length L and cross section A:

$$I = E/(L/\rho A) \tag{4.19}$$

The similarity between equations 4.18 and 4.19 illustrates the analogies between magnetic and electric quantities:

Electric current I = Magnetic flux ϕ (Wb)
Electromotive force E = Magnetomotive force nI (amp-turns)
Electric resistance R = Magnetic resistance R_m (1/H)

Drawing the analog resistance network is invaluable in analyzing magnetic circuits with various materials, shapes, and air gaps in the flux path, and the method will be used extensively in this book. Another term for magnetic resistance is reluctance.

Two air gaps are treated like two reluctances (resistors) in series, while stray flux paralleling the gap is treated as a parallel reluctance (similar to parallel resistors).

Also used are magnetic permeance P, which is magnetic conductivity equal to the reciprocal of resistance. P values are particularly handy for parallel combinations.

EFFICIENCY OF READ HEADS

The media flux flows ideally through the read head as shown in Fig. 4.22, left. However, a portion of the flux links across the front gap and does not contribute to the useful flux through the winding.

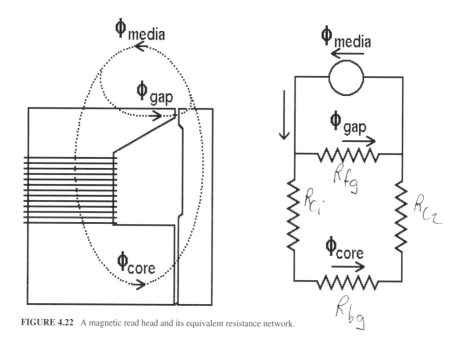

FIGURE 4.22 A magnetic read head and its equivalent resistance network.

The total flux is:

$$\phi_{media} = \phi_{core} + \phi_{gap}, \text{ or}$$

$$\phi_{core} = \phi_{media} - \phi_{gap}$$

Across the gap there exists a magnetic potential:

$$\begin{aligned} V_m &= \phi_{gap} \times R_{fg} \\ &= \phi_{core} \times (R_{bg} + R_{c1} + R_{c2}) \\ &= \phi_{media} \times (R_{fg}/(R_{bg} + R_{c1} + R_{c2})) \end{aligned}$$

We can now find the core flux and determine the read efficiency:

$$\eta = \phi_{core}/\phi_{media} \text{ or}$$

$$\eta = R_{fg}/\Sigma R_m \tag{4.20}$$

where:

$$\Sigma R_m = R_{fg} + R_{bg} + R_{c1} + R_{c2}$$
$$= \text{sum of all reluctances going once around the circuit.}$$

R_{fg} should include the stray fields at the gaps (mandatory in modern microhead structures). It is generally safe to ignore the stray fields at the back gap due to its short length and large area.

The front gap stray fields are discussed in Chapter 8. The formulas provide the permeances of the various sections of the stray fields; the total magnetic resistance is equal to the reciprocal of the sum of the permeances.

A quick examination of the expression for the efficiency tells us how to make a good head:

1. Make the front gap resistance large:

   1.1. Make the gap length L_{fg} large; this is not possible due to short-wavelength losses, as illustrated in Fig. 4.23. When the gap length equals two bit lengths, then the flux through the head core is zero at all times. A gap length equal to one bit length is optimum. With a packing density of 10,000 FRI (flux reversals per inch), the bit length is 100 μin. Therefore, L_{fg} should be no longer than 100 μin.

   1.2. Make the gap depth small. This is possible for heads used in high-speed disk drives, where the head assembly is flying over the disk surface. Then there is no wear, and the gap depth is as small as 2 μm (80 μin). This gap depth would be too shallow for in-contact applications, such as tape and floppy disks. One must allow 25–75 μm (1–3 mils) for wear. (See further in several sections of Chapter 18.)

2. Make the back gap reluctance small, i.e., short length and large area. The inclusion of a back gap is necessary in most head designs for the reason of assembly. A back gap also makes life easier in the production by increasing yield (see Chapter 9).

3. Make the core reluctance small, i.e., use short, stubby core halves. Use high-permeability material.

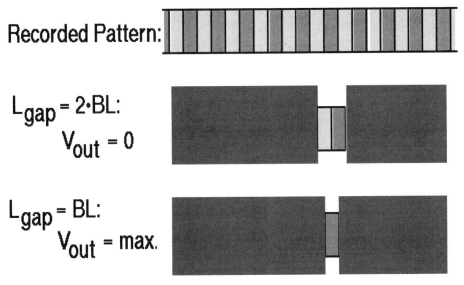

FIGURE 4.23 The playback (read) head gap length must not exceed the bit length.

FIELD STRENGTH IN THE WRITE HEAD GAP

Recording of a tape or disk is done when the media moves through the stray field from the gap in the write head. We need to know this field so we can determine the locations where the field strength exceeds the coercive force of the individual particles, and hence may change their magnetization (interaction fields should be included; see Chapter 19 on ac-bias recording).

The field strength components H_x and H_y were determined by Karlqvist (1954), and are listed in Fig. 4.24 where also the field lines are shown. The vector field strength $H = \mathbf{i}H_x + \mathbf{j}H_y$ can now be calculated, with known values of H_g, g, x, and y.

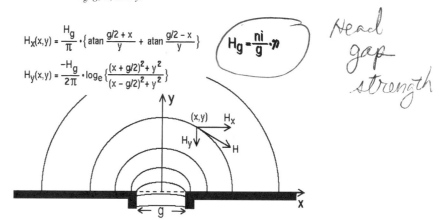

$$H_x(x,y) = \frac{H_g}{\pi} \cdot \left\{ \text{atan}\, \frac{g/2 + x}{y} + \text{atan}\, \frac{g/2 - x}{y} \right\}$$

$$H_y(x,y) = \frac{-H_g}{2\pi} \cdot \log_e \left\{ \frac{(x + g/2)^2 + y^2}{(x - g/2)^2 + y^2} \right\}$$

$$H_g = \frac{ni}{g} \cdot \eta$$

Head gap strength

FIGURE 4.24   Equations for the magnetic field strength in front of a gap (*Karlqvist*).

We need to determine the value of H_g, the field inside the gap, and will use the magnetic circuit analogy:

We have:

$$H_g = B_g/\mu_o$$

where:
Also

$$B_g = \phi_g/A_g$$

$$\phi_g = nI/\Sigma R_m$$

Substituting

$$H_g = \phi_g/\mu_o A_g$$
$$= nI/\Sigma R_m \mu_o A_g$$

or, after reduction:

$$H_g = (ni/g) * \eta \tag{4.21}$$

The appearance of the read efficiency factor accounts for the reciprocity property of magnetic heads. It serves equally well in the write or read process.

We need to multiply with two additional terms. First, the gap flux is only a fraction of the total front gap flux, as we interpreted it in Fig. 4.22. A portion of the flux goes through air (stray) and contributes to the formation of H_g. Consider that the total stray flux reluctance of the front gap is summed up in a term R_{stray}. Then we can easily find that the gap field needs to be multiplied with the additional term of a write efficiency:

$$\eta_w = R_{stray}/(R_{stray} + R_{fg}) \tag{4.22}$$

Write heads also have lower apparent efficiency than read heads due to flux leakage around the coil (insufficient coupling) and to shielding that may be next to the core. An efficiency η_{coupl} can be calculated for the flux from nI that reaches and goes through the front (stray plus gap). This matter is discussed in Chapter 8; see specifically Fig. 8.31. Also see Chapter 10, under write head efficiency.

For now we will use a term coupling efficiency η_{coupl}, and for the deep-gap field H_g for use in Karlqvist's formulas we find:

$$H_g = (ni/g) \tag{4.23}$$

We may compare this with the magnitude of the field H in the center of a solenoid L meters long with n turns: $H = nI/L$. This makes the formula for H_g easy to remember: The field strength H_g is equal to that of a very short solenoid with n turns, placed in the gap of length g, and multiplied with the product of the three efficiencies.

The true write head efficiency is defined as the ratio between the magnetomotive force in the gap and the input magnetomotive force ni.

HEAD IMPEDANCE

The impedance of a magnetic write or read head is made up of three components: the wire resistance R, the loss resistance R_{loss}, and the inductance L (here we use the same symbol as for the gap length, but that appears better than using a lowercase l for length; that is often mistaken for a 1 (one)).

The inductance of an electric circuit is a measure of the magnetic flux that links the circuit when a current I flows in that circuit:

$$
\begin{aligned}
L &= n\phi/I \\
&= n(nI/\Sigma R_m)/I, \text{ or}
\end{aligned}
\tag{4.24}
$$

$$L = n^2/\Sigma R_m$$

The wire resistance and the equivalent loss resistance will be discussed in Chapter 10.

BIBLIOGRAPHY

Joseph, R.I., and E. Schloemann. May 1965. Demagnetizing Field in Nonellipsoidal Bodies. *Jour. Appl. Phys.*, 36 (5): 1579–1593.

Karlqvist, O. June 1954. Calculation of the Magnetic Field in the Ferromagnetic Layer of a Magnetic Drum. *Trans. Royal Inst. Tech.* 86: 3–27.

Mallinson, J.C. Sept. 1966. Demagnetization Theory for Longitudinal Recording. *IEEE Trans. Magn.* MAG-2 (3): 233–236.

Osbron, J.A. July 1945. Demagnetizing Factors of the General Ellipsoid. *Physical Review* 67 (11 and 12): 351–357.

Smaller, P. Sept. 1966. An Experimental Study of Short Wavelength Recording Phenomena. *IEEE Trans. Magn.* MAG-2 (3): 242–247.

Westmijze, W.K. June 1954. Studies in Magnetic Recording, I: Introduction. *Philips Res. Rep.* 8 (3): 148–157.

---

1. Alnico magnets are made from minute particles (with inherent high coercivity) that are sintered together. Magnetization changes therefore occur by rotations.

CHAPTER 5
FUNDAMENTAL RECORDING THEORY

The magnetic recording process converts an electric current signal into an equivalent magnetization in the coating of a magnetic tape or disk. This is done with a transducer (called a record or write head) that transforms the electrical signal into a magnetic field through which the coating passes. The result in the coating is a magnetic remanence proportional to the field.

We will often use the term "write" rather than "record," although the two words are identical in meaning. "Write" is used in the digital disk and tape industry, while "record" is used among audio and instrumentation engineers.

There is a difference between the recording of analog and digital signals. Figure 5.1 illustrates an analog signal recording. The signal may represent voice, music, video, or analog data. It can be a pure sine wave, and the recorded wavelength λ is determined by $\lambda = v/f$ = media velocity divided by the signal frequency.

Figure 5.2 shows a recorded digital signal. Here the media magnetization changes from $+M_{sat}$ to $-M_{sat}$ in accord with the medium velocity and the wavelength of a rectangular signal waveform, representing bits. This is a sequence of zeros and ones that represent some digitized signal originating in music, speech, measured data, or the contents of the memory in a computer.

Throughout the application of a positive or negative value of the digital signal, the media is magnetized to positive or negative saturation. The length of the shortest saturated portion is the bit length, and is related to data rate and media velocity as $BL = v/(max.\ data\ rate)$. Wavelength λ and bit length BL are related by:

$$\lambda = 2\,BL$$

Figure 5.3 shows the recorded magnetization patterns inside the media for a thin disk media ($\sim 0.1\ \mu m$) and a thick tape coating ($\sim 2\ \mu m$). The disk coating is fully saturated, while the tape in this example is only partly magnetized. This incomplete magnetization is called partial penetration recording.

Notice the transition between bits. In this chapter we will seek to understand and, moreover, determine the length of this transition zone that we will designate Δx. It should be as short as possible to allow for many Δx along a given length of the recorded track, typically one inch. A short Δx will also assure a high read voltage by making

$$\frac{d\phi}{dt}$$

high ($dt = \Delta x/(velocity\ of\ media)$).

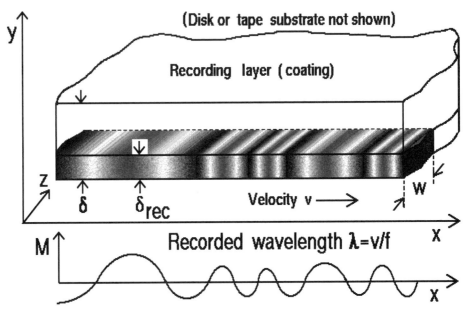

FIGURE 5.1 Recording a sine-wave signal.

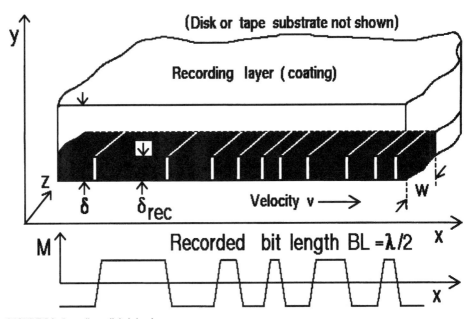

FIGURE 5.2 Recording a digital signal.

The magnetization varies gradually from negative saturation in one bit to positive saturation in the next bit. Δx is defined as the length over which the magnetization changes 50 percent, from $-0.5M_s$ to $+0.5M_s$.

These boundaries are outlined by two black lines in the illustrations. The gray levels were adjusted so each represents a change of 10 percent from positive to negative full saturation. The transi-

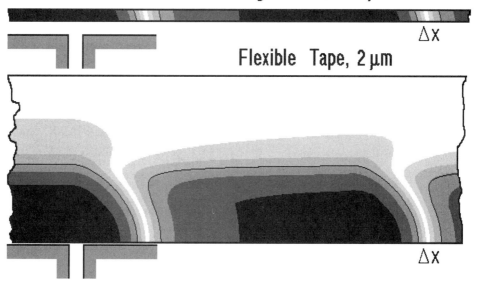

FIGURE 5.3 Typical dimensions of disk and tape media.

tion includes four gray levels, plus the white band, plus another four gray bands, totalling nine bands of each 10 percent. Black is used for saturation at either end.

A measure of the true packing density is

$$fpi_{max} = 1''/(BL + \Delta x)$$

The ultimate high density is achieved by letting BL decrease toward zero.

APPROACH TO MODELLING

A write head consists of a ring core with a short gap, from which the stray flux forms the recording field. Many disk and tape drives use thin film heads that have fields that are much the same as those from conventional ring core heads. Figure 5.4 shows the general field pattern.

The figure also sets the stage for the geometrical parameters involved: gap length g, coating thickness δ, recorded depth δ_{rec} ($\delta_{rec} \le \delta$), and the spacing d between coating and head. We will also need to know the values of the coating coercivity H_r, the switching field distribution Δh_r, and the deep-gap field H_g. Finally, we assume that the field and the coating magnetization do not change in the z-direction and our analysis is thus limited to the x-y plane. This assumption will need modification for very narrow track widths (see Chapter 19 on advanced recording).

Several observations can now be made for a coating moving through the field H, which for the time being is constant. Considering just a small volume $\Delta vol = dx \times dy \times w$, where w is the track width, we find:

• Each Δvol undergoes a magnetizing field that increases to some maximum strength and then decreases to zero.

- The volumes nearest the head "see" the strongest field: The magnetization throughout the coating is not uniform.
- The direction of the field changes, or rotates, during the pass through it. The overall magnetization is not necessarily longitudinal.
- If the coating is a distance from the head, the field strength is less and a recording loss occurs.

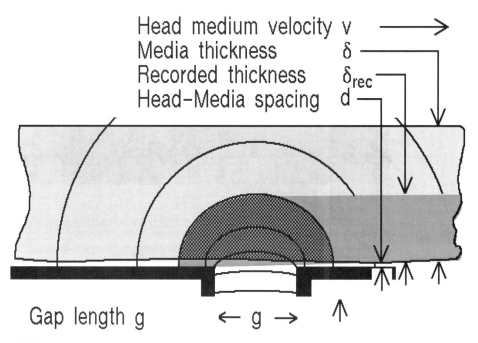

FIGURE 5.4 Dimensions associated with the write (record) process.

We will start our exploration of the write process by considering a simplified model with a very thin coating so the field strength from the write head is the same in the front and the back of the coating, and that only longitudinal magnetization occurs. This will follow the earliest explanation of the write process, used until the mid-sixties. It is of educational value, but remains a weak model of the write process.

The switching field distribution is assumed zero, which results in a zero-length transition zone between written bits. We will next improve the model by introducing a nonzero switching field distribution, plus full vector magnetization—that is, longitudinal plus perpendicular magnetizations in a thick coating. We can then make estimates of the length of the transition zone Δx between two bits, and will learn that the length of the write gap has negligible influence upon the length of Δx, and hence the packing density (see Chapter 19).

The method of adding a high-frequency ac-bias signal to the write signal is briefly introduced to familiarize the reader with the idea of a recording zone. This is the region where a Δvol moves away from the gap and the field strength becomes less than the switching field strength; then the particles' magnetizations will stop flipping (following the ac-bias field frequency), and are "frozen" into a final magnetization. The length of this zone is identical to the length of the transition zone. For details, see Chapter 20 on ac-bias recording.

We complete our exploration of the write process by considering the transition zone to be a small line pole, stretching from side to side of the track; its field should be added to the field from the gap as a demagnetizing field (see formula 4.3). Further details of the write (record) process can be found in Chapter 19.

MEDIA MAGNETIZATION

We have experienced a remarkable increase in the amount of information that can be stored on a given length of tape. Computer tape and disk systems have gone from 800 to between 50,000 and 100,000 bpi (bits-per-inch). Instrumentation recording is now extending to an upper frequency of 4 MHz (4,000,000 Hz) at 120 IPS (inches per second tape speed). Audio has gone from 15,000 Hz at 30 IPS to 20,000 Hz at 1⅞ IPS!

This means that the smallest bit size or wavelength is shorter than the coating thickness. We are then dealing with a remanence that is perpendicular rather than longitudinal, in spite of the particle's orientation. The idea of perpendicular remanence was supported as early as 1963, when two engineers at Philips, Tjaden and Leyten, carried out a large-scale recording experiment.

The magnetization pattern inside a coating is not visible by the Bitter method used in ordinary observations of domain patterns; the pictures that develop when iron powder is laid down on a side-cut of a magnetized coating will, at best, show closed flux patterns. An example is Iwasaki's observation in 1975, which led to the discovery of circular magnetization and then to his suggestion, in 1977, of using a perpendicular mode at high packing densities.

Tjaden and Leyten fabricated a 5000:1 scale model of the coating and the head for use in a recording experiment. The two-inch-thick coatings were, after recording, cut into small blocks. Magnitude and direction of magnetization in each block were measured and plotted. One of the results is shown in Fig. 5.5, clearly indicating variations in magnetization through the thickness of the coating.

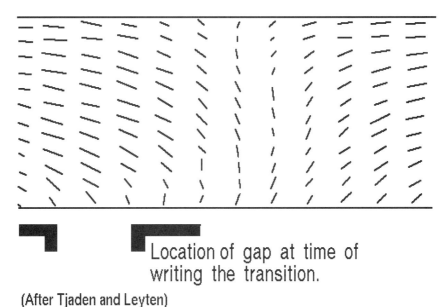

Location of gap at time of writing the transition.

(After Tjaden and Leyten)

FIGURE 5.5 Magnetization in and around a recorded transition (*Tjaden and Leyten*).

Perpendicular magnetizations appear to be preferred for short bit lengths. When the bit size (or half wavelength) equals the coating thickness, the demagnetization factor is ½ for the longitudinal or perpendicular remanence. For shorter wavelengths the perpendicular demagnetization is less than the longitudinal. A simple dipole model shows minimum energy for perpendicular remanence at very short wavelengths.

It would appear that oriented, acicular particles have little remanence to offer in the perpendicular direction, but each of them may be considered a chain of single domains (Jacobs-Beans model), where indeed the magnetization could be "fanned-out" toward perpendicular remanence, as shown earlier in Fig. 3.62.

The perpendicular magnetization components require some amount of phase equalization upon playback. This principle has long been used in instrumentation recorders, and recently in other units. We would expect this to be so when the recorded wavelength becomes shorter than the coating thickness. Examples of this, for a coating thickness of $\delta = 0.2$ mils, are as follows:

<table>
<tr><td>Instrumentation:</td><td>$v = 120$ IPS, $\lambda/2 < \delta : f > 300{,}000$ Hz</td></tr>
<tr><td>Audio :</td><td>$v = 1\frac{7}{8}$ IPS, $\lambda/2 < \delta : f > 4700$ Hz</td></tr>
<tr><td>Computer disk :</td><td>$v = 500$ IPS, $BL < \delta$: rate $> 2{,}500{,}000$ bps</td></tr>
</table>

where $\lambda =$ wavelength and bps $=$ bits per second.

The reader can now understand why we should not consider longitudinal recording only. It is, after all, the high packing density associated with perpendicular magnetization that is the pacing item in modern recording equipment. The end results are smaller disks and tape reels, or recorders with longer playing time.

RECORDING ON A THIN MEDIUM

Direct recording implies that a signal voltage is amplified and converted into a current through the record head winding. No ac bias is applied. The tape passing over the record head is assumed to be neutral, e.g., completely erased to a zero remanence. A thin coating implies that the field strength in front of the coating is equal to the field strength in the back, as depicted in Fig. 5.6.

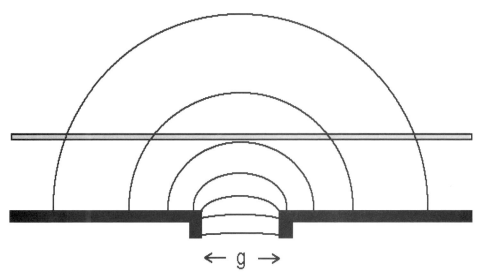

FIGURE 5.6   Recording on a very thin coating; the field strength in the back of the coating equals that in the front.

The initial magnetization curve (from a nonmagnetic state to saturation) is shown in Fig. 5.7. The quality of a recording depends upon the linearity between the field strength in the record gap and the magnetization (remanence M_r) left in the coating. A doubling of the field strength (below saturation) should result in a doubling of the remanence.

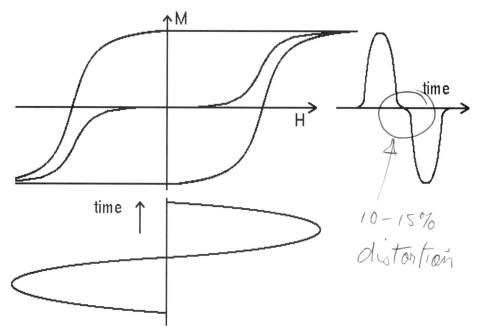

10 – 15%
distortion

FIGURE 5.7 Remanent magnetization in a coating, recorded directly. Notice zero-crossing distortion.

The pure waveform of the record current (and field) has become a highly distorted remanence signal, which is useless for any quality recording and can be used only in digital recorders, where the signal consists of ones and zeros that are written as positive or negative saturation levels.

The remanence curve around H_r is quite linear. If this region is used, then the remanence on the tape will be a linear function of the input signal. This was recognized by Valdemar Poulsen, and his friend and associate, Professor P.O. Pedersen, and they added a dc-bias current to the signal current, as shown in Fig. 5.8. The resultant remanence has a cleaner alternating waveform, with a superimposed dc remanence.

The presence of dc theoretically does not present a problem, because the induced read voltage is proportional to the differential of the flux, but in practice the story is different. Figure 5.10 shows a coating with two imperfections: an agglomerate of particles, and a void. This causes a variation in the dc remanence, with resulting noise spikes in the read voltage. Any nonuniformity of the coating will produce noise under dc magnetized conditions. The noise level is easily 20 dB or more above that produced by a properly ac-erased coating, and this method of biasing was abandoned.

This method of dc magnetizing a coating instead found its proper home in the Quality Control Department, for evaluation of the coating uniformity of tapes and disks.

The dc-bias method can be improved by writing on a premagnetized coating, as shown in Fig. 5.9. The coating passes over a permanent magnet that saturates the coating into the negative region, erases old information, and leaves it with a remanence equal to $-M_{rs}$ before reaching the record head. Biasing again to $+H_r$ provides a larger linear region for the transfer function as well as higher sensitivity ("amplification").

The dc bias can be adjusted so the dc-level remanence is zero. The bias adjustment is critical, though, and a change of a few percent of the magnetic properties of the coating coercivity (and/or coating thickness) will throw the carefully adjusted working point off. In Chapter 12 we will learn that the temperature coefficient of the coating's coercivity is 2–3 Oe/C, and a temperature change may thus throw the dc bias point out of adjustment.

Digital recording uses all three methods (no bias, dc bias with virgin tape, and dc bias with pre-saturated [dc-erased] tape). The explanations above of each write process are more qualitative than

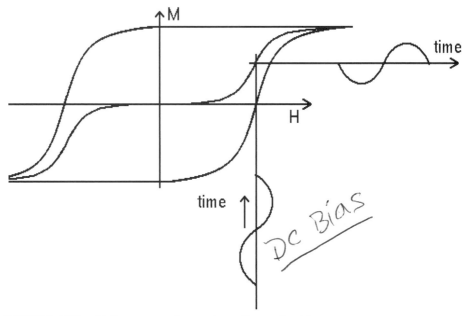

FIGURE 5.8 Addition of dc bias moves recording operation to a linear portion of the remanence curve.

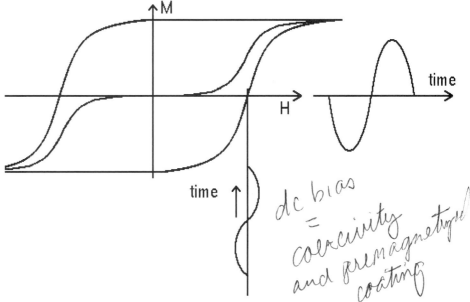

FIGURE 5.9 Linear operation is also achieved by using a bias field equal to the coercivity and an opposite premagnetized (possibly erased) coating.

quantitative. They give us no answer about the length of the transition zone between bits, and they are quite useless if we wish to explore partial penetration recordings, or the overwrite process. (Partial penetration recording is advantageous in that it provides short transition zones, as shown later in this chapter, where overwrite of old information is discussed.)

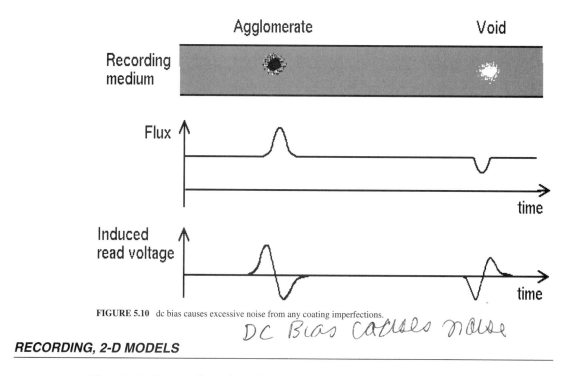

FIGURE 5.10 dc bias causes excessive noise from any coating imperfections.

DC Bias causes noise

RECORDING, 2-D MODELS

The criterion for recording information onto a media coating is quite simple: If the field magnitude $H(x,y)$ at the element Δvol is greater than the switching field (coercivity) h_{ci}, then the particles within Δvol will switch magnetization in accordance with the field.

On the other hand, when Δvol has moved away from the strong field immediately above the gap, and the field $H(x,y)$ is less than h_{ci}, then the particles will no longer follow the field; their magnetizations have been frozen at the point where $H(x,y)$ became just less than h_{ci}.

A "freezing" zone therefore exists at the trailing edge of the recording head gap; this is where transitions in digital recordings are formed. The freezing zone also corresponds to the ac-bias recording zone, shown later in this chapter.

The freezing zones will therefore be sheaths with banana-shaped cross sections in the x-y plane, as shown in Fig. 5.11. A recording model based on this idea was introduced in the early sixties by C.D. Mee and B.B. Bauer.

They used a simple model where only longitudinal recording was considered, and also assumed that all particles have identical coercivities, i.e., the *SFD* was set equal to zero. Contours of constant field strength $H_x(x,y)$ are shown in Fig. 5.11 and the pattern of a recording shown in Fig. 5.12. The model postulates a transition zone of length zero, and therefore predicts an infinite number for the FPI, flux-reversals-per-inch.

The contours are circles that go through the gap corners, centered on the gap centerline. The 3-D picture shows the shapes of transition zones at small field strength (near the gap) and at higher field strengths, away from the gap. The shapes helped corner the nicknames "bubble" or "cylinder" theory.

In a real recording medium there are particles (tapes or floppies) or grains (rigid disk and ME-tape) within a spread from low to high coercivities. *SFD* (switching field distribution) is a measure of this spread.

The *SFD* is obtained by finding the difference between the H-values that correspond to the 25 percent and 75 percent points on the *MH*-loop, as in Fig. 5.13. The value of *SFD* is found by dividing this difference ΔH with the mean value H_r, and is then assigned the symbol Δh_r. (Values of *SFD* obtained in this fashion are very close to those obtained from another expression for $SFD = 1 - S^*$, where S^* is the coercivity squareness factor; see further in Chapter 12.)

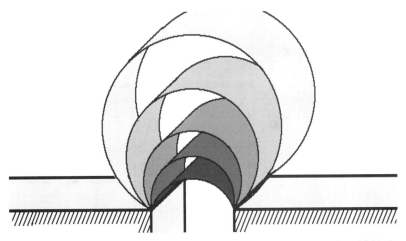

FIGURE 5.11 The first 2-D recording model (Mee and Bauer). Contours are of constant horizontal field values.

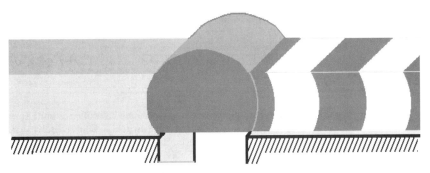

FIGURE 5.12 The recorded magnetization pattern using the "cylinder" model (Figure 5.11).

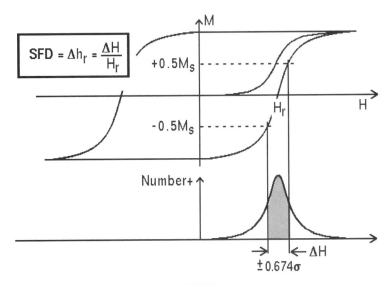

FIGURE 5.13 The definition of a switching field *SFD*.

The *SFD* accounts for exactly 50 percent of the particles. If their distribution is normal (Gaussian), then their mean deviation will correspond to $\pm 0.674\sigma$, where σ is the standard deviation. If we had chosen *SFD* points that correspond to a standard deviation $\sigma = \pm 1$, then 68.26 percent of the particles/grains would be represented by the *SFD*.

The transition length Δx between two bits is found by using the field contours $H_x(x,y)_{\text{const}}$ corresponding to $H_r \times (1 \pm \Delta h_r/2)$. This is illustrated in Fig. 5.14.

As the coating moves to the right, out of the write field, the right transition zone remains magnetized as shown, while the zone written on the left side of the head field is erased as it enters the region of higher field strength above the gap. In general, we do not care what happens on the ingoing side of the recording field; the magnetization in that region is always changed when it crosses the higher field above the gap.

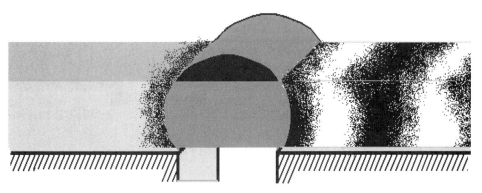

FIGURE 5.14 Magnetization changes gradually across transitions when *SFD* is included in the cylinder model.

USING Δh_r, VECTOR MAGNETIZATION

The cylinder model served well for all longitudinal-only models, but it became apparent in the late 1960s and early 1970s that the perpendicular magnetization component should be included. One method would be to consider a longitudinal and a perpendicular write process separately, and then combine the results to a final transition. We will go on by including the perpendicular field component $H_y(x,y)$ and form the field magnitude $H = [H_x^2(x,y) + H_y^2(x,y)]^{1/2}$.

Let us for a moment recall Fig. 3.64, which showed little directional change in a particulate coating's coercivity. In other words, its switching properties are primarily related to the field's strength, not to its direction nor to its anisotropy (which may not be true for solid, metallic coatings). We can, therefore, use the total field magnitudes as criteria for finding the inside and outside surfaces of the transition zone. A chart for contours of constant field strength $[H_x^2 + H_y^2]^{1/2} = \text{constant})$ was made by Westmijze (see ref.), and is reproduced in Figs. 5.15, 5.16, and 5.17. They are very useful illustrations for budgetary estimates of write conditions and resulting transition zones. Figure 5.15 is intended for coating thicknesses up to 3 times the gap length; Fig. 5.16 is for a thicker coating, and Fig. 5.17 for a thinner one.

Notice that the charts have horizontal lines spaced one or more gap lengths away from the head interface, i.e., lines at g, $2g$, $3g$ etc. One will start with the knowledge of the gap length g (often dictated by the read resolution, that is $g = 0.6 - 0.9 \times BL$, where BL = shortest recorded bit length). The coating thickness δ and the head/coating distance d allow for drawing the front and back of the coating in proper scale to the gap length by using the horizontal, evenly spaced lines. Next the write level is set, with a 100-percent level defined as that level where the contour corresponding to H_r is tangent to the coating back side.

Next, observe the numbers marked v. They represent the field strength on a given contour, relative to the deep-gap field H_g. This is the field we calculated in formula 4.21. We can write $H = v \times H_g$.

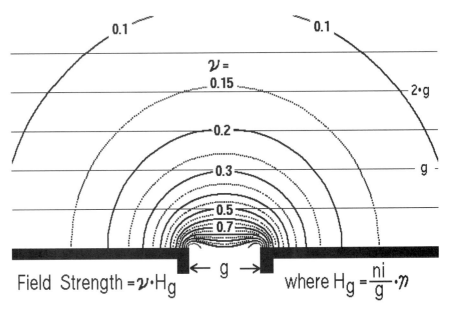

FIGURE 5.15 Constant field strength contours in front of a gap (*Westmijze*).

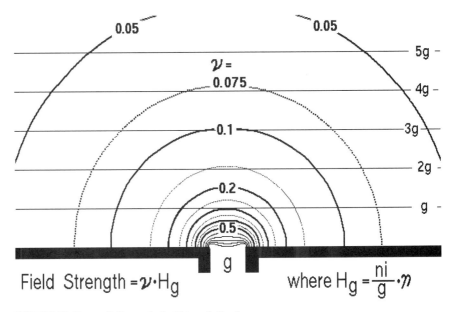

FIGURE 5.16 Constant field strengths for thick media (tape).

Finally note that most of the contours are almost circles, quite identical to the actual field lines from Fig. 4.24. Discrepancies are only found near the gap corners. That is of little concern, because most write fields have a strength that places the write zone at contour $v = 0.6$ or smaller. From the foregoing we can conclude that the field directions that are tangents to the field lines in Fig. 4.24 are also near tangents to the magnitude contours in Figs. 5.15 through 5.17.

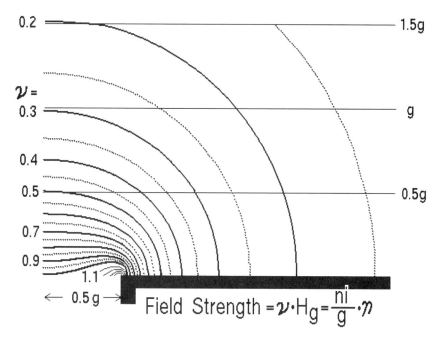

FIGURE 5.17 Constant field strengths for thin media (disk).

The transition zone Δx is determined by locating the contours that correspond to the fields of strengths equal to H_r $(1 + \text{or} - \Delta h_r/2)$, or simpler by using the contours for $\nu \times (1 + \text{or} - \Delta h_r/2)$, where the value of ν corresponds to H_r/Hg. This will in general require a visual interpolation. The length of the transition zone is defined as the dimension Δx in the very front of the coating.

Example 5.1, Tape System: Determine the length of the transition zone that results from writing bits on a coating with an $SFD = \Delta h_r$ of 0.40. The coating has a thickness of $\delta = 5$ μm (200 μin), the head/media spacing is $d = 0.1$ μm, and the gap length is $g = 1.7$ μm. The write level is 100 percent. Also determine the maximum density.

Solution: In Fig. 5.18 the coating is outlined by lines that are spaced $3g$ and $g/17$ away from the head surface. The recording zone contours are found by interpolation: The center contour that is tangent to the back side has $\nu = 0.105$, and the lines outlining Δx are therefore located at 0.105 ± 20 percent, or $\nu = 0.126$ and $\nu = 0.084$.

The actual length of Δx is determined by using the gap length dimension g as a scale = 1.7 μm. We find $\Delta x = 1.2 \times g = 2.04$ μm = 82 μin. The maximum packing density, where transition follows transition, is (1,000,000 μin/in)/(82 μin/flux reversal) = 12,200 fpi.

Example 5.2, Rigid Disk: Determine the length (and shape) of the transition zone in a write process onto a 0.05 μm (2 μin) thick coating, spaced 0.05 μm (2 μin) away from the head, that has a 0.3 μm (12 μin) long gap. The write level is set at 140 percent, and the SFD is 0.3. Also, determine the required gap field strength if H_r is 1600 Oe.

Solution: A 100 percent write level requires a gap field of $H_g = H_r/\nu = 1600/0.63 = 2540$ Oe. At 150 percent write level the gap field must therefore be $1.5 \times 2540 = 3810$ Oe. The corresponding contour is found at $\nu = 1600/3810 = 0.42$. The transition zone is outlined by $\nu = 0.42 \times (1 \pm 0.15) = 0.42 \pm 0.063 = 0.357$ and 0.486.

The transition zone is drawn in Fig. 5.19. Δx is approximately 0.073 μm (2.9 μin), and the direction of the magnetization varies between 45 degrees in the back, and 65 degrees in the front.

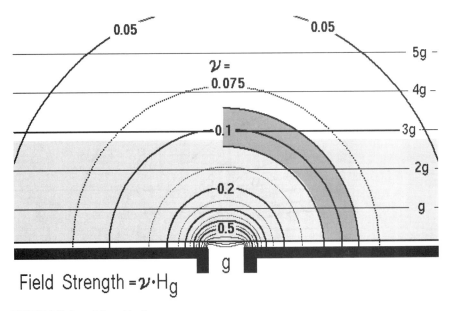

FIGURE 5.18 Recorded transition in a tape.

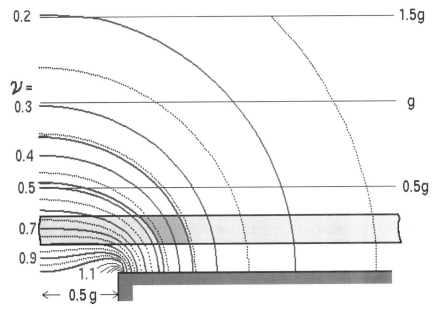

FIGURE 5.19 Recorded transition in a disk.

SETTING THE RECORD LEVEL

The two examples show that the absolute value of the medium's coercivity has no direct influence on the length of the transition zone Δx. (H_r and M_r will occur later in a formula for the transition length that includes demagnetization.) In example 5.1, the gap field H_g is set so the field strength at the magnetic coating's back equals H_r. We will define this level as the 100-percent record level.

It follows the nearly circular field magnitude curve that has a radius $R = d + \delta$. This is indeed very correct if all particles (grains) are alike and do not interact ($SFD = 0$, $\Delta H = 0$). Other record/write levels are referenced to the 100-percent level. A 50-percent level will write nearly halfway into the coating in example 5.1, while nothing will be recorded in example 5.2. The definition of record level correlates with conventional adjustment of the write current level.

Inclusion of SFD will change the conditions along the contour of value H_r. Exactly 50 percent of the particles along that contour are affected, more on the inside and less on the outside.

Consider a very thin medium: When $i_{\mathrm{write}} = 100\%$, then half the particles/grains are affected. Erasure of a recorded magnetization will be only 6 dB. If $S^* = 0.7$ (i.e., $SFD = 1 - 0.7 = 0.3$), one can generate a small table:

Write level	Erasure	Particles affected:
100%	6 dB	50%
122%	12 dB	75%
139%	28 dB	96%
152%	40 dB	99%

That was the reason for setting the write current in example 5.2 to 150 percent: to ensure sufficient overwrite (see further information on overwrite in Chapters 8 and 19).

Let us evaluate the effects of a variation in the SFD. That a larger SFD causes a longer Δx is obvious from the discussions above. But the pictorial presentations we made in Figs. 5.12, 5.18, and 5.19 leave the reader with an overly simplistic impression of an abrupt transition zone.

Let us see what happens when we write at a 70-percent level with SFDs that vary from 0 to 40 percent of H_r. This is illustrated in Fig. 5.20. The 0 percent SFD gives the ideal transition (although it follows the curvature of the field contour, rather than being straight and perpendicular to the coating surface). The not recorded portion of the coating (in this case 30 percent, from the back) is shown as a checkered pattern, indicating that there are equally many particles magnetized at either polarity; therefore, that portion of the coating does not contribute to the read flux.

As the SFD is increased from 0 to 10, 20, and 40 percent, the picture becomes rather grim. At this point it should be mentioned that a modern, high-quality medium has an SFD in the vicinity of 30 percent—which does not look too good, does it? The visual impression is supported by a program

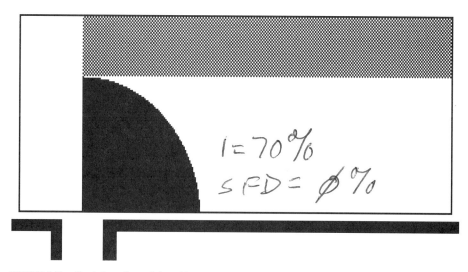

FIGURE 5.20a Simulations of recorded transitions: SFD is 0.

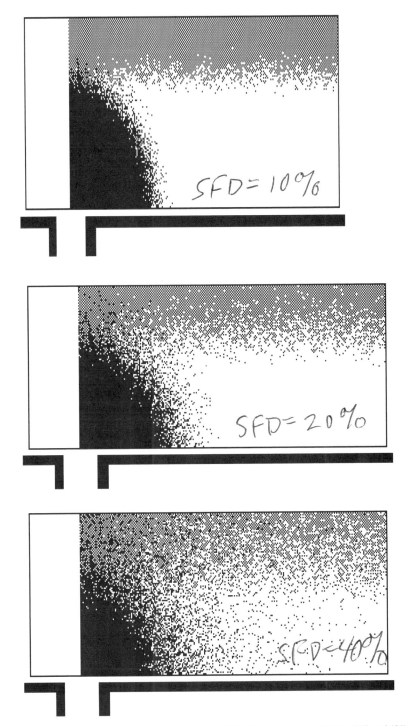

FIGURE 5.20b   Simulations of recorded transitions: SFD is 10% in top picture, 20% in middle and 40% in bottom.

subroutine that keeps track of the magnetized levels in the front of the coating, and the results are plotted in Fig. 5.21.

Two horizontal lines are drawn, corresponding to $SFD = \Delta h_r$ (and the 50 percent switched particles' measure). The magnetization transitions are indeed seen to follow arctangent like curves. For a 40-percent SFD we read that $\Delta x = 54$ μin. Had the level been 100 percent (as in example 5.1), then $\Delta x = 77$ μin; that compares with 82 μin by the graphical method. For a given medium the transition zone length will be proportional to the write (record) level.

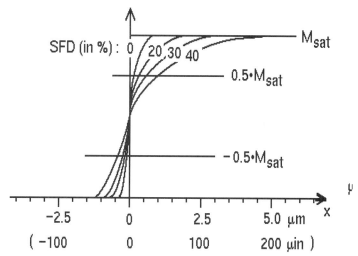

FIGURE 5.21   The change in magnetization M along the front of the recorded transition in Figure 5.20.

It is also necessary to appraise what happens when transitions are written close to one another. The program provided a pictorial answer, shown in Fig. 5.22. The medium has thickness δ, the bit length is $BL = \delta/4$, and the write level is 70 percent (that is, underwrite).

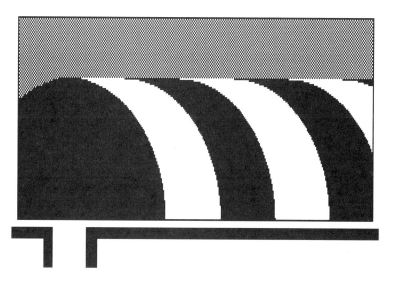

FIGURE 5.22a   Simulation of recorded transitions: SFD is ϕ.

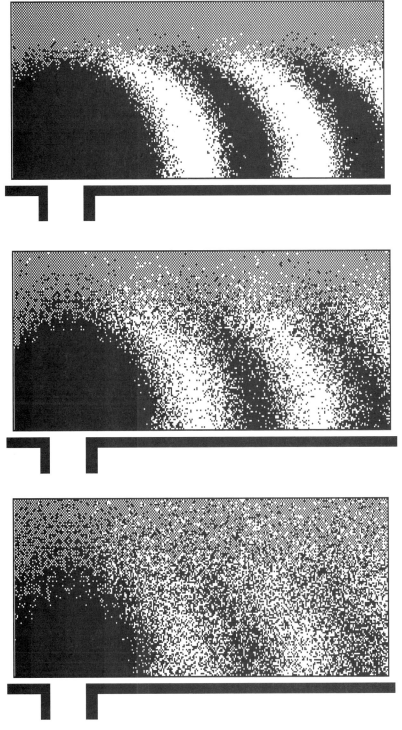

FIGURE 5.22 b Simulation of recorded transitions: SFD = 10, 20, and 40%.

The resolution suffers with increasing *SFD*, and the overall amplitude decreases. This loss is discussed at the end of this chapter, after an introduction to ac-bias recording. The interested reader will find a detailed analysis of the write process in Chapter 19.

The model used in the record simulation in the third edition of this book was rather simple: At some point (x,y) the total head field $H(x,y)$ is calculated in magnitude. The coercivity of the small assembly of particles/grains at the point is estimated from a Gaussian distribution and, if smaller than $H(x,y)$, then its magnetization is switched to a polarity set by the program. Otherwise it is not switched. It is assumed that the assemblies initially are at zero magnetization.

This model did not take into account what happens when the small assemblies at first were magnetized coming into the head field, from the left. And what happens if the media is premagnetized to some level? The latter occurs during overwrite or dc erasure.

Let us first tackle the writing of a transition on an ac-erased medium. The magnetization loop and the magnetization distribution is shown in Fig. 5.23. The maximum magnetization is set to 1 (M_s) and -1 ($-M_s$). Each distribution shown, therefore, has areas of $-\frac{1}{2}$ and $+\frac{1}{2}$, resulting in a net magnetization of $M = 0$.

A small element Δvol now enters the head field (Fig. 5.4) and is exposed to a maximum field H_{max} when crossing the gap center line. This changes the magnetization to M_1, Fig. 5.23 middle. A portion B of the negative distribution is switched to positive, where a similar number B of particles already is positive. The net magnetization therefore changes to $M = B + B = 2B$.

We now move the assembly to the right of the gap and at some time t the field is switched in polarity. Its strength at that time is H_t and its magnitude is less than (or possibly equal to) H_{max}. A portion $2C$ is now switched from positive to negative magnetization, and the resulting magnetization $M = 2B - 4C$ (Fig. 5.23). We will shortly return to the computation of M.

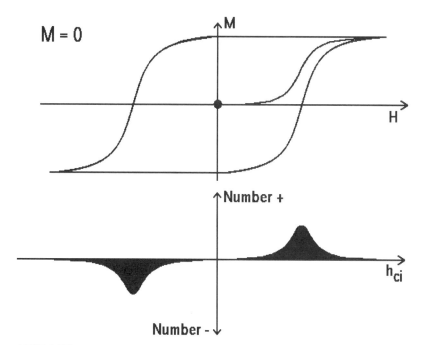

FIGURE 5.23 The change in magnetization for a field polarity change of a value less than saturation, in an ac-erased medium.

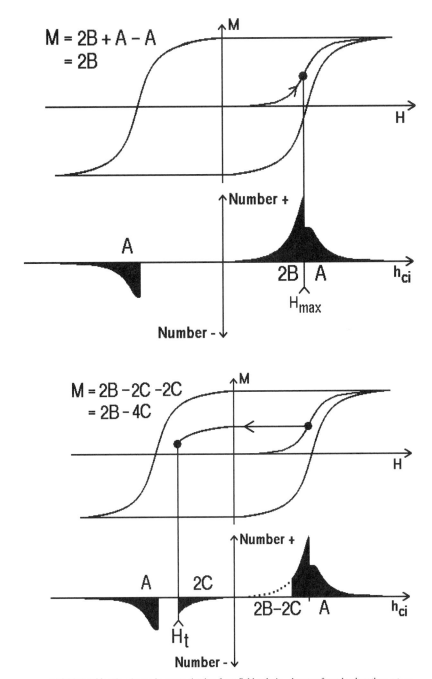

FIGURE 5.23 The change in magnetization for a field polarity change of a value less than saturation, in an ac-erased medium (*Continued*).

If the media is premagnetized to, for example, $-M_s$, then a different but similar situation exists. Figure 5.24 (top) shows a magnetization of $M = -1$ $(-M_{sat})$.

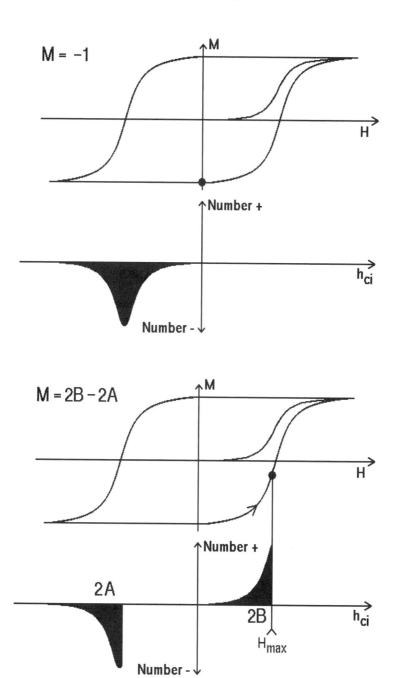

FIGURE 5.24 The change in magnetization for a field polarity change of a value less than saturation, in a negatively magnetized medium.

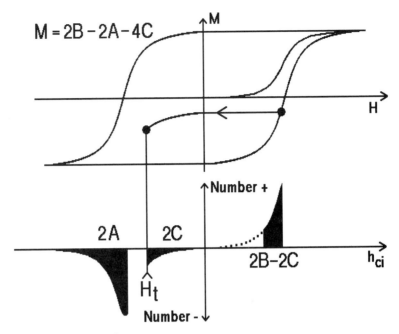

FIGURE 5.24 (*Continued*).

When it experiences the maximum field H_{max} above the gap, the magnetization changes by an amount $2B$, being switched from negative to positive. That leaves the negative magnetization at a value of $-2A$ and the difference results in a new net magnetization of $M = 2B - 2A$.

The assembly now moves to smaller field values that do not change the magnetization unless the field polarity is switched. Then a $2C$ quantity of particles is switched from positive to negative magnetization, and the resulting magnetization becomes $M = 2B - 2A - 4C$.

A similar argument can be made for a medium that was dc erased to a positive magnetization of $+1$ ($+M_s$). We will find $M = 2B + 2A - 4C$.

The results for the three cases can be summarized for writing a transition that changes the magnetization from a negative bit to a positive bit:

$$M = 2B + P \times 2A - 4C \tag{5.1}$$

where:
$P = -1$ for negatively premagnetized media.
$P = 0$ for ac-erased media
$P = +1$ for positively premagnetized media.

The computation of M is very difficult if one tries to use a numerical approximation for the MH-loops involved, but very straightforward using Gaussian distributions and setting the nominal value of $H_r = 1$:

$$2B = \text{Prob}(0 \leq h_{ci} \leq H_{max})$$
$$2A = 1 - 2B$$
$$2C = \text{Prob}(0 \leq h_{ci} \leq H_t)$$

These probabilities are equal to the areas under the distribution curves. The numerical results are found by looking up the corresponding values of the complementary error function (erfc), which is tabulated in most books on probability. There is no simple formula that allows the calculation of erfc.

It was mentioned earlier that the arctangent function has served well to describe the transition properties—that is, the complementary error function. A good approximation is:

$$\text{ercf}(x) = 0.5 + \arctan(1.25\ x)/2.62 \quad (-3 < x < +3)$$

This distribution function has its mean at 0. We wish to move it to +1 and truncate its values to fall between 0 and +2. This results in:

$$M = 0.5 + \arctan(3.75\ (x-1))/2.62$$

This M has values of -0.5 and $+0.5$ at $x = \pm 0.225$, i.e., a span of 0.45. This corresponds to $SFD = 0.45$, and to accommodate any SFD value we write:

$$M = 0.5 + \arctan((0.45/SFD)\ 3.75\ (x-1))/2.62$$

It finally becomes necessary to scale the results so we obtain $M = 0$ at $x = 0$ and $M = 1$ at $x = 2$ ($\approx 2\ M_s$):

$$M = 0.5 + N/D \tag{5.2}$$

where:
$N = \arctan((0.45/SFD)\ 3.75\ (x-1))/2.62$
$D = (1 + 0.45 - SFD)/237$

The reader will find formula 5.2 handy for simulating magnetization curves in numerical computations. It was used in the computations of A, B, and C in formula 5.1.

Some results of simulating the writing of transitions will now be presented. The formulas listed above were used in a program named MathVISION, designed for the Amiga computer (Seven Seas Software, PO Box 1451, Port Townsend, WA 98368; 206-385-1956).

The program can represent an XY-coordinate system on the screen with pixel values calculated from a formula like 5.2, where H is obtained from Karlqvist equations. Different values will be assigned different colors that can be assigned to bands of values. In the following examples each band represents a 20-percent change in value, with a white band assigned to value between -0.1 and $+0.1$ (corresponding to M equal to zero $\pm$ 10%). Separated by four bands come the extreme values of -1 and $+1$, representing saturation magnetizations and both shown as black in the following illustrations.

Writing with a level of 100 percent and $SFD = 0$ results in Fig. 5.25a, showing a circular, thin transition zone. When the SFD is given values of 10, 30, and 50 percent, Figs. 5.25b through 5.25d result. The front transition zone dimension Δx is outlined by two thin black lines. The figures show the gradual change in magnetization, and measurement of Δx reveals that it is proportional to the SFD value.

Additional computer runs for other SFD values resulted in calculated Δx values:

SFD fraction	Δx **Fraction of write depth**
0.1	0.117
0.2	0.212
0.3	0.308
0.4	0.387
0.5	0.465
0.6	0.555

A good approximation is (in contact)

$$\Delta x \approx SFD_{\text{fraction}} \times write\ level \tag{5.3}$$

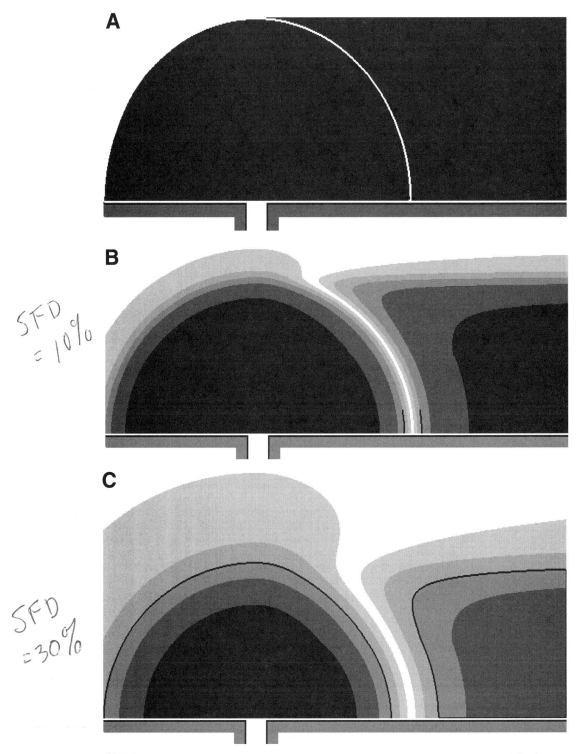

FIGURE 5.25 Simulations of recorded transitions: *SFD* is 0, 10, 30, and 50%. Transitions are written in the hard direction.

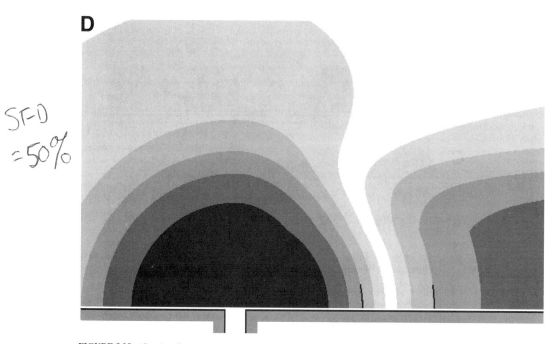

SFD
=50%

FIGURE 5.25 (*Continued*) .

There is a great spread in *SFD* values for various media, and they depend on the particulate/granular material characteristics as well as upon its squareness *S* (or *SQ*). The following table shows some recently published values:

Material	SFD
BaFeO	0.1
γFe_2O_3	0.3
Metal (MP)	0.5

Further results are shown in Figs. 5.26a through 5.26d, where the *SFD* is held constant at 0.3 and the *write level* changed. Tabulating we find:

Write level fraction	Δx fraction of write depth
0.25	0.076
0.50	0.154
1.00	0.304
1.50	0.450

Again, these computations are in good agreement with formula 5.3.

FIGURE 5.26 Simulations of recorded transitions in a medium with *SFD* = 0.3. Write levels are 25%, 50%, 100%, and 150%.

B

C

D

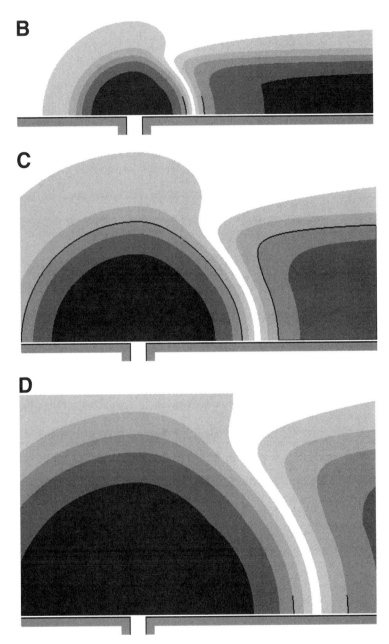

FIGURE 5.26 (*Continued*).

COMPARISON OF Δx *WITH THE TRANSITION PARAMETER* a

The transition zone length is traditionally determined from a model originated by Comstock and Williams. The magnetization is assumed to change after an arctangent function:

$$M(x) = (2M_r/\pi)\tan^{-1}[(x - x_0)/a].$$

The magnetization changes from $-M_r/2$ to $+M_r/2$ when we set $(x - x_0) = -$ or $+ a$. This coincides with the definition of *SFD* and the associated Δx:

$$\Delta x = 2a$$

The transition parameter a is derived in several places in the literature. The earlier expression from Comstock and Williams paper has been modified by Middleton to include a thick medium:

$$a \approx [(M_r/\pi H_r) \times \delta(d+\delta/2)]^{1/2} \tag{5.4}$$

(M_r and H_r are in kA/m)

Example 5.2 revisited: A CoPt alloy disk coating has M_r/H_r that ranges from 7 to 16. With $\delta = d = 0.05$ µm we calculate that $2a$ ranges from 0.182 to 0.276 µm. The graphic determination gave $\Delta x = 0.073$ µm.

The simple graphic approach tends to be overly optimistic. In Chapter 19 we will recompute Δx with an added demagnetization field. The results will show an increase of 2 to 2.5 times the graph value. This would bring the Δx value of 0.073 µm to 1.5 to 1.8 µm, very close to the calculated values.

The graphical approach has its strength in its intuitive presentation.

ERASURE AND OVERWRITE

The program used in the prior section is readily modified to illustrate the effect of an erase head. An ideally saturated coating is shown in Fig. 5.27, top. It is again a side view of the magnetized coating, but now with ideal, straight transitions. If this coating, having a switching field distribution of *SFD* = 0, passes over a write head (or erase head) with a field strength halfway through the coating, H_r that equals the coercivity halfway, then exactly half the coating is erased and rendered nonmagnetic.

In real life the coating has an *SFD* around 30 percent, for which the middle of the coating is only partially erased as shown in Fig. 5.28. And even when the field is increased so the field strength at the back of the coating equals the coercivity, then there is clearly residual magnetization from the prior recording, shown in Fig. 5.29, bottom. The overwrite is never quite complete (more on this topic in Chapter 19).

A

FIGURE 5.27   Top: "ideal" magnetization in a coating. A 50% write level erases less than 50% of the magnetization, and a 100% write level erases only approximately 90%.

B

(handwritten note: Low coercivity particles on back side)

C

(handwritten note: SFD = 30%)

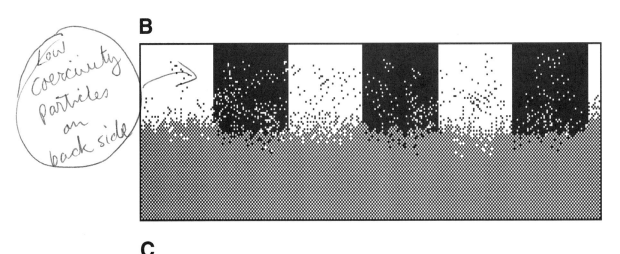
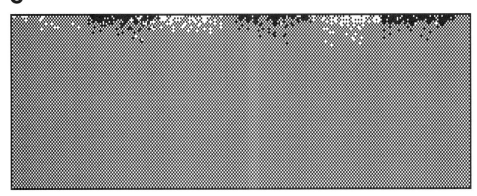

FIGURE 5.27   *(Continued).*

A further increase in field strength now results in apparent complete erasure, but there is yet residual magnetization, although it is many dB below saturation.

This analysis does not include the tendency of the erase signal, or new data signal when overwriting, to re-record the old recorded information. Both may act as ac-bias signals that greatly enhance the sensitivity of the record process (see Chapter 8, *Head Fields*). It is now useful to introduce ac-bias in its elementary form.

RECORDING WITH AC BIAS

In the late twenties, it was discovered that the addition of a high-frequency signal to a sound or data signal greatly improved the quality of magnetic recording. This technique is called ac-bias recording and is used in all the audio and instrumentation recorders of today. The action of ac bias is not easily explained.

An elementary understanding is obtained by studying the effect upon a magnetic material by a decaying ac field. Figure 5.28 illustrates how a magnetic material is demagnetized by an alternating field that slowly decays to zero. Any initial remanence M_r is gradually reduced through each cycle of the field and finally reaches a value of zero. The figure shows only a few cycles to illustrate the principle; in reality hundreds of cycles are required for the decaying signal. It is generally recommended to apply at least 300 smoothly decaying cycles in order to achieve a good, quiet erasure. This lasts 5 seconds at 60 Hz.

The alternating field that erases the tape or disk should leave the media in a zero-magnetization state. This is shown in Fig. 5.29 where the negative and positive magnetization distributions show equal values equal to half the total saturation magnetization.

This is the principle used in erasing a tape, where the tape is subjected to a strong alternating field that slowly decays to zero.

If a dc field is present during the field's decay, then the alternating cycles will leave the tape strongly magnetized. If the dc field, representing a signal to be recorded, varies in strength, then the remanence will vary accordingly. This remanence will be left where the coating leaves the record head field; both field magnitudes decay with distance (the ac field as well as the slower varying signal field).

Figures 5.28 and 5.29 do not predict that this will happen. The theory needs to include interaction fields between particles/grains, and this is done by introducing a Preisach diagram that represents the media magnetization distribution with regard to not only switching field distribution but also interaction field distribution. A full treatment of this topic is found in Chapter 19. (It will also show that without interaction fields ac bias will not work! Hence Figs. 5.28 and 5.29 fail to show a dc remanence when a dc field is present during the decay of the alternating signal.)

For a certain value of the superimposed ac-bias field the recording process is very linear, with a high signal-to-noise ratio. Therefore, high-quality recordings are obtained by adding an ac-bias field to the signal field; the frequency of the ac bias must be several times higher than the highest audio (or data) frequency to be recorded to avoid beat frequency appearing in the recording.

The ac-bias field also further reduces the tape background noise, because the alternating field causes the tape to leave the record head in a neutral condition when no signal is present. Tape coating irregularities will still cause noise, but to a much lesser degree than when recording with dc bias,

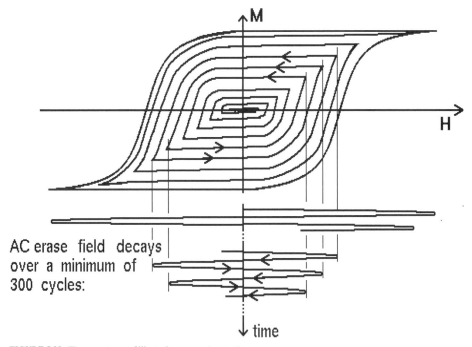

FIGURE 5.28   The correct way of illustrating erasure (see text).

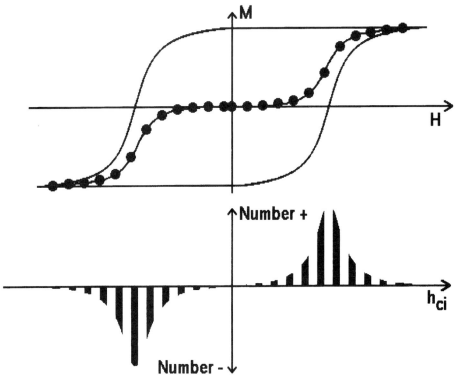

FIGURE 5.29 Erasure.

where the tape is recorded essentially as one long magnet in the absence of a signal field. Any coating irregularities show up as a noise output caused by a change in the external tape flux.

A distorted ac-bias waveform will cause excessive tape noise because the distortion very likely will contain a dc component. The better recorders use a push-pull oscillator to develop bias, which produces an ac field that is largely free of any dc component and second harmonic distortion. The amplitude of the ac-bias field plays a major role in the quality of a recording. Each particular tape (i.e., coercivity, formulation, and/or tape thickness) requires a certain bias level for optimum performance. The following general rules apply:

- A bias level that is too low will result in a noisy and highly distorted recording with excessive high-frequency response; this condition is called underbias.

- An overly high bias level will result in a quiet recording, but with a noticeable drop in the high-frequency level. This condition is called overbias, in which the bias field is so strong that the short wavelength resolution suffers.

RECORDING ZONE; RECORDING LOSSES

The freezing of a particle's magnetization occurs when it crosses the head field line that has a strength equal to its coercivity h_{ci}. This leads us to the concept of a recording zone which, speaking in terms of field lines, has a width of $SFD \times H_r$ around the line corresponding to a strength equal to H_r. This is illustrated in Fig. 5.30, where the recording zone is shown for different bias levels. The recording zone therefore has the same geometry as the transition zone discussed earlier (re. Koester, et al., 1984, Wilson, 1975).

The bottom figure shows a recording where the recording zone reaches through the magnetic coating of the tape. When the tape approaches the gap, it enters a magnetic field of increasing strength, and as it passes over the gap, most particles "flip" back and forth under the influence of a saturating bias field. When the coating passes the trailing edge of the gap, it goes through the recording zone and becomes permanently magnetized in accordance with the presence of a signal field H_{signal}.

The length of the recording zone determines the shortest wavelength that can be recorded on the tape for a particular tape speed. If the polarity of the data field changes 180 degrees as the tape moves through the recording zone, the net recording will be zero. This null occurs when the recorded wavelength equals the length of the recording zone; this is not a very well-defined zone, and will be discussed further in chapter 19.

The length of the recording zone in a typical cassette audio recorder is normally between 2 and 5 μm (80 and 200 μin) for underbias and overbias settings respectively. The band edge losses (highest frequencies), due to the recording zone length, range from a few dB to 10–20 dB. They are caused by the *SFD*, and are not a result of bias or self-erasure (whatever that may be). The loss due to *SFD* is (in dB):

$$Loss_{\text{SFD}} = 20 \times (1 - S^*) \times i_{\text{write}}/i_{100\%} \times (\delta + d)/BL$$
$$= 40 \times (1 - S^*) \times (i_{\text{write}}/i_{100\%}) \times (\delta + d)/\lambda$$

Example: $\delta = 5$ μm, $d = 0.1$ μm, $i_{\text{write}} = 1.1 \times i_{100\%}$ (normal bias), $S^* = 0.7$, $\lambda = 1\frac{7}{8}''/20{,}000$ (*v/f*) $= 2.38$ μm ($= 95$ μin). Loss is $40 \times 0.3 \times 1.1 \times (5.1)/(2.38) = 28.3$ dB.

Modern tapes with a narrow range of switching fields reduce this loss. Reduced thickness will provide the same result and tapes with a magnetic coating thickness on the order of 0.5 μm ($= 20$ μin) are under development. Due to reduced long-wavelength output, these tapes are strictly intended for storing digital information.

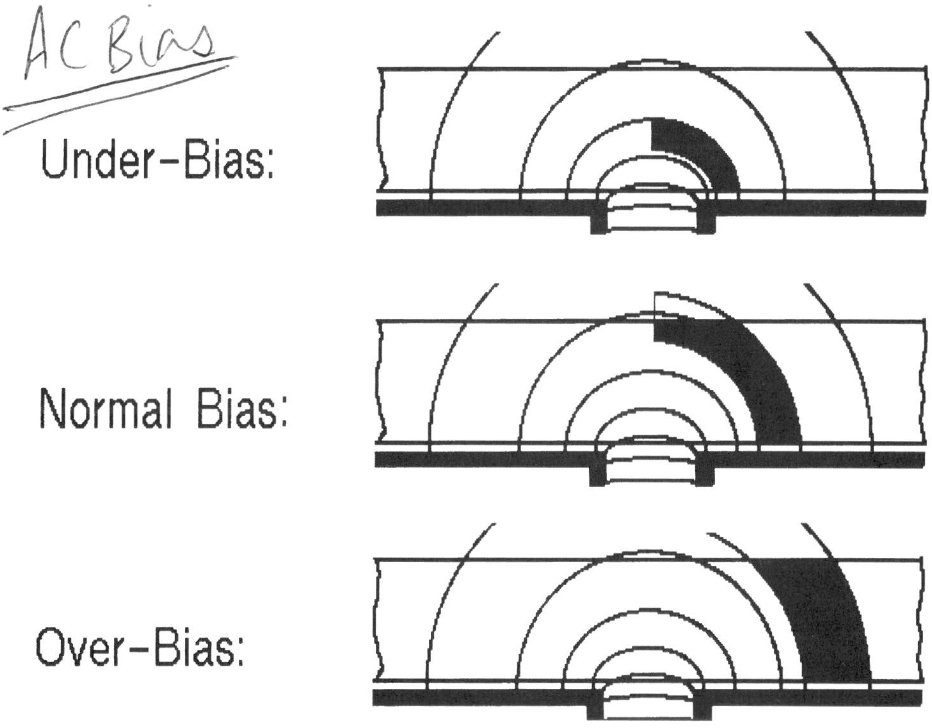

FIGURE 5.30   "Freezing" zones, ac-bias.

Another loss during recording occurs when the record head is contaminated by debris and the tape is therefore spaced away from the recording head. Then only a portion of the coating is magnetized, and the loss may be analyzed using the earlier discussed models for the recording process. A general formula for this loss as a function of the spacing is not available (as it is in playback, see next chapter).

In order to reduce the recording losses it is necessary, therefore, to have a recording zone as short as possible. This can be achieved by reducing the bias level, as shown in Fig. 5.30. But as we reduce the bias level, the tape is no longer recorded throughout its full thickness, the level at long wavelengths is reduced, and the signal is thus highly distorted.

During the past years considerable effort has been devoted to the design of record heads with a short recording zone; some success has been reported with pole shaping (thin film heads), multiple poles (the cross field head), very-high-frequency eddy current field shaping, and perpendicular pole heads on a magnetic coating with a highly permeable undercoat. Simultaneously, work continues in developing tape and disk coatings with smaller *SFD*. The interested reader is referred to later chapters in this book and their literature listings.

MAGNETIZATION PRODUCED BY A SINE WAVE SIGNAL RECORDING

Figure 5.31 illustrates how the remanent recorded flux level increases to a maximum at $i_{100\%}$, where it would remain if the coating was isotropic. The different squareness in the x and y directions (lon-

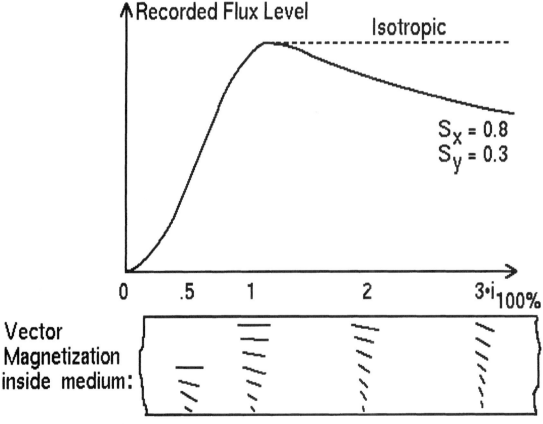

FIGURE 5.31 Flux level vs. write current.

Write level: 50 %

Write level: 100 %

Write level: 175 %

FIGURE 5.32 Magnetization patterns.

gitudinal and perpendicular) causes an internal vector magnetization that tends toward the perpendicular field direction that prevails at high write levels (Bertram 1982).

This reduces the net magnetization level. Also demagnetization and *SFD* losses reduce the remanent flux. A reduction in flux level is also observed at high write current levels for digital signals. Figure 5.32 shows how the magnetization level is reduced when the magnetization becomes perpendicular at high levels. Demagnetization also increases.

Note also the longer transition zones. They will produce less read voltage in inductive heads, because:

- The flux magnitude change becomes smaller from bit to bit.
- The rate of change $d\phi/dt = \Delta\phi/\Delta t$ is reduced when the length $\Delta x = 2a$ becomes longer ($\Delta t = \Delta x \times v$).

These matters have less significance when the disk magnetic is very thin. But as bit lengths become smaller, their thicknesses are no longer vanishing. It was recently commented that magnetic media should all be treated as thick in future models (Middleton, 1992).

REFERENCES

Bauer, B.B., and C.D. Mee. Jan 1961. A New Model for Magnetic Recording. *IRE Trans. Audio* AU-9 (1): 139–145.

Bertram, H.N. 1994. *Theory of Magnetic Recording*. Cambridge, UK, Cambridge University Press, 356 pages.

Bertram, H.N. May 1984. Geometric Effects in the Magnetic Recording Process. *IEEE Trans. Magn.* MAG-20 (3): 468–478.

Bertram, H.N., and R. Niedermeyer. Nov. 1982. The Effect of Spacing On Demagnetization in Magnetic Recording, *IEEE Trans. Magn.* MAG-18 (6): 1206–09.

Comstock, R.L., and M.L. Williams, 1971. An Analytical Model of the Write Process in Digital Magnetic Recording. AIP Conf. Proc., Vol. 5 (1): 738–42.

Dinnis, A.K., B.K. Middleton and J.J. Miles. 1993. Theory of longitudinal digital magnetic recording on thick media. Jour. Magn. and Magn. Matls. 120 (1993): 149–153.

Iwasaki, S.I., and K. Takemura. Nov. 1975. An Analysis for the Circular Mode of Magnetization in Short Wavelength Recording. *IEEE Trans. Magn.* MAG-11 (5): 1173–1176.

Iwasaki, S.I., and Y. Nakamura. July 1977. An Analysis for the Magnetization Mode for High Density Magnetic Recording. *IEEE Trans. Magn.* MAG-13 (5): 1272–1278.

Koester, E. Jan. 1984. Recommendation of a Simple and Universally Applicable Method for Measuring the Switching Field Distribution of Magnetic Recording Media. *IEEE Trans. Magn.* MAG-20 (1): 81–83.

Koester, E., H. Jakusch, and U. Kullman. Nov. 1981. Switching Field Distribution and A.C.Bias Recording Parameters. *IEEE Trans. Magn.* MAG-17 (6): 2550–2552.

Mee, C.D. Jan 1962. Applications and Limitations of the New Magnetic Recording Model. *IRE Trans. Audio.* AU-10 (1): 161–164.

Middleton, B.K. 1987. Recording and Reproducing Processes, Ch. 2 in *Magnetic Recording*, Vol. 1, Ed. by Mee and Daniel, McGraw-Hill, pp. 22–97.

Suzuki, T. Sept. 1992. Orientation and Angular Dependence of Magnetic Properties for BaFe Tapes. *IEEE Trans. Magn.* MAG-28 (5): 2388–90.

Tjaden, D.L.A., and J. Leyten. Sept. 1963. A 5000:1 Scale Model of the Magnetic Recording Process. *Philips Technical Review* 25 (11/12): 319–329.

Wilson, D.M. Sept. 1975. Effects of Switching Field Distributions and Coercivity on Magnetic Recording Properties. *IEEE Trans. Magn.* MAG-11 (5): 1200–1202.

BIBLIOGRAPHY

Daniel, E.D., and P.E. Axon, and W.T Frost. Jan. 1957. A Survey of Factors Limiting the Performance of Magnetic Recording Systems. *Jour. AES*, 5 (1): 42–52.

Hoagland, A.S. Jan. 1958. High-Resolution Magnetic Recording Structures. *IBM Jour. Res. & Devel.* 2 (1): 91 ff.

Westmijze, W.K. July 1953, Studies on Magnetic Recording: pt. III. The Recording Process. *Philips Res. Repts.* 8: 245–255.

Westmijze, W.K. Sept. 1953. The Principle of the Magnetic Recording and Reproduction of Sound. *Philips Technical Review* 15 (3): 84–95.

CHAPTER 6
PLAYBACK (READ) THEORY

A recorded tape is, during playback, moved across a read (also known as reproduce or playback) head, shown in Fig. 6.1. The head's core is fabricated from a magnetically soft material with high permeability, and the magnetic flux from the tape links through this core and induces a voltage in its winding.

We can easily find the signal voltage from the playback of a sinusoidal signal recorded on a very thin coating in contact with the read head. The read voltage e is, from Faraday's law:

$$\begin{aligned} e &= -nd\phi/dt \\ &= -n\Phi_m\omega\cos\omega t \\ &= -n\Phi_m(2\pi f)\cos(2\pi x/\lambda) \end{aligned} \tag{6.1}$$

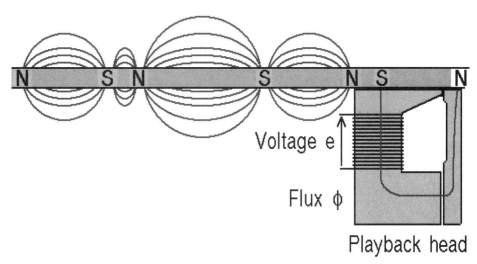

FIGURE 6.1 Playback (read) of recorded data.

where

n = number of turns on the core

$\omega = 2\pi f$

f = signal frequency in Hz

Φ_m = peak flux level (Wb)

$\phi = \Phi_m \sin \omega t$

$\quad = \Phi_m \sin(2\pi x/\lambda)$

x = position along coating ($x = vt$)

λ = recorded signal wave length (m)

v = head-to-medium speed in m/sec

Hence the voltage increases proportional to $1/\lambda$, or the frequency f, at a rate of 6 dB/octave. This is only observed for low values of the signal frequency. The measured output voltage curve is the one showed lowest in Fig. 6.2. The difference at higher frequencies between a theoretical 6 dB/octave rise, and the actual voltage is caused by several losses. In this chapter we will examine these playback losses:

• Spacing loss

• Coating thickness loss

• Gap length loss

• Gap alignment loss

• Reproduce core loss

The read, or playback, voltage from complicated magnetization patterns can only be calculated by means of a computer. We have shown a formula for the read voltage only for purely longitudinal magnetization recorded on a very thin coating with sine or cosine wave patterns. By writing the flux variations along the coating as $\phi = \Phi_m \times \sin(2\pi x/\lambda)$, and substituting this flux into Faraday's formula, we arrive at formula 6.1, which is the loss-free playback voltage.

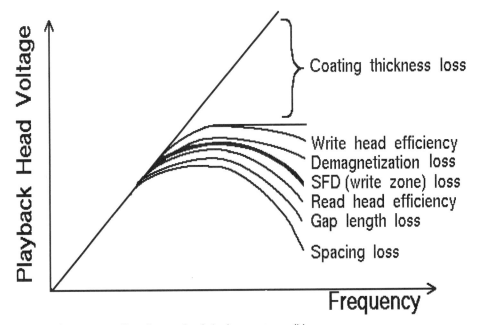

FIGURE 6.2   A summary of losses in magnetic write/read systems, tape or disk.

The voltage is, to no surprise, proportional to the number of turns and the flux level on the tape. It is also proportional to the recorded frequency and would therefore result in a very unbalanced reproduction of the input signal. For example, the higher frequency content of an input signal would be read out at a higher voltage than would lower frequencies. Some electronic equalization will be necessary to restore the signal, a topic we will discuss in Chapter 22.

FLUX LEVEL THROUGH HEAD CORE; THICK COATING

The flux from a recorded tape is rarely purely sinusoidal, and is always comprised of a mix of longitudinal and perpendicular magnetizations. It is also apparent that equal magnetizations in front and back of a coating result in different flux levels at the read head. How does the flux actually link to and through the head? An answer to this question is no easy task when the coating thickness is comparable to the gap length.

Take for example a simple, small magnetic dipole, placed in front of the read head. When alone, the field strength from a dipole is readily calculated, but the presence of the core greatly modifies the field pattern. Elaborate calculations using magnetic field imaging and mapping have been carried out, giving excellent results in agreement with measurements, but with a loss of understanding during the mathematical derivations. Moreover, the reader is reminded of Fig. 5.5, showing how the total coating magnetization can be thought of as being a large number of dipoles, each contributing to flux through the head.

If, however, we can find a method for calculating the flux into the head core from just one such dipole element, then we can sum up the fluxes from all elements in the vicinity of the head to account for the total flux. We will therefore consider the simple system shown in Fig. 6.3. Recall that a magnetic dipole can be characterized by its magnetic moment $m = pl$, and that this is equivalent to the magnetic dipole moment $j = iAn$ from a small coil (formulas 3.4 and 3.15). Let us therefore replace the small magnetic dipole with the coil dipole.

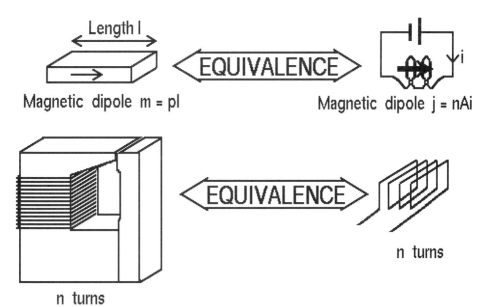

FIGURE 6.3 The playback principle can be simplified by replacing the permanent-magnet bit with its equivalent coil dipole moment.

A magnetic head can likewise be modelled by a small coil, of length equal to the gap length, and with n turns. Let us therefore replace the head with such a coil. We have now reduced the problem to that of finding the flux coupling from one coil into another nearby coil. Finding an answer to this problem may be straightforward, and we must then substitute the magnetic dipole moment back into the expression we derive.

The magnetic circuit of two coupled coils is an item we all come across. Take for example the small power supply you connect to $110\ V_{ac}$, which then delivers $6\ V_{dc}$ to your transistor radio or tape recorder. Inside the power supply is a small transformer that changes the $110\ V_{ac}$ into $6\ V_{ac}$ (which then is rectified to provide the $6\ V_{dc}$). The transformer is an example of two closely coupled coils.

One may also connect $6\ V_{ac}$ backward through the transformer and obtain $110\ V_{ac}$. This property of being able to act in both directions is called reciprocity. We call the device a reciprocal transducer; this property is found in many devices, for example piezoelectric crystals. With properly mounted electrodes you will find that an applied voltage will bend the crystal. If you, however, bend the crystal, then a voltage is produced between the electrodes. Another example is the magnetostrictive transducers used in ultrasonic cleaning tanks, and in underwater sound producing/receiving devices. The loudspeaker in an intercom system is reciprocal; it will produce sound as well as act as a microphone.

A magnetic ring core head will record as well as read: It is a reciprocal transducer. This fact will allow us to proceed with finding the read flux. Let us consider two coils, L_1 and L_2, with a mutual coupling M (see Fig. 6.4). A current i_1 through L_1 produces a flux:

$$\phi_2 = M \times i_1 \tag{6.2}$$

through L_2; and a current i_2 through L_2 produces flux

$$\phi_1 = M \times i_2 \tag{6.3}$$

through L_1. When comparing Fig. 6.3 with 6.4 we see that the answer we seek is formula 6.2, flux through the head. The current i_1 is part of the dipole moment, and the unknown is therefore M.

The coupled coils we are interested in are shown in Fig. 6.5. The left and right drawings correspond to those in Fig. 6.4.

By expressing (6.3) in words we have: At the location of coil 1 a field with flux ϕ_1 is produced by the current i_2, and its magnitude is $M \times i_2$. This brings to mind the record head field that is produced outside L_2 when energized. Karlqvist found the formulas (Fig. 4.24) for the field produced at points (x,y) in front of a record head, fed with the current i_2.

We can now find M, by first finding the flux and then isolating i_1; the rest will be M. We are again considering a two-dimensional field pattern, assuming a wide track. ϕ must therefore be found for both longitudinal and for vertical components. We will solve for ϕ_x, and can write:

$$\begin{aligned}
\phi_{x1} &= B_{x1}A_1 \\
&= H_{x1}\mu_o A_1 \\
&= (H_{x1}/i_2)i_2\mu_o A_1 \\
&= H_{sx1}i_2\mu_o A_1
\end{aligned} \tag{6.4}$$

Comparing formulas 6.3 and 6.4 we see then the mutual coupling coefficient equals $H_{sx1}\mu_o A_1$. $H_{sx} = H_{xs}\mu_o A_1$ is called the sensing function for the longitudinal magnetization. It has dimensions m^{-1}, similar to an antenna. We obtain our intermediate result by substituting into formula 6.2:

$$\begin{aligned}
\phi_{x2} &= M \times i_1 \\
&= H_{sx}\mu_o A_1 i_1
\end{aligned}$$

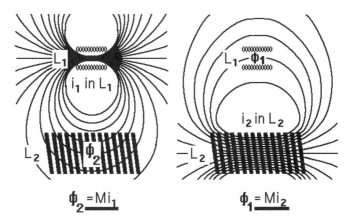

FIGURE 6.4 Mutual coupling between two coils.

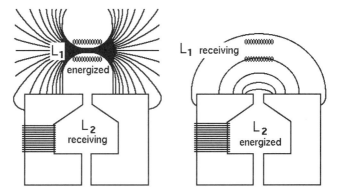

FIGURE 6.5 Mutual couplings. Left: from dipole moment to read head. Right: from write head to location of dipole.

The dipole moment for a single turn loop equals $j = Ai$, and we may now further substitute m_{x1} for $j = A_1 i_1$:

$$\phi_{x2} = H_{sx}\mu_o m_{x1}$$

The result is a net head flux that equals the product of the sensing field H_{sx} and the magnetization $\mu_o m_x$. A small volume ($\Delta vol = dxdyw$ in Fig. 5.4) has magnetization

$$\Delta m_x = S_x M_{sat}\Delta vol$$

where S_x is the longitudinal squareness and M_{sat} the saturation magnetization. Demagnetization is not included. The flux from longitudinal magnetization is:

$$\Delta\phi_x = H_{sx}\mu_o\Delta m_x \ \text{(Wb)}$$

A similar derivation for flux from any perpendicular magnetization is:

$$\Delta\phi_y = H_{sy}\mu_o m_y \text{ (Wb)}$$

The total flux from the small element Δvol with magnetization $\Delta m = \Delta m_x + \Delta m_y$ is therefore (formula 6.5a):

$$\Delta\phi = H_{sx}\mu_o\Delta m_x + H_{sy}\mu_o\Delta m_y \tag{6.5a}$$

We recognize this expression for the flux as the dot product of the sensing function H_s and the coating's magnetization $\mu_o\Delta m$. Both are vectors.

Formula (6.5a) is convenient for numerical computations. Another way of writing the dot product is:

$$\Delta\phi = H_s\mu_o\Delta m \cos(\beta) \tag{6.5b}$$

In vector notation the read flux is:

$$\Delta\phi = \mu_o \boldsymbol{H_s} \boldsymbol{\cdot} \boldsymbol{\Delta_m} \tag{6.5c}$$

Formula 6.5b is convenient for quick estimates: β is the angle between the direction of the magnetization and the sensing function. Figure 6.6 illustrates a simple example. The magnetization in a coating is often tilted as shown. m_1 forms an angle of 90° with the sensing function direction and its contribution to the flux is therefore zero (cos 90° = 0). m_2 is parallel to the sensing field, and its flux contribution is therefore equal to $H_s\mu_o m_2$.

The total flux is now obtained by a double summation (integration), one from minus infinity to plus infinity to account for all magnetizations along a layer in the tape, and the other from d to $d+\delta_{rec}$,

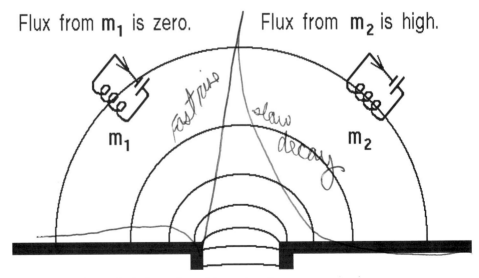

Flux from $\mathbf{m_1}$ is zero.   **Flux from $\mathbf{m_2}$ is high.**

m_1 m_2

FIGURE 6.6   The flux contribution from m_1 is zero; it is at maximum from m_2.

where d is the spacing between the head and tape surfaces and δ_{rec} the recorded thickness of the coating. The last summation accounts for all the layers in the coating.

A few comments are in order. It is difficult to envision the reciprocal sensing function, but an analog example will help. Envision a VHF antenna being used by a radio amateur for transmitting (see Fig. 6.7). The antenna is made from a single dipole with added elements for directivity. This will focus the transmitted energy into a narrow beam, which is directed toward the geographical location where the receiver is. When the operator switches to listening, he disconnects the antenna from the transmitter and connects it to the input of the receiver. This assures a higher sensitivity toward the point where he expects to receive a reply. The sensing function does have the same pattern as the transmitted energy pattern. The same holds true for the write/read (record/playback) heads.

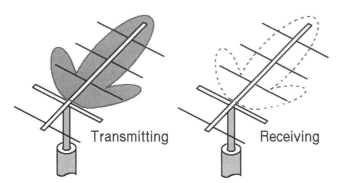

FIGURE 6.7 An antenna has directionality in both operating modes.

Note also the dimensions in the formulas above. The field strength from the head gap is measured in A/m (Oe), using the expressions in Fig. 4.24; H_g equals $\eta ni/g$, where η is the head efficiency, ni the ampere turns, and g is the gap length. The definition of the sensing function is field strength divided by a unit current of 1 Amp, which gives the sensing function the unit of m^{-1}. When using Fig. 4.24 to find values of the sensing components (x and y), use $H_g' = \eta n/g$ when figuring the sensing function.

The unit for the dipole magnetization is Am^2. Here we multiply with μ_o, which has a unit of H/m = Vs/Am, and we obtain Wbm. Further multiplication of the magnetization with the sensing function of unit m^{-1} results in units of Wb, which is correct for flux.

The playback waveforms depend upon the magnetization in the coating. A longitudinal magnetization will result in read waveforms shown in Fig. 6.8. Perpendicular magnetization results in the waveforms in Fig. 6.9.

A single transition in a longitudinal coating is represented by two opposite magnetizations, both lying in the longitudinal direction. The reader may wish to trace the head coil on a small piece of paper and slide it past the two upper coils that represent the magnetization on each side of the transition. It is then easy to observe how the flux varies, and a differentiation of this waveform provides the familiar bell-shaped voltage waveform. The same analysis can be applied for the perpendicular magnetization, where the coils now are parallel to each other with their axes perpendicular to the head coil. The voltage waveform is now S-shaped.

In a longitudinally oriented coating, the longitudinal magnetization is typically 80 to 90 percent of M_s while the level of the perpendicular component is only 30 percent of M_s. This is shown in Fig. 6.10, in which the voltage-versus-time curve has a flat start (possibly with undershoot), then a fast rise, followed by a slow decay. We will return to this topic in Chapter 21.

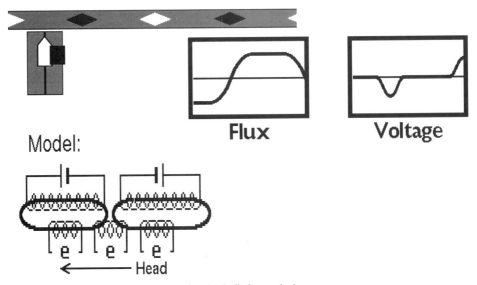

FIGURE 6.8   Read flux and voltage waveforms from longitudinal magnetization.

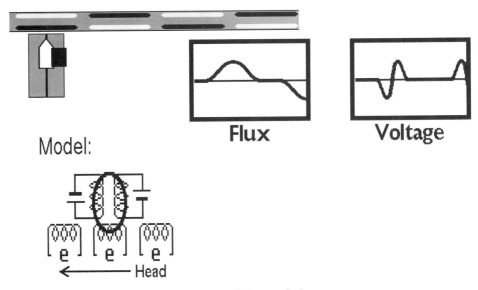

FIGURE 6.9   Read flux and voltage waveforms from perpendicular magnetization.

TOTAL PLAYBACK LOSSES

The read voltage can be calculated for a magnetization pattern that is sinusoidal, longitudinal, and uniform through the thickness. The result is, after a lengthy integration, a general formula for the reproduce voltage, shown in Fig. 6.11. The last four terms are playback losses, which we will discuss in separate sections.

The peak voltage is found by inserting numbers in the first portion of the formula.

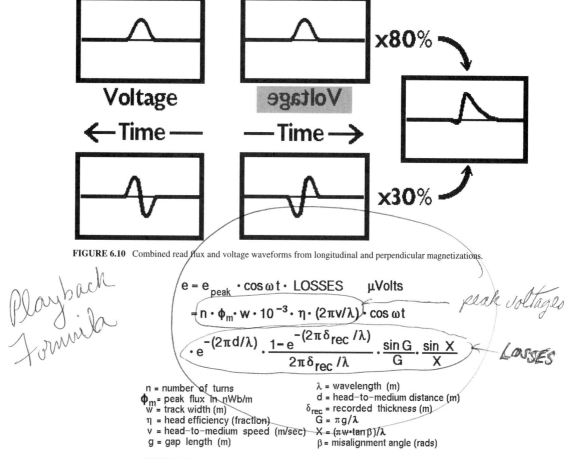

FIGURE 6.10 Combined read flux and voltage waveforms from longitudinal and perpendicular magnetizations.

Playback Formula

$$e = e_{peak} \cdot \cos\omega t \cdot \text{LOSSES} \qquad \mu\text{Volts}$$

$$= n \cdot \phi_m \cdot w \cdot 10^{-3} \cdot \eta \cdot (2\pi v/\lambda) \cdot \cos\omega t$$

peak voltage

$$\cdot e^{-(2\pi d/\lambda)} \cdot \frac{1 - e^{-(2\pi\delta_{rec}/\lambda)}}{2\pi\delta_{rec}/\lambda} \cdot \frac{\sin G}{G} \cdot \frac{\sin X}{X}$$

LOSSES

n = number of turns
ϕ_m = peak flux in nWb/m
w = track width (m)
η = head efficiency (fraction)
v = head–to–medium speed (m/sec)
g = gap length (m)

λ = wavelength (m)
d = head–to–medium distance (m)
δ_{rec} = recorded thickness (m)
G = πg/λ
X = (π·w·tan β)/λ
β = misalignment angle (rads)

FIGURE 6.11 The read voltage formula for sinusoidal signals in recording and playback.

Example 6.1: Find the loss-free peak voltage at 1000 Hz from the head in an audiocassette player. The tape is recorded within its full thickness of 4 μm, at a peak level of 160 nWb/m. The track width is 0.5 mm (20 mils), the head read efficiency is 70 percent, and the number of turns is 280.

Answer: The flux level is $\Phi_m \times (\delta_{rec}/\delta) \times w = 160 \times (4/4) \times 0.5 \times 10^{-3} = 0.08$ nWb. The peak voltage is

$$e_p = 280 \times 0.08 \times 10^{-3} \times 0.7 \times 2\pi \times 1000$$
$$= 98.5 \text{ mV}_{peak}$$

SPACING LOSS

This loss was first described by R.L. Wallace, Jr. (see ref.), and its magnitude is generally referred to as the "Wallace Loss":

$$\text{Spacing loss} = 55 \ d/\lambda \text{ dB} \tag{6.7}$$

This makes it quite simple to appraise: if the spacing d equals 20 percent of the wavelength then the loss is 11 dB; for a wavelength of 125 μin (15,000 Hz at 1⅞ IPS), the spacing equals 25 μin (much less than the diameter of a human hair— about the size of the particles in cigarette smoke).

The spacing loss is explained by an example in Fig. 6.12: the flux from dipole m_1 produces 2.5 times less flux than the dipole m_2, even though they are of the same magnitude. The further away from the head the dipoles are, the smaller their flux contributions. The spacing loss is therefore proportional to the spacing d.

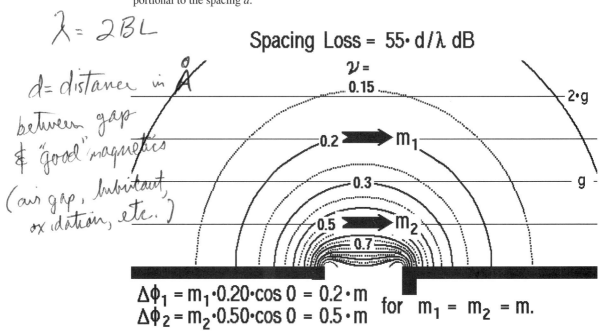

$$\lambda = 2BL$$

d = distance in Å between gap & "good" magnetics (air gap, lubricant, oxidation, etc.)

Spacing Loss = 55· d /λ dB

$$\Delta\phi_1 = m_1 \cdot 0.20 \cdot \cos 0 = 0.2 \cdot m$$
$$\Delta\phi_2 = m_2 \cdot 0.50 \cdot \cos 0 = 0.5 \cdot m \quad \text{for} \quad m_1 = m_2 = m.$$

FIGURE 6.12 Spacing loss reduces the flux contribution from magnetization spaced away from the head surface.

The inverse proportionality to λ is found in flux calculations from the sum of magnetizations "seen" by the head field. At short wavelengths the flux contributions along a thin layer in the coating tend to cancel each other, as shown by the flux contribution due to the top row of dipoles in Fig. 6.13.

The spacing loss in tape recorders and floppy disk drives is mainly due to coating surface roughness, provided the heads are clean and smooth without any debris formed on their surfaces. (Another cause for spacing loss may be coldworking of the core surface, which causes a loss of permeability, which in turn will act as a spacing loss. New heads should be broken in with a mildly abrasive disk or tape.) The quality of modern computer disks and audio, video, and instrumentation tapes assures an effective coating roughness in the range of 0.010–0.002 μin (0.25–0.05 μm).

The spacing losses in rigid disk drives are, of course, controlled by the spacing between the flying head and the disk surface. This dimension has in recent years decreased to a couple of μin.

The Wallace spacing loss assumes a coating permeability of μ_r equal to one. It actually is somewhere between 2 and 4 and will affect the sensing function in a way that increases the spacing loss (see Chapter 21 for a detailed discussion). The use of Wallace's formula in flying height measurements, therefore, may not be valid.

The spacing loss in perpendicular read/write systems differs from the Wallace law when a soft magnetic underlayer below the coating is used. When a single-pole head is used, a loss in the order of 100 d/λ dB is typical, while there is not a specific rule for the use of a conventional ring core head (ref. Iwasaki, Speliotis, and Yamamoto).

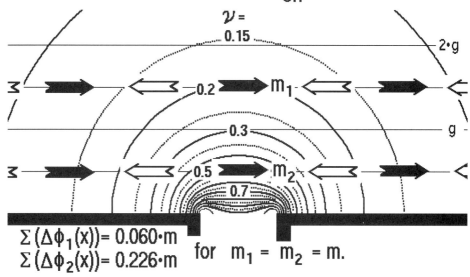

FIGURE 6.13 Spacing and phase cancellation create a read deficiency for magnetizations in the back of a media coating.

COATING THICKNESS LOSS

Coating thickness loss shows up as a reduction in external flux from the tape. The name is unfortunate, because the loss is not a loss in magnetization, as the name implies. Having just discussed the spacing loss we can explain the thickness loss as a spacing loss for the various layers of the coating, with those furthest away from the head suffering the greatest spacing loss (see Fig. 6.13).

Figure 6.14 shows the coating thickness losses in analog recording. For normal or overbiased recordings, the flux levels at long wavelengths ($\lambda > 2\pi\delta_{rec}$) is proportional to the coating thickness. At short wavelength the coating thickness does not matter.

An example will serve to illustrate this. Figure 6.15 shows the magnetization pattern in a tape recorded with a square wave. There are 12 layers in the coating, and the vector magnetization for each small dipole in the array is in the direction it "saw" when it left the recording zone.

The computed flux levels from each element in Fig. 6.15 clearly show how the layers away from the head contribute very little to the total flux. The layer nearest the head provides about 50 percent of the total flux, the next two layers the rest. Layers further away from the head provide fluxes that cancel each other, an effect that increases with decreasing wavelength.

This picture fits a formula that states that the thickness of the layer that contributes 75 percent of the flux is (see Eldridge et al., ref):

$$\delta_{eff} = 0.22\lambda \tag{6.8a}$$

or

$$\delta_{eff} = 0.44BL \tag{6.8b}$$

We now have three coating thicknesses to deal with: the physical coating thickness δ, the recorded thickness δ_{rec}, and now the effective coating thickness δ_{eff}. They are related by the following inequality:

$$\delta \geq \delta_{rec} \geq \delta_{eff} \tag{6.8c}$$

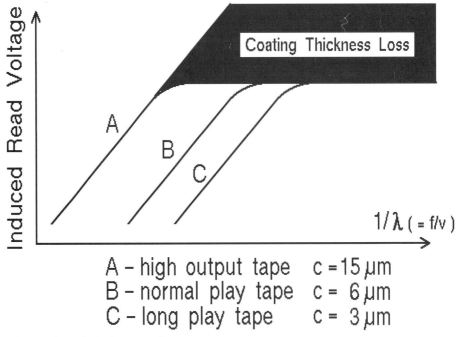

FIGURE 6.14 Spacing loss in tape recordings.

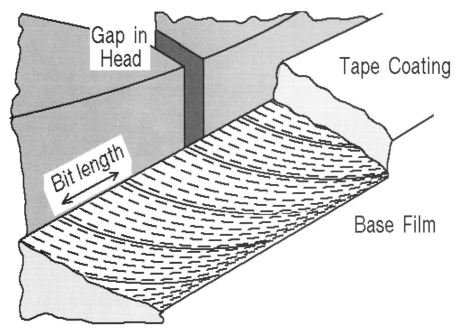

FIGURE 6.15 Recording of a square-wave signal on an isotropic, thick coating leaves vector magnetization as shown.

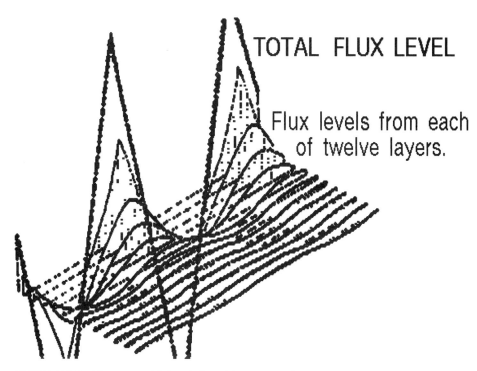

TOTAL FLUX LEVEL

Flux levels from each
of twelve layers.

FIGURE 6.16   Read flux comes mainly from the front of the recorded coating.

A coating thickness greater than δ_{eff} adds no output, but increases the demagnetization losses in longitudinal recording. (But a thick coating is preferable in the perpendicular record mode. Why?)

REPRODUCE GAP LOSS

The reproduce gap loss is caused by the finite length of the front gap in the head. A limit is reached when the recorded wavelength is equal to the gap length: The flux contributions from the two opposite magnetized half wave lengths cancel, and the induced voltage is zero.

$$\text{Gap Loss} = 20 \log(\sin x/x) \text{ dB} \qquad (6.9)$$

where $x = \pi g/\lambda$

The length g of the gap in the $\sin x/x$-type expression for the loss is slightly longer than the measured physical gap length. An accurate loss is determined by using $1.12 \times g$ instead of g, or better: Measure the magnetic gap length by recording a number of frequencies and determining the one where a null occurs. Modern thin-film heads have multipliers different from 1.12 (see ref. Westmijze, and Chapter 8). The bandwidth of a recorder is seldom extended beyond the frequency corresponding to a wavelength twice the effective gap length (f = tape speed in IPS/(2 × gap length in inches)). Otherwise, high-frequency equalization will be excessive and the noise level too high. Disk drives use gap lengths ranging from 0.5 times the BL for high resolution to 1.0 times the BL for high output. Figure 6.17 shows the loss curve as a function of $x = \pi g/\lambda$.

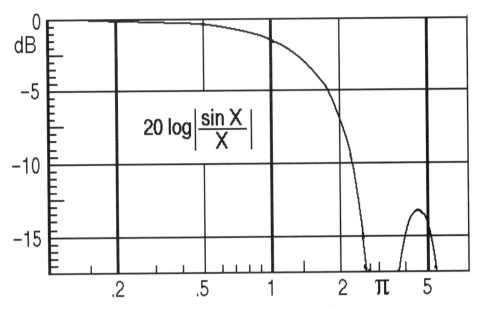

FIGURE 6.17   Loss due to finite gap length or misalignment of the head to the recorded pattern.

GAP ALIGNMENT LOSS; AZIMUTH RECORDING

A loss, similar to the gap length loss, occurs if the reproduce head gap is not parallel to the recording heap gap pattern. This is named alignment loss, and is a $\sin x/x$-type loss, like the gap loss:

$$\text{Alignment loss} = 20 \log(\sin x/x) \text{ dB} \qquad (6.10)$$

where $x = \pi w \tan(\beta)/\lambda$

Track width w and the misalignment angle β between the gap and transitions recorded on the track are shown in Fig. 6.18.

Most recorders have their reproduce heads mounted on a plate that allows for correct adjustment of the alignment angle. It should be noted that the alignment is less critical when the track width is narrow.

Home video recorders (all formats) use alignment loss as a guard against reading information from adjacent tracks. The write/read heads are mounted on a drum, 180 degrees apart, each laying down (or reading) a track that has the data for half a picture frame. Two such frames make up a complete interlaced picture. (See the chapter on video applications.)

The tape area is fully utilized by laying down the tracks with ±7 degrees of misalignment, as shown in Fig. 6.18. When playing back the videotape, adjacent tracks will produce less output due to severe misalignment loss. The alignment loss is not like an ordinary lowpass filter, but rather like a comb filter. Figure 6.19 shows the incomplete isolation between tracks in a helical-scan home VCR.

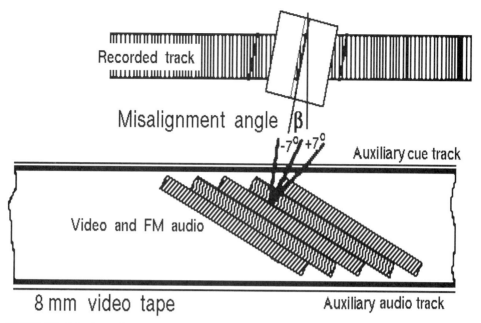

FIGURE 6.18 Azimuth recording: No spacing is required between the alternating slanted tracks on the videotape.

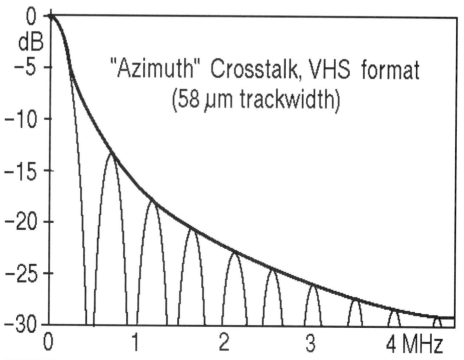

FIGURE 6.19 Incomplete isolation between tracks in azimuth recording.

The color signal is centered near 0.6 MHz and is, therefore, poorly attenuated. The reader may have observed this as a color-bleeding, in particular from saturated shades of red.

The isolation afforded by azimuth recording is further expanded in the newer units with digital sound capabilities; the sound is recorded with two additional heads, also 180 degrees apart, with misalignment angles of ±15 degrees. The sound is recorded first, in depth, with a low carrier frequency. Next, the video is recorded by partial-depth recording so it does not erase (overwrite) the digital audio. Filtering upon playback separates the signals. Note that the terms "first" and "next" apply to the sequence of heads on the head drum.

LOSSES DEPEND UPON DIMENSIONS

In the foregoing sections, several losses have been discussed and explained for the playback process. Note that they all are functions of the recorded wavelength, *not* the signal frequency! (This is true also for the losses during the record process.) Playback losses are always analyzed by way of the wavelength (or bit length) in relation to the spacing, the recorded depth of the coating, the gap length, and the head alignment.

Additional losses do occur in write and read heads at higher frequencies, and these will be covered in Chapter 7.

RESPONSE TO A TRANSITION

The foregoing discussion has pertained to playback or readout of sinusoidal recorded signals. The results are applicable to any other recorded signal which can be described by its Fourier components. The losses are calculated at the frequencies (wavelengths) of interest.

The pulse response from isolated transitions in digital signal recordings can be characterized by their duration, or width, at the halfway maximum voltage. This dimension is called *PW50*, and has a unit of length (after Middleton 1966):

$$PW50 = [g^2 + 4(d + a)^2]^{1/2} \tag{6.11}$$

where g = gap length
  d = head-to-medium distance
  a = transition parameter = $\Delta x/2$

This expression is exact for thin media and a long-pole head, and is a good approximation for TFH if the poles are not too short. For a thick medium with arctangent transitions and no phase change with depth,

$$PW50 = [g^2 + 4(d+a)(d+a+\delta)]^{1/2} \tag{6.12}$$

where δ = medium thickness (Middleton and Miles, 1993).

The read pulse has a shape that best is reproduced by the Lorentzian form (re. Mallinson, 1993):

$$e(x) \approx 1/[\ 1 + (2x/PW50)^2] \tag{6.13}$$

REFERENCES

Alstad, J. Sept. 1973. A Novel Technique for Measuring Head-Tape Spacing. *IEEE Trans. Magn.*, MAG-9 (3): 327–329.

Dinnis, A.K., B.K. Middleton and J.J. Miles. 1993. Theory of longitudinal digital magnetic recording on thick media. *Jour. of Magn. and Magn. Matls.* 120: 149–153.

Geurst, J. A. Nov. 1963. The Reciprocity Principle in the Theory of Magnetic Recording. *Proc. IEEE* 51 (11): 1573–1577.

Lindholm, D. Mar. 1978. Spacing Losses in Finite Track Width Reproducing Systems. *IEEE Trans. Magn.* MAG-14 (2): 55–59.

Mallinson, J.C. 1993. *The Foundations of Magnetic Recording.* 2nd Ed. Academic Press. 217 pages.

Wallace Jr., R.L. Oct. 1951. The Reproduction of Magnetically Recorded Signals. *The Bell System Tech. Jour.* 30 (10): 1145–1173.

Westmijze, W.K. 1953. Studies on Magnetic Recording: Pt. II. Field Configuration Around the Gap and Gap-length Formula. *Philips Res. Repts.* (8): 161–183.

CHAPTER 7
MATERIALS FOR MAGNETIC HEADS

The magnetic heads in a tape recorder or disk drive are the focal points of recording and playback of signals. They have one thing in common: They both have a magnetic core that either guides a concentrated field for the purpose of recording (and/or erasing) or senses the magnetic flux from a recorded tape or disk.

A traditional core structure is shown in Fig. 7.1, left. Thin laminations of a high-permeability material are stacked and cemented together, and each half of the core is provided with a winding. The shallow front gap either generates the recording field or collects the tape flux; the deep back gap offers a minimum reluctance to the flux lines through the core. (This gap, in general, is present because of methods of fabrication of the winding and for reasons of assembly techniques. It does not affect the record or reproduce process.)

Magnetic cores are also made from ferrites, or a combination of a ferrite core and a wear-resistant pole piece made from a magnetic iron-aluminum alloy such as Alfenol, Alfesil, Sendust, Spin-alloy, or Vacodur 5. These heads are called hard-tipped heads; see Fig. 7.1. A new amorphous iron-nickel

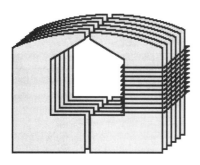

Laminated Mu–Metal head.

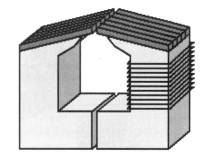

Ferrite core with laminated Fe–Al tip plates.

FIGURE 7.1 Tape head cores made from laminated NiFe (left) or from solid ferrite with laminated high-wear pole tips of AlFe alloy (right).

alloy named Metglass or Vitrovac has recently been introduced in some applications, in addition to other high-wear-resistance materials.

Cores are found in all kinds of shapes and sizes, ranging from about one inch in overall diameter to the size of the head of a pin. Figure 7.2 shows all ferrite heads for floppy drives. Figure 7.3 shows the small slider assemblies for rigid disk drive heads.

Materials for head cores are investigated in three recent books (Chen 1986; Snelling 1988; Boll 1977) and in the classic by R. Bozorth. The latter was reprinted in 1993 by IEEE. Early descriptions of heads, materials for heads, and fabrication are found in two papers by Kornei (1953 and 1954). A chapter in Mee's and Daniel's book covers heads, and a sizeable book by Ciureneau and Gavrila (1987) is recommended for the specialist.

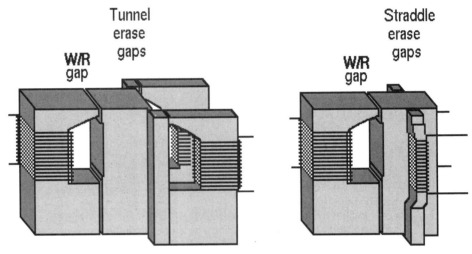

FIGURE 7.2 All ferrite heads for floppy disks. Erase cores are provided to trim erase new tracks.

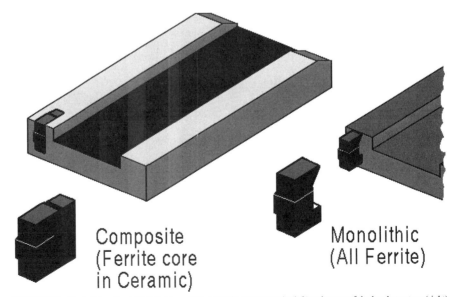

FIGURE 7.3 Head sliders for rigid disk drives. Slider may be nonmagnetic (left) or be part of the head structure (right).

The reader may wonder: Why is there such a large variety of head designs? This and the following three chapters will examine the many performance criteria placed on a magnetic head, and where trade-offs among them are necessary. The most important ones are:

- Frequency range
- Wavelength response
- Signal sensitivity
- Losses (and noise)
- Useful life
- Track configurations
- Size
- Ease of fabrication
- Materials and processes
- Overall performance versus cost

Most of these can be analyzed by a study of the magnetic circuit that represents the magnetic core with its coil. The analysis will depend on the head's use, whether for erasing, recording (writing), or playback (reading). The fabrication may be piecewise assembly of materials or a vacuum deposition of thin film cores, insulators, and conductive windings as integral parts.

Many head assemblies contain several tracks and consideration must be given to crosstalk. This consists of unwanted signals coupled from adjacent tracks into a core; it may be stray flux from recorded tracks or from transformer coupling between core windings.

Modern disk drives apply single-channel write/read heads. There are several designs: Ferrite and Ferrite MIG heads are shown in Fig. 7.3 and inductive thin-film heads in Fig. 7.4, with details shown in Fig. 7.5. Ferrite is often combined with an MR (magnetoresistive) element, where the inductive portion performs recording and the MR section is reading only; see Fig. 7.6.

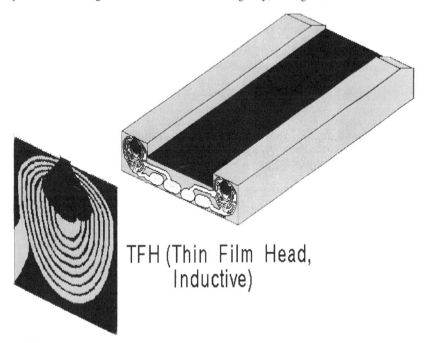

TFH (Thin Film Head, Inductive)

FIGURE 7.4 Slider for hard disk drive with thin-film heads. Only one head is used, on top of a disk or on the underside.

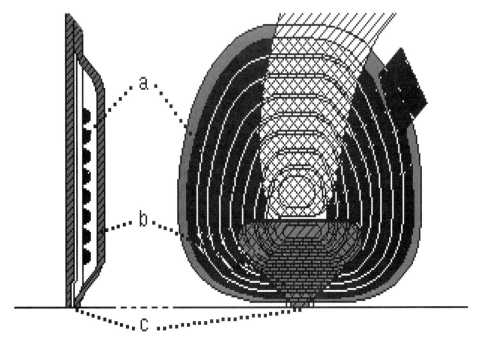

FIGURE 7.5   Copy of original patent drawing (*Jones at IBM, 1972*).

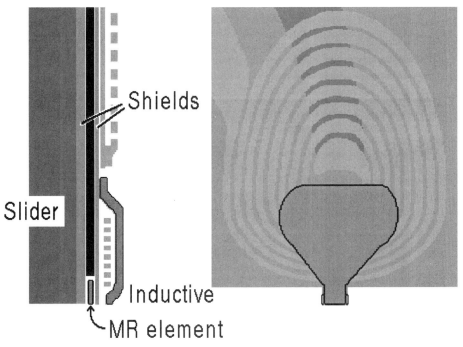

FIGURE 7.6   Combined TFH/MR head.

The optimized designs and performances for various applications are deferred to Chapter 10 in order that we may first learn how many modern heads are built, and why certain construction techniques are used. Examples will reflect applications in the following areas: computer data storage on disk and tape, instrumentation, audio, and video. I hope that this arrangement of discussing head materials first, then the vital topic of fields in Chapter 8, and then construction techniques will be valuable when the book is used as a reference. This should also provide a good background for the design guides in Chapter 10.

In this chapter we will examine the magnetic properties of several materials, striving for the optimum magnetic performance of a head. That spawns a requirement for high permeability.

An important practical concern is also how long a tape head will last for a given initial cost. High sensitivity may result in head designs with a very shallow gap depth, in the range of 0.2–3 mils (5–7.5 μm). The abrasive action of the magnetic tape coating may wear out such heads rather quickly, and attention must be paid to the mechanical wear properties of the materials.

The electromagnetic performance of a head is appraised by its efficiency, impedance, and noise levels, all over a specified frequency range. We will review the efficiency and inductance (from Chapter 4), and apply the formulas to material selection. We will discover that a rather high resistive component of the head's impedance may appear at high frequencies, and by examining its cause we find restrictions on the material selection. Eddy currents are induced in the cores, and they must be reduced by laminating the core or by using a high-resistivity material—a ceramic named ferrite.

It also becomes apparent that a small size is desirable for optimum efficiency, and that borrowing techniques from the semiconductor industry has in recent years given us a new head, the so-called thin-film head.

Upon examining materials we also find some whose properties are field-dependent, such as changing resistance value in a varying field. Such materials are used in flux-sensing heads, for instance the magnetoresistive heads. Elements for MR (magnetoresistive) heads will be discussed at the end of this chapter.

MAGNETIC CIRCUITS

Magnetic heads in use today are ring-shaped cores of a flux conducting magnetic material, provided with a winding and a small gap in the core. This gap interrupts the magnetic circuit and is used in record heads to provide a concentrated recording field. In the reproduce head it collects the available flux from a recorded track, and the core links the flux through the winding where it produces a voltage.

The majority of all heads utilize a magnetic core similar to the examples in Fig. 7.1. They may differ in size and geometry but are essentially alike in construction, whether for digital data, instrumentation, or audio and video recording or playback.

Common core materials are laminated iron-nickel or ferrite, with occasional use of a special alloy for long-wear-life applications.

We can, for each of the cores shown, sketch a magnetic circuit from the flux path, while breaking down the more complex structure into simple geometric blocks for each of which we can calculate the magnetic resistance. This was done in Chapter 4 (see Fig. 4.22), and from an elementary circuit analysis we found expressions for the efficiency, record current required, and the self inductance of the winding:

$$\text{Efficiency } \eta = R_{\text{fg}}/\Sigma R_{\text{m}} \tag{4.26}$$

$$\text{Current } i_{\text{write}} = H_{\text{g}}g/n\eta \tag{4.29a}$$

$$\text{Inductance } L = n^2/\Sigma R_{\text{m}} \tag{4.30}$$

where:

$R_{fg} = g/\mu_o D_g w$

g = gap length in meters

D_g = gap depth in meters

w = track width in meters

$\mu_o = 4\pi 10^{-7}$ Henry/meter

ΣR_m = sum of all magnetic resistances, tracing the circuit once

H_g = Deep gap field required, as shown in Figs. 5.15 through 5.17

n = number on turns in winding

OPTIMIZE EFFICIENCY

The magnetic circuit determines the efficiency of the head core, in addition to the value of L. A low value is achieved in a head core design where the flux path length through the core is short and the cross-sectional area is large. So we want a small but wide core. The permeability μ_r of the core material should be as high as possible, and traditionally, a designer will select an iron-nickel alloy, known by trade names such as Permalloy or Mu-metal. This should ensure the best efficiency.

A low write current or a high read voltage is obtained for a large number of turns, but it is limited due to restrictions on the inductance value. It must be small enough so the head impedance has a resonance above the highest signal frequency (or bias frequency). And the time constant of head inductance and amplifier impedance must be smaller than the rise time of the pulse edges in pulse recording; there is possibly a desired time constant for each application.

Assume a prototype head has been built, the core is a laminated iron-nickel material, and testing is underway. This may be a bench test where a current-carrying wire is placed in front of the head to simulate the playback of a constant flux recording. The current source is high-impedance to assure a constant current from low to high frequencies. The induced voltage should ideally increase 6 dB/octave, but above a few kHz a slower rise is observed and we wish to know why.

Another measurement is made to verify the calculated inductance. It may be quite correct at low frequencies, but decreases toward higher frequencies, and the impedance bridge reveals that the resistive component of the winding resistance apparently increases quite dramatically.

EDDY CURRENT LOSSES

The effects of increased resistance values are named core losses and have their origin in circulating currents in the core. Figure 7.7 shows how a magnetic field penetrates a sheet of conductive material. When the field changes value, as in an ac field, then circulating currents I_e are induced inside the sheet, in accordance with Lenz's law.

This law is quite like Le Chatelier's law, which states that all systems at rest will react to an outside change by an action that opposes the change. Lenz's law applies specifically to the induced currents, making them have direction so they create a field that is in opposition to the field that creates them.

If the ac field varies fast enough (higher frequencies), then the induced fields may completely oppose the outside field, an effect used in shielding. The effectiveness of such shielding is not reduced by drilling a number of small holes in the sheet, since the circular currents still are operative. Hence, we get good shielding effect of conductive screens, such as Faraday shields. An example is the door of a microwave oven, where holes permit the user to look at the destruction of a good filet mignon, but the field cannot radiate out.

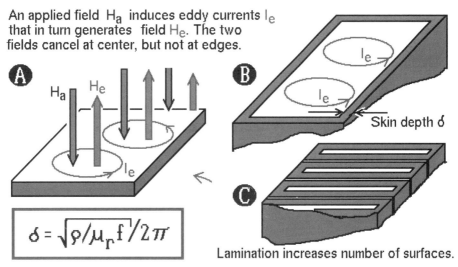

An applied field H_a induces eddy currents I_e that in turn generates field H_e. The two fields cancel at center, but not at edges.

Ⓐ H_a H_e I_e

Ⓑ I_e I_e Skin depth δ

$$\delta = \sqrt{\rho/\mu_r f}/2\pi$$

Ⓒ

Lamination increases number of surfaces.

FIGURE 7.7 Eddy currents induced by incoming field H_a prevents field flow in the core's interior.

When the eddy currents are strong enough, no magnetic field goes through the center section of the sheet. Because the eddy currents increase with frequency, we will, with increasing frequency, have the flux concentrated at the edges of the sheet. This will still hold true when we next increase the thickness of the sheet so it, in the limit, becomes a rectangular bar. It is then logical to talk about a surface flux that only penetrates into the bar to a certain depth, the skin depth.

This skin depth is small for materials with high permeability μ_r and small resistivity ρ, and it does, as we have seen, decrease with higher frequencies. We could therefore, in our expression for the magnetic resistance of the core (part of ΣR_m), substitute the area A with an effective area A', that decreases in size with increasing signal frequency; see Fig. 7.8. The area for the flux may be increased by dividing the core into laminations of thicknesses equal to twice the skin depth at the frequency of interest (see Fig. 7.9).

The skin depth can be calculated from (after Bozorth):

$$\delta = \frac{(\pi/2)}{2\pi} \times \sqrt{\rho/\mu_r f} \text{ cm} \tag{7.1}$$

where:

ρ = resistivity in $\mu\Omega$cm
μ_r = relative permeability (dc)
f = frequency in kHz

This formula would allow the calculation of the area available for the flux, at each frequency we desire. The use of these values in the expressions for the efficiency and the inductance would indeed show a decrease in the value of both as we raise the frequency.

It would not provide an answer to the increase in resistance. Upon studying the literature (Bozorth, Peterson, and Wrathall, Lee) we learn that a complex value for the permeability must be used:

$$\mu = \mu' - j\mu'' \tag{7.2}$$

where j is the complex number $\sqrt{-1}$.

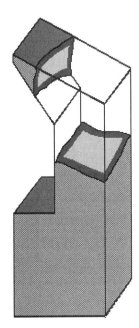

FIGURE 7.8 Eddy currents reduce effective core area for flux to flow.

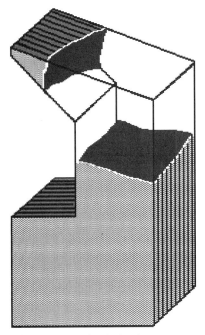

FIGURE 7.9 Use of laminations in core construction recovers area for magnetic flux.

Use of this expression is in agreement with the real world, and will give us correct results in our later analysis of the head impedance, losses, and the resistance-type noise generated in the head.

Do note that the complex permeability is not a true material constant; it depends not only upon the material, but also its shape and the signal frequency. An expression for μ is found from the references listed above, and is given in Fig. 7.10.

The parameter K is:

$$K = 2\pi d_{cm} \times \sqrt{\mu_{rdc} f_k / \rho} \qquad (7.3)$$

where:

d_{cm} = lamination thickness in cm
μ_{rdc} = relative permeability at dc
f_k = frequency in kHz
ρ = resistivity in $\mu\Omega$cm

The expressions in Fig. 7.10 are formidable, but rather easy to handle today with a small subroutine in a computer program. The expressions do simplify, at high frequencies, to:

$$\mu = \mu_0 \mu_r \times (1-j)/K \qquad (7.4)$$

The complex permeability therefore has a phase angle of 45 degrees at high frequencies, and decreases at a rate of 3 dB/octave (K is proportional to the square root of the frequency; see formula 7.3).

Figure 7.10 shows that the permeability is mostly real and equal to μ_r up to a value of $K = 1$, after which it falls and becomes complex. When $K = 1$ the skin depth equals the lamination thickness, and the core has a permeability equal to the low frequency value. The total core area is available for the flux; see Fig. 7.9. At higher frequencies the flux is carried by less than the full lamination thickness, which is equivalent to it having a smaller permeability.

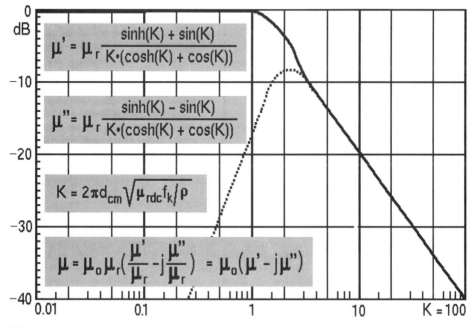

FIGURE 7.10 Real and imaginary permeabilities as a function of K (proportional to the square root of the frequency).

There is a full 180 degrees of phase shift between the surface flux and the flux at a depth of δ. The latter opposes the surface flux, as we explained earlier when using the Lenz law to explain eddy current losses. The significance of the complex permeability is that at high frequencies it reduces the inductance L and also appears as a loss at the head winding terminals. A toroidal inductor without an airgap has an impedance:

$$
\begin{aligned}
Z &= R_{wdg} + j\omega L \\
&= R_{wdg} + j\omega n^2/\Sigma Rm \\
&= R_{wdg} + j\omega n^2 \mu A_c/L_c \\
&= R_{wdg} + j\omega n^2 (A_c/L_c) \times (\mu' - j\mu'') \\
&= R_{wdg} + j\omega n^2 \mu' (A_c/L_c) - (-1)\mu'' \omega n^2 (A_c/L_c) \\
&= R_{wdg} + \mu'' \omega n^2 (A_c/L_c) + j\omega L \mu'/\mu \\
&= R_{wdg} + R' + j\omega L'
\end{aligned}
\tag{7.5}
$$

(Note that no subscript is used for the permeabilities, which means they represent the absolute values, i.e., $\mu = \mu_o \mu_r$, $\mu' = \mu_o \mu_r'$ etc.)

The net effect is an added resistance value R' and an inductance value L' that decreases as μ'/μ. The total magnetic resistance in a head assembly is due to core sections and air gaps; only the core sections include the μ-term to be substituted with $\mu' - j\mu''$. This complicates the computations.

The net effect is the same, possibly diluted by the presence of the air gaps; the inductance value decreases at higher frequencies (μ'/μ), and the head's impedance has an added resistive component that traces the curve for μ'' versus frequency, i.e., it rises 6 dB/octave for $K < 1$, and decreases 3 dB/octave for $K > 1$. The term ω is a steady 6 dB/octave rise, and the overall pattern of R' is therefore an impressive 12 dB/octave rise when $K < 1$, and a modest 3 dB/octave rise when $K > 1$. The changes are shown in Fig. 7.11.

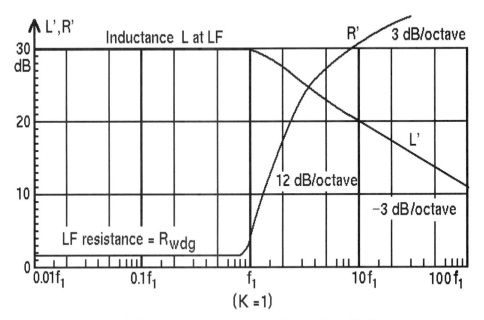

FIGURE 7.11 The reduction in inductance and increase in resistance of head impedance at high frequencies.

SELECTION OF CORE MATERIAL

We have discussed the magnetic circuit and the high frequency effects of eddy currents, and can now apply our gained knowledge toward selection of a suitable magnetic material. It should have the following properties:

For efficiency:	High permeability μ_r
For low hysteresis loss:	Low coercivity H_c
For low remanence:	Low coercivity H_c
For low eddy current loss:	High resistivity ρ
	Low permeability μ_r and thin lamination d

To this we should add ease of machining or forming, and good durability against mechanical wear (which is somewhat equivalent to a reasonably high hardness).

The required high permeability for achieving a good efficiency at low frequencies overrides the conflicting low permeability for small eddy current losses. And it is resolved by using cores made from thin laminations, or from ferrites with high resistivities.

Table 7.1 is a chart of materials currently used for fabricating magnetic heads. We will now discuss their various merits.

TABLE 7.1 Materials

Material	μ_r–init –	H_C Oe	B_{sat} kG	σ $\mu\Omega$cm	ε μm/m /°C	T_C °C	Composition
Permalloy	20,000	0.05	8.7	25	13.2	460	Ni79Fe17Mo4
HyMu80	40,000	0.02	7.5	25	12.9	460	Ni80Fe15MoMnSi
Tough Perm	40,000	0.02	5.0	90	–	280	NiFeTi
Sendust	10,000	0.06	10.0	90	15	500	Fe85Al6Si9
Vacodur 16	8,000	0.06	8.0	145	15	350	FeAl
MetGlass	20,000	0.01	5.5	130	12.5	250	FeCo
MnZnFerrite	4,000	0.1	5.5	10^7	11.1	170	MnOZnOFeO
NiZnFerrite	3,000	0.2	3.0	10^{10}	9	1–200	NiOZnOFeO
FeRuGaSi[1]	1,500	0.5	14	130	14.0		Fe68Ru8Ga7Si17
Fe/FeCrB[2]	3,000	0.5	19				Fe73Cr7B20
FeN[3]	>1,000	10	28				Fe16N2

[1] Laminated: 2μm/100 nm, after Ash, et al.
[2] Multilayer: 10 nm/3 nm,. after Dirne, et al.
[3] After Kolmoto.

METALLIC CORE MATERIALS

A high permeability is achieved when the walls between magnetic domains are easily moved, and when few hindrances are in their way. From this we can conclude that the materials should be uniform, isotropic, and free from impurities.

We will also find that the core manufacturing processes may produce stresses from machining, bending, pressing (from epoxies), and from polishing and lapping the pole tip interface. The material should be insensitive with regard to magnetostriction. Stress in a magnetostrictive material will produce an ordering of the domains and, therefore, a change in permeability. And a magnetic field will produce a similar ordering of the domains, with resulting stress leading to expansion or contraction of the material. This property is used in magnetostrictive transducers such as, for instance, sonar transducers and ultrasonic cleaning devices.

IRON-NICKEL ALLOYS

A material with very low crystalline anisotropy and low magnetostriction is the iron-nickel alloy with about 80 percent Ni. The variation of λ_{100} and λ_{111} (magnetostriction in crystal direction (100) and (111), and of the crystal anisotropy K, is shown in Fig. 7.12. They are all near zero at around 80 percent Ni. That is the composition of 78 Permalloy, a popular soft magnetic material. Other material compositions have been analyzed and are described by Hall (see reference).

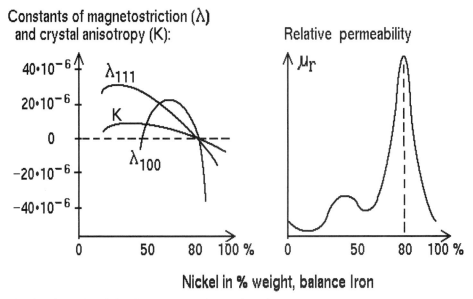

FIGURE 7.12   Selection of NiFe alloy in order to obtain zero anisotropies.

When other elements are added to an Fe-Ni base, several useful alloys result. The addition of 4 to 5 percent molybdenum increases the initial permeability and about triples the electrical resistivity (so eddy current losses are lower). These alloys are called "4-79 Permalloy" and "Supermalloy."

Mu-Metal is obtained by the addition of 5 percent copper, and often 2 percent chromium. This material is easier to roll into thin sheets, and has by now expanded into a small family of alloys.

HyMu-80 has high μ_r, low losses, and is easy to photoetch into laminations. (Stamping laminations for cores is really a thing of the past.) HyMu-80 Mark II is a new alloy with excellent high-frequency properties.

HyMu-800 alloys have smaller grains and also smaller domain sizes. It is available in grades A, B, and C. Grade A is manufactured in sheets down to 0.001 inch thickness and has the best electrical performance. Grade B has a minute (and carefully controlled) amount of abrasive particles embedded uniformly throughout the material, thus providing hardness for a high resistance to wear from the tape moving across the head. Grade C is employed mostly for deep drawn cans for shielding and for head housings.

The HyMu-800-B has a lower permeability than HyMu-800-A due to the added impurities, which are rolled-in aluminum oxides. While this provides for a tenfold increase in wear resistance, it also hinders domain wall movement. Its initial permeability is less than 10,000, which compares with 70,000 for HyMu-800-A. Calculated values of μ' and μ'' are shown in Fig. 7.13.

We can, at this point, temper our striving for the highest permeability, because materials with this property are very sensitive to any form of stress and coldworking. This occurs in three steps of head manufacturing:

1. Stacking and gluing together of thin laminations to obtain a final thickness equal to the track width.

2. Inserting and clamping the finished cores into the head block (where they, in general, are held into place by epoxies).

3. Lapping of pole tip interfaces.

All three processes lead to a severe degradation of the permeability in fabrication of standard core heads as well as in thin film heads. It is found that a permeability of at best 2000–5000 is retained after the stacking and gluing of laminations (McKnight 1977). The lacquers or epoxies used in bonding the laminations together (and providing for insulation) shrink during use, and result in compression stresses in the laminations.

Now, a material like HyMu-800-B has already been "downgraded" by the addition of impurities (the hard abrasives), and therefore, is not degraded much further by the additional processing stresses. This is clearly evident in Fig. 7.14, which shows that there is little difference between HyMu-80 and HyMu-800-B after stacking and embedding the laminations (Bendson 1976).

Another form of strengthening the material consists of adding titanium and niobium to the nickel-iron alloy. This results in an interstitial molecular structure. These materials are known as Tough-Permalloy and Recovac (ref. Miyazaki 1972, Pfeifer and Radeloff 1980).

Figure 7.14 provides another interesting piece of information. When we, in Chapter 10, go through the design and performance calculations of heads, we will find that the nonmagnetic gaps in the magnetic circuits offset the benefit of a very high permeability at low frequencies. It doesn't matter much whether it is 5000 or 100,000. We will find that we should concentrate on selecting a thin lamination and a material that will keep up a reasonable permeability at high frequencies. Here we note, in Fig. 7.14, that HyMu-80 and HyMu-800-B behave almost identically after encapsulation. Also, lowering the permeability makes the K values small and extends the head's operating range.

There are two additional materials that play a prominent role in modern magnetic heads: the iron-aluminum alloys and the amorphous metallic glasses.

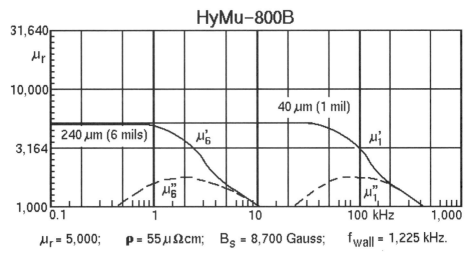

FIGURE 7.13 Permeability spectra for HyMu-800B alloy.

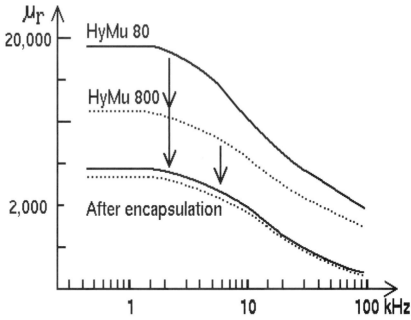

FIGURE 7.14   Coldworking (stress and pressure) reduces core permeability.

IRON-ALUMINUM ALLOYS

The first Fe-Al alloy was Alfenol (84 percent Fe, 16 percent Al), a hard alloy with modest permeability and high saturation density. Unfortunately, it is very difficult to work; rolling it into thin sheets tends to make it buckle, and therefore makes the stacking of etched laminations quite difficult.

Sendust (85 percent Fe, 6 percent Al, 9 percent Si, also named Alfesil 5) is also a hard alloy, but very brittle and impossible to roll into sheets. Spinalloy is another name for this alloy. It has higher permeability than Alfenol and is often employed in tip plates on a regular permalloy core or better ferrite cores, as shown in Fig. 7.1. The material has been tried in ordinary head fabrication (Tsukagoshi 1981; Tsuya 1981; and Matsumoto et al. 1981). Vacodur has properties similar to Alfenol, and appears easier to process as evidenced by its use in heads by European manufacturers.

Why all these names for essentially the same material? AlfeNOL was first developed by Naval Ordnance Laboratory, SENDust came from the university of Sendai, SPINalloy is a product from Spin Physics, and VACodur is from Vacuumschmelze AG.

METALLIC GLASSES

Glassy metals are soft magnetic materials that are prepared in continuous ribbon form by rapid quenching directly from the melt (Raskin and Davis 1981; Liebermann 1979). The alloy solidifies so fast that grains do not form; the structure is amorphous.

This brings about some unique characteristics, such as very high strength and low energy losses. The alloys that produce soft magnetic materials are made from iron and nickel plus nonmetallic substances such as boron and silicon. Trade names are MetGlass (Allied Corp.) and Vitrovac (Vacuumschmelze). The ribbons are typically 50 mm wide, and are cast at a rate of 30 meters per second. The cooling rate is about one million degrees centigrade per second.

The initial permeability for Fe40Ni40B20 is 10,000, the losses are comparable to the FeNi alloys, and M_{sat} is 10 kgauss (O'Handley 1974). Also, it is much more forgiving to handling than the NiFe alloys; it can be bent and twisted without appreciable loss in magnetic performance.

DEPOSITED LAMINATIONS, MULTILAYERS, AND MATERIALS WITH HIGH B_{sat}

Laminated cores for very high frequencies require laminations too thin for ordinary fabrication. Deposition techniques are required to make laminations of thickness 2 nm (0.08 mils), which is ten times thinner than ordinary core lamination material.

One recent application was a 150-MHz (300 Mbps) helical scan recorder (Ash et al. 1990). The measured permeability at 150 MHz was 200 (μ') for a 10-layer film, while a single-layer film had permeability of 300 (μ'), as shown in Fig. 7.15. Theory predicts a somewhat higher μ'. Poor insulation between layers, stress, and coupling between layers may play a role. Eddy-current damping associated with magnetization change (wall motions and rotations) may play a role (Yuan and Bertram 1993).

Multilayer films have also provided high-frequency cores. These multilayers are not laminated with insulation. In microcrystalline multilayers, Fe layers are used to obtain a high M_{sat}. The interlayers, which limit the Fe grain size, are used to obtain good soft magnetic properties (Dirne et al. 1991). RF-diode sputtering was used to deposit a 10 nm/3 nm Fe/FeCrB multilayer film with B_{sat} = 1.9 T (19,000 Wb/m²), H_c = 40 A/m (0.5 Oe) and $\mu' > 2000$ at 10 MHz.

Such multilayers are used in heads of trackwidth 10 µm (0.4 mils) and smaller. Nonmagnetic ceramics provide the outsides of a sandwich construction similar to the video head shown in Fig. 7.15.

Other multilayer films are Fe(N)/NiFeCo(N) of thicknesses (20–125 nm)/(5–66 nm) and a total number of layers from 9 to 75 (Jones et al. 1993). These multilayer films possess high saturation flux densities to write on future very-high-coercivity media. They also have excellent high frequency performances.

Other films are CoHfTaPd (Ohkubo et al. 1993), FeIrGaSi, FeRhGaSi (Seltser and Jagielinski, 1993) and FeTaN (Okumura et al. 1993). The trend is toward iron nitride compounds. More than 20 kinds of alloy systems with high saturation induction of more than 10 kG have been reported (Kohmoto 1991).

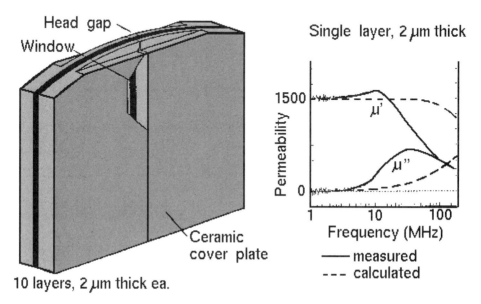

FIGURE 7.15 150-MHz video head with 2-mm laminations. Measured permeabilities differ from theory (*Ash et al.*).

FERRITES

Ferrites are magnetic ceramics that have come to play an important role in magnetic heads (Néel 1948; Snelling 1988). Early after their discovery in the forties they were employed in erase heads because of low losses. Ferrite is a sintered material, and the first ferrites had a high porosity (5 percent at that time), which precluded their application for heads requiring precise gap dimensions.

These magnetically soft ferrites are generally formed as $MOFe_2O_3$, where MO is a bivalent metal oxide, such as manganese oxide (MnO) or zinc oxide (ZnO) mixed with the Fe_2O_3 powder. They are solid solutions and are sintered together. Their outstanding features are high resistivity with reasonably good magnetic properties. This means that they can operate with virtually no eddy current loss at high frequencies, and thus are superior to laminated permalloy cores. There is no loss in effective area for the flux (see Fig. 7.16).

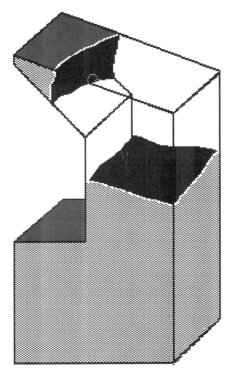

FIGURE 7.16 Ferrite cores with high resistivity have no eddy current losses, and full core area available for flux.

Efforts in the sixties toward reducing the porosity succeeded by hot pressing the ferrites during final sintering (Suguya 1968; Monforte 1971; LaGrange 1972). The results were materials with 0.1 percent porosity or less. Earlier high-density materials were also hot pressed, but at lower temperatures. These materials were primarily used for audio heads.

The details of ferrite fabrication are proprietary to the manufacturers, but the basic processing steps are similar:

1. A powder of ferric oxide (Fe_2O_3) is mixed with NiO or MnO (or NiO + ZiO, or MnO + ZiO) in distilled water. They are thoroughly mixed in a ball mill to produce smaller particle sizes and a uniform mixture.

2. After milling, the mixture is dried and pressed into loose blocks, which are presintered at 900–1100 C. This produces a partial formation of ferrite.

$$MO + Fe_2O_3 \rightarrow MOFe_2O_3.$$

3. This material is ground again to promote mixing of any unreacted oxides and to achieve further reduction in particle size, to about 0.5–1.0 μm. (Smaller sizes may be obtained by a method where an aqueous solution of Fe, Zi, and Ni hydrated sulphates are dehydrated in a spray drier at 250 degrees C. Decomposition at 800 degrees C in air produces particle sizes of 0.1–0.2 μm.)

4. The powder is mixed with an organic binder and formed into its final shape, which may be slabs about $1 \times 2 \times 3$ inches, or 1"–2" thick disks of 6" diameter.

The final sintering has several process variables that to a great extent determine the characteristics of the final product: these variables include temperature, pressure, atmosphere, and time. Their combined effects on grain size are shown in Fig. 7.17. This shows the proportionality of final sizes to temperature, pressure, and time. A small grain size and absence of pores provide a ferrite with higher resistance to wear and chipping. This conflicts with the desire to have a high permeability, which results from large grain sizes as shown in Fig. 7.18 (Snelling 1988).

The temperature cycle extends from one to many hours and must be carefully controlled (Withop 1978). The atmosphere is even more critical, because it has a pronounced effect upon permeability and saturation magnetization.

The improved technique of applying pressure during the final sintering originated in ceramic tool manufacturing in Japan—a rather popular skill there. The pressure is in the order of 500 kg/cm^2 and may be applied in one or two directions, or may be isostatic. These hot-pressed ferrites are successfully applied in heads (Rigby, Kehr, and Meldrum 1984).

It is also possible to grow single-crystal ferrites with magnetic properties similar to the poly-crystalline material. They often exhibit better wear characteristics, but are also fragile. The single crystal material is used for audio and video heads. One drawback of the single crystal material is anisotropy in regard to permeability and wear characteristics. The permeability varies as much as 2:1, and wear rates 3:1. Another factor is the limited range of ferrite compositions from which single crystals can be grown (Mizushima 1971).

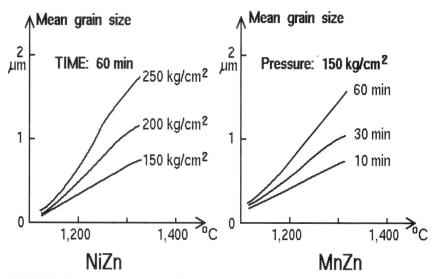

FIGURE 7.17 Grain sizes in ferrites are controlled by processing temperature and pressure.

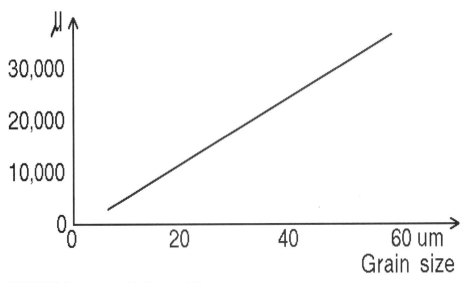

FIGURE 7.18 Large grains result in high permeability.

The magnetic properties of ferrites are heavily influenced by the starting powder composition, and the final sintering process. Manufacturers do not list the exact composition of their ferrites, calling them merely MnZn or NiZn types, with a letter/number specifying the various grades.

The MnZn ferrites have initial permeabilities of 2000–5000, coercivities of less than 1.0 Oe, and resistivities of greater than 10^6 $\mu\Omega$cm (Brissette 1983; Takama, 1979).

NiZn ferrites are designed for very-high-frequency operations, with initial permeability of 100–200 and resistivity in the order of 10^{10} $\mu\Omega$cm.

The properties of both ferrites are heavily influenced by the percent content of zinc. As the nickel content is decreased from 50 to 30 percent of the MO content, then both saturation flux density and coercivity drop, while the permeability increases. (It is thus possible to tailor-make a ferrite.) Also, the higher the content of zinc, the lower the Curie temperature.

Low Curie temperatures T_c are typical for ferrites; above this temperature the spontaneous magnetization disappears (Fig. 7.19). Composition plays a major role for the Curie temperature, intrinsic magnetization M, and the permeability μ_r; the latter is shown to have a very large variation with temperature. This, for instance, makes the use of ferrites impossible in video heads, where the heat from head-coating friction may exceed T_c.

The permeability of ferrites decreases toward higher frequencies. This is due to a slowdown in the rotation of the magnetization, or the spin's directions. Ferrites therefore exhibit a permeability spectrum with a drop-off at the spin frequency:

$$f_s = 1.87 \times B_{rs}/\mu_{rdc} \text{ MHz} \tag{7.6}$$

where B_{rs} = remanence after saturation in gauss and μ_{rdc} = relative permeability at dc.

This spectrum is shown in Fig. 7.20 for three values of the permeability. A high spin frequency is obtained by sacrificing permeability, which will decrease the head's efficiency somewhat.

The value of the remanent saturation flux density should be high, which also concurs with a desired high value for the material's application in a recording head.

Ferrites are always used as solid cores, and eddy currents may reduce the effective permeability for thick cores at high frequencies. Eddy currents in MnZn can only be disregarded at low frequencies or when the core is very small. Modern high-resistivity ferrites have, to some extent, eliminated this problem (Brissette 1983).

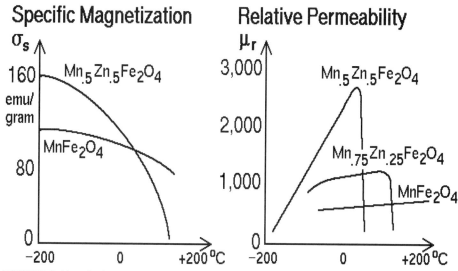

FIGURE 7.19 Magnetization (left) decreases with temperature. Permeability also changes with temperature.

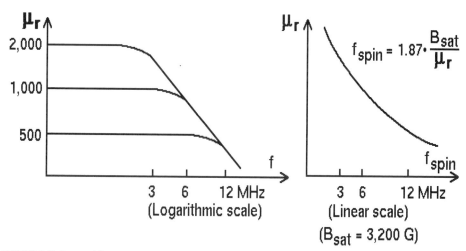

FIGURE 7.20 Increased frequency response is obtained at lower permeabilities.

All curves shown in this chapter are guides only and accurate calculations of head performance requires use of data from the materials data sheet. It is always best to make one's own measurements on a ring core sample of the ferrite (and on a sample that is free from machining stresses).

THIN-FILM HEADS

We will now have a final discussion on the high-frequency permeability in metals. We have learned that a high permeability requires easy domain wall movements and that a certain energy is associated with these movements, much like moving a spring-loaded mass back and forth. An equation for the wall movement will, as a matter of fact, be identical to the equation of motion for a spring-loaded mass.

The wall movements will decrease in amplitude above a certain resonance frequency, called the wall relaxation frequency f_w, and the permeability, thereafter, decreases. A detailed discussion is beyond the scope of this book, but we will quote the results from which we can determine the cut-off frequencies (Boll 1960):

$$f_w = 0.84 \times B_{rs}/\mu_{rdc} \text{ MHz} \tag{7.7}$$

where B_{rs} = remanence after saturation in gauss and μ_{rdc} = relative permeability at dc.

This expression is similar to (7.6) for the resonance for magnetization rotation (see under ferrites). For Supermalloy we find $f_w = 0.07$ MHz, and for Mu-metal $f_w = 0.2$ MHz. In the practical case of a laminated core with final permeability below 5000, the frequencies are in the 1–2 MHz range.

The significance of this frequency is that it does not pay to reduce the lamination thickness below a certain limit. (It should be emphasized, though, that f_w is a function of the material; it is not influenced by the lamination thickness.)

We first realized a decrease in effective permeability due to eddy current losses, and then we established a relationship between the lamination thickness, permeability, conductivity, and a frequency f, where eddy currents begin reducing the effective permeability.

This relationship fails when the lamination thickness is below 3 μm (0.12 mils). The reason for this is the earlier mentioned wall resonance frequency f_s. If we eliminate the frequency f in our previous calculation, substituting the factor 0.84 then we find:

$$2\delta_{min} = \sqrt{\rho/B_{rs}\mu_r} \tag{7.8}$$

The lamination with the very high permeability has a wall resonance frequency around 100 kHz and the permeability falls off faster than eddy currents alone would cause.

The essence of this discussion was to point out that a material with very high permeability is detrimental to the overall performance of a high-frequency head. Also, it does not pay to make cores from laminations that are thinner than 3 μm = 0.12 mils. Now, such thin laminations are virtually impossible to work with in a piecewise assembled head, but they are easily deposited in thin film heads. And we have, from the above, learned that little, if anything, may be gained by deposition of laminations so thin that eddy current limitations are nonexistent; the wall and magnetization rotation resonance frequency limitations remain.

We will in Chapter 10 show the efficiency of several heads, and will find that it is almost as high for materials with modest permeability (2–5000) as for very-high-permeability (100,000) materials.

DEPOSITED FILMS

Very small head structures can be manufactured using the deposition and etch techniques developed by the semiconductor industry. Most thin films used today are again the NiFe alloy composition with 82 percent Ni to make the magnetostriction near zero; see Fig. 7.12.

There are limits to the smallness of the magnetic films, though. We already learned that problems occur when the thickness is reduced below 3 μm (formula (7.8)). It is generally found that the permeability decreases and the coercivity increases as the films are made thinner (Feng and Thompson 1977). A double layer magnetic film appears to improve matters (Herd 1979).

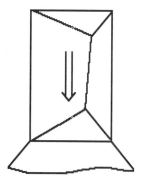 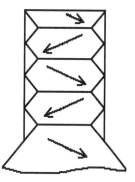

Wall resonance:

$$f_w = 0.84 * \frac{B_{rs}}{\mu_r} \text{ MHz}$$

Spin resonance:

$$f_s = 1.87 * \frac{B_{rs}}{\mu_r} \text{ MHz}$$

FIGURE 7.21 Magnetization in TFH pole tips is controlled by domain wall motion (left), magnetization rotation (right), or a mixture thereof (*Boll*).

It has been observed that distinct walls form in thin films (Kryder et al. 1980). Figure 7.21 illustrates how the domains will form if the film is isotropic, i.e., its properties are independent of direction. The central wall will move from one side to the other during flux reversals, which is a slow process. The film's permeability is improved by depositing an anisotropic film, so the magnetization reversal takes place through rotation, which is a fast process (Wells and Savoy 1981; Jones 1979).

Domain walls in thin films differ from walls in bulk material. The magnetization in bulk material rotates in a 3-D pattern as shown in Fig. 3.49, called a Bloch wall. In thin films the rotation may only occur in the 2-D plane of the film. These walls are named Néel walls.

The magnetization in the thin film segments that make up a TFH should ideally change very smoothly under influence of a field. This may be from a write current in the head coil, or from the read flux off the media.

Shifts in domain wall positions are hindered by imperfections and anisotropies introduced by stress. Total freedom from magnetostriction in every grain direction is not possible. Stresses will further change as the head heats up from the write current or cools off later on.

The domain walls will be shifted minute distances when the field in the head is comparable to the domain wall coercivities (Williams and Lambert 1989). Movements will be in sudden jumps, and are observed as wiggles on the trailing edge of read pulses.

The irreversible wall displacements are also associated with the abundance of shallow pinning centers in the Permalloy films (Klaassen and van Peppen 1989). The jumps also occur as a delayed relaxation after writing, associated with the rapid cooling of the head after current turn-off (Klaassen and van Peppen 1989). The noise spikes are known as popcorn noise and can occur during the servo read phase between writing from one sector to the next. This will obviously reduce servo performance (Klaassen and van Peppen 1994).

Similar Barkhausen noises occur in MIG ferrite heads (Schaeffer et al. 1993). Optimum design requires fairly large-grain-size ferrites, as smaller grains lead to reduced permeability. Also controlled grain orientation is needed, with very careful processing.

The elimination of walls in thin films was accomplished by laminating the film (Lazzari et al. 1971) and demonstrated in TFH structures (Slonczewski et al.). Its effectiveness is illustrated in Fig. 7.22. Two thin-film layers have several domains (a) until they are in near contact (c). Only one domain remains in each film when the spacing between them is on the order of 10 nm. This is due to magnetostatic coupling without exchange force interaction.

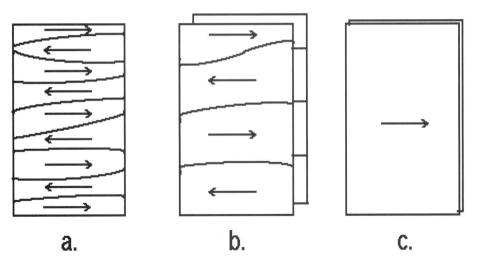

a.   **b.**   **c.**

FIGURE 7.22   Magnetostatically coupled films have few domains (*Lazzari*).

FLUX-SENSING MATERIALS

We have so far considered the read signal from a tape or disk as the signal that is generated by flux changes in an inductor (the head core plus winding). Other means exist for reading the flux from a tape recording, such as the incorporation of a flux-sensitive material, or generation of second harmonics in a magnetic modulator.

The semiconductor Indium-Antimonide will generate a voltage when magnetic flux lines pass through it; a dc-bias current is required. The voltage is proportional to the bias current I and a material constant K. The current is limited by the element's heat capacity and heat sink arrangement.

The element should be thin and is in practice placed in one of the gaps in a conventional head structure. One practical layout (after Camras) has the element placed in the front gap. The back gap is made large to concentrate the flux in the front gap. This assembly is called a Hall element.

It is correct that this Hall element will respond to a steady (dc) flux, but this is often projected to the erroneous statement that a Hall Head will read very long wavelengths. The core structure is still responsible for gathering the flux from the tape and is limited in long wavelength response. An improved response may be obtained by deleting the back gap, but even this structure will gather less and less flux as the wavelength increases. Flux-sensitive heads do not, therefore, respond well to dc during playback.

MAGNETORESISTANCE (MR) AND NiFe

The electrical resistivity of certain iron-nickel alloys will decrease under the influence of an external magnetic field. The FeNi element can be a thin plate ($t = 50$ nm) that is in near-contact with a recorded track. The field from the recorded transitions, in particular the perpendicular component H_y, will change the element's resistance R. If a constant current i circulates through the resistor, then a voltage $i \times R$ is developed. Any change in R due to changing field will deliver the read signal as the change in $i \times R$.

The application of the FeNi elements property was proposed in 1971 by Hunt. The exact nature of magnetoresistance is obscure and very difficult to analyze. Magnetism is created by the electron spins, and in materials like FeNi all generated magnetons are lined up within domains due to the exchange force. If the magnetons are affected by an external field (like compass needles), then the electron spins are reacting. And the consequence is always a reduction in electrical resistance.

A fundamental analysis is shown in Figs. 7.23 and 7.24 (Thompson et al., 1975). Figure 7.25 shows a numerical example where the resistance value drops one ohm, irrespective of field polarity. The signal response is nonlinear, and the loss of polarity makes recovery of data impossible. It becomes necessary to bias the element to a midway point of resistance change. In the example in Fig. 7.25 that value is 49.5 Ω, and now the signal polarity has been recovered in a read signal of ±10 mV_{peak}.

M
θ
I
H'y
H'y
θ
H0
H0 = Hk + Hd = Hk +N·Ms
Hk = Anisotropy field
sin θ = H'y / H0
I= = I·cosθ
I⊥ = I·sin θ
E=
E⊥
θ
E
I=
I⊥
θ
I

FIGURE 7.23 Definitions of MR head parameters.

$$
\begin{aligned}
E &= E_{=}\cos\theta + E_{\perp}\sin\theta \\
&= \rho_{=} I_{=} \cos\theta + \rho_{\perp} I_{\perp}\sin\theta \\
&= \rho_{=} I\cos^2\theta + \rho_{\perp} I\sin^2\theta \\
&= I\cdot(\rho_{=}\cos^2\theta + \rho_{\perp}\cdot(1 - \cos^2\theta)) \\
&= I\cdot(\rho_{\perp} + (\rho_{=} - \rho_{\perp})\cdot\cos^2\theta) \\
&= I\cdot\rho_{\perp}\cdot\left(1 + \frac{\rho_{=} - \rho_{\perp}}{\rho_{\perp}}\cdot\cos^2\theta\right) \\
&= I\cdot\rho_{\perp}\cdot(1 + \Delta\rho/\rho_{\perp}\cdot(1 - \sin^2\theta)) \\
&= I\cdot\rho_{\perp} + I\cdot\Delta\rho\cdot(1 - (H'_y/H_0)^2)
\end{aligned}
$$

Let $H'_y = H_{bias} + H_{data}$, and discard squared terms:

$$
\boxed{Eac = 2\cdot I\cdot R_0\cdot(\Delta\rho/R_0)\cdot(H_b/H_0^2)\cdot Hdata}
$$

FIGURE 7.24 Simplified MR head analysis (*Hunt, D.A. Thompson*).

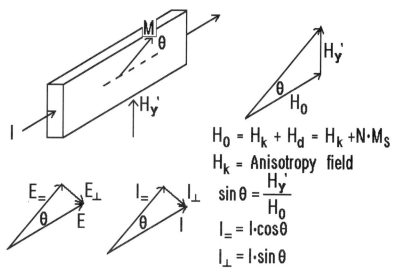

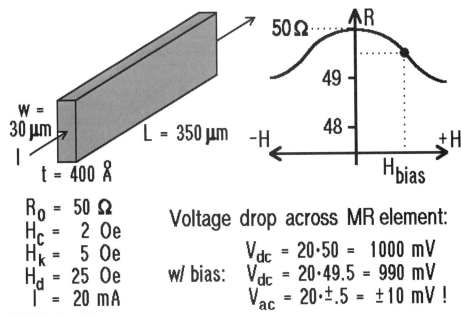

$W = 30\ \mu m$

$L = 350\ \mu m$

I

$t = 400\ Å$

$R_0 = 50\ \Omega$
$H_c = 2\ Oe$
$H_k = 5\ Oe$
$H_d = 25\ Oe$
$I = 20\ mA$

Voltage drop across MR element:

$$V_{dc} = 20 \cdot 50 = 1000\ mV$$

w/ bias: $V_{dc} = 20 \cdot 49.5 = 990\ mV$

$V_{ac} = 20 \cdot {\pm}.5 = \pm 10\ mV$!

FIGURE 7.25   An MR element, practical example.

This is an impressive value of the read signal, particularly when compared with the read voltages from inductive read heads. The number of turns in a TFH is limited at high data rates due to resonance with cable capacitance, amplifier capacitance, and self-capacitance. Figure 7.26 compares the output from an unshielded MR head output with an inductive head (Smith et al. 1987).

The necessary bias field can be provided in several different ways. They have pros and cons that are discussed in the literature. A tutorial paper by Shelledy and Nix (1992) is recommended. We will now outline the primary bias methods.

One obvious method is to place a permanent magnet next to the MR element (Bajorek and Thompson 1975), shown in Fig. 7.27. A certain spacing between magnet and element is necessary to avoid saturation of the MR element and Barkhausen noise.

An electric current produces a field and can be used for bias. A Ti element is part of the MR sensor (Fig. 7.28) and it carries most of the sensor current, the balance being the sense current for the MR element. A push-pull configuration shown in Fig. 7.29 improves the total linearity (Shelledy and Brock 1975). The sense and bias currents can also run separately, as shown in Fig. 7.30. Note that small cross sections warrant limitation of currents.

Two MR elements can also function such that their sense currents mutually bias the other element. Their output voltages are connected to a differential amplifier. Any heat spike generated by, for example, a tape passing over will cause the same polarity voltage change due to the temperature coefficient of resistance (Shelledy and Cheatham 1976).

There are two biasing schemes that have withstood the test of time. One is the SAL, Soft Adjacent Layer. Here a high-permeability film (SAL) is located close to an MR element; see Fig. 7.32. The sense current in the MR element will magnetize the SAL, which in turn generates a magnetic bias field for the MR element (Jeffers et al. 1985).

The other method conducts a current for biasing and sensing at a 45° angle to the element. This changes the element's R versus H characteristics so it becomes linear around $H = 0$; see Fig. 7.33. The magnetic field from a recorded track is conducted to the element via high-permeability flux guides (Kuijk et al. 1975; Gestel et al. 1977). These heads have successfully been designed into Philips' DCC recorders for digital sound recording and playback.

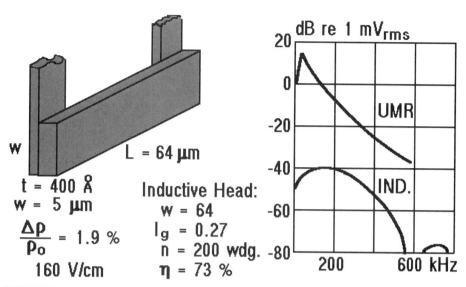

$$L = 64 \ \mu m$$

$$t = 400 \ \overset{\circ}{A}$$
$$w = 5 \ \mu m$$

Inductive Head:
$$w = 64$$

$$\frac{\Delta \rho}{\rho_o} = 1.9 \ \%$$

$$l_g = 0.27$$
$$n = 200 \ wdg.$$

$$160 \ V/cm$$

$$\eta = 73 \ \%$$

FIGURE 7.26   Response of unshielded MR heads versus ordinary inductive TFH (*Smith, et al., 1987*).

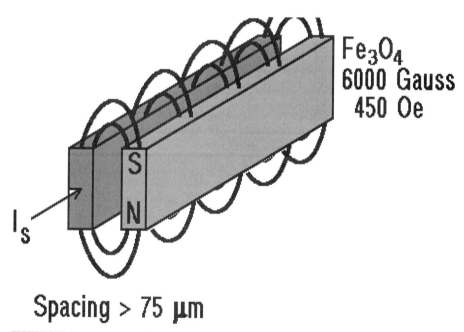

$$Fe_3O_4$$
6000 Gauss
450 Oe

$$I_S$$

Spacing > 75 μm

FIGURE 7.27   Permanent-magnet bias of an MR element (*Bajorek patent, 1975*).

A recent innovation is the Dual MR element head, shown in Fig. 7.34 (Smith et al. 1992). Two identical elements are connected in parallel, and the sensing/biasing current divide evenly between the two. One element is therefore biased up, the other one down. When an external, uniform field is present, then one element will decrease its resistance value further (to 99 Ω) while the other increases (to 101 Ω).

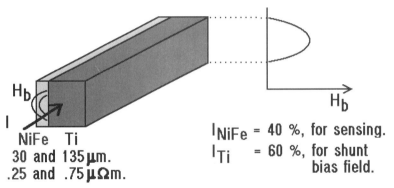

I_{NiFe} = 40 %, for sensing.
I_{Ti} = 60 %, for shunt
bias field.

NiFe   Ti
30 and 135 μm.
.25 and .75 μΩm.

FIGURE 7.28 Shunt bias of an MR element.

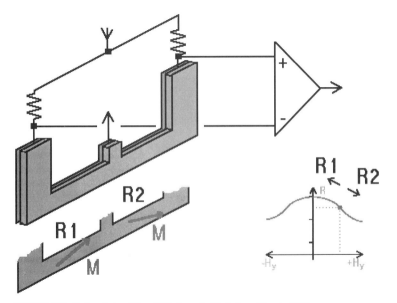

FIGURE 7.29 Push-pull shunt-biased MR elements (*Shelledy and Brock , 1975*).

If a field differential is present, as when the sensor is located exactly over a transition, then both elements' resistances will increase or decrease in resistance values. The net parallel resistance will then change dramatically in value. The DMR head has smaller maximum voltage than the comparable SAL head, but has extended response (Fig. 7.35).

MR heads are very sophisticated transducers, in spite of their apparent simplicity. The reader will have noticed that most bias configurations, etc., were conceived in the mid-1970s. No hardware has used MR heads until the early eighties, where they found use in the IBM 3480 tape drives. Now MR tape heads are used in QIC drives; see Fig. 7.36. Disk drives started using MR heads in the late eighties, again under the leadership of IBM. Today they are finally emerging in many disk drive designs.

The problems have been many. One was corrosion of the small 50-nm-thick elements containing Fe. Magnetically, the element must also be configured as a single domain. Both ends of the element are exposed and subject to simultaneous contact with the conductive thin-film media, which would short the element. This required the development of an insulating overcoat. An interesting solution to this problem, and to the one of diminishing track width, is the orthogonal MR head shown in Fig. 7.37 (Wang et al. 1993).

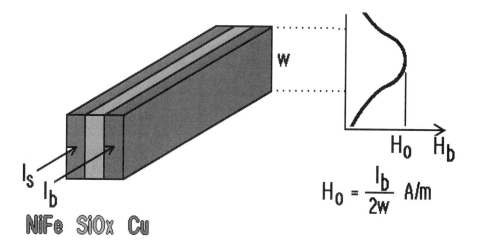

$$H_0 = \frac{I_b}{2w} \ A/m$$

Note current limit: A current density of $0.5 \cdot 10^{10}$ Am^2 causes a temperature rise of 10° C. (Example: I_b = 100 mA.)

FIGURE 7.30   Insulated shunt-biased MR element.

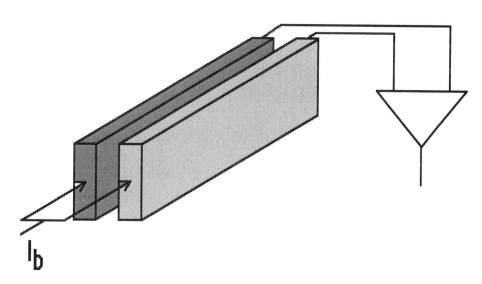

Insensitive to heat spikes.

FIGURE 7.31   Mutually biased, coupled MR elements (*Shelledy and Cheatham*).

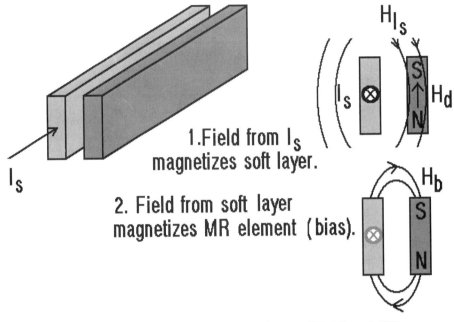

1.Field from I_s magnetizes soft layer.

2. Field from soft layer magnetizes MR element (bias).

FIGURE 7.32 A soft adjacent layer (SAL) -biased MR element (*Beaulieu patent 1975, Jeffers et al., 1975*).

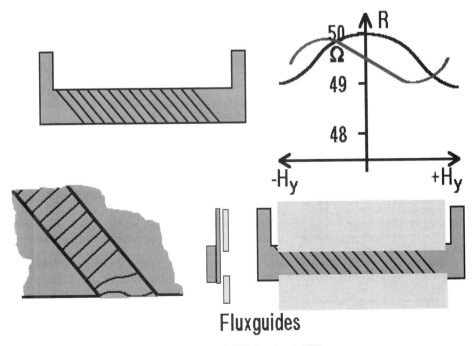

Fluxguides

FIGURE 7.33 Barber Pole biased MR element (*Kuijk et al. 1975; Gestel et al., 1977*).

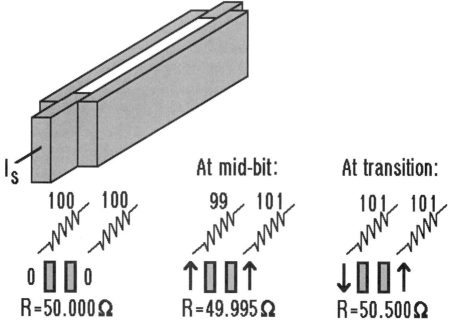

FIGURE 7.34 Dual-element MR head, DMR (*Smith, 1992*).

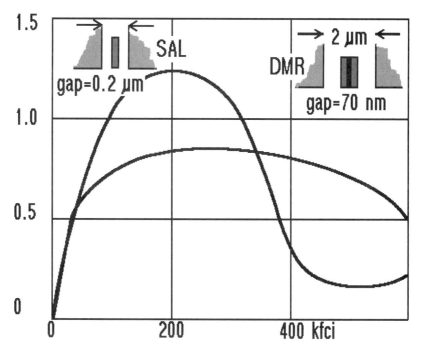

FIGURE 7.35 Comparison of SAL and DMR heads, both shielded.

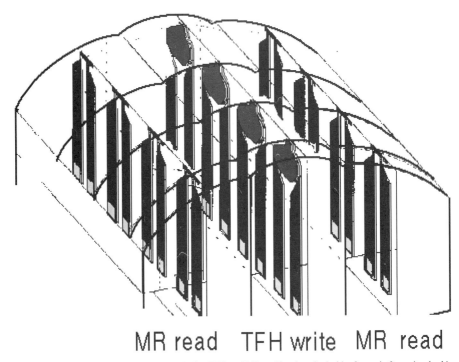

MR read   TFH write   MR read

FIGURE 7.36 QIC tape head with center-inductive TFHs and MR read heads on both sides for read-after-write checking.

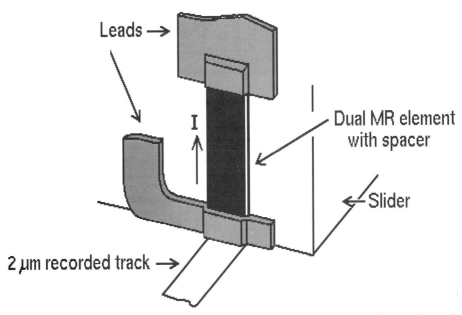

FIGURE 7.37 An orthogonal MR head offers narrow-track performance. The bottom connection can be at ground potential (disk potential) and the signal voltage not shorted (*Wang et al., 1993*).

GIANT MR HEADS

The maximum resistance change in the FeNi element is a couple of percent, as indicated in Fig. 7.25. In 1988 a group of researchers in France reported magnetoresistance change in the order of fifty percent in Fe/Cr ultra-thin-film multilayers. This was the first observation of giant magnetoresistance. It appeared to be a very general phenomenon, occurring regularly when thin (1–3 nm) magnetic layers of transition metals (Fe, Co, Ni) are separated by thin nonmagnetic metal (Cr, Cu, Ag, Au) layers (White 1994).

Also from France came the first NiFe/Ag multilayer GMR head (Mouchot et al.). IBM has unveiled the future promise of GMR by reporting an NiFe/Cu/Co sensor that produces an output of 1000 mV/mm from a disk with 1.25 memu/cm^2 at a flying height of 1.5 μin (Tsang et al., 1994).

MATERIALS FOR HOUSINGS AND SLIDERS: PLASTICS, EPOXIES

Several materials besides magnetics make up a finished head assembly (exclusive of suspension, cables, and connectors). Both tape and disk heads need a housing for the magnetic cores that will provide support and shielding. A tape head housing will also provide the contact/wear surface for the tape while a disk head slider provides air bearing/landing surface(s). Suitable materials are:

Aluminum	(ϵ = 22 μm/m/C)
Brass	(ϵ = 20 μm/m/C)
Stainless Steel	(ϵ = 9.2 μm/m/C)
MnNi Oxide	(ϵ = 12.5 μm/m/C)
Al_2O_2/TiC (3M 851-D)	(ϵ = 11.1 μm/m/C)
Al_2O_3/TiC (general)	(ϵ = 7.8 μm/m/C)

The coefficient of expansion ϵ, measured in micro-meters per meter per Centigrade degree, is important because it should match that of the magnetic core material. That is why the 3M Company "stretched" their aluminum-oxide/titanium-carbide material to match MnZn ferrites.

MnNi oxide has been applied in making very small laminated video heads (Ash et al. 1990). This material has excellent wear resistance. In a low-speed tape application, its wear rate was only 10^{-4} Å/foot. MnZn is used in some MR tape heads to serve as a shield as well as a wear surface, because also MnZn wears very slowly.

Epoxies, adhesives, and a host of different materials are used in head fabrication. Some materials are listed in this chapter's references. Others, as well as some process descriptions, are found in Chapter 15's bibliography for assorted papers and in Chapter 16's references on Testing, Materials, and Processes.

REFERENCES

GENERAL INTEREST

Bozorth, R.M. 1993 (Orig. 1951)). *Ferromagnetism.* IEEE Press New York, 992 pp.

Chen, C-W. 1986. *Magnetism and Metallurgy of Soft Magnetic Materials.*1986. Dover, New York. 571 pages.

Ciureanu, P., and H. Gavrila. 1987. *Magnetic Heads for Digital Recording.* Elsevier, Amsterdam 714 pp.

Jones Jr., R.E., and C.D. Mee. 1987. Recording Heads, in *Magnetic Recording.* Vol. 1. Ed. by C.D. Mee and E.D. Daniel. New York. McGraw-Hill, pp. 244–336.

Kornei, O. July 1953. Structure and Performance of Magnetic Transducer Heads. *Journal of Audio Engineering Society* 1 (3).

Kornei, O. March 1954. Survey of Flux-Responsive Magnetic Reproducing Heads. *Journal of Audio Engineering Society* 2 (3): 86 ff.

EDDY CURRENTS

Boll, R. 1960. Metalliche Magnetwerkstoffe bei hohen Frequenzen. *Technische Informationsblätter*, Vacuumschmelze AG, Germany, 8 pages.

Boll, R. Aug. 1960. Über Bezichnungen dem Schaltkoeffizienten und der Grenzfrequenz ferromagnetischer Werkstoffe. *Zeits. für angew. Physik* 12 (8): 364–370.

Boll, R. May 1960. Wirbelstrom- und Spinrelaxationsverluste in dünnen Metallbandern bei Frequenzen bis zu etwa 1 MHz. *Zeits. für angew. Physik* 12 (8): 212–223.

Lee, E.W. 1958. Eddy-Current Losses in Thin Ferromagnetic Sheets. *Proc. IEE* C105: 337–342.

Peterson, E., and L.R. Wrathall. Dec. 1977. Eddy Currents in Composite Laminations. *Journal Audio Engineering Society* 25 (12): 1026-1032; Original: Feb. 1936. *Proc. IRE* 24 (2): 275–286.

Yuan, S.W., and H.N. Bertram. Nov. 1993. Eddy Current Damping of Thin Film Domain Walls. *IEEE Trans. Magn.* 29 (6): 2515–17.

METALLIC MATERIALS AND LAMINATIONS

Ash, K.P., D. Wachenschwanz, C. Brucker, J. Olson, M. Trcka, and T. Jagielinski. Nov. 1990. A Magnetic Head for 150 MHz, High Density Recording. *IEEE Trans. Magn.* MAG-26 (6): 2960–65.

Bendson, S. 1976. New Wear-Resistant Permalloy for Optimum High-Frequency Permeability and Machinability. *Journal of Audio Engineering Society* 24: 562–567.

Boll, R. 1977. *Soft Magnetic Materials*. Siemens/Heyden. 349 pages.

Dirne, F.W.A., H.J. de Wit, C.H.M. Witmer, P. Boher, Ph. Houdy, G.H. Rotman, and J.J.M. Ruigrok. Nov. 1991. Soft-Magnetic Multilayers for Video Recording Heads. *IEEE Trans. Magn.* 27 (6): 4882–84.

Feng, J.S.Y., and D.A. Thompson. Sept. 1977. Permeability of Narrow Permalloy Stripes. *IEEE Trans. Magn..* MAG-13 (5): 1521–1523.

Hall, R. 1959. Single Crystal Anisotropy and Magnetostriction Constants of Several Ferromagnetic Materials Including NiFe, SiFe, AlFe, CoNi and CoFe. *Jour. Appl. Phys.* 38: 816–819.

Herd, S.R. Nov. 1979. Magnetization Reversal in Narrow Strips of NiFe Thin Films. *IEEE Trans. Magn.* MAG-15 (6): 1824-1826.

Jones, R. Nov. 1979. Domain Effects in the Thin Film Head. *IEEE Trans. Magn.* MAG-15 (6): 1619–1621.

Kryder, M. Jan. 1980. Magnetic Properties and Domain Structures in Narrow NiFe Stripes. *IEEE Trans. Magn.* MAG-16 (1): 99–103.

Liebermann, H.H. Nov. 1979. Manufacture of Amorphous Alloy Ribbons. *IEEE Trans. Magn.* MAG-15 (6): 1393–1397.

Masumoto, H. Ota, and T. Yamamoto. Dec. 1981. Ribbon Sendust Magnetic Tape Heads. *Journal of Audio Engineering Society* 29 (12): 867–872.

McKnight, J. May 1977. The Permeability of Laminations for Magnetic Recording and Reproducing Heads. *Journal of Audio Engineering Society,* Preprint no. 1265.

Miyazaki, T. Sept. 1972. New Magnetic Alloys for Magnetic Recording Heads. *IEEE Trans. Magn.* MAG-8 (3): 501–502.

O'Handley, R. March 1979. Low-Field Magnetic Properties of Wide Fe-Ni Metallic Glass Strips. *IEEE Trans. Magn* MAG-15 (2): 970–972.

Preifer, F., and C. Radeloff. Apr. 1980. Soft Magnetic Ni-Fe and Co-Fe Alloys-Some Physical and Metallurgical Aspects. *Jour. of Magnetism and Magnetic Materials* 19 (1-3): 190–207.

Raskin, D., and L. A. Davis. Nov. 1981. Metallic Glasses: A Magnetic Alternative. *IEEE Spectrum*, pp. 28–33.

Tsukagoshi. T., H. Ito, K. Ogasawara, T. Namiki, S. Uasuda, Y. Masumoto, H. Ota, and T. Yamamoto. Dec. 1981. Ribbon Sendust Magnetic Tape Heads. *Jour. Audio Engr. Soc.* 29 (12): 867–872.

Tsuya, N. Nov. 1981. Magnetic Recording Head Using Ribbon-Sendust. *IEEE Trans. Magn.* MAG-17 (6): 3111–3113.

Wells, O. and R. Savoy. May 1981. Magnetic Domains in Thin-Film Recording Heads as Observed in the SEM by a Lock-In Technique. *IEEE Trans. Magn.* MAG-17 (3): 1253–1261.

THIN FILMS

Klaassen, K.B., and J.C.L. van Peppen. March 1994. Magnetic Instability of Thin-Film Recording Heads. *IEEE Trans. Magn.* MAG-30 (2): 375–80.

Klaassen, K.B., and J.C.L. van Peppen. Sept. 1989. Delayed Relaxation in Thin-Film Heads. *IEEE Trans. Magn.* MAG-25 (5): 3212–14.

Klaassen, K.B., and J.C.L. van Peppen. Sept. 1989. Irreversible Wall Motion in Inductive Recording Heads. *IEEE Trans. Magn.* MAG-25 (5): 3209–11.

Lazzari, J.P., and J. Melnick. March 1971. Integrated Magnetic Recording Heads. *IEEE Trans. Magn.* MAG-7 (1): 146–150.

Slonczewski, J.C., B. Petek, and B.E. Argyle. May 1988. Micromagnetics of Laminated Permalloy Films. *IEEE Trans. Magn.* MAG-24 (3): 2045–54.

Williams, M.L. and S.E. Lambert. Sept. 1989. Film-Head Pulse Distortion due to Microvariation of Domain Wall Energy. *IEEE Trans. Magn.* MAG-25 (5): 3206–08.

HIGH-MOMENT THIN FILMS

Jones Jr., R.E., J. Lo, and J.L. Williams. Nov. 1993. Magnetic Properties of RF Sputtered Fe(N)/NiFeCo(N) Films. *IEEE Trans. Magn.* MAG-29 (6): 3072–74.

Kohmoto, O. July 1991. Recent Development of Thin Film Materials for Magnetic Heads. *IEEE Trans. Magn.* 27 (4): 3640–47.

Ohkubu, K., Y. Arimoto, T. Fumoto, and H. Yamasaki. Nov. 1993. Magnetic Properties and Domain Structure of High Bs Co-Based Amorphous Films. *IEEE Trans. Magn.* MAG-29 (6): 2530–32.

Okumura, T., A. Osaka, N. Ishiwata, Y. Takeshima, and H. Urai. Nov. 1993. High Frequency Read/Write Characteristics for Laminated Fe-Ta-N Heads. *IEEE Trans. Magn.* 29 (6): 3843–45.

Zeltser, A.M., and T.M. Jagielinski. Nov. 1993. Soft Magnetic Properties of FeXGaSi (X=Ir, Rh) Films. *IEEE Trans. Magn.* MAG-29 (6): 3508–3532.

FERRITES (CERAMICS)

Brissette, Leo. 1983. MnZn Ferrites extend Performance. *Electronic Design.*

Lagrange, A. Sept. 1972. Preparation and Properties of Hot-Pressed Ni-Zn Ferrites for Magnetic Head Application. *IEEE Trans. Magn.* MAG-8 (3): 494–497.

Mizushima, M. Sept. 1971. Mn-Zn Single Crystal Ferrite as a Video-Head Material. *IEEE Trans. Magn.* MAG-7 (3): 342–345.

Monforte, F. Sept. 1971. Pressure Sintering of MnZn and NiZn Ferrites. *IEEE Trans. Magn.* MAG-7 (3): 345–350.

Néel, L. March 1948. Propriétés magnetiques des ferrites: ferrimagnétisme et antiferromagnétisme. *Ann. Phys.* Paris. pp.137–98.

Rigby, E.B., W.D. Kehr, and C. B. Meldrum. Sept. 1984. Preparation of Coprecipitated NiZn Ferrite. *IEEE Trans. Magn.* MAG-20 (5): 1506–1508.

Snelling, E.C. 1988. *Soft Ferrites, Properties and Applications.* Butterwirths, London, 2nd ed., 366 pp.

Snelling, E.C., and A.D. Giles. 1983. *Ferrites for Inductors and Transformers.* Research Studies Press Ltd. (Wiley), 167 pages.

Sugaya, H. Sept. 1968. Newly Developed Hot-Pressed Ferrite Head. *IEEE Trans. Magn.* MAG-4 (3): 295–301.

Takama, E. Nov. 1979. New Mn-Zn Ferrite Fabricated by Hot Isostatic Pressing. *IEEE Trans. Magn.* MAG-15 (6): 1858–1860.

Withop, A. Sept. 1978. Manganese-Zinc Ferrite Processing, Properties and Recording Performance. *IEEE Trans. Magn.* MAG-14 (5): 439–441.

MAGNETORESISTIVE MATERIALS

Bajorek, C., and D.A. Thompson. Sept. 1975. Permanent Magnet Films for Biasing of Magnetoresistive Transducers. *IEEE Trans. Magn.* MAG-11 (5): 1209–11.

Druyvesteyn, W.W.F., J.A.C. van Oogen, L. Postma, E.L.M. Raemakers, J.J.M. Ruigrok, and J. de Wilde. Nov. 1981. Magnetoresistive Heads. *IEEE Trans. Magn.* MAG-17 (6): 2884–2889.

Hunt, R. March 1971. A Magnetoresistive Readout Transducer. *IEEE Trans. Magn.* MAG-7 (1):150–54.

Jeffers, F. Nov. 1979. Magnetoresistive Transducer with Canted Easy Axis. *IEEE Trans. Magn.* MAG-15 (6): 1628–1630.

Jeffers, F., J. Freeman, R. Toussaint, N. Smith, D. Wachenschwanz, S. Shtrikman, and W. Doyle. Sept. 1985. Soft-Adjacent-Layer Self-Biased Magnetoresistive Heads in High-Density Recording. *IEEE Trans. Magn.* MAG-21 (5): 1563–65.

Kuijk, K., W.J. van Gestel, and F.W. Gorter. Sept. 1975. The Barber Pole, A Linear Magnetoresistive Head. *IEEE Trans. Magn.* MAG-11 (5): 1215–17.

Mouchot, J., P. Gerard, and B. Rodmacq. Nov. 1993. Magnetoresistive Sensors Based on Ni81Fe19/ Ag Multilayers. *IEEE Trans. Magn.* MAG-29 (6): 2732–34.

Shelledy, F. and G. Brock. Sept. 1975. A Linear Self-Biased Magnetoresistive Head. *IEEE Trans. Magn.* MAG-11 (5): 1206–08.

Shelledy, F.B., and J.L. Nix. Sept. 1992. Magnetoresistive Heads for Magnetic Tape and Disk Recording. *IEEE Trans. Magn.* MAG-28 (5):2283–88.

Shelledy, F.B., and S.D. Cheatham. July 1976. Suppression of Thermally Induced Pulses in Magnetoresistive Heads. *IEEE Conf. Proc.* No. 35, pp. 251–260.

Smith, N., and D. Wachenschwanz. Sept. 1987. Magnetoresistive Heads and the Reciprocity Principle. *IEEE Trans. Magn.* MAG-23 (5): 2494–96.

Smith, N., J. Freeman, P. Koeppe, and T. Carr. Sept. 1992. Dual Magnetoresistive Head for Very High Density Recording. *IEEE Trans. Magn.* MAG-28 (5): 2292–94.

Thompson, D.A., L.T. Romankiw, and A.F. Mayadas. July 1975. Thin Film Magnetoresistors in Memory, Storage, and Related Applications. *IEEE Trans. Magn.* MAG-11 (4): 1039–50.

Tsang, C., R. Fontana, T. Lin, D. Heim, V. Speriosu, B. Gurney, and M. Williams, Nov. 1994. Design, Fabrication & Testing of Spin-Valve Read Heads for High Density Recording. *IEEE Trans. Magn.* MAG-30(b): 3801–06.

van Gestel, W.J., F.W. Gorter, and K.E. Kuijk. March 1977. Read-out of a Magnetic Tape by Magnetoresistance Effect. *Philips Tech. Rev.* 37 (2/3): 42–50.

Wang, P-K., M. Krounbi, D.E. Heim, and R. Lee. Nov. 1993. Sensitivity of Orthogonal Magnetoresistive Heads. *IEEE Trans. Magn.* MAG-29 (6): 3820–22.

White, R.L. March 1994. Giant Magnetoresistance Materials and Their Potential as Read Head Sensors. *IEEE Trans. Magn.* MAG-30 (2): 346–352.

OTHER MATERIALS

Chandrasekar, S., and B. Bhushan. Jan. 1990. Friction and Wear of Ceramics for Magnetic Recording Applications. *Jour. Tribology* 112 (1): 1–16.

Chu, M.Y., B. Bhushan, and L.C. Dejonghe. Nov. 1992. Processing of Diamond Alumina Composites for Low Wear Applications. *Jour. Matl. Res.* 7 (11): 3010–18.

Fisk, G.E. April 1990. Ceramic Materials for Magnetic Heads. *Amer. Ceramic Soc. Bul.* 69 (4): 696–702.

Wada, T. Aug. 1991. Magnetic Head Slider Materials. *Jour. Japan Soc. Tribo.* 36 (8): 603–08.

CHAPTER 8
FIELDS FROM MAGNETIC HEADS

The function of the record head is to create a permanent magnetization pattern on a magnetic tape or disk moving past. The recording process is performed by the magnetic field from the gap in a magnetic core, energized by a current through its winding.

Upon playback, or read-out, the head performs a sensing function whereby its field pattern senses any magnetization moving past the gap in the head and converts the detected magnetization into flux through the core; this flux will, in turn, induce a voltage in the winding.

Investigation of the field in front of the magnetic head is therefore playing an increasingly important role in recording technology. This is spurred by the arrival of perpendicular write/read modes and the requirements for high-density tracks. We wish to answer the following questions:

1. How does the field strength vary on the trailing side of the gap in the write head? It must decay fast for the best writing of short transitions and short wavelengths.

2. What is the direction of the field on the trailing edge of the write gap? Longitudinal, perpendicular, or both?

3. How much wider than the head core is the field? This will ultimately set the limit for the smallest trackwidth, for writing as well as reading.

We need to explore the fields from various head configurations, and in this chapter we will discuss several ways of doing this. We will start by describing some past and present head designs, and their fields. Plots of these fields can be made freehand, aided by measurements or models. Ample references are supplied.

With our knowledge of fields we can continue to investigate how wide a track is. It will always be wider than the core width dimension, due to stray fields on the sides, and it poses an upper limit to the number of tracks that can be written parallel to each other. This limitation will also apply to the read process. In this connection the reader is reminded that the design of magnetic fields with sharply defined patterns is as difficult as designing electric circuits for use in salt water. Electric circuits enjoy a difference in conductivity between copper and air of more that 10^{10}, while magnetic circuits have magnetic conductances that only differ by 10^2 to 10^4.

We are also working with dimensions that preclude using the design techniques from antennas, where multiple elements aid in focusing the field. Any useful directivity will require an operating frequency of 10^{14} Hz, which happens to be near the frequency of light.

The content of this chapter leaves the inventor-to-be of novel heads with the knowledge of cleverly applied materials and processes, such as the phasing of multiple pole pieces; that's why this chapter on fields has been made quite detailed.

FIELDS FROM MAGNETIC HEADS

The field lines from a gap in a ring core head are semicircular patterns outside the recording gap. This pattern has been studied extensively in the past years because it, together with the medium magnetic characteristics, determines the recording resolution. The direct recording process without bias was introduced in Chapter 5, and will be further treated in Chapter 19. It will then become clear that a desired feature of the recording field is that the field strength decreases rapidly with distance from its maximum value over the center of the gap.

Figure 8.1 illustrates the field patterns associated with the conventional ring core head (so-called in spite of its rectangular shape, because its original form was a round core shape). We are first interested in the field in and around the front gap. The field into the coating, including that from both corners, does the recording. A picture of the field pattern can be developed by employing a couple of magnetic poles (cut from thick Mu-metal lamination stock or iron plate), energized by a dc current through a suitably located winding; see Fig. 8.2. Place a stiff piece of paper on top of the magnet, with iron powder evenly sprinkled on top of the card. Now gently tap the card, and the powder will move about to form the patterns shown in Fig. 8.3.

The field on the side and in back of the front gap lowers the head's efficiency; the stray flux at the back gap has little bearing on the head's performance and will not be discussed further. The leakage flux around the coil decreases the head's efficiency when applied as a write head. This leakage is increased when shielding around the core, or between cores in a multitrack head, is employed.

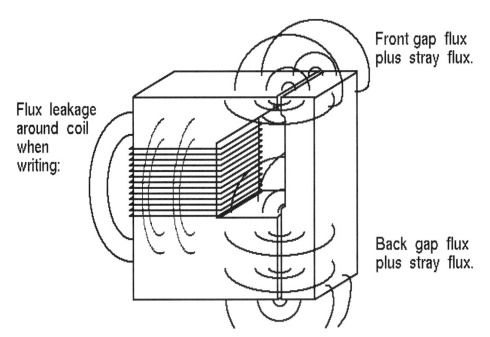

Front gap flux plus stray flux.

Flux leakage around coil when writing:

Back gap flux plus stray flux.

FIGURE 8.1   Magnetic flux paths in and around a magnetic head core.

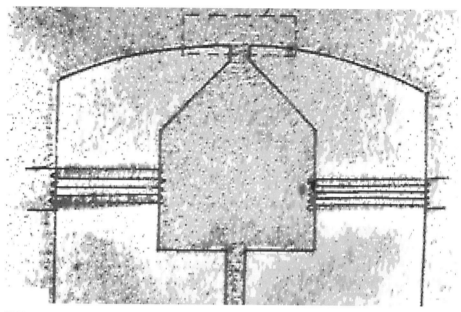

FIGURE 8.2 Magnetic flux patterns shown by iron filings on cardboard placed on top of a magnet structure. The magnet is energized by a current through the windings. Detail in the gap region is shown in Figure 8.3.

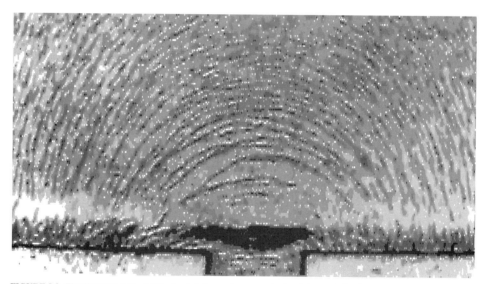

FIGURE 8.3 Flux lines follow the field in front of the gap in a magnetic circuit (magnetic head).

Mathematical expressions for the various fields are often needed in calculations of recording performance. The most commonly used equations for the field in front of the write gap are from Karlqvist, and the expressions for the longitudinal component H_x and perpendicular component H_y are expressed in Fig. 4.24. In the past only the longitudinal (also called horizontal) component was considered, justified by the fact that the tape coating particles were longitudinally oriented and demagnetization losses were not too great.

Today's packing densities are very high, and the assumption of a purely longitudinal recording scheme is not valid. This matter was covered in Chapter 5, and much work has revealed the presence of strong perpendicular components in recordings.

There is, for the time being, no unified treatment of the recording process that expresses itself in a few useful formulas. All subsequent discussions in this book, therefore, will rely on the reader's understanding of the interaction between a medium's switching field distribution and the write head field pattern. We will use the curves of constant, total field strength, rather than those of constant longitudinal or perpendicular field components.

We will further assume a value of unity for the coating's relative permeability; this appears justified in the write process, where the coating is magnetized to saturation.

HIGH-GRADIENT HEADS

Improvements in the field gradient over that of the stray field from a gap are possible. Attempts to do this have not been great successes, mainly due to difficulties in the manufacturing processes.

The crossfield head, created by Camras (1964), uses superposition of two ac-bias fields with pole pieces and their strengths arranged in a manner that produces a higher gradient. This is illustrated in Fig. 8.4. This head has (to the author's knowledge) only been used by the early Roberts (Akai) and Tandberg recorders.

Another design uses the limitation of field penetration due to eddy currents. This head, called the driven-gap-record head, has the bias current running in a conductor wedged between two pole pieces of highly permeable materials. If the bias frequency is high enough (above 50 MHz), then the field is confined as shown in Fig. 8.5. The gradient can be several times higher than that for a conventional head (Johnson and Jorgensen 1966).

The driven gap head has been used to record data with a rate of 60 Mbps, utilizing a bias frequency of 150 MHz (ref. Krey, 1973). Multichannel applications were difficult due to lack of uniformity between tracks; this could be a problem of the past if modern microcircuit techniques are used in the manufacture, whereby the conductor geometry is closely controlled.

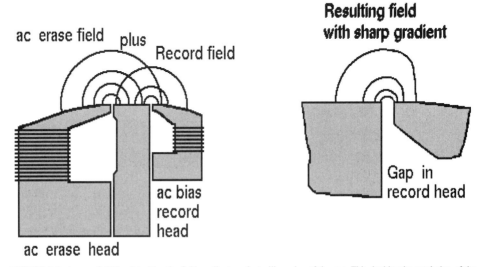

FIGURE 8.4 A cross-field head doubles the field gradient on the trailing edge of the gap. This doubles the resolution of the recording (*Camras, 1964*).

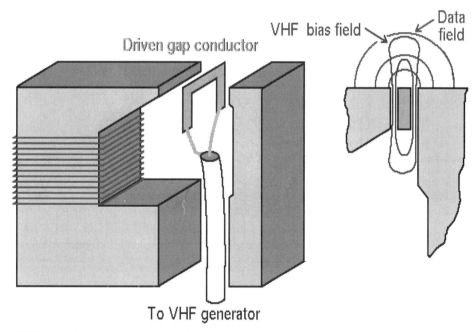

FIGURE 8.5 Driven-gap head with very-high-frequency ac bias improves the field gradient and resolution (*Johnson and Jorgensen, 1966*).

The bias current is about 2 A for a coating with $H_c = 300$ Oe. The difficulty in providing this current at UHF was readily solved by connecting the gap conductor as the termination in a quarter-wavelength coax cable.

A variety of the driven-gap head has been employed in studio recorders and duplicators (K. Johnson 1965). A conductive shim was wedged between the two front pole pieces and shorted to the head-housing block. VHF bias applied to the ordinary winding will induce circulating currents in the shim, with such polarity that the total field was focused.

Figure 8.6 shows a 3D-picture of the field magnitudes in front of the standard ring core gap. The highest field gradient is found at the head surface, and it then diminishes with increasing distance from the head surface. The field strength variation away from the gap is bell-shaped, of Lorentzian form.

The high-gradient heads described above are, because of their field patterns, also perpendicular recording heads (the H_y field being greater than the H_x field component). Experiments with heads having perpendicular fields continue to develop a more practical and efficient way to assist the perpendicular tape remanence. We shall return to a discussion thereof after we have learned more about computing and plotting head fields.

COMPUTATIONS AND PLOTS OF FIELDS

A good qualitative method of establishing the shape of magnetic fields has been demonstrated in Fig. 8.3, using iron powder. A good source for the powder is the local hobby store: ask for Iron, Powdered Fe, item no. 721 from Perfect Parts Company, Baltimore, MD 21224.

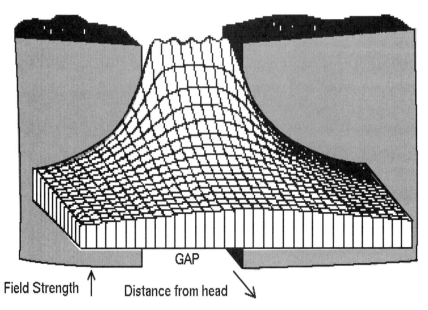

Field Strength ↑   Distance from head ↘

FIGURE 8.6 The field strength and gradient decrease as the distance to the head increases.

HOW TO SKETCH FIELD PATTERNS

Quite accurate field patterns may be drawn by following a few simple rules. From high school physics you will recall that electric field lines in a simple capacitor (Fig. 8.7) run straight from plate to plate, and so it is for the magnetic field lines between two magnetic poles. Outside these uniform regions the field lines are curved, as shown.

The primary aids in drawing field lines are shown as broken lines, which are the potential lines. Here they are drawn freehand, but they may be determined by an electric model, as will be described later.

If for a moment we look at the magnetic field lines, we see that the distances between them increase as we move away from the gap between the poles. This simply means that the field strength decreases in the same proportion. The square pattern of field and potential lines between the poles divides the space up into reluctance elements, which all have the same value of reluctance (reluctance equals magnetic resistance).

Assume that the pole pieces extend w meters into the paper. The volume defined by the two sheaths following adjacent field lines, and of width w meters, is called a flux tube. It is a characteristic of the field line drawing that all flux tubes carry the same amount of flux; the magnetic potential difference equals $2M$, and the flux in each tube equals $2M/\Sigma R_m$ ($= 2M/4R_m$ in Fig. 8.7 because there are 4 tubes in series, each of magnetic resistance R_m). This is quite clear for the flux tubes in the uniform region between the poles.

Outside the poles the flux tubes become longer, but the distance between field lines increases so that the area of each tube also increases. This geometry change is such that the reluctance of each tube remains unchanged. This leads to the requirement that the elements dx by dy should remain square. That is not possible, so the requirement is relaxed to the following rules:

1. Start the field drawing by using squares in a region where the field is uniform (as between the pole pieces in the front gap of a magnetic head).

2. Extend the potential lines into adjacent regions.

3. Draw the field lines so they form distorted square pillow patterns with the potential lines.

4. Step three must be done in such a way that the lines are perpendicular to each other where they intersect; this requirement overrides the requirement of the pillow shape. A good check is to also draw a pillow's diagonals: they must be perpendicular to one another.

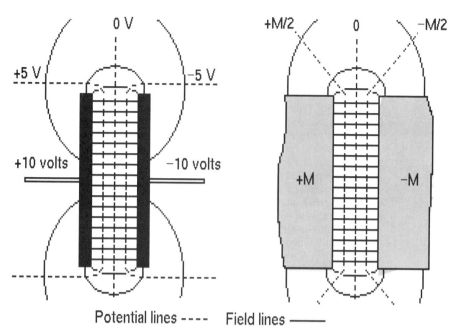

Potential lines - - - -   Field lines ————

FIGURE 8.7 Field and potential analogies between electric (capacitor) and magnetic (poles) circuits.

These rules are summarized in Fig. 8.8.

This graphic method does require a good estimate of the patterns of the potential lines, and that ability only comes with practice and experience. There is a simple method for the determination of points on potential lines, using the analogy that exists between electric and magnetic fields.

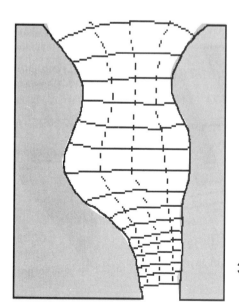

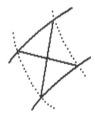

1. Angles between field lines and equipotential lines must all be ninety degrees.

2. Diagonals must be perpendicular to each other.

3. Field lines must be perpendicular to magnetic surfaces.

FIGURE 8.8 Rules for drawing a. potential lines, and b. field lines.

PLOT OF ELECTRIC POTENTIALS AND FIELDS

This method is also known from high school physics, using so-called Teledeltos paper, an electrically conducting paper. The only problem is that such paper does not exist any more; the author has searched in vain. An alternative is to use an electrolytic tank model (see references), but that tends to get a bit messy.

Another alternative is a different conductive paper, available from EDO Western Corporation, Salt Lake City, Utah 84115, part no. 22250-3. It is a black paper stock, coated with white wax, and is used in level recorders where a conductive stylus cuts through the wax and finds its proper position by sensing a voltage on the paper.

The procedure for making a model is now quite simple. Prepare a base by gluing $\frac{1}{16}$" artist cardboard onto one or more pieces of balsa wood. Next glue the conductive paper onto the cardboard using, for example, 3M Co.'s Super 77 Spray Adhesive.

Now use a soft pencil to trace the pole patterns onto the paper. Next scrape the wax overcoat away inside each pole pattern, exposing the black conductive paper in $\frac{3}{32}$" wide stripes. Use the flat side of a screwdriver so as not to cut or scratch the carbon paper.

The electrodes for the pole pieces themselves are best made from a 0.005" flexible brass stock (from K and S Engineering Co., Chicago, Ill 60638). Cut the pieces to exact size so they fit with the center lines of the scraped patterns.

A good contact is essential on the outside edges of the pole patterns and may be achieved by either beveling the edges of the brass pieces or holding them down against the paper with thumbtacks. The author found better results by cutting a jagged edge with a scissor; see Fig. 8.9. (Conductive paint proved of little value due to cracking, e.g., open circuit.)

The model is now finished. Connect a battery (or power supply) to opposite pole electrodes, and a probe with a pinpoint as shown in Fig. 8.9. Trace the paper for points with the same voltages; this will give the pattern for one potential line. Continue on and trace points of different voltages to find different potential lines. It is a good idea to select voltages that are $V_{battery}/10$ volts apart.

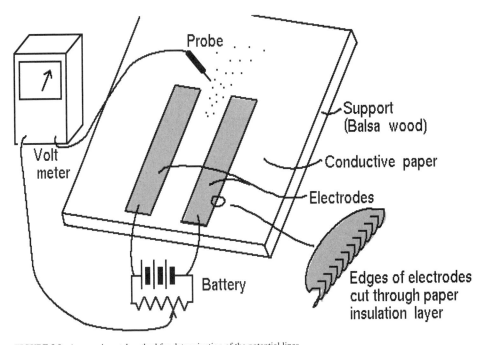

FIGURE 8.9 An experimental method for determination of the potential lines.

Finally trace the points of equal potential with a pencil, and the potential line plot is complete. To now add field lines to this plot, we must use the reluctance method. Since this method involves trial and error, use a copy machine and do the rest of the work on the obtained copy (e.g., starting with squares in an area where the potential lines are parallel and evenly spaced, working into the more difficult area). With training, an accuracy of 5 to 10 percent for the finished field pattern plot is possible.

When outlining the pole pieces, care must be taken to use the correct scaling. For magnetic heads a side view is generally used, and this may be scaled up without restrictions, provided the thickness of the core (track width) does not change. Changing the thickness (as in micro-composite heads) requires that the scaling correlates with the magnetic resistance of the various elements of the circuit. Round structures are particularly tricky, as for example the linear motors (voice coil motors) for head positioning in large disk drives.

COMPUTATIONS OF FIELDS AND FORMULAS FOR H_x, H_y and H_z

Electromagnetic field theory is required for detailed analysis of the field from magnetic heads. Approximations are necessary during development of formulas like Karlqvist's. This has in particular been true for the development of expressions for H_x and H_y field components around TFH poles and the single pole MR or perpendicular head. And some of the latter two may have magnetic shielding that will influence performance.

The earliest analytic expressions were done by Potter, et al. (1971, 1975). With the aid of computers, more elaborate and more accurate expressions became possible. The derivations by Szczech (1979, 1981), and Szczech, et al. (1982, 1987) are in wide use today. Their expressions for H_x, H_y, and H_z use several constants that are obtained from a large scale model.

Another field solution for H_z and H_y from a TFH is presented in a paper by Bertero, et al. (1993). Computational analysis was done by Megory-Cohen and Howell (1988). One of their results is shown in Figs. 8.10 and 8.11. It is a good example that verifies the rules from Fig. 8.8. It also shows

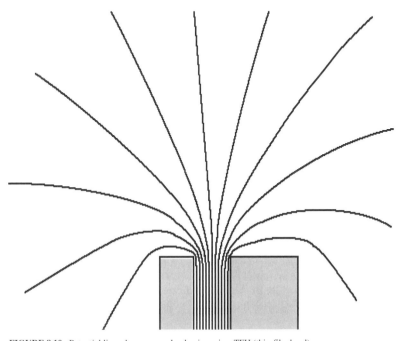

FIGURE 8.10 Potential lines drawn around pole pieces in a TFH (thin-film head).

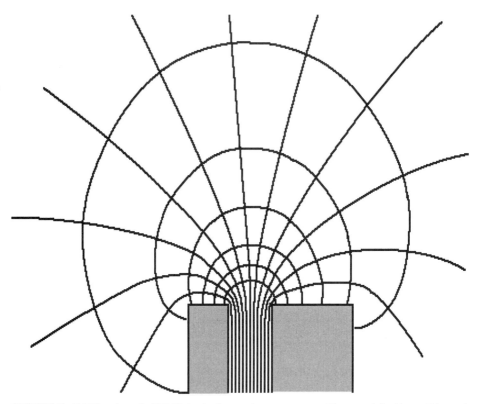

FIGURE 8.11   Field lines drawn for TFH. Notice the downward curvature near outside corners (*after Megory-Cohen and Howell, 1988*).

how the field lines curve in toward the head in the regions outside the pole pieces. A longitudinal medium magnetization will, at those positions, produce a negative flux contribution. The angle between the magnetization and the sensing pattern is greater that 90 degrees, and their dot product (formula 6.5b) is therefore negative. We therefore have undershoots in the read voltage from a single transition; see Fig. 8.12.

The length p of the pole piece will affect the timing of the undershoot. If p is large, then the undershoot occurs further away from the pulse center. This is verified in Fig. 8.13.

One may further speculate that a difference in the length of the individual pole pieces may play a role in the head's quality. The pole tip region is now defined as shown in Fig. 8.14. Aoi et al. found optimum resolution for $p_t/p_L = 1.2$ and maximum field strength H_z for $p_t/p_L = 1.4$; see Fig. 8.15. Chi (1985) found, on the other hand, that $p_t/p_L = 0.8$ reduce amplitude variations and peak shifts in a tribit pattern. No doubt that timing, density and choice of modulation/detection scheme will be affected by the p_t/p_L ratio.

The overall length of the TFH pole face length also affects performance as shown in Figs. 8.16 and 8.17 (After Ohura, et al., 1987).

The only real way to get rid of the influence of the undershoots is to eliminate them. Ion milling has been used to trim TFH pole pieces. The original geometry can be changed in the two geometries shown in Fig. 8.18. Both offer significant reduction in the negative pre- and post-pulses (Yoshida, et al., 1993). The remaining problem is their relative difficulty of fabrication.

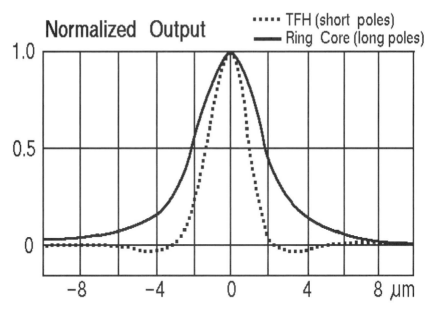

FIGURE 8.12   Read pulses from a single transition from a long-pole head (ring core) and from a short-pole head (TFH).

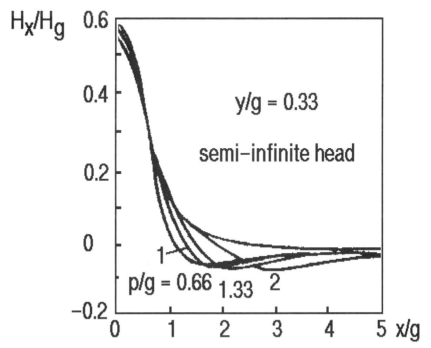

FIGURE 8.13   Variation in response with the change in ratio between pole-to-gap lengths.

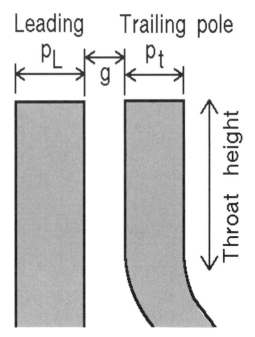

FIGURE 8.14 Geometry of the pole tip region in a TFH.

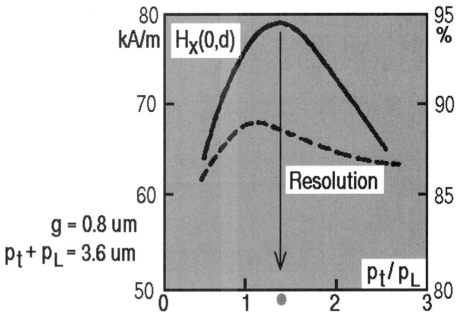

FIGURE 8.15 Write field and resolution versus the ratio of the pole tip lengths (film thicknesses) (*Aoi et al., 1982*).

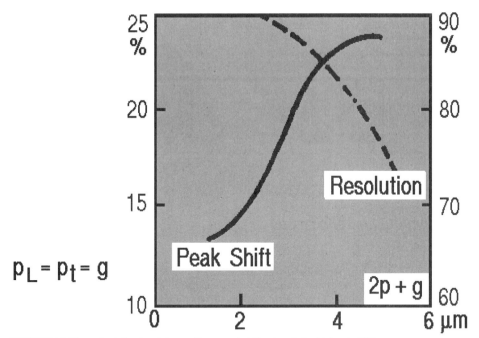

FIGURE 8.16   The total pole face length has an effect on peak shift and resolution (*Aoi et al., 1982*).

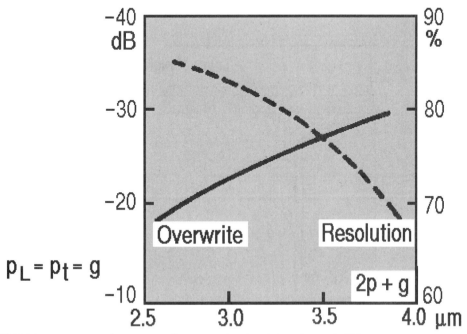

FIGURE 8.17   The total pole face length has effect on overwrite and resolution (*Ohura et al., 1987*).

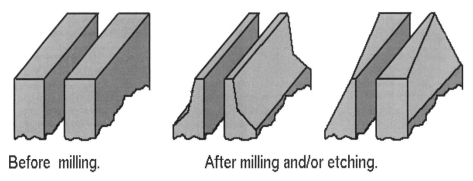

Before milling.   **After milling and/or etching.**

FIGURE 8.18 Pole edge elimination (*Yoshida et al., 1993*).

PERPENDICULAR HEAD FIELDS

Since 1980, several individuals and groups have worked to develop a more practical and efficient write head to emphasize the perpendicular remanence. We know that this remanence pattern has little or no demagnetization when the bit length is shorter than the recorded thickness, and will therefore dominate at very short wavelength recordings.

Iwasaki first promoted perpendicular recording in the late 1970s, and started experimental work using a single-pole head, also called the probe head. Now, because single poles do not exist, the counter pole was placed on the backside of the coating. This presents a very practical problem; the spacing between poles would have to be quite large to accommodate a disk or tape, and this would seriously impair the field gradient.

A solution was quickly found by providing the coating with a high permeability underlayer such that the field lines from the pole would seek a path through the underlayer, and return some distance from the pole. Practical ways of doing this are illustrated in Fig. 8.19.

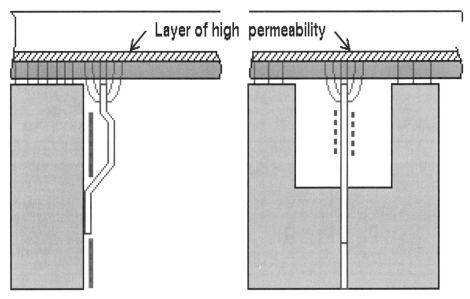

Layer of high permeability

FIGURE 8.19 A side view of the cross section of two perpendicular heads.

A standard ring core or thin-film head (see Chapter 10) will do quite well in recording onto a perpendicular coating. And there is quite a debate about the merits of the fields from the perpendicular heads (Mallinson and Bertram (1984), Minuhin (1984)). Minuhin claims that the characteristics of the probe models are inferior to those of the ring core head; he also shows that the resolution of the probe head is reduced if a coating with soft magnetic underlayer is used. And thus the debate goes on.

RESPONSE OF PROBE HEADS

Perpendicular and MR heads both use a single pole for scanning a recorded track. Figure 8.20 shows the equipotential lines found with the method from Fig. 8.9, and with field lines traced after the rules in Fig. 8.8. Near the pole the field lines look like the equipotential lines in Fig. 8.11, and vice versa.

The response of these head structures were examined by Davies and Middleton (1975) that found useful expressions for *PW50*:

$$PW50_{unshielded} = [t^2 + 4(d + a)(d + a + 2\delta)]^{1/2} \tag{8.1}$$

$$PW50_{shielded} = [(t + g)^2 + 4(d + a)(d + a + 2\delta)]^{1/2} \tag{8.2}$$

Druyvesteyn et al. (1981) further analyzed and summarized the read voltages from sinusoidal recordings; see Figs. 8.21 and 8.22. The pole thickness is t.

We recognize a couple of terms in both formulas: the spacing loss ($\exp - (2\pi d/\lambda)$) and the slightly more complicated thickness loss (δ). For the unshielded MR, one additional term is present: a sensor height loss. It means that less and less of the sensor height h is used when the wavelength becomes shorter.

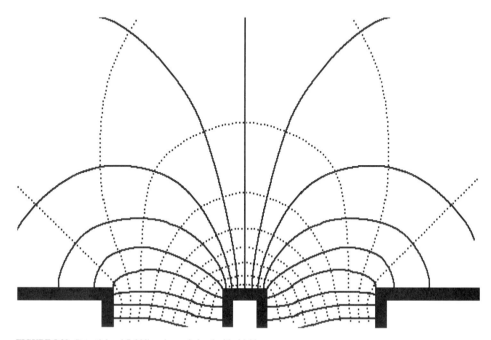

FIGURE 8.20 Potential and field lines in a pole head with shields.

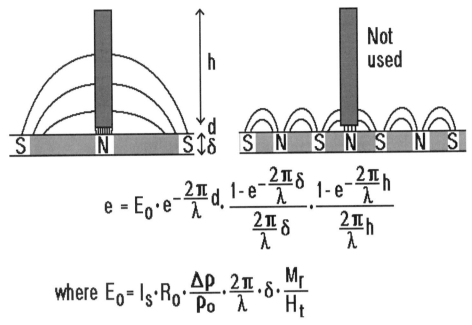

$$e = E_0 \cdot e^{-\frac{2\pi}{\lambda}d} \cdot \frac{1-e^{-\frac{2\pi}{\lambda}\delta}}{\frac{2\pi}{\lambda}\delta} \cdot \frac{1-e^{-\frac{2\pi}{\lambda}h}}{\frac{2\pi}{\lambda}h}$$

$$\text{where } E_0 = I_s \cdot R_0 \cdot \frac{\Delta\rho}{\rho_0} \cdot \frac{2\pi}{\lambda} \cdot \delta \cdot \frac{M_r}{H_t}$$

FIGURE 8.21 Output voltage and PW50 for an unshielded MR head (*Druyvesteyn et al., 1981; Davies and Middleton, 1975*).

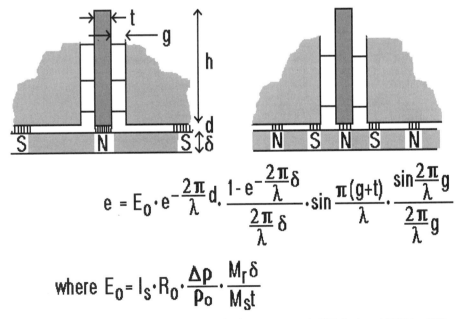

$$e = E_0 \cdot e^{-\frac{2\pi}{\lambda}d} \cdot \frac{1-e^{-\frac{2\pi}{\lambda}\delta}}{\frac{2\pi}{\lambda}\delta} \cdot \sin\frac{\pi(g+t)}{\lambda} \cdot \frac{\sin\frac{2\pi}{\lambda}g}{\frac{2\pi}{\lambda}g}$$

$$\text{where } E_0 = I_s \cdot R_0 \cdot \frac{\Delta\rho}{\rho_0} \cdot \frac{M_r\delta}{M_s t}$$

FIGURE 8.22 Output voltage and PW50 for shielded MR head (*Druyvesteyn et al., 1981; Davies and Middleton, 1975*).

This sensor height loss does not appear in the read voltage for the shielded MR. The shields carry the flux back into the entire gap, irrespective of wavelength (in absence of eddy currents). But now the gap length loss has appeared, plus a straight sinusoidal loss term that means a dropoff toward dc and some short wavelength.

Field equations for the perpendicular field components are found in paper by Szczech (1982).

SIDE WRITE/READ EFFECTS (NARROW TRACKS)

The quest for increasing storage capacity of magnetic tapes and disks has naturally led to narrower tracks. We have so far disregarded any variations of the field strength in the z-direction, perpendicular to the plane of the paper. The recorded track will be slightly wider than the core.

Assume that the recorded track width equals the width in the coating where the field strength equals or exceeds the coercivity H_c of the medium. The final track width can then be determined from Fig. 8.23, showing the total write field that includes: the ordinarily assumed field configuration (top), the side fields, and the wedges connecting the two.

The amount of extra trackwidth on both sides has been analyzed and calculated by Y. Ichiyama. His results are shown in Fig. 8.24, giving all operating conditions plus a graph for finding the answer. It covers also a tapered core, which has a wider track than a core with parallel sides. Tapered cores may not be used when the track width decreases below 100 μm (4 mils), because a certain amount of core body must remain just for core handling (and higher efficiency).

Another effect of the side field is the associated side-sensing function, which equals the head field divided by a unit current. This may lead to crosstalk from adjacent tracks, and sets the limit for how many tracks can be placed parallel to each other. Each recording system has its own crosstalk requirements, and the designer is referred to the papers listed in the references and further readings.

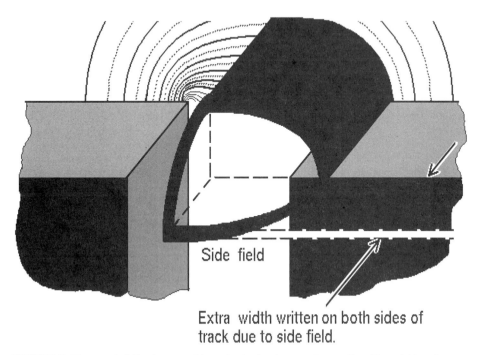

Side field

Extra width written on both sides of track due to side field.

FIGURE 8.23 The magnetic field at the corner of the gap in a head reaches out and causes side-writing and side-reading.

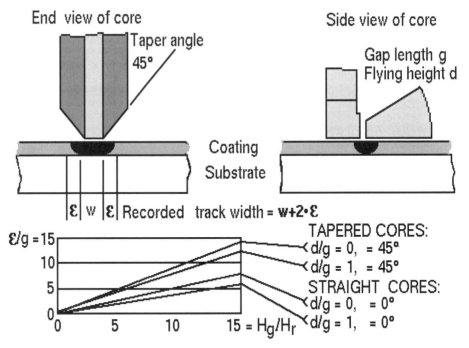

FIGURE 8.24   Written track is 2*e* wider than mechanical track width *w* (*Ichiyama*).

One valuable clue is found in work done by van Herk. An early assumption was that the crosstalk, or side sensitivity, varied similarly to the Wallace's spacing loss of $55d/\lambda$ dB. This applied to a very narrow track running off the core, as shown in Fig. 8.25. Van Herk found that the loss actually approaches $55d/\lambda$ dB plus another 6 dB. For very small spacings the loss is more like $110d/\lambda$ dB.

More recent studies used a large scale model of the TFH (Heim and Monson, 1987). They found a magnetization in the side writing that was different from the magnetization in the center of transitions. Recent micromagnetic modeling is described in Chapter 19 (J.G. Zhu). The side writing and reading is a complex matter that must include the media's anisotropies.

When recording surface area is at a premium, as for example in home video recorders, a clever solution applies. It was introduced by Sony, and is made by purposely misaligning the azimuth of two video heads, which are then placed 180 degrees apart on the head drum. The heads are offset in azimuth by plus and minus 7 degrees, and are synchronized upon playback to read their own tracks laid down earlier (see Fig. 6.18).

There is now a misalignment error of 14 degrees to the adjacent track; this causes a near-complete loss in picking up crosstalk, and the tracks can therefore be placed next to each other without the use of a guard band. The insulation between tracks is not perfect, in particular at long wavelengths. This presents a problem in color-bleeding (red) in VHS systems due to the location of the color carrier at a low frequency, i.e., long wavelength.

A couple of disk drives have dual actuators to increase the data handling speed. This system easily allows for an azimuth system where one actuator uses even tracks and the other uses odd tracks. Then azimuth recording should be feasible. A look at the side-written data in Figs. 19.9, 19.10, and 19.11 may explain why this has not happened: It is not a clean azimuth recording that results.

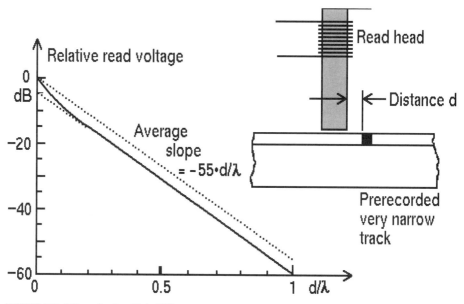

FIGURE 8.25 Side reading (*van Herk, 1978*).

THE ERASE PROCESS AND OVERWRITE

The principle of erasure of magnetic tapes was illustrated in Fig. 5.29: The tape is exposed to a decaying alternating field in which the initial value of the field is great enough to saturate the tape (typically five times the coercivity). This procedure leaves the tape in a state of net zero magnetization.

The erasure is in practice implemented by passing the tape over a magnetic head with a strong ac field in the gap (erase head), or by placing the reel of tape in a device with a strong magnetic field that slowly decays to zero (bulk degaussers; see Chapter 14).

Heads with an alternating and decaying field can be made from permanent magnets in arrangements, as shown in Fig. 8.26. The wrap angle, where the tape moves away from the head, is critical and may need adjustment for a particular tape. The last field "seen" by the tape should be of such magnitude that it, combined with the preceding (and opposite) field magnitudes, leaves the tape with $M_r = 0$. (The tape magnetization will follow diminishing minor loops.)

These permanent magnet erase heads require no power, but must be moved away from the tape when not used. A record head with a fairly long gap and an ac current serves well as an erase head. It does not provide complete erasure of a high-level recording. At the point where the field strength of the ac erase field has decayed to approximately the coercive force H_c of the tape there exists, in reality, a recording zone. Any foreign field that may be present will therefore be recorded onto the tape.

McKnight found that for a given erasing system operating at a given erasing current, and erasing a given wavelength, the signal is erased to a certain percentage of the original signal; this is almost independent of the original signal level.

Short-wavelength recordings are more easily erased than those of long wavelengths. Examples of erasures are shown in Fig. 8.27. The higher degree of erasure for shorter wavelengths appears to be related to reduced re-recording; the available flux from short wavelengths is small compared with the flux from long wavelengths, assuming identical recording levels.

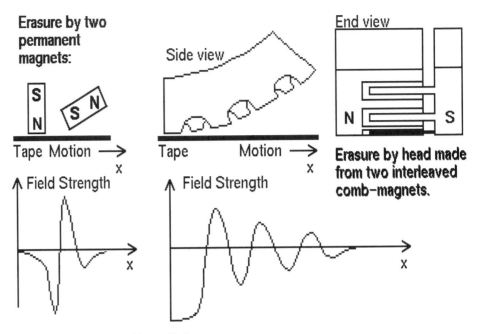

FIGURE 8.26 Simple erasure of the recording by permanent magnets.

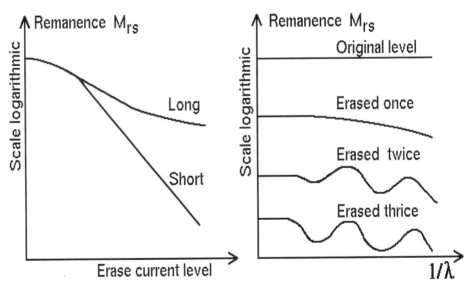

FIGURE 8.27 Erasure as a function of the recorded wavelength (left) and the number of passes over the erase head gap (right) (*McKnight*).

It also appears likely that the fixed percentage of erasure in one erasure following another is related to the re-recording phenomenon. But there is no explanation for the increased difficulty in erasing a tape recorded with a higher bias level than another.

It appears that there are two sources for the field that creates the re-recording. Keep in mind that the data field strength in an ac-bias recording is much smaller than the level required for a nonbias recording. A fraction of an Oe will suffice, if added to the ac-bias field near the recording zone.

The erase head has a sensing function of its own, and will "see" magnetizations before their passage through the erase field. This sensing results in a flux through the core (as a read head), and will cause re-recording thereof. Remember that any ring core-type head acts as a read as well as a write head, at any time, as illustrated in Fig. 8.28. The two processes are independent of each other, possibly with some coupling due to nonlinearities in the head permeability due to high write levels.

The other source for the re-recording field occurs primarily in digital recordings, where difficulties arise in erasing while writing new information. Actually, the write field acts as an erase field for the old data. The situation is again like that shown in Fig. 8.28, but often aggravated by the fact that the bit length of the old data is quite close to the length of the write field. The erasure of the write field is not complete, and it has been estimated that the remaining magnetization is sufficiently strong to act as a source for re-recording of itself (presentation by J.U. Lemke, January 1986 THIC meeting). Hence there should be some peculiar cases of very poor erasure (in-phase) or very good erasure (out-of-phase), which indeed occur with FM and MFM codes. This pattern dependency is explored in recent studies (Dee et al., and Sugaya et al., 1993; Jeffers et al., 1992).

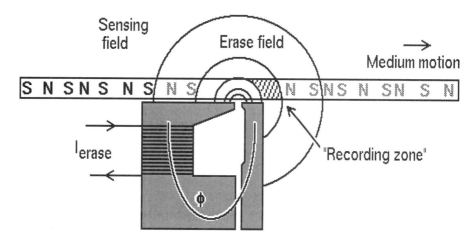

$$\text{TOTAL flux } \phi \text{ through core} = \text{Flux for } I_{\text{erase}} \text{ } \underline{\text{PLUS}} \text{ Sensed magnetization}$$
$$= n \cdot I_{\text{erase}} / \Sigma R_m \text{ } \underline{\text{PLUS}} \text{ } \Sigma \Delta \phi_{\text{left}} + \Sigma \Delta \phi_{\text{front}}$$

FIGURE 8.28 Unerased magnetization is sensed by the head sensing function, converted into flux, and then rerecorded at the "recording zone" formed by the new signal being recorded.

One should also bear in mind that the interaction field among particles (in tapes) or grains (in thin-film disk media) is not a fixed, spatial distribution. Particles will largely be surrounded with a field that is changed when a recording is made. This will make a next recording "see" a different magnetization pattern. And some particles may not switch so readily; keep in mind the fairly broad distribution of particle coercivities even for materials with small SFD.

There appear to be three contributing factors to poor overwrite:

• Re-recording of sensed, not-yet-erased old data.

• Re-recording of fields from not-erased old data, mostly in back of the coating.

• High coercivity particles retain the old data recording.

Interactions between two or all of these can explain the patterns observed. The patterns were already apparent during erasure with a head; see Fig. 8.27. They do not appear after erasure with a bulk degausser, which is a large magnet that produces a field several times larger than the tape's coercivity, i.e., its mean value of coercivities.

Repeat erasures can be accomplished during a tape pass if two (or more) erase gaps are incorporated in the erase head. They must belong to separate core structures, spaced away from each other, as shown in Fig. 8.29. This is accomplished with a double center leg that isolates the re-recording fluxes to their respective head sections.

The frequency of the erase current must be so high that each tape section undergoes several hundred decaying field cycles as it leaves the center of the erase gap. If the tape speed is 7.5 IPS and the erase gap length is 40 mils (1 mm), the frequency should be above 40 kHz.

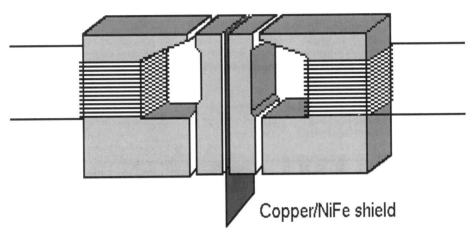

Copper/NiFe shield

FIGURE 8.29   A double-gap erase head eliminates the flux from the sensing function reaching the "recording zone."

STRAY FLUX COMPUTATIONS

Magnetic circuits control the flux as poorly as an electric circuit immersed in salt water controls the current flow. While the difference in electric conductivity is infinite between air and conductive wires (at dc), the difference in magnetic conductivity is at most 100,000, often only 3000 and decreasing at higher frequencies because of eddy currents.

We should, therefore, include the stray flux conductance through the air, shunting the front as well as back gaps in a magnetic head structure. This is illustrated in Fig. 8.30, where all elements making up the front gap stray flux are identified and characterized. The various permeances were originally found in a classic book by Roters.

When calculating the magnetic resistances of so many parallel elements it is easier to calculate the sum of the permeances (magnetic conductances), and then find the magnetic resistance from the reciprocal of the sum of permeances.

The effect of stray fluxes cannot be neglected for narrow track cores, especially at high frequencies. The net effects of stray flux will affect calculations of head performance.

The formulas for the various permeances come from integrations of small flux tubes, summing their contributions to obtain the total permeance of the element. This is quite simple when, as an example, we deal with the permeance of the front gap of dimensions g long, w wide, and D_g deep. We divide the volume up into cubes with sides equal to 1 mm.

Magnetic resistance of a flux tube (also called a magnetic field cell) with cross section $dxdy$ is:

$$R_m = length/\mu_o \times area$$
$$= dx/\mu_o dy\ w \tag{8.3}$$

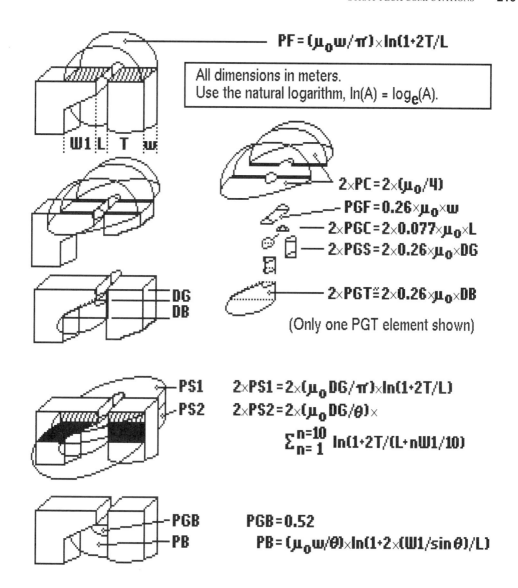

$$PF = (\mu_0 w/\pi) \times \ln(1 + 2T/L$$

All dimensions in meters.
Use the natural logarithm, $\ln(A) = \log_e(A)$.

$W1 \quad L \quad T \quad w$

$2 \times PC = 2 \times (\mu_0/4)$

$PGF = 0.26 \times \mu_0 \times w$

$2 \times PGC = 2 \times 0.077 \times \mu_0 \times L$

$2 \times PGS = 2 \times 0.26 \times \mu_0 \times DG$

DG
DB

$2 \times PGT \doteq 2 \times 0.26 \times \mu_0 \times DB$

(Only one PGT element shown)

$PS1$
$PS2$

$2 \times PS1 = 2 \times (\mu_0 DG/\pi) \times \ln(1 + 2T/L)$

$2 \times PS2 = 2 \times (\mu_0 DG/\theta) \times$

$\sum_{n=1}^{n=10} \ln(1 + 2T/(L + nW1/10)$

PGB
PB

$PGB = 0.52$

$PB = (\mu_0 w/\theta) \times \ln(1 + 2 \times (W1/\sin\theta)/L)$

Total Permeance P of Front Gap Stray Fields equals the sum of all above, and $R_m = 1/P$.

FIGURE 8.30 Magnetic permeances (conductivities) around the gap in a magnetic head (*Roters*).

For a cube with sides 1 mm we find:

$$R_m \mu = 10^{13}/4\pi \; 1/H \tag{8.4}$$

and the permeance is $4\pi 10^{-13}$ Hy. We treat these elements as admittances, adding them when they are in parallel, and dividing by N when N elements are in series. The permeance of the gap is thus found by multiplying by wD_g, and dividing by g. A numerical example will clarify:

$$w = 125 \ \mu m = 5 \ mils$$

$$D_g = 50 \ \mu m = 2 \ mils$$

$$g = 2 \ \mu m = 80 \ \mu in$$

$$PG = (125 \times 50/2) \ 4\pi 10^{-13} \ Hy$$
$$= 3.93 \times 10^{-9} \ Hy$$

hence

$$R_{gap} = 1/PG = 255 \times 10^6 \ Hy^{-1}$$
$$= 255 \ 1/\mu H$$

The formulas in Fig. 8.30 provide for all stray flux around the gap in a magnetic head. We will use them in Chapter 10.

FLUX LINKAGE

Figure 8.31 illustrates two impediments to a full coupling of the flux from the write current into the gap area of the write head.

First, not all of the field generated by the write head coil produces core flux. A few field lines from turns in the middle of the winding do not even get into the core. This inefficiency is termed flux linkage.

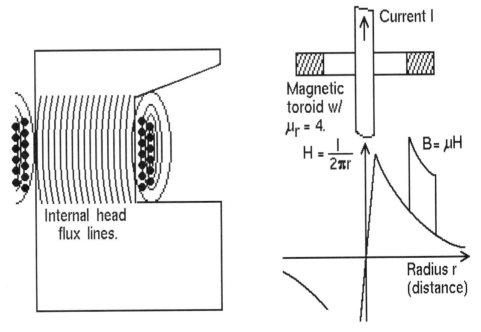

FIGURE 8.31 Incomplete coupling of the flux into a magnetic circuit (core). Flux linkage < 100%.

It is quite clear from the illustration on the right, where the field from a single conductor couples flux into a toroid-shaped core, that flux linkage varies with the separation between conductor and core. Because the field strength is higher on the inside of the core than on the outside, and because H is inversely proportional to the distance from the wire center, the highest flux is achieved for a small-diameter core.

When placing the winding onto the write head core, care must be taken in packing the windings as close to the core as possible to keep the flux linkage close to 100 percent. It is very difficult to arrive at a number for the flux linkage efficiency.

FLUX LEAKAGE

Flux leakage occurs when a portion of the flux strays from the core before reaching the front gap. This causes a loss that can be characterized as a coupling efficiency:

When shields are placed near cores (as in multitrack heads) the leakage flux increases, as shown in Fig. 8.32. These shields serve two purposes. First, they reduce the transformer crosstalk between adjacent cores. Second, two sets of write and read heads are often used, and the channels are interleaved, e.g., one set of heads handles all odd-numbered tracks, the other all even-numbered tracks. Then shields serve the additional function of short-circuiting the flux from the interleaved tracks moving past the read head.

An inside core with shields on both sides has higher leakage than an outside core; this will additionally cause track-to-track nonuniformity (sensitivity), and it is common practice to install shields outside the outer cores also.

SHIELDING

The last topic that belongs in this chapter on head fields is the shielding of heads. This is more of an art than a scientifically studied topic; we know more about what does not work (Lopez et al. (1981),

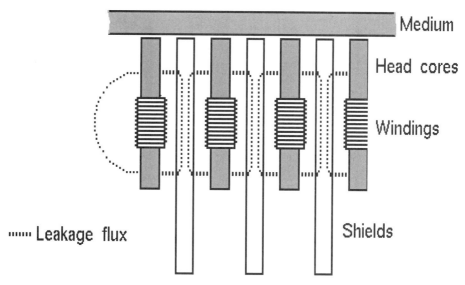

FIGURE 8.32 Wound cores couple to shields in a multitrack head.

Jorgensen (1975)) than we do about predicting what should work. Apparently field strength levels of more than 5 Oe affect ac-bias recordings as well as the writing of digital data; in the latter case the digital data field acts as an ac-bias source for the recording of the foreign field. The situation is aggravated when overwriting on old data. The susceptibility to foreign fields has been reported as low as 3 Oe for a modern TFH. This disagrees with the 10 to 30 Oe field reportedly required to destroy the BER from a TFH in a stray vertical field (Smith, et al., 1988).

In a shielded MR sensor for hard disk applications, the sensitivity to external fields is affected by the shape of the shields. Tall shields are more sensitive to axial fields; asymmetrically shaped shields can increase the sensitivity to radial fields by converting them to axial fields at the sensor. Moderate fields can sometimes induce small, systematic changes into the small-signal spatial sensitivity (Feng et al., 1991).

During playback, "hum" pickup from power transformers and motors may occur, and only shielding of the head can reduce this pickup. Audio heads are often built into housings of soft magnetic materials, as shown in Fig. 8.33.

The magnetics involved with shielding is treated by Sasada et al. (1988) and by Mager (1970), with some application to magnetic heads. A practical shield is shown in Fig. 8.33, top right, where a cover is closed after the tape has been loaded. The housing itself is made from annealed nickel-iron or shielding material from PerfectionMica. A sandwiched layer of copper has also been used in the past.

Deflection of fields (humbugging) is a last resort to prevent foreign flux from entering the heads. This is achieved by placing pieces of permalloy in the head's vicinity to deflect hum fields.

Another problem facing the equipment designer is radiation of fields from the heads to nearby electronic equipment, and indeed, from the write head to the read head. Insufficient attenuation of the radiated write field often causes a very poor signal-to-noise ratio when attempts are made to measure the read signal while writing.

A small slab of ferrite often helps (see Fig. 8.33, lower right). It is a good choice to use the type used for antennas in transistor radios. Further reduction is possible by installing a Faraday shield between the heads. It has a small slot open for the tape to pass through and opens up only for loading and unloading of tapes.

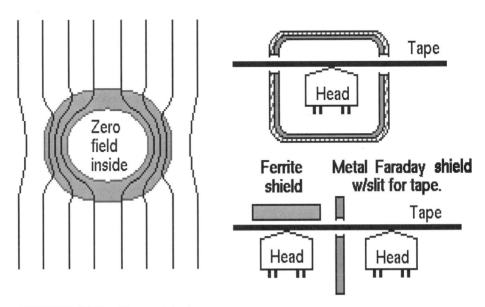

FIGURE 8.33 Shielding of the magnetic heads.

REFERENCES

HEAD FIELDS

Aoi, H., M. Saitoh, T. Tamura, M. Ohura, H. Tssuchiya, and M. Hayashi. Nov. 1982. Pole-Tip Design of High Density Recording in Thin Film Heads. *IEEE Trans. Magn.* MAG-18 (6): 1137–39.

Bertero, A.B., H.N. Bertram, and D.M. Barnett. Jan. 1993. Fields and Transforms for Thin Film Heads. *IEEE Trans. Magn.* MAG-29 (1):67–76.

Camras, M. May 1964. An X-Field Micro-Gap Head for High Density Magnetic Recording. *IEEE Trans. Audio* AU-12: 41–52.

Chi, C.S. Sept. 1985. A thin-film head design technique to correct read pulse asymmetry. *IEEE Trans. Magn.* MAG-21 (5): 1569–71.

Davies, A.V., and B.K. Middleton. Nov. 1975. The resolution of vertical magneto-resistive readout head. *IEEE Trans. Magn.* MAG-11 (6): 1689–91.

Druyvesteyn, W.W.F., J.A.C. van Ooijen, L. Postma, E.L.M. Raemakers, J.J.M. Ruigrok, and J. de Wilde. Nov. 1981. Magnetoresistive Heads. *IEEE Trans. Magn.* MAG-17 (6): 2884–2889.

Gopal, S.H., and A.V, Pohm. Sept. 1986. Models for M-R Head Analysis. *IEEE Trans. Magn.* MAG-22 (5): 845–46.

Johnson, K.O., and D.P. Gregg. Oct. 1965. Transient Response and Phase Equalization in Magnetic Recorders. *Jour. AES* 13 (10): 323–30.

Johnson, W., and F. Jorgensen. Aug. 1966. A New Analog Magnetic Recording Technique. *ITC Conf. Proc.* Vol. 2, pp. 414–430.

Krey, K., and A. Lieberman. Fall 1973. Principles and Head Characteristics in VHF-Recording. *ITC Conf. Proc.* Vol. 9, pp. 49–59.

Mallinson, J.C., and H. N. Bertram. Sept. 1984. On the Characteristics of Pole-Keeper Head Fields. *IEEE Trans. Magn.* MAG-20 (5): 721–724.

Megory-Cohen, I., and T.D. Howell. May 1988. Exact Field Calculations for Asymmetrical Finite-Pole-Tip Ring Heads. *IEEE Trans. Magn.* MAG-24 (3): 2074–80.

Minuhin, V.B. July 1985. Characteristics of Ideal Probe Heads with Ideal Windings in the Presence of a Permeable Media Underlayer. *IEEE Trans. Magn.* MAG-21 (4): 1289–1294.

Minuhin, V. May 1984. Comparison of Sensitivity Functions for Ideal Probe and Ring-Type Heads. *IEEE Trans. Magn.* MAG-20 (3): 488–494.

Minuhin, V.B. April 1985. Field of Probe Heads with Ideal Winding. *Jour. Appl. Phys.* 7 (1): 4006–4009.

Ohura, M., Y. Tsuji, S. Kuwatsuka, M. Hanazono, K. Kawakami, and M. Saitoh. April 1987, Design of High Density Thin-film Heads for Particulate Rigid Disks. *Jour. Appl. Phys.* 61 (8-Part IIB): 4182–84.

Potter, R.I., R.J. Schmulian, and K. Hartman. Sept. 1971. Fringe Field and Readback Voltage Computations for Finite Pole-Tip Length Recording Heads. *IEEE Trans. Magn.* MAG-7 (3): 689–95.

Potter, R.. Jan. 1975. Analytic Expression for the Fringe Field of Finite Pole-Tip Length Recording Heads. *IEEE Trans. Magn.* MAG-11 (1): 80–81.

Potter, R.I. Sept. 1974. Digital Magnetic Recording Theory. *IEEE Trans. Magn.* MAG-10 (3): 502–508.

Szczech, T.F., M. Steinback, and M. Jodeit, Jr. Jan. 1982. Equations for the Field Components of a Perpendicular Magnetic Head. *IEEE Trans. Magn.* MAG-18 (1): 229–232.

Szczech, T.J. Nov. 1981. Exact Solution for the Field of a Perpendicular Head. *IEEE Trans. Magn.* MAG-17 (6): 3117–3119.

Szczech, T.J. Sept. 1979. Analytic Expressions for Field Components of Nonsymmetrical Finite Pole Tip Length Magnetic Head Based on Measurements on Large-Scale Models. *IEEE Trans. Magn.* MAG-15 (5): 1319–22.

Szczech, T.J., and P.R. Iverson. Sept. 1987. Improvement of the Coefficients in Field Equations for Thin-Film Recording Heads. *IEEE Trans. Magn.* MAG-23 (5): 3066–67.

Yoshida, M., M. Sakai, K. Fukuda, N. Yamanaka, T. Koyanagi, and M. Matsuzaki. Nov. 1993. Edge Eliminated Head. *IEEE Trans. Magn.* MAG-29 (6): 3637–39.

SIDE WRITE/READ FROM ADJACENT TRACKS

Heim, D.E., and J.E. Monson. Jan. 1987. Measurements of Side Writing on a Large-Scale Recording Model. *IEEE Trans. Magn.* MAG-23 (1): 198–200.

Ichiyama, Y. Sept. 1977. Analytic Expressions for the Side Fringe Field of Narrow Track Heads. *IEEE Trans. Magn.* MAG-13 (5): 1688–1689.

van Herk, A. April 1978. Side-Fringing Response of Magnetic Reproducing Heads. *Jour. AES* 26 (4): 209–211.

van Herk, A. Jan. 1980. Measurement of Side-Write, -Erase, and -Read Behavior of Conventional Narrow Track Disk Heads. *IEEE Trans. Magn.*, MAG-16 (1): 114–119.

van Herk, A. Nov. 1977. Analytical Expressions for Side Fringing Response and Crosstalk with Finite Head and Track Widths. *IEEE Trans. Magn.* MAG-13, (6): 1764–1766.

ERASE HEADS AND OVERWRITE

Dee, R.H., K.S. Franzel and Joe Jurneke. Nov. 1993. Pattern Dependent Overwrite on Thick Media. *IEEE Trans. Magn.* MAG-29 (6): 4059–61.

Jeffers, F., and J. Bero. Dec. 1991. Record head saturation and overwrite performance of a BaFe floppy disk. *IEEE Trans. Magn.* MAG-27 (6): 4885–4887.

Lemke, J.U. Jan. 1986. Re-recordings during Overwrite. THIC Meeting.

McKnight, J.G. Oct. 1963. Erasure of Magnetic Tape. *Jour. of AES* 11 (10): 223–233.

Sugaya, H., T. Nakagawa, and T. Arai. Nov. 1993. Re-recording in Overwrite Recording (Re-recording Mechanism by Contact Printing Method). *IEEE Trans. Magn.* MAG-29 (6): 4080–82.

Wachenschwanz, D., and F. Jeffers. Sept. 1985. Overwrite as a function of record gap length. *IEEE Trans. Magn.* MAG-21 (5): 1380–82.

LEAKAGE FLUX AND SHIELDING

Feng, J.S., J. Tippner, B.G. Kinney, J.H. Lee, R.I. Smith, and C. Chue. Nov. 1991. Effects of Uniform Magnetic Fields on Shielded MR Sensors. *IEEE Trans. Magn.* MAG-27 (6): 4701–03.

Hammond, M.A. Jan 1955. Leakage Flux and Surface Polarity in Iron Ring Stampings. *IEEE Proceedings* 43 (1): 138–147.

Jorgensen, F. Fall 1975. The Influence of an Ambient Magnetic Field on Magnetic Tape Recorders. *ITC Conf. Proc.* Vol. 11, pp. 373–390.

Lopez, O., T. Lam, and R. Stromsta. July 1981. Effects of Magnetic Fields on Flexible Disk Drive Performance. *IEEE Trans. Magn.* MAG-17 (4): 1417–1422.

Mager, A. March 1970. Magnetic Shields. *IEEE Trans. Magn.* MAG-6 (1): 67–75.

Roters, H.C. 1941. *Electromagnetic Devices*, John Wiley & Sons, 561 pages.

Sasada, I., S. Kubo, and K. Harada. Oct. 1988. Effective Shielding for Low-Level Magnetic Fields. *Jour. Appl. Phys.* Vol. 64, pp. 5696–8.

Smith, N., D. Wachenschwanz, and F. Jeffers. Nov. 1988. The Effect of Stray Magnetic Field on Thin Film Record Heads. *IEEE Trans. Magn.* MAG-24 (6): 2835–37.

CHAPTER 9
MANUFACTURE OF
MAGNETIC HEAD ASSEMBLIES

A large variety of magnetic heads for recording and playback are available from a number of manufacturers. Figure 9.1 shows just a fraction of tape heads used in a variety of applications. Disk heads are shown later. Magnetic heads for all applications have evolved from the large, single-track ring core in the Magnetophone to very small disk drive heads that are smaller than the head on a pin.

Beneath the surface of these heads lies numerous hours of ultraprecise manufacturing, machining, and assembly. The finished head must meet a set of rigid standards covering tracks widths, spacing dimensions, and other mechanical tolerances. An important mechanical tolerance is the maximum deviation of each track from being in-line (skew). Record or reproduce sensitivities must be within certain limits, for all tracks. Frequency response (efficiency, gap length, losses) must be uniform and alike for all tracks. And the assembly must withstand a hostile environment of tape abrasivity and, often, temperature extremes. The temperature requirement dictates that core material, bonding epoxies, and head housing materials have essentially the same coefficient of thermal expansion.

The fabrication processes for multichannel metal core and ferrite core heads are pretty well established, and are described first in this chapter. The novel approach of using semiconductor fabrication technology in production of thin film heads (TFH) is covered later.

The chapter ends with a discussion of specifications, some methods for the test and measurement of magnetic head performance, and a design test guide.

HEADS WITH LAMINATED METAL CORES AND SOLID FERRITE CORES

Many heads have cores that are made from laminated Ni-Fe alloys, or from solid ferrite. So-called hard-tipped heads have ferrite cores with wear-resistant pole tips made from Alfenol, Sendust, Spinalloy, or Vacodur. These are four trade names for the alloy AlFeSi.

Laminations are fabricated by photochemical milling (or etching) (Boll 1965). Figure 9.2 shows an etched NiFe sheet with a large number of core sections. These sheets are stacked to a total thickness equal to the track width; see Fig. 9.3.

FIGURE 9.1   Multitrack tape head assemblies (*courtesy of Saki Magnetics, Inc.*).

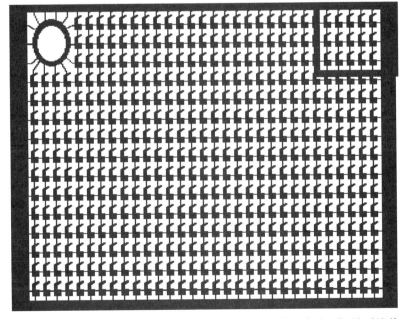

FIGURE 9.2   An etched NiFe sheet for head cores. The sheet contains single laminations for 16-×-31 half-cores plus a permeability check ring (upper left corner).

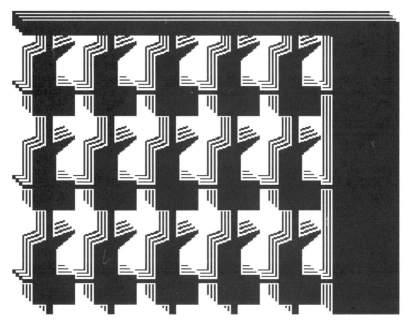

FIGURE 9.3 Four lamination sheets make half-cores with trackwidth 4 × lamination thickness plus 3 × photoresist or some other insulator between sheets.

Any coldworking of a soft magnetic alloy will reduce its permeability drastically; there are no magnetic materials completely free from magnetostriction. If the sheets for etching are bent with a radius of only 10-20 cm, a large reduction in permeability takes place. And a pressure of 981×10^3 N/m^2 (= 142 lbf/in^2 = 10 kgf/cm^2) will reduce the permeability of a well-annealed Mumetal by a factor of ten (Boll 1956). The manufacturer must be aware of these limitations when stacking the cores for bonding under pressure, and later during application of epoxies to keep them in place in the head housings (Liechti 1985). It is good practice to include a sample ring in each lamination sheet; see Fig. 9.2, top left. This ring can be tested for permeability of the affiliated cores. A simple check of the inductance of a few turns will do; the derivation of formula 7.5 shows that the inductance value is directly proportional to $\mu_r = \mu/\mu_o$.

The coil windings are either wound on a separate bobbin and slipped onto a core leg (when the core geometry permits it) or they are placed directly on the core, which then is provided with an insulation of oilpaper, teflon, or tape. Figure 9.4 shows four core halves provided with coils. One problem in the design of multitrack heads is finding room for the winding. Shielding between tracks must be provided for and should, in interleaved assemblies, have a thickness equal to the track width. This leaves only a distance equal to the track spacing for winding plus coil. A so-called half-shell has machined slots and additional room for windings; see Fig. 9.5.

Grounding of the cores and the shields is essential for low-noise operation. Not only does this reduce pickup of interfering electric fields, but it also prevents the buildup of electrostatic charges from the moving tape (making the device an unintentional Van de Graaff generator). Some manufacturers mix graphite or other conductive powder in the bonding epoxies to eliminate buildup of any electrostatic charge.

The next critical step in the process is the lapping of the head-halves. One four channel head-half is shown in Fig. 9.6. They should be artificially aged (by way of temperature cycling) to be free of stresses prior to the finish lapping process. The finished surface is checked with an optical flat placed against it, in monochromatic light, and should be within as little as one quarter of a light band for narrow gap heads (one light band is, for example, in orange light equal to 0.3 μm (12 μin), because the wavelength of orange is 6000 angstrom = 0.6 μm = 24 μin).

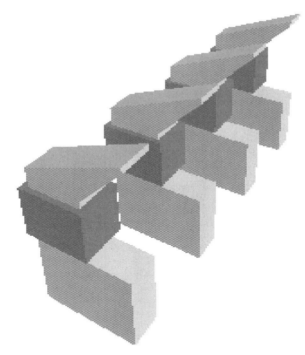

FIGURE 9.4   Four half-cores with windings.

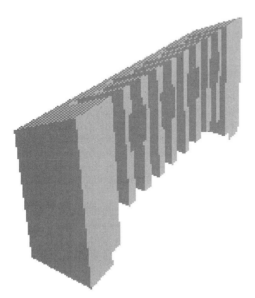

FIGURE 9.5   Half-shells for four half-cores. Note the milled slots for core-halves as well as for shields (inserted after the head has been assembled).

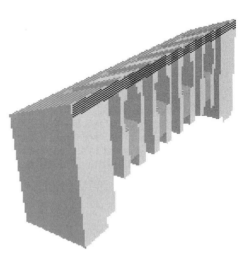

FIGURE 9.6   Half-cores inserted into the half-shell. The interface surface to other half-shell has been lapped, polished, and provided with a thin layer of SiOx for front gap length.

During the grinding and final lapping of the mating surfaces of the head halves, great care must be exercised. The ultimate tensile strength of Mumetal is quoted to be 5.5×10^8 N/m^2, and during grinding and lapping this force must occur in the minute areas where material is removed (Rabino-wicz 1968).

The net result is that the outside surface layers of the interfacing poletips have very low permeability, which will appear as a gap that is longer than its physical dimensions. This is not to be confused with the fact that the magnetic gap of a head is 10–20 percent longer than its mechanical length; coldworking will add even more to these percentages. Only by careful and patient lapping methods can these layers be so thin as to be insignificant.

Gap spacers of various materials are used to control the length of the front gap in the head assembly. Cleaved mica has been used extensively, while modern narrow gap heads have a silicon-monoxide layer for gap spacers. It is generally placed on a head-half by vacuum deposition techniques; see Fig. 9.6. This method can also be applied for spacers of conductive materials.

A finished half-shell is shown in Fig. 9.7, and two such halves are then mated and held together with screws or epoxy, or both; see Fig. 9.8. The laminations must meet and match so one set is a mirror image of the other; otherwise, poor high-frequency response results (remember the flux flows through the lamination surfaces at high frequencies, and should not flow into the middle of a mating lamination).

FIGURE 9.7   Finished half-shell for a 21-channel head assembly *(Courtesy of Omutec, a division of Odetics Corp).*

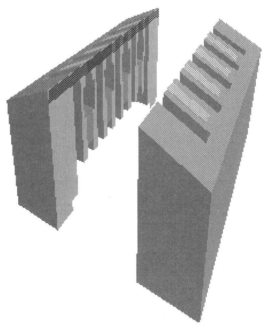

FIGURE 9.8   Two half-shells make a full head stack.

The final step in the assembly of a multichannel head is the insertion of shields; see Fig. 9.9. These must be full shields to prevent eddy currents from the windings' fields in setting up a write field or sensing function where two shield halves otherwise would meet, at the gap line. Multitrack recorders often have two interleaved head stacks for write as well as for read, and shields are typically located at alternating tracks. A shield joined at the gap line would be detrimental to partial recording or reading from these tracks. This would be due to crosstalk between the neighbor cores and the two half-shields.

Final assembly involves the pouring of an epoxy to hold the components together. Care must be taken at all steps to prevent excessive stresses that may affect permeability and coercivity (Le Floc'h et al. 1981; Liechte 1985).

Finishing the assembly is done by contouring the head surface by grinding and then lapping it to its final dimensions. A green lapping tape is run over the head (Fig. 9.10) for a short period, at low tape speeds and preferably cooled with denatured alcohol. There are various (some proprietary) means of checking the grinding so a well-defined final gap depth is achieved. One method is a simple check of the inductance values of sample tracks.

For long-life applications, the gap must be about 4 mils deep (or provided with hard tips; see next section). Higher efficiencies are achieved in short gap heads for wideband applications by cutting the gap depth to 2 mils.

Overall shielding of the head assembly is a commonly used technique in audio recorders, where 60- or 50-Hz power-line hum from the power transformer, and from motors, is a source of noise inside the signal frequency range (20–20,000 Hz). Shielding can be provided for by a magnetically soft shell around the head (see Fig. 8.33); the material may be drawn, annealed mumetal (MuShield 1980), or a Co-Netic material that can be cut and bent within wide ranges without losing its permeability (Perfection Mica Corp. 1980).

Heads with Mumetal cores will, in certain applications, wear out too fast (see Chapter 18), and it has become standard practice to provide head types with wear-resistant pole tips (instrumentation, video, and some audio heads). The tip material is one of the hard iron-aluminum alloys of many names (for variation in composition): Alfesil, Sendust, Spinalloy, Vacodur, etc. The tips may be solid or laminated, and they are, as a step in the processing, placed in a tip plate.

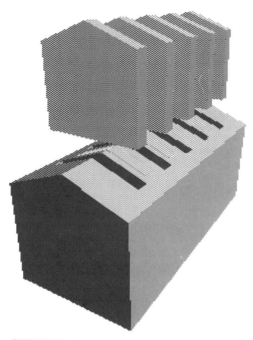

FIGURE 9.9 The final insertion of shields prior to epoxy bonding of all head components and the wire/solder connection to a connector (beneath the head; not shown).

FIGURE 9.10 Mild abrasive tape is used in the final contour lapping of a finished head assembly.

The tip plate is placed against the head block and bonded to it so the tips make intimate contact with the cores. The magnetic reluctance in the gap between the tip and the core has a generally negligible effect on the net value of the head efficiency. But the hard tip must be laminated for good high-frequency performance.

HEADS WITH FERRITE CORES

Ferrite materials offer a more straightforward manufacturing technique, with fewer steps than for laminated core heads. Single-track as well as multitrack heads are made from ferrite bars ground to profiles, lapped, gapped, and bonded together with glass. Several ferrite pieces in various states of machining are shown in Fig. 9.11.

FIGURE 9.11 Ferrite pieces in various states of machining *(courtesy of Tranetics, Inc).*

The bonding process uses various glasses with coefficients of expansion equal to those of the selected ferrites, and there are glasses with high as well as low melting temperatures. Occasionally both are used in the same head; a high-temperature glass for the gap and a low-temperature glass for bonding the head into the head housing (Fig. 9.17). Recent papers discuss the properties and applications of such glasses (Kijima et al. 1990; Mizuno and Ikeda 1992; Mizuno et al. 1992).

Machining and grinding of ferrites requires experience and skill. Too fast and too deep a cut (high removal rate) can produce stresses that affect the magnetic properties. Final lapping operations are just as critical. When material is removed by the cutting and smearing action of the particles in the abrasives, then tensile stresses in the surface result. During the motion of the scratching particle the material is deformed not only beneath the indenter, but also before, around, and behind it (Broese van Gruenau 1975; Okada et al. 1993).

The result is a dead layer (called the "Beilby layer") where the stresses greatly reduce the material's permeability via the mechanism of magnetostriction. The thickness of this layer in HIP materials (hot-pressed ferrites) may be estimated by (Wada 1980):

$$\text{Thickness} \approx 4 \times R_z + 50 \text{ angstrom} \tag{9.1}$$

where R_z equals the surface roughness in angstroms (mean value is assumed; the reference is not clear).

The stress inside the material, just below the surface, reaches values equal to the ultimate compressive strength of the material, and the stresses perpendicular to the direction of grinding are about 40 percent larger than those parallel to the direction of grinding (Knowles 1975). Influence of lapping conditions were investigated by Chandrasekar, et al., (1992).

The thickness of the dead layer can be determined by a straightforward experiment, illustrated in Fig. 9.12. A ferrite sample is provided in the form of a toroid. Wind a suitable number of turns (say 10) onto the core, and measure the low frequency inductance value L'. Next cut the toroid in half, and assemble (with a light pressure) to form a closed inductor again (Fig. 9.12, right). Now measure the inductance value L", and find the thickness of the layer destroyed by the cutting plus the effective air-gap due to surface roughness from formula 9.2 in Fig. 9.12.

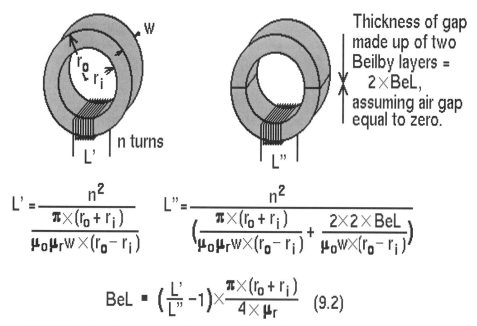

$$L' = \frac{n^2}{\dfrac{\pi \times (r_0 + r_i)}{\mu_0 \mu_r w \times (r_0 - r_i)}}$$

$$L'' = \frac{n^2}{\left(\dfrac{\pi \times (r_0 + r_i)}{\mu_0 \mu_r w \times (r_0 - r_i)} + \dfrac{2 \times 2 \times BeL}{\mu_0 w \times (r_0 - r_i)} \right)}$$

$$BeL \approx \left(\frac{L'}{L''} - 1 \right) \times \frac{\pi \times (r_0 + r_i)}{4 \times \mu_r} \quad (9.2)$$

FIGURE 9.12   Method for measurement of the thickness of a Beilby layer (see text).

Next lap and polish the surfaces to be flat with a light band, and measure L" again; it should be a higher value, indicating a reduction in the thickness of the dead layer. Ultimately, etch the surfaces with cold concentrated hydrochloric acid (HCl—use extreme care!), and it should be possible to achieve a very small value of BeL, the Beilby layer. The etching removes the stressed surface material quite uniformly, and with a ferrite with closed porosity no HCl penetrates below the surface (Knowles, 1975).

Now apply your usual grinding method, measure the resulting BeL, and repeat for the lapping and polishing processes. This method will aid in developing the optimum processing.

Another concern with ferrites is crystal pullout. Such voids in the gap edges will deteriorate the head's short-wavelength resolution. Voids in the surface may collect dirt and dust and may cause "crashing" of flying heads, scratches in tape, and generation of dropouts. These problems are eliminated by using single-crystal ferrite cores (Tanji et al. 1985; Ichinose and Aronof 1990). It is unfortunate that selection of optimum crystal direction is different for high efficiency and for low wear rate.

The NiZn ferrites are reportedly of higher stability and better intergrain bonding, and therefore are easier to machine with high precision and little chipping. They do, on the other hand, appear to have higher magnetic losses from machining than do MnZn ferrites.

Wear resistance of the two ferrite types are inadequately described in the literature. NiZn ferrites wear little and are therefore used in modern MR heads as shields while at the same time providing a wear-resistant head contour. A large variety of head configurations are made today, as the following descriptions and illustrations will show.

Single-track heads for disk use have cores that are diced, ground, and sliced from ferrite as shown in Fig. 9.13. More complicated structures have been developed in the past few years. For diskette use, write/read heads with tunnel-erase and straddle-erase cores were shown in Fig. 7.2. Although the latter version trims the written track closer to the write/read gap, the first appears to have become industry standard (Butsch 1981); on track width, trimming, TPI etc., see Chapter 10.

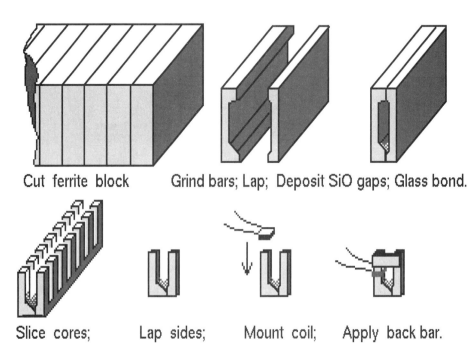

Cut ferrite block   Grind bars; Lap; Deposit SiO gaps; Glass bond.

Slice cores;   Lap sides;   Mount coil;   Apply back bar.

FIGURE 9.13   Glass bonding, slicing, dicing, and lapping of 4- to 7-mils-wide (thick) cores for early disk drive heads.

Heads for rigid disk applications are shown in Fig. 7.3. They are required to fly over the surface of the disk at a distance of 2–5 μin, and operate at frequencies in the MHz range. The technique of flying a head over a moving surface is covered in Chapter 18 (Head/Media Interface), and the design of very high frequency heads is addressed in Chapter 10.

Some video ferrite core and core assemblies are shown in Figs. 9.14, 9.15, and 9.16. Their assembly often involves the use of glass as a bonding agent, rather than epoxies (which tend to change dimensions with time, humidity, absorption, etc.). Glasses match the ferrites quite well in terms of coefficient of expansion, and are easy to apply and melt to form a meniscus for holding core halves together (see Fig. 9.17). A high-melting-temperature glass is used in the gap region, while the glass used for holding the core in the slider (or housing) has a lower melting temperature. The glasses react with the ferrite and may partly destroy the permeability (Freitag 1980). The glass should not contain lead, even though it provides a low melting temperature (Rigby 1984).

An interesting phenomenon may occur in ferrites: standing waves. The propagation speed of an electromagnetic wave inside the material is much lower than the speed in air due to the material's dielectric constant and its permeability: $v = v_{air}/(\in_r \mu_r)^{1/2}$. Resonances (standing waves) may occur at 1 MHz when the smallest dimension is in the millimeter range. The velocity of the wavefront (group velocity) is only $v_c = \omega\delta$ where $\omega = 2\pi f$ and δ = skin depth (Deschamps 1983).

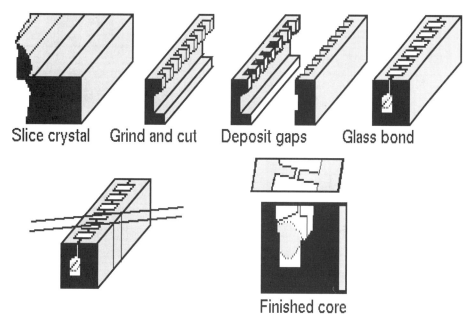

FIGURE 9.14 Fabrication steps for making video heads for VTR with azimuth recording.

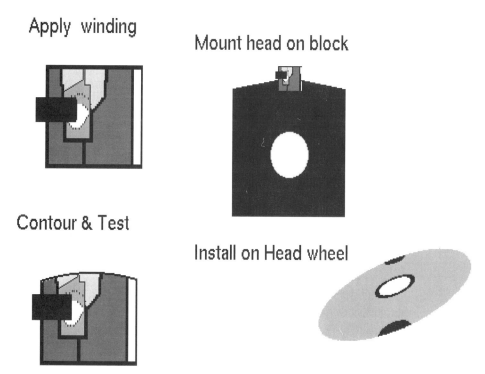

FIGURE 9.15 Final assembly of video heads on head assembly drum.

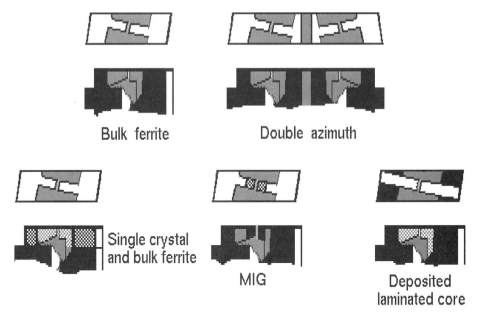

Bulk ferrite   Double azimuth

Single crystal and bulk ferrite   MIG   Deposited laminated core

FIGURE 9.16 Configurations of five video heads.

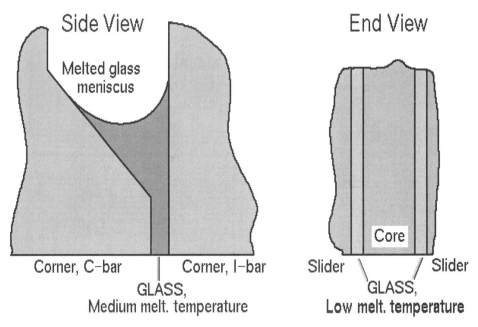

Side View

Melted glass meniscus

Corner, C–bar   Corner, I–bar

GLASS,
Medium melt. temperature

End View

Core

Slider   Slider

GLASS,
Low melt. temperature

FIGURE 9.17 Glass bonding of ferrite cores (left), and of finished cores into a slot in the head pad.

MIG (Metal-In-Gap) heads have deposited high-B_{sat} material on the pole pieces; see Fig. 9.18 (Coughlin 1985). This allows for a short gap for reading high-FCI tracks while still able to produce a strong enough gap field while writing on high coercivity disks or tapes (Jeffers 1982). See further in Chapter 10.

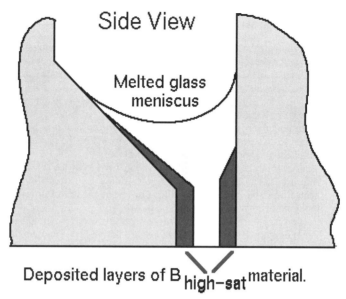

FIGURE 9.18 Glass bonding of a MIG head.

For extended-bandwidth heads (toward 50 MHz and beyond) neither conventional lamination heads nor MIG heads are adequate. New models have been fabricated where the core structure is a very thin lamination construction. These laminations are deposited and have a thickness of 2 μin = .05 μm = 500 Å. Their performance is illustrated in Fig. 9.19 (laminated).

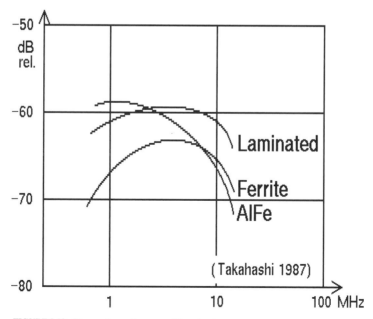

FIGURE 9.19 Comparative performance of three heads.

THIN-FILM HEADS

We can, from the preceding section, see that the manufacturing of multichannel heads becomes more difficult as the track density and the upper frequency limit both increase. This is in particular a problem in disk files for computer storage. There are units that use only one head assembly with one data channel. Rigid disks have thousands of concentric tracks, and the head must therefore move back and forth to select a given track. This movement requires precision mechanical actuators and it takes time to settle down at a track.

In computers time is precious, and a one-head-per-track assembly could eliminate the time required to change tracks. A partial solution would be the movement of a multichannel head (shorter travel equals shorter time) or several interlaced heads.

A new fabrication technology was borrowed from the semiconductor industry to fabricate thin-film heads. The obvious advantages were low cost, track-to-track uniformity, and production repeatability. The substrate is a silicon or ceramic wafer of dimensions up to six by six inches, and thousand of heads are fabricated onto each.

Two thin-film head configurations were proposed by Lazzarri in 1970–71 (Lazzarri and Melnick 1971; Lazzarri 1973). One was the vertical design that is in use today; the other was a horizontal layout that has some appealing features, but never has made it into production. Difficulties with gap fabrication appear to be the major stumbling block. The vertical design was quickly put to use (Chynoweth et al. 1972; Valstyn and Shew 1973).

Thin-film heads offer several advantages over bulk heads, such as those fabricated from solid ferrite or laminated materials. They were introduced in the 1970s and applied since the early 1980s, first in the IBM 3370/3380 disk drives.

The first single-turn thin-film heads performed well, except that they suffered from insufficient read voltages from the single turn. A multiturn head was formed by plane-parallel conductors, electrically insulated by dielectric layers. The magnetic circuit was also multilayered, which enhanced the formation of a single large domain and magnetization changes accomplished by magnetization rotation. The manufacturing process of these heads was vacuum evaporation through metallic masks of thin magnetic, dielectric, and conductor films. This multilayer fabrication had many difficulties.

The first commercial 8-turn head was proposed by Jones (1975) and applied by IBM in disk drives. The coil is a spiral shape, as shown in the by-now classic drawing of a TFH, Fig. 7.5. (NOTE the change in scale between the cross-section view and the head-on view.) This is the head configuration that is covered in the remainder of this chapter.

The TFH (thin-film head) is a good example of the general trend in magnetic storage devices: Smaller is better. The current overall size of a TFH is about one-tenth of a millimeter (= four one-thousands of an inch). This size allows for batch fabrication of thousand of heads at once and adapting technology from semiconductor fabrication.

These small heads are advantageous for other reasons:

- The small size permits very high densities in terms of bpi (bits per inch along the track) and tpi (the number of parallel tracks per inch).

- The switching behavior of the few magnetic domains can be controlled by material selection and processing controls. Coherent magnetization rotations are enhanced over the slower domain wall motions.

- The inductance of the tiny coil/magnetics is also small and permits very high-frequency operations.

- The isolated readback pulse is narrower with steeper fronts than the comparable ring core heads (ferrite). A small negative pre- and post-pulse may at first appear detrimental, but each can be used with the proper codes and signalling techniques. The TFH provides a higher field gradient than the comparable ring head, and therefore provides higher densities.

- The magnetics in the small TFH offers better high-frequency response. The film thicknesses are in the order of 1 micron, and eddy currents do not appear before a frequency of many MHz. This is in contrast to core materials for ring heads that are much bulkier, and therefore exhibit eddy current losses at fairly low frequencies.

The horizontal head has seen a revival and may be a factor in the future (Springer 1988). The head structure is typically deposited/etched into a well in a ceramic pad; see Fig. 9.37. The interested reader can consult references (Lazzarri and Deroux-Dauphin 1989; Chapman 1989; Umesaki et al. 1991).

A review of TFH technology is found in a paper by Oshiki and Hamasaki (1990), and for additional reading the book by Ciureanu and Gavrila (1990) is highly recommended.

TFH FABRICATION

The flat structure of the multilayer head shown in Fig. 7.6 requires sequential depositions and pattern constructions of alternating layers of magnetic films, insulators, and conductors on top of a strong and stable substrate. It has been fortunate that the pioneers of TFH fabrication have been able to lean on technologies used in semiconductor device fabrication. The advances in photolithography are an example; they allow the fabrication of very small structures, such as coils and pole pieces.

Numerous methods exist for the deposition of one material on another (Bhushan and Gupta 1991). The insulating layers for thin-film heads are aluminum oxides (for planar, strong surfaces), photoresist-deposited or remaining after an etch operation, and cured. The magnetic films are NiFe alloys, possibly with addition of N (nitride) for high saturation flux density. The substrate that serves as support for the deposited layers is a ceramic (Aluminum oxide/titanium carbide). The substrate ends up as the flying head structure with the heads remaining on the ends (see Fig. 9.26).

The deposition methods in use today have evolved over the past two decades. There are, for our purpose, four methods available:

- Evaporation
- Electroless deposition
- Electrodeposition
- Sputtering

In evaporation, a material is heated above its melting point and the evaporating fumes will condense on any surface they encounter. The coating's adhesion to the substrate is weak, and makes further processing of additional layers difficult. Further, the method is incapable of preserving the stoichiometry of an alloy, with a resulting film that is quite different from the melted bulk material. Evaporation has only been used for applying a gap spacer in tape heads.

In electroless deposition, a continuous coating of metal or alloy is deposited on a substrate that is immersed in an electroless bath solution. The method has not been reported used in head fabrication, while rigid disks with CoNi alloys have been made.

Electrodeposition and sputtering are both applicable for producing thin-film heads. Electrodeposition has evolved as the method for economical and efficient TFH fabrication. The opposite holds true for fabrication of rigid disks, where sputtering clearly holds the lead.

The first method (electrodeposition) is referred to as the wet process, and two papers are recommended for details: Romankiw and Simon (1975) and Bischoff (1990). Sputtering (and associated process steps) is the dry process; a good tutorial is the paper by Hanazano et al., (1987). For thin-film fabrication in general, the reader is referred to Vossen and Kern's book, Thin Film Processes II (1991).

Both the wet and the dry method involve design and fabrication of masks to define the flat patterns of NiFe or coil/conductors. Each mask will typically contain patterns for hundreds of heads, and will have to be moved (stepped) about to cover an entire 4"-×-4" wafer holding up to 6000 heads. The application (spin coating) of photoresist must be carefully controlled. Several papers on processes etc. are found in the bibliography listing at the chapter's end. These are, for instance: A booklet from Kodak (1971), which describes the photoresist technology and wafer alignment in papers by Farrow et al. (1991), and by Yuan and Strojwas (1991).

The general sequence in fabricating an inductive thin-film head is shown in Figs. 9.20 through 9.26. For simplicity, a dry deposition process is shown, and the figure's text tells the story.

Figure 9.27 shows the differences between the dry and wet processes. In both processes a carefully prepared substrate is used. Part of the preparation is the application, by RF sputtering, of an aluminum oxide layer of thickness 15 μm. Its surface has an orange-peel texture, and the wafer is lapped and polished.

The dry process continues with sputter deposition of a 2–4 μm layer of NiFe. Next, photoresist is applied and exposed so it masks etching in the desired pattern. After the etching of the pattern, the remaining photoresist is dissolved and removed, leaving a trace of pole structure of coil (if Cu was processed).

The few steps outlined in Fig. 9.27 are repeated for each layer of the head structure. The insulating layers that are needed between coil and NiFe, and possibly between coils in a multilayer coil head, is often cured positive photoresist (Bischoff 1990).

Such sandwich construction require careful choice of materials and processes in order to ensure similarity in coefficients of expansion and good layer-to-layer adhesion. The adhesion between metal films and many substrates can be improved by the initial sputtering of a few hundred Å of Cr. This layer is not really continuous, but forms small islands of Cr that are points of good adhesion.

The wet process requires more processing steps, as shown in Fig. 9.27. The substrate must be made conductive in order to do electroplating. The first step is, therefore, the sputtering of a thin metal layer.

We wish to deposit a designed pattern of NiFe or Cu through openings in a photoresist layer. The openings are part of the mask required for exposure of a thin, continuous layer of photoresist. If we only deposit the desired geometries, we will find that the current distribution in the electrolyte will be quite nonuniform, as will be the deposit. We will therefore employ a "window frame" process where the desired geometries are outlined by a 10 μm-wide photoresist trace (Romankiw and Manor 1974). Now we will also plate outside the desired geometry. These islands will have to be etched away separately, but the result is a very uniform current density across the entire wafer, and this leads to uniform thickness and well-defined geometry of the deposits.

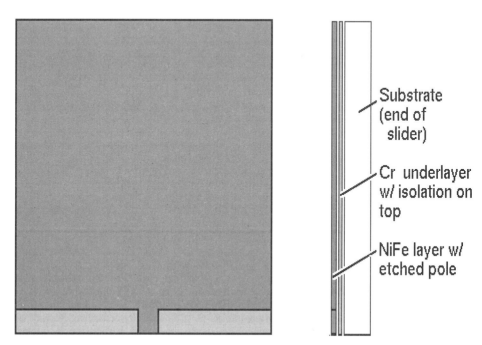

FIGURE 9.20   A few hundred angstroms of chromium is flashed onto a wafer surface to assist in adhesion for the NiFe film that is deposited next. Photoresist is coated on top of the NiFe film; it will form the head's bottom magnetic film plus its leading pole piece.

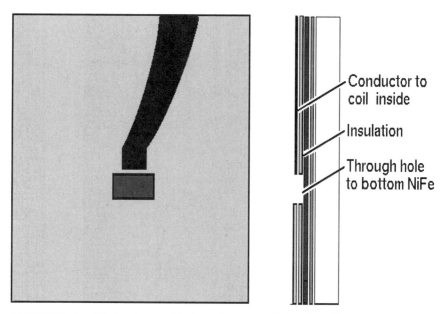

FIGURE 9.21 A mask is aligned on top of the photoresist and exposed. An etchant will remove the exposed resist, and another etchant will next remove the unprotected NiFe so the leading pole is formed (at the bottom). The unexposed photoresist may be left on top of the NiFe film, and a layer of insulator material deposited on top. Next a layer of copper is deposited to become the conductor to the center of the future coil. The wafer is again covered with photoresist, and a pattern for the conductor exposed.

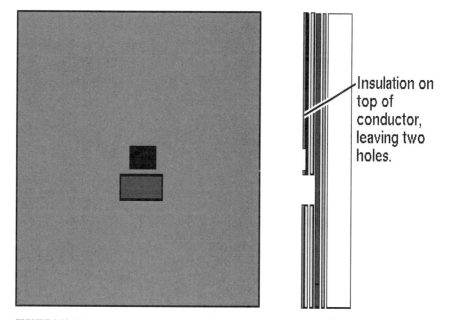

FIGURE 9.22 The exposed photoresist and underlaying copper are etched away and the center conductor formed. Note the preparation of a hole down to the NiFe film for later contact to the top NiFe pole piece. An insulating material is deposited on the wafer, with the exception of the bottom portion of the conductor and the hole for later NiFe contact to the bottom film.

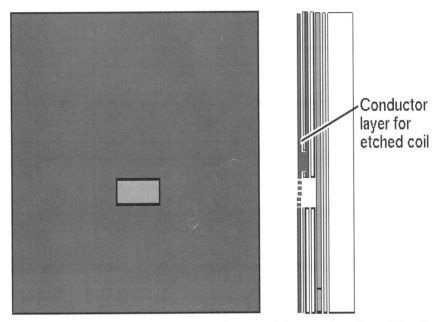

FIGURE 9.23 Copper is deposited, making contact at the center to the bottom conductor. Photoresist is again coated onto the wafer.

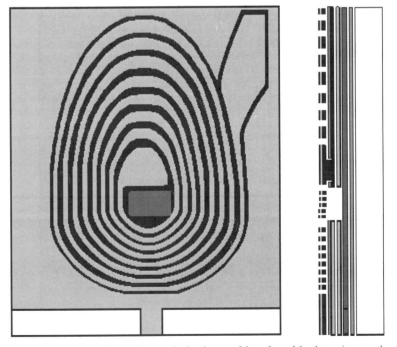

FIGURE 9.24 A mask with the coil pattern is placed on top of the wafer, and the photoresist exposed. After etching, the coil remains.

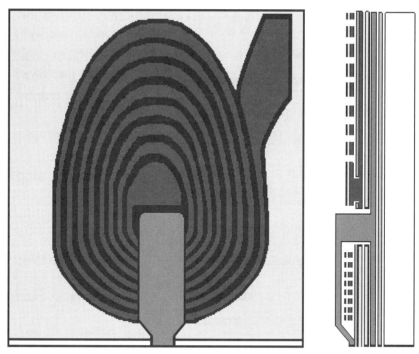

FIGURE 9.25 A BeCu mask with a pattern like the final top pole is placed into position as shown. Insulating material is deposited, and a slightly smaller mask placed for final deposition of the top (trailing) pole piece. Contact is made through the hole to the bottom NiFe layer.

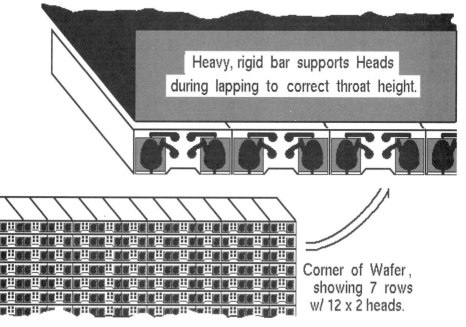

FIGURE 9.26 The wafer is cut into bars, and each bar with as many as 20 heads is lapped to provide a final throat height.

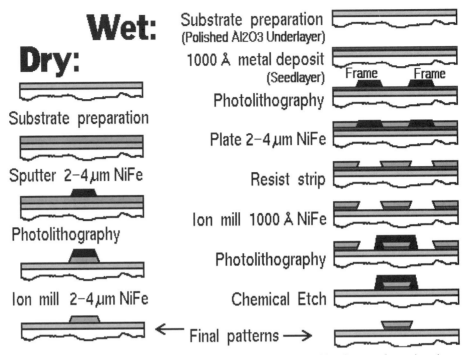

FIGURE 9-27 Processing steps in the dry and the wet fabrication processes for deposition of patterns of magnetics and conductors in a TFH (*Bischoff*).

The window frame is shown in Fig. 9.27, right. After electroplating and stripping the resist, we have the deposited pattern and the islands between patterns. Also, everything is shorted by the remaining thin metal deposit. The latter is first removed from between the patterns by ion milling (or back sputtering). Next, we cover the desired pattern with photoresist through a photolithography process. The unwanted islands are etched away, the photoresist is removed, and we have a final pattern similar to the one we made with the dry process.

The wet process clearly demands many more steps but does, all in all, provide the most economical way of fabricating thin-film heads. There are over 140 process steps, which can include 18 or more individual masking layers. And a great number of processes must be monitored and controlled. Sputtering may be used for high-saturation type NiFeXY alloys, where X and Y are additional components, such as N (nitride).

There will be many processing variables to control. Let us consider the properties of a NiFe film, specifically its resistivity. The film may be used in the MR section of the thin-film head (see Fig. 7.6). We desire a film with high change in resistivity; see Fig. 9.28. The percentage change approaches 4 as the Ni content approaches 90% (the rest is Fe). This alloy is not suited for use in our MR head, because the magnetostriction is quite high at that composition. The film must conform to the standard 82% Ni film used in all TFHs.

The MR element should also be reasonably thin, to provide high resolution. Again, we find limits: Figure 9.29 shows an increase in the resistivity as the film becomes thin, and Fig. 9.30 shows a dramatic decrease in the MR effect for very thin films. A compromise is struck by using MR films of several hundred angstroms thickness.

The properties of three evaporated and two sputtered films are shown as examples in Figs. 9.31 through 9.35. Substrate temperature clearly affects B_{sat}, becoming lower as the temperature increases above 275° C. H_c is best (lowest) at 310° C and permeability seems to simply increase with temperature.

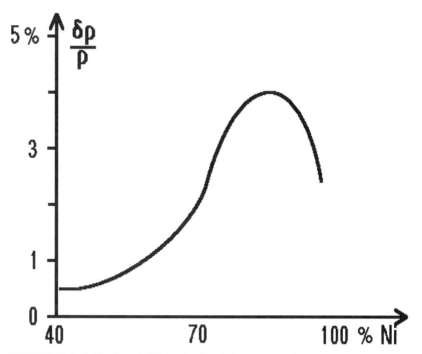

FIGURE 9.28 Resistivity change in NiFe is a function of Ni content (*Bozorth; McGuire and Potter, 1975*).

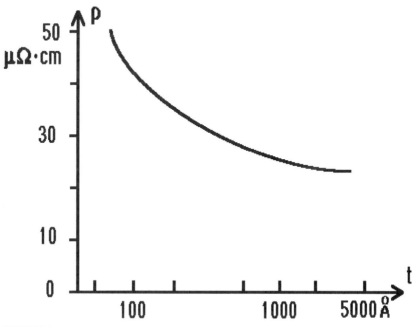

FIGURE 9.29 The resistivity value of a NiFe film is a function of film thickness.

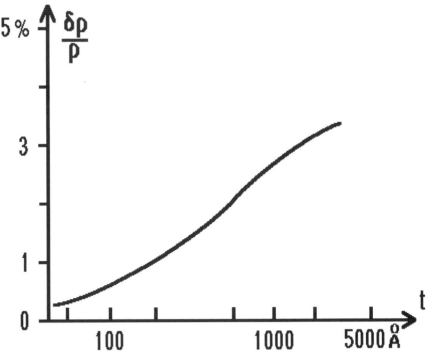

FIGURE 9.30   The resistivity change of a NiFe film is a function of film thickness (*Thompson et. al., 1975*).

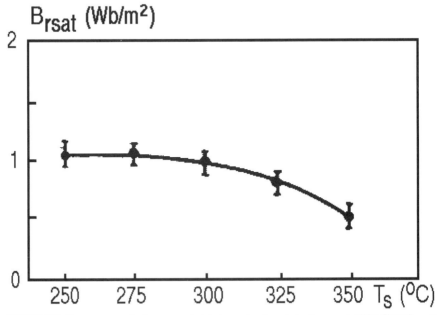

FIGURE 9.31   Remanent magnetization versus substrate temperature for anisotropic evaporated NiFe films (*Ciurenau and Gavrila, 1990*).

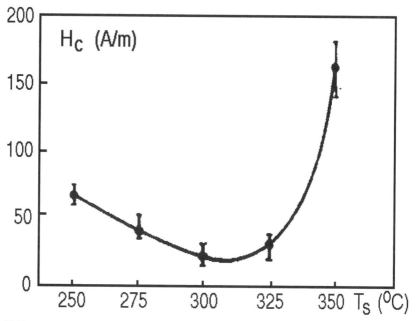

FIGURE 9.32 Coercivity versus substrate temperature for anisotropic evaporated NiFe films (*Ciurenau and Gavrila, 1990*).

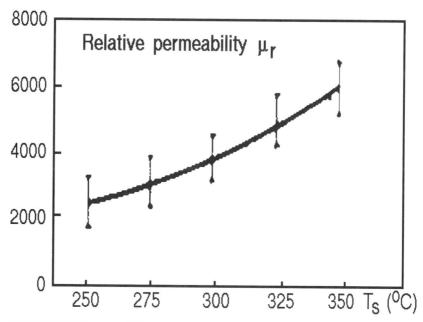

FIGURE 9.33 Relative permeability versus substrate temperature for anisotropic evaporated NiFe films (*Ciurenau and Gavrila, 1990*).

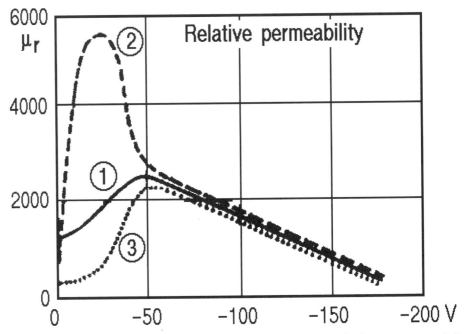

FIGURE 9.34 Permeability of sputtered Mu-metal layers versus negative bias voltage for three argon pressures: 1: 2 mTorr; 2: 5 mTorr; 3: 10 mTorr (*Kao and Kasiraj, 1991*).

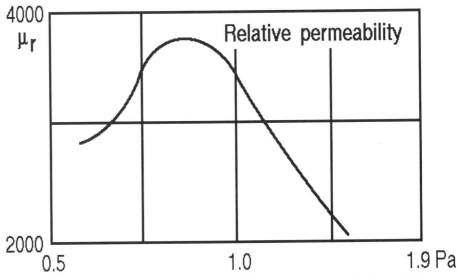

FIGURE 9.35 Permeability of sputtered Mu-metal multilayered films versus Argon pressure (*Potzlberger, 1984*).

Sputtering conditions will likewise influence the final magnetic properties. The sputtering process itself is covered in Chapter 13, because its primary use within magnetic recording is for fabrication of rigid disks.

PRODUCTION TOLERANCES

One of the major difficulties in the fabrication of magnetic heads is achieving satisfactory production yields. The processes are sequential, and final quality can rarely be tested until the assembly is completed. Head manufacturers have therefore implemented stringent quality-control steps throughout production. Needless to say, a good portion of the fabrication is done in a clean room environment. Minute dust particles are known to have ruined production yields.

We recall the expression for a head's efficiency, which is a proportionality factor in its sensitivity to record or playback:

$$\text{Efficiency } \eta = R_{\text{fg}}/\Sigma R_{\text{m}} \tag{4.20}$$

where R_{fg} equals the reluctance of the front gap, in parallel with the stray fields around the front gap, and ΣR_{m} is the sum of the magnetic reluctances going once around the equivalent circuit of the head.

The gap reluctances are controlled by the spacings L_{fg} and L_{bg} between the two core halves, the depths of the gaps (front gap: d_{fg}, back gap: d_{bg}), and the track width w. The core reluctance is determined primarily by the relative permeability of the core material, and is subject to gross changes during the fabrication process (deposition or etching of metallic laminations, or machining of ferrites; pressure from insertion and epoxy bonding; and lapping operations).

The acceptable production tolerances can be determined if we differentiate the expression for the efficiency η with respect to R_{fg} or $R_{\text{c}} = R_{\text{c1}} + R_{\text{c2}}$, the total core reluctance:

$$\delta\eta/\delta R_{\text{fg}} = (\Sigma R_{\text{m}} - R_{\text{fg}})/(\Sigma R_{\text{m}})^2$$
$$= (1 - \eta)/\Sigma R_{\text{m}}$$

or

$$\delta\eta/\eta = (1 - \eta) \times (\delta R_{\text{fg}}/R_{\text{fg}}) \tag{9.1}$$

Also,

$$\delta\eta/\eta R_{\text{c}} = -R_{\text{fg}}/(\Sigma R_{\text{m}})^2$$

or

$$\delta\eta/\eta = -R_{\text{c}} \times (\delta R_{\text{c}}/R_{\text{c}})/\Sigma R_{\text{m}} \tag{9.2}$$

These last two equations enable the head designer to work out acceptable production tolerances.

Example 9.1: If the nominal value of η is 0.8 for a recording head with:

$$R_{\text{fg}} = 30 \ \mu\text{H}^{-1}$$
$$R_{\text{c}} = \ \ 2 \ \mu\text{H}^{-1}$$
$$\Sigma R_{\text{m}} = 37.5 \ \mu\text{H}^{-1}$$
$$R_{\text{bg}} = \ \ 3.5 \ \mu\text{H}^{-1}$$

then we find, for an allowed + or − 0.5 dB variation (= 5.9 percent) in η:

$$\delta\eta/\eta = \pm(1 - 0.8) \times (\delta R_{fg}/R_{fg}) \tag{9.1}$$

or

$$\delta R_{fg}/R_{fg} \leq 0.295$$

or 30 percent variation in the gap reluctance. If we applied the tolerance to variations in the core reluctance we find:

$$\delta\eta/\eta = \pm(2/37.5) \times (\delta R_c/R_c)$$

or

$$0.06 \geq \pm(2/37.5) \times \delta R_c/R_c$$

or

$$\delta R_c/R_c \leq 1.125$$

or a 112 percent variation (in, say, the permeability of the core material).

If the gap length was changed to one-third of the original length, then R_{fg} would become $10 \ \mu H^{-1}$, $R_{bg} = 1.2 \ \mu H^{-1}$ and $\eta = 0.66$; this would be the case where the same core structure was used for a short-gap head.

The permissible tolerances are now:

$$\delta R_{fg}/R_{fg} \leq 18 \text{ percent,}$$

or

$$\delta R_c/R_c \leq 46 \text{ percent.}$$

This example illustrates the difficulty in achieving a high yield with short gap, multitrack head assemblies. The shorter gap lengths must be much better controlled (from say 2.25 μm (90 μin) ±0.7 μm (27 μin) to 0.75 μm (30 μin) ±0.135 μm (5.4 μin)!). And similar reasoning holds for the core permeability of nominal value of, say, 2000, from a range of $\mu_r = 4240$ to 944 to a tighter range of $\mu_r = 2920$ to 1370.

Here lies the reason why a head assembly with n tracks is not priced at n times the cost of a single-track head, but rather n^q times, where $q > 1$.

Accurate modelling of heads requires the inclusion of stray fields and eddy-current losses. Formulas 9.1 and 9.2 are instructive in making budgetary estimates of allowable tolerances in production, but a far more accurate insight is gained by making "what-if" questions to a computer program that models the head performance based on all relevant input parameters.

The quality of the gap definition is also more demanding to produce in short gap heads for tape systems. Irregularities in the gap line cause an additional loss for the write/read signals at short wavelengths (Mallinson). Assume that the irregularities deviate from a straight line in a random fashion with variance σ^2 (The variance σ^2 is found from $\sigma^2 = \Sigma x^2/N - X^2$, where x are the measured gap lengths at N locations, and $X = \Sigma x/N$ is the arithmetic mean value). This results in an additional loss in head output voltage:

$$\text{Irregularity loss} = A_t = 170 \times (\sigma/\lambda)^2 \text{ dB} \tag{9.3}$$

where λ is the recorded wavelength. This result may be remembered by its similarity to Wallace's spacing loss formula $54.6 \times d/\lambda$ dB. This loss is not the same as gap scatter, which refers to the potential difference in gap alignments in a multitrack head.

Example 9.2: If a tape was recorded with a head having random irregularities with variance σ_1^2 and reproduced with a head with variance σ_2^2, then:

$$A_t = 170 \times (\sigma_1^2 + \sigma_2^2)/\lambda^2 \quad \text{dB} \tag{9.4}$$

Answer: The deviations from a straight line at 10 locations along the gap were measured to be 12, 7, 3, 8 , –9, 5, –14, 10, 4, and 7 μin. The variance σ^2 is 69, and the irregularity loss for $\lambda = 0.75$ μm (30 μin for a 4 MHz recording at $v = 120$ IPS) is then $170 \times 69/30^2 = 13$ dB. The standard deviation for this gap is 0.2 μm (8.3 μin).

Head designs can be modelled to any level of accuracy. These models will assist in setting fabrication dimensions and material parameters for heads. These may be laminated, ferrite, TFH or MR. Figure 9.36 shows how four parameters vary with frequency with core permeability as parameter. μ_r varies from 100 to 2000 in 9 steps. The head is a horizontal TFH with a single layer coil in a milled cavity and pole structure in center; see Fig. 9.37 (Springer 1988). The changes in each parameter can be judged from this model, and verified by small scale experimental head fabrication.

Statistical process control is of vital importance for success in making heads, in particular thin-film heads where thousands of heads are processed concurrently on their supporting wafer. Powerful data becomes available when production deviations are analyzed together with performance data. Documented models of the recording process have been incorporated in Monte Carlo simulations and can be used to examine the importance of thin film geometry control (Williams 1990.) Some of the critical TFH parameters were:

Variable (all in μm)	Mean	Std.Dev. 1990	Std.Dev. 1995
Flying Height	0.17	0.01	0.0035
Throat Height	3.0	0.40	0.20
Pole Thickness	3.0	0.07	0.07
Gap Length	0.50	0.03	0.005
P2W	13.0	0.33	0.10

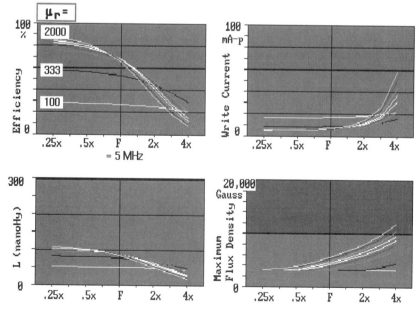

FIGURE 9.36 Computer simulation of thin-film head characteristics as function of changes in film permeability. Head construction is shown in Figure 9.37.

FIGURE 9.37 Early development of the horizontal TFH. The main body of the MetGlass and top NiFe layers join at the outside perimeter of an etched cavity wherein the coil is placed (insulated). A gap is formed where the top layer meets the center post (*Springer, 1988*).

This data set clearly shows the reduced tolerances made in a five-year span.

Of interest are the weight of the effects that different parameters have on the timing window $Tw10$, which is defined as the sum of the noise-induced jitter, the pattern-induced bit shifts (ISI = intersymbol interference), and the write-induced shift (from overwrite).

The individual contributions of fly height and other parameters were found to be (for 1990 data):

Parameter	$Tw10$ Std.Dev.
Fly Height	0.86 nsec
Throat Height	0.43 nsec
Gap Length	0.21 nsec
Pole Thickness	0.12 nsec
Track Width	0.12 nsec
All others	0.08 nsec

SPECIFICATIONS

Table 9.1 is a comprehensive list of things not to forget on a specification drawing; it is advisable to check all of the items listed, and to examine which additional items should be included.

TABLE 9.1 Comprehensive Listing of Magnetic Head Specifications

Signal	Medium
Digital: Recording code	Material, oxide? film?
Bit rate, f-upper	H_c, M_{rs}, SFD, S.
Analog: Data bandwidth	Head-medium speed
ac-bias frequency	Head-medium spacing

Core geometry	Core material
Track width	Relative permeability
Gap length & depth	H_c, M_{rs}, resistivity
Core dimensions	How well does it demagnetize?

Electrical	Head geometry
Self-capacitance	Track location(s)
Losses	Gap scatter
Load impedance	Write-read spacing
Number turns	Surface contour
Efficiency vs. frequency	Flying attitude
Impedance vs. frequency	Mounting method
Signal & noise signals	Outline dimensions
SNR	Azimuth: tilt
Crosstalk	
Other interface specs.	

Environment
Temperature
Humidity, vibrations
External magnetic fields
RFI

Note: The items listed may not all be applicable, and additional items may be included.

MEASUREMENTS AND TESTS

Techniques for measurements of the effective gap length, head core losses, self-capacitance, and impedance are important aspects of magnetic recording. Additional tests of head performance such as required write current, overwrite capabilities, read response, resolution, PW50, etc. are typically performed on head test units made by specialized companies (for a listing consult the IDEMA membership directory).

MEASUREMENT OF EFFECTIVE GAP LENGTH

Because the magnetic gap lengths are always larger than the mechanical gap lengths, an error can be induced in a design by measuring the gap lengths under a microscope and using this dimension to establish the gap losses. The gap length can be determined accurately only by measuring the wavelength λ_o, where the induced voltage from the playback head goes through a null; the gap length is then calculated from the gap loss function. This results in:

$$\text{Effective gap length} = \lambda_o \tag{9.5}$$

Because this null normally is beyond the frequency range of the recorder (in a properly designed recorder, by a factor of 2), the easiest measurement is undertaken by connecting the head leads directly to a signal generator. (The higher frequencies required for this measurement makes the use of high-frequency bias questionable because of the generation of beat notes.) The level of the record current from the sine-wave generator should be of the same magnitude as the bias current normally used, which essentially means the tape is recorded to saturation. A series of frequencies are recorded and played back; then a curve can be plotted and the null frequency interpolated.

In a more powerful method, a sweep signal generator is used to provide the record signal. The sweep rate is set to 30 per second. The upper and lower frequencies are selected to cover the recorder's range and up to a frequency beyond the gap null frequency (experiment may be necessary). An amplitude detection circuit detects the read signal, which is fed to the vertical input on any oscilloscope. The scope's sweep rate is synchronized to the generator's sweep rate. A dip in the output will show where the first gap null is located (compare with Fig. 6.17).

Figure 9.38 shows an example in which the tape speed was 120 inches per second; the null at 5.5 MHz corresponds to an effective gap length of 22 μin = 0.55 μm.

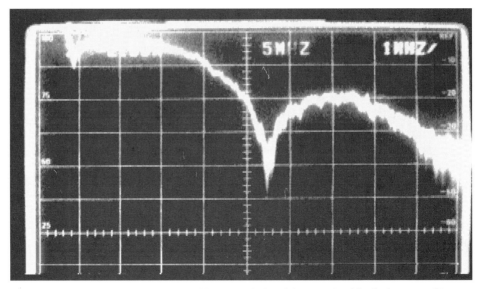

FIGURE 9.38   The sweep frequency test method facilitates determination of the true gap length in a head *(courtesy of Tranetics, Inc.)*.

MEASUREMENTS OF HEAD CORE LOSSES

If the flux level through the core in a magnetic head (without losses) is held constant with frequency, the induced voltage will ideally rise 6 dB per octave. The constant flux can readily be provided as shown in Fig. 9.39 by a thin, straight wire placed in front of, and parallel with, the gap. The departure from the straight 6 dB per octave line is evident, and is a measure of head losses.

The constant current is obtained by connecting the wire to a sine-wave generator in series with an induction-free resistance of a value equal to the recommended termination for the generator. Connecting a voltmeter across the resistor provides a method to keep the current constant, because most

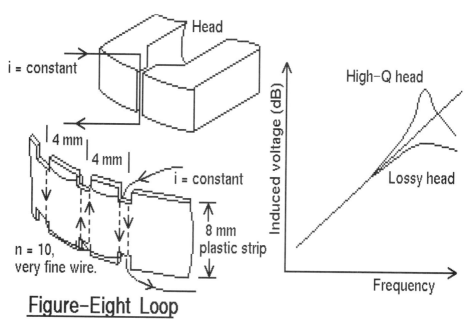

Figure-Eight Loop

FIGURE 9.39 Measurement of head core losses. Signal is injected into the head by a single current-carrying wire in front of the gap, or through a figure-eight loop.

signal generators require slight readjustments of the output as the frequency is changed. Another voltmeter (high impedance) is connected to the head output terminals to measure the induced open-circuit voltage. Losses in the core reduce the induced voltage at high frequencies, and these losses are represented by the distance between the 6-dB-per-octave line and the measured curve in Fig. 9.39. The frequency range of the head will also be evident, because the head inductance and its self-capacitance will cause it to resonate at a certain frequency f_o.

Instead of using a straight wire, it may be advantageous to fabricate a small figure-eight loop, as shown in the same figure. It can be wound on a thin strip of plastic or celluloid, for example, and can easily be positioned in front of the head similar to the path followed by a magnetic tape. The loop should be pressed lightly against the head in such a position that the induced voltage is maximum, and if possible, tape it down so it will not move during the measurements.

MEASUREMENT OF HEAD IMPEDANCE AND SELF-CAPACITANCE

The head impedance varies with frequency and needs to be plotted for the electronic circuit designer (write and read circuits). These measurements are also required for incoming inspection, and are readily done using a commercial impedance bridge, such as Hewlett-Packard's HP 4291A.

A simple test is outlined in Fig. 9.40. It requires only a sine wave generator, a frequency counter, and an oscilloscope. The self-capacitance C_e is determined by substitution, as shown in Fig. 9.40 bottom. The true inductance L_{true} is always smaller than the measured inductance L_{meas}. If the value of the tuning capacitor is C_1, then

$$L_{true} = L_{meas}{}^x C_1/(C_1 + C_e) \tag{9.6}$$

Measurements of output versus write current, of overwrite (erasure of old data), of resolution, and of noise are covered in Chapter 10. Mechanical properties such as flying height and attidue are covered in Chapter 18.

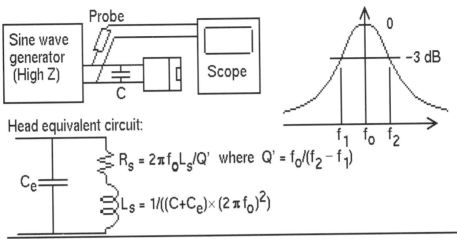

Head equivalent circuit:

$$R_s = 2\pi f_0 L_s / Q' \quad \text{where} \quad Q' = f_0/(f_2 - f_1)$$

$$L_s = 1/((C+C_e) \times (2\pi f_0)^2)$$

Determine C_e by first resonating the head with an additional capacitor C_1; Note resonance frequency f'. Now add C_2 and note new resonance frequency f". Then $C_e = [f''^2/(f'^2 - f''^2)] \times (C_2 - C_1)$.

FIGURE 9.40 Measurements of head impedance and its RLC-equivalent components.

DESIGN CHECKLIST FOR TAPE HEADS

This chapter concludes with the useful Table 9.2 of potential problems in the application of magnetic write/read tape heads. Some items are also applicable to disk heads.

TABLE 9.2 Design Test for Digital Recording (courtesy of Nortronics)

Symptom	Probable cause	Solution
Low output	1. Poor tape contact	1. Check tape wrap angle and tension and/or face finish of head. Check pressure pad and look for tape wear pattern on face of head to see that gaps are being contacted.
	2. Excessive write current	2. Use write current = 150% of saturation current.
	3. Very low write current	3. Same as above.
	4. Tape speed too slow	4. Adjust speed to correct value.
	5. Poor tape (worn oxide)	5. Replace tape.
	6. Write and read coil connections reversed	6. Connect write head to writer and read head reversed to read amplifier.
	7. Read coil open	7. Replace head.
	8. Open leads (poor connections)	8. Replace leads and/or clean connections.
	9. Read load excessive	9. A load impedance on read head of less than 10K ohms/200 pF will reduce output somewhat depending on head impedance, frequency and load impedance.
	10. Write density too high	10. Use density for which output test is specified.
	11. Poor tape guiding	11. Adjust azimuth of head.
	12. Dirt on tape	12. Clean tape **and** head.

High output	1. Tape speed too fast	1. Adjust speed to correct value.
	2. No load on read coil	2. Read head should have 10K ohm/200 pF for comparison with spec. Higher impedance will increase Eo a little.
	3. Resonant condition: Resonant Frequency = $\dfrac{1}{2\pi\ \sqrt{LC}}$ Where C is distributed C of head + leads + electronics	3. If correct density and tape speed results in approaching resonant frequency, then lower inductance head should be used.
	4. Write density is too low	4. Use density for which output test is specified.
High write current	1. Poor tape contact	1. Check tape wrap angle and tension and/or face finish of head. Check pressure pad and look for tape wear pattern on face of head to see that gaps are being contacted.
	2. Tape on backwards on reel to reel deck	2. Turn oxide side of tape towards the head.
	3. 0-Peak level used instead of peak-peak	3. Write current is specified in ma. peak-peak.
	4. Rise and fall time set improperly (too slow)	4. Set rise and fall times per spec.
	5. Shorted write coil	5. Replace head.
	6. Write density too low	6. Use density for which saturation test is specified.
	7. Dirt on tape	7. Clean tape and head.
Low write current	1. Open write coil	1. Replace head.
	2. Write and read coil connections reversed	2. Connect write head to writer and read head to read amplifier.
	3. Write density too high	3. Use density for which saturation test is specified.
	4. Defective writer or incorrect current measuring set-up	4. Check waveshape and amplitude of write current and current measuring set-up.
High percentage of crossfeed	1. High write current	1. Use write current = 150% of saturation current
	2. Rise and fall times set too fast	2. Set rise and fall times per spec (20% of pulse length).
	3. Poor tape contact	3. Check tape wrap angle and tension and/or face finish of head. Check pressure pad and look for tape wear pattern on face of head to see that gaps are being contacted.
	4. Head not grounded	4. Make ground connection to case of head.
	5. Write and read coil connections reversed	5. Connect write head to writer and read head to read amplifier.
	6. Tape speed too slow	6. Adjust speed to correct value.
	7. No load on read coil	7. Loading head will reduce fast rise spikes due to instrumentation ground loops and crossfeed from fast rise.
	8. MU metal shield in cassette cartridge	8. Remove screening shield from cartridge.
	9. HY MU 800 or ferrite near face of head	9. Remove material from close proximity to head.
	10. Improper shielding on read/write cables	10. Separate read and write cables; read cable must be well shielded and shield grounded.
	11. Electro-magnetic radiation from motors, relays and electronics	11. Head must be shielded from strong magnetic fields.

TABLE 9.2   Design Test for Digital Recording (courtesy of Nortronics) (Continued)

Symptom	Probable cause	Solution
Distorted output	1. Shorted coil	1. Replace head.
	2. Shorted coil to ground	2. Replace head.
	3. Defective writer or incorrect current measuring set-up	3. Check waveshape and amplitude of write current and current measuring set-up.
	4. Defective read circuit	4. Check loading of read amplifier.
	5. Magnetic radiation into head	5. Head must be shielded from strong magnetic fields.
	6. Excessive secondary or ghost pulses	6. Check for excessive tape wrap angle (10° in and 10° out is sufficient) and for excessive write current (use 150% of saturation current).
	7. Low write current	7. Use 150% of saturation current.
	8. Incomplete erasure of old data	8. Use 150% of saturation current.
Low pulse resolution	1. High write current	1. Use write current = 150% of saturation current.
	2. Poor tape contact	2. Check tape wrap angle and tension and/or face finish of head. Check pressure pad and look for tape wear pattern on face of head to see that gaps are being contacted.

REFERENCES

LAMINATED CORES AND HALF-SHELL CONSTRUCTION

Boll, R. 1956. Einbettung weichmagnetischer Werkstoffe in Kunststoffe. *Technische Informationsblatter M7.* Vacuumschmelze Aktiengesellschaft, Hanau, West Germany. 5 pages.

Boll, R. June 1965. Fotochemisch hergestellte Lamellen und Praezisionsteile aus weichmagnetischen Legierungen und anderen Sonderwerkstoffen. *Feinwerktechnik* 69 (6): 241–246.

Floc'h, M. Le, J. Loaëc, H. Pascard, and A. Globus. Nov. 1981. Effect of Pressure on Soft Magnetic Materials. *IEEE Trans. Magn.* MAG-17 (6): 3129–34.

Liechti, K.M. Sept. 1985. Residual Stresses in Plastically Encapsulated Microelectronic Devices. Experimental Mechanics 25 (3): 226–231.

Rabinowicz, E. June 1968. Polishing. *Scientific American* 236 (6): 91–99.

Staff. 1980. MuShield Magnetic Shields and Shielding Materials. MuShield Co. Brochure. 8 pages.

Staff. 1980. Netic and Co-Netic Magnetic Shielding Manual. Perfection Mica Co., 1322 Elston Ave, Chicago 22, Ill. 32 pages.

FERRITE CORE HEADS

Broese van Groenou, A. Sept. 1975. Grinding of Ferrites, Some Mechanical and Magnetic Aspects. *IEEE Trans. Magn.* MAG-11 (5): 1446–1451.

Butsch, O. Nov. 1981. Refining head design in high-density minifloppies. *Mini-micro Systems* 14 (11): 221–224.

Chandrasekar, S., K. Kokini, and B. Bhushan. July 1990. Influence of Abrasive Properties on Residual Stresses in Lapped Ferrite and Alumina. *Jour. Amer. Ceramic Soc.* 73 (7): 1907–1911.

Coughlin, T.M. Sept. 1985. Sendust Films for High Temperature Processing. *IEEE Trans. Magn.* MAG-21 (5): 1897–1899.

Deschamps, R.G. May 1983. Mise en evidence des resonances electromagnetiques dimensionnelles dans des circuits magnetiques et en particulier dans les pots ferrites. *l'onde electrique* 63 (5): 46–50.

Freitag, W., P. Mee, and R. Petersen. Sept. 1980. Glass/Ferrite Interactions and Corrosion of Gap Glasses in Recording Heads. *IEEE Trans. Magn.* MAG-16 (5): 876–878.

Ichinose, M., and M. Aronof. Nov. 1990. Single-Crystal Ferrite Technology for Monolithic Disk Heads. *IEEE Trans. Magn.* MAG-26 (6): 2972–77.

Jeffers, F. Nov. 1982. Metal-In Gap Record Head. *IEEE Trans. Magn.* MAG-18 (6): 1146–1148.

Kijima, T., H. Sakaguchi, Y. Satou, T. Fujine, T. Tanaka, T. Hara, A. Enomoto, K. Himeshima, and T. Kimura. Dec. 1990. Development of Bonding Low-Melting Glass for Magnetic Head. *Sharp Tech. Jour.* No. 47, pp. 57–63.

Knowles, J. Jan. 1975. The Origin of the Increase in Magnetic Loss Induced by Machining Ferrites. *IEEE Trans. Magn.* MAG-11 (1): 44–49.

Mizuno, Y., and M. Ikeda. Jan. 1992. Preparation and Properties of Bonding Glass for Amorphous Magnetic Head. *Jour. Ceramic Soc. Japan* 100 (1): 84–87.

Mizuno, Y., M. Ikeda, and A. Yoshida. Dec. 1992. Application of Tellurite Bonding Glasses to Magnetic Heads. *Jour. Matl. Sci. Letters* 11 (24): 1653–1656.

Okada, A., Y. Yamamoto, T. Yoshiie, I. Ishida, K. Hamada, and E. Hirota. April 1993, Surface Layer of Mechanically Polished Mn-Zn Ferrite Single Crystals. *Materials Transactions Jim* 34 (4): 343–350.

Rigby, E.B. Sept. 1984. Diffusion Bonding of NiZn Ferrite and Nonmagnetic Materials. *IEEE Trans. Magn.* MAG-20 (5): 1503–1505.

Tanji, S., S. Matsuzawa, N. Wakatsuki, and S. Soejima. Sept. 1985. A Magnetic Head of Mn-Zn Ferrite Single Crystal Produced by Solid-Solid Reaction. *IEEE Trans. Magn.* MAG-21 (5): 1542–1544.

Wada, T. Sept. 1980. An Improvement of Ferrite Substrate. *IEEE Trans. Magn.* MAG-16 (5): 884–886.

THIN-FILM HEADS: EVOLUTION, CONFIGURATIONS

Chapman, D.W. Sept. 1989. A New Approach to Making Thin-Film Head-Slider Devices. *IEEE Trans. Magn.* MAG-25 (5): 3686–88.

Chynoweth, W., J. Jordan, and W. Kayser. 1972. PEDRO - A Transducer-Per-Track Recording System with Batch-Fabricated Magnetic Film Read/Write Transducers. *Honeywell Computer Journal* 7 (5): 103–117.

Ciureanu, P., and H. Gavrila, 1990. *Magnetic Heads for Digital Recording*, Elsevier. 714 pages. Chapter 6 "Thin-Film Inductive Heads", pp. 455–567.

Jones, Jr., R.E. 1975. IBM 3370 Film Head Design and Fabrication. IBM Publication: *Disk Storage Development*, pp. 3–5.

Lazzarri, J.P., and P. Deroux-Dauphin. Sept. 1989. A New Thin-Film Head Generation IC Head. *IEEE Trans. Magn.* MAG-25 (5): 3190–3193.

Lazzari, J.P., and I. Melnick. March 1971. Integrated Magnetic Recording Head. *IEEE Trans. Magn.* MAG-7 (1): 146–150.

Lazzari, J.P. Sept. 1973. Integrated Magnetic Recording Heads Applications. *IEEE Trans. Magn.* MAG-9 (3): 322–26.

Oshiki, M., and S. Hamasaki. 1990. Thin-Film Head Technology, *Fujitsu Sci. Tech. Jour.* 26 (4): 353–64.

Springer, G., Personal communication, fall 1988.

Umesaki, M., Y. Ohdoi, H. Hata, K. Yabushita, K. Morikawa, H. Kishi, S. Horibata, and H. Shibata. Nov. 1991. A New Horizontal Thin-Film Head. *IEEE Trans. Magn.* MAG-27 (6)): 4933–35.

Valstyn, E. and L. Shew. Sept. 1973. Performance of Single-Turn Film Heads. *IEEE Trans. Magn.* MAG-9 (3): 317–21.

THIN-FILM HEADS: FABRICATION

Bhushan, B., and B.K. Gupta. 1991. *Handbook of Tribology*. McGraw-Hill, Inc.

Bischoff, P.G. June 1990. Electrochemical Deposition Requirements for Fabricating Thin-Film Recording Heads. *Magnetic Materials, Processes, and Devices, Electrochemical Society, Inc., Proc.* 90 (8): 221–232.

Hanazano, M., S. Narishige, S. Hara, K. Mitsuoka, K. Kawakami, Y. Sugita, S. Kuwatsuka, T. Kobayashi, M. Ohura, and Y. Tsuji. April 1987. Design and fabrication of thin-film heads based on a dry process. Jour. Appl. Phys. 61 (8): 4157–4162.

Kao, A.S., and P. Kasiraj. Nov. 1991. Effect of Magnetic Annealing on Plated Permalloy and Domain Configurations in Thin-Film Inductive Head. *IEEE Trans. Magn.* MAG-27 (6): 4452–57.

Mallinson, J. March 1969. Gap Irregularity Effects in Tape Recording. *IEEE Trans. Magn.* MAG-5 (1): 71.

Potzlberger, H.W. Sept. 1984. Magnetron Sputtering of Permalloy for Thin-Film Heads. *IEEE Trans. Magn.* MAG-20 (5): 851–53.

Romankiw, L. and P. Simon Jan. 1975. Batch Fabrication of Thin-Film Magnetic Recording Heads: A Literature Review and Process Description for Vertical Single Turn. *IEEE Trans. Magn.* MAG-11 (1): 50–54.

Romankiw, L.T., and B. Manor. Dec 1974. Elimination of Undercut in Anodically Active Metal During Chemical Etching. U.S. Patent 3,853,714.

Vossen, J.L., and W. Kern. 1991. *Thin Film Processes II*. 866 pages. Academic Press, Inc.

Williams, E.M. Nov. 1990. Monte Carlo Simulation of Thin-Film Read-Write Performance. *IEEE Trans. Magn.* MAG-26 (6):3022–26.

BIBLIOGRAPHY

FABRICATION, GENERAL

Bonnie, G.P. Sept. 1989. Inductive Thin-Film Heads Directions for Development and Manufacturing in the 1990's. *DISKCON Tech. Conf.* 27 Viewgraphs.

Farrow, R.C., S.D. Berger, J.M. Gibson, J.A. Liddle, J.S. Kraus, R.M. Camarda, H.A. Huggins. Nov. 1991. Alignment and Registration Schemes for Projection Electron Lithography. *Jour. Vacuum Sci. & Tech.-B* 9 (6): 3582–3585.

Fontana Jr., R.E. June 1990. Magnetic Thin–Film Heads, A Review on Processing Issues, Magnetic Materials, Processes, and Devices. *Electrochemical Society, Inc., Proc.* 90 (8): 205–219.

Kawabe, T., M. Fuyama, S. Narishige, and Y. Sugita. Nov. 1991. Fabrication of Thin Film Inductive Heads with Top Core Separated Structure. *IEEE Trans. Magn.* MAG-27 (6): 4936–38.

Staff. Aug. 1982. IC Process Technology. *HP Journal* 33 (8): 36 pages.

Tsang, C. and R. Fontana. Nov. 1982. Fabrication and Wafer Testing of Barber-Pole and Exchange-Biased Narrow-Track MR Sensors. *IEEE Trans. Magn.* MAG-18 (6): 1149–51.

Vossen, J.L., and W. Kern. 1978. *Thin-Film Processes*. 564 pages.

Yuan, C.M., and A.J. Strojwas. May 1991. Modeling Optical Equipment for Wafer Alignment and Line-Width Measurement. *IEEE Trans. Semicond. Manuf.* 4 (2): 99–110.

Zieren, V., G. Somers, J. Ruigrok, M. de Jongh, A. van Straalen, W. Folkerts, E. Draaisma, F. Pronk, and T. Mitchell. Nov. 1993. Design and Fabrication of Thin-Film Heads for the Digital Compact Cassette Audio System. *IEEE Trans. Magn.* MAG-29 (6): 3064–68.

GENERAL METHODS OF DEPOSITION

Deshpandey, C.V., and R.F. Bunshah. 1991. Evaporation Processes, in *Thin-Film Processes II*, pp. 79–132.

Freeman, J.D. May 1991. Effect of Deposition Conditions on the Properties of Thin Permalloy Film. *Jour. Vacuum Sci. & Tech.*, 9 (3): 421–425.

Komaki, K. Oct. 1990. Domain Structure and Permeability of Electrodeposited NiFe Strip Films. *Magnetic Materials, Processes, and Devices, 92/10, Electrochemical Society, Inc., Proc.* 92 (10): 245–253.

Liao, S.H., and S.E. Anderson. Jan. 1993. Domain Characteristics and Conformal Deposition of NiFe Thin Films. *Jour. Electrochem. Soc.*, 140 (1): 208–211.

Narayan, P.B., and S.C. Herrera. Oct. 1992. Permalloy Plating Imperfections - Effect of Contamination. *Magnetic Materials, Processes, and Devices. Electrochemical Society, Inc.*, Proc. 92 (10)): 355–366.

Poupon, G., T. Braisaz, and P. Deroux Dauphin. June 1990. Microelectrodeposition and Domain Structure of Ni Fe Alloys in I.C. Head. *Magnetic Materials, Processes, and Devices, Electrochemical Society, Inc. Proc.* 90 (8): 267–278.

Wagner, U. and A. Zilk. May 1982. Selective Microelectrodeposition of Ni-Fe Patterns. *IEEE Trans. Magn.* MAG-18 (3): 877–79.

SPUTTERING

Berghof, W. and H.H. Gatzen. Sept. 1980. Sputter Deposited Thin-Film Multilayer Head. *IEEE Trans. Magn.* MAG-16 (5): 782–84.

Chen, G., C. Leu, J.M. Sivertsen, and J.H. Judy. Sept. 1985. Sputter-Induced Composition Effects on the Magnetic and Microstructure of NiFeCuMo Permalloy Thin Films. *IEEE Trans. Magn.* MAG-21 (5): 1939–41.

Jahnes, C.V., M.A. Russak, B. Petek, and E. Klokholm. July 1992. Ion Beam Sputter Deposited Permalloy Thin Films. *IEEE Trans. Magn.* MAG-28 (4): 1904–10.

Krusch, K. Sept. 1986. Sputter Parameters and Magnetic Properties of Permalloy for Thin Film Heads. *IEEE Trans. Magn.* MAG-22 (5): 626–28.

Lo, J., C. Hwang, T.C. Huang, and R. Campbell. Sept. 1987. Near-Zero Magnetostriction NiFe Films Deposited by Ion Beam Sputtering. *IEEE Trans. Magn.* MAG-23 (5): 3065–67.

Narishige, S., K. Mitsuoka, and Y. Sugita. March 1992. Crystal Structure and Magnetic Properties of Permalloy Films Sputtered by Mixed Ar-N2 Gases. *IEEE Trans. Magn.* MAG-28 (2): 990–93.

Parsons, R. 1991. Sputter Deposition Processes, in *Thin Film Processes II*, pp. 177–208.

APPLICATIONS OF ION-BEAM AND LASERS

Clampitt, R., P.W. Mingay, and S.T. Davies. Oct. 1991. Micromachining with Focused Ion Beams. *Sensors and Actuators* 25 (1-3): 15–20.

Cohen, U. Oct. 1992. A Self-Aligned Ion Beam Pole Trimming Process for TFH. *Magnetic Materials, Processes, and Devices, Electrochemical Society, Inc., Proc.* 92 (10): 211–216.

Cohen, U. Oct. 1992. Selective Wet Chemical Etching of the Coil Seed-Layer in TFH Fabrication. *Magnetic Materials, Processes, and Devices, Electrochemical Society, Inc., Proc.* 92 (10): 217–221.

Das, S.C., and J. Khan. June 1990. Applications of Laser Etching in Thin Film Head Fabrication. *Magnetic Materials, Processes, and Devices, Electrochemical Society, Inc., Proc.* 90 (8): 279–288.

Das, S.C., and J. Khan. June 1990. New Bonding Pad Process for Thin-Film Heads Using Selective Laser Etching. *Jour. Appl. Phys.* 67 (12): 4860–2.

Fawcett, S.C., and T.A. Dow. July 1992. Influence of Wheel Speed on Surface Finish and Chip Geometry in Precision Contour Grinding. *Precision Engineering* 14 (3): 160–167.

Kasai, T., and T. Karakidoy. May 1992. Tribology from a Viewpoint of Ultraprecision Polishing Technology. *Jour. Japan Soc. Tribo.* 37 (5): 357–362.

Kinoshita, K., K. Yamada, and H. Matsutera. Nov. 1991. Reactive Ion Etching of Fe-Si-Al Alloy for Thin Film Head. *IEEE Trans. Magn.* MAG-27 (6): 4888–90.

Lutwyche, M.I. March 1992. The Resolution of Electron Beam Lithography. *Microelectronic Engr.* 17 (1-4): 17–20.

Nakanishi, T., K. Kogure, T. Toshima, and K. Yanagisawa. Sept. 1980. Floating Thin-Film Head Fabricated by Ion Etching Method. *IEEE Trans. Magn.* MAG-16 (5): 785–87.

Okano, H., T. Yamazaki, and Y. Horiike. March 1983. High-Rate Reactive Ion Etching Technology. *Toshiba Review* No. 143, pp. 31–35.

Orloff, J. Oct. 1991. Focused Ion Beams. *Sci. American* 265 (4): 96–101.

Prewett, P.D. March 1993. Focused Ion Beams Microfabrication Methods and Applications. *Vacuum* 44 (3-4)): 345–351.

Young, R.J. March 1993. Micro-Machining Using a Focused Ion Beam. *Vacuum* 44 (3–4): 353–356.

MEASUREMENTS

Anderson, N.C. and R.E. Jones. Nov. 1981. Substrate Testing of Film Heads. *IEEE Trans. Magn.* MAG-17 (6): 2896–98.

Gieraltowski, J., J. Loaëc, and H. Le Gall. Sept. 1989. Non-Destructive AC Permeability Measurement of Full Wafer Soft Magnetic Films. *IEEE Trans. Magn.* MAG-25 (5): 4219–21.

Kawakami, K., M. Suda, M. Aihara, H. Fukuoka, Y. Hagiwara, and K. Takeshita. April 1987. Electrical Detection of End Point in Polishing Process of Thin-Film Heads. *Jour. Appl. Phys.* 61 (8): Part IIB, pp. 4163–66.

Kirk, C.P. June 1987. Design of an Automated Optical Microscope for Measuring the Critical Dimensions of Magnetic Recording Heads. *Optical Engr.* 26 (6): 507–12.

Sonnenfeld, R. Sept. 1990. Fly Height, Pitch, and Crown Measurements of Hard-disk Sliders by Capacitance Stripe. *IEEE Trans. Magn.* MAG-28 (5): 2545–47.

Yip, Y., M.H. Vos, M. Lu, M.P. Dugas, and J.H. Judy. Nov. 1988. In-Situ Measurements of Permeabilty During Rotational Magnetic Annealing of Co Zr Nb. *IEEE Trans. Magn.* MAG-24 (6): 3072–74.

CHAPTER 10
DESIGN AND PERFORMANCE OF MAGNETIC HEADS

The performance of magnetic heads can be computed from the basic model of a resistance network, and the results can inform us about things like the write head's recording efficiency, current requirements, and impedance. We can also get information about the possibility of signal distortion, and whether the head core may become permanently magnetized or not. Other items are pole-tip saturation and potential heating of very small write heads.

For read heads we need to know about their sensitivity, efficiency, and noise. For both type of heads, susceptibility to external fields need to be evaluated.

Keeping the book's title of a "handbook" in mind, it would have been nice to have one large table that lists all head types and their performances. This proved to be impossible, and a different approach is used: The chapter will discuss the design and performance of several heads, designed to operate in ranges covering from audio frequencies up to 100 MHz. The information may then be applied to other heads, with proper interpolation.

The head designer may wish to learn from interdisciplinary engineering tasks; magnetic heads are inductive transducers and are related to other inductors and transformers, where books by Zinke (1982) and Grossner (1983) are recommended. This chapter will refer to numerous papers on various aspects of head performance. We will discuss traditional multichannel heads with cores made from laminated NiFe or ferrite materials. Inductive thin film heads have deposited windings for write and read operations, while magnetoresistive thin film heads use a small NiFe element for sensing the field from a recorded track.

Typical head core lamination thicknesses are:

Ni-Fe (Permalloy)	50 μm (2 mils)
Ni-Fe (Permalloy)	25 μm (1 mils)
Hard Ni-Fe ("Tough-malloy")	15 μm (0.6 mils)

Ferrites are almost exclusively MnZn because of their low high-frequency losses. NiZn is used as shielding material (MR heads) and for head interface for low wear.

The materials properties were covered in Chapter 7. Heads with AlFe(Si) pole tips will behave as if the cores were made entirely from the ferrite used for the body of the core; the magnetic reluctance between a pole piece and the core is vanishing compared to the other head reluctances.

DESIGN PROCEDURE

The design of heads has clearly followed an evolutionary path where current designs are modified to new requirements, and so on. There are numerous considerations, often with conflicting demands on the design, and trade-offs are made.

It is a good starting point to provide answers to all applicable items in Table 9.1. The desired packing density for a given application will dictate the geometry of the track width and gap length, and the design can now proceed by adopting a core geometry that appears compatible therewith; this may be an existing core design.

The head's performance can now be computed in as much detail as desired and the design optimized. This chapter will show how, and will illustrate the methods with computed results. This was done in part by a spreadsheet program, and in part by Head/Media Design programs by the author. A detailed head analysis is found later in this chapter in the section Saturation. It is a lengthy procedure, but serves as an example for the interested engineer. Further detailed head analysis procedures are found in a book by Ruigrok (1990).

The computed performances are evaluated in light of the desired specifications, and changes may be made. The next step is the fabrication of a few engineering model heads, upon which measurements are made and compared with the computed results. Discrepancies should be appraised and understood.

The modelling of a head rests on many formulas and certain approximations, in particular with regard to stray fields and eddy current losses. The first are computed from discrete approximations or by using finite-element methods. The eddy currents cause a drop in permeability toward high frequencies (and the upsurge of an imaginary component μ''); these changes are readily computed, but neglect the fact that flux will run on the surface of the core, NOT in its interior, at high frequencies.

An example to keep in mind is the way two head halves with laminated cores match. If the core sections do not mate as mirror images of each other, a poor performance will result; the flux coming up along a lamination surface will cross the gap and then bounce head-on into the misaligned lamination in the other core half, and then spread out. The result is an irregular field in and around a gap in a misaligned laminated head. (We have no good way of observing this field irregularity.)

The results from computations based on head modelling may not agree completely with measurements on the prototype heads. Understanding the discrepancies and repairing the model is an important effort to provide for closer agreements between computations and measurements. The use of a model for trend analysis is nevertheless always excellent.

The sections in this chapter need not be read in sequence for the design or examination of a head; rather, scan through it and use what is appropriate.

SELECTING UPPER FREQUENCY, GAP LENGTH, AND DEPTH

The upper operating frequency f_u of a head is limited by the resonance between its inductance L and the sum of the capacitances of the coil (C_e), the cable, and the amplifier. The value of L is equal to the number of turns squared, divided by the magnetic impedance Z_m of the head structure, including the leakage field around the coil (Z_m is the complex value of the reluctance R_m that results at frequencies where μ_r becomes complex).

For audio and instrumentation applications, the selection of f_u is straightforward: for playback, it must be higher than the highest data frequency, possibly by 50 percent (to keep the self-resonance and phase shifts away). In the record mode, f_u must equal or exceed the ac-bias operating frequency (60–100 kHz for audio, higher for instrumentation; typically 4–5 times the highest data frequency). The temptation to locate the resonance at the bias frequency to achieve a peaking effect should be avoided: First of all, heads actually seldom have a quality factor Q exceeding two to five at bias frequencies. There will be variations in production values of L, and the head impedance may further change with head wear.

In digital applications the data rate DR and encoding technique dictates f_u. FM-encoding results in $f_u = DR$, while MFM only requires $f_u = 0.5 \times DR$. This is for the fundamental frequency only, and should be multiplied by three if it is desired to include the third harmonic. It is assumed that the data rate DR is the rate after error-correction coding.

With an established speed between the head and the tape/disk, the resulting packing density in BPI and bit length BL can now be calculated:

$$BPI = 2 \times f_u/speed \qquad (10.1)$$

$$\text{and } BL = 1,000,000/BPI \text{ } \mu\text{in.} \qquad (10.2)$$

The gap length g is determined by:

$$g \approx 0.6 \times BL \text{ for high resolution}$$
$$\approx 0.9 \times BL \text{ for high output.} \qquad (10.3)$$

The mechanical gap length should be specified shorter, by two factors:

1. Correct for the effective gap length (see formula (10.14)), which is longer than the mechanical length and depends on the media permeability. In general, $g = 1.15 \times L_{fg}$ (L_{fg} = mechanical length).

2. Correct for dead layer (Beilby layer) on the surface of machined ferrites. Subtract four times the surface roughness per pole face (see formula (9.1)).

We can also establish the depth of the gap:

$$D_{gap} = 4 \text{ mils for long life}$$
$$= 2 \text{ mils for high output}$$
$$= 0.1 \text{ mils (no contact)}$$

Let us for a moment digress to the matter of units. We are currently in a transition from μin, mils, and inches to μm, mm, and meters. Both cgs/inches and MKS (SI) units are in current use, with the following conversions:

$$1 \text{ } \mu\text{m} = 40 \text{ } \mu\text{in}$$

$$1 \text{ } \mu\text{in} = 25.4 \times 10^{-9}\text{m}$$

$$1 \text{ mil} = 25.4 \times 10^{-6}\text{m}$$

SELECTING TRACKWIDTH

Reduced trackwidth is one factor that has made possible the large amount of data that can be stored on magnetic tapes and disks. Forty years ago a good quality music recording required a speed of 15 IPS, with a trackwidth equal to the tape width of 6.25 mm (0.25 inches). Today a digital hi-fi recording can be contained within a small cassette, with a playing time of 2 hours (Philips DCC).

Three considerations must be made when the track width is reduced. One, reduction of the track width itself reduces the signal-to-noise ratio by 3 dB for each halving of the width. Secondly, the amplitude can become modulated by mistracking; i.e., the track might not be moving past the read head with perfect registration, but might be wavering back and forth. Thirdly, the last action may also cause signals from an adjacent track to be picked up, which we classify as noise, because it is an unwanted signal.

The problem of mistracking, with resulting signal amplitude modulation and added noise from side reading, is particularly troublesome when using a tape or disk made from PET-film (PolyEthylene Terephthalate, with tradenames such as Hostaphan, Mylar, and Terylene; for properties, see Chapter 11 on materials for tapes and disks). The dimensions of a PET film change drastically with changes in temperature and/or humidity:

$$\Delta L = 1.1 \times 10^{-5} \text{ m/m/percent RH}$$

$$\Delta L = 1.5 \times 10^{-5} \text{ m/m/degree F}$$

Example 10.1: A 5.25-inch diskette drive is to operate at temperatures ranging from 50 to 100 degrees F; add 20 degrees of temperature increase inside the drive. The humidity may vary over an 80-percent range. Estimate track width and track center spacing.

Answer: The radius of an outer track is 6.35 cm = 2.5 inches, and can change by $(50 + 20) \times 1.5 \times 10^{-5} \times 6.25 = 0.0066$ cm, plus $80 \times 1.1 \times 10^{-5} \times 6.25 = 0.0056$ cm, totalling 0.0121 cm = 121 μm = ±61 μm. We must design for the worst-case conditions, and use the full peak value of 121 μm. A track width of 121 μm = 4.8 mils could fail totally.

We must now add centering and head positioning tolerances, and find:

Max. error (temperature)	66 μm
Max. error (humidity)	56 μm
Center hole (oversize)	12 μm
Disk spindle (undersize)	12 μm
Head positioning	12 μm
Total, no compensation	158 μm
Temperature compensation	−33 μm
Humidity compensation	−28 μm
Total change, p-p	97 μm

Without compensation, the error would be 158 μm = 6.3 mils, calling for a track width of 0.32 mm minimum. The compensation is accomplished by building the drive transport and head arm assembly to match the PET film coefficients of expansion versus temperature. The humidity is compensated partly by using a track index scale made from PET material, closely matching the disk. The main problem in this approach lies in the anisotropic residual stress gradients that exist in PET film; when exposed to temperature and/or humidity, a circular track will distort into a pattern that resembles a peanut shell (Greenberg 1977; Brock 1983).

From the above example we see that a track width of 320 μm can write a track that may be ±80 μm off center line when read later on. This calls for a minimum width allocation of $320 + 2 \times 80 =$ approximately 500 μm per track, and a minimum of 500 μm = 19.7 mils between track center lines. The resulting track density is 1000/19.7 = 50 TPI. The standard is 48 TPI, and a typical trackwidth is 300 μm = 11.8 mils.

From the preceding discussion, we can establish that the minimum track width and track-to-track center distance should be:

$$\text{Track width} \approx 2 \times (\text{Total change})_{pp}$$

$$\text{Center to center} \approx 4 \times (\text{Total change})_{pp}$$

Reductions in the read amplitude modulation can be made by trimming the written track with two narrow erase gaps, as was shown in Fig. 7.2. This also decreases the side-reading from the adjacent tracks, and it is possible to optimize the write/read and erase gaps for a given system. This should ultimately be done in a way that optimizes the ratio of signal to noise from adjacent and previously written tracks (Edelman 1985).

There is only one method for a drastic reduction of the tracking errors listed above, and that is a system where a prerecorded pattern is recorded on a disk during formatting. The best solution has been the floptical disk drive, where an optical servo track is impressed onto the diskette surface to provide a reference track. The servo system is therefore optical while the recording method remains magnetic.

Modern thin-film heads have special tracking considerations because the write head is a wide inductive head and the read head is a narrower MR element. Write current and frequency affects the erase bands during overwrite and subsequent tracking (Lin et al. 1989). The offtrack behavior has been analyzed and design formulas generated for optimum trackwidths of the two head sections (Bonyhard and Lee 1990; Lee and Bonyhard 1990).

Traditionally, inductive thin-film heads have been fabricated with a bottom pole (leading) somewhat wider than the top pole, mainly due to alignment tolerances in printing the top pole mask through which the pole is subsequently plated. The overlap of the bottom pole causes writing of poor-quality side information. A track density increase of 7 to 9% can be obtained by trimming the inductive head poles (Cain et al. 1994).

SELECTING CORE SIZE AND MATERIAL

The choice of core size and material is a most frustrating task if the designer has had little experience. Some guidance is necessary and is best obtained by considering head designs similar to the the one at hand. Good sources are manufacturers' catalogs and recent papers on head performances (trade journals and IEEE transactions on magnetics).

A decision is made for a start design, and computations will reveal how close the performance agrees with the specifications. The design is modified until the agreement is satisfactory (say within ±20 percent, or ±2 dB). Verifications of the design are carried out by taking measurements from a few engineering models.

The permeability spectra for typical materials are shown in Figs. 10.1 and 10.2. Note that the relative permeability μ_r of MuMetal is listed as only 2000, while manufacturers' data sheets show much higher values. The value of 2000 reflects the permeability that realistically exists in a head after finished fabrication (see Fig. 7.14).

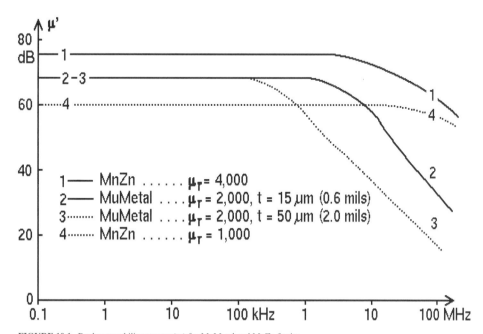

FIGURE 10.1 Real permeability spectra (μ_r) for MuMetal and MnZn ferrite.

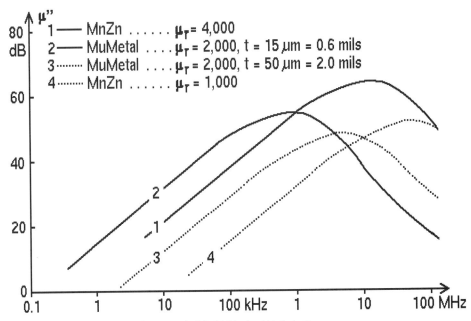

FIGURE 10.2 Imaginary permeability spectra (μ_{im}) for MuMetal and MnZn ferrite.

ELECTROMAGNETIC DESIGN

In Chapter 4 we arrived at an expression for the "deep-gap" field H_g in the front gap of a ring core head (formula 4.21). It should be multiplied with three efficiencies: coupling of the field from the coil into the core (Fig. 10.3), the ratio of the flux running through the gap to the total core flux, and finally the efficiency when the head operates in the write mode.

WRITE AND READ HEAD EFFICIENCY

The write efficiency may appear different than the read efficiency, and will indeed be so if the write efficiency is defined as the ratio of the flux in the gap to the flux generated in the core by the write current. The error is caused by the two entirely different magnetic impedance levels (reluctances) for the two fluxes.

Write efficiency must instead be defined as the ratio between the ampere-turns generated in the write gap to the ampere-turns fed into the head winding (Schelor 1986). This efficiency is exactly equal in value to the earlier-defined read efficiency. The reason is, of course, nested in the head's reciprocity. One may carry out a calculation of the write efficiency based on the coupling efficiency, etc., and verify that this number is equal to the more simply calculated read efficiency.

The field in front of the write gap depends on the gap field strength; this matter was discussed in Chapter 5, and the designer can use Figs. 5.15, 5.16, or 5.17 to determine the required field H_g. To do this he must know the desired write level, the coercivity of the magnetic media, and the dimensions (gap length, media thickness, and spacing from the head surface). We will now finalize the expression for the coupling efficiency so we can determine the required write current for a prescribed gap field.

The exact value of R_s is difficult to calculate. The stray flux around the coil is largest when a short coil is used. A short as well as a long coil develops a magnetomotive force ni. The short coil makes a short distance in the core for the magnetic potential; that makes the effective value of R_s small, and

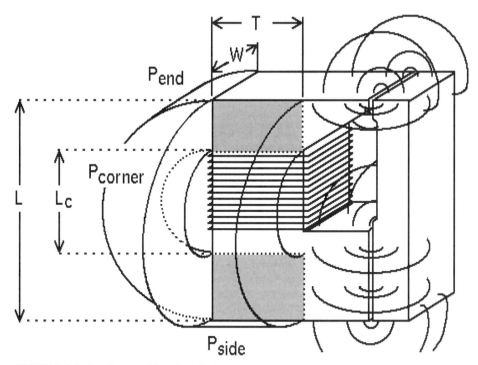

FIGURE 10.3 Leakage flux around the coil on a head core.

the coupling efficiency is also small. On the other hand, a long coil that is split between the two core halves results in a large coupling efficiency.

It is assumed that the coil is wound gradually from one end to the other so the potential is across the entire coil length. If wound back and forth and ending up with the last turn located on top of the first, then the coil's electromagnetic length is very short: The shunt reluctance will be small, while at the same time the self-capacitance C_e will be large; both factors are detrimental to an optimum design.

If we assume a gradual buildup of the magnetomotive force from one coil end to the other, then a simple algorithm for the computation of R_s is shown in Figs. 10.3 and 10.4. In the book by Ruigrok (1990) a detailed finite element analysis of a video head is carried out.

The analog circuit for a write head is shown in Fig. 10.4, where the shunt reluctance R_s around the coil is added. The inclusion of this element plus details of the stray fields around the gap are mandatory in computations of the performance of small, high-frequency heads. Further details on R_s are necessary if the core is near magnetic shielding or housing.

For the required write current, we find:

$$i_{\text{write}} = H_g \times L_g/(n \times \eta_r \times \eta_w \times \eta_{\text{coupl}}) \tag{10.4}$$

where η_r and η_w are found as (4.20) and (4.22). η_{coupl} is derived from the insertion loss of R_s in Fig. 10.4:

$$\begin{aligned}\eta_{\text{coupl}} &= R_s/(R_s + R_{\text{Ctop}} + R_{\text{fg}}' + R_I + R_{\text{bg}}' + R_{\text{Cbottom}}) \\ &= R_s/(R_s + \Sigma R_{\text{ms}})\end{aligned} \tag{10.5}$$

R_{fg}' and R_{bg}' are the gap reluctances, including stray fluxes. The magnetic resistance of the core elements are relatively easily determined, and the stray fluxes are determined by the formulas in Fig. 8.30. (NOTE: As a check, you may compute the write efficiency as ratio between magnetomotive forces in the gap versus the head winding).

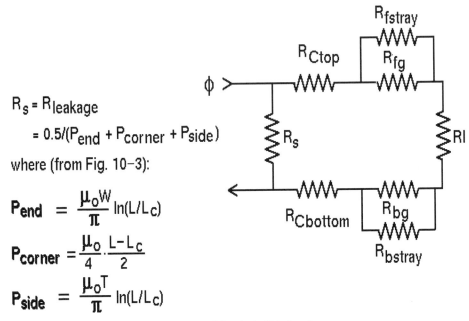

$R_s = R_{leakage}$

$= 0.5/(P_{end} + P_{corner} + P_{side})$

where (from Fig. 10-3):

$$P_{end} = \frac{\mu_0 W}{\pi} \ln(L/L_c)$$

$$P_{corner} = \frac{\mu_0}{4} \cdot \frac{L-L_c}{2}$$

$$P_{side} = \frac{\mu_0 T}{\pi} \ln(L/L_c)$$

FIGURE 10.4 Magnetic resistance (reluctance) network for a head with leakage flux.

A finer division of the core into more sections may be necessary in order to achieve sufficient accuracy; this will also preserve a feel for the magnitudes of the various elements (visible if programmed into the cells of a spreadsheet). This method follow the classic work described in Roter's book (see Chapter 8) and several papers cover the details, the first including crosstalk in a multitrack head assembly (Sansom 1976; McKnight 1979). More recent papers use the full power of finite difference and finite element computations (Wood et al. 1985; Visser et al. 1985; Katz 1980).

Figure 10.5 shows the computed values of read efficiencies (formula (4.20), using R_{fg}' (which includes stray fields) and the total, complex value of ΣR_m. Phase shifts are associated with the drop-offs at high frequencies, and the read circuitry may use a conventional RLC-circuit equalizer to correct for the variations in magnitude and phase with frequency. The read voltage can be calculated by using the formula in Fig. 6.11.

The efficiencies of thin-film heads (TFH) are more complex to compute due to the distributed nature of the magnetics and air reluctances. A TFH may have from one- to four-layer coils in order to maximize the number of turns and hence maximize the read voltage. A two-layer coil is shown in Fig. 10.6. The space needed by the layers increases the distance between magnetic bottom and yoke films, and so also increases the detrimental shunt reluctance between these two films. The magnetic films are double-layered in order to minimize the number of domains formed; see Fig. 7.22. This layering technique has no effect on eddy current losses, in contrast to the conventional lamination methods.

The early analysis of the efficiency of a TFH assumed a single conductor surrounded by a NiFe film; see Fig. 10.7. The flux was compared to the current in a simple transmission line model (Paton 1971; Jones 1978; and Heim 1986). Some computed results are shown in Fig. 10.7 (Katz 1978; Miura et al. 1978; Yeh 1982).

A reluctance model has certain advantages and is represented by the ladder network shown in Fig. 10.8. The magnetic film reluctances are named $ZF(N)$ and $ZB(N)$; they are complex values due to the complex permeabilities (Schelor 1986 and Jorgensen 1988). A reluctance model computation for an eight-turn head results in much higher values of head efficiencies than the transmission line model gives for a single turn in Fig. 10.9 (Ciureanu 1986). This discrepancy remains in another example comparing an eight-turn with a nine-turn TFH; see Fig. 10.10.

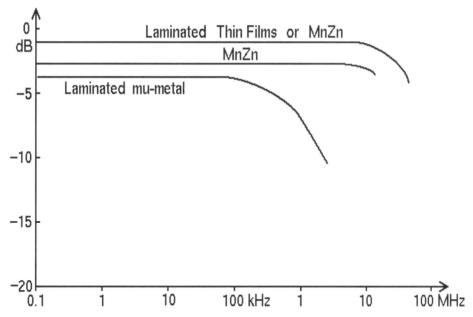

FIGURE 10.5 Efficiency of heads falls off at high frequencies.

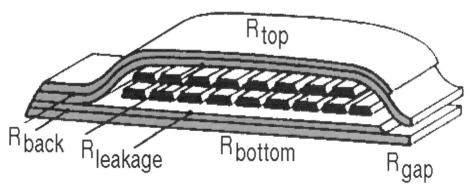

FIGURE 10.6 Two-coil TFH. Double lamination layers are used to control domain structure.

    In both figures, the variable parameter was the coil region. For a large coil we can expect less efficiency due to more stray flux between the film layers, and Fig. 10.9 verifies this. The opposite will happen if we increase the film spacing; see Fig. 10.10.

    Permeability of the NiFe films should be high in order to achieve high efficiency. Figure 10.11 shows the variation in efficiency for the relative permeability range of 0 to 3000. Values of 500 to 1000 appear adequate, while little is gained by using higher values. We might even wish to limit the value in order to have adequate high-frequency response for domain wall movements and rotations (see Fig. 7.21).

    The efficiency of a ring-core type head is highly dependent on the gap depth. This is not true for a TFH, as the computations in Fig. 10.12 show. Throat height (gap depth) control in a TFH is nevertheless very important in order to ensure adequate overwrite.

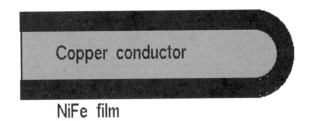

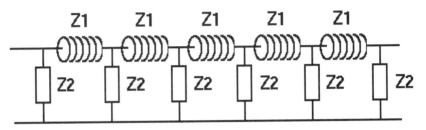

For an ordinary transmission line: $Z1 = R_{series} + j\omega L$

$$Z2 = G_{shunt} + 1/j\omega C$$

FIGURE 10.7 The transmission-line model for a single-turn TFH. Current elements at the gap will contribute most to the flux in front of the gap. The flux is reduced when the conductor's dimension away from the gap increases (*after Yeh, 1982*).

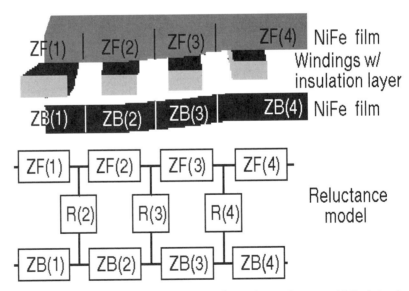

FIGURE 10.8 A multiturn TFH can be analyzed by a discrete element reluctance model. Conductors 1 through 4 are shown between two NiFe films, and the network below shows the reluctances of the top film ($ZF(N)$) and the bottom film ($ZB(N)$) and the shunts between them $R(N)$. $ZF(\ )$ and $ZB(\ )$ are complex, due to eddy current losses.

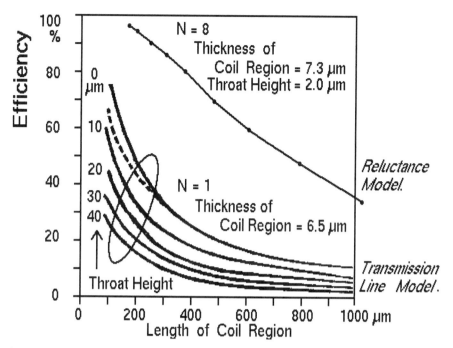

FIGURE 10.9 Computed efficiencies for a single-turn head (*N* = 1, *after Ciureanu* 1986), and an eight-turn head (*after Jorgensen*).

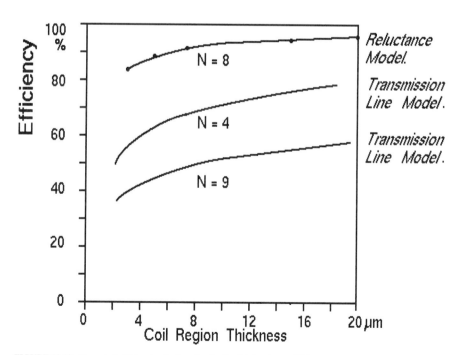

FIGURE 10.10 Computed efficiencies for three heads: *N* = 4 (*after Jones, 1978*), *N* = 9 (*after Ciureanu, 1986*) and *N* = 8 (*Jorgensen 1988*).

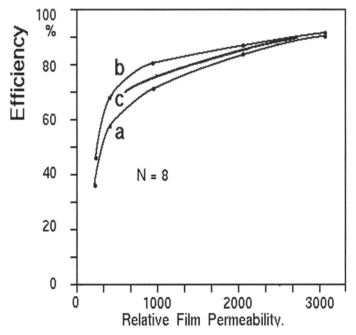

FIGURE 10.11 Efficiencies versus thin film permeability. (a) *after Schelor, 1986*, (b) *after Katz, 1986* and (c) *after Jorgensen, 1988.*

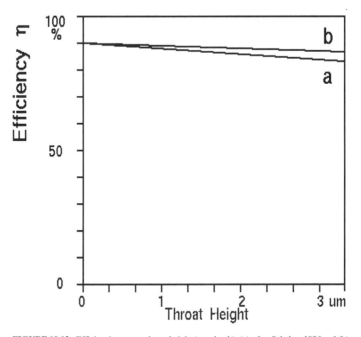

FIGURE 10.12 Efficiencies versus throat height (gap depth). (a) *after Schelor, 1986* and (b) *after Jorgensen, 1988.*

The influence of the top pole (the yoke section of the thin-film head) has been examined and is generally shown to be optimum for the shape shown in Figs. 7.5 and 7.6 (Akoh et al. 1990). Better performance in terms of magnetic domain and wall behavior is achieved with a straight rather than heart-shaped yoke; see Fig. 9.25 (Ed Williams, Read-Rite).

HEAD IMPEDANCE

The head impedance is determined from

$$Z = R_{wdg} + j\omega L$$
$$= R_{wdg} + j\;\omega n^2/\Sigma R_{ms}$$

which is the same formula used earlier for the impedance of a toroid, leading to formula (7.5). The core magnetic resistance ΣR_{ms} is now made up of the two core halves, the front and back gaps, stray fluxes, and leakage fluxes (in parallel with the core impedances at the location of the coil). Its value is complex, and varies with frequency. Computations are fairly straightforward, once the head core structure has been modeled and the low-frequency values of the core reluctances established.

The high-frequency reluctances of the core of length L and cross sectional area A are complex values because the permeability is complex: $\mu = \mu' - j\mu''$. The detail computations are as follows:

$$R_m = L/\mu A \text{ (at low frequencies)}$$

$$\begin{aligned}
Z_m &= R_m \text{ at high frequencies} \\
&= L/(\mu' - j\mu'')A \\
&= L \times (\mu' + j\mu'')/\{(\mu' + j\mu'')(\mu' - j\mu'')A\} \\
&= L \times (\mu' + j\mu'')/\{(\mu'^2 + \mu''^2)A\} \\
&= (L/\mu A)\mu(\mu' + j\mu'')/(\mu'^2 + \mu''^2) \\
&= R_m[\mu\mu'/(\mu'^2 + \mu''^2) + j\mu\mu''/(\mu'^2 + \mu''^2)] \\
&= Z_{mreal} + jZ_{mimag}
\end{aligned} \tag{10.6}$$

Z_m is computed for all elements in the head model, and added to give the total magnetic impedance. Let us designate that value $Z_M = Z_{Mreal} + jZ_{Mimag}$.

The head's impedance is, with a winding resistance of R_{wdg},

$$\begin{aligned}
Z &= R_{wdg} + j\omega L \\
&= R_{wdg} + j\omega n^2/Z_M \\
&= R_{wdg} + j\omega n^2/(Z_{Mreal} + jZ_{Mimag}) \\
&= R_{wdg} + j\omega n^2 \times (Z_{Mreal} - jZ_{Mimag})/(Z_{Mreal}^2 + Z_{Mimag}^2) \\
&= R_{wdg} + \omega n^2 \times Z_{Mimag}/(Z_{Mreal}^2 + Z_{Mimag}^2) + \\
&\quad j\omega n^2\;\omega\;Z_{Mreal}/(Z_{Mreal}^2 + Z_{Mimag}^2) \\
&= R_{wdg} + R' + jL'
\end{aligned} \tag{10.7}$$

where R' is proportional to μ'' and L' is proportional to μ'.

Computed values of L' and R' are shown in Figs. 10.13 and 10.14. R' represents the eddy current losses. Notice the rapid increase in the value of R' once it has sprung up from values below one ohm. It originates in Z_{Mimag}, which is proportional to μ'', which increases in proportion to the frequency (see Fig. 7.10).

Z_{Mimag} is further multiplied in formula (10.7) by $\omega = 2\pi f$. The result is a resistance value proportional to f^2, and it therefore increases 12 dB/octave ($20 \log_{10} 2^2 = 12$).

Inductance variations for thin-film heads are shown in Figs. 10.15 and 10.16. The small change in throat height eliminates the possibility of controlling throat height lapping by measuring the head inductance. This method is used with great success for ring-core type heads. Lapping of thin-film heads is now done for a row of heads at a time, as shown in Fig. 9.26, and the progression of lapping is monitored and controlled by the changes in a number of resistors located along the bar.

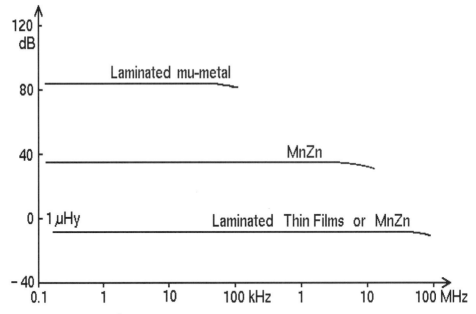

FIGURE 10.13   Inductance L' versus frequency for a low-, medium-, and high-frequency head.

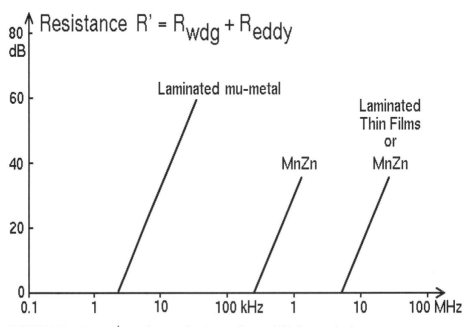

FIGURE 10.14   Resistance R' versus frequency for a low-, medium-, and high-frequency head.

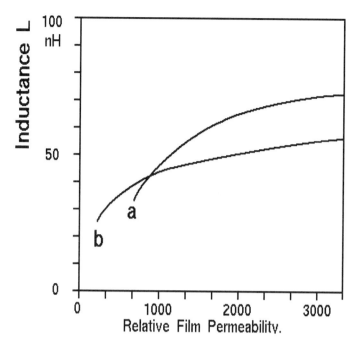

FIGURE 10.15 Inductances versus film permeability for two heads: (a) *after Katz, 1978,* (b) *after Jorgensen, 1988.*

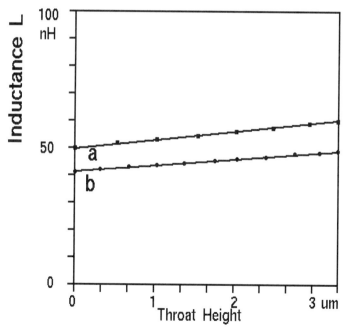

FIGURE 10.16 Inductances versus throat height for two heads: (a) *after Schelor, 1986,* (b) *after Jorgensen, 1988.*

To the total impedance we must add the distributed capacitance of the winding C_e and the capacitances of the cabling and the amplifier impedance. The number of turns n is always made as high as possible in order to minimize the required write current and to maximize the induced read-head voltage.

When the number of turns have been determined, then a wire size is selected that will fit into the available winding space. This normally results in a very fine wire (No. 40 or smaller) and R_{wdg} can now be calculated. Its value adds to the R' values above, and the coil quality Q can be determined.

The ratio $\omega L'/(R_{wdg} + R')$ is the coil quality Q' of the head impedance. Q' is quite low when compared with RF-coils (radio frequency coils). They will be lower in the finished head due to coupled losses from shields (expand R_s to include shields or a metal housing, if any is present), and dielectric losses (unpredictable magnitude).

Very low values of Q (less than three) will reduce the read voltage to less than the induced voltage, and simultaneously introduce a phase lag. For $Q = 1$ the loss is 3 dB and the phase angle –45 degrees.

This loss is part of the often observed 5½-dB slope of the read head voltage versus frequency curve, at low frequencies; the slope should theoretically be 6 dB/octave.

WRITE HEAD CORE LOSSES

The head core losses at high frequencies such as the bias frequency in ac-bias recorders are in the milliwatt range. They are calculated from:

$$P = i^2(R' + R_{wdg})$$

The losses are often just estimated as $V \times i$, where V is the winding terminal voltage and i the head current. Additional losses are hysteresis losses, which are covered later in this chapter under erase heads. The two losses may together produce excessive heat in small core structures that are held embedded in epoxies of low heat-transfer capability.

READ HEAD NOISE

There are two sources of noise in magnetic heads. First, there is Barkhausen noise, caused by jumps in the magnetic domain wall movements. The second source is resistance noise. The thermal agitation in a resistor produces a noise voltage that is:
where

$$V_{noise} = \sqrt{4kTBR} \text{ volts} \tag{10.8}$$

k = Boltzman's constant = 1.38×10^{-23} J/°K
T = Temperature in degree Kelvin = 273° + °C
B = Bandwidth in Hertz
R = Resistance value in ohms

The thermal noise in thin-film inductive heads is discussed in a recent paper (Klaassen and van Peppen 1992).

Barkhausen noise has been reported in connection with the switching of thin heads (see Fig. 7.21). Its contribution to the overall recorder noise level does otherwise remain undetected, although it is present in other devices (Bittel 1969). It appears to be one of the noise sources in magnetoresistive heads.

Recent studies have taught us more about the domain behaviors in thin-film recording heads (Klaassen and van Peppen, 1990 and 1991).

One mechanism that causes popcorn noise is believed to be associated with pinning and later release of domain walls. The pinning may occur during the writing, where the head is also heated

(Monson 1984). This changes the magnetostriction in such a way that the domains stay in the positions and shapes as determined during the write cycle. The anisotropy associated with the magnetostriction will change when the head cools and the domains snap back to their normal positions. This snap causes a change in magnetization and hence produces a noise spike. Because it is so irregular and happens just after writing, it has the nickname "popcorn noise." It is particularly annoying because it may occur at the instant the head is reading servo-wedge information. Futher details on this noise-after-write is found in a paper by Liu et al. (1992).

Ferrites in MIG heads may have unstable isolated transition responses that can be attributed to unfavorable magnetization directions in the crystallites (Klaassen and van Peppen 1992). The domain behavior depends strongly on the stress state and surface treatment of a ferrite head (Schäfer et al. 1992), and asymmetry in the read waveform is traceable to complexities in multiple domains (Takayama et al. 1992).

The noise voltage from the equivalent loss resistance R', on the other hand, can be of sufficient level to impair the overall signal-to-noise ratio of a tape recorder. This is, in particular, true of high-frequency wideband recorders and video recorders. The computation of the noise voltage is straightforward once R' has been determined.

The SNR improves about 10 dB by using 2-mil rather than 6-mil laminations in audio heads. Further improvement is made by using ferrite cores (Byers 1971).

Some heads have been reported to be microphonic; i.e., they produce a crackling noise when a blank tape moves across their surface. These effects may relate to residual magnetostriction and/or domain wall pinnings (Watanabe 1974).

POLE TIP SATURATION

Modern recording materials with high coercivity, in conjunction with short-gap-length heads, have created a new problem for head designers. The field in the gap of the recording head must be higher than in the past, and the danger of saturating the core material becomes real. A brief example will illustrate the problem.

Example 10.2: A 1-μm gap length head is used for recording and playback in a cassette tape machine. Determine the record gap field when using a 5-μm-thick coating of a cobalt treated gamma ferric oxide tape with coercivity 600 Oe.

Answer: Assume an overbias setting that corresponds to a write level of 120 percent. From Fig. 5.16 we find that $\nu = 0.05$, i.e., $H_g = 600/0.05 = 12,000$ Oe. This corresponds to a gap flux density of 12,000 gauss. A ferrite core cannot be used due to low saturation flux density, nor can Mu-metal; only an Fe-Al alloy will handle the high flux level (see Table 7.1).

The situation in this example did happen a few years ago when high-coercivity tapes were introduced into the audio cassette market. A decision was also made to make the 3.5-inch diskette from a 550 Oe material. In both cases a short read gap was required for adequate resolution, but using the same head for recording would result in saturation of the core pole tips during the record or write process.

A solution was simple in the audio field: adding a separate record head with longer gap. This increased the value of ν and H_g could be lowered. The reader will recall that the length of the write gap has little influence on the resolution, with the exception of situations where the recording is very shallow, i.e., ν is large.

The diskette problem was solved by making the coating thinner: $\delta = 2$ μm. That also increases the ν value. At the same time demagnetization of short bits is reduced. As a matter of fact, with packing densities of 5 KFCI, the coating need only be $\delta_{eff} = 0.44 \times BL = 0.445$ μm = 2.2 μm or less (formula 6.8b).

The saturation process is complex. The permeability decreases at high flux densities (Fig. 10.17), which in turn increases the magnetic resistance of the head core. When the permeability becomes very low, then the head loses its ability to form a proper head field. It has been established that the corners saturate first, as shown in Fig. 10.18, and this occurs when the flux density in the gap approaches half the value of the material's saturation flux density (Suzuki and Iwasaki 1972; Shibaya and Fukuda 1977; Szczech et al. 1978).

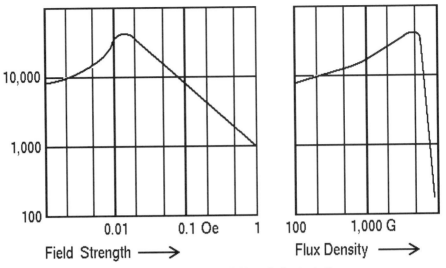

FIGURE 10.17 Permeability μ_r varies with the magnetizing field H or the flux density B.

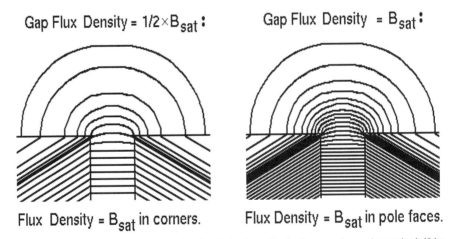

FIGURE 10.18 The gap corners start saturating when the head core flux density approaches a value equal to half the saturation value B_{sat}. The number of flux lines in the very corner is twice the average value of flux. The pole interfaces saturate when the flux level reaches B_{sat} at the gap interface.

The write resolution does not suffer much thereby (Thornley and Bertram 1978; Fujiwara 1979). This is partly so because the recording or writing takes place at some distance behind the trailing edge of the write gap.

When the gap flux is increased, then the pole faces will at some point saturate and no further increase in gap flux is possible. Any attempt to drive more flux through the head will only result in an increase in the flux going through the stray flux elements around the gap, not through the gap. After that level is reached, the recorded magnetization does not change. This is observed as a sudden leveling of the read-voltage-vs.-write-current curve. There are only two alternatives: Use a higher M_{sat} material for the head core or for pole tips (MIG), or use a longer gap for recording, i.e., separate heads are required.

Saturation does not happen like a switch was included in the equivalent circuit for the head. The transition is smoothed by the stray fluxes. This is emphasized by the fact that heads will experience a gradual increase in the pole tip reluctance. The relative permeability μ_r is one at the pole tip surface when saturation starts.

Pole-tip saturation occurs when making recordings on high-coercivity media with heads having small gap lengths. The result is a recorded level that is less than full saturation. The onset of saturation causes a rounding of the corners (edges) of the field at the head poles, as shown in Fig. 10.18. Complete saturation of the pole pieces (shown right) occurs at a field strength that is double the one causing corner saturation (Bertram and Steele 1976).

Pole-tip saturation causes reduced permeability at and near the pole interface. The magnetic resistance of the head is increased, and more current is required to produce the gap flux. Stray fluxes will play a larger role when the head core permeability drops and a further increase in drive current is needed.

The reduced relative permeability (approaching a value of 1 at the gap) also causes a reduction in the field gradient. At the same time, the effective magnetic gap becomes slightly longer, and a likewise slightly higher field is produced in front of the gap.

The result is a very complex deterioration of the write head performance at the onset of saturation:

1. Pole tip permeability drops.

2. Effect of stray fields increases.

3. Field strength near the gap increases only slightly.

4. Field gradient near gap decreases.

The main degradation of the head performance is the inability to provide enough field strength to make the recording. Higher write currents do not generate higher gap fields after the head is saturated in the pole region.

The following section will deal with a solution to item 1 (MIG heads), and provides insight into items 2, 3, and 4. A numerical example will be used to illustrate the underlying theory and give the reader a good insight in what "goes on."

THE HEAD RELUCTANCE MODEL—A TUTORIAL HEAD ANALYSIS

The analysis of magnetic heads is easy when one applies a resistance network where the elements represent air gaps and core sections. A core sample is shown in Fig. 10.19 and its equivalent network in Fig. 10.20, lower left. The magnetic resistors, also called reluctances, are calculated from

$$R_m = length/(permeability \times area)$$

Notice the step in core cross section width in our example. This requires than the reluctances for the I and the C cores are split into at least two elements, 1 and 2.

One additional reluctance, R_{pole}, is included for portions of the C-core. The reluctance of the tapered pole is quite large near the gap, where the area for the flux is small and the permeability low. Stray fluxes are left out of the example for simplicity. The variations in flux density and permeability are shown in Fig. 10.20.

We will first determine the reluctance dR of the small element of length dx and area $w \times d(x)$. The total reluctance is found by integration from $x = 0$ to $x = l_{pole}$.

$$dR = l/\mu A$$
$$= dx/(\mu_o \times \mu_r(x) \times d(x) \times w)$$

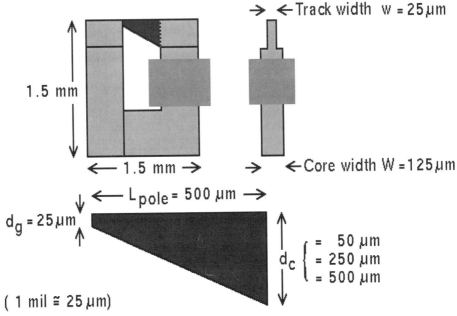

FIGURE 10.19 Ferrite core dimensions for the example in the text. Track width = 25 μm = 1 mil. Core width = 125 μm = 5 mils.

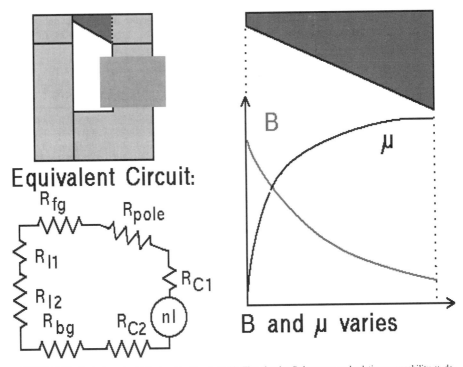

FIGURE 10.20 Equivalent circuit for head in Figure 10.19. Flux density B decreases and relative permeability μ_r decreases away from the gap region.

where $\mu_r(x)$ and $d(x)$ are functions of x.

The dimension $d(x)$ is found from the geometry in Fig. 10.21:

$$d(x) = d_g + x \times (d_p - d_g)/l_{pole} \quad (x > 0)$$
$$= d_g(1 + xK) \tag{10.9}$$

where $K = (d_p - d_g)/(l_{pole} \times d_g) \quad (d_p > d_g)$.

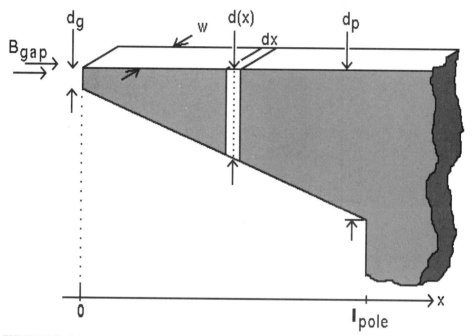

FIGURE 10.21 Pole tip geometry for computation of its reluctance and permeability variation.

$\mu(x)$ is found by using a well-known model for the initial magnetization curve, shown in Fig. 10.22 (Fischer and Moser 1956):

$$B = B_{sat}/(1 + H_c/H)$$

Solving for $\mu = B/H$ we find:

$$\mu = \mu_{init} \times (1 - B/B_{sat})$$

where the maximum value of B is

$$B_{max} = B_{sat} \times (\mu_{init} - 1)/\mu_{init}.$$

This is a straight line that goes from $(\mu,B) = (\mu_{init},0)$ to $(\mu,B) = (1,B_{sat})$; see Fig. 10.23. A curve measured for a MnZn ferrite is shown for comparison (Fig. 3.54). The straight line for the model of μ is below this curve, and therefore conservative.

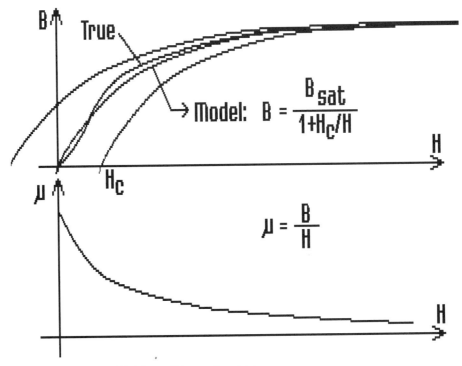

FIGURE 10.22   Classic model of B versus H in magnetic materials.

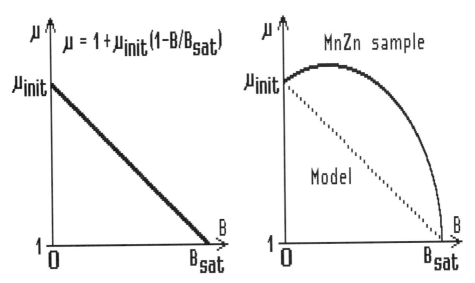

FIGURE 10.23   Model of permeability μ versus flux density B.

The variation in μ is now found by substituting for B:

$$B = B_{gap} \times d_g/d(x).$$

B decreases with the distance x from the gap. Substituting for $d(x)$ (10.9):

$$
\begin{aligned}
B &= B_{gap} \times d_g/(d_g(1 + xK)) \\
&= B_{gap}/(1 + xK)
\end{aligned}
\tag{10.10}
$$

We now introduce a number S to represent the level of flux density in the pole tip, and therefore also in the gap:

$$S = B_{gap}/B_{sat} \tag{10.11}$$

Substituting (10.11) into (10.10) and then into the expression for μ results in (after reduction):

$$\mu(x) = \mu_{init} \times (1 - S/(1 + xK)) \tag{10.12}$$

Now we substitute $d(x)$ (10.9) and $\mu(x)$ (10.12) into dR and find:

$$dR = dx/\{\mu_o \times w \times d_g \times \mu_{init} \times (1 - S+xK)\}$$

The total pole reluctance R_{pole} is found by integration of dR from $x = 0$ to $x = l_{pole}$. The result is:

$$R_{pole} = \{l_{pole}/(\mu_0\mu_{init} \times w \times (d_p - d_g))\} \times \log_e\{1 + (d_p - d_g)/((1 - S) \times d_g)\} \tag{10.13}$$

When d_p approaches d_g, the first term goes toward infinity while $\log_e\{1 + (d_p - d_g)/((1 - S) \times d_g)\}$ becomes zero. This is similar to the conflict in evaluating $\sin X / X$ when X goes to zero. Replace the natural log term in formula (10.13), i.e., $\log_e(1 + X) \rightarrow X$, and the terms $(d_p - d_g)$ cancel. The formula now equals the value of the tapered pole with $d_p = d_g$, divided by $(1 - S)$. Because μ_r is between μ_{init} and 1, the maximum value of S becomes

$$S_{max} = (\mu_{init} - 1)/\mu_{init} \tag{10.14}$$

This extreme case then corresponds to $\mu_{init} = 1$, and correlates with the maximum value of the flux density B.

Several computations are made for the reluctances of various pole-tip configurations. The gap front depth is equal to 25 μm (1 mil) while the dimension d_p is selected equal to 50, 250, and 500 μm. These values correspond to a taper ratio of 1:2, 1:10, and 1:20 of the pole piece.

The air gap reluctances in the head shown in Fig. 10.19 are (in Hy^{-1}):

$$R_{fg} = 636.0 \times 10^6$$

$$R_{bg} = 2.5 \times 10^6$$

The core reluctances in Fig. 10.21 are, with an initial permeability of 1000,

$$R_{I1} = 34.0 \times 10^6$$

$$R_{I2} = 11.5 \times 10^6$$

$$R_{C1} = 25.5 \times 10^6$$

$$R_{C2} = 17.8 \times 10^6$$

The six fixed reluctances add up to a total of 727.3×10^6 Hy^{-1}. Their sum is shown in Figs. 10.24 and 10.25 as the bottom of the bars. The top of the bars show the computed values of the reluctances of the tapered pole piece at several flux density levels.

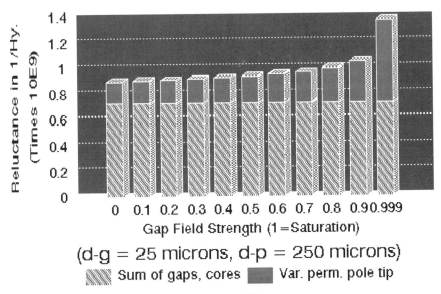

FIGURE 10.24 Head model reluctances for a pole taper ratio of 10:1.

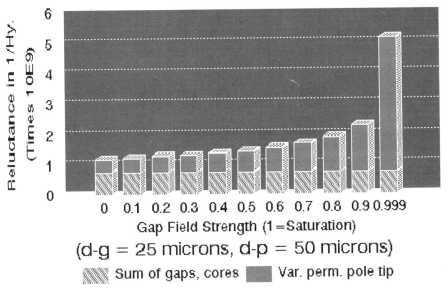

FIGURE 10.25 Head model reluctances for a pole taper ratio of 2:1.

The effect of saturation is pronounced in the core with a taper of 2:1; see Fig. 10.25. This reflects in the amount of current required to drive the head, shown in Fig. 10.26. More current is required near saturation when the taper is small. This is shown in Fig. 10.27.

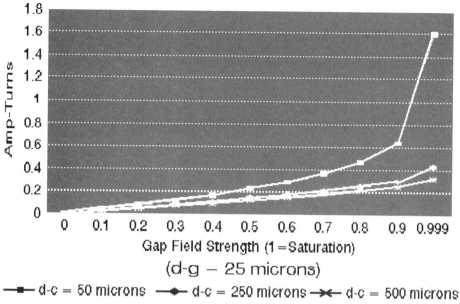

FIGURE 10.26 Drive currents required for taper ratios of 2:1 (d–p = 50 μm), 10:1 (d–p = 250 μm) and 20:1 (d–p = 500 μm).

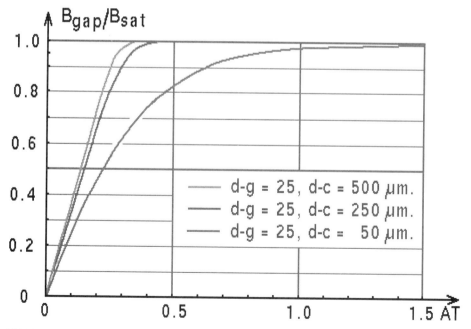

FIGURE 10.27 Variation in gap flux densities for three tapers of head core.

An insight in the magnetic changes brought about by saturation can be illustrated in a different way. Formula 10.13 can first be used to first calculate the pole-tip reluctance at saturation ($S = S_{max}$). It can next be used to solve for the location of x-values that will divide the pole tip into ten equal reluctances. The result is shown in Fig. 10.28 and reveals that half the pole tip reluctance is located a few gap lengths away from the gap. This dilution of the magnetics near the gap is also illustrated in Fig. 10.29, and the reduced field gradient in Fig. 10.30. Formula 10.12 permits computations of the permeability variations through the pole piece, and may be valuable in appraising the diluted field pattern that exists in front of a saturated head. It is recognized that this problem has been approached from more advanced theoretical considerations (Bertram and Steele 1976), but the reluctance model has advantages in providing more insight into the saturation process.

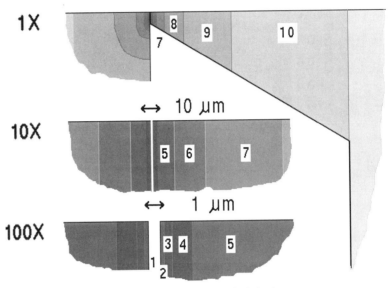

FIGURE 10.28   Gradual saturation in pole piece of saturated ferrite head.

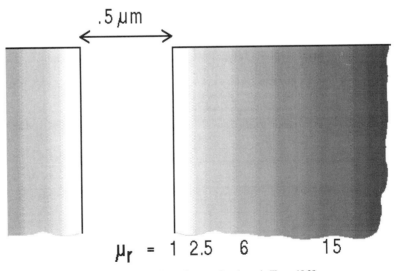

FIGURE 10.29   Reduced permeabilities due to the saturation shown in Figure 10.28.

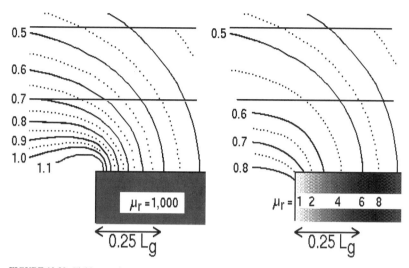

FIGURE 10.30 Field strength contours near a normal and a saturated head gap.

The flux level inside the tip is at a level almost equal to saturation, and the permeability is much lower than it is for small flux levels. Hence the pole-tip reluctance is high, and the efficiency low. The gap flux will therefore level off gradually when saturation is approached. The effect is enhanced by the stray flux entering the pole piece behind the gap interface, where it may cause further saturation (Valstyn and Packard 1986)

This results in a lowering of the write efficiency, and the gap field will remain almost constant over a range of drive currents. The net effect is a broadening of the specification for the drive current, and a write current response as shown in Fig. 10.31 (Valstyn and Packard 1986).

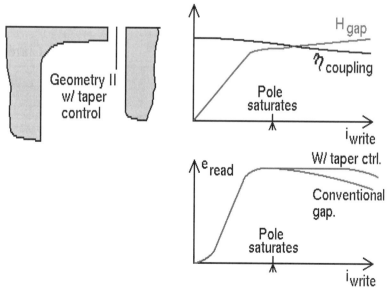

FIGURE 10.31 Reluctance control through excessive pole piece taper. This method allows for large variations in drive currents without much change in the operating characteristics of head.

METAL-IN-GAP (MIG) HEADS

The pole-tip saturation problem can be overcome by using a write gap construction with deposited metal pole pieces; see Fig. 9.18. The metal may be AlFeSi with M_{sat} equal to 12,000 gauss, and the improved performance of these heads is evident from Fig. 10.32 (Jeffers et al. 1982). MIG heads were analyzed by Ruigrok (1984).

Also, the efficiency of a reproduce head can be improved by using a metal gap spacer, which at high frequencies increases the front gap reluctance due to eddy currents. Improvements of several dB have been reported (McKnight 1979).

Eddy currents may also play a role in the MIG head, but the author has found no references treating this aspect. A pseudo-effect appears to cause small pre- and post-pulses in the read response. They are caused by the magnetic gaps formed between the metal pole pieces and the Beilby layer in the ferrite pole pieces. Ion beam milling (see Chapter 13) of the ferrite pole pieces prior to the metal sputtering is required to clean the ferrite.

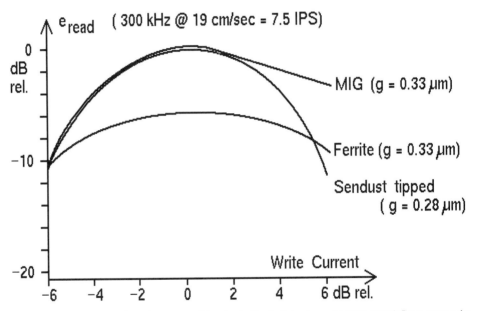

FIGURE 10.32 The comparative write response of three heads. The ferrite head cannot produce enough flux to saturate the tape (maximum output occurs at −6 dB). The MIG head improves the recorded level, but cannot overwrite adequately. A Sendust-tipped head gives maximum output and has the characteristic parabola shape (*after Jeffers, 1982*).

SATURATION IN THIN-FILM HEADS

The gap region in a thin film head is shown in Fig. 10.33 for three angles of the throat section that conducts the flux down to the gap. Notice that the thickness of this deposited (or plated) section is inverse proportional to the angle. So the steep throat denoted a carries a high flux density compared to case c (Tagami and Nishimoto 1980). The selected slope is a compromise between saturation and efficiency.

A typical design is shown in Fig. 10.34. During writing the throat is saturated, and the flux flow limited. The gap field H_g is proportional to the flux density B_g which in turn is inversely proportional to the area available for the flux, i.e. the track width multiplied by the throat height. A certain gap field must be reached for adequate overwrite and proper write performance, and the throat height must be carefully reduced to achieve this. The variation in H_g versus throat height is shown.

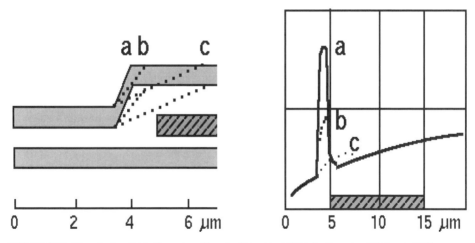

FIGURE 10.33 Magnetization distribution varies with angle of throat in a TFH (*after Tagami and Nishimoto, 1980*).

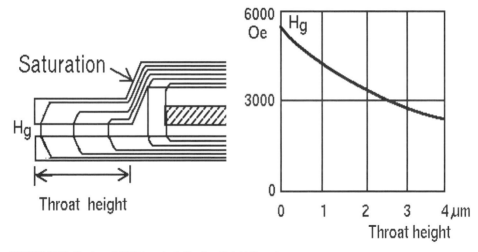

FIGURE 10.34 Front gap field H_g is controlled by throat height dimension.

BACK GAP BENEFITS

The low-frequency efficiency of a magnetic head is highest when the back gap reluctance has a zero value. This is achieved by using a back gap with a large area and a short gap length. There are nevertheless a couple of good reasons for designing a long back gap into magnetic heads: In audio record heads, the effects of nonlinearities of the permeability are rendered "small," while the heads become less prone to become permanently magnetized. These topics are discussed next.

DISTORTION IN INSTRUMENTATION AND AUDIO HEADS

The magnetic reluctance of a core is inversely proportional to the permeability μ_r, which depends on the strength of the magnetizing field H or flux density B. The relationship between μ_r and H, and also

μ_r and B, was shown in Fig. 10.17 for a typical metallic core material. The permeability increases with H (or B) up to a maximum value μ_{rmax}, whereafter it reduces dramatically as B approaches saturation.

This change in μ_r will change the values of the read efficiency η and therefore also the write current requirements. The increase in μ_r with flux density increases write efficiency; this means that the peak levels of a sine wave drive current will produce too-high flux levels (peaking), and the recording will contain third harmonic distortion components. (NOTE: This record-head distortion is not the third harmonic distortion measured on ac-biased recorders; it may be part of it, but should not be, with a properly designed record head.)

Attempts have been made to calculate this distortion using Bozorth's results for changes in μ_r at low field strengths (Rayleigh loops). That approach is limited by:

- Approximations
- Different induction levels exist in the various parts of the core ($B = \phi/area$, and the core area varies from tip to back gap). Eddy currents (skin depth) will affect the flux densities across the core cross sections.
- Superposition of data and bias signals in an ac-bias recording is difficult to handle.

A reasonable approach is instead to evaluate what percentage change occurs in the record-current sensitivity when the permeability changes, say, 10 percent. The result of a series of calculations is shown in Fig. 10.35, and illustrates that a large back gap is beneficial in reducing the effect of change in μ_r. (This technique is used in audio record heads.)

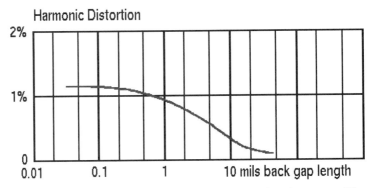

FIGURE 10.35 Percent change in record efficiency for a 10-percent change in core permeability, as a function of the length of the back gap in an instrumentation head.

PERMANENT MAGNETIZATION IN CORES

The core halves in a magnetic head form a magnetic circuit that is capable of maintaining a remanent magnetization M_{rs}'. If no gaps were present, the level of remanence could equal the maximum remanence for the core material, M_{rsat}.

The presence of even very small air gaps causes demagnetization, which lowers the level of remanence. In Chapter 4, example 4.1, we derived an expression for the maximum remanence in a magnetic head core with air gaps (from Fig. 4.8):

$$M_{rs}' = H_r/N$$

where $N = (L_{fg} + L_{bg})/(Length\ of\ core\ plus\ gaps)$. When no back gap is present (highest efficiency), the demagnetization is small and current pulses (or turn-off current after saturation recording) may leave the core in a highly magnetized state. It may be high enough to cause partial erasure of a recorded track moving past the gap.

If the ac current in biased recording is turned off properly (slow decay), then the head will automatically degauss. Reproduce heads have shorter gaps than record heads, and are therefore more prone to become permanently magnetized.

A long back gap in any head has its price: reduced efficiency. This is shown in Fig. 10.36 for the write current requirements. A short back gap requires less current at low frequencies, but a larger amount of boost at high frequencies. A long back gap can just as well be used, requiring a current registering only 3 dB, because the write amplifier has to deliver the higher current at high frequencies anyhow. The drop in read level for long back gaps may not be acceptable because the midband signal-to-noise ratio suffers thereby (Fig. 10.37).

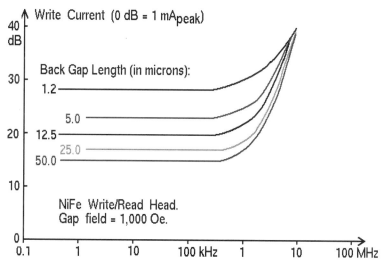

FIGURE 10.36 Typical write current levels as function of frequency, for five different values of the back gap length.

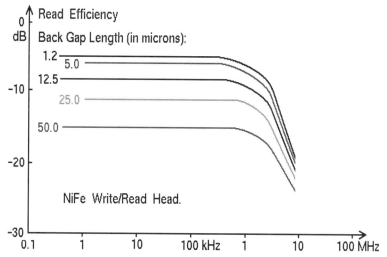

FIGURE 10.37 Efficiencies plotted as function of frequency, for five different values of the back gap length.

EFFECTS OF CORE GEOMETRY

The wavelength resolution in magnetic recording is bounded by two mechanical dimensions: The length of the gap in the core will limit the short wavelength resolution while the overall core length will limit the long wavelength output.

The interactions between a head core and adjacent cores and recorded tracks are covered later in this chapter under Multitrack Heads.

EFFECT OF OVERALL CORE SIZE

Low-frequency recordings result in wavelengths that are comparable in size to the overall length of the head interface. This will result in a higher flux density through the body of the core (and hence the winding), and the net effect is an increase in the head's efficiency at low frequencies. (Neither this phenomenon, nor the undulations listed later, occur in record heads.) The equivalent diagram of the head core gradually changes for increasing wavelengths as shown in Fig. 10.38. The core reluctance is divided into 3 portions for clarity. The two "sliders" on the core tip reluctances move outwards with increasing wavelengths, hence increasing the efficiency.

This increase in output at long wavelengths is called secondary-gap effect (Fritzsch 1968).

Another irregularity, undulations, occur in the same long-wavelength range. This is most easily explained by examination of the flux through the entire core structure, as shown for $\lambda = L$ and $\lambda = L/2$ in Fig. 10.39, where L = the length of the head interface. The flux through the cores from the half wavelength magnets "hanging" over the edge of the core are in phase with the main flux when $\lambda = L$, but out-of-phase when $\lambda = L/2$.

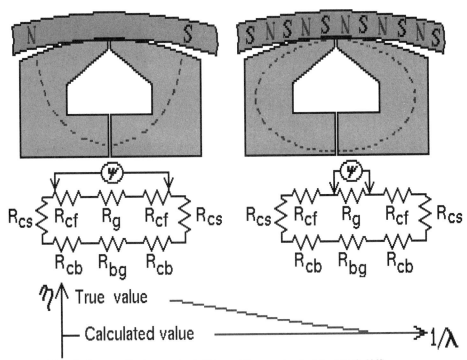

FIGURE 10.38   Head contour effect increases head efficiency at long wavelengths (*after Fritsch, 1968*).

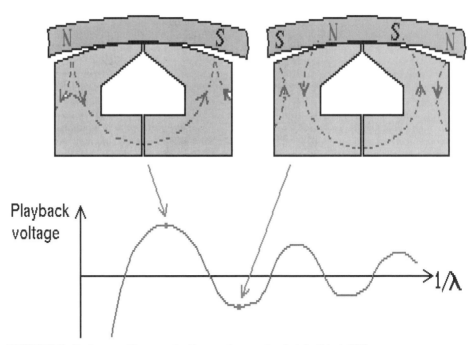

FIGURE 10.39 Head contour effect creates head bumps at long wavelengths (*after Fritsch, 1968*).

The total effect is a response function (called the Spiegelfunktion) that is a superposition of the secondary gap effect and the undulation. (A slang term is "head bumps.") The latter are lessened by avoiding tape-head contact at the corners of the core, shown in Fig. 10.40a. The undulations are modified if the core is inside a shield can, which provides a flux return path for the overhanging magnets (Fig. 10.40b). They can also be suppressed by using asymmetrical core halves (Fig. 10.40c). Some computer write/read heads are of this construction, and may exhibit phase distortion because the reproduce system is asymmetrical.

The undulation effects are so different from one head design to another (including the effect of trackwidth) that no general formula can be given for their magnitudes.

FINITE WIDTH OF CORE (TRACK WIDTH)

A cylindrical head of infinite width was the base of Westmijze's studies. If the width is reduced toward zero then the core becomes a round disk, and this is a better approximation to the real world of narrow-track heads.

Geurst (1965) investigated the behavior of such a narrow head. He found that there were much stronger fluctuations for the disk head, and that wavelengths as small as one-tenth of the diameter were still detectable. The two heads (a long cylinder and a short disk) are shown in Fig. 10.41. The associated head bumps are shown for a wrap angle of $2 \times 5°$ (solid lines) and $0°$ (broken lines).

There is obviously a higher degree of wavelength interference in the disk-shaped head. The patterns of the fluctuations reflect the sensing function H_s (see Chapter 6). Because the field strengths at sharp edges are quite high, strong fluctuations in H_s can also be expected, such as at the edges of the

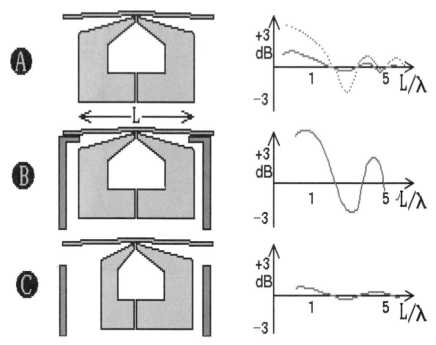

FIGURE 10.40   Head contours are affected by the presence of a shield and associated geometries (*after Baker, 1977*).

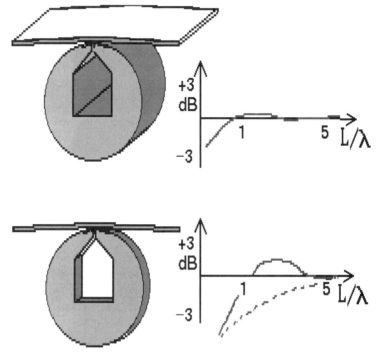

FIGURE 10.41   Head contours are affected by trackwidth (*after Geurst, 1965*).

disk head and at the edges of the core where the tape enters or leaves, and also at the corners of shielding housing (Fig. 10.40b) (Baker 1977 and Lindholm 1980).

The finite pole lengths in thin-film heads act as spatial equalizers, equivalent in effect to a pulse-slimming equalizer. Asymmetry of the pole lengths (film thicknesses) can be used to compensate the phase shift in the media magnetization (Kakehi et al. 1982, Aoi et al. 1982, and Singh and Bischoff 1985). Modern mathematical analysis has been made by Menuhin (1990). Figures 8.14 through 8.17 illustrate the effects of the pole geometries.

SIDE-READING EFFECTS (CROSSTALK)

The sensing field exists in the entire space around the read gap, as shown in Figs. 8.1 and 8.23. The question is: "What does the sensing field see from adjacent tracks on the tape or disk?". Any signals picked up in this way are named Crosstalk, and are part of the overall recorder system noise. A portion of this noise is residual old data from incomplete overwrite (Fig. 10-42).

Eldridge and Baaba (1962) investigated this Crosstalk and found that it behaved quite like the well-known Wallace spacing loss when a head is spaced away from a media surface. They found a Crosstalk figure of $65 \times d/\lambda$ dB, where d is the lateral distance from the side of the head side to an adjacent track.

Luitjens and van Herk (1982) calculated the Crosstalk using a three-dimensional field. Their results are comparable and are shown in Fig. 10.43. Two results can be deduced from these graphs: 1) There exists a registration loss when the tracking between the reproduce head and the recorded track varies. This may occur when recording and playback is done on two different drives, and temperature/humidity conditions alter the tracking. 2) Another loss is the read Crosstalk from an adjacent track (when center-of-head to center-of-track distance is greater than the track width). The Crosstalk (equal to the reduced output from an adjacent track due to spacing d) was shown in Fig. 8.25.

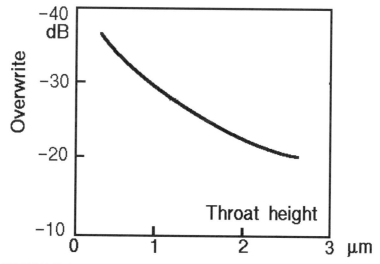

FIGURE 10.42 Overwrite is inversely proportional to throat height.

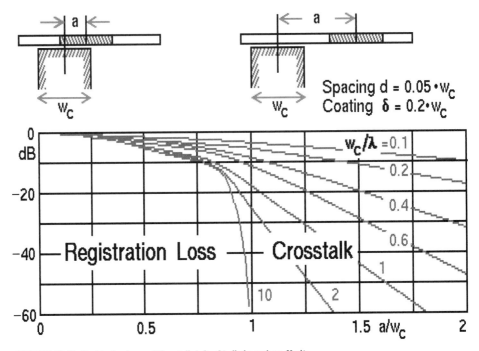

FIGURE 10.43 Registration loss and Crosstalk (*after Lindholm and van Herk*).

Another result was derived by van Herk for the increased output a playback head will produce when it reads a track that is wider than the core. This side-fringing response is strongest at long wavelengths (low frequencies) and must be considered when a multitrack tape reproducer is calibrated with full-track standard tapes (Melis and Nijholt 1978).

EFFECTS OF COATING PERMEABILITY

The classical expression for the loss of signal that increases with decreasing wavelength is:

$$Loss = \sin(X)/X$$

where $X = \pi g/\lambda$

When the wavelength λ equals the length g, the output is zero.

GAP LENGTH

The length g is not equal to the mechanical gap L_{fg} of the front gap in the reproduce head. The $\sin(X)/X$ formula was first derived for optical (motion picture) playback, where g was the slit width of the optical scanning head. In late 1940, Daniel and Axon found that the measured gap-length response did not agree with the gap loss formula. The magnetic gap length appeared to be 1.15 times the mechanical length.

In 1953, Westmijze published his now-famous study of the magnetic recording process, and verified that the semi-infinite head has an "effective" gap length that is about 10 percent longer than the mechanical gap. The tape permeability was assumed equal to one.

Siakkou (1969, 1974) investigated the effect of tapes with isotropic permeability and found that g in our previous formula should be substituted with

$$g = L_{fg} \times \sqrt{(1.5\mu_{rt} + 1)/(\mu_{rt} + 1)} \tag{10.14}$$

where μ_{rt} is the relative permeability of the tape. When $\mu_{rt} = 1$, $g = 1.12 \times L_{fg}$, which agrees with Westmijze's work. For $\mu_{rt} = 2$, we find $g = 1.15 \times L_{fg}$, and at the limit (for μ_{rt} very large) we find $g = 1.22 \times L_{fg}$.

SPACING LOSS AND COATING THICKNESS LOSS

The general playback theory assumes that the tape or disk coating has a relative permeability equal to one. If it is greater than one, then the sensing function $H_s(x,y)$ in our previous formula (see Chapter 6) will be different from the sensing field derived from Karlqvist's expressions. The resulting flux ϕ through the playback core will thus be modified and the result of the reciprocity calculations in Chapter 6 will likewise be modified. Westmijze's evaluation was based on a tape coating with uniform permeability in all directions (isotropic). He concluded that the reproduce flux was reduced at short wavelengths for μ_r. Siakkou arrived at essentially the same result: The mathematical derivation led to a modification of the factor δ_{rec} in the formula in Fig. 6.11, namely that δ_{rec} should be multiplied with μ_{rt}.

The result thereof is a coating thickness loss that starts at a frequency that is lowered by a factor of μ_{rt}; Fig. 10.44 illustrates the phenomenon. The sensing lines (corresponding to field lines) sensing the magnetization in the back of the coating "think" that the magnetization is further away, i.e., the coating is thicker.

Bertram (1978) considered a coating with anisotropic permeability; μ_{rx} in the direction of tape motion and μ_{ry} perpendicular to it. The numerical values for a well-oriented tape are typically $\mu_{rx} = 1$ to 2, and $\mu_{ry} = 3$ to 5; their values are determined as the tangent to the hysteresis loop where it crosses the M-axis, at M_{rsat}, minus one.

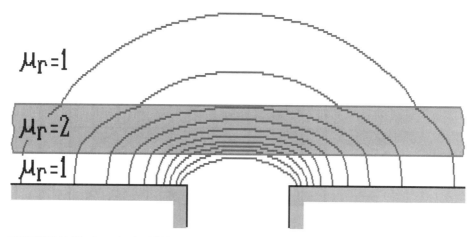

FIGURE 10.44 The change in a head field pattern due to media permeability μ_r = two.

Bertram concluded that the anisotropic permeability reduces the coating thickness loss with 3–5 dB in ac-biased recordings, and about 2 dB in unbiased recording (digital), for a media with $\mu_{ry}/\mu_{rx} = 4$. He also found that the multiplier 55.4 in the spacing loss formula should be 110, or double the traditional Wallace loss, for small values of the spacing (note slope in Fig. 8.25).

The increased output due to $\sqrt{\mu_{ry}/\mu_{rx}}$ and the increased loss due to $\sqrt{\mu_{ry} \times \mu_{rx}}$ may offset each other; they are difficult to observe and separate in actual measurements using modern tapes with high anisotropy and small spacings due to surface finish.

CROSSTALK IN MULTITRACK HEADS AND TRANSFORMER COUPLING BETWEEN CORES

Several head cores placed close to each other will interfere by transformer and capacitive coupling (Fig. 10.45). This form of crosstalk can be greatly reduced by insertion of shields (Mu-metal-Copper-Mu-metal) between the cores. It is essential that full shields are used. If they are split then the gap line between them may not only behave as a secondary recording gap, but the shielding efficiency is lowered. The secondary recording may be on top on an "in-between track" when staggered head assemblies are used. The level of crosstalk can be calculated from an analysis of the stray flux patterns between the structures shown in Fig. 10.45 (Tanaka 1984).

Thin-film heads do not always offer an easy way of inserting shielding; reduction of crosstalk has been accomplished by insertion of feedback from a record track to the disturbed read track. Multitrack TFHs are discussed in a paper by Heijman et al. (1988).

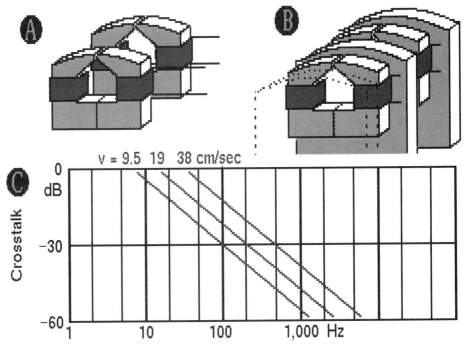

FIGURE 10.45   Crosstalk in magnetic heads. (a) Transformer coupling between channels via stray flux and capacitive coupling. (b) Reduction of crosstalk by shielding. (c) Typical crosstalk versus frequency for adjacent tracks in an IRIG head assembly.

Crosstalk during recording results in permanent noise on recorded tracks. Read-while-Write crosstalk is a noise source that can be reduced by individual shielding of the record and playback head assemblies. The placement of a ferrite block immediately in front of the record gaps will reduce this crosstalk.

SIDE-READING EFFECTS

During playback of interleaved recordings, the fringing field of the playback gap picks up flux from adjacent channels (in addition to transformer coupling in the head assembly itself). This effect is particularly pronounced at long wavelengths. It is good practice to make the width of the shield between tracks equal (or greater) than the width of the tracks that ride over them.

Figure 10.45 serves as a yardstick for predicting crosstalk. The sloped lines are smooth approximations of measured values of crosstalk in a 31-track system, using direct electronics for adjacent tracks with a separation of 0.80 mm (.030") center-to-center. They point out crosstalk as a problem when low frequencies are recorded and played back at high tape speeds.

Crosstalk can be reduced by using FM record techniques (see Chapter 27), and/or using a slow tape speed. Critical data may also be separated by recording them on tracks that are not located immediately next to each other. The use of ac bias reduces the required data currents, and therefore also reduces the read-while-writing crossfeed (assuming the higher-frequency ac-bias signal is filtered away).

ERASE HEADS

The construction and performance of erase heads were covered in Chapter 8 (see Figs. 8.26 through 8.29). The text also treated the problem in writing new data on top of old data, i.e., supposedly erasing the old data with the field from the new write process. An audio erase head design is discussed by Sawada and Yoneda (1988).

Just how well this is done is measured by the overwrite ratio, which is defined as the ratio of an original $1f$ (or half-band-edge signal) to that remaining after being overwritten by a $2f$ (band-edge signal). Overwrite of 30 dB is considered adequate for proper performance (Wachenschwanz and Jeffers 1985.)

Measurements of overwrite require knowledge of the original read level of the $1f$ signal. A better method appears to be the overwrite method that specifies the ratio of residual $1f$ signal to new $2f$ signal. A recommended method is to measure the residual spectral noise level to the $1f$ or $2f$ levels, in conjunction with the signal's resolution, i.e., the ratio of the $2f$ read voltage to the $1f$ read voltage; it should be no more than 80 percent (2 dB).

CORE LOSSES

Two losses account for the power that is required to drive an erase head (and similarly, by smaller amounts, write heads for high-coercivity media). These are eddy currents and hysteresis losses; the total power loss is:

$$
\begin{aligned}
P_{\text{total}} &= P_{\text{eddy}} + P_{\text{hysteresis}} \\
&= kf^2 + W_h f \text{ watts/m}^3
\end{aligned}
\tag{10.15}
$$

where:

$\quad W_h$ = area of hysteresis loop in watt $\times$ sec/m^3
$\quad$ k = material constant
$\quad f$ = frequency in Hz

The eddy currents, kf^2, can also be written:

$$P_{eddy} = 10^{-4}\pi^2 d^2 B_o^2 f^2 / 6\sigma \text{ watts/m}^3 \tag{10.16}$$

where:
   d = lamination thickness in cm
   B_o = flux density amplitude in gauss
   f = frequency in Hz
   σ = resistivity in $\mu\Omega$-cm

Example 10.3:   A MuMetal erase core operates at 50 kHz at a flux density level of 5000 gauss. The core is made from 2-mils lamination; the total core volume is 1 cm^3. Find P_{total}.

Answer: From Table 7.1 we find the resistivity σ = 25 $\mu\Omega$-cm and H_c = 0.02 Oe = 1.6 A/m. The flux density level is 5000 gauss = 0.5 Wb/m^2, so the total area of the rectangular BH-loop is

$$W_h = 1.6 \times 0.5 \times 4 = 3.2 \text{ Ws/m}^3$$

The lamination thickness is d = 0.0051 cm. The eddy-current loss calculates to

$$P_{eddy} = 10.4 \times 10^6 \text{ watts/m}^3 \times 10^{-6} \text{ m}^3$$
$$= 10.4 \text{ watts}$$

and the hysteresis loss is

$$P_{hysteresis} = 3.2 \times 5 \times 10^4 \times 10^{-6}$$
$$= 1.6 \text{ watts}$$

The total power loss in the MuMetal erase core is 12 watts.

The losses can be reduced significantly by using a NiZn ferrite core. The head core losses cause a temperature rise in the core, and the material should therefore be reasonably well-connected to a heat sink through a heat-conducting epoxy (or direct contact to metal parts, if possible). Ferrite materials must also have a reasonably high Curie temperature, and the design must not have flux levels that exceed the M_{sat} levels. The heating of erase heads caused designers of early audio equipment to incorporate an interlock so the heads were inoperative unless the tape was moving and thus cooling the heads!

An analysis as shown in the example above should be carried out for most write heads that work at high write flux levels (Monson et al. 1984). A final note on the circuitry to drive erase heads: The current waveform must be absolutely symmetrical, i.e., free from even harmonic signals and dc currents. Otherwise the result is erased tapes with high residual noise levels. Push-pull circuits are recommended. The basic oscillator may operate with the erase head in a tank circuit in its oscillator, while a frequency doubler delivers a synchronous bias signal to the record-head bias drivers. It is not advisable to use separate erase and bias oscillators without some form of synchronization; otherwise, the result may be a set of recorded beat signals between the two oscillators.

RADIO FREQUENCY INTERFERENCE FROM WRITE HEADS

The write current signal contains VHF signals that can radiate from the head; the head cores are dipole antennas, although so small that their efficiency as antennas is very poor. If the radiation levels are troublesome, shielding of the heads may become necessary (see Fig. 8.33).

Shielding of radio frequencies can be successful if an aluminum housing is used around the head structure, and by using a head cable that is a two-conductor overall shield, properly grounded (i.e., ground the shield in one or both ends; see Morrison's book for details).

A multichannel recorder with ac bias can successfully suppress radiation by alternating the phase of the bias signals to a row of write cores.

SUSCEPTIBILITY TO EXTERNAL FIELDS

Shielding of heads for attenuation of outside fields may also be necessary. The earth field itself has caused many problems by linking through the write head and appearing as an unwanted dc level mixed in with the write field. The results were (and are) noisy and distorted recordings. A particularly grave situation may occur when heads are demagnetized with an ac degausser; the result has often been a permed (dc-magnetized) head, with subsequent noise and distortion problems.

Another example is a spacecraft recorder placed in the vicinity of three orthogonal (perdendicular to one another) electromagnets that has been proposed used for space craft attitude control. Applying power to one or more of the magnets will bring about a torque, in relation to the Earth's field, and thus slowly turn the spacecraft. The power is generated from solar panels and is therefore not dispensed as ordinary propellant fuel for attitude boosters.

Figure 10.46 shows the concept, and Fig. 10.47 illustrates how an external field can link through a recorder and be collected by the heads, with the ill results reported above. This was modelled in a large coil (1 meter in diameter), and the results revealed the following field strength limits versus performance impairment (Jorgensen 1975):

- $H > 5$ Oersteds: Write function affected
- $H > 25$ Oersteds: Loss of write function
- $H > 30$ Oersteds: Loss of read while write
- $H > 35$ Oersteds: Total failure, any mode

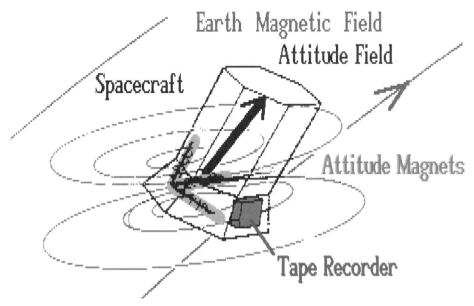

FIGURE 10.46 Magnetic attitude correction for spacecraft may affect head operation in an onboard tape recorder (see text).

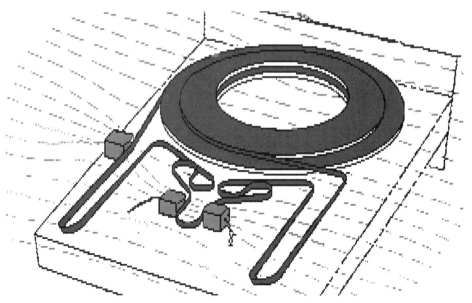

FIGURE 10.47   Flux linkage from external fields through a magnetic tape unit.

The 5-Oersted limitation is the same value as determined by Lopez and Stromsta (1981). This presents a major design problem in shielding the heads from magnets in permanent-magnet position motors in disk drives. The actuator may have a voice-coil motor with a field of several thousand Oe, located one to two inches away from the disk edge; the field from the motor must at that point be less than 5 Oersteds (= 400 A/m). Some thin-film heads do not tolerate more than 3 Oe of ambient field strength.

The new horizontal thin-film heads, as shown in Figs. 9.37 and 10.56, are shielded by way of their construction (Springer and Jorgensen 1985; Lazzari 1993).

VARIATIONS IN PERMEABILITY (PRODUCTION TOLERANCES)

It was mentioned in Chapter 7, under Metallic Core Materials, that coldworking of the core material during head fabrication would significantly lower the relative permeability. The effect thereof on the head's efficiency is shown in Figs. 10.48 through 10.50.

The read efficiency at low frequencies will become lower as the permeability drops in value. This is not necessarily the case at high frequencies, because the eddy currents are reduced when the permeability decreases (formulas 7.1 and 7.3). The change in permeability spectra is shown in Fig. 10.48, and is reflected in the efficiency curves in Fig. 10.49.

The write-current requirements increase when the permeability decreases; see Fig. 10.50.

CHANGES IN HEAD PARAMETERS WITH WEAR

The front gap reluctance in heads increases when the heads wear. The read efficiency increases with wear, in particular at high frequencies. This is an interesting situation, where an item gets better as it wears out. An operator should be alert to the subtle improvement in a recorder's amplitude versus frequency response; it may be just before the head wears out. When it does, the performance will nosedive. A write head will, in general, require less and less drive current as it wears out.

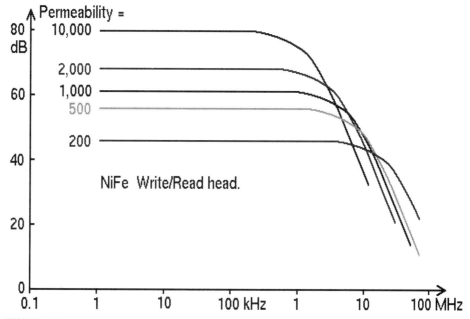

FIGURE 10.48 Variation in relative permeability versus frequency.

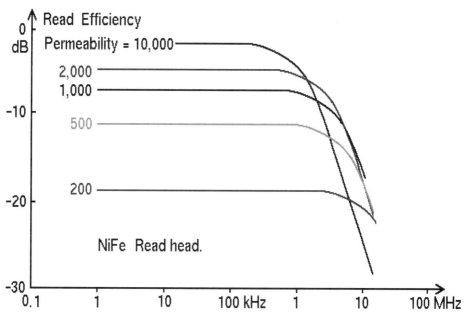

FIGURE 10.49 Variation in head efficiency versus frequency, for five different permeabilities. Low permeability assures extended response.

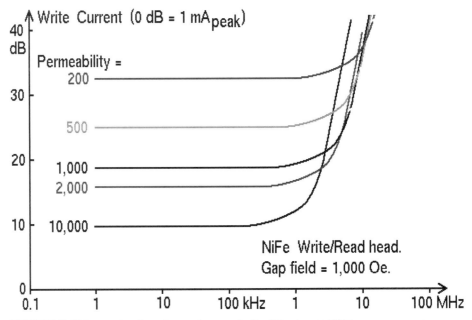

FIGURE 10.50 Write current requirements versus frequency, for five different permeabilities.

TOWARD HIGHER DATA RATES AND STORAGE DENSITIES

Higher data rates will demand better high-frequency performance of heads (Wood et al. 1990). Thin-film heads are no exceptions. They exhibit eddy-current losses as shown in Fig. 10.51 (Re et al. 1985). The flux flows on the inside of the film structure, in a thickness that becomes smaller as the frequency increases.

The result is a phase shift in the field just outside the pole pieces, as shown in Fig. 10.52 (Re et al. 1985). Recent investigation shows the phase shift to increase with increasing frequency, ranging from 30° to 70° at 100 MHz (Shi and Kryder 1993).

One solution lies in making heads with cores from very thin laminations (Makino et al. 1993; Cheng 1993). The latter are fabricated by sputtering methods, which also allow for materials with several elements (such as NiFeN) that also have a very high B_{sat}.

The flying height of the disk head assemblies has steadily decreased, and is now really called the glide height. The evolution is shown in Fig. 10.53. And with the reduced spacing follows higher densities, as shown in Fig. 10.54 (Williams, 1993).

New heads will be innovated. Recently three such candidates have appeared. They are:

1. A single-crystal ferrite MIG head that is bonded to the side of a slider, as shown in Fig. 10.55 (Sano et al. 1993). This technique may also be applied to a thin laminated core structure.

2. A horizontal head has been under developments for the past decade, and one version is shown in Fig. 10.56 (Lazzari 1993).

3. And lastly there is an intriguing head called the Diamond head, in which the flux loops twice through the coil, providing higher output without additional turns; see Fig. 10.57 (Mallary and Ramaswamy 1993).

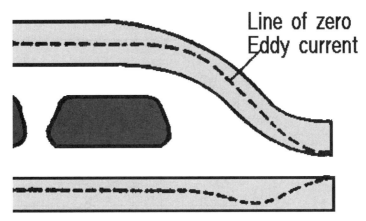

FIGURE 10.51 Eddy currents are not produced symmetrically near the gap in a TFH due to the field from the winding near the gap (*after Re at al., 1985*).

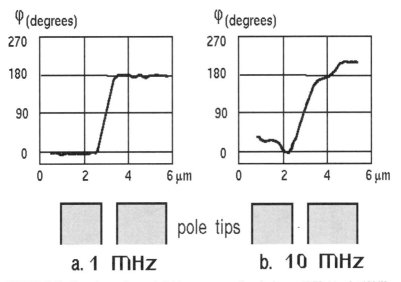

FIGURE 10.52 Phase change of magnetic field response across the pole pieces at 1 MHz (a) and at 10 MHz (b) (*after Re et al., 1985*).

REFERENCES

GENERAL INTEREST

Grossner, N.R. 1983. *Transformers for Electronic Circuits*. New York. McGraw-Hill, 467 pages.

Ruigrok, J.J.M. 1990. *Short-Wavelength Magnetic Recording*. Elsevier Advanced Technology, Oxford, UK. 564 pages.

Zinke, O., and H. Seither. 1982. *Widerstaende, Kondensatoren, Spulen und ihre Werkstoffe*. Berlin. Springer Verlag, 350 pages.

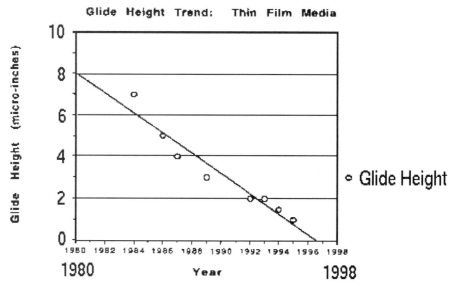

FIGURE 10.53   The recent trend is toward in-contact recording in disk drives (*after Chen, 1993 Head/Media Conference, Las Vegas, NV*).

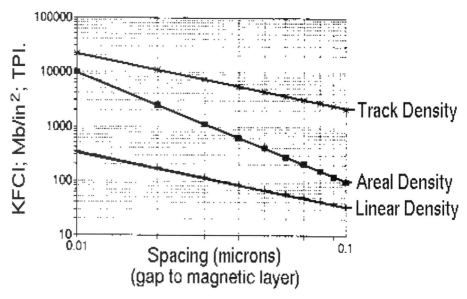

FIGURE 10.54   Storage densities versus spacing (*after Williams, 1993*).

TRACK WIDTH

Bonyhard, P.I., and J.K. Lee. Sept. 1990. Magnetoresistive Read Magnetic Recording Head Offtrack Performance Assessment. *IEEE Trans. Magn.* MAG-26 (5): 2448–50.

Brock, G.W. May 1983. Instability of Flexible Magnetic Recording Media. *Symposium on Magn. Media Manuf.(MMIS).* paper no. A-2, 22 pages.

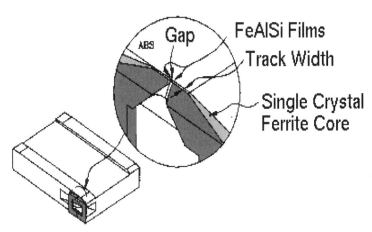

FIGURE 10.55   A slider with a side-core (*after Sano et al., 1993*)

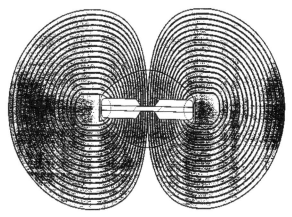

FIGURE 10.56   The coil in a planar thin film head (*after Lazzari, 1993*)

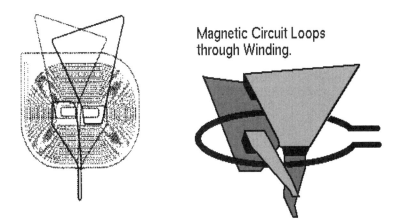

FIGURE 10.57   A diamond-head doubles the flux through the winding (*after Mallary and Ramaswamy, 1993*).

Cain, W.C., D.K. Thayamballi, and Vea, M.P. March 1994. The Effects of Pole Geometry on Recording Performance of Narrow Track Film Heads. *IEEE Trans. Magn.* MAG-30 (2): 275–280.

Edelman, H., and M. Covault. Sept. 1985. Design of Magnetic Recording Heads for High Track Densities. *IEEE Trans. Magn.* MAG-21 (6): 2583–2587.

Greenberg, H.J., R.L. Stephens, and F. E. Talke. Nov. 1977. Dimensional Stability of Floppy Disks. *IEEE Trans. Magn.* MAG-13 (6): 1397–1399.

Lee, J.K., and P.I. Bonyhard. Sept. 1990. A Track Density Model for Magnetoresistive Heads Considering Erase Bands. *IEEE Trans. Magn.* MAG-26 (5): 2475–77.

Lin, T., J.A. Christner, T.B. Mitchell, J-S. Gau, and P.K. George. Jan. 1989. Effects of Current and Frequency on Write, Read, and Erase Widths for Thin-Film Inductive and Magnetoresistive Heads. *IEEE Trans. Magn.* MAG-25 (1): 710–715.

ELECTROMAGNETIC DESIGN AND EFFICIENCY

Atoh, S., M. Tagagishi, and T. Hiroshima. Nov. 1990. Study of the Relation Between Yoke Shape and Performance of Thin-Film Heads. *IEEE Trans. Magn.* MAG-26 (5): 1674–76.

Ciureanu, P.C. Jan 1986. Efficiency and Inductance Analysis of a Three-Region Thin-Film Head. *IEEE Trans. Magn.* MAG-22 (1): 26–32.

Heim, D.E. Feb. 1986. Flux Propagation in Thin-Film Magnetic Structures. *J. Appl. Phys.* 59 (3): 864–872.

Jones Jr., R.E. Sept. 1978. Analysis of the Efficiency and Inductance of Multiturn Thin-Film Magnetic Recording Heads", *IEEE Trans. Magn.,* Vol. MAG-14, No. 5, pp. 509–11.

Jorgensen, F., *Head/Media Performance Design/Evaluation program*, Vers. 2, Danvik, 1988.

Katz, E. Nov. 1980. Numerical Analysis of Ferrite Recording Heads with Complex Permeability. *IEEE Trans. Magn.* MAG-16 (6): 1404–1409.

Katz, E.R. Sept. 1978. Finite Element Analysis of the Vertical Multi-Turn Thin-Film Head. *IEEE Trans. Magn.* MAG-14 (5): 506–08.

McKnight, J. March 1979. Magnetic Design of Tape Recorder Heads. *Journal Audio Engineering Society.* 27 (3): 106–120.

Miura, Y., S. Kawakami, and S. Sakai. Sept. 1978. An Analysis of the Write Performance on Thin Film Head. *IEEE Trans. Magn.* MAG-14 (5): 512–514.

Paton, A. Dec. 1971. Analysis of the Efficiency of Thin-Film Magnetic Heads. *Jour. Appl. Phys.* 42 (13): 5868–70.

Sansom, D. Sept. 1976. Recording Head Design Calculations. *IEEE Trans. Magn.* MAG-12 (3): pp. 230–233.

Schelor, J.R. 1986. *A Reluctance Model for Thin Film Magnetic Recording Heads.* A Thesis—UCSB (Univ. of Calif., Santa Barbara), 49 pages.

Visser, E.G., and L.R.M. Van Rijn, J.J.F. Maas. July 1985. An Improved Measurement of the Absolute Efficiency of Magnetic Heads by Saturating the Gap Field. *IEEE Trans. Magn.* MAG-21 (4): 1283–1288.

Wood, R., D. Lindholm, and R. Haag. Sept. 1985. On the Bandwidth of Magnetic Record/Reproduce Heads. *IEEE Trans. Magn.* MAG-21 (5): 1566–1568.

Yeh, N. Jan. 1982. Analysis of Thin Film Heads with a Generalized Transmission Line Model. *IEEE Trans. Magn.* MAG-18 (1): 233–237.

HEAD NOISE

Bittel, H. Sept. 1969. Noise of Ferromagnetic Materials. *IEEE Trans. Magn.* MAG-5 (3): 359–365.

Byers, R.A. June 1971. Theoretical Signal-to-Noise Ratio for Magnetic Tape Heads. *IEEE Trans. Magn.* MAG-7 (2): 254–259.

Klaassen, K.B., and J.C.L. van Peppen. April 1991. Field-driven domain-wall jumps in thin-film Heads. *Jour. Appl. Phys.* 69 (8): 5417–19.

Klaassen, K.B., and J.C.L. van Peppen. Sept. 1990. Barkhausen Noise in Thin-Film Recording Heads. *IEEE Trans. Magn.* MAG-26 (5): 1697–99.

Klaassen, K.B., and J.C.L. van Peppen. Sept. 1992. Noise in Thin-Film Inductive Heads. *IEEE Trans. Magn.* MAG-28 (5): 2097–99.

Klaassen, K.B., and J.C.L. van Peppen. Sept. 1992. Ferrite Head Instability. *IEEE Trans. Magn.* MAG-28 (5): 2641–43.

Liu, F.H., P. Ryan, X. Shi and M.H. Kryder. Sept. 1992. Correlation between Noise-After-Write and Magnetization Dynamics in Thin Film Heads. *IEEE Trans. Magn.* MAG-28 (5): 2100–02.

Monson, J.E., K.P. Ash, R.E. Jones Jr., and D.E. Heim. Sept. 1984. Self-Heating Effects in Thin-Film Heads. *IEEE Trans. Magn.* MAG-20 (5): 845–47.

Schäfer, R., B.E. Argyle, and P.L. Trouilloud. Sept. 1992. Domain Studies in Single-Crystal Ferrite MIG Heads with Image-Enhanced Wide-Field Kerr Microscopy. *IEEE Trans. Magn.* MAG-28 (5): 2644–46.

Takayama, S., K. Sueoka, H. Setoh, R. Schäfer, B.E. Argyle, and P.L. Trouilloud. Sept. 1992. A Study of MIG Head Readout Waveform Asymmetry, Using Magnetic Force and Kerr Microscopy. *IEEE Trans. Magn.* MAG-28 (5): 2647–49.

Watanabe, H. Sept. 1974. Noise Analysis of Ferrite Head in Audio Tape Recording. *IEEE Trans. Magn.* MAG-10 (3): 903–906.

POLE TIP SATURATION AND MIG

Bertram, N.H., and C.W. Steele. Nov. 1976. Pole Tip Saturation in Magnetic Recording Heads. *IEEE Trans. Magn.* MAG-12 (6): 702–706.

Fischer, J., and H. Moser. 1956. Die Nachbieldung von Magnetisieringskurven durch einfache algebraische oder Tranzendente Funktionen. *Archiv für Elektrotechnik* Vol. 42 pp. 286–299.

Fujiwara, T. May 1979. Record Head Saturation in AC Bias Recording. *IEEE Trans. Magn.* MAG-15 (5): 1046–49.

Jeffers, F., R.J. McClure, W.W. French, and N.J. Griffith. Nov. 1982. Metal-In Gap Record Head. *IEEE Trans. Magn.* MAG-18 (6): 1146–1148.

McKnight, J.G.. March 1979. How the Magnetic Characteristics of a Magnetic Tape Head are Affected by Gap Length and a Conductive Spacer. *Journal Audio Engineering Society.* 27 (3): 106–120.

Ruigrok, J.J.M. Sept. 1984. Analysis of Metal-In-Gap Heads. *IEEE Trans. Magn.* MAG-20 (5): 872–874.

Shibaya, H., and I. Fukuda. Sept. 1977. The Effect of the B_s of Recording Head Cores on the Magnetization of High Coercivity Media. *IEEE Trans. Magn.* MAG-13 (3): 1005–1008.

Suzuki, T., and S. Iwasaki Sept. 1972. An Analysis of Magnetic Recording Head Fields Using Vector Potential. *IEEE Trans. Magn.* MAG-8 (3): 536–537.

Szczech, T.J., E.F. Wollack, and D.B. Richards. July 1978. A Technique for Measuring Pole Tip Saturation of Low Inductance Heads. *IEEE Trans. Magn.* MAG-14 (4): 197–200.

Tagami, K., and K. Nishimoto. Sept. 1980. Write Field Analysis of Multiturn Thin Film Heads. *IEEE Trans. Magn.* MAG-16 (5): 791–93.

Thornley, R.F.M., and H.N. Bertram. Sept. 1978. The Effect of Pole Tip Saturation on the Performance of a Recording Head. *IEEE Trans. Magn.* MAG-14 (5): 430–432.

Valstyn, E.P., and E. Packard. Sept. 1986. Optimization of Ferrite Heads for Thin Media. *IEEE Trans. Magn.* MAG-22 (5): 847–49.

EFFECTS OF CORE GEOMETRY

Aoi, H., M. Saitoh, T. Tamura, M. Ohura, H. Tsuchiga and M. Hayashi. Nov. 1982. Pole-Tip Design of High Density Recording Thin Film Heads. *IEEE Trans. Magn.* MAG-18 (6): 1137–1139.

Baker, B. Sept. 1977. Long Wavelength Response of Shielded Elliptical Reproduce Heads. *IEEE Trans. Magn.* MAG-13 (3): 1009–1012.

Eldridge. D.F., and A. Baaba. Feb. 1962. The Effect of Track Width in Magnetic Recording. *IRE Trans. Audio* AU-9 (1): 10–15.

Fritzsch, K. Dec. 1968. Long-Wavelength Response of Magnetic Reproducing Heads. *IEEE Trans. Audio and Electroacoustics* AU-16 (4): 486–494.

Geurst, J.A. 1965. Theoretical Analysis of the Influence of Track Width on the Harmonic Response of Magnetic Reproducing Heads. *Philips Res. Repts.* 20 (?): 633–657.

Kakehi, A., M. Oshiki, T. Aikawa, M. Sasaki, and T. Kozai. Nov. 1982. A Thin Film Head for High Density Recording. *IEEE Trans. Magn.* MAG-18 (6): 1131–1133.

Lindholm, D. Sept. 1980. Secondary Gap Effect in Narrow and Wide Track Reproduce Heads. *IEEE Trans. Magn.* MAG-16 (5): 893–895.

Luitjens, S. and A. Van Herk. Nov. 1982. A Discussion on the Crosstalk in Longitudinal and Perpendicular Recording. *IEEE Trans. Magn.* MAG-18 (6): 1804–1812.

Melis, J., and B. Nijholt. April 1978. A Comparison of Measured and Calculated Fringing Response of Multitrack Magnetic Reproducers. *Journal Audio Engineering Society* 26 (4): 212–216.

Minuhin, V.B. June 1990. Field of Thin Film Head of Finite Dimensions. *Jour. Appl. Phys.* 67 (12): 4866–68.

Minuhin, V.B. May 1990. Wavelength Response on Ring-Type Head with Arbitraty Gap Depth. *IEEE Trans. Magn.* MAG-26 (3): 1213–16.

Singh, A., and P. Bischoff. Sept. 1985. Optimization of Thin Film Heads for Resolution, Peak Shift and Overwrite. *IEEE Trans. Magn.* MAG-21 (5): 1572–1574.

MULTITRACK HEADS

Heijman, M.G.J., J.H.W. Kuntzel, and G.H.J. Somers. Dec. 1988, Multi-track Magnetic Heads in Thin-Film Technology. *Philips Tech. Review* 44 (6): 169–187.

Tanaka, K. Jan. 1984. Some Considerations on Crosstalk in Multihead Magnetic Digital Recording. *IEEE Trans. Magn.* MAG-20 (1): 160–165.

EFFECTS OF COATING PERMEABILITY

Bertram, H. May 1978. Anisotropic Reversible Permeability Effects in the Magnetic Reproduce Process. *IEEE Trans. Magn.* MAG-14 (3): 111–118.

Siakkou, M. 1974. Playback of Magnetic Recordings with a Reproduce head with Finite Gap Length and Tapes of Various Permeabilities (in German). *Zeitschrift elektr. Inform. und Energietechnik* 4 (?): 311–316.

Siakkou, M. Dec. 1969. Influence of Coating Permeability in Thin-Film Pulse Recording. *IEEE Trans. Magn.* Vol. MAG-5 (4): 891–895.

ERASE HEADS AND OVERWRITE

Sawada, T., and K. Yoneda. Sept. 1988. AC Erase Head for Cassette Recorder. *IEEE Trans. Magn.* MAG-21 (5): 2104–10.

Wachenschwanz, D., and F. Jeffers. Sept. 1985. Overwrite as a Function of Record Gap Length. *IEEE Trans. Magn.* MAG-21 (5): 1380–1382.

RADIO FREQUENCY INTERFERENCE

Jorgensen, F. 1975. The Influence of an Ambient Magnetic Field on Magnetic Tape Recorders. *Intl. Telemetry Conf. Proc.* Vol. 11, pp. 372–390.

Lopez, O., and R. Stromsta. July ,1981. Effects of Magnetic Fields on Flexible Disk Drive Performance. *IEEE Trans. Magn.* MAG-17 (4): 1417–1422.

Morrison, Ralph. 1986. *Grounding and Shielding Techniques in Instrumentation.* Wiley-Interscience 172 pages.

Springer, G., and F. Jorgensen. Sept. 1985. A Novel Magnetic Transducer for Magnetographic Printing. *IEEE Trans. Magn.* MAG 21 (5): 1548–1550.

FUTURE DEVELOPMENTS

Cheng, P. March 1993. MIG Head Developments/Laminated MIG. *Components Technology Review*, Singapore. 12 pages.

Lazzari, J.P. Nov. 1993. Planar Silicon Heads Review. *Head/Media Technology Review*, Las Vegas. 22 viewgraphs.

Makino, S., S. Shinkai, Y. Takeshima, T. Nakamura, M. Yabuta, M. Kitamura, N. Ishiwata, S. Tuboi, and H. Urai. Nov. 1993. High Density Recording FeTaN Laminated Hard Disk Heads. *IEEE Trans. Magn.* MAG-29 (8): 3882–84.

Mallary, M.L., and S. Ramaswamy. Nov. 1993. A New Thin Film Head which Doubles the Flux through the Coil. *IEEE Trans. Magn.* MAG-29 (6): 3832–36.

Re, M.E., D.N. Shenton, and M.H. Kryder. Sept. 1985. Magnetic Switching Characteristics at the Pole Tips of Thin Film Heads. *IEEE Trans. Magn.* MAG-21 (5): 1575–77.

Roscamp, T. 1987. *Thin Film Disk Head User's Manual.*

Sano, A., M. Egawa, M. Nitta, K. Takayanagi, T. Matsushita, T. Fujita, E. Packard, V. Nguyen, R. Limb, J. Frost, and T. Watanabe. Nov. 1993. A Low Inductance Metal-In-Gap Head Using A Side-Core Concept. *IEEE Trans. Magn.* MAG-29 (6): 3888–3890.

Shi, X., and M.H. Kryder. Nov. 1993. Phase Lag in the High-Frequency Response of Thin-Film-Recording Heads. *IEEE Trans. Magn.* MAG-29 (6): 3855–57.

Williams, E. Nov. 1993. Proximity Recording. *Head/Media Technology Review*, Las Vegas. 15 viewgraphs.

Wood, R., M. Williams, and J. Hong. Nov. 1990. Considerations for High Data Rate Recording with Thin-Film Heads. *IEEE Trans. Magn.* MAG-26 (6): 2954–59.

BIBLIOGRAPHY

GENERAL INTEREST

Autino, E., J.P. Lazzari, and C. Pisella. Sept. 1992. Compatibility of Silicon Planar Heads with Conventional Thin Film Heads in Hard Disk Drives. *IEEE Trans. Magn.* MAG-28 (5):: 2124–25.

Brittenham, S. Nov. 1990. Characterisation of Magnetic Recording Heads for Drive-Specific Applications. *IEEE Trans. Magn.* MAG-26 (6): 2989–94.

C.18 Koshimoto, Y., T. Mikazuki, and T. Ohkubo. Jan. 1988. Magnetic Head Design for Large-Capacity Fast-Access Magnetic Disk Storage. *Review of the Elec. Comm. Labs., Nippon T&T* 36 (1): 97–102.

Coutellier, J.M., H. Magna, and X. Pirot. Sept. 1992. A 384 Track Fixed Head. *IEEE Trans. Magn.* MAG-28 (5): 2653–55.

Dohmen, G.M., E.A. Draaisma, A. van Herk, and T. Wielinga. Nov. 1990. Thin Film Tape Heads, Analog and Digital Applications. *IEEE Trans. Magn.* MAG-26 (6): 2983–88.

Kitamura, M., S. Makino, Y. Takeshima, T. Sugimura, K. Minami, Y. Motomura, H. Tamai, and H. Urai. Nov. 1991. Laminated Hard Disk Head for High Density Magnetic Recording. *IEEE Trans. Magn.* MAG-27 (6): 4945–47.

Maillot, C., and F. Maurice. Sept. 1992. The Kerr Head: a Multitrack Fixed Active Head. *IEEE Trans. Magn.* MAG-28 (5): 2656–58.

Muller, M.W., R.S. Indeck, E.S. Murdock, and R. Omes. May 1990. Track edge fluctuations. *Jour. Appl. Phys.* 67 (9): 4683–85.

Romankiw, L.T. Sept. 1990. Thin Film Inductive Heads; From One to Thirty One Turns. *Magn. Matls., Proc. and Devices,* El. Chem. Soc. Vol. 90-8 pp. 39–53.

Yoneoka, S., T. Ohwe, and Y. Mizoshita. 1990. Flying Head Assemblies. *Fujitsu Sci. & Tech. Jour.* 26 (4): 404–14.

TRACK WIDTH, MULTITRACK

Doved, M.M., J.K. Spong, J.H. Eaton, and D.A. Thompson. Sept. 1992. Microtrack Profiling Technique for Narrow Track Tape Heads. *IEEE Trans. Magn.* MAG-28 (5): 2304–06.

Gibson, G.A., S. Schultz, T. Carr, and T. Jagielinski. Sept. 1992. Spatial Mapping of the Sensitivity Function of Magnetic Recording Heads Using a Magnetic Force Microscope as a Local Flux Applicator. *IEEE Trans. Magn.* MAG-28 (5): 2310–11.

Heijman, M.G.J., J.H.W. Kuntzel, and G.H.J. Somers. 1988. Multi-Track Magnetic Heads in Thin-Film Technology. *Philips Tech. Rev.* 44 (6): 169–78.

Komoda, T., T. Kira, R. Minakata, T. Nakamura, K. Nakai, and H. Deguchi. Sept. 1993. Combined Type Thin Film Head for DCC. *IEEE Trans. Magn.* MAG-29 (6): 3826–28.

Lambert, S.E., and M.L. Williams. Nov. 1988. Recording Head Characterization Using 1 micron Wide Discrete Tracks. *IEEE Trans. Magn.* MAG-24 (6): 2832–34.

ELECTROMAGNETIC DESIGN AND EFFICIENCY

Muramatsu, T., T. Kira, A. Miyamoto, and T. Sasada. March 1993. Thin Film Magnetic Head. *Sharp Tech. Jour.* No. 55 pp. 5–11.

Trouilloud, P.L., B. Petek, and B.E. Argyle. Nov. 1992. Flux Propagation in Inductive Thin-Film Heads. *Magn. Matls., Proc., and Devices*, Electrochem. Soc. Vol. 2 92-10 pp. 177–201.

POLE TIP SATURATION AND MIG

Huang, H.L., T.Y. Lee, and Y.T. Huang. Sept. 1992. Saturation Effects on Read/Write Characteristics of Asymmetric Ring Head. *IEEE Trans. Magn.* MAG-28 (5): 2650–52.

Iwata, H., K. Noguchi, S. Suwabe, and T. Nishiyama. Sept. 1990. MIG Mini Composite Head Using Single Crystal Mn-Zn Ferrite. *IEEE Trans. Magn.* MAG-26 (5): 2394–96.

Jeffers, F., and J. Bero. Nov. 1991. Record Head Saturation and Overwrite Performance of a Bafe Floppy Disk. *IEEE Trans. Magn.* MAG-27 (6): 4885–87.

Koeppe, P.V., and M.H. Kryder. Sept. 1989. Direct Observation of the Dynamic Action in Metal-In-Gap Recording Heads. *IEEE Trans. Magn.* MAG-25 (5): 3701–3703.

Nishikawa, M., and A. Kirihara. Sept. 1986. The Effect of Write Head Saturation on the Characteristics of AC Bias. *IEEE Trans. Magn.* MAG-22 (5): 1310–14.

Wachenschwanz, D., and H.N. Bertram. Nov. 1991. Modeling the Effect of Head Saturation in High Density Tape Recording. *IEEE Trans. Magn.* MAG-27 (6): 4981–83.

MAGNETORESISTIVE ELEMENTS

Bonyhard, P.I. Nov. 1990. Design Issues for Practical Rigid Disk Magnetoresistive Heads. *IEEE Trans. Magn.* MAG-26 (6): 3001–3003.

Cole, R.W., R.I. Potter, C.C. Lin, K.L. Deckert, and E.P. Valstyn. Nov. 1974. Numerical Analysis of the Shielded Magnetoresistive Head. *IBM Jour. Res. Develop.* 18, pp. 551–555.

Collins, A.J., and R.M. Jones. 1979. Review of Un-shielded Magnetoresistive Heads. *Intl. Conf. Video and Data IERE* Publ. No. 43, pp. 1–17.

Dee, R.H., and R.F.M. Thornley. Nov. 1991. Thermal Effects in Shielded MR Heads for Tape Applications. *IEEE Trans. Magn.* MAG-27 (6): 4704–05.

Dovek, M.M., D.J. Seagle, T.J. Beaulieu, E.R. Christensen, and R.E. Fontana, Jr. Sept. 1992. Performance Comparison of Unshielded and Shielded MR Heads for Digital Tape Recording. *IEEE Trans. Magn.* MAG-28 (5): 2301–03.

Jeffers, F., J. Freeman, R. Toussaint, N. Smith, D. Wachenschwanz, S. Shtrikman, and D. Woyle. Sept. 1985. Soft-Adjacent-Layer Self-Biased Magnetorestive Heads in High Density Recording. *IEEE Trans. Magn.* MAG-21 (5): 1563–1565.

Jones, Jr., R.E. Nov. 1991. An Unshielded Horizontal Dual-Element MR Sensor. *IEEE Trans. Magn.* MAG-27 (6): 4687–89.

Markham, D., and F. Jeffers. Aug. 1990. Magnetoresistive Head Technology. *Magn. Matls., Proc., and Devices*, Electrochem. Soc. Vol. 90-8 pp. 185–205.

O'Connor, F.B. Shelledy, and D.E. Heim. Sept. 1985. Mathematical Model of a Magnetoresistive Read Head for a Magnetic Tape Drive. *IEEE Trans. Magn.* MAG-21 (5): 1560–1562.

Ruigrok, J.J.M. 1981. Analytical Description of Thin-Film Yoke Magnetoresistive Heads. *Philips Jour. Res.* 36 (4-5-6): pp. 289–310.

Shelledy, F.B., and J.L. Nix. Sept. 1992. Magnetoresistive Heads for Magnetic Tape and Disk Recording. *IEEE Trans. Magn.* MAG-28 (5):2283–88.

Simmons, R., B. Jackson, M. Covault, C. Wacken, and J. Rausch. Sept. 1983. Design and Peak Shift Characterization of a Magnetoresistive Head Thin Film Media System. *IEEE Trans. Magn.* MAG-19 (5): 1737–1739.

Smith, N., D.R. Smith, and S. Shtrikman. Sept. 1992. Analysis of a Dual Magnetoresistive Head. *IEEE Trans. Magn.* MAG-28 (5): 2295–97.

van Gestel, W.J., F.W. Gorter, and K.E. Kuijk. March 1977. Read-out of a Magnetic Tape by Magnetoresistance Effect. *Philips Tech. Rev.* 37 (2/3): 42–50.

Vinal, A.W. Sept. 1984. Considerations for Applying Solid State Sensors to High Density Magnetic Disk Recording. *IEEE Trans. Magn.* MAG-20 (5): 681–686.

HEADS FOR PERPENDICULAR RECORDING

Hamilton, H., R. Anderson, and K. Goodson. Oct. 1991. Integrated Read/Write Head/Flexure/Conductor Structure for Contact Perpendicular Recording. *Magn. Matls., Proc., and Devices*, Electrochem. Soc. Vol. 92-10 pp. 161–170.

Hokkyo, J. 1991. High-Sensitivity Single Pole Head for Perpendicular Magnetic Recording. Ch. 3 in *Perpendicular Magnetic Recording*, Ohmsha Ltd., Tokyo, Ed. by S. Iwasaki and J. Hokkyo, 211 pages, pp. 43–63.

Zhu, J.G., and H.N. Bertram. Jan. 1987. Comparison of Thin Film and Ring Heads for Perpendicular Recording. *IEEE Trans. Magn.* MAG-23 (1): 177–179.

VTR APPLICATIONS

Ash, K.P., D. Wachenschwanz, G. Brucker, J. Olson, M. Trcka, and T. Jagielinski. Nov. 1990. A Magnetic Head for 150 MHz, High Density Recording. *IEEE Trans. Magn.* MAG-26 (6): 2960–65.

Dirne, F.W.A., and J.J.M. Ruigrok. Sept. 1990. Sputtered Sandwich Heads for High-Density Digital Video Recording. *IEEE Trans. Magn.* MAG-26 (5): 1683–85.

Himeshima, K., A. Enomoto, T. Hara, T. Fujine, T. Kijima, Y. Sato, F. Takahashi, A. Kitaya, and T. Kimura. Dec. 1991. Thin Film Laminated Head for S-VHS VTR. *Sharp Tech. Jour.* No. 51 pp. 41–47.

Luitjens, S.B., G.J. van den Enden, and H.A.J. Cramer. April 1990. A Way to Assess the Performance of Heads and Tapes for Digital Video Recording. *Intl. Conf. Video, Audio and Data Rec.*, IEE Publ. No. 319 pp. 43–49.

Miura, M., M. Saito, N. Arai, K. Konishi, H.O. Nagatomo, T. Tsuchiya, and H. Takahashi. Nov. 1991. Thin Film Magnetic Head for High Definition VCR. *IEEE Trans. Magn*. MAG-27 (6): 4948–50.

Okumura, T., A. Osaka, N. Ishiwata, M. Kitamura, and H. Urai. Sept. 1992. Read/Write Characteristics for Laminated High Moment Fe-Ta-N Film Heads for HDTV VTR. *IEEE Trans. Magn*. MAG-28 (5): 2121–23.

Ono, H., T. Okamoto, N. Kaku, K. Ogiro, S. Ozaki, and F. Scott. Aug. 1991. A New Drum and Heads for the D-2 Digital VTR. *SMPTE Journal* 98 (8): 596–600.

Saito, M., M. Miura, Y. Shibayama, H. Nagatomo, S. Saito, and S. Imai. Sept. 1990. A Fabrication Process of Thin Film Head for VCR. *IEEE Trans. Magn*. MAG-26 (5): 1686–88.

CHAPTER 11
TAPE AND DISK MATERIALS

The ultimate quality of a recording system is determined by the recording medium. This is a most intriguing magnetic/chemical product, in several forms and shapes, and one that is evaluated by its electromagnetic properties as well as mechanical uniformity, strength, and tolerances.

Valdemar Poulsen used piano wire for his first Telegraphone, but it was only a few years later when his associate, P. O. Pedersen, filed a patent on a metal ribbon with an electrochemically deposited magnetic coating. The later Stille recorder used a ¼" wide steel tape. Experiments with magnetic particles coated onto a paper base and homogeneous, cast plastic tapes were tried in Germany in the late twenties.

The first coated plastic magnetic tape was conceived 50 years ago by Fritz Pfleumer in Germany. It was fabricated by BASF in 1934 and it used a black iron oxide Fe_3O_4 in a binder, coated onto a cellophane base. The Fe_3O_4 is an iron ferrite that has good magnetic properties for recording, but with severe drawbacks in the form of instability and susceptibility to print-through. The latter is a copy effect whereby recordings from adjacent tape layers imprint on the layer between them (see Chapter 14). The brown iron oxide γFe_2O_3 was soon introduced, and (much refined) remains the workhorse of the recording and storage industry today.

The base film was changed from paper to PVC (polyvinyl chloride) in 1944, and later to today's PET (polyethylene terephthalate), or polyester, or Mylar (tradename of duPont). We have also seen new particles emerge: chromium dioxide (CrO_2), cobalt-doped or coated iron oxides, and recently iron powders (high energy tapes, MP) and barium ferrite particles (perpendicular magnetization media). There are also particles with very high coercivities for credit cards and the like, where erasure-resistant recordings are wanted.

Another recording material exists in plated disks. Platings on flexible films were tried but they were not successful. They were incompatible with in-contact recordings, where the metal had to be in moving contact with metal interfaced heads, and they were further very sensitive to mechanical damages (nicks, creases). Corrosion was also a problem. Efforts have paid off and metal film tapes are available today (ME, metal evaporated), for camcorders and 8 mm data systems.

This chapter will first survey and discuss materials for both particulate and plated media. Their magnetic properties, measurements, demagnetization, and noise are discussed in Chapter 12. Chapter 13 consists of a description of the manufacture of tapes as well as disks, while the performances of the end products are covered in Chapter 14.

PARTICULATE MEDIA COMPONENTS

It is customary in the media industry (tapes and disks) to classify the magnetic powders by their intrinsic properties, such as the specific magnetization σ_s (in EMU/cc), and the mean value H_{ci} of intrinsic coercivities h_{ci}. The unit EMU is equal to 4π Maxwell-centimeter (or $4\pi \times 10^{-10}$ Weber-meter). Hence 1 EMU/cc $= 4\pi \times 10^{-4}$ Wb/m^2.

It should be noted that σ_s, which corresponds to the remanence (or retentivity) of the final coating, is insensitive to the particles' internal structure, while h_{ci} is structure sensitive.

MAGNETIZATION OF A POWDER MAGNET

The powder pack (the media coating) will have a smaller saturation magnetization M_{sat} due to a volume occupancy or packing fraction p that is less than one, typically in the range of 0.35 to 0.40. The coercivity H_{ci} depends largely on the particle sizes and shapes, as we shall see. Both M_{sat} and H_{ci} are, therefore, extrinsic properties of the coating. The intrinsic value σ_s is the absolute maximum magnetization inside a solid sample of the material (and therefore the magnetization in each domain, or single-domain particle).

The level of magnetization in a saturated, nonoriented tape coating is:

$$B_{rsat} = \mu_o M_{rsat}$$
$$= 0.5 \times \mu_o M_{sat}$$
$$= 0.5 \times p \times \sigma_s$$

where p is the packing fraction of particles in the coating ($0.0 < p < 1.0$). The theoretical maximum of a packing fraction of one is, of course, unrealistic; its maximum value is about 0.4. A powder of cubic particles has a packing fraction of approximately 0.17 when loosely packed, and 0.25 when compacted. It can increase from 0.35 to 0.40 when compressed under high pressure, and this is also the maximum for a tape coating. Higher values result in a weak coating, leading to oxide shedding, debris formation on heads and guides, and subsequent signal drop-outs.

The value of M_{rsat} can be increased by orienting the particles in the direction of the anticipated magnetization. In the past, this has always been longitudinal for audio, instrumentation, and computer tapes, and transverse for two-inch broadcast videotapes. The measure of orientation is the squareness S, which is defined as the ratio M_{rsat}/M_{sat}, both measured on the MH-loop. S ranges from 0.5 to a theoretical maximum of 1.

Replacing the factor 0.5 with S, and converting to Wb/m^2 results in:

$$B_{rsat} = S \times p \times \sigma_s \text{ EMU/cc}$$
$$= S \times p \times \sigma_s \times 4\pi 10^{-4} \text{ Wb/m}^2$$

It has become common practice to quote the flux level from a coating in nWb/m, and the end user can then multiply this figure with the recorded trackwidth in meters to obtain the flux from that track. Thus we arrive at the following formula for the flux level (with mixed units for practical reasons):

$$\phi_{rsat} = 0.4\pi \times S \times p \times \sigma_s \times \delta \text{ nWb/m} \qquad (11.1)$$

where:

$S = M_{rsat}/M_{sat} =$ the squareness
$p =$ packing fraction ($0.35 - 0.45$)
$\sigma_s =$ specific magnetization in EMU/cc
$\delta =$ coating thickness in μm

A high-output medium obtains its high flux level by maximizing the four items in formula 11.1.

This technique has long prevailed in the audio and instrumentation field, where a very high signal-to-noise ratio is necessary (see Chapters 27 and 28). The reader will probably encounter BH- and MH-loops with units from cgs or SI, and conversions at times become a chore. We had earlier:

$$B_{rsat} = \mu_o M_{rsat} \qquad [B \text{ in Wb/m}^2, M \text{ in A/m}]$$
$$= 4\pi 10^{-7} M_{rsat}$$
$$= 4\pi 10^{-4} M_{rsat} \quad [M \text{ now in kA/m}]$$
$$= 4\pi M_{rsat} \qquad [B \text{ in gauss, } M \text{ in kA/m}]$$

From this we find:

$$M_r \text{ [kA/m]} \leftrightarrow 4\pi M_r \text{ [gauss]}$$

$$\text{or } M_r \text{ [gauss]} \leftrightarrow M_r/4\pi \text{ [kA/m]}$$

$$\text{and } H \text{ [kA /m]} \leftrightarrow 4\pi H \text{ [Oe]}$$

$$\text{or } H \text{ [Oe]} \leftrightarrow H/4\pi \text{ [kA/m]}$$

Media for data storage, such as 3380, 3390, and QIC cartridge tapes, aim at very high packing densities of bits along a track. We will examine which magnetic properties will enhance this.

A thin coating is desired when pulses of opposite polarity are recorded. Between bits there exist transition zones, and it is shown in Chapter 19 that the length Δx of this zone is inversely proportional to the coating thickness (for equal levels of overwrite).

Demagnetization in all recordings is decreased when a thin coating is used, or when the recording is made only partially into the coating. The latter principle is used in wideband instrumentation recorders, and is used for analog data as well as in high-density digital recording (HDDR, QIC) This does have the same effect as that of decreasing the coating thickness (less demagnetization). No mechanical difference can be observed between a thick and a thin coating except for the fact that a thin-coat tape is more pliable, provides better head-to-tape contact, and affords longer playing time. An increase in coercive force H_{ci} will reduce the effect of demagnetization. The *BH*-loops for two coatings are shown in Fig. 11.1. They differ only in coercive force, and the advantage of the higher value of H_{ci} is clearly evident by the higher values of both open-circuit remanence and short-circuit remanence (during contact with a high-permeability reproduce head).

Final improvements can be achieved by orientation of the particles, which results in a more square loop (Fig. 11.1, right). The analysis of pulse recording in Chapter 19 will further show that the steep slope of the sides of the hysteresis loop will enhance short transition zones; the slope corresponds to the switching field distribution. They are made steep not only by orientation of the particles, but also by using a magnetic powder material that has a narrow distribution range of single-particle coercivities. The latter is also beneficial in anhysteretic ac-bias recordings.

The mean coercive force H_{ci} of an assembly of particles depends on the material as well as the particle size. Large particles will have smaller shape anisotropy and may each have more than one domain. The net result is low coercivity, as shown in Fig. 11.2. Conversely, the assembly of very small and (in particular) nonacicular particles leads to nonmagnetic behavior, also called superparamagnetic behavior, below a critical size that happens to coincide closely with the thickness of the walls in that material.

The coating's coercivity is, in other words, controlled by the particle sizes. As the size is reduced, it is typically found that the coercivity increases, goes through a maximum, and then tends to go toward zero. The behavior changes simultaneously from multidomain to single-domain when the particle size is below a critical diameter; that diameter is not well-defined. Predictions of the critical sizes (and volumes and shapes) for single-domain and superparamagnetic transitions are difficult to make with any accuracy, as the theory is not well enough understood.

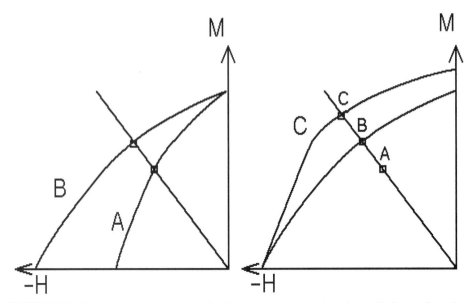

FIGURE 11.1   The level of magnetization in a recorded medium is reduced to level A due to demagnetization (see chapter 4). An increase in medium coercivity will cause the remanent level to increase to B. Further improvement in remanent level is possible by orientation of the magnetic particles into the direction of head-media motion, to level C.

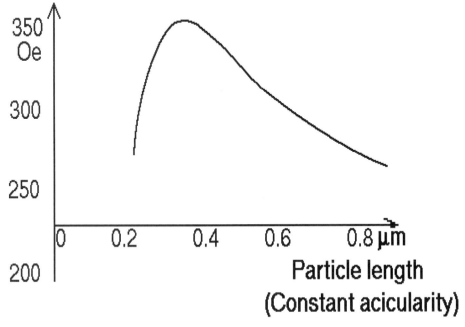

FIGURE 11.2   Coercivity of γFe_2O_3 particles as function of size (*after Daniel, 1972*).

Our knowledge of single-particle magnetization is limited, and gets worse when we put several of them together. The added complexity is due to the interaction fields between the particles. The idealized view that a particle is a perfect ellipsoid in shape and consists of a single domain is shattered by the observation that they actually are of irregular shape, polycrystalline and with cubic anisotropy. Further, a high proportion of particles are found to be multiple; that is, they comprise several particles lying side by side. Most particles in recording tapes are of this type (Knowles 1981).

The single particle's coercivity can be estimated. For a needle-shaped particle with predominant shape anisotropy and with a ratio of length to diameter greater than 10, we have:

$$h_{ci} = (N_d - N_l) \times M_{sat} \tag{11.2}$$

where N_d is the demagnetization factor along the short axis d and N_l is the demagnetization factor along the long axis l. A spherical particle would thus have $h_{ci} = 0$ because $N_d = N_l$ for l/d = one. The particle coercivity is then controlled by its crystalline anisotropy. Such particles were once made by by Pathe in France and used in coatings named "Isomax" (Tradename of Spin Physics). The final coercivity was controlled by a cobalt treatment that brings about an equiaxis anisotropy; i.e., a coating made from these particles will magnetize equally well in the longitudinal and perpendicular mode. The commercial use of Isomax never materialized, due to the classic cobalt problems: excessive sensitivity to temperature and stress.

Calculations of the total coating coercivity based on a single particle's coercivity is also difficult because of particle interactions, as was shown in Fig. 3.25. It has been suggested that the coercivity is best represented by:

$$H_c = (h_{ci} - H_i)^2 / h_{ci} \tag{11.3}$$

where h_{ci} is the coercivity of an individual particle and H_i is the interaction field applied to a given particle by its neighbors.

When crystal anisotropy prevails, then the coercivity is expected to be independent of the packing fraction p. For particles with shape anisotropy, the inverse is true and the coercivity H_{ci} decreases with increasing p:

$$H_c = h_{ci} \times (1 - p) \tag{11.4}$$

When $p = 1$, all particles are everywhere in contact; shape anisotropy is lost and the coercivity becomes zero—if other forms of anisotropy are absent.

The ultimate behavior of coatings must be determined by extensive experimental measurements, while work continues on the theoretical aspects; see "Behavior of an Assembly of Particles," Chapter 12.

It was mentioned earlier that particles should have a narrow range of coercivities. Noting Fig. 11.2, this means a narrow range of particle sizes and shapes. Typical particle size distribution for modern particles are shown in Fig. 11.3 (Koester 1993). It is desirable to reduce their sizes because lower noise and higher coercivity results. The magnetization of a particle becomes unstable with time for very small sizes, placing a lower limit on their dimensions (Sharrock 1990).

The instability, appearing as a decay in coercivity, is caused by thermal fluctuation, and is treated further in Chapter 12, Effects of Time. The very-short-time coercivity prevails during recording (time equal to the field-rise time for the transition in a digital signal). It is possible to estimate an optimum particle size where the coercivity at time of recording is twice the value after long-time storage. The result is a particle volume of 3×10^{-17} cm^3 for an oxide particle. This implies, for a ten-to-one length-to-diameter ratio, a length of 0.156 μm and a diameter of 0.0156 μm (Sharrock 1990). This dimension is indicated in Fig. 11.4 with an arrow.

The lower limit for particle sizes due to complete instability due to thermal fluctuations are indicated by vertical lines (Koester 1993). The size distribution for a given particle should not include such small particles; they can cause instability and print-through.

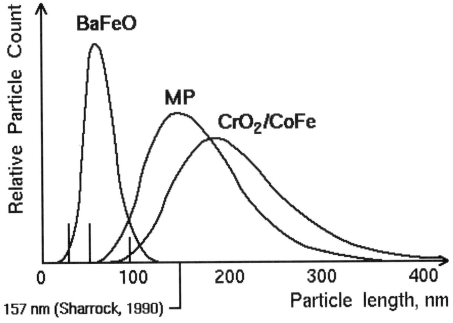

FIGURE 11.3 Particle size distribution of modern particles (*after Koester, 1993*).

FIGURE 11.4 Gamma iron oxide (γFe_2O_3) particles.

Another particle size effect is the loss in recorded signal resolution, as was earlier indicated in Fig. 3.62. The recorded bit length is limited to about 1.5 times the mean particle length (Sharrock 1990). Koester estimates the current packing density limits for tapes as

13,000 fc/mm = 330,000 fci for BaFeO
5000 fc/mm = 127,000 fci for MP
4000 fc/mm = 102,000 fci for CrO_2, CoFe,

where CoFe is cobalt-treated γFe_2O_3.

FIGURE 11.5   Chromium dioxide (CrO_2) particles.

FIGURE 11.6   Metal (Fe) particles.

The reduced particle size requirement reflects in another characteristic of oxide powders: their specific surface area (named *BET* in the industry). The *BET* for CoFe has increased from an average of 20 m^2/g in the early 1970s to currently 30 m^2/g, while the maximum *BET* in use today has reached 45 m^2/g (Corradi 1992). A surface area of 40 m^2/g corresponds to a particle volume of 2×10^{-17} cm^3.

Many of the conclusions we have arrived at concerning recording performance have been verified. Pulse width has been found to be proportional to H_c^{-x}, where x is between ½ and 1; the remanence has no influence on the pulse width (Bate and Alstad, 1969).

The signal amplitude was, in three investigations, found to be proportional to $(H_c \mu_o M_{sat})^{0.5}$, $H_c^{0.15}$ $\times (\mu_o M_{sat})^{0.85} \times \delta^{0.15}$, or plainly $(\mu_o M_{sat})^{1.0}$ (Bate and Alstad, 1969).

The pulse playback phenomenon called peak-shift (see Chapter 21) could not be traced to any particular properties of the magnetic media. This problem is complex and involves the position and phase of the "in-depth of recording."

We can, in summary, set goals for selection of a magnetic particle material:

- The coercivity should be high, consistent with compatibility and record (write) capabilities. A high coercivity provides maximum utilization of the material's magnetization and packing density. It also reduces the effects of demagnetization of recorded signals.

- The specific magnetization should be high in order to achieve a high playback (read) signal.

- The individual particles should be small. This leads to increased coercivity and reduced noise.

- The particles shall, when feasible, be oriented. This increases the overall signal output and improves resolution (steeper sides of the hysteresis loop).

The matter of selecting a coating thickness will be a trade-off decision. Audio tapes for studio use will employ a fairly thick coating (10 μm = 400 μin) and offset the short wavelength losses from demagnetization by running the tape at a speed higher than normally required, i.e., 76 or 38 cm/sec (30 or 15 IPS) instead of 19 cm/sec (7½ IPS).

In high-density digital storage recording, the signal-to-noise ratio of a thick coating can easily be relaxed, and a thin coating is chosen for its better resolution. As a matter of fact, 75 percent of the output from a recording is contributed by a coating layer of thickness $0.22 \times \lambda$, or $0.44 \times BL$ (from Chapter 6)—and the coating should be no thicker.

MAGNETIC PARTICLES

Several oxide materials used for tape production are listed in Table 11.1. We will briefly discuss each of them. A recent paper describes these particles, fabrication, and application in detail (Sharrock 1989). Typical characteristics of oxides and particles are listed in Table 11.1.

TABLE 11.1 Characteristics of magnetic pigments used in modern tape

	Particle length μm	Aspect ratio	Magnetization		Coercivity		Surf. area m^2/g	SFD Δh_r	Curie temp. °C
			Wb/m^2	EMU/cc	kA/m	Oe			
γFe$_2$O$_3$	0.2	5:1	0.44	350	22–34	420	15–30	0.2–0.3	600
CO-γFe$_2$O$_3$	0.2	6:1	0.48	380	30–75	940	20–35	0.2–0.45	700
CrO$_2$	0.2	10:1	0.50	400	30–75	950	18–55	0.25–0.3	125
Fe	0.15	10:1	1.4*	1100*	56–176	2200	20–60	0.4–0.6	770
BaFeO	0.05	0.02 μm thick	0.40	320	56–240	3000	20–25	0.05–0.2	350

*For overcoated, stable particles use only 50 to 80% of these values due to reduced magnetic particle volume.

FERRIC OXIDE (Fe₃O₄)

This was the particle used earliest in magnetic tape production, in the 1930s. It had a specific magnetization that essentially is the same as the other iron oxide materials, but the low value of $H_c = 90$ Oe lowered the useable remanence to about 25 percent of the saturation remanence. The tapes suffered from aftereffects (loss of level, in particular at short wavelengths) and heavy print-through, and the material was soon superseded by gamma ferric oxide.

GAMMA FERRIC OXIDE (γFe₂O₃)

This was at one time the most used particle. It is inexpensive and uniform, and there exists considerable experience in its manufacture and use. Today the cobalt-treated (adsorbed) particles are preferred due to their higher coercivity.

These particles are considerably shorter than one micron (40 μin) (see Fig. 11.4). They are acicular and have a length-to-width ratio of approximately six to one. The oxide particles are manufactured by first dissolving iron in an appropriate acid. The resulting particles, called alpha particles, have the elongated shape of the finished magnetic oxide, but are nonmagnetic.

To continue in the manufacturing process, the particles are heated and the oxide is reduced to the Fe₃O₄ form in an atmosphere of hydrogen or natural gas. Then the particles are reoxidized under a controlled temperature and the final particles, in the form of gamma ferric oxide, are obtained:

$$(FeO)OH \quad \rightarrow \quad \alpha Fe_2O_3(red) \quad \rightarrow \quad Fe_3O_4(black) \quad \rightarrow \quad \gamma Fe_2O_3(brown) \quad (11.5)$$

process:	Solution	Reduction	Oxidation
temp:	120°C	350–500°C	250–330°C

Small amounts of Zn or Cr salt may be added to improve particle stability and dispersion properties.

The (FeO)OH and αFe₂O₃ powders were used as pigmentation in the paint industry, and that is how some companies got into the tape business (I.G. Farben, now BASF; and C.K. Williams, now ISK Magnetics, Inc.). This, of course, ties in with their knowledge of applying binders (paint) to the base films.

The shape and size of the final particles are largely determined by the preparation of the (FeO)OH-Goethite. Agglomeration and sintering of the particles occur at the high-temperature steps of dehydration and reduction, which are often combined into one step. A small particle size is advantageous, because the coating then has better short-wavelength response (less demagnetization because of higher coercivity) and less noise (more particles per coating volume). The lower noise from smaller particles has been verified by measurement (Podolsky 1981).

A new type of particle, named NP (nonpolar), has an ellipsoidal shape and therefore no internal demagnetization from an irregular shape (Corradi et al. 1984). The morphology of these particles allows a much better packing and homogeneous distribution of the particles, providing for a higher remanence and lower noise. It is also suggested that they may provide for a perpendicular recording surface, in addition to the Ba ferrite particles to be described further on. The NP particles are also free from porosities. They have found use in high-grade audio tapes for studio master recorders.

COBALT-SUBSTITUTED (DOPED) IRON OXIDE

It has long been known that cobalt-doped iron oxide particles can be fabricated with a large range of coercivities; $H_c = 150 - 800$ Oe. In these particles, up to four percent of the ferric ions have been replaced by cobalt ions.

It has also been known that the cobalt-doped particles in a tape suffer from the temperature dependence of the coercivity and the loss of remanence after storage at elevated temperatures. This is due to temperature sensitivity in the crystal anisotropy. A user might adjust storage conditions to live with these disadvantages, but will be faced with repeated signal losses each time the tape is played. This is due to a high magnetostriction constant, and the signal would literally be erased after a large number of passes (see Chapter 14).

ISOTROPIC IRON OXIDE

An iron oxide particle named "Isomax" was evaluated in the early seventies by Spin Physics. The particles were shaped like rice, and were essentially isotropic. By cobalt treatment it was possible to introduce anisotropy so that the nonoriented coating had a high remanence for longitudinal as well as perpendicular magnetization. Thus Isomax delivered the best of both worlds.

It was unfortunate that a serious flaw, excessive thermal and magnetostrictive instabilities, prevented their use for ordinary tapes. The *MH*-loop has a certain squareness, whether measured in the plane of the coating (longitudinally) or perpendicular thereto.

COBALT-ADSORBED (SURFACE) IRON OXIDE

These particles are surface-modified iron oxides, obtained by producing a reaction of the Fe_3O_4 or γFe_2O_3 particles, with a solution containing Co_2+. Their size and shape are therefore that of γFe_2O_3. The resulting material consists of particles with a thin surface layer of cobalt/iron oxide. The depth and composition of the layer determine both the coercive force and its temperature dependence.

The process control over the final material coercivity is a great advantage. New tape products should not only be superior to their predecessors, but also compatible with existing equipment (Umeki et al. 1974).

CHROMIUM DIOXIDE

Chromium dioxide was introduced in the late 1960s, and recording equipment was provided with a switch for normal (γFe_2O_3) and high (CrO_2) bias currents to the record head. One cobalt-adsorbed particle tape was introduced as "Avilyn," and the coercivity was adjusted to match the high-bias setting.

Chromium dioxide was developed during DuPont's investigations into compounds formed at intense heats and pressures. It soon became evident that chromium dioxide possessed magnetic properties superior to iron oxides, and could be made in a needle shape well-suited to magnetic recording applications, being (at that time) 0.8 µm long and with a length-to-diameter ratio (aspect ratio) of ten to one. Commercial production of CrO_2 began in 1967.

One advantage is that the particles can be made of uniform size and free of dendrites. They are readily dispersed, easily oriented, and give excellent high-density (short-wavelength) recording performance. Coercivities range from 450 to 670 Oe, and the particle aspect ratio can be controlled over a wide range (Chen et al. 1984).

Their disadvantages are the cost (three times that of ferric oxides but about the same as co-adsorbed ferric oxide) and a somewhat higher wear rate (see Chapter 18). Wear is often overrated, and can be judged as a trade-off for better performance (1000 hours of head life instead of 2000; that is still a lot of tape playing).

Modern CrO_2 particles are made in a range of sizes, obtained by doping with antimony trioxide (Sb_2O_3). A high doping level (approximately one percent) results in a higher H_c and smaller, more uniform particles as in Fig. 11.5 (Braginskij 1981). The popularity of CrO_2 is evident from the recent choice by IBM to use these particles for the new 3480 digital tape system.

CrO_2 particles are often rated better than CoFe in some aspects of recording performance, such as print-through and erasure. Its low Curie temperature also allows its use in mass videotape duplication by heat-contact-transfer (Auweter et al. 1990). Recently, CrO_2 has been improved to a point where videotapes with an SFD of 0.17 have been produced (Veitch et al. 1993).

METAL PARTICLES

Metal particles (pure Fe) have a magnetization that is almost three times that of ferric oxide, and they can be made smaller than the oxide particles without the onset of paramagnetism.

The coercivity is high, which presents a two-fold problem: lack of compatibility with existing equipment, and a potential danger of pole tip saturation in recording heads. On the other hand, a high coercivity is desirable for high-density recordings. The coating procedure for a metal-particle tape presents its own set of problems. The particles are highly reactive with the atmosphere, and will rust or even burn when exposed to air. They will need a protective coating to prevent contact with air.

Most modern tapes are made with polyurethane binders, which have high transmission rates for water vapor. A protective and permanent coating must therefore be used for the Fe particles. Indications are that some of the coating problems are being solved (Chubacki and Tamagawa 1984). The durability is approaching that of conventional tapes; the SNR is about 10 dB above oxide tapes for the new 8 mm video camera.

Modern Fe particles used in MP (metal particle) tapes are overcoated with a 10–12 Å layer of aluminum oxide and silicon oxide that protects the particle very well. Tapes made by brand name manufacturers since 1990 are stable. The cost is less magnetization, about 50% volume lost to the protective overcoat. This problem will increase for a higher-density tape where the particles must be made smaller, and the volume occupied by the protection layer becomes relatively larger. Special protective binder formulations are also under development and evaluation, with some success (Liang and Nikles 1993). These binders must still fulfill the many mechanical specifications for flexible tapes.

BARIUM FERRITE

Fine barium ferrite particles has been developed during the past few years, and show great promise as a particulate perpendicular recording medium. The particles, about 0.08 μm (3.2 μin) in average diameter, are thin hexagonal platelets with easy magnetization axes normal to their plane (Fig. 11.7). The addition of Co and Ti makes the coercivity controllable over a wide range without significant reduction in magnetization; see Fig. 11.8 (Kubo et al. 1982). This allows for the design and optimization of tapes for use with ferrite or metal head recording capability (Isshiki et al. 1985). (Ferrite heads have tendencies to saturate when writing on high-coercivity tapes or disks; see Chapter 10.)

Recent data verify the excellent properties of Ba-ferrite, as shown in Fig. 11.9 which illustrates packing densities approaching 80,000 fr/cm (200,000 FRPI) (Fujiwara 1985). The signal envelope has less amplitude modulation due to a more uniform particle dispersion in the coating, and the SNR is 6 dB higher than a low-noise γFe_2O_3 tape.

Orientation of BaFeO was until recently a problem, because the platelets would tend to stack due to magnetic attraction. The surface treatment with a 6–8 Å thick nonmagnetic layer reduces the attractive forces (which vary with the distance cubed), making dispersion and orientation much easier.

Longitudinally oriented BaFeO coatings on rigid disks have been tested and found equal in performance to the current metal film disks (Speliotis et al. 1993). Complex business decisions forestalled the application in manufactured disk drives. Application in the QIC (quarter-inch cartridge) drive market is developing much more satisfactorily, as BaFeO offers the high-density tape needed for cartridge capacities above 1 GB.

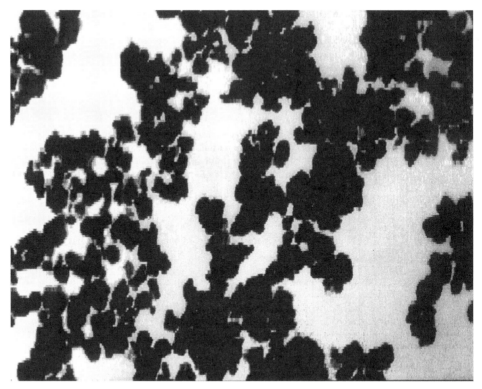

FIGURE 11.7   Barium ferrite (BaFeO) particles. Average particle diameter = 300–400 Å (*courtesy of 3M Co.*).

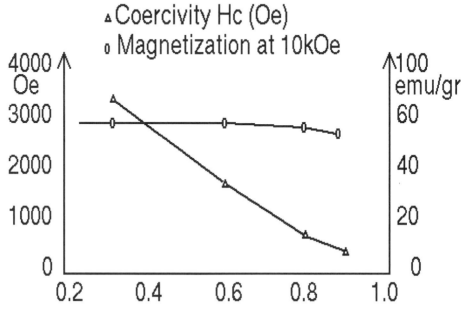

FIGURE 11.8   Co and Ti substitution elements control coercivity and magnetization for BaFeO (*after Kubpo et al,. 1982*).

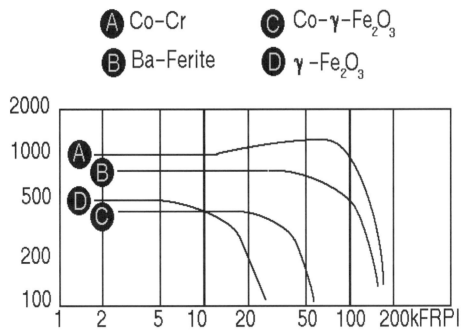

FIGURE 11.9 Density response curves for barium ferrite (b) flexible disk compared with CoCr (a) sputtered perpendicular, CoγFe$_2$O$_3$ (c) and γFe$_2$O$_3$ (d) (*after Fujiware, 1985*).

PARTICLES WITH VERY HIGH COERCIVITY

A considerable effort was made during the seventies to develop rare-earth-cobalt materials for high-energy permanent magnets. The high coercivity of these materials is well suited for secure credit-card applications. Barium ferrites have also been used and require a recording head made from an iron-cobalt alloy. BaFe$_{12}$O$_9$ and SmCo$_5$ particle tapes have been compared and both were found suitable for erasure-resistant applications (Fayling and Bendson 1978).

THIN MAGNETIC FILMS

Metallic coatings on circular aluminum substrates are in widespread use in all hard disk memories. The coatings are thin (0.03–0.1 μm, or 0.12–4 μin.), but the output is comparable or better than the thicker oxide coatings because the packing factor is 1 in thin films, and there is no intrinsic demagnetization.

Film coatings on flexible media remain in development in order to solve the problems of corrosion and excessive head wear. The wear is in particular a problem when thin-film tapes are used in conjunction with metal heads, rather than ferrites. And all ferrite heads remain a bit of a mystery; one head manufacturer has excellent results with head life, while another has problems with wear and crystal pull-outs. There is also the matter of using a stable, long-lasting lubricant, without which the contact between the head and the metal film would be like a bearing without grease. Recent efforts to manufacture such tapes are described by Feuerstein and Mayr (1984).

The films can be deposited by several methods that, together with a large range of materials, have been evaluated since the late 1970s. Some of the earliest coatings were made from cobalt phosphorus (CoP) or cobalt-nickel (CoNi) by vacuum deposition, chemical deposition, or electrical deposition. Modern coatings are deposited by sputtering, and a fairly large number of materials have been evaluated, with many claims to superiority over particulate oxide coatings.

Ferrite films have been made with a coercivity of 40 kA/m (500 Oe), and remanences B_{rs} ranging from 0.05 to 0.26 Wb/m^2 (500 to 2600 gauss) (Ianagaki et al. 1976). Rare-earth films ($Fe_{1-x-y}Tb_xGd_y$) show promise (Desserre and Jeanniot 1983).

Cobalt and cobalt-chromium are both good candidates for perpendicular thin-film recording surfaces made by electroplating (Chen and Martin 1979) or by RF (radio frequency) sputtering (Sagoi et al. 1984). Both materials can be made in a columnar structure that enhances the perpendicular magnetization mode; see Fig. 11.10. This should enhance their ability to carry a high packing density, in contrast to some of the ordinary films that suffer from resolution limitation due to the formation of zigzag transitions; see Fig. 3.72 (Chen and Martin 1979).

A couple of ternary films were evaluated: CoPtCr and CoNiCr (Sanders et al. 1989). They found a variation in coercivity with film thickness that pointed toward today's thin disks, with $\delta = 2$–400 Å (see Fig. 11.11).

The materials have today sorted themselves into several ternary alloys in extended commercial use: CoPtCr, CoPtNi, CoTaCr, and CoTaNi. Next-generation thin-film coatings are heading toward CoPtCrTa and CoPtCrX, where X = Ni, B, or Si (Johnson 1992 and Procker 1993). One investigation of CoCrPtB/Cr was reported by Paik et al. (1992).

When a thin film is specified CoCrPtB/Cr, the last /Cr implies that the ternary film is deposited on an underlayer of Cr. This underlayer is particularly important because its grain structure controls the initial film formation during sputtering. The grain structure should be well-textured in order to enhance a "grain-to-grain epitaxy" (Hsu et al. 1990).

The granular nature of thin films is a model that looks like that of particulate media. It can be modelled and used to explore optimum film structures in order to achieve high data densities along with low noise (Zhu 1992 and El-Hilo et al. 1991). Multilayer films has thus been found experimentally and by modelling to have lower noise—but of course, at a higher price. Is it worth it in large-volume manufacture?

Several graphs showing the annual improvements in storage density were shown in Fig. 2.4, and a point indicated an IBM demo. That was the demonstration of a disk drive with capacity of 10 Mb/in^2. The media to use for such densities is discussed in a paper by Yogi and Nguyen (1993).

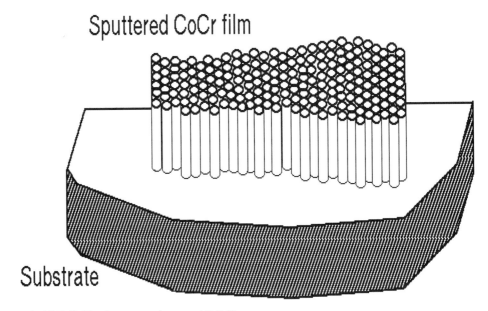

FIGURE 11.10   The microstructure of a sputtered CoCr film.

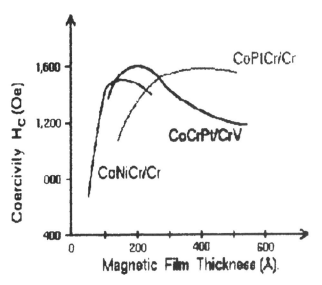

FIGURE 11.11 Coercivities of deposited thin films as function of film thickness (*after Sanders, et al., 1989*).

FLEXIBLE MEDIA SUBSTRATES

Several methods have been employed to support the recording surface. Tapes and flexible disks use plastic films; in the early days, this was a cellulose triacetate material. Today's tapes and diskettes use a PET film (polyethylene terephthalate), often called Mylar (a tradename of DuPont). Rigid disks use an aluminum, glass, or ceramic disk substrate. Cellulose triacetate was abandoned because of its low tensile strength, which could often cause a tape to break in the event of careless tape recorder operation. Nevertheless, the break was always clean and easy to splice; Mylar will stretch to about twice its normal length before breaking, ruining that portion and making splicing a questionable operation.

PET films

The surface of PET (polyethylene terephthalate) is inferior to that of PVC. PET is one of the many long-molecule materials where cross-linking causes uneven surfaces, primarily due to lumping. Efforts continue to develop new base films by modifying the PET and by mixing polymers—a very difficult process due to intermolecular interactions and difficulties in building a uniform molecular structure from the different long chains (Braginskij 1981).

High-temperature application requires special films, such as a polyimide film named Kapton (a tradename of DuPont). Polyimides maintain their physical characteristics up to 400°C and do not support combustion (Ochsner 1986). Tapes in flight data recorders on board commercial airliners are, therefore, made with a Kapton-based film.

Polyimide substrates are candidates for flexible disks for perpendicular recording. Sputter metallizing of cobalt chromium is most effectively done at a temperature of approximately 200°C, where most other films would have failed completely.

MANUFACTURE OF PET FILM

PET belongs to a group of quite complicated polymer esters made from terephthalic acid and ethylene glycol (Mark 1957 and 1968; Bhushan 1992). The molecular weight is between 15,000 and 30,000, and the raw material exists in the form of granules of a mean size 4 by 4 by 2.5 mm (Braginskij 1981). Fillers are added to the polymer in order to control the surface structure of the finished film (Goerlitz and Ito 1993). During a stretching process in the film production, the filler particles protrude from the otherwise completely smooth film and form the topography.

The creation of a continuous film is started by thoroughly drying the granules so that the water content is below 0.01 percent, at a temperature between 150 and 180°C. Any excess humidity will result in film defects (small bubbles and blisters).

The granules are next fed into one end of a screw conveyor, where they are melted at a temperature of 280 to 290°C and form a liquid syrup-like substance (Fig. 11.12). This liquid is forced through an extrusion head to form a liquid sheet that immediately is cooled by contact with a cooling drum. It is important that the liquid sheet is extruded under a uniform pressure, because pulsations result in thickness variations of the PET film. This uniform extrusion pressure can be provided by a double-screw conveyor, or a single conveyor followed by a gear pump.

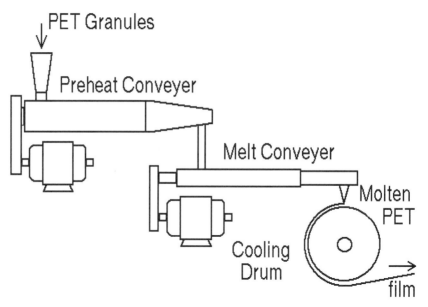

FIGURE 11.12   Screw conveyers for fabrication of PET films.

The PET may, at the moment of extrusion, crystallize into spherulites; this is prevented by rapid cooling when it touches a drum that has a cooling water system built into its surface. Good contact between the PET and the drum is required, and can be obtained by an air stream pushing onto the PET film, coating the drum with glycol, or even the use of electrostatic adhesion (Braginskij 1981).

The fabricated film is next stretched in order to increase its strength. This must be done in the machine direction, which corresponds to the motion of the film, as well as in the direction of its width. If done in the machine direction only, the result is an alignment of the molecules leading to strength in just that direction only. The temperature and the speed with which the film is stretched will influence its tensile strength.

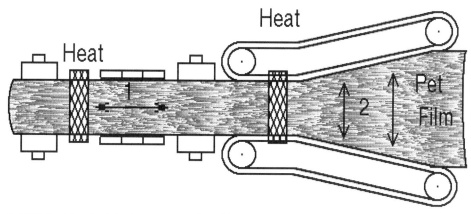

FIGURE 11.13 Longitudinal and transverse stretching of PET film.

The process takes place in a piece of equipment where the film is heated, stretched lengthwise and then sideways, and finally fixed (cross-linking the long molecules) and cooled (Fig. 11.13). This process is referred to as tensilizing, and is done at a speed of 6–60 m/minute. The machinery is a rather large structure, from 30 to 50 meters in length, 3–6 meters wide, and 2–3 meters tall.

The finished PET film varies in width from 0.6 to 4 meters, and is trimmed from its edges and wound onto drums; care must be taken to avoid dust (which may originate from the slitting operation). Any small particles wound within a pack cause deformation of the film, and can cause dropouts.

Table 11.2 summarizes the properties of modern base films. The values are averages of published data and should be used for budgetary design purposes only. Comparative stress-strain curves are shown in Fig. 11.14. In a tensilized (stressed) PET film there are residual stresses that will deform and buckle the film if it is heated.

The long-chained molecule links are, furthermore, such that the films have properties that differ in directions away from the machine direction (length of web). The modulus and thermal expansion coefficient are affected, and may vary as much as 2:1 and 4:1, respectively, in various directions.

TABLE 11.2 Characteristics of flexible base films used for magnetic tapes floppy disks

	Unit	PET	PEN	Kapton™	ARAMID	PBO
Density	g/cm^2	1.395	1.355	1.420	1.500	1.54
Melting temperature	C	263	272	None	None	None
Glass transition temp.	C	68	113	350	280	None
Young's modulus	kg/mm^2	500–850	650–1400	300	1–2000	5000
Tensile strength	kg/mm^2	25	30	18	50	60
Tensile elongation	%	150	95	70	60	1–2
Long term temp.	C	120	155	230	180	300, >
Heat shrinkage*	%	5–10	1.5	0.1	0.1	<0.1
Coeff. of thermal exp.	10^{-6}/C	15	13	20	15	?
Coeff. of hydrosc. exp.	10^{-6}/%RH	10	10	20	18	-7×10^{-6}
Moisture absorption	%	0.4	0.4	2.9	1.5	<1%

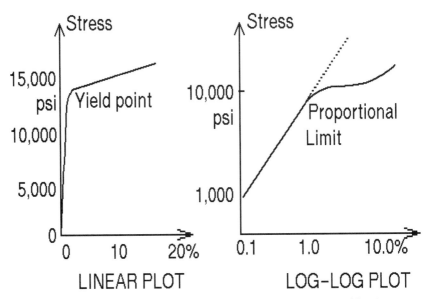

FIGURE 11.14   Stress-strain plots for PET film (*courtesy E.F. Cuddihy, JPL, Pasadena, California*).

This leads to problems when diskettes are stamped out from a coated PET film. Changes in temperature cause different degrees of expansion in different directions, and a round track may end up being oval. Sufficient trackwidth must be used to avoid loss of contact between the head and the deformed track; this limits the TPI.

PET films are inherently very smooth, which is desirable for a recording substrate, but leads to problems in the handling of the film (difficult to wind, blocking of surfaces). It is, therefore, general practice to roughen the surface by embossing, sandblasting, chemical etching, and so on. It is also possible to produce the rougher surface by altering the polymer by incorporating inorganic additives either grown or injected into the polymer prior to film extrusion. Silicon dioxides, titanium dioxide, calcium carbonate, etc. are often used as fillers (Goerlitz and Ito 1993; Heffelfinger 1986).

After the biaxial stretching, the result is a bumpy surface. The surface roughness is often graded by the use of a measurement of the arithmetic mean deviation from the center line, called the center line average (*CLA*) or R_a. PET for videotapes for ½ inch VCR has $R_a = 0.035$ μm (= 1.4 μin), while $R_a = 0.005$ μm (= 0.2 μin) for higher density ME media (Heffelfinger 1986).

Tapes are often backcoated in order to provide a rougher surface for better tracking. The backcoat often contains carbon-black that simultaneously lowers the tape's resistivity and thereby prevents electrostatic buildup.

OTHER BASE FILMS

The earlier mentioned Polyimide film (Kapton, tradename by DuPont) is expensive and is mainly applied in tapes for flight recorders. A stronger film is a thermoplastic film called PEN; it has 20% higher mechanical strength and thermal stability. The polymer processing is similar to PET, and existing tape production technology can be used. PEN can, therefore, replace PET for long-play versions of existing tape formats, at a slight increase in cost.

A new high-performance substrate, PBO, was announced in 1992 (Perettie). Its tensile strength is 2½ times that of PET, and tensile elongation 1–2% versus 150% for PET. Influences of temperature and humidity are negligible. For comparison of the media, see Table 11.2.

The PBO (Polybenzoxazole) is a rigid-rod polymer being developed for numerous applications (Perettie et al. 1993). Tape substrates have been made in thicknesses down to 2 μm. PBO will withstand 500–600°C for short periods and continuous use to 350°C. It is therefore a good candidate for a base film for metal-sputtered ME tape. Commercial development of PBO films is not underway due to negative financial marketing prospects. Other, similar substrates may appear. One negative aspect of the steadily increasing data density per area is that less and less tape or disk is required!

RIGID DISK SUBSTRATES

The definitive material for rigid disk substrates is aluminum, ranging in thicknesses from 6 to 0.6 mm; brass has been used, but is a thing of the past. Aluminum is produced by the electrolytic reduction of aluminum oxide dissolved in a molten cryolite bath. The result can be 99.9 percent pure aluminum, with iron and silicon impurity elements. A large cast ingot is the start of operations involving rolling and annealing to relieve stresses.

Pure aluminum is very soft, so magnesium is therefore added (three to four percent, by weight). Magnesium dissolves in aluminum, providing a solid solution that has high structural strength. A small amount of manganese is added to absorb the iron impurities into intermetallic compounds. The manganese improves corrosion resistance, which is enhanced by the further addition of chromium (Westerman 1986).

Circular blanks are stamped or cut from the finished aluminum sheets. They must be perfectly flat; flatness may be achieved by additional machining on a lathe, followed by a polishing operation. Disks turned with a diamond tool have the smoothest surfaces.

A nickel alloy of 10 μm approximate thickness is electroplated onto the aluminum disk. This provides a layer that covers up faults in the aluminum surface, and it is polished to a smoothness of about 50 Å prior to texturing. This type of substrate remains the king in disk making and applications, in spite of several new glass/ceramic materials.

Rigid disks have been used in various applications since the 1950s. The first Winchester-class fixed disk drive with brown oxide coatings was introduced in 1972. It was IBM's 3340, a 60 MB disk drive the size of a washing machine. The disk substrate was plain aluminum.

Plating technology was tried in the manufacture of disks with a thin metal film. First, the aluminum substrate is nickel plated, then the magnetic layer is applied by immersing the substrate in a solution containing a high-coercivity cobalt alloy. The key problem in these disk were surface asperities, or raised areas on the surface. They resulted from the growth of crystals in the plating process.

Plated disk had another problem in corrosion. The plating process limits the selection of cobalt alloys that can be used for disks to only highly corrosive cobalt phosphorous (Payne 1990). The deposition technology therefore shifted to sputtering, which eliminated asperity and corrosion problems. Other materials than aluminum can be used for a substrate to sputter onto, such as an insulator, because the sputtering can be assisted by an RF (radio-frequency) power supply, which permits a nonconductive substrate.

The final disk surface should ideally be free of any voids or asperities and have a perfect mirror finish, with surface smoothness less than 0.006 μm (0.25 μin). This is necessary in order to fly the head at distances of 0.05 μm (2 μin) or less (quasi-contact).

Such a smooth surface presents a stiction problem when the head is parked, landing on a mirror-finish surface. A molecular adhesion between the head glider and disk surface will gradually develop, and increased bonding will result after some time (see Chapter 18). The bond may become so strong that it will hinder startup of disk rotation.

It has therefore become standard practice to texture the disk surface after its original coating and finish. This will produce small asperities upon which the head gliders will rest. Granted, the unit area pressure at those points will increase, but the net effect is a lowered coefficient of friction, and less tendency to freeze a drive by head-media stiction.

The controlled asperities can be produced by a random buffing of the surface or by controlled turning with a diamond tool. In modern disks these processes to roughen the surface are only done on the inner section of the disks, leaving a perfectly smooth surface for the large, outside area wherein writing and reading of data take place (zone texture).

Aluminum-magnesium has over time been the chosen substrate for rigid disks. There are other qualified candidates, such as glass, ceramics, silicon carbides, and carbon. But established disk drive designs cannot readily be switched to use a substrate that has a different coefficient of thermal expansion from the original design. The potential for new substrates are in novel disk drive designs (Payne 1990; McLaughlin 1986). Hence, these newer substrates are found in small disk drives for laptop computers.

In a review of the requirements of a substrate, we find some key specifications:

- Young's modulus needs to be high so that the disks can be made thinner. This is very important today in order to cram many disks into a small volume. Current substrate thicknesses are:

Diameter 130 mm	t = 1.25 mm
Diameter 95 mm	t = 0.80 mm
Diameter 65 mm	t = 0.64 mm
Future, small ceramic	t → 0.25 mm

- Young's modulus must be high to reduce flutter at high rpm. The typical value is 70 to 90 GPa for Al alloy, the NiP overcoat, and a variety of glass and ceramic disks. The mechanical structure of the substrate will also affect the magnitude and nature of clamping errors when the disk is mounted on the spindle motor (Mann, D. 1994).

- The substrate must exceed a certain hardness to prevent damage during CSS and head slap against the disk. Aluminum is only rated at 70 kg/mm^2, while it is 450 for NiP; hence the overcoat of the latter. Glasses and Canasite have values of 600.

- Substrates for very high density and low head flying heights must be super smooth, and the head must land in a designated area only on the zone-textured disk. The uneven surface caused by texturing has until recently been ignored in noise considerations, but it has been found that disks with substrate texture exhibit significant modulation noise as compared to highly polished disks (Wu et al. 1994).

Glass or ceramic disks certainly satisfy the main requirements listed above. There are a number of other factors to consider. It has been shown that the magnetics of a well known film (CoCrTa) are very different on a glass substrate than the conventional AlMg/NiP substrate (Duan et al. 1994).

Recent findings also point to improvements that may be obtained by coating a glass substrate with additional thin reactive films before the sputtering of the magnetics (Tang et al. 1994). General evaluations of glass substrate disks (Mann, D. 1994; IDEMA Symposium 1993; and H.C. Tsai 1993) and flying characteristics (Tsai, H.C. et al. 1991) must enter the designer's evaluation before selecting a media/substrate combination for a drive.

REFERENCES

PARTICULATE MEDIA

Auweter, H., R. Feser, H. Jakusch, M.W. Muller, N. Muller, E. Schwab, and R.J. Veitch. Jan. 1990. Chromium Dioxide Particles for Magnetic Recording. *IEEE Trans. Magn.* MAG-26 (1): 66–68.

Bate, G., and J.K. Alstad. Dec. 1969. A Critical Review of Magnetic Recording Materials. *IEEE Trans. Magn.* MAG-5 (4): 821–839.

Braginskij, G.I., and E.N. Timoteev. 1981. *Technologie der Magnetbandherstellung*, Akademie-Verlag, DDR—1080 Berlin, Leipziger Strasse 3–4, 320 pages. Chapter 4.

Chen, H.Y., K.M. Hiller, J.E. Hudson, C.J.A. Westenbroek. Jan. 1984. Advances in Properties and Manufacturing of Chromium Dioxide. *IEEE Trans. Magn.* MAG-20 (1): 24–26.

Chubacki, R., and N. Tamagawa. Jan. 1984. Characteristics and Applications of Metal Tape. *IEEE Trans. Magn.* MAG-20 (1): 45–47.

Corradi, A.R. Sept. 1992. Towards the Ultimate Oxide Particle. Presented at the *First Intl. Conf. on BaFeO*, Kalamata, Greece. 10 pages.

Corradi, A.R., S.J. Andress, J.E. French, G. Bottoni, D. Candolfo, A. Cecchetti, and F. Masoli. Jan. 1984. Magnetic Properties of New (NP) Hydrothermal Particles. *IEEE Trans. Magn.* MAG-20 (1): 33–38.

Daniel, E.D. March 1972. Tape Noise in Audio Recording. *Jour. AES*, 20 (2): 92–99.

Fayling, R.E., and S.A. Bendson. Sept. 1978. Magnetic Recording Properties of SmCo5. *IEEE Trans. Magn.* MAG-14 (5): 752–755.

Fujiwara, T. Sept. 1985. Barium Ferrite Media for Perpendicular Recording. *IEEE Trans. Magn.* MAG-21 (5): 1480–1485.

Isshiki, M., T. Suzuki, T. Ito, T. Ido, and T. Fujiwara. Sept. 1985. Relations between Coercivity and Recording Performances for Ba-Ferrite Particulate Perpendicular Media. *IEEE Trans. Magn.* MAG-21 (5): 1486–1488.

Knowles, J.E. Nov. 1981. Magnetic Properties of Individual Acicular Particles. *IEEE Trans. Magn.* MAG-17 (6): 3008–3013.

Köster, E. 1993. Trends in magnetic recording media. *Jour. Magn. and Magn. Matls.* 120 pp. 1–10.

Kubo, O., T. Ido, and H. Yokoyama. Nov. 1982. Properties of Ba-Ferrite Particles for Perpendicular Magnetic Recording Media. *IEEE Trans. Magn.* MAG-18 (6): 1122–1125.

Liang, J-L., and D.E. Nikler. Nov. 1993. Amine-Quinone Polyurethanes as Binders for Metal Particle Tape. *IEEE Trans. Magn.* MAG-29 (6): 3649–51.

Podolsky, G. Nov. 1981. Relationship of Gamma-Fe_2O_3 Audio Tape Properties to Particle Size. *IEEE Trans. Magn.* MAG-17 (6): 3032–3034.

Sharrock, M.P. Jan. 1990. Time-Dependent Magnetic Phenomena and Particle-Size Effects in Recording Media. *IEEE Trans. Magn.* MAG-26 (1): 193–197.

Sharrock, M.P. Nov. 1989. Particulate Magnetic Recording Media: A Review. MAG-25 (6): 4374–89.

Speliotis, D., P. Judge, W. Lynch, J. Burgage, and R. Keirsted. 1993. Extremely Narrow Switching Fields in Oriented Ba-Ferrite Particulate Media. *Jour. Magn. and Magn. Matls.* 120 pp. 172–176. See also *IEEE Trans. Magn.* MAG-29 (6): 3625–27.)

Umeki, S., S. Saitoh, and Y. Imaoka. Sept. 1974. A New High Coercive Particle for Recording Tape. *IEEE Trans. Magn.* MAG-10 (2): 655–657.

Veitch, R.J., E. Held, H. Jakusch, and R. Körner. Nov. 1993. Chromium Dioxide Recording Tape with an Extremely Narrow Switching Field Distribution. *IEEE Trans. Magn.* MAG-29 (6): 3637–39.

DEPOSITED THIN FILMS

Chen T., and R.M. Martin. Nov. 1979. The Physical Limits of High Density Recording in Metallic Magnetic Thin Film Media. *IEEE Trans. Magn.* MAG-15 (6): 1444–1446.

Desserre, J., and D. Jeanniot. Nov. 1983. Rare Earth—Transition Metal Alloys : Another Way for Perpendicular Recording. *IEEE Trans. Magn.* MAG-19 (6): 1647–1649.

El-Hilo, M., K. O'Grady, R.W. Cantrell, I.L. Sanders, M.M. Yang, and J.K. Howard. Nov. 1991. Interaction Effects in Multi-Layer Thin Film Media. *IEEE Trans. Magn.* MAG-27 (6): 5061–63.

Feuerstein, A., and M. Mayr. Jan. 1984. High Vacuum Evaporation of Ferromagnetic Materials—A New Production Technology for Magnetic Tapes. *IEEE Trans. Magn.* MAG-20 (1): 51–56.

Hsu, Y., J.M. Sivertsen, and J.H. Judy. Sept. 1990. Texture Formation and Magnetic Properties of RF Sputtered CoCrTa/Cr Longitudinal Thin Films. *IEEE Trans. Magn.* MAG-26 (5): 1599–1601.

Inagaki, N., S. Hattori, Y. Ishii, A. Terada, and H. Katsuraki. Nov. 1976. Ferrite Thin Films for High Recording Density. *IEEE Trans. Magn.*, Nov. 1976, Vol. MAG-12, No. 6, pp.785–787.

Johnson, K.E. 1992. Fabrication of Low Noise Thin-Film Media. *Noise in Digital Magnetic Recording* ed. by Arnoldussen, T.C., and L.L. Nunnelley, World Scientific, Singapore, pp. 7–63.

Paik, C.R., I. Suzuki, N. Tani, M. Ishikawa, Y. Ota, and K. Nakamura. Sept. 1992. Magnetic Properties and Noise Characteristics of High Coercivity CoCrPtB/Cr Media. *IEEE Trans. Magn.* MAG-28 (5): 3084–86.

Procker, L. Nov. 1993. Media Technology—Evolution or Revolution. *Head/Media Technology Review*, Las Vegas.

Sagoi, M., R. Hishikawa, and T. Suzuki. Sept. 1984. Film Structure and Magnetic Properties for Co-Cr Sputtered Films. *IEEE Trans. Magn.* MAG-20 (5): 2019–2024.

Sanders, I.L., T. Yogi, J.K. Howard, S.E. Lambert, G.L. Gorman, and C. Hwang. Sept. 1989. Magnetic and Recording Characteristics of very thin metal-film media. *IEEE Trans. Magn.* MAG-25 (5): 3869–71.

Yogi, T. and T.A. Nguyen. Jan. 1993. Ultra High Density Media: Gigabit and Beyond. *IEEE Trans. Magn.* MAG-29 (1): 307–316.

Zhu, J-G. Sept. 1992. Modelling of Multilayer Thin Film Recording Media. *IEEE Trans. Magn.* MAG-28 (5): 3267–69.

FLEXIBLE SUBSTRATES

Bhushan, B. 1992. Physical and Chemical Properties of PET Substrate and Coated Magnetic Media. Chapter 2 in *Mechanics and Reliability of Flexible Magnetic Media*, Springer-Verlag, pp. 85–163.

Braginskij, G.I., and E.N. Timoteev, *Technologie der Magnetbandherstellung*, Akademie-Verlag, DDR—1080 Berlin, Leipziger Strasse 3-4, 1981, 320 pages, Chapter 3.

Campbell, R.W. May 1986. Biaxially Oriented Polu(Phenylene Sulfide) Film. *SMART Symposium.* Paper No. WS 1-D-1, 16 pages.

Goerlitz, W. and A. Ito. 1993. Substrates for flexible magnetic recording media: The role of base films for modern performance requirements. *Jour. Magn. and Magn. Matls.* 120, pp. 76–82.

Heffelfinger, C.J. May 1986. Improved Performance Polyester Films. *SMART Symposium.* Paper No. WS 1-C-1, 26 pages.

Mark, H.F. 1968. Giant Molecules. *Time-Life Books*, 200 pages.

Mark, H.F. Sept. 1957. Giant Molecules. *Scient. Amer.* 197 (3): 81–89.

Ochsner, J.P. May 1986. Polyimide Films for Recording Substrates. *SMART Symposium.* Paper No. WS 1-D-2, 17 pages.

Perettie, D. Oct. 1992. High-performance, flexible substrates look promising for recording media. *NML Bits* 2 (4): 1–6.

Perettie, D., W. Hwang, T. McCarthy, P. Pierini, D. Speliotis, J. Judy, and Q. Chen. 1993. A high-performance, flexible substrate for thin-film media. *Jour. Magn. and Magn. Matls.* 120, pp. 334–37.

RIGID SUBSTRATES

Duan, S., B. Zhang, C. Gao, G.C. Rauch, J.L. Pressesky, and A. Schwartz. Nov. 1994. A Study of Magnetic Recording Media on Glass Substrates. *IEEE Trans. Magn.* MAG-30 (6): 3966–68.

Mann, D. Aug. 1994. Substrate Flatness and Runout Velocity Acceleration Research. *Head/Media Research Journal*, Vol. 1, 120 pages.

McLaughlin, H.J. May 1986. Problems in Non-Aluminum Substrates for Rigid Disk Media. *SMART Symposium.* Paper No. WS 1-B-3, 22 pages.

Payne, H. Nov. 1990. Substrate Advances Critical to Small Drive Design. *Computer Techn. Review*, 10 (14): 49–55.

Sept. 1993. *IDEMA 2nd Symposium on Alternative Substrates*. 230 pages.

Tang, X., B. Reed, R. Zubeck, D. Hollars and K. Goodson. Nov. 1994. High Coercivity and Low Noise Media Using Glass Substrate. *IEEE Trans. Magn.* MAG-30 (6): 3963–65.

Tsai, H-c. Jan. 1993. Advantage and Challenge of Nonmetallic Substrates for Rigid Disk Applications. *IEEE Trans. Magn.* MAG-29 (1): 241–46.

Tsai. H-c., and A. Eltoukhy. Nov. 1991. *IEEE Trans. Magn.* MAG-27 (6): 5142–44.

Westerman, E.J. May 1986. Improvements and Problems in Aluminum Alloys for Magnetic Media Substrates. *SMART Symposium.* Paper No. WS 1-A-1, 27 pages.

Wu, E.Y., J.V. Peske, and D.C. Palmer. Nov. 1994. Texture-Induced Noise and Its Impact on Magnetic Recording Performance. *IEEE Trans. Magn.* MAG-30 (6): 3996–98.

BIBLIOGRAPHY

GENERAL INTEREST

Bate, G. 1980. Recording Materials, Chapter 7, Ferromagnetic Materials, Vol. 2, North-Holland, 126 pages.

Craig, D.J.(Editor). 1975. *Magnetic Oxides*, John Wiley and Sons, 2 volumes, total 798 pages.

Granum, F., and A. Nishimura. July 1979. Modern Developments in Magnetic Tape. Intl. Conf. Video and Data 79, IERE Conf. Proc. No. 43, pp. 49–61.

Koester, E. 1995. "Recording Media", in *Magnetic Recording*, Ed. by C.D. Mee and E.D. Daniel, McGraw-Hill, pp. TBD.

Monson, J.E. 1995. "Recording Measurements", in *Magnetic Recording*, Ed. by C.D. Mee and E.D. Daniel, McGraw-Hill, pp. TBD.

Sakai, H., K. Hanawa, and K. Aoyagi. Nov. 1992. Preparation and Magnetic Properties of Barium Ferrite Fine Particles by the Coprecipitation Salt-Catalysis Method. *IEEE Trans. Magn.* MAG-28 (6): 3355–62.

Sharrock, M.P. Jan. 1990. Anisotropy and switching behaviour of recording media; Comparison of Barium Ferrite and acicular particles. *IEEE Trans. Magn.* MAG-26 (1): 225–227.

Speliotis, D.E. Sept. 1987. Distinctive Characteristics of Barium Ferrite Media. *IEEE Trans. Magn.* MAG-23 (1): 25–28.

Speriosu, V.S., D.A. Herman, I.L. Sanders, and T. Yogi. Nov. 1990. Magnetic Thin Films in Recording Industry. *IBM Jour. Res. & Dev.* 34 (6): 884–902.

Suzuki, M. 1991. Description of Particulate Assemblies. *Powder Technology Handbook*, pp. 73–88.

Yamamori, K., T. Tanaka, and T. Jitosho. Nov. 1991. Recording Characteristics for Highly Oriented Ba-Ferrite Flexible Disks. *IEEE Trans. Magn.* MAG-27 (6): 4970–62.

SUBSTRATES

Kaempf. G., H. Loewer, and M.W. Witman. Oct. 1987. Polymers as Substrates and Media for Data Storage. *Polymer Engr. and Sci.* 27 (19): 1421–35.

Lebourvellec, G., J. Beautemps, and J.P. Jarry. Jan. 1990. Stretching of Pet Films Under Constant Load. 1. Kinetics of Deformation. *Jour. Appl. Poly. Sci.* 39 (2): 319–28.

Moore, G.R., and D.E. Kline. 1984. *Properties and Processing of Polymers for Engineers.* Soc. of Plastic Engrs., Prentice-Hall, 209 pages.

Rodriquez, F. 1970. *Principles of Polymer Systems*. McGraw-Hill, 560 pages.

CHAPTER 12
MAGNETIC PROPERTIES OF TAPES AND DISKS

Magnetic recording surfaces are made from small magnetic particles or from deposited films. We discussed the basic magnetic properties of these materials in Chapter 11, and will now describe how they behave when finished into a coating for writing and reading information. We are particularly interested in those properties that relate to the performance of the write/read processes we examined in Chapters 4, 5, and 6.

We can briefly summarize how the shape and the magnitudes from the *BH*-, or better, *MH*-loop relate to performance:

- The current required for writing data onto a recording surface is proportional to the coating coercivity H_c.
- Long-wavelength read output is proportional to the value of B_{rsat} (= $\mu_o \times M_{rsat}$).
- Short-wavelength read output is reduced by demagnetization; this loss can be reduced by using a coating with a high value of coercivity H_c.
- The read output in digital recording is 75 percent of its possible maximum when the coating thickness equals $0.44 \times BL$ (bit length). A thicker coat produces only a small improvement in signal level, but increases demagnetization losses (which really is a function of the bit geometry) and peak shifts. The optimum coating thickness is therefore approximately 50 percent of the longest bit length (or 25 percent of the longest wavelength).
- Resolution (and packing density) is improved by having a *BH*- (*MH*-) loop with steep sides. This results if the particles have a low *SFD* (switching field distribution), and are well-oriented in the direction of the magnetization.

The coating's hysteresis loop gives us all essential information to evaluate the write/read performance. We will start the chapter by outlining two methods for the measurements of the magnetization curve, the *MH* meter and the VSM (vibrating sample magnetometer). Either instrument will serve as long as it is properly calibrated, but there can be discrepancies between results from the two instruments. This issue has recently been resolved, and the cause will be described (time effects in materials magnetism). The *BH*-loop for a given material will vary in accordance with the particle loading, orientation, etc. Knowledge of the exact behavior is important for the optimum design of a media coating, as well as for production quality control. It is also desirable to know the magnetization loop of a single particle, and experimental data are reported to illustrate this.

The chapter concludes with a review of the important issue of noise, from a particulate and also from thin-film media.

MEASUREMENTS OF THE MH-LOOP

The coercivity H_c represents a tape's recording sensitivity and its resistance to change, and is an important specification. It can be measured on a *MH* meter, VSM, or the newer Kerr optical instruments; some researchers also use torque meters.

MH METER

A coated tape or disk sample, which may be several layers thick, is placed inside a large field coil in the *MH*-meter (Fig. 12.1) (Newman 1978). It is magnetized by the 60 Hz field H_a, and the change in flux ϕ is sensed by the small coil around the sample. This coil also picks up the field H_a, but this component is cancelled by the voltage induced by H_a in a series connected balance coil.

The induced signal $d\phi/dt$ is integrated, amplified, and connected to the vertical deflection on an oscilloscope. A small resistor in the bottom of the field coil provides a voltage proportional to H_a, which provides the horizontal deflection. Thus the *MH*-loop is displayed on the scope, and may be photographed or otherwise plotted for reading the magnitude of H_{ci} and M_{rsat}. The *BH*-loop is obtained when the voltage from the compensating coil is shorted. From the *BH*-loop we determine H_c and B_{rsat}.

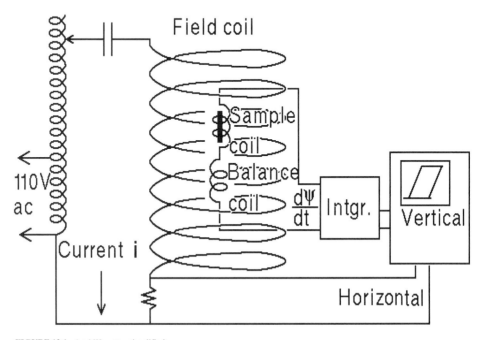

FIGURE 12.1 An *MH*-meter, simplified.

VIBRATING SAMPLE MAGNETOMETER (VSM)

The VSM instrument (Fig. 12.2) operates by measuring the magnetization in a sample by mechanical oscillation of the sample in a gradient sensing field provided by two coils. The motion of the sample can be in the same direction of the field (Newman 1978), or perpendicular thereto (Foner 1959).

The sensing mechanism of the coils is much like the sensing field from a head; it is passive, but senses magnetization. The sample's magnetization may be some remanent state from an earlier magnetization process, or it can be brought about by a magnetic dc field (which itself it not sensed).

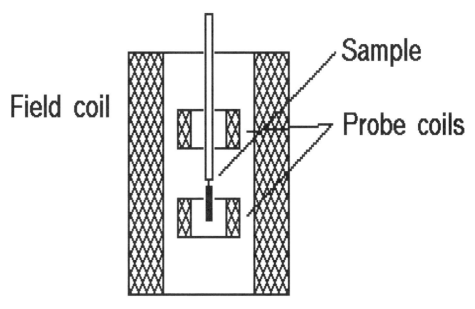

FIGURE 12.2   A VSM, simplified. In this configuration the field is parallel to the sample motion.

MAGNETO-OPTIC KERR MEASUREMENT

A polarized light beam will undergo a slight phase change when reflected off a mirror-finished surface of a magnetized material sample. This is the principle in magneto-optical recording, briefly addressed in Chapter 30. The reflected beam carries information about the magnetics of the material and will, after signal processing, produce the material's *MH*-loop, from which several parameters like H_r, S, and SFD can be determined (Krafft et al. 1986; Josephs 1989.)

Figure 12.3 shows the principle in the magneto-optical instrument. It can measure magnetic properties in a nondestructive fashion anywhere on a thin-film disk surface. Figures 12.17 and 12.18 are examples. Also, thin-film magnetics in TFH fabrication processes can be measured.

The ring magnet in Fig. 12.3 can produce a field of up to 8000 Oe, and so it is possible to measure high-coercivity materials. An *MH*-loop is readily measured in a second and a half.

SYMMETRICAL READ HEAD

There is finally a "poor man's" method to determine the absolute flux level; it requires a symmetrical playback head, where the geometries of the front and back gaps are identical (Fig. 12.4). The core

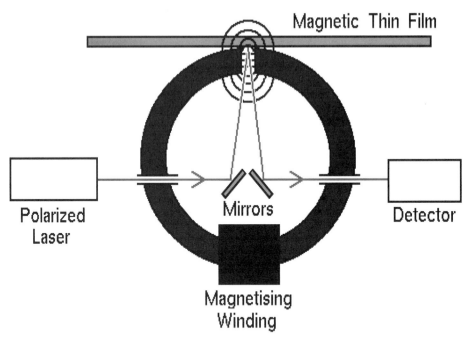

FIGURE 12.3   Magneto-optical Kerr *MH*-meter (*after R.M. Joseph, 1989*).

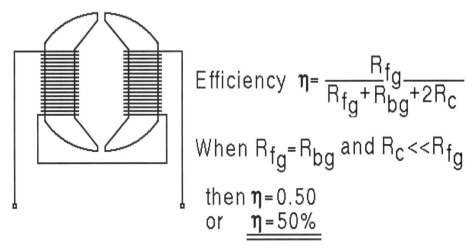

FIGURE 12.4   Symmetrical read head with efficiency equal to 50.0 percent.

halves are made from high-grade laminated mu-metal in order to assure that the core reluctance is at least 1000 times less than the gap reluctances at long wavelengths. The flux is determined from:

$$e = n \times \eta \times d\phi/dt \ (= E \times \cos\omega t)$$
$$= n \times 0.5 \times \omega \times \phi_m \times \cos\omega t$$

where $\omega = 2\pi f$ and $\eta = 0.5$.

Rewriting yields:

$$\phi_m = E/(0.5 \times \omega \times n)$$

Dividing with the trackwidth w (in m) gives:

$$\phi_m = E/(0.5 \times \omega \times n \times w) \text{ [Wb/m]}$$

and additional division with the coating thickness δ (in m) gives:

$$B_{rs} = E/(0.5 \times \omega \times n \times w \times \delta) \text{ (Wb/m}^2, \text{ volts)} \tag{12.1}$$

SAMPLE SIZE

The samples are usually small pieces of tape, a few inches long for the *MH*-meter, and small squares for the VSM. Several layers may be used to provide increased signal output. The sample preparation is more difficult for disks, which actually have to be cut up so a square centimeter sample can be provided. An aluminum substrate will cause eddy current losses in the *MH*-meter, and results may be erroneous (Newman 1978). Errors will also prevail in the VSM if the substrate is conductive and the field gradient is not constant (Kerchner et al. 1984).

Measurements in the plane of the coating seldom require any correction for demagnetization. For a circular tape coating sample of diameter one cm and coating thickness 5 μm = 0.0005 cm the diagonal demagnetization factor is zero (longitudinal or transverse), while it is almost one in the perpendicular direction.

CORRECTION FOR DEMAGNETIZATION IN SAMPLE

It is necessary to correct the *BH*-loops for the quite large demagnetization factor involved in perpendicular measurements. An approximation for the demagnetization factors, diagonally (N_x and N_y) and perpendicularly (N_z) are:

$$N_x = N_y = 1.13 \times \delta/a \tag{12.2}$$

$$N_z = 1 - 2.26 \times \delta/a \tag{12.3}$$

where δ is the sample thickness and a is the sample diameter. (Bate 1978; the formulas are changed to MKS values for N. The nominal maximum is 1.)

Bate did find that the perpendicular properties for a longitudinally oriented coating could be determined from the transverse properties (diagonally, 90 degrees to the longitudinal direction); this eliminates the need for correction of demagnetization. However, this will not apply to a perpendicularly oriented coating without the cylindrical symmetry of the angular distribution.

The correction for demagnetization is normally done after completion of the measurement. This will only correct for the component of the demagnetization field that is parallel to the applied field. It is possible to correct for the total demagnetizing field by varying the applied field strength and simultaneously rotating the sample (Bernards and Cramer 1991). The method was applied to samples of ME tape and a CoCr layer, and the intrinsic magnetic properties of the two were very similar.

COERCIVITY, REMANENT FLUX, AND SFD

We will now examine longitudinal, transverse, and perpendicular properties.

LONGITUDINAL AND TRANSVERSE *MH*-LOOP

The level of magnetization in a tape or disk coating may be determined by formula 11.1, and it is proportional to σ_s = a particle's specific magnetization, which depends on the particle material used. It is also proportional to the particle loading, or volume fraction p, the coating thickness δ, and the squareness ratio S. The latter three factors must therefore be included in a comparison of various tapes.

The largest variation will be found in the coating thickness, which is 15 μm (600 μin) for high-output audio tapes and early computer tapes, 6 to 3 μm (240 to 120 μin) for most other tapes, and 3 to ½ μm (120 to 20 μin) for high-resolution tapes and disks. ϕ_{rsat}, which is proportional to the coating thickness, can vary widely without being an indicator of the tape's or disk's merit, as long as the medium provides an adequate playback or read voltage.

The remanent magnetization of all tape and disk products includes the influence of particle orientation. Perfectly aligned, ideal particles should possess no remanence in directions perpendicular to the alignment direction. This corresponds to a squareness ratio of one, which is never achieved. The net result is a transverse remanence that is less than the longitudinal remanence. It can be measured in properly oriented samples with a *MH*-meter or VSM.

Another factor, the orientation factor, is introduced to illustrate how good the orientation is. The orientation factor is simply the ratio between the saturation remanence in the longitudinal direction and the saturation remanence transverse thereto. Both are measured in the coating plane with perpendicular (sometimes called vertical) magnetization, at ninety degrees to the coating plane.

The transverse remanence should equal the longitudinal remanence in flexible disk products to avoid cyclic patterns in the read signal. This ideal situation is found in a peripheral orientation, as in a rigid disk. An orientation in the coating plane is inherent in the extrusion of the wet coating onto the substrate for flexible disks, and any further alignment is discouraged by mechanical or electromagnetic agitation. This can, for instance, be achieved by a plane rotating orienter field. The squareness of a diskette cannot, therefore, exceed 0.7.

Figure 12.5 shows the magnetization loops for a longitudinally oriented coating; it is highly desirable for the designer of a write/read system to have both loops available from the media manufacturer so he or she can make a fair estimate of the written magnetization in a coating (see Chapters 19 and 20).

PERPENDICULAR PROPERTIES

The curvature of the record-head field lines will tend to generate a curved magnetization pattern in the coating, being perpendicular at the coating surface for in-contact recordings. The remanence pattern will be modified by demagnetization, and by the longitudinal alignment of the particles. The perpendicular remanence will be in the order of X percent of the longitudinal, where X is the ratio between perpendicular and longitudinal saturation remanence (at long wavelengths). Pure perpendicular remanence may prove advantageous at very high packing densities.

THE REMANENCE *MH*-LOOP

The graphs have so far provided data for M that result while the sample is exposed to a certain field strength. These are named hysteresis loops.

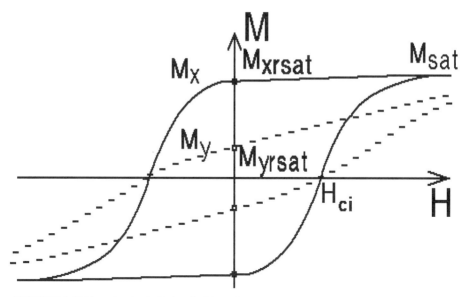

FIGURE 12.5 *MH*-loops for a longitudinal media ($M_{xrsat} > M_{yrsat}$).

In recording, we are ultimately interested in the magnetization that remains after field exposure, the remanence. The magnetization will, after field removal, move along a recoil line. This was used as a final correction to demagnetization losses discussed in Chapter 4 (see Fig. 4.12).

The measurement of the remanence MH_r-loop is easily done by applying the VSM or magneto-optical method. The sample is exposed to a range of field strengths that each decrease to zero, and the resulting residual flux, i.e., magnetization, is measured and entered in the data bank (Speliotis 1990). A plot of a remanence loop MH_r is shown in Fig. 12.6.

SWITCHING FIELD DISTRIBUTION; PREISACH'S DIAGRAM

Several methods are used for the measurement of the switching field distribution *SFD*, a quantity that was introduced in Chapter 5. It is a measure of a particulate coating's ability to record short transitions; i.e., a small *SFD* results in high resolution (for ac-bias recordings as well).

The recommended definition of *SFD* is the magnetizing field range wherein 50 percent of the coating's particles will switch magnetization. The *SFD* is then named Δh_r, and is derived from the remanence MH_r-loop as shown in Fig. 12.7.

Another way to determine the *SFD* is made by drawing the tangent to the *MH*-loop at $M = 0$ (step A in Fig. 12.8). Next draw a line through point $(0,M_{rs})$, parallel to the *H*-axis (step B). The distance from the intersection between the first drawn tangent and the M-axis equals $S^* \times H_{ci}$ (step C). S^* is then $S^* H_{ci}/H_{ci}$, and *SFD* is:

$$SFD = 1 - S^*$$

Values of $1 - S^*$ have been found to closely match the Δh_r values for the same coatings; it is therefore recommended as an easy method to determine the *SFD* (Kneller and Koester 1977).

It has also been suggested that the *MH*-loop be differentiated, and the width of the differentiated curve measured at the level that is 50 percent of its maximum value. This method correlates with the

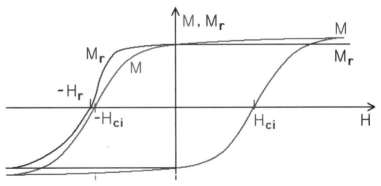

FIGURE 12.6   An *MH* hysteresis loop and remanent *MH*$_r$-loops.

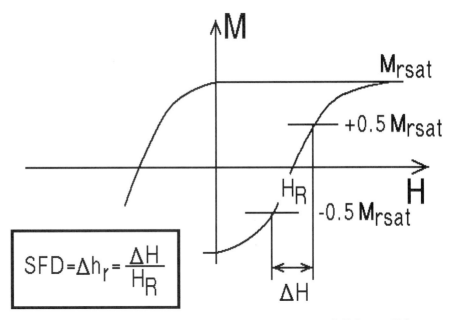

FIGURE 12.7   The definition of the switching field distribution SFD = ηh_r, and how to find it from an *MH*$_r$-loop.

PW50 measurements of read signals from a single transition; *PW50* (in time, for example in μsec-onds) equals the pulse width at the 50-percent amplitude points). The *SFD* measurement on the *MH*-loop results in a value that does not correspond well with Δh_r.

A combined display of the *SFD* plus the interacting fields is possible by using a so-called Preisach diagram. A series of ac-bias measurements are needed to get the data points (see Chapter 19), and a three-dimensional plot of the distribution of the particles' coercivities (h_{ci}) and their interacting fields (ΔH) will result in an image like the one shown in Fig. 20.2. The coating sample is an audio tape. The two humps at low field levels are interesting: they tell us that the coating contains some large parti-cles (low coercivities h_{ci}) that are subject to some large interacting fields.

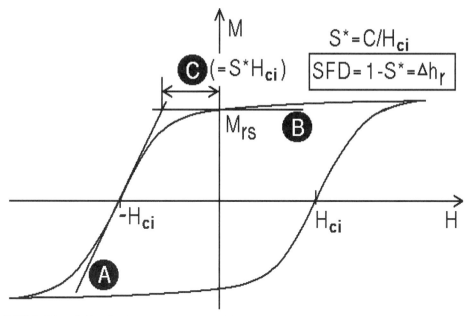

FIGURE 12.8 Definition of S^* for an MH_r-loop, and its relationship to ηh_r.

FACTORS INFLUENCING THE MAGNETIC PROPERTIES

The most obvious factors that influence the magnetic behavior of a magnetic coating is the orientation of particles, or actually the preferred direction of magnetization. The ratio between the extreme remanence values is the so-called orientation factor $= M_{rx}(\infty)/M_{rz}(\infty)$.

ORIENTATION

All media is made with the highest possible orientation on the magnetics into the direction of the media motion. This assures highest read output. A measure of this is the squareness:

$$S = M_{rx}(0)/M_{rx}(\infty)$$

PARTICLE SIZE

Particle sizes are made as small as possible, for two reasons: First, small particles assure the highest signal resolution; this principle is equivalent to that of photographic film. The finer the grains, the sharper the picture. Second, small particles assure low noise, as will be explained.

These guidelines have been verified by Podolsky (1981). Both sensitivity and saturated output increase for decreasing particle volume at a short wavelength, while the opposite is true for the long-wavelength output.

PACKING FACTOR *p*

The variation of coercivity with packing factor *p* is generally expressed as:

$$H_{ci}(p) = h_{ci}(1 - p) \tag{12.4}$$

where h_{ci} is the coercivity of an isolated particle. Equation (12.4) does not usually agree with experiments, and a novel computational approach (Monte Carlo treatment) combined with better understanding of the individual particles' magnetization switching modes (explained later in this chapter) provides better agreement (Knowles 1985).

MEASUREMENTS ON SINGLE PARTICLES

When particles are suspended in a liquid, one can observe their behavior through a microscope and do experiments on them; for instance, by turning them with a magnetic field (Scholten 1975). One approach has led to a fairly accurate way of finding the coercivities of these essentially single particles (Knowles 1980).

The particles are kept in a very dilute suspension, using a lacquer base with a viscosity of 100 P or more. This suspension was then sucked into a small glass tube, and an individual particle observed with the microscope; see Fig. 12.9. The particle was aligned by a continuously applied field of a few Oersted, and a large pulsed field was then applied in the opposite direction. If this pulsed field was larger than the coercive force h_{ci} of the particle, then the magnetization was reversed and the particle subsequently rotated through 180°. By a simple extension of the method, a remanent loop for the particle can be obtained. A histogram of many test results shows the coercivities for a typical tape sample; this curve corresponds to the distribution shown to the right in Fig. 3.66. These data permit the construction of the tape's overall magnetization loop, exactly as was done in Fig. 3.69.

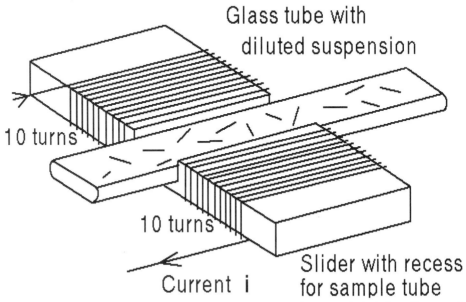

Glass tube with diluted suspension

10 turns

10 turns

Slider with recess for sample tube

Current i

FIGURE 12.9 Observation of the rotation of single particles in a dilute suspension; remagnetization is done with an external, pulsed field (*Knowles, 1980*).

Work continues in order to get a better idea about the magnetic behavior of particles, from the Stoner-Wohlfarth single-domain model to the Jacob-Beans multidomain particle with various magnetization modes—buckling, fanning, etc. (Knowles 1980, 1984). A recent investigation used the Foucault mode of Lorentz microscopy to detect the polarity of the field from a single-domain particle (Salling et al. 1991).

BEHAVIOR OF AN ASSEMBLY OF PARTICLES

All the measurements and associated experiments listed and discussed above will direct our design efforts toward a better particulate medium. This can today be supplemented with computations. A three-dimensional model has recently been used to present some impressive results by computing the characteristics of particulate media (Vos et al. 1993).

The model consists of a particle assembly with a few hundred particles, and includes particle shape effects (oblate or prolate spheroids), particle anisotropy distribution (*SFD*), particle position/orientation variation, and magnetostatic particle interactions. The total magnetization of the assembly is computed dynamically, integrating the Landau-Lifshitz-Gilbert differential equation in combination with a self-consistent iterative approach. The model generates results based on fundamental physics and does not require experimental inputs.

Figure 12.10 illustrates a couple of particle configurations, namely prolate (γFe_2O_3, CrO_2) and oblate (BaFeO). The computed hysteresis loops for representatives of these two particle types are shown in Fig. 12.11. Future work is planned to obtain more data and correlate these with measurements.

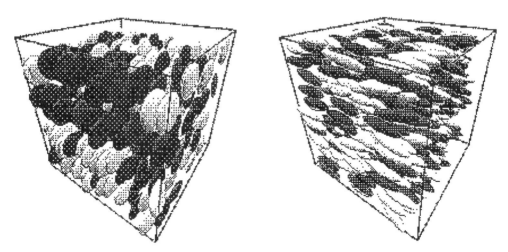

FIGURE 12.10 Particle assemblies for the computer simulation of magnetic properties. The left illustration shows a prolate particle assembly with $p = 20\%$, while the right illustration is an oblate particle configuration, also with $p = 20\%$ (*after Vos et al., 1993*).

AC-BIAS SENSITIVITY

Among other things, ac-bias greatly increases the record current sensitivity. This can be expressed as an amplification factor much like the alpha of a transistor, and is named the anhysteretic susceptibility χ_{ar}. It will be further characterized in Chapter 20.

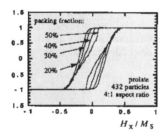 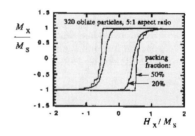

FIGURE 12.11 Computer hysteresis loops for prolate particle assemblies (left) and oblate particles assemblies (*after Vos et al., 1993*).

Kneller and Köster (1977) have found that χ_{ar}'s relation to the static magnetic parameters can be expressed as:

$$\chi_{ar} = p(1-p)M_{sat}^2/(2(1-S^2)H_r^2) \qquad (12.5)$$

where:

M_{sat} = saturation magnetization
p = packing fraction of magnetic particles
S = remanence to saturation ratio (= squareness)
H_r = remanence coercivity ($\approx H_c$)

Typical values for χ_{ar} are in the range of 15 to 25.

EFFECTS OF TIME

The coercivity of a magnetic recording material is an important parameter; it determines the required field strength (and, by extension, the required write current) for writing, and plays a role in the de-magnetization process (such as transition broadening) during long-term storage. It has been found that measurements that produce rapid changes of magnetization will yield a higher value of coercivity than those that operate on a longer time scale (Sharrock and McKinney 1981).

The two types of measurements that were described earlier in this chapter will show such a discrepancy: The *MH*-meter operates at 60 Hz (or 50 Hz) and the magnetization time is therefore less than $1/100$ of a second. The VSM, in contrast, operates at a magnetization time of 10 seconds or more.

The difference in coercivity values is due to the thermally assisted nature of the magnetic switching process. Figure 12.12 shows the decay for two materials. The one at the top is cobalt-treated Co-γFe_2O_3; at bottom is ordinary γFe_2O_3. The formula shows also that the difference between short- and long-term coercivities becomes more pronounced with time.

Sharrock and McKinney (1981) give an example where they use 10^{-2} seconds as an appropriate value of τ for the *MH*-meter, and 10 seconds for the VSM. The values for H_c are 322 Oe for the *MH meter*, and 288 Oe with the VSM. Signal amplitudes in magnetic tapes are also observed to decay with time; this decay is a function of the writing density, increasing with wavelength to 7.3 percent in 1 hour at 4 kfc/cm in one case (Kloepper et al. 1984).

Both phenomena are explained by time decays of magnetization, M, which are attributed to a thermal-fluctuation aftereffect proposed by Neel (1972), involving irreversible switching of interacting particles.

The temperature does also play a role, as the presence of T in the formula for H_c in Fig. 12.12 shows; its exact role is also the target of recent investigations (Oseroff et al. 1985).

The time dependence has received more attention recently as it is evident that the short-time coercivity is somewhat higher than the coercivity as measured with a VSM (Sharrock 1984 and

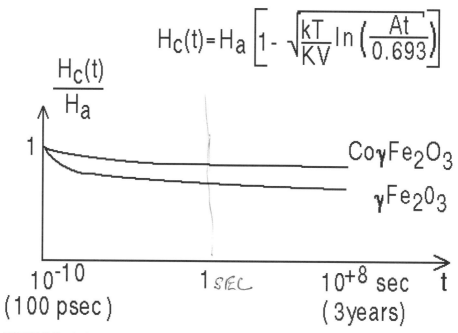

FIGURE 12.12 The time decay of coercivity (*after Sharrock and McKinley. 1981*).

1990; de Witte et al. 1993; Doyle et al. 1993). It is surprising that very little is known about this relationship for thin films, with the exception of one report (Lyberatos et al. 1990). Work does indicate that an even higher increase in H_c will occur at very short writing time, i.e., high data rates (W.D. Doyle 1995).

EFFECTS OF TEMPERATURE

Magnetization in general will decrease with increasing temperature, and eventually will cease to be present at and above the material's Curie temperature. It will recur when the temperature is lowered.

Figure 12.13 illustrates the variations in coercivities and remanences for modern magnetic particles (Kubo 1982; Fujiwara 1985). These changes may cause errors in exact measurements of write and read experiments, and also play a role in the exchange of recorded data (crossplay). An example will serve well to illustrate what might happen.

Example 12.1: Figure 12.14 shows the changed write (record) conditions that occur due to a change in temperature only, everything else remaining constant. The coating has a coercivity of $H_r = 300$ Oe at $T = 0°C$, and the writing is about 85 percent into the coating, and Δx equals $0.40 \times g$.

At the higher temperature $T = 50°C$, the write level has increased to 100 percent, and Δx has increased to $0.55 \times g$. Therefore, a lower resolution should result. This is confirmed upon read-out, but the higher level of magnetization is not realized into a higher flux because the remanence is also lowered at the higher temperature.

Read-out of the data at $0°C$ results in a higher flux level, but this may be masked by the longer Δx, i.e., smaller $d\phi/dt$! Any write/read experiments should be done at a constant temperature, which means that several hours should be allowed for equipment warmup.

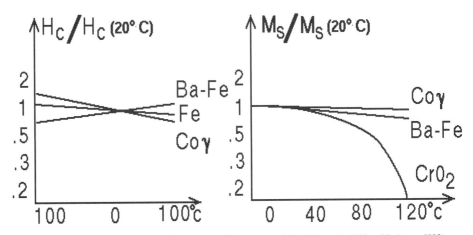

FIGURE 12.13   Changes in coercivities and remanences with temperature (*after Kubo et al., 1982 and Fujiwara, 1985*).

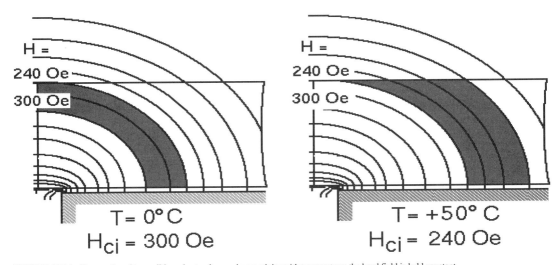

FIGURE 12.14   Changes in write conditions due to changes in coercivity with temperature; the head field is held constant

MIXTURE OF TWO OR MORE PARTICLE TYPES

Makers of audio tapes have sought an improvement of overall response by making a double-coated tape: First, a thick coating of standard coercivity is laid down, whereafter it is overcoated with a high-coercivity material. The bias current can now be adjusted for overbias of the thick undercoat; the record zone will still be quite short for recording the short wavelengths in the top coat. And only the top coat magnetization contributes to the short-wavelength output during read. (Remember: 75 percent of read flux comes from a media thickness of $0.22 \times$ wavelength.)

The situation is illustrated in Fig. 12.15. This technique improves the frequency response by providing the best of two worlds: a large signal output at long wavelengths due to the thick coating, and a good short-wavelength response due to a short recording zone.

The major shortcoming is a double transient response for a single transition. This causes audio signals from string instruments to have a muddled sound during the attack time (see Chapter 21 for

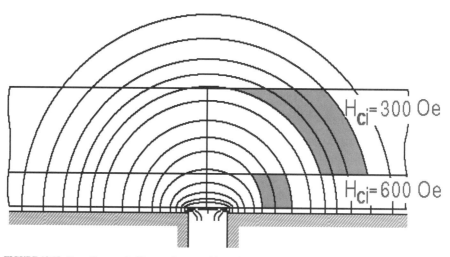

FIGURE 12.15 Recording on a double-coated tape provides optimum bias adjustment for long and short wavelengths simultaneously. Frequency response is very good, but transient response is very poor.

computation). The same problems will occur for the so-called multimodal media, which also will have two (or more) recording zones with transient responses centered around the average value of the coercivities. The *MH*-loops for a bimodal and a trimodal media, where the particles are mixed within one coating layer, are shown in Fig. 12.16 (Manly 1976).

Recent double-coat tapes provide improved response. A normal undercoat (1.5 μm) is first laid down, followed by a thin overcoat (0.5 μm). Both layers have the same coercivity in order to avoid the problem shown in Fig. 12.15. But the top layer is a higher-grade oxide with smaller *SFD*. Another possibility is that the first layer is nonmagnetic and merely enables the process of coating a very thin film (Fuji film data sheets).

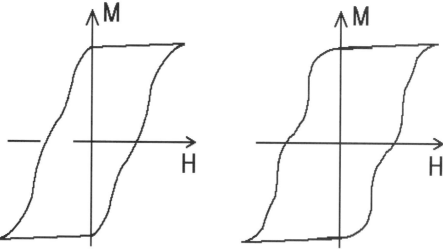

FIGURE 12.16 *MH*-loops for bi- and trimodal media (*after Manly, 1976*).

SPATIAL VARIATIONS IN MAGNETIC PROPERTIES

A medium's magnetic characteristics are not static, precise numbers; there is variation in the values for coercivity, magnetization, and so forth. When measured on the ordinary *MH* or VSM instruments, average numbers result. The Kerr magneto-optical instrument shown in Fig. 12.3 can provide detailed information about the magnetics around the entire disk surface. Figures 12.17 and 12.18 show the ranges of H_c and M_{rt}, and display a ±10 percent variation.

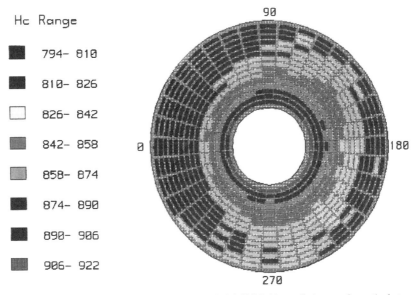

FIGURE 12.17   A range of coercivities measured on typical rigid disk drive media (*courtesy Innovative Instrumentation, Inc.*).

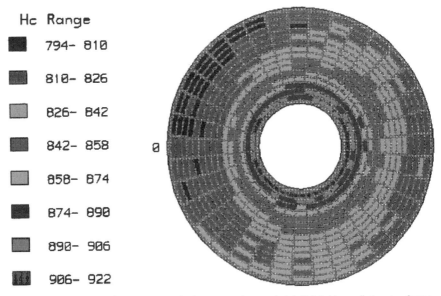

FIGURE 12.18   A range of remanent magnetizations measured on a typical rigid disk drive media (*courtesy Innovative Instrumentation, Inc.*).

NOISE FROM PARTICULATE MEDIA

Noise in magnetic recording has several sources, such as the electronic circuit's noise, head-loss resistance noise, man-made noise (radio interference), and ultimately noise from the tape or disk coating itself.

Its origin is best understood if we at first envision a highly diluted coating where a few individual particles are well separated from each other. Each particle is a permanent magnet with a field strength that decreases as the inverse cube to square of the distance (near field).

The fields from these individual particles are sensed and summed by the reproduce head (Fig. 12.19). This random noise voltage will, on a statistical basis, increase in proportion to the square root of their number n per unit of volume:

$$E_{noise} = K \times \sqrt{n}$$

where K is a proportionality factor.

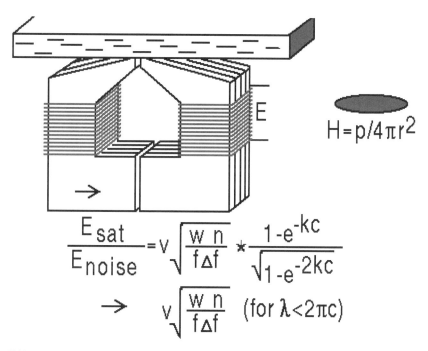

$$H = p/4\pi r^2$$

$$\frac{E_{sat}}{E_{noise}} = v\sqrt{\frac{w\,n}{f\Delta f}} * \frac{1-e^{-kc}}{\sqrt{1-e^{-2kc}}}$$

$$\rightarrow \quad v\sqrt{\frac{w\,n}{f\Delta f}} \quad (\text{for } \lambda < 2\pi c)$$

FIGURE 12.19 Noise voltage generated by separate particles in a coating (*after Mallinson, 1969*). v = media speed, w = track width, f = signal frequency, Δf = bandwidth observed, k = wavenumber = $2\pi f/v = 2\pi/\eta$, c = coating thickness, and $n = N/V$ - particle density (N = the number of particles in volume V).

A small number of particles will, when recorded, exhibit a sum remanence (in the direction of magnetization) that is proportional to the level of the recording field. Their number will, in an ac-bias recording (anhysteretic), be directly proportional to the record current, and their induced voltage is:

$$E_{signal} = K \times n \times i_w$$

where i_w is the write current amplitude and K is unchanged.

The ratio between the two voltages relates the signal-to-noise ratio:

$$\begin{aligned} SNR &= E_{signal}/E_{noise} \\ &= i_w \times n/\sqrt{n} \\ &= i_w \times \sqrt{n} \end{aligned}$$

(12.6)

This simple relationship shows that the signal-to-noise ratio is proportional to the square root of the number of particles per volume unit. We derived this relation in a grossly simplified way, and we should compare this result with the detailed analysis that resulted in the expression shown in Fig. 12.19 (Mallinson 1969).

These results show that the signal-to-noise ratio is proportional to tape velocity and the square roots of the track width and the particle density. In the latter fact lies the reasoning for today's quieter tapes, as compared with early low-density iron-oxide particle tapes. The new metal particle tapes have very small particle sizes (with attendant higher density) and, therefore, low noise characteristics.

The noise of a recorder system is often measured with a ½- or ⅓-octave band-pass analyzer. This allows for a closer examination of the noise spectrum. The electronics noise, for instance, is not flat, but boosted at both ends of the spectrum as a result of playback equalization. Erased tape noise is often slightly above that of the electronics and will differ according to tape type.

Studies into the particulate media noise mechanism remains intriguing and subject to several investigations (Mallinson 1991; Nunnelley et al. 1987; Thurlings 1985 and 1982).

AC-BIAS-INDUCED NOISE

A noticeable rise in the playback noise level occurs when a tape has been recorded with ac-bias (and zero data record current).

This matter was investigated by Ragle and Smaller (1965), who suggested that the increased noise was a result of "recording of particle interaction fields under the influence of the bias." A reduction of the bias noise occurs only when the tape is separated from the head. When the tape is in contact with the highly permeable head core, magnetic images of the flux in the tape form in the head core material. The particle noise flux from coating and image combine, in the worst case, to double the mean square noise flux in the coating, giving an increase of 3 dB in the root mean square value.

This increase would be temporary if it were not for the presence of the bias field. Daniel has offered the theory that the effect of the bias is to cause the reinforced noise flux to be anhysteretically recorded on the coating as it moves away from the gap. (This is essentially the "recording" process earlier suggested.)

An experiment verified the theory: a length of tape was cut from a reel, then interleaved with a portion of the remaining tape on the reel, so that coating-to-coating contact was made. The composite reel was bulk erased, the interleaved length of tape was removed, and the reel played back on a recorder: It was found that the noise spectrum was identical to the bias noise otherwise produced by the same tape. This confirmed the theory of rerecording of magnetic images of the noise (Daniel 1972).

DC NOISE

The average number of particles seen by the playback head will vary slightly as the tape moves by, and a random noise voltage is generated.

If, as an experiment, we feed a dc current through the recording head, we will observe a large increase in the noise voltage. This voltage is proportional to a varying magnetization along the tape and is related to a variation in the number of particles magnetized by the dc current.

When the dc current is so strong that the coating is completely saturated, the noise voltage is directly proportional to the variation in the number of particles sensed by the playback head. Occasional voltage impulses will indicate the presence of particle agglomerates or voids. If the dc record current is reduced to a value somewhat below the saturation value, the noise level will, in general, increase at low frequencies. This noise reflects discrete projections (drop-ins) or asperities (drop-outs) in the tape surface. The noise spectrum of drop-outs, which cause the tape to form a tent with circumference d, is (Daniel 1972):

$$\Delta e = f^{1.5} \times e^{-\pi df/v} \tag{12.7}$$

where Δe is the noise in ⅓-octave bands df, f is the center frequency, and v the tape velocity. From such measurements (see upper curve in Fig. 12.20) d has been determined to be about 7 mils.

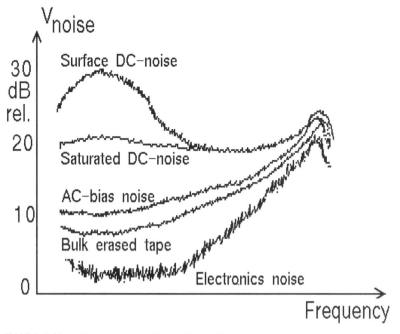

FIGURE 12.20 Noise spectra measured in an audio recorder/reproducer (*Daniel, 1972*).

A general conclusion to be drawn is that a uniform dispersion is essential in the preparation of low-noise tape. Surface treatment after coating may result in improved short-wavelength recording, but will have no effect on any imperfection already cast into the coating.

A note on standardization: the surface dc noise can be measured with a dc current that results in a maximum noise level. Or a value can be used that equals the effective level of a low-frequency ac current that results in a recorded level 10 dB below saturation (NAB, CCIR).

MODULATION NOISE

The playback of a nondistorted sine wave recording does not sound quite as good as its direct transmission to the loudspeaker. When the playback sine wave is displayed on an oscilloscope, it is also observed that its peaks cannot become sharply focused. This noise signal riding on top of the recorded signal is called modulation noise, and is similar in nature to the noise that arises when the record current is a dc current (Fig. 12.21).

The signal-to-noise ratio in digital recording is the ratio of the peak pulse voltage output to the dc-record noise voltage. It is the noise that is present during the peak of a pulse that determines the bit-error-rate in a pulse peak detection circuit.

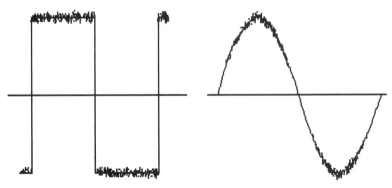

FIGURE 12.21 Modulation noise.

NOISE FROM THIN-FILM MEDIA

Thin-film media has taken on a very important role for the hard disk drives for computer data storage. Thin films provide a higher read voltage than their oxide-coated relatives, with satisfactory SNR. The thin-film disk would be the sole choice, were it not for problems with head crashes, contamination, and so on.

An ideal film would also be free from noise. The read head would sense alternating magnetizations separated by perfectly straight walls. In reality, the transitions are controlled by Neel spikes as shown earlier in Fig. 3.72.

Figure 12.22 is an early result of noise measurements (Belk et al. 1975). The noise spectra shown have been scrutinized, and the reader may find this data set referenced in several papers on noise in magnetic recording.

A substantial reduction of all types of noise was discovered when the CoCr layer was divided into thinner layers separated by nonmagnetic layers (Murdock et al. 1990). The separation between magnetic layers must be great enough to break the magnetic exchange coupling with just the magnetostatic coupling left. An average transition now exists, composed of individual transitions. They will

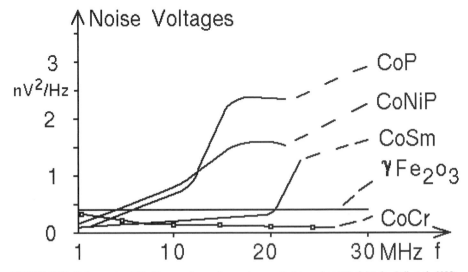

FIGURE 12.22 Noise spectra of thin film recording surfaces and one particulate surface (γFe_2O_3) (*after Belk et al., 1985*).

produce a signal flux that is proportional to the total magnetization from all n layers. The noise flux will only add as random contributions, i.e., proportional to $n^{1/2}$. For the signal-to-noise ratio we find an improvement of $n/n^{1/2} = n^{1/2}$.

Having three layers improves the SNR by 4.5 dB, but these disks become more expensive when several layers of magnetics and insulators are to be sputtered. A 25-Å Cr separation layer is adequate to decouple the two magnetic films (Min et al. 1991).

Further investigations showed that the segregation into small magnetic islands surrounded by nonmagnetic material offers a reduction in noise. This brings the analysis of thin-film noise mechanisms to an analysis similar to the one done for particulate media (Bertram and Che 1993). And the obvious conclusion is to make the number of islands large; i.e., make the islands small and break their exchange coupling. The model we arrive at is shown in Fig. 12.23 (Zhu 1992). A lucid account of noise mechanisms is found in Arnoldussen and Nunnelley's book from 1992.

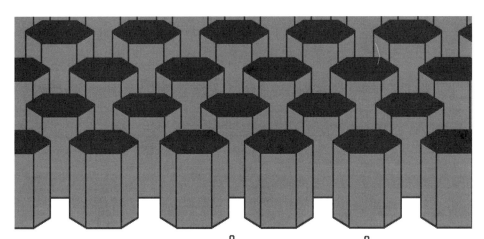

Grain Size: Height: 300 Å, Diameter 300 Å
No exchange coupling.
Magnetostatic Interaction constant $h_m = M_s/H_k = 0.2$.

FIGURE 12.23 A micromagnetic model for the computation of thin film media characteristics (*after Zhu, 1992*).

REFERENCES

MEASUREMENTS OF THE *BH*-LOOP

Bate, G. Sept. 1978. The Cylindrical Symmetry of the Angular Distribution of Particles in Magnetic Tapes. *IEEE Trans. Magn.* MAG-14 (5): 869–870.

Bernards, J.P.C., and H.A.J. Cramer. Nov. 1991. Vector Magnetization of Recording Media; A New Method to Compensate for Demagnetising Fields. *IEEE Trans. Magn.* MAG-27 (6): 4873–75.

Foner, Simon. July 1959. Versatile and Sensitive Vibrating Sample Magnetometer. *Review Sci. Instr.* 30 (7): 548–557.

Josephs, R.M. 1989. United States Patent No. 4,816,761.

Kerchner, H.R., S.T. Sekula, and J.R. Thompson. April 1984. Errors due to Field Nonuniformity in Vibrating Sample Magnetometry. *Rev. of Sci. Instr.*, 55 (4): 533–536.

Krafft, C.S., R.M. Josephs, and D.S. Crompton. Sept. 1986. Magneto-Optic Kerr Effect Hysteresis Loop Measurements on Particulate Recording Media. *IEEE Trans. Magn.* MAG-22 (5): 662–664.

Monson, J.E. 1987. Recording Measurements. *Magnetic Recording*, Ed. by Daniel and Mee, Vol. 1, pp. 376–426. See also 1995 edition

Newman, J.J. July 1978. Magnetic Measurements for Digital Magnetic Recording. *IEEE Trans. Magn.* MAG-14 (4): 154–159.

Newman, J.J. Nov. 1978. Correction to "Magnet Measurements for Digital Magnetic Recording. *IEEE Trans. Magn.* MAG-14 (6): 1187–1188.

Speliotis, D.E. 1990. Magnetometry of Magnetic Recording Material. *Proc. Electrochem. Soc.* 90-8, pp. 57–84.

FACTORS INFLUENCING MAGNETIC PROPERTIES

Kneller, E., and E. Koester. Sept. 1977. Relation between Anhysteretic and Static Magnetic Tape Parameters. *IEEE Trans. Magn.* MAG-13 (5): 1388–1390.

Knowles, J.E. Nov. 1985. Packing Factor and Coercivity in Tapes: A Monte Carlo Treatment. *IEEE Trans. Magn.* MAG-21 (6): 2576–2582.

Manly, W.A. Nov. 1976. Multimodal Media for Magnetic Recording. *IEEE Trans. Magn.* MAG-12 (6): 764–766.

Podolsky, G. Nov. 1981. Relationship of γFe_2O_3 Audio Tape Properties to Particle Size. *IEEE Trans. Magn.* MAG-17 (6): 3032–3034.

Vos, M.J., R.L. Brott, J-G. Zhu, and L.W. Carlson. Nov. 1993. Computed Hysteresis Behaviour and Interaction Effects in Spheroidal Particle Assemblies. *IEEE Trans. Magn.* MAG-29 (6): 3652–57.

MEASUREMENTS ON SINGLE PARTICLES

Knowles, J.E. Jan. 1984. The Measurement of the Anisotropy Field of Single "Tape" Particles. *IEEE Trans. Magn.* MAG-20 (1): 84–86.

Knowles, J.E. Jan. 1980. Magnetic Measurements on Single Acicular Particles of $GammaFe_2O_3$. *IEEE Trans. Magn.* MAG-16 (1): 62–67.

Salling, C., S. Schultz, I. McFadyen and M. Ozaki. Sept. 1991. Measuring the coercivity of individual sub-micron ferromagnetic particles by Lorentz microscopy. *IEEE Trans. Magn.* MAG-27 (6): 5184–86.

Scholten, P.C. Nov. 1975. Magnetic Measurements on Particles in Suspension. *IEEE Trans. Magn.* MAG-11 (5): 1400–1403.

TIME AND TEMPERATURE EFFECTS

de Witte, A.M., M. El-Hilo, K. O'Grady, and R.W. Chantrell. 1993. Sweep rate measurements of coercivity in particulate recording media. *Jour. Magn. and Magn. Matl.* Vol. 120, pp. 184–86.

Doyle, W.D., 1995. Personal communication and paper for Intermag '95.

Doyle, W.D., L. He, and P.J. Flanders. Nov. 1993. Measurement of the Switching Speed Limit in High Coercivity Magnetic Media. *IEEE Trans. Magn.* MAG-29 (6): 3634–36.

Fujiwara, T. Sept. 1985. Barium Ferrite Media for Perpendicular Recording. *IEEE Trans. Magn.* MAG-21 (5): 1480–1485.

Kloepper, R.M., B. Finkelstein, and D.P. Braunstein. Sept. 1984. Time Decay of Magnetization in Particulate Media. *IEEE Trans. Magn.* MAG-20 (5): 757–759.

Kubo, O., T. Ido, and H. Yokoyama. Nov. 1982. Properties of Ba-Ferrite Particles for Perpendicular Magnetic Recording Media. *IEEE Trans. Magn.* MAG-18 (6): 1122–1124.

Lyberatos, A., R.W. Chantrell, and A. Hoare. Jan. 1990. Calculation of Time Dependence in Thin Films. *IEEE Trans. Magn.* MAG-26 (1): 222–224.

Néel, L. 1972. Theorie du Trainage Magnetique des Substances Massives dans le Domaine de Rayleigh, as reported by B. D. Cullity in *Introduction to Magnetic Materials*, Addison-Wesley, page 472, 1972.

Oseroff, S.B., D. Clark, S.Shultz, and S. Shtrikman. Sept. 1985. Temperature Dependence of the Time Decay of the Magnetization in Particulate Media. *IEEE Trans. Magn.* MAG-21 (5): 1495–1496.

Sharrock, M.P. Jan. 1990. Time-Dependent Magnetic Phenomena and Particle-Size Effects in Recording Media. *IEEE Trans. Magn.* MAG-26 (1): 193–197.

Sharrock, M.P. Sept. 1984. Particle-Size Effects on the Switching Behaviour of Uniaxial and Multiaxial Magnetic Recording Materials. *IEEE Trans. Magn.* MAG-20 (5): 745–756.

Sharrock, M.P., and J.T. McKinney. Nov. 1981. Kinetic Effects in Coercivity Measurements. *IEEE Trans. Magn.* MAG-17 (6): 3020–3022.

NOISE FROM MEDIA

Arnoldussen, T.C., and L.L. Nunnelley, editors. 1992. *Noise in Digital Magnetic Recording*. World Scientific, Singapore, 280 pages.

Belk, N.R., P.K. George, and G.S. Mowry. Sept. 1985. Noise in High Performance Magnetic Recording Media. *IEEE Trans. Magn.* MAG-21 (5): 1350–1355.

Bertram, H.N., and X.D. Che. Jan. 1993. General Analysis of Noise in Recorded Transitions in Thin Film Recording Media. *IEEE Trans. Magn.* MAG-29 (1): 201–208.

Daniel, E.D. March 1972. Tape Noise in Audio Recording. *Jour. AES*, 20 (2): 92–99.

Mallinson, J.C. July 1991. A New Theory of Recording Media Noise. *IEEE Trans. Magn.* MAG-26 (5): 3519–31.

Mallinson, J.C. Sept. 1969. Maximum Signal-to-Noise Ratio of a Tape Recorder. *IEEE Trans. Magn.* MAG-5 (3): 182–186.

Min, T., J.G. Zhu, and J. Judy. Nov. 1991. Effects of Inter-layer Magnetic Interactions in Multilayered CoCrTa/Cr Thin Film Media. *IEEE Trans. Magn.* MAG-27 (6): 5058–60.

Murdock, E.S., B.R. Natarajan, and R.G. Walmsley. Sept. 1990. Noise Properties of Multilayered Co-Alloy Magnetic Recording Media. *IEEE Trans. Magn.* MAG-26 (5): 2700–05.

Nunnelley, L.L., D.E. Heim, and T.C. Arnoldussen. March 1987. Flux Noise in Particulate Media: Measurement and Interpretation. *IEEE Trans. Magn.* MAG-23 (2): 1767–75.

Ragle, H.U., and P. Smaller. March 1965. An Investigation of High-Frequency Bias-Induced Tape Noise. *IEEE Trans. Magn.* MAG-1, pp. 105–110.

Thurlings, L.F.G. April 1982. *Studies on Noise in Magnetic Recording on Particulate Media*. Ph.D. thesis, Holland (Philips). 84 pages.

Thurlings, L.F.G. Jan 1985. Basic Properties of AC noise. *IEEE Trans. Magn.* MAG-21 (1): 36–40.

Zhu, J-G., 1992. Micromagnetic modelling of thin film recording media. In *Noise in Digital Magnetic Recording*. World Scientific, Singapore, 280 pages.

CHAPTER 13
MANUFACTURE OF MAGNETIC TAPES AND DISKS

A large number of firms worldwide manufacture many types of magnetic tapes and disks. These products are all of similar construction, although there may be subtle differences between the different brands and different types of tapes.

The basic requirements for a magnetic tape are as follows: The coating should be completely uniform and have a perfectly flat surface, which ensures good contact with the recording and playback heads. The combination of the coating and film base should be completely pliable and at the same time possess adequate mechanical strength to prevent stretching or breaking of the tape. It should also be completely insensitive to storage duration, temperature changes, and humidity changes.

The vast majority of flexible media coatings are composed of a paint with magnetic particles instead of color pigmentation. These coatings are used for all diskettes, and for virtually all types of tapes—audio, video, computer, and instrumentation.

In the past, rigid-disk media also used a particulate coating, but solid metal films have been developed that provide higher outputs and better SNR (signal-to-noise ratio). Such are the disks found in all modern hard disk drives. Thin metal films are now also coated onto flexible tapes, called ME-tapes (metal evaporated). A sophisticated and lasting lubrication is required, in particular where the transport uses metal heads.

A modern magnetic tape consists of a 25-μm (one mil) plastic film with a coating of magnetic particles in a binder, with a thickness of from 2.5 to 10 μm (0.1 to 0.4 mils); diskettes are made on a thicker base film, typically 75 μm (3 mils). Figure 13.1 shows the cross-section of a magnetic tape as viewed through an electron microscope. The coating consists of a binder with a large number of magnetic particles. The binder is, in essence, the cement that holds the magnetic particles together when the dispersion has been applied to the base and the solvents have dried. The base film is commonly a polyethylene terephthalate (PET) material; fabrication and properties of PET are covered in Chapter 11.

The materials and processes used in making flexible particulate media and solid metal film will be covered in this chapter. Literature is very scarce because detailed chemical formulations and processes are trade secrets. A general outline of magnetic tape and disk manufacturing is possible, however. Considerations in the development of a magnetic tape coating are well summarized in papers by Sischka (1973) and by Daniel and Naumann (1971).

FIGURE 13.1 An SEM photo of a cross-section of a coating for a particulate magnetic medium (tape, floppy, or credit card) (*courtesy Memorex Corp.*).

FABRICATION OF PARTICULATE MEDIA, TAPES, AND DISKS

The following issues are addressed in the fabrication of particulate media.

BINDER INGREDIENTS

Typical ingredients for the binder holding the magnetic particles are shown in Fig. 13.2. The chemical formulation of the binder material depends largely on the material used for the base film, because the binder must adhere very strongly to it. Because most tapes today have plastic bases, we find that the binder dispersions usually contain the following components:

BINDER MATERIAL

A binder must satisfy several basic requirements. It must be totally soluble in a solvent so that the magnetic particles can be uniformly dispersed and remain so after the binder has solidified again as a coating. It must then also possess the necessary adhesion to the substrate, and have strength and elasticity.

Early tapes were made on a base film of cellulose acetate with binders of polyvinyl chloride plastic, and later cellulose nitrate or polyvinyl chloride. Both were closely related to the base film and displayed excellent adhesion and flexibility, but were sensitive to light and heat.

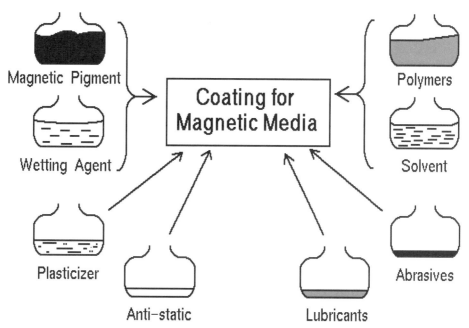

FIGURE 13.2 Ingredients for a particulate magnetic coating.

The most commonly used binders today for use on PET film are polyurethanes, which range from thermoplastic to thermoset or polyvinyl (like "Saran," which is thermoplastic). A thermoplastic material can repeatedly be heated and thereby softened to take a new shape; this plays a role in the general plastic industry, because scrap material can be re-used. A thermoset material is "fixed" during a heating cycle, and can no longer be made soft.

One important consideration in the selection of the binder material is the amount of solvent required to dissolve the material. All coating plants have strict requirements as to the recovery of solvents, rather than letting them escape into the atmosphere where they contribute to air pollution. The recovery process is costly and energy-consuming.

A remarkable solution would be a binder that merely needed water rather than an organic solvent. This would require a binder that could later be treated to become non-water-absorbent.

SOLVENTS

A solvent plays an indispensable role in dissolving the binder material. It must do so in order that the binder becomes a liquid into which the particles are dispersed and which can readily be applied to the base film. Two or more solvents are generally included in dispersions.

The process of dissolving the binder can be explained as a breaking of the crosslinking bonds between the binder molecules. The cause is predominantly due to the thermal motion of solvent molecules that cause a weakening of the intermolecular forces. If these forces between the solvent molecules equal those between the binder molecules, then the latter becomes dissolved. But if the forces between the solvent molecules are greater than between solvent and binder molecules, then no solution takes place; this is also the case when the forces between the binder molecules are greater than between solvent and binder molecules.

The solution process is quite complicated, and requires much experimentation to get the optimum formulation and process. This includes the drying process, where the entire amount of solvent is removed from the binder. It must leave a stress-free polymer structure where crosslinking recurs in an orderly fashion, leaving the coating well-bonded to the substrate.

PLASTICIZER (SOFTENER)

Using solvent alone to prepare the binder would leave a very hard and brittle film on the substrate. Polymers have a glass temperature below which the molecular network is rigid, with no allowance for translations or rotations between molecules. For nitrocellulose, for example, this temperature is 160° C. A lowering to 0° C is achieved by adding 20 percent Rizinusoil (Braginskij).

The plasticizer molecules must work their way into the polymer structure to be effective. The amount used may equal or exceed that of the binder, and the material must be such that it does not wander about and, for example, settle on the PET backing of a tape; this is possible in a wound reel of tape.

WETTING AGENTS

All pigments need to be surface-treated before media production to improve their "wettability," which enhances their uniform dispersion in the binder. The particles used in the coating are magnetic, and will therefore attract each other. They also tend to agglomerate into lumps; it is the purpose of the wetting agent to break up such agglomerates and provide for the best possible dispersion of the particles. Examples of agglomeration are shown in Fig. 5.10. Such agglomerates deteriorate the signal-to-noise ratio of a recording.

Small particles do also have electrically active surfaces, and the charges hold a film of air which, in effect, is bonded to the surface. This film of air must be replaced with a film of solvent containing the binder. A substantial amount of work is necessary to do this, because the air film is firmly bonded to the pigment surface. This work is done by a milling device in which the dry oxide is placed, and then small amounts of a wetting agent are added incrementally. The particles will begin to form balls that eventually coalesce into a continuous high-viscosity mass.

The wetting agent is often lecithin—not the commercial liquid type used in the food industry, but rather an oil-free type now prepared for the media producers (Lueck 1983). Long-chain fatty-acid amines or polyglycerides are also used.

The wetting process is quite complex and beyond the scope of this book (Braginskij pp.179–194). It is nonetheless worth noting the very large area that the wetting agents must work on and that must be uniformly coated with the binder. An appreciation thereof is achieved if the reader considers the surface area of one cubic centimeter; it equals 6 square centimeters.

Now divide the cube into 8 smaller ones. This increases the area by 6 square centimeters to a total of 12. Divide once more, now into 8-×-8 cubes, and the area increases to 24 cm^2. If n is the number of times we divide the cube, then the area is 6×2^n cm^2, while the length of one cube edge is $1/2n$ cm. Set n = 16 and you will find a cube with side length = 0.156 μm (which is about the average size of magnetic particles used). The corresponding area has grown to 39.2 square meters! Compare this with the specific surface area of particles in Table 11.1.

ANTI-STATIC AGENTS

Magnetic tapes are inherently good insulators, and may attain very high electric potentials that will only slowly bleed off. This necessitates tape transports that have grounded guides and heads, or they may act as Van de Graaff generators. Electrostatic charges on a tape not only cause noise when they discharge through a head assembly, but will most certainly attract airborne dust particles that cause dropouts.

Carbon is added to dispersions to reduce the electrical resistance and thereby prevent buildup of electrostatic charges. The carbon can be in the form of graphite or soot. Both will weaken the mechanical strength of the coating, and yet they must occupy a fair amount of the binder volume to be effective (10 to 25 weight percent of the binder, or 2 to 6 weight percent of the total coating).

This binder weakening is avoided if the carbon is applied as a back coat. Polyethylene and polypropylene are both excellent binders for soot. The back coat technique brings two benefits; any differential expansion (and hence cupping) of the tape is compensated, and the danger of layer-to-layer slippage on the reels is greatly reduced. This reduces the danger of cinching.

Carbon powder originates in the soot from gas flames burning with an insufficient oxygen supply. The carbon particles are recovered by placing a water-cooled baffle in the flame, and later scraping off the soot. Another method produces graphite from thermal decomposition of gas in a preheated chamber; the end product is acetylene graphite, if acetylene is decomposed.

Graphite is not an amorphous structure, as was once believed; modern electron microscopes have revealed a microcrystalline structure, held together by the Van der Waals forces and chemical bindings. The conductivity of polymers is affected when the distance between the graphite particles decreases below 10 nm (0.01 μm = 0.4 μin). This occurs when the graphite particles occupy 10 to 15 percent of the polymer volume, or two to three percent of the total binder volume. It is also found that the conductivity is highest for small graphite particles. Graphite is named carbon black in the media industry. The particles are considerably finer than the magnetic particles (Burgess 1986):

Fineness:	Particle diameter:	Surface area:
Low	0.09 μm	24 m^2/g
Medium	0.03 μm	88 m^2/g
High	0.01 μm	259 m^2/g

Processing techniques and associated parameters are described in Burgess' paper, and in Braginskij, pages 194–201.

LUBRICANTS

To overcome stickiness and scrape-flutter problems, a lubricant is generally added to the binder material. Great care is exerted in the selection and the amount applied, because it may transfer to the backside of the base material. This would result in poor friction characteristics between the capstan and the rubber puck. Lubricants should be selected and applied so they become an integral part of the polymer structure.

There are many candidates for the lubricant. In modern tapes and disks with very smooth surfaces, the lubrication problem is twofold. The primary function is to lubricate the binder-to-head interface, the latter being ferrite/ceramics or metal; this problem is traditional. The next function is to prevent, if possible, the buildup of binder polymer molecules onto the head surface ("clear varnish;" see Chapter 31). The frictional properties of a polymer-to-polymer interface invite stick-slip behavior. This affects the media motion. It is a degenerate mode, where the polymer buildup increases until spacing losses disrupt operations.

The lubricants are traditionally added to the coating mix and become part of the binder system, acting at times also as a plasticizer. These materials are fatty acids (Mihalik 1983). Another approach places the lubricant on the surface of the dried coating. The most frequently used lubricants for this method are the perfluoro-polyethers (Bagatta et al. 1984). These materials do not interact with most plastic materials and are practically insoluble in most organic solvents. They are stable up to 260°C, and are very efficient in reducing the dynamic coefficient of friction and the wear rate of tested polymer materials in sliding contact against head surfaces (tested by Fulmer Research, England; Bagatta 1986).

The method of application to the surfaces varies: For rigid disks a spray system is used, while for flexible disks sometimes the dip coating is preferred. For tapes a roll-coating machine or extrusion die is used.

ABRASIVE AGENTS

Only recently were abrasive powders introduced into binders to remove any polymer buildup on the heads. The material is typically aluminum oxide, and is included in a very sparse amount to avoid excessive head wear. (In some audiocassettes, a few inches of the leaders are mildly abrasive to remove any buildup on the heads; head cleaning with a Q-tip has thus been eliminated.)

Properly dispersed alumina particles provide a gentle cleansing action with little wear or damage to the head. The average particle size is 0.4 to 0.6 µm (16–24 µin.). The variance is controlled to eliminate very large particles that could cause damage, a scratch in the head surface, or a head crash in a flying-head application. The limit for the maximum particle size varies from five to less than one micron (200 to 40 µin).

The amounts of alumina used vary from one producer to another; general guidelines are (Crowe and Arvidson 1986):

Rigid disks:	0–1 percent
Flexible disks:	0.5–2 percent
Computer tapes:	0.5–2 percent
Video/audio tapes:	1–3 percent

OTHER ADDITIVES

A host of other additives for binders are cited in the literature, each contributing to an improvement in one or another characteristic of a magnetic media. An example is oleic acid, a fatty acid still used in some applications. It serves as an effective plasticizer, but may form a thin film on the media surface that most likely is accompanied by stick-slip friction (Mihalik 1983).

Other additives modify the basic binder polymer. Crosslinking (thermoset) is enhanced by isocyanate, and the result is less layer-to-layer adhesion, better scratch resistance, better cohesive strength, and more resistance to solvents than polymers that are cross-linked without it.

The entire formulation issue is complex, and again reflects the tradeoff decision process in magnetic recording as a whole. The reader is referred to papers by Williams and Markusch on polyurethane coatings (1986), Mihalik on additives (1983), and Brown et al. on E-beam curable coatings (1983).

DISPERSION OF THE MAGNETIC PARTICLES

The dispersion process distributes the oxide particles uniformly throughout the binder. There must be a sufficient amount of binder available to thoroughly disperse the particles, and the process must be one of true dispersion, not forcible milling. The particle agglomerates must be broken up by the shearing forces in the agitated binder liquid, not by contact or collisions between agglomerates.

The destruction of an agglomerate is illustrated in Fig. 13.3, simplified. The two spheres will move as shown under the influence of the shear force τ, produced by agitation of the binder liquid. The parameter K is (Braginskij, p. 208):

$$K = 6\pi R \tau / F_a \tag{13.1}$$

where:

R = radius of spheres
τ = shear force
F_a = force between spheres

The force between the spheres is predominantly magnetic, and varies in inverse proportion to the distance squared for the far field, and to the distance cubed in the near field. Hence this force is large

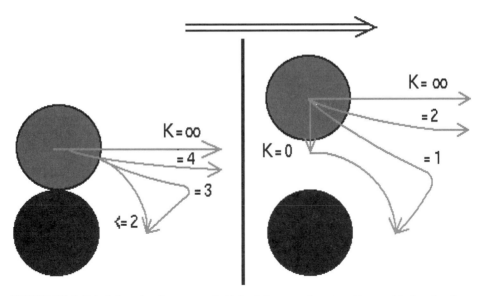

FIGURE 13.3 Sufficiently large shear forces are required to break down an agglomerate and to prevent the formation of new ones (*after Braginskij*).

for very small particles such as pure iron (which, in addition, has higher magnetization). Clearly K must exceed a value of three for particles in contact, and less for particles not in contact (from Fig. 13.3). (The exact number is irrelevant in this discussion; the formula was introduced to characterize the action of breaking up agglomerates.)

The shear force must exceed a certain value, or no dispersion will take place. It is further advantageous to continuously alter the direction of the shear forces, thereby increasing the probability of dispersing all agglomerates.

A dispersion improves with milling time. In a metal particle (MP) dispersion, changes in several properties can be measured: an increase in squareness, orientation ratio, and magnetic susceptibility, and a decrease in coercivity as functions of milling time (Mayo et al. 1992). Similar trends exist for other pigments and may be used as a measure of quality control of dispersion and milling time. Figure 13.4 shows the changes versus milling time for γFe_2O_3 particles (after Dasgupta 1984).

It is also possible to draw a small sample of the dispersion and expose it to a small 60-Hz ac field, and then a much higher-intensity dc field. The response (amplitude and phase) of the sample to these field exposures can be related to the quality of the dispersion (Jung et al. 1993). Another method involves the use of a capillary viscometer (Gooch 1989).

THE DISSOLVER

A traditional dissolver is shown in Fig. 13.5. It is used to produce the shear that will dissolve the solid binder material parts into the solvent, and often to disperse the magnetic pigments. The rotating blades and their speed, plus the dimensions of the container, are carefully designed to produce a laminar flow. The rotational speed of the dissolver disk must reach a certain speed so the necessary shear forces are produced; otherwise, the machine will merely act as a stirring or milling device for the solvent/binder mix.

The process can be monitored by observing the energy consumed by the motor drive. It will increase rapidly during the first few minutes and decrease during the next ten or so minutes. This means that the dissolving process is completed within 15 to 20 minutes.

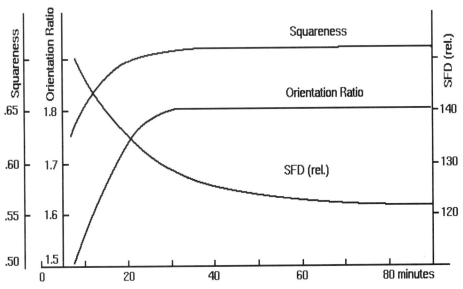

FIGURE 13.4   Magnetic properties of a γFe_2O_3 particulate dispersion (slurry) versus milling time (*after Dasgupta, 1984*).

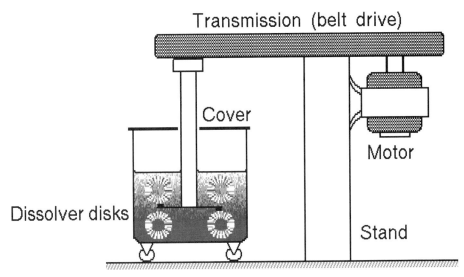

FIGURE 13.5   Traditional dissolver with stirrer (dissolver disk).

The geometry of the rotating blade is, of course, of paramount importance for the process. Its efficiency is improved by providing its perimeter with small toothlike structures.

THE KNEADER

The viscosity range is limited in the dissolver; it will only move the material in the immediate vicinity of the stirrer. Many modern formulas call for a high viscosity mix, which can be handled by the kneader (shown in Fig. 13.6). It is also becoming popular in the prewetting process of the magnetic pigments. They are loaded dry into the kneader and the wetting agent is added incrementally. Small balls start forming, and they eventually coalesce into one large ball, which becomes the premix to which binder and the other additives are added (Missbach, Lueck 1983).

FIGURE 13.6   A twenty-liter horizontal media mill. The slurry agitation is done by eccentrically mounted disks on a rotating shaft (*courtesy Netzsch Inc.*).

THE BALL MILL

The oldest and most common dispersion method is a ball mill, which is simply a jar or a bell that is partially filled with a dispersing media of either metallic or ceramic balls, cylinders, rods, or a random shape (as in sand). The time of the milling process may last from a few hours to several days, and it is dependent upon the chemical composition of the binder and the particles used. Too long a milling time has been found to break down particles. The process is therefore aimed at shorter milling cycles.

The horizontal ball mill consists of a slowly rotating metal or porcelain drum, provided with a cooling jacket. The drum is partially filled with the milling media, and their motions are greatly influenced by the rpm of the drum; they change from those of an avalanche to those of a waterfall, as shown in Fig. 13.7.

The size of the balls range from a couple of millimeters in diameter down to slightly less than one millimeter. The optimum size depends on the operating conditions of the mills, the rpm, the mix viscosity, and so on. It is also important that the dispersion takes place by generating liquid shear forces in the small volumes that exist near the points where the balls either touch or are very close to each other. The action must not develop into one where too many particles break up into smaller particles with assemblies that are superparamagnetic.

A vertical mill is shown in Fig. 13.8. It completely encloses the mix, which is advantageous, and the shear forces are produced by the rotating disk(s). These are available with different cross sections and different geometries, such as a worm or a set of eccentric rings. The latter can now become a horizontal mill again.

FILTRATION

Lumps, aggregates, agglomerates, and downright foreign particles cannot be tolerated in the mix for the coating process. The mix must pass through one or more filters. Excessively large particles are prevented from passage, either at the filter surface or deeper down in its structure.

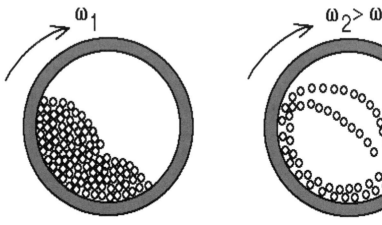

Avalance motion: Waterfall motion:

ω_1 $\omega_2 > \omega_1$

FIGURE 13.7   Ball mill, horizontal.

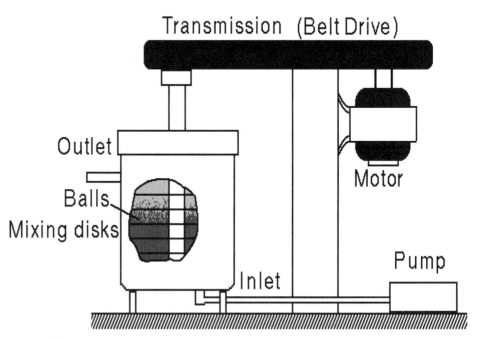

Transmission (Belt Drive)

Outlet

Balls

Mixing disks

Inlet

Motor

Pump

FIGURE 13.8   Ball mill, vertical.

Sintered metal filters were once used, but have now been replaced by pleated polypropylene filters with an absolute rating in the 2.5 to 5 micron range. The polypropylene yarn is graded in size as well as closeness in structure, as shown in Fig. 13.9. This provides the deep filter action.

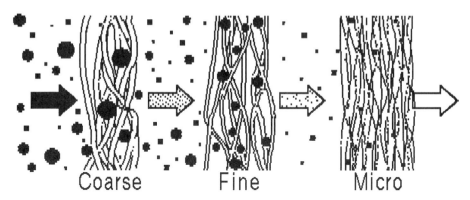

Coarse   Fine   Micro

FIGURE 13.9   A graded filter, using polypropylene yarn (*after Missbach, 1983*).

COATING OF FLEXIBLE PARTICULATE MEDIA

After the milling process, the material is fed to a coater, which applies the coating onto the PET film. A complete coating line is shown in Fig. 13.10, while the coating process is shown simplified in Figs. 13.11 and 13.12. The coating machinery is really a giant tape transport. Its function is to feed a cleaned and stress-relieved PET film at a constant speed to a station where the wet coating is applied. A following station orients the magnetic particles in the coating, and the coating must then be dried.

The drying is done in one or more zones in a ventilated duct, and heating of the coating causes it to cure into a hard yet pliable surface. The surface finish is at this point fairly rough, and therefore dull to the eye; this will cause spacing losses later on, so the surface is improved by treatment in a calender station. Here the coated web loops through several nips existing between heated, polished steel-to-cotton rollers, pushed together with a very high force. The result is a high-gloss finish coating. Care must be exercised so that the coating does not become over-stressed, which can cause a number of particles to break into the smaller, potentially superparamagnetic particle assemblies.

The finished web is wound onto a roll that may proceed to "slitting," where it is cut into the planned tape widths, or it may go into storage for further aging (crosslinking). Another destination for the web may be back to the front of the coating line to be coated on the backside, either with carbon black for a high-quality video or instrumentation tape, or with another magnetic coat so it may later be stamped out into double-sided diskettes.

The design and construction of such a coating line have much in common with those of paper production lines and instrumentation tape transports. Web guidance, tension controls, servo sensors, and servo controls operate within 10 milliseconds in a modern coating line; that requires a servo system with a 50-Hz bandwidth (Landskroener 1983; Braginskij 1981).

METHODS OF COATING

Several coating processes are used, and are roughly divided into two categories, as shown in Figs. 13.11 and 13.12. The direct gravure process uses a gravure roll (a cylinder with etched cells) that picks up the coating from a pan or a pressure-fed applicator. Excess material is wiped off with a doc-

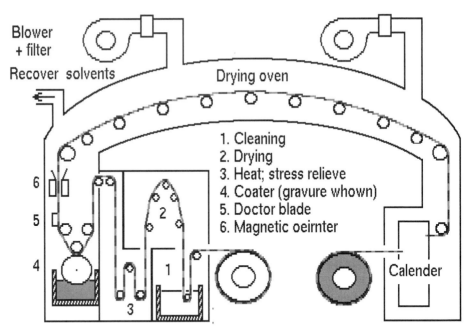

FIGURE 13.10   A production line and drying oven for thermoset particulate magnetic tape coatings.

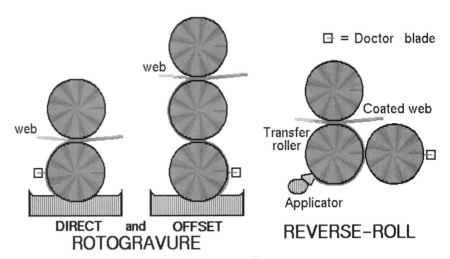

FIGURE 13.11   Nipped coating processes: gravure and reverse-roll coaters.

tor blade before the coating material is transferred to the PET film in the nip between the gravure roll and the backing roll; the backing roll has a resilient surface.

The pattern of the cells in the surface of the gravure roll is imprinted on the PET film so the coating film is discontinuous (split). This requires a post-coating smoothing that, for instance, can be done by a PET sheet that wipes the coated surface. An offset gravure coater tends to reduce the split

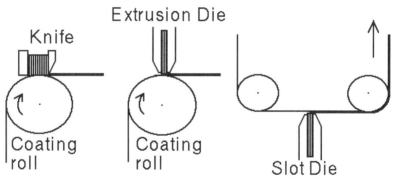

FIGURE 13.12 Flow coating processes.

film patterns. Further improvements are possible by using reverse gravure systems, where the applicator roll moves in the direction opposite the backing roll.

A very smooth coated surface can be made with the reverse-roll coater, which has numerous configurations; one of these is shown in Fig. 13.11. The wet coating is applied through a pressure-fed applicator, and excess material is removed with a metering roll, which in turn is cleaned with a doctor blade. The coat is transferred, or "peeled off," in the nip between the coating roll and the backing roll.

The other coating process category in Fig. 13.12 includes the so-called unnipped systems, which apply the coating in a more direct fashion to the PET film. The knife coater is the veteran, in use since the 1940s. The gap between the knife blade and the film determines the coating thickness, and is set by adjusting the backing roll-to-knife distance. This will necessarily include the film thickness—and the variations thereof; the latter variations can be eliminated from the metering process by the extrusion die or slot die methods.

ORIENTATION OF PARTICLES

Immediately after coating, while the binder is still wet, the coated web undergoes orientation in a magnetic field of 1000 to 5000 Oe in strength, whereby the magnetic particles are aligned. This provides (as mentioned earlier) for increased output at long wavelengths. The effectiveness of orientation is generally referred to as the squareness ratio, which is the ratio between the remanent saturation magnetization M_{rsat} and the saturation magnetization M_{sat}. For randomly-oriented particles, the squareness ratio is 0.5, and for ideally and perfectly oriented particles the ratio is equal to 1.0, providing for a potential 6-dB increase in long wavelength output. Practical values for the squareness ratio normally fall around 0.70 to 0.90, which gives an increase of 3 to 5 dB.

Most tapes have the magnetic field applied in the direction of tape travel. One exception is the two-inch-wide tape used in transverse scanning video recorders, and another is the web used for diskettes. In the latter a uniform distribution of the particle orientation is made over 360 degrees by a proprietary, rotating magnetic field. This results in a stable signal amplitude when the finished disk rotates through one rotation.

The mechanism of rotating particles in the viscous coat has been examined (Newman and Yarbrough 1969) and applied to actual coating processes (Newman 1978). This work, plus the analysis of orienting magnet design (Bate and Dunn 1980), set guidelines for the design of orienting magnets.

High orientation ratios for metal particles require a very high field strength. Permanent magnets are relatively cheap and safe for orienting magnetic particles, but they cannot obtain a high orientation ratio in iron particles because of deorientation that occurs after leaving the gap of the magnets (Bate and Dunn 1980). An electromagnet provides a good orientation, but requires high electric power and poses the risk of igniting an explosion. A combination permanent magnet (400 kA/m or 5 kOe) followed by an electromagnet of lesser strength appears a good solution (Peng et al. 1992).

DRYING AND CURING THE COAT

After coating and orienting, the tape enters a drying oven, which normally is over a hundred feet long, where all the solvents are evaporated and removed with the aid of heat and airflow. The drying oven is typically designed as shown in Fig. 13.10, divided into zones. The solvents are removed (and recovered), and the degree of cure of the coating is a function of time as well as temperature; together they dictate the speed of the web.

There is a rapid removal of solvents in the first zone, yet the speed of this process must be limited or a crust will form on top of the otherwise wet coating. This will hinder the ongoing extraction of solvents from the interior of the coating, and a nonuniform coating structure results.

In the following zone, the coating starts to solidify and further solvent removal now causes a complicated, slightly porous coating to form. The coating is simultaneously cured by heat, causing crosslinking between the polymer molecules.

A total removal of the solvents is not always possible within the time of the normal drying process. It is advantageous to store and age the coated web for anywhere from one day to one month, a time that also would allow for complete curing of the binder.

CALENDERING, FINISHING, AND SLITTING TAPES

The web continues to the final treatment called calendering, which compacts the binder and smooths its surface. The machinery for this process is another loan from the paper industry; one configuration is shown in Fig. 13.13.

The calendering causes a plastic deformation of the coating. The pressure must be above a certain value to prevent a mere elastic deformation. The plastic deformation depends on factors such as binder polymer, pigment loading, and solvent retention. The deformation can be quite large in a fresh coating with high level of solvent. Under those conditions, however, it does not result in a lower final value for the surface roughness after calendering under optimal conditions. The optimal results could be achieved much more easily, however (less pressure, fewer nips) (Brondijk et al. 1987).

Further finishing treatment depends on the intended use of the tape. A burnishing process is very much like the first few passes over a head, and will remove any high points on the coating surface. This procedure is common for all high-packing-density tapes.

The web is finally slit into the finished widths, which may be from one-sixth of an inch for audio uses to up to 2 inches for video uses; see Fig. 13.14. This final slitting process must be carefully controlled, because any width variation will cause skew (improper tracking of the tape). Also, the debris materials from the slitting action must be completely removed (with vacuum), or they will cause further problems when they become attached to or embedded in the tape coating.

The slitting action is very rough on the knives. They are essentially cutting into a plastic containing abrasive particles, which shortens the lifetime of the blades and changes their dimensions. This will be reflected in the tolerances that can be achieved.

The slitting machine is normally separate from the coating line. It has its own web feed and guidance mechanism, friction drive for the web and the slit tapes, and tape-up hubs for the tapes. The operation on the web is a mix of knife/scissor operations, and typical slitters are shown in Fig. 13.15. Each version has its own advantages and disadvantages in terms of slitting tolerances, quality of cut, lifetime before sharpening/regrinding, and so forth.

Alternating Cotton—Steel rollers.

FIGURE 13.13   Calender machinery where a web with a dull finish enters the pressure of alternating, heated cotton and steel rollers. after treatment, the coated tape surface is glossy and smooth.

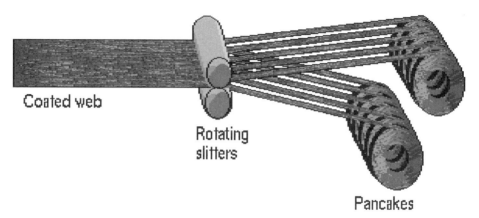

FIGURE 13.14   The slitting of finished web into tape pancakes.

The speed of the slitting machine ranges from 80 to 200 m/min (52.5 to 131 IPS). Any faster speed would lead to problems with windage when winding the slit tapes onto the reel hubs; air drawn with the tape will take time to spread away when the tape is laid down on the reel pack, and a loose pack could result. This would make subsequent handling of the wound "pancakes" very difficult, and it is often common practice to use a pack wheel for each reel hub to firmly lay down the slit tape.

The speed of the circumference of the rotating knives is slightly higher than the web speed to assure a clean cut; too high a speed would only cause excessive wear and shorten the life of the knives. A lesser speed would result in a very poor edge.

The slitting operation results in reels of tape, commonly called pancakes, which often need further processing. During the winding onto the pancakes the slit tape may pass over a cleaning blade. Then a tissue plus vacuum system removes all dirt and dust. The finished tape product may stay on the hubs, which are provided with flanges then sealed and packaged. The pancakes may also go on to a rewinder station for the loading of tapes into data, video, or audiocassettes and cartridges. The pancakes may also have prerecorded music or video duplicated onto them prior to loading into cassettes.

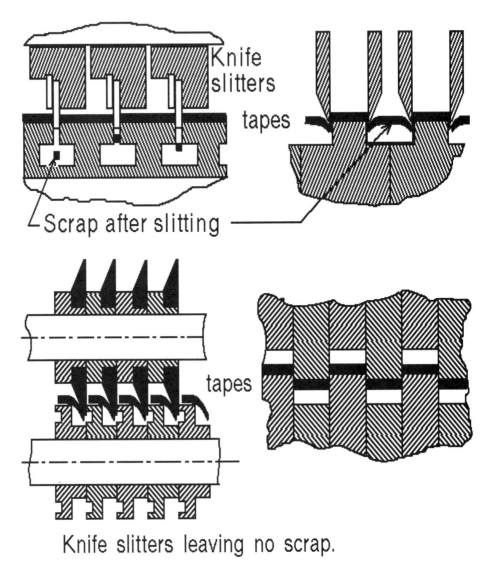

FIGURE 13.15 A symmetrical knife slitter that leaves scrap. Also shown: An asymmetrical knife slitter and scissors slitter without waste.

STAMPING, FINISHING, AND PACKAGING FLEXIBLE DISKS

Flexible disks with particle coatings are manufactured using the same principles as magnetic tapes. There are a couple of differences, though. The particles are not oriented in the machine direction, but as randomly as possible in the plane of the coating. And the total thickness of flexible disks are typically 75 μm (3 mils).

While calendering is a satisfactory surface treatment for many tapes, one finds that flexible disks must undergo a polishing or buffing process before their surface is smooth enough to minimize spacing losses and reduce head wear. This is done by clamping the disk against a rotating (but firm) support and placing the disk into contact with a buffing wheel.

The disk is placed in a protective jacket with an inside liner that protects both sides of the disk while collecting dust and foreign particles that inadvertently enter the jacket. The inside (fiber) liner can be chosen from a large variety of materials such as rayon, polypropylene, polyesters, and copolymers (Ostrowski 1983).

COATING PARTICULATE RIGID-DISK MEDIA

A particulate rigid disk has a high-quality aluminum base of thickness ⅛" to ¼". It is highly polished and then cleaned by dipping in acid, alkaline, and water, and then scrubbed by an ultrasonic process.

The coating is applied by either spraying or spinning, or both. The centrifugal forces on the wet coating cause it to spread out in a spiral fashion. The speed is highest at the periphery of the disk and the final coating tends to be rather thick there. This is, of course, advantageous for high-density recording, but limits the useful area of a disk to a 2"-wide outer band.

The coated disks are placed in ovens, where the coating is hardened and bonded to the aluminum disk. A final step is polishing the disk surface, using proprietary methods. The final coating thickness for high-density disks is about 0.5 to 1.25 μm (20 to 50 μin).

Each disk is tested for coating defects such as voids, improper dispersion of oxide, and wavy or sloping edges. Some surfaces are checked for hardness and smoothness. The disk is then placed into a single disk tester to check magnetic properties (output level and noise) and potential defects (dropouts and dropins). Finally, a head is made to fly at one half the regular height, and this burns off any protruding chunks of oxide. This is called burnishing.

FABRICATION OF THIN-FILM MEDIA, TAPES, AND DISKS

The issues that concern thin-film media fabrication are as follows.

THIN-FILM COATING OF FLEXIBLE MEDIA

Coating a continuous magnetic film on a flexible tape has been under development for the past two decades. Excellent performance has been achieved for the write/read signals, and many mechanical problems have been resolved.

Several methods for deposition of the magnetic film have been evaluated, and the result has been the selection of CoNi for the alloy and evaporation for the method of deposition. The two components have similar vapor pressure, and the composition is well-preserved in the film even during long deposition runs.

The alloy is in the range CoNi (80/20–70/30), where magnetostriction is small, i.e., there is no stress dependence of magnetic properties. Layer thicknesses are 50 to 150 nm. Coating machinery is outlined in a paper by Feuerstein and Mayr (1984).

Early problems were in the area of exposing the base film to high deposition temperatures followed by differential expansion between the metal film and the base, causing cupping. Contact problems are also evident when the metal film is run over a metal head. Ferrite heads have been shown to be a partial solution, and surface lubrication helps, but abrasion characteristics are worse than for polymer-coated tapes.

A stable 10–30 Å thick lubricant layer is required to be placed on the metal coating to prevent gouging between metal and head (Lee 1990). Too thin or thick a lubricant layer does not work, and these problems caused Ampex and BASF to cancel their ME tape developments while Sony continued (Hokkyo et al. 1993). This is not surprising, because Sony has products that need the best tapes, such as the HI-8 camcorder. The lubrication stays on Sony's ME tapes because of a patented oxidation treatment of the metal surface before the application of lubricant.

There is also the environmental stability question of corrosion. This will always be more difficult for metal film tapes than for particulate tapes, because the tapes are subject to bending shear and other stresses as they move through a transport. The stresses alternate from compressive to expansive, and may in time cause cracks in the metal surface; the result is a severe permanent dropout, plus noise. One suggested method for corrosion protection applies a thin plasma polymer film over the metal (Griesser 1989).

Figure 13.16 shows a vacuum coating station that produces a high-grade ME (metal evaporated) tape (Feuerstein and Mayr 1984). The angle of deposition is controlled by the location of the slot through which the film is produced. The columnar structure of the film is tilted, and makes recordings in opposite directions different. This phenomenon was clarified by Ouchi et al. (1986) and verified and expanded by Krijnen et al. (1988).

The magnetic anisotropy matches the head field direction on the ingoing side when the tape moves as shown in example A, Fig. 13.17. The field component in the anisotropy direction then rapidly decreases as the tape moves over the gap. This seems to result in a short transition and a small recording demagnetization; the reverse direction (B) has a large recording demagnetization (Hokkyo et al. 1993). The response curves for read voltage versus write current are shown as A and B in Fig. 13.17; the dotted line shows the response for an ordinary MP tape. The result is a product that has excellent short-wavelength response and an output that exceeds MP tapes by 8–9 dB and HI-8-MP by 4 dB.

Also, sputtering (described in next section) has been employed in the making of tapes, such as a tri-layer VTR tape for the perpendicular mode (Numazawa et al. 1993). The polyimide base film was 10 μm thick, and the magnetic layers were (from base film and up): 0.2 μm $Ni_{80}Fe_{20}$ ($\mu_r = 1700$), 0.03 μm $Co_{67}Cr_{33}$, and 0.1 μm $Co_{79}Cr_{21}$.

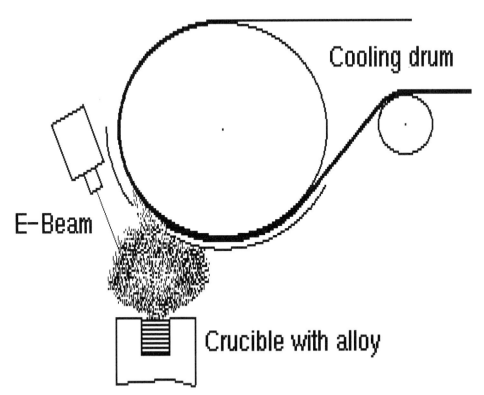

FIGURE 13.16 A coater for metal evaporated (ME) tape (*after Feuerstein, 1984*).

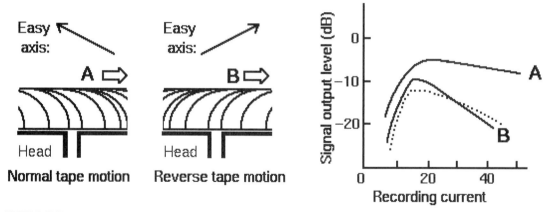

FIGURE 13.17 Anisotropy in ME tape coated on a line as shown in Figure 13.16. The response in the forward direction is many dBs above that in the opposite direction (*after Hokkyo et al., 1993*).

Stresses between the metal film and the base film can cause severe cupping of the tape, and should be prevented by (for instance) the application of a back coat. A select sputtering process has achieved the same result by sputtering $Co_{80}Cr_{17}Ta_3$ in Kr (Krypton) on a thin PEN (polyethylene napthalate) film using a technique called FTS (facing targets sputtering) (Akiyama et al. 1991).

THIN-FILM COATING OF RIGID DISKS

Early disks were made with metallic coatings of CoP and CoNiP on aluminum substrates. The film thickness ranged from 0.1 to 0.25 μm (4 to 10 μin). A thin oxide coating would have very low magnetization, while a metallic film of comparable thickness would have magnetic moments comparable to that of thick oxide coatings. The volume factor is 1.0, and there is no internal demagnetization because $BL/\Delta x \gg 1$ (see formula 4.14). It is further possible to obtain a wide range of coercivities (4 to 120 kA/m, 50 to 1500 Oe) in metal films.

The films can be deposited by several methods, such as electrochemical deposition, electrolytic deposition, and vacuum evaporation. Plated media have reached a rather sophisticated level of perfection (Suganuma et al. 1982; Garrison 1983).

Most composite materials cannot be deposited by evaporation, due to the different vapor pressures of the ingredients. Evaporation has only survived in the production of ME-tapes, as described above. Today sputtering techniques are used. Here the molecules to be deposited are literally knocked off the target surface, and subsequently travel away to become deposited on the substrate, i.e., the disk (Judy 1983).

Figure 13.18 illustrates the deposition process. Molecules of the coating material (target) are dislodged and ejected from the solid surface due to bombardment by energetic particles. The latter are usually positive ions of a heavy inert gas, such as Argon. The action is somewhat similar to the strike of a billiard ball against a tightly packed group of the other balls, as at the opening of a game.

The target molecules now travel right toward the substrate under the energy that was transferred during the Argon molecules' impact. They gradually build up a film that is a true replica of the target composition. To achieve this two things must occur: There must be freedom from interference from air molecules, and there must be some means of introducing the energetic ions.

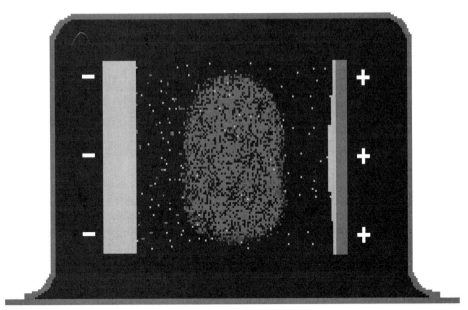

Argon ions knock off target molecules.

FIGURE 13.18   Basic sputtering system.

The target, substrate, and holders are placed inside a chamber from where the air is pumped out. The pressure is typically brought down to 10^{-6} torr (1 torr = 1 mm mercury = 133 pa = 133 N/m². 1 atmosphere = 760 torr). The mean free path length between colliding air molecules is then greater than 20–30 cm.

Next, ions must be provided. This is done in many installations by backfilling the chamber with Argon to a pressure of 10^{-3} torr. A plasma is next formed by a dc diode discharge: A negative voltage is applied to the target, so that an electric field exists between the target and the substrate. The latter may be grounded or connected to a positive voltage.

A free electron may at first occur by some means such as a cosmic ray or a UV photon striking the cathode (target). That electron will be accelerated toward the anode (substrate) and will gain sufficient energy to ionize one of the argon atoms while simultaneously creating a second electron. Each will be accelerated: the ion toward the cathode (target) and the electron toward the anode (substrate). The first electron is still attracted toward the anode (substrate).

Additional ionizing collisions may occur. The ions are accelerated to the cathode (target) and may strike it at high energy, causing emission of secondary electrons. These additional electrons can cause further ionization and a breakdown occurs (after Rossnagel 1991). A spectacular, purple-colored glowing discharge appears and the sputtering process is underway.

The ions strike the target and dislodge layer upon layer of molecules, which travel to the substrate to build up the thin film. Their travel energy is partly due to the momentum exchange with the ions, partly to the applied anode (substrate) field. The sputtering process is a high-energy process when compared to some of the other deposition methods. Cooling (water) of the target is necessary. Deposition rates are in the 100–1000 Å/min range.

The energy and current density of the bombarding ions can be controlled by ion beam sputtering (secondary ion beam deposition). Figure 13.19 shows a single ion beam sputtering system where a separate ion source provides bombarding ions to provide the coating process. A second ion beam can

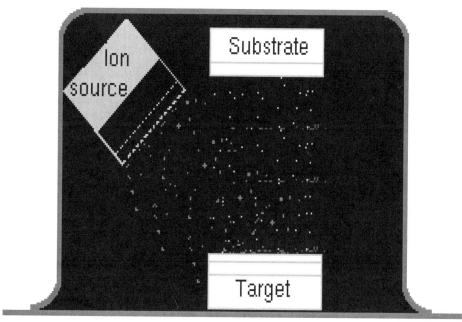

FIGURE 13.19 Basic ion beam deposition system.

assist by bombarding the growing layer on the substrate with relatively energetic ions. This method produces very hard coatings with excellent adhesion.

Ion-beam sputtering was used to produce a film of CoCr that has the easy axis (the preferred direction of magnetization, i.e., anisotropy) perpendicular to the plane of the substrate and, therefore, enhances perpendicular recordings (Gill and Yamashita 1984).

Deposition rates can be increased by Magnetron sputtering, where a magnet system is placed just behind the target. Its field forces free travelling electrons to remain in the plasma and thus increase the electron efficiency.

Sputtering methods do also allow the deposition of insulating materials. An RF (radio-frequency) voltage is superimposed on the cathode (target) and serves to discharge any buildup of dc voltage from the ion bombardment. RF-sputtering is hence used for deposition of insulators as well as oxide and ferrite films.

Details of the sputtering processes are found in recent literature (Rossnagel 1991; Parsons 1991; Bhushan and Gupta 1991). Methods for production of sputtered disks were described by Moore et al. (1985) and Drennan et al. (1985), and more recently by Ishida and Seki (1991) and Johnson (1992).

Production methods are described in papers by Johnson et al. (1993), and by Cord et al. (1993). The first paper considers the manufacturability concerns regarding the making of a low-noise media. This requires minimization of strong exchanged intergranular coupling forces. This can be achieved by multilayer construction (Glijer et al. 1994) or by the inclusion of tantalum (Ta), which reduces media noise. Ta is reported to assist in the creation of very small grains of mean diameter 80–100 Å; see Fig. 13.20 (Sin et al. 1994). Many other items can affect sputter deposition, and a few selected papers are listed in the bibliography.

The typical thin-film disk now consists of the layers shown in Fig. 13.21. Production lines for sputtering are shown in Figs. 13.22 and 13.23.

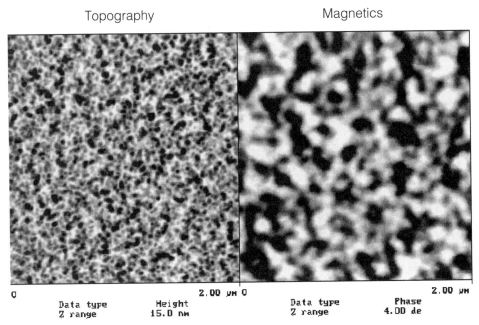

FIGURE 13.20   Experimental small-grain thin film coatings for disks (*courtesy of Ken Babcock, Digital Instruments, Inc., 1995*).

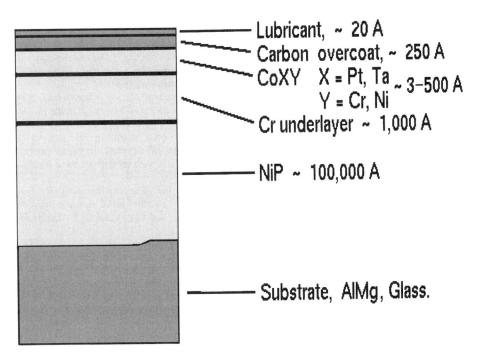

FIGURE 13.21   Typical construction of a thin-film magnetic disk.

FIGURE 13.22 Typical design of an in-line system for production of thin-film magnetic disks (*courtesy of Leybold*).

FIGURE 13.23 A single-disk sputtering station (*courtesy of Balzers Process Systems, Hanau, Germany*).

TEXTURING, OVERCOATING, AND LUBRICATION OF THIN-FILM RIGID DISKS

The AlMg disk substrate is diamond turned or fine ground, and a 10-μm layer of NiP is plated on for strength. The next step is a process whereby the surface is provided with a great number of minute grooves. The surface is now rougher and has a smaller coefficient of friction against the head, in some cases about half the value of a very smooth surface. This reduces the forces required to start the disk spinning, and the risk of stiction damage is reduced.

The grooves can be patterned like on a phonograph record. This can result when the substrate is turned on a lathe with a diamond tool. This circular pattern emphasizes the orientation ratio for longitudinal recording along a track. The patterns may also be radial rosettes, or circumferential with cross-hatching. The latter patterns have low stiction (Bhushan 1993). The finished substrate is burnished to remove high points.

The next step is cleaning and plating with a 0.1 μm Cr underlayer. This layer serves as a seed layer for the crystal growth of the subsequently sputtered magnetic film.

A protective layer is placed on top of the magnetics. Many materials have been tried, such as carbon, Rhodium, ZiO_2, Zi-alloy nitrides, and SiO_2. By suitable RF sputtering techniques, a 150–250 Å thick diamond-like carbon film is formed that has the desirable properties of an overcoat: strength, corrosion resistance, low friction, and stability (Angus 1992; Latev et al. 1991; Khan et al. 1988; Kurokawa et al. 1987).

The final manufacturing step is the application of a lubricant. Perfluoropolyethers have proven best for most applications. A layer of thickness 20–40 Å reduces wear rates by 100 to 1000 times.

QUALITY CONTROL METHODS

Substrates need to be tested before production, and several methods are useful for PET films (thickness, surface roughness, strength, and so on—see Persoon 1983), and for aluminum disks (Morrison and Brar 1983).

In-process measurements have been evaluated, and some have been introduced into production as a means of quality control. Some of the measurement techniques are described in the next chapter, and we will look briefly at some of the sample preparations.

The liquid coating material for tapes can be sampled in a small capillary tube. A solid dispersion is obtained by immersing the sample in liquid nitrogen, and orientation is obtained by applying a field of 160 kA/m (= 2000 Oe) during the solidification. Measurements are next made on the vibrating sample magnetometer (VSM) and can be used to examine and follow the effects of milling, solvent changes, viscosity, stability etc. (Fisher et al. 1982; further developed by Sumiya et al. 1984). Measuring instruments and techniques have been developed that allow in-process testing, from the dry magnetic particles to the finished, calendered product (Steinberg 1983).

The most modern tool added to inspection of magnetic recording media is the magnetic force microscope, which is discussed briefly in Chapter 14 (Fig. 14.10), with examples shown in Figs. 14.11 through 14.14.

A typical disk surface is shown in Fig. 13.24. The picture to the left shows the surface topography; note the height variation of 90 nm between dark and light. The dimensions of the shown square are 3×3 μm, and the small "grains" may be mechanical grains similar to those shown in Fig. 13.20; or, they may be grains in the carbon overcoat. The magnetics show domains ranging from 70 to 175 nm in diameter.

Figure 13.25 shows the very edge of a track, with random domains in the unwritten region to the side and transitions made by orienting the domains using the head field. There is clearly roughness in the transitions on the order of the domain size, causing noise.

Topography: Magnetics:

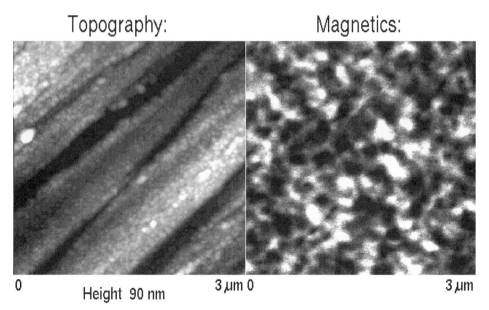

0   Height 90 nm   3 μm 0   3 μm

FIGURE 13.24 MFM pictures of small grains (left, light-to-dark = 90 nm) and small magnetic clusters (right) in a modern thin-film magnetic disk (*courtesy of Ken Babcock, Digital Instruments, Inc., 1995*).

Topography: Magnetics:

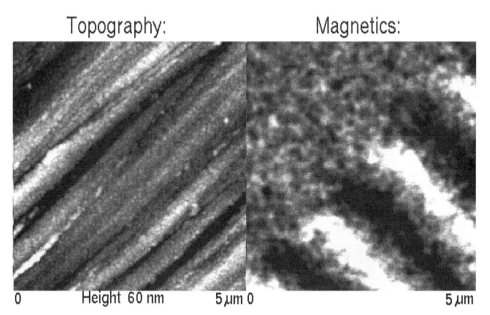

0   Height 60 nm   5 μm 0   5 μm

FIGURE 13.25 MFM pictures of grains (left, light-to-dark = 60 nm) and recorded track on thin-film disk. Note the reduced resolution due to magnetic clusters (*courtesy of Ken Babcock, Digital Instruments, Inc., 1995*).

REFERENCES

PARTICULATE MEDIA BINDER INGREDIENTS AND DISPERSION

Bagatta, P.U. May 1986. Concepts of Advanced Lubricant Systems for All Forms of Magnetic Media. *SMART Symposium*, paper no. WS 3-C-1, 39 pages.

Bagatta, U., A.R. Corradi, L. Flabbi, and L. Salvioli. Jan. 1984. Lubrication of Tapes with Fluorocarbon (Fomblin) Oils. *IEEE Trans. Magn.* MAG-20 (1): 16–18.

Braginskij, G.I., and E.N. Timoteev. 1981. *Technologie der Magnetbandherstellung*, Akademie-Verlag, DDR-1080 Berlin, 320 pages.

Brown, W.H., R.E. Ansel, L. Laskin, and S.R. Schmid. May 1983. Radiation Curable Coatings. *SYMPOSIUM Mag. Media 83*, paper no. D-3, 69 pages.

Burgess, K.A. May 1986. Carbon Black: Effect on Electrical Conductivity in Polymer Systems. *SMART Symposium*, paper no. E-3, 32 pages.

Crowe, J.T., and D.B. Arvidson, Jr. May 1986. The Role of Abrasive Additions in Magnetic Media. *SMART Symposium*, paper no. E-4, 13 pages.

Daniel, E.D., and K.E. Naumann. Nov. 1971. Audio Cassette Chromium Dioxide Tape. *Jour. AES* 19 (11): 822–28.

Dasgupta, S. Jan. 1984. Characteristics of Magnetic Dispersions: Rheological, Mechanical and Magnetic Properties. *IEEE Trans. Magn.* MAG-20 (1): 7–12.

Gooch, J.W. 1989. Rheological Characterization of high solids magnetic dispersions. In *Polymers in Information Storage Technology*. Edited by K.L. Mittal, Plenum Press, 1991, 457 pages, pp. 273–290.

Jung, C., S. Raghavan, and M.C.A. Mathur. Nov. 1993. Magnetic Probing of Dispersion Quality of Cobalt Modified Iron Oxide Particle Inks. *IEEE Trans. Magn.* MAG-29 (6): 3643–45.

Lueck, L.B. May 1983. The Wetting of Magnetic Pigment. *SYMPOSIUM Mag. Media 83*, paper no. D-1, 21 pages.

Mayo, P.I., K. O'Grady, and P.C. Hobby. Sept. 1992. Magnetic Characterization of Metal Particle Pigment Dispersion. *IEEE Trans. Magn.* MAG-28 (5): 2374–76.

Mihalik, R.S. May 1983. Binder/Additive Relationships. *SYMPOSIUM Mag. Media 83*, paper no. E-2, 37 pages.

Missbach, F.S. May 1983. Premilling, Milling, and Filtration Methology in Magnetic Media Dispersion Preparation. *SYMPOSIUM Mag. Media 83*, paper no. B-1, 46 pages.

Sischka, F.J. Oct. 1973. Development of an Advanced Fe_2O_3 Cassette Tape Coating. *Jour. AES* 21 (10): 789–808.

Williams, J.L., and P.H. Markusch. May 1986. Polyurethane Coatings. *SMART Symposium*, paper no. D-2, 34 pages.

COATING OF PARTICULATE MEDIA

Landskroener, P.A. May 1983. Conventional Coating Methods. *SYMPOSIUM Mag. Media 83*, paper no. B-2, 38 pages.

PARTICULATE ORIENTATION

Bate, G., and L.P. Dunn. Sept. 1980. On the Design of Magnets for the Orientation of Particles in Tapes. *IEEE Trans. Magn.* MAG-16 (5): 1123–1125.

Newman, J.J. Sept. 1978. Orientation of Magnetic Particle Assemblies. *IEEE Trans. Magn.* MAG-14 (5): 866–868.

Newman, J.J., and R. Yarbrough. Sept. 1969. Theory of the Motions of a Fine Magnetic Particle in a Newtonian Fluid. *IEEE Trans. Magn.* MAG-5 (3): 320–324.

Peng, W.G., S.S. Wong, Y.S. Lin, and C.D. Wu. Sept. 1992. Magnetic Orientation of Acicular Iron Particle in Recording Tape. *IEEE Trans. Magn.* MAG-28 (5): 2377–79.

SLITTING AND FINISH

Brondijk, J.J., P.E. Wierenga, E.E. Feekes, and W.J.J.M. Sprangers. Jan. 1987. Roughness and Deformation Aspects in Calendering of Particulate Magnetic Tape. *IEEE Trans. Magn.* MAG-23 (1): 146–49.

Ostrowski, H.S. May 1983. Nonwoven Liners for Floppy Disks and Related Jacket Manufacturing Factors. *Symposium Magn. Media 83*, paper no. F-2, 33 pages.

THIN-FILM MEDIA: METAL EVAPORATED TAPES (ME)

Akiyama, S., S. Nakagawa, and M. Naoe. Nov. 1991. Deposition of stress-free CoCr alloy thin films on thin tape substrate by Kr ion sputtering. *IEEE Trans. Magn.* MAG-27 (6): 4751–53.

Feuerstein, A., and M. Mayr. Jan. 1984. High Vacuum Evaporation of ferromagnetic materials—a new production technology for magnetic tapes. *IEEE Trans. Magn.* MAG-20 (1): 51–56.

Griesser, H.J. 1989. Plasma polymer films for corrosion protection of cobalt-nickel 80:20 magnetic thin films. In *Polymers in Information Storage Technology*. Edited by K.L. Mittal, Plenum Press, 1991, 457 pages, pp. 351–372.

Hokkyo, J., T. Suzuki, K. Chiba, K. Sato, Y. Arisaka, T. Sasaki, and Y. Ebine. Jan. 1993. Co-Ni obliquely evaporated tape and its magnetic and recording characteristics. *Jour. Magnetism and Magn. Matl.* 120 ((1-3): 281–285.

Krijnen, G., S.B. Luitjens, R.W. de Bie, and J.C. Lodder. March 1988. Correlation between anisotropy direction and pulse shape for metal evaporated tape. *IEEE Trans. Magn.* MAG-24 (2): 1817–19.

Lee, T-H.D. Jan. 1990. Selection of Lubricants for Metal Evaporated Tape. *IEEE Trans. Magn.* MAG-26 (1): 171–73.

Numazawa, J., K. Kuga, and H. Yoshimoto. Nov. 1993. Thin-Film Tape Media with Multilayers by Sputtering. *IEEE Trans. Magn.* MAG-29 (6): 3058–63.

Ouchi, N., H. Yoshida, K. Shinohara, and A. Tomago. Sept. 1986. Analysis due to vector magnetic field for recording characteristic of metal evaporated tape. *IEEE Trans. Magn.* MAG-22 (5): 385–87.

SPUTTERED RIGID DISKS

Angus, J.C. Aug. 1992. Diamond and Diamond-Like Films. *Thin Solid Films* 1 (4): 126–133.

Bhushan, B., 1993. Magnetic Slider/Rigid Disk Substrate Materials and Disk Texturing Techniques—Status and Future Outlook. *Adv. Info. Storage Syst.* Vol. 5 pp. 175–209.

Bhushan, B., and B.K. Gupta. 1991. Sputtering (Sect. 9.3), in *Handbook of Tribology*—Materials, Coatings and Surface Treatments. McGraw-Hill.

Cord, B., S. Schulz, and K.H. Schuller. Jan. 1993. Single-disk process technology for magnetic media. *Jour. of Magnetism and Magn. Matls.* 120 (1–3): 330–333.

Drennan, G.A., R.J. Lawson, and M.B. Jacobson. Nov. 1985. In-Line Sputtering Deposition System for Thin-Film Disc Fabrication. *HP Journal* 36 (11): 21–25.

Garrison, M.C. May 1983. Plated Media Manufacturing Methods. *SYMPOSIUM Mag. Media 83*, paper no. MMS-G, 23 pages.

Gill, H.S., and T. Yamashita. Sept. 1984. The Growth Characteristics of Ion-beam Sputtered CoCr Films on Isolation Layers. *IEEE Trans. Magn.* MAG-20 (5): 776–778.

Glijer, P., J.M. Sivertsen, and J.H. Judy. Nov. 1994. Advanced Multilayer Thin Films for Ultra-High Density Magnetic Recording Media. *IEEE Trans. Magn.* MAG-30 (6): 3957–59.

Ishida, S., and K. Seki. Feb. 1991. Thin-Film Disk Technology. *Fujitsu Sci. Tech. Jour.* 26 (4): 337–352.

Johnson, K.E. 1992. Fabrication of Low-Noise Thin-Film Media. In *Noise in Digital Magnetic Recording*, ed. by Arnoldussen, T.C., and L.L. Nunnelley, World Scientific, Singapore, 280 pages, pp. 7–63.

Johnson, K.E., J.B. Mahlke, K.J. Schulz, and A.C. Wahl. Jan. 1993. Fabrication of Low Noise Media—A Manufacturing Perspective. *IEEE Trans. Magn.* MAG-29 (1-Pt. 1): 215–222.

Judy, J. May 1983. Sputtered Perpendicularly Oriented Metal Coatings. *SYMPOSIUM Mag. Media 83* paper no. F-1, 39 pages.

Khan, M.R., N. Heiman, R.D. Fisher, S. Smith, M. Smallden, G.F. Hughes, K. Veirs, B. Marchon, D.F. Ogletree, M. Salmeron, and W. Siekhaus. Nov. 1988. Carbon Overcoat and the Process Dependence on Its Microstructure and Wear Characteristics. *IEEE Trans. Magn.* MAG-24 (6): 2647–49.

Kurokawa, H., T. Mitani, and T. Yonezawa. Sept. 1987. Application of Diamond Like Carbon Films to Metallic Thin-Film Magnetic Media. *IEEE Trans. Magn.* MAG-23 (5): 2410–12.

Latev, D., V. Dorfman, and B. Pypkin. Aug. 1991. Application of Diamond Polymer Films in Hard Disk Technology. *Surf. & Coat.* 47 (1-3): 308–314.

Moore Jr, G.E., R.S. Seymour, and D.R. Bloomquist. Nov. 1985. Manufacturing Thin-Film Discs. *HP Journal* 36 (11): 34–35.

Parsons, R. 1991. Sputter Deposition Processes (sect. II-4), in *Thin Film Processes II*, ed. by Vossen, J.L., and W. Kern, Academic Press, pp. 177–208.

Rossnagel, S.M. 1991. Glow Discharge Plasmas and Sources for Etching and Deposition (sect. II-1), in *Thin Film Processes II*, ed. by Vossen, J.L., and W. Kern, Academic Press, pp. 11–76.

Sin, K., J.M. Sivertsen, and J.H. Judy. Nov. 1994. Effect of Ta on the Structure, Magnetic properties, and Recording Performance of CoCrTaPt/Cr Thin Film Media. *IEEE Trans. Magn.* MAG-30 (6): 4008–10.

Suganuma, Y., H. Tanaka, M. Yanagisawa, F. Goto, and S. Hatano. Nov. 1982. Production Process and High Density Recording Characteristics of Plated Disks. *IEEE Trans. Magn.* MAG-18 (6): 1215–1221.

QUALITY CONTROL IN THIN-FILM PRODUCTION

Fisher, R.D., L.P. Davis, and R.A. Cutler. Nov. 1982. Magnetic Characteristics of Gamma-Fe$_2$O$_3$ Dispersions. *IEEE Trans. Magn.* MAG-18 (6): 1098–1110.

Morrison, J.R., and A.S. Brar. May 1983. Disk Substrate Requirements for Future High Areal Density. *SYMPOSIUM Mag. Media 83*, paper no. A-3, 45 pages.

Persoon, A.H. May 1983. Characterization of Parameters for Developing Quality Control Programs for Magnetic Media. *SYMPOSIUM Mag. Media 83*, paper no. TMM-1, 18 pages.

Steinberg, G. May 1983.Testing Dispersions and Magnetic Coatings. *SYMPOSIUM Mag. Media 83*, paper no. TMM-2, 43 pages.

Sumiya, K., N. Hirayama, F. Hayama, and T. Matsumoto. Sept. 1984. Determination of Dispersibility and Stability of Magnetic Paint by Rotation—Vibration Method, *IEEE Trans. Magn.* MAG-20 (5): 745–747.

BIBLIOGRAPHY

PARTICULATE MEDIA

Bourgin, P., and F. Bouquerel, 1993. Winding Flexible Media: A Global Approach. *Adv. Info. Storage Syst.* vol. 5, pp. 493– 512.

Coyle, D.J., C.W. Macosko, and L.E. Scriven. Feb. 1990. The Fluid Dynamics of Reverse Roll Coating. *AICHE Journal* 35 (2): 161–174.

Edwards, M.F., and M.R. Baker. 1992. A Review of Liquid Mixing Equipment. In *Mixing in the Process Industries*, 2nd. ed., pp. 118–136.

Kwon, T.M., M.S. Jhon and T.E Karis. 1992. Dispersion Quality of Rod-Like Fe$_2$O$_3$ and CrO$_2$ and Plate-Like Ba-Ferrite Suspensions for Magnetic Recording. *Adv. Info. Storage Syst.* Vol. 4 pp. 87–101.

Kwon, T.M., M.S. Jhon, and T.E. Karis. Feb. 1992. A Device for Measuring the Concentration and Dispersion Quality of Magnetic Particle Suspensions. *IEEE Trans. Instr. and Meas.* 41 (1): 10–16.

Mathur, M.C.A., S. Raghavan, and C. Jung. 1991. Dispersion Quality of Magnetic Inks: A Review. *Adv. Info. Storage Syst.* Vol. 1 pp. 337–351.

Okuyama, K. 1991. Agglomeration (Coagulation). In *Powder Technology Handbook*, pp. 293–305.

Wolf, I., and T. Neuman. Nov. 1991. Recording at High Volumetric Packing Densities. *SMPTE Journal* 98 (7): 515–519.

FACTORS AFFECTING THIN-FILM MEDIA FABRICATION

Doerner, M.F., T. Yogi, D.S. Parker, S. Lambert, B. Hermsmeier, O.C. Allegranza, and T. Nguyen. Nov. 1993. Composition Effects in High Density CoPtCr Media. *IEEE Trans. Magn.* MAG-29 (6): 3667–69.

Ishikawa, M., K. Terao, M. Hashimoto, N. Tani, Y. Ota, and K. Nakamura. Sept. 1990. Effects of Thin Cr Film Thickness on CoNiCr/Cr Sputtered Hard Disk. *IEEE Trans. Magn.* MAG-26 (5): 1602–04.

Johnson, K.E. Nov. 1992. Underlayer Epitaxial Effects on Magnetic Thin Films Used in Rigid Disk Recording. *Magnetic Matls. Proc., and Devices.* Electrochem. Soc. Inc. Proc. Vol. 92-10, pp. 27–37.

Khan, M.R., R.D. Fisher. N. Heiman, and J. Pressesky. Nov. 1988. Effects of Nitrogen, Oxygen and Air on the Magnetic Properties of Sputtered Materials for Thin-Film Recording Disks. *IEEE Trans. Magn.* MAG-24 (6): 2985–87.

Lu, M., J.H. Judy, and J.M. Sivertsen. Sept. 1990. Effects of RF Bias on Texture, Magnetics, and Recording Properties of RF Sputtered CoCr/Cr Longitudinal Thin-Film Media. *IEEE Trans. Magn.* MAG-26 (5): 1581–83.

Mahvan, N., E. Ziera, and A. Eltoukhy. Nov. 1993. Oxidation of Seed-Layer for Improved Magnetic & Recording Performance of Thin-Film Rigid Discs. *IEEE Trans. Magn.* MAG-29 (6): 3694–93.

Okumura, Y., H. Morita, H. Fujimori, X.B. Yanf, and I. Endo. Nov. 1993. Segregation Structure of Bias-Sputtered CoCr/Cr and CoCrTa/Cr Films. *IEEE Trans. Magn.* MAG-29 (6): 3144–46.

Onishi, Y., H. Matsumora, T. Hase, H. Hayashi, M. Sato, and K. Katsumoto. Sept. 1989. Substrate Effects on the Magnetic Characteristics of Sputtered Media. *IEEE Trans. Magn.* MAG-25 (5): 3887–3889.

PERPENDICULAR MEDIA

Ouchi, K. 1991. Co-Cr Perpendicular Recording Medium. Ch. 7 in *Perpendicular Magnetic Recording.* pp. 109–129.

Sugita, R. 1991. Co-Cr Perpendicular Recording Medium by Vacuum Deposition. Ch. 9 in *Perpendicular Magnetic Recording.* pp. 143–157.

Uchiyama, Y., H. Sato, and Y. Kitamoto. Sept. 1992. Effect of Substrate Temperature on Magnetic and Microstructure Properties of sputtered CoCr Films with Perpendicular Magnetic Anisotropy. *IEEE Trans. Magn.* MAG-28 (5): 2010–17.

CHAPTER 14
PROPERTIES OF MAGNETIC TAPES AND DISKS

In selecting a magnetic media many factors should be considered, such as application, data density and rate, storage capacity (or playing time), compatibility, cost, and quality.

This chapter's update reflects the substantial shift in application of magnetic recording from analog to digital. The home reel-to-reel tape recorder has all but disappeared, along with the desire to record the FM stereo broadcast of fine music. Today the CD and its player serves the audiophile. A tape cassette recorder/player is still there, but mostly just for duplication or playback. The professional amateur now uses Philips' DCC (Digital Compact Cassette) or Sony's MO (magneto-optical) disk to record new material.

The same trend is obvious in video recording, which today is done digitally at the studio level. This will sift into home use when HDTV with digital picture compression matures. Add thereto the large growth in libraries storing information on tapes (3490 cartridges, 8 mm, DAT, QIC) and very high storage volumes on digital D-1, D-2, D-3, and D-5 cartridges, a technology transfer from digital video.

Applications used to divide tapes into four user groups: audio, instrumentation, video, and computer/data. Digital recordings are replacing analog methods, and formats and technologies are transferred from one application area to another.

The situation is shown in Table 14.1, which tabulates the main tape configurations in use. Floppies in the 3½" size use coatings similar to those of tapes. Rigid disk drives use aluminum, glass, or ceramic disks coated with a thin magnetic film, plus a protective overcoat, plus lubrication.

Certain media properties are important in any application. They are:

Properties of substrates:	Flatness (tape, disk)
Properties of coatings:	Smoothness (tape)
	Overcoat protection (disk)
	Lubrication (tape, disk)
	Durability (tape)
	Abrasion (tape)
Basic Magnetics (tape, disk):	Write level
	Response
	Short-wavelength response
	Eraseability
	Overwrite
	Dropouts
	Storage stability

The quality of ac-bias recordings is further influenced by:

- Record level (normal, audio, and distortion)
- Frequency response vs. data and bias levels
- Long wavelength response and drop-ins
- Dead layer
- Print-through
- Contact duplication

TABLE 14.1 Tape Applications and Configurations ca.1995

Analog or digital:	Longitudinal tape						Helical-scan recorders					
	Audio		Instrument		Comp.	Audio	Instr.	Composite		Component		Data
	A	D	A	D	D	D	D	A	D	A	D	D
Tape width:												
4 mm	•	DCC				DAT						DAT
6.25 mm	•		•									QIC
8 mm		PCM						8MM				8MM
12.7 mm	•		•		•	•	VHS	VHS	D–3	S–VHS		3480 & 3490
19 mm			•		•			C	D–2		D–1	
25.4 mm	•	•	•	•								
		(studio)			•							
50.8 mm	•	•										

Magnetic coatings: gamma—Fe_2O_3, CoFe, CrO_2,BaFeO, MP. Particle coating thicknesses range from 15 µm (Analog audio, studio) through 5 µm (standard audio) to 2 µm (video, QIC) and recently 0.5–0.1 µm (Fuji). ME tapes have 0.1 µm film thickness.

Reels, cassettes and cartridges are used to store tapes.

MECHANICAL PROPERTIES OF SUBSTRATES

There are a number of relevant parameters for substrates for tapes, flexible disks, and rigid disks, and they are as follows:

TAPES

The base film for magnetic tapes and floppy disks today is almost exclusively PET or polyester (an oriented polyethylene terephthalate). In the early days, cellulose acetate was used, but was later succeeded by PVC (polyvinyl chloride). The PCV film was not as strong as PET, nor would it withstand temperatures approaching 70° C. It did have a better surface than the first PET films, and the acetate films had the advantage of breaking clean rather than stretching under heavy loads, such as occurred during the start and stop operations in early recorders. A clean break could be spliced, while a recording on a stretched tape was lost.

PET combines the strength required of a thin base film with the limpness needed for intimate head-to-tape contact. Details about the PET and other base films are covered in detail in pages 85 through 294 in *Mechanics and Reliability of Flexible Media* (Bhushan 1992).

Although polyester base materials are twice as expensive as either of the two earlier base films, it is employed exclusively in the manufacture of precision magnetic tape for audio, computers, video, and instrumentation recorders. The strength of the PET film can be increased further by prestretching it. The stretching orients the long-chain molecules in the film in the direction of the stretch. The disadvantage of such prestretched or tensilized polyester films is that they have a memory; if they are reheated beyond a certain point, they will shrink back close to their original size with consequent distortion of the recorded material.

The strength of a tape is controlled by the base film. The coating's mechanical properties alone are difficult to measure. A combination of a tensile tester and a computer has made it possible to separate the coating's contribution to the tape's Youngs' modulus (Schaake et al. 1987). A number of other tape/coating properties, like secant modulus, elasticity, break boundaries, etc., can easily be determined.

Other types of distortion of the finished tape product are cupping, curling, and layer-to-layer adhesion. Cupping is illustrated in Fig. 14.1, left. Two tape samples are placed on a planar supporting surface. The cupped tape will stand straight out, while the better tape will bend in a smooth arc. Cupping is generally found in inferior tapes and is due to improper manufacturing, or differences between coefficients of thermal or hygroscopic expansions of coating and base film.

Curling results in a twist in a free-hanging length of tape. This is again due to residual stress gradients in the PET film, possibly aggravated by a bad slitting process. Curl is detected by unwinding a few feet of tape, holding the reel up and looking down along the free-hanging length of tape. It should be a flat sheet, as illustrated in Fig. 14.1, center.

Layer-to-layer adhesion is detected by observing how easily the tape unwinds. Hold the reel vertically and slowly turn it to unwind the tape; it should unwind freely and smoothly; see Fig. 14.1, right. Any jerky motion caused by sticking to the tape pack will cause flutter in many recorders. This defect is particularly common after prolonged storage of a tape.

The dimensional stability of PET films was commonly neglected in the past due to the overshadowing problem of slitting tolerances. Tapes are slit to a final width with a tolerance of +.000" to –.0008" in the very best case. This almost 1 mil of uncertainty sets an ultimate limit for the number of tracks that can be successfully recorded on one machine and played back on another; in reality, differences in temperature and humidity cause additional dimensional changes.

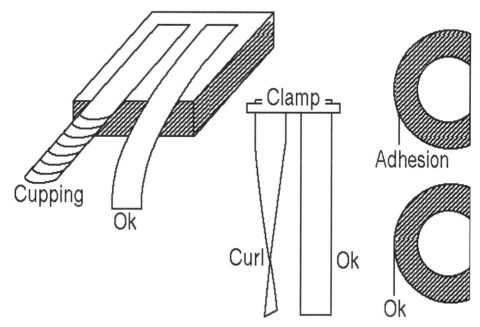

FIGURE 14.1 Tests of magnetic tape for cupping, curl, and layer-to-layer adhesion.

FLEXIBLE DISKS

A similar tolerance applies to floppy disks for the accuracy of the center hole. We must add to it the dimensional changes that take place in a PET disk. There are relaxations of built-in stresses created during the manufacturing process, plus temperature and humidity changes. How much and when they occur is difficult to predict. Furthermore, PET film properties vary across the web with the orientation of the polymers.

All the dimensional changes due to temperature, humidity, time, and centrifugal forces may change an initially circular track into an elliptic or otherwise noncircular track. This will obviously limit the number of tracks per inch of radius of the disk.

These changes were discussed in the section on track width selection in Chapter 10. There are methods for the determination of dimensional changes in a disk, such as recording a pair of adjacent tracks followed by playback and comparison of the data (Behr and Osborn 1981); users of such methods must be aware of errors involved that may be substantial (Izraelev 1983).

RIGID DISKS

The requirements for aluminum disk blanks were covered in Chapter 11. The surface characteristics can be evaluated by a profilometer, or optically (which does not disturb the surface). The surface properties of the substrate will reflect through the magnetic overcoat, whether it is particulate or metallic. This is not just in the form of asperities or the overall smoothness, but it affects the magnetic properties of film as well (Thompson and Mee 1984).

The thickness of a rigid disk is steadily decreasing to shrink both height and volume of a disk drive (to fit into laptops). The required degree of flatness increases at the same time. The disk may warp when clamped to the spindle motor shaft.

A substrate can be tested for flatness in laser interferometry equipment (see Fig. 14.7 for the principle of operation). This should be done before and during manufacture. Well-known instrument makers are WYKO and ZYGO. The flatness measurements produce pictures that are maps of the disk height variations; see Fig. 14.2.

Of equal interest are plots of variations around a track; i.e., for a 360° revolution of the disk. A solid spin stand with precision air bearings is used to test disks and heads, as shown in Fig. 14.3. Figure 14.4 shows the result of a fluttering glass disk where head runout as well as disk runout is shown (Mann 1994). The amplitude of the plot is the acceleration of the velocity, which is a critical parameter for the head flying attitude (i.e., how well the servo can keep it on-track). The numbers plotted are called RVA (runout velocity acceleration).

MECHANICAL PROPERTIES OF COATINGS

Coatings also have a number of relevant mechanical properties that must be taken into account.

SURFACE PROPERTIES OF TAPES AND FLEXIBLE DISKS

The magnetic coating on a tape is fabricated from magnetic particles dispersed in a suitable binder. The overall coating thickness and the surface finish of the coating bear a significant influence on the quality of recordings. For optimum recordings, different coating thicknesses require different amounts of high-frequency bias currents, and a recorder is therefore tuned up for a particular tape that should be the only tape used for future recordings. The difference between a thick coating and thin coating is illustrated in Fig. 14.5. The thicker coating gives a higher output at low frequencies, but at the price of low output at high frequencies (overbias).

Disk pressed against optical flat produces Moire patterns.

FIGURE 14.2 The Moire patterns of a warped disk (*courtesy of WYKO Corp., 1995*).

FIGURE 14.3 Disk testing on a spin stand (*courtesy of Teletrac Inc., 1995*).

Digital storage response is measured by resolution which is the ratio between the outputs at the frequencies $2f$ and $1f$, expressed in percentage (Fig. 14.6). Here, as above, the outcome depends on the media thickness, write current, packing density, and so forth, and no resolution rating can be assigned to media without qualifications. $2f$ and $1f$ are the highest and the lowest signal frequencies in the encoded data (see chapter 24), originating in the FM code.

The surface characteristics have a significant bearing on the high-frequency performance of a magnetic tape. Any spacing between the magnetic heads and the tape surface causes a reduction of high-frequency (short-wavelength) response, and a rough tape surface does, in effect, act as a spacing between the magnetic coating and the head. In the manufacture of magnetic tape it is, therefore,

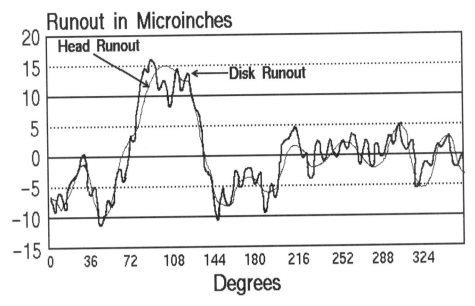

FIGURE 14.4 Head runout and disk runout (*courtesy of Don Mann Assoc., 1995*).

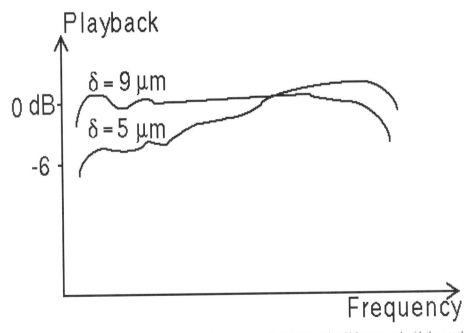

FIGURE 14.5 Amplitude versus frequency response for two tapes with different coating thicknesses, each with the record level optimized at long wavelengths.

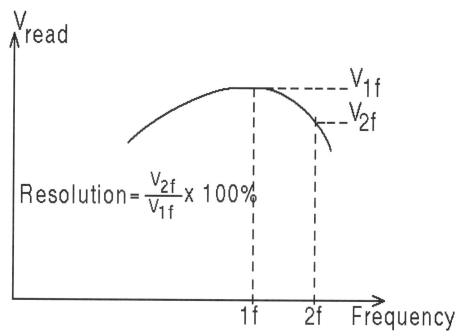

FIGURE 14.6 Resolution in digital recording.

common practice to surface-treat the tape to obtain a finished product that has as smooth a surface as practical. Tapes are calendered and floppies are, in addition, polished in the circular band assigned to data storage (both sides).

Surface flatness can be evaluated by means of a profilometer. This instrument moves a very small diamond needle across the tape surface and any motions of the needle (like a phonograph pickup) are amplified, and the output waveform plotted on a paper chart (Fig. 14.7). Although the diamond stylus may deform the tape surface under the high unit area pressure, it is still a useful instrument in comparing various tapes. Photographs of two tape surfaces are shown in Fig. 14.8.

Another method of gauging the surface flatness of a recording media results when an MR-head is placed in contact with the moving surface, because localized heat from friction between the surface asperities and the MR-element results in a voltage output.

A superior technique has been employed in recent years by using laser interferometry (Wahl et al. 1983; Perry et al. 1984; Robinson et al. 1984). The instrumentation is shown simplified in Fig. 14.7, and two scans of a tape surface are shown in Fig. 14.8 (Robinson et al. 1985). A laser beam, frequency shifted by an acousto-optic cell, is focused on a moving surface and reflected. Surface measurements of the sample are derived from the phase shifts of the carrier frequency; these shifts are created by interference.

This measuring technique has allowed for collection of some very interesting data pertaining to the SNR of video tapes (Robinson et al. 1985; Wierenga et al. 1985). The recorded RF-signal amplitude varied in inverse proportion to the surface roughness, which is to be expected from the spacing loss rule. This, in turn, affected the SNR of the chrominance signal in the ½-inch VTRs with color-under; see Fig. 14.9. The chrominance signal is discussed in Chapter 29.

Making a tape perfectly smooth does have a price, by causing an almost molecular-level adhesion to the head surfaces. This phenomenon is chemical/mechanical in nature. The effect demonstrates itself in two ways: gap smear, or "varnish." Gap smear, which is particularly noticeable in microgap

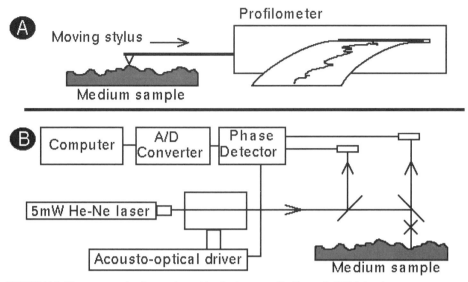

FIGURE 14.7 Measurements of surface roughness: (a) stylus instrument (Profilometer), (b) light interference meter.

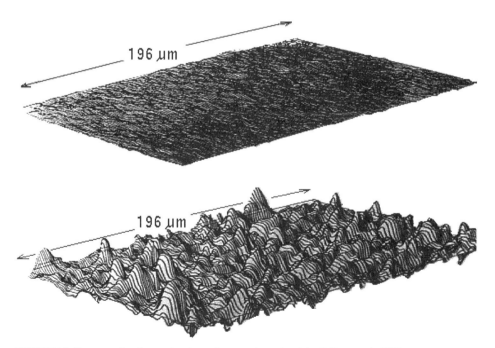

FIGURE 14.8 The topography of a smooth and a rough tape coating surface (*after Robinson et al., 1985*).

heads, will cause the core lamination material to flow across the gap and, in essence, short circuit it magnetically. The varnish phenomenon is a microscopic buildup of a clear film on the magnetic heads, which may cause separations between the tape surface and the head surface. This varnish may come from resins in the coating material, and should be eliminated by abrasive coating additives such as alumina powder.

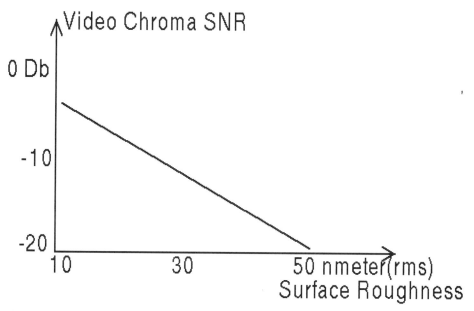

FIGURE 14.9 Video chrominance SNR for a color-under system, as a function of the tape surface roughness (*after Robinson et al., 1985*).

A tape must further exhibit good temperature stability when used in video recording. In a transverse-scan video recorder, the relative speed between the head and the tape is in the order of 500 to 1500 IPS, and the localized heat generated by friction can be very high and cause a breakdown of the coating. This is particularly pronounced in the helical-scan video recorders with still-frame capability; that is, when the tape motion is stopped while the scanning head rotates and thereby produces a single-frame picture of the video information. This generates locally concentrated high temperatures, and special binder formulations are therefore used for video tapes.

In recent years it has become common to apply a second coating to the backside of many varieties of tapes. This coat is a binder containing carbon, and serves to lower the resistivity of the tape and thereby prevent the annoying pop noises associated with electrostatic discharges. This coat is sometimes called carbon black.

The surface finish of this coat is matte, and it provides the additional advantage of increasing friction among tape layers and hence reducing the danger of tape cinching and slipping during fast starts and stops in the winding mode. It may also be beneficial due to the stronger grip of the tape between capstan and pressure puck. Durability of the coating polymer binder increases with the yield stress and strength of repetition fatigue of the coating (Sumiya et al. 1989).

There is finally the question of lubrication of tape surfaces. Early lubricants were fluorocarbons (Au-Yeung 1983), while other, organic lubricants were used (Wright and Tobin 1975; Nakamura et al. 1984). Today, lauric acids are used in tapes (Aoyama and Kishimoto 1991). It has been shown to be present in three states: adsorption on the surface of magnetic particles, dissolution in the binder, and the free state at the surface. The latter controls the runnability of the tape.

It is quite difficult to apply a lasting lubrication to ME tapes: Less than 15 Å thickness is insufficient, and thicker than 40 Å is too much. Only ME tapes with 15- to 40-Å-thick lubrication withstood the 5 minutes still-frame test in a VTR. Two lubricants have shown similar results: stearic acid and perfluoroethers (Lee 1990).

A protective layer has also been tried to improve a perpendicular CoCr thin-film ME tape. A 100- to 500-Å overcoat of FeMoB was sputtered onto a 1500-Å-thick CoCr film; FeMoB is wear and corrosion resistant and has a low coefficient of friction (Akiyama et al. 1991).

SURFACE PROPERTIES OF RIGID DISKS AND OVERCOATS

Disks are overcoated with two thin layers: diamond-like carbon (≈ 250 Å) and a lubricant (≈ 20 Å). The carbon layer provides corrosion resistance and damage protection from impacts. The carbon film must therefore be hard and offer toughness against fracture. These are the properties that are intended by sputtering the carbon so that it becomes diamond-like. The film must, in addition, provide good adhesion to the magnetic film layer, plus compatibility with the entire disk-making process.

A tool for examination of carbon overcoats (and other surfaces) is the STM (Scanning Tunneling Microscope, Marchon et al. Nov. 1991; Wickramasinghe 1989; and Hansma et al. 1988).

Topography as well as recorded tracks on a disk can be observed by MFM, *magnetic force microscopy* (Gomez et al. 1993; Burke et al. 1994). MFM is an application of scanning probe microscopy that provides imaging of magnetic fields near a sample surface.

In its simplest form, MFM detects static cantilever deflection that occurs when a magnetic field exerts a force on a tip coated with magnetic material; see Fig. 14.10. This method was used in obtaining Figs. 3.11 and 3.12. The MFM sensitivity can be enhanced by oscillating the cantilever near its resonance frequency. When the tip encounters a magnetic force gradient, the effective spring constant, and hence the resonance frequency, is shifted (Babcock et al. 1994. Patent Pending).

Measurements are taken in two passes over each scan line. On the first pass, topographical information is recorded by letting the oscillating cantilever lightly tap the surface. On the second pass, the tip is lifted to a user-selected separation (typically 20–200 nm) between the tip and local surface topography. By using the stored topographical data instead of the standard feedback, the separation remains constant without sensing the surface.

Two-pass measurements are taken for every scan line, producing separate topographic and magnetic image field images. Shown in Fig. 14.11 are a pair of images of a thin-film disk. The topographical image on the left shows disk roughness. The right image shows magnetic force gradients as sensed. In this example, the magnetics consist of a track that was written, then later overwritten along its center line by a narrower head. In Chapter 30 a similar set of images shows recorded bits on a magneto-optical disk.

Figure 14.12 and 14.13 show a 2-D and 3-D MFM picture of recorded disk tracks at 4300 tpi and user density 74,800 bpi. That is an areal density of 322 Mb/in². Figure 14.14 shows a very-high-density recording made by actually recording bits with the MFM probe tip (to be published; K. Babcock, Digital Instruments Inc., and M. Dugas, Advanced Research Corp. 1995).

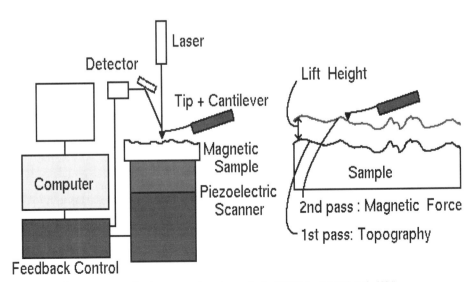

FIGURE 14.10 Measurements of surface topography and magnetics by MFM (*after Babcock et al., 1994*).

Topography: Magnetics:

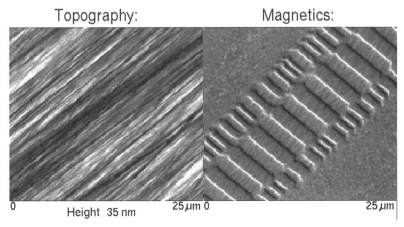

0   Height 35 nm   25 μm 0   25 μm

FIGURE 14.11 The topography and magnetics of rigid CoCr disk medium. Texture peak-to-valley = 35 nm. Magnetics shows one wide track recorded (diagonally), overwritten by a narrow track recording (*courtesy of Digital Instruments Inc., 1995*).

Topography: Magnetics:

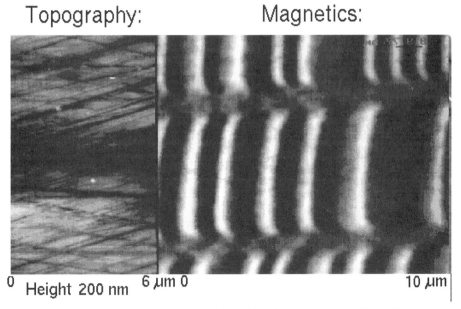

0   Height 200 nm   6 μm 0   10 μm

FIGURE 14.12 The topography and magnetics of a rigid CoCr disk medium. Texture peak-to-valley = 200 nm. Magnetics show tracks recorded at density of 4,300 tpi, with user data density of 74,800 bpi (*courtesy of Advanced Research Lab., 1995*).

Friction between a head slider and a carbon-overcoated disk tends to build up during repeated passes. This suggests that frictional behaviour may be related to chemical changes on the disk surface (Yang et al. 1991)

The carbon film structure, friction, and other properties need to be measured and monitored during development and production. Raman spectroscopy has been identified as a measure of film examination (Ager III 1993). The method helped evaluate the difference between sputtered amorphous

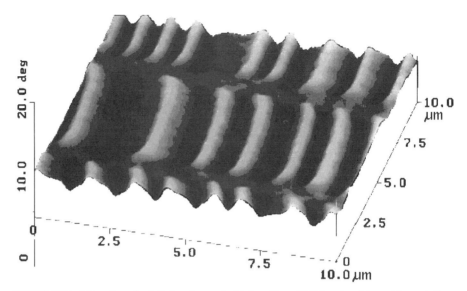

FIGURE 14.13 A three-dimensional display of magnetic data from Figure 14.12 (*courtesy Advanced Research Corp., 1995*).

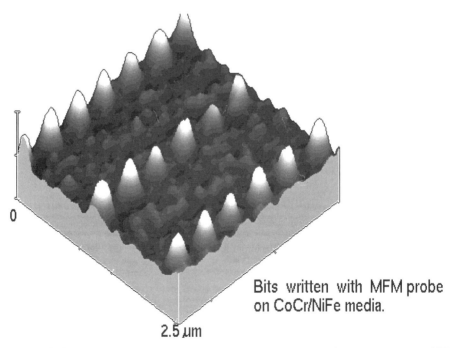

Bits written with MFM probe on CoCr/NiFe media.

FIGURE 14.14 The magnetics of bits recorded with the tip of an MFM probe (*courtesy of Digital Instruments, Inc., 1995*).

and hydrogenated amorphous carbon films. The latter films are harder, and have superior wear resistance and frictional properties (Agarwal et al. 1993.) The experimental results suggest that tribomechanical wear is the dominant wear mode in the amorphous carbon films, whereas abrasive wear occurs with the hydrogenated amorphous carbon.

Wear resistance of thin-film media is needed during the brief periods of head-to-media contact, i.e., when disk rotation starts and stops. Perfluoroether lubricants are used because they will chemically bond to the carbon overcoat, while the top layer of lubricant molecules were found to physically bond to each other. It has been shown that the chemically bonded monolayer is not removed during accelerated lubricant spin-off testing (Merchant et al. 1990).

The effects of lubricant film thickness, disk surface topography, sliding speed, viscosity, and durability have been investigated (Streator et al. 1991 [2]). In many cases, a sharp increase in dynamic and static friction coefficients occur when a certain film thickness is exceeded. For a very smooth disk (R_a = 2.0 nm), the increase occurred when the lubricant thickness increased beyond 5 nm. On a rougher disk, the lubricant could be thicker.

DURABILITY AND ABRASIVITY OF TAPES AND FLEXIBLE DISKS

Certain applications require a coating binder that will endure a large number of passes over a head. Examples are computer tape drives, in-contact floppy disks, and satellite telemetry recorders. Still-framing in video recorders adds the additional environment of frictional heat, which decomposes most binders.

It is in general required that binders shall meet the following requirements:

Computer drive tapes:	> 20,000 passes
Floppy disks:	> $3 \times 10^{+6}$ revolutions
Video tapes:	> 1 hour still framing

An often-highlighted feature of helical-scan video recorders is their still-frame capability. In other words, with the tape motion stopped, the rotating head assembly will scan a particular portion of the recorded material and display it on the TV screen. Very high flash temperatures are generated in the contact area between the rapidly moving head and the stopped tape (possibly several hundred degrees C). This action may wear out the tape, or cause the formation of debris on the video head (called head clogging). Debris is a general term used for oxide and/or binder buildup on magnetic heads and tape guides. Some tapes have a still-frame capability of only a few seconds and are obviously rated poor, while others will last in excess of one hour. The still-frame test is consequently used as a figure of merit for a videotape.

Still-frame life of a tape is strongly affected by the surface micro roughness (Tomago et al. 1985); CoNi films can be optimized in this respect for their use in the 8 mm cameras. Wear and durability of disks have been investigated by Tereda et al. (1985).

It is not recommended to use several different tape types in a given camcorder (Patton and Bhushan 1995). Each tape type has its specific wear characteristics, elastic modulus, and interface chemistry. Thus the rotating heads' contours will change, and will never be properly "seated" for the next tape if it is a different type or brand. It was found that particulate tapes (MP and BaFeO) were compatible with each other in a tape drive, but they were not compatible with thick ME tape. It appears that a change to a different tape should be initialized by running one or two cartridges through the camcorcer in order to properly "seat" the heads.

Two recent papers cover the wear mechanisms of metal-evaporated and particulate tapes in VTRs (videotape recorders) (Osaki et al., 1990, 1992). The papers give a good insight into the wear problems, and the reader interested in the details of the head/tape relationship is referred to Chapter 18.

ELECTROMAGNETIC PROPERTIES

The magnetic recording surfaces of tapes or disks are subject to a series of measurements to determine how well they may perform in a recording system. The chore has become almost overwhelming, with a fairly large number of tests made in order to fully characterize a media.

The list of test data should be compared against some standard values so the tapes or disks can be graded, and a proper selection made. There is today no complete set of standards, and the choice in the end is made for brand A's media because it has certain critical properties that are superior to brand B's media, and because it has a known performance history.

This situation is not the fault of the standards groups, but it is really a result of the rapid developments among media manufacturers. The past decade has seen more new and better products coming on the market than during the entire history of magnetic recording.

Direct digital recording requires a record current that will produce a field that is greater than the coercivity H_r of the coating. The write current is normalized to a value i_{95} that produces 95 percent of the maximum read level, and is then adjusted upward to $2.1 \times i_{95}$ for low-bit-density recordings, or $1.8 \times i_{95}$ for high bit density. It may be even less for higher-density recordings.

Updating of digital recordings is made by overwriting on top of old data, and a typical criterion is that old data must be reduced below –30 dB in reference to the new data; see Fig. 14.15. This results in a method for adjusting the normal operating level in digital recording:

1. Write data at 1f.

2. Overwrite the 1f data with a new 2f signal, and increase the write current until the old 1f signal is 30 dB below the new 2f data (or better: until the SNR is better than 30 dB). Repeat step 1 with the new write current, and verify step 2; if necessary, readjust the write current, and repeat steps 1 and 2.

3. Use the write current from step 2 to write 1f and 2f data. Measure the corresponding peak-to-peak voltages V_{1f} and V_{2f}. The resolution can now be calculated: Resolution $= 100 \times V_{2f}/V_{1f}$ percent.

4. When adjusting i_{write} for a disk system do steps 1, 2 and 3 at an inner and an outer track.

Video recording has the simplest procedure of all: Adjust the record current for maximum RF-signal output during playback. At this point, evaluate the SNR of the tape.

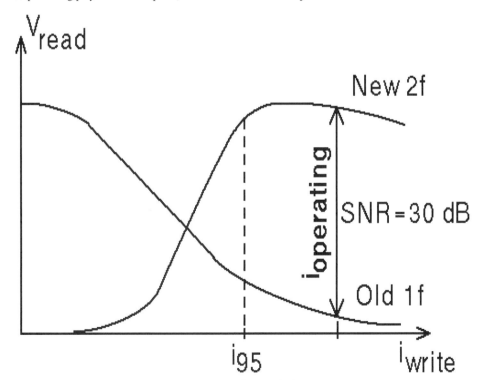

FIGURE 14.15   Amplitudes of new 2f and old 1f signals, as a function of write current.

RESOLUTION AND AMPLITUDE VERSUS FREQUENCY RESPONSE

A media for direct digital recording should retain the shortest possible transition zones while providing an adequate signal output. This requires a high value for the remanence M_{rs}, a high value for the coercivity H_r, and furthermore a thin coating for small demagnetization ($BL >$ coating thickness). The reduced output from a thin coat can be compensated for by increasing the remanence. This results in a small degradation in resolution, and a thin metallic coating may provide the optimum future coating for disks.

A narrow range of individual particle coercivities is also desirable, providing shorter transition zones at flux reversals in digital recording and providing a shorter recording zone in ac-bias recording. This coercivity range is commonly referred to as switching field distribution, SFD. It is the width in Oersteds of the derivative of the MH-curve in the switching region around H_r. See Fig. 12.7 for details.

SHORT WAVELENGTH RESPONSE

High densities of data on disk or tape are possible if the following conditions are satisfied:

1. Transition zones Δx are short. This requires:
    1.1. A medium with small $SFD = 1 - S^* = \Delta h_r$.
    1.2. A thin medium.
    1.3. In-contact recording.
    1.4. Small write current (only sufficient for overwrite).

2. In-contact record/playback, or very low flying height.

3. No demagnetization loss, i.e.:
    3.1. Partial-penetration recording
    3.2. High coercivity H_r

A combination of the listed requirements has resulted in an ongoing reduction of the wavelength on tapes, as shown in Fig. 14.16. Part of this success is result of increased coercivities, as shown in Fig. 14.17. Note also the increased outputs—an impressive 24 dB from γFe_2O_3 to ME tapes.

The reader can find a brief description of the effect of surface roughness on video color SNR in this chapter's section on the surface properties of tapes and flexible disks. A graph of the changing roughness in tapes made since 1955 shows an interesting correlation to the wavelengths recorded: Smoother tapes are required for short-wavelength recordings (see Fig. 14.18.)

This is understandable for analog signals. At short wavelengths, only a small thickness of the recorded coating contributes to the reads signal: $\delta_{eff} = .22 \times \lambda$. If the average roughness R_a is ten times less than the flux-contributing coating, then the amplitude variation (noise) level is -20 dB; see Fig. 14.19. The required smoothness for a density of 50 kbpi should therefore be:

$$R_a \leq BL/22.5 = .25/22.5 \ \mu m, \text{ or}$$

$$R_a = 11 \ nm$$

This is exactly the predicted smoothness for a 40–80 kbpi tape for a QIC application (Smith 1989). Not only tapes but also disks will have smoother surfaces. Disks are normally textured over their entire recording surface in order to prevent stiction between slider and disk (see Chapter 18). Texturing will, in modern disk drives, be done only in a small circular area near the disk center for a landing zone for the head slider. The rest of the surfaces will have a surface roughness R_a in the order of a couple of nanometers. The read spacing loss may change from $55 \times d/\lambda$ dB to $110 \times d/\lambda$ dB. This could be caused by a media relative permeability $\mu_r = 2$.

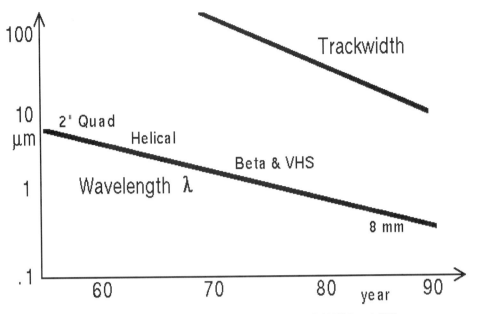

FIGURE 14.16   The decrease in trackwidth and operational wavelengths for the period 1955 through 1990.

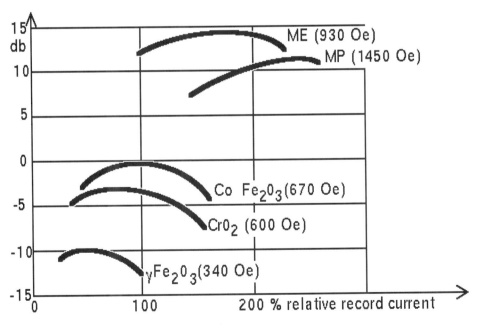

FIGURE 14.17   Comparative write/read level curves for five tapes.

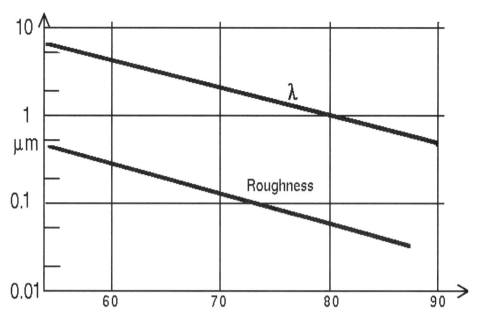

FIGURE 14.18   The decrease in wavelengths (from Figure 14.16) and reported tape surface roughness for satisfactory operation at these wavelengths. The evolution is shown for 1955 through 1990.

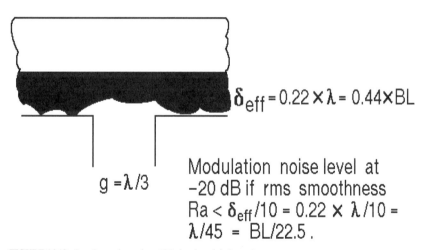

FIGURE 14.19   Roughness determines AM-signal modulation noise.

A spacing loss also occurs during writing. For ME tapes, the net write-and-read loss at a spacing of d becomes (Schildberg et al. 1993):

$$\text{Total spacing loss} = 110 \times d/\lambda \text{ dB} \tag{14.1}$$

It was also deduced that *PW50* will be larger due to the spacing:

$$PW50' \cong PW50 + 4d \tag{14.2}$$

and the drop in peak height is:

$$L = 1/(1 + 4d/PW50) \tag{14.3}$$

Formula (14.3) is valid only at low densities, where no pulse interference takes place.

If contact between an ideal head and media is perfect, one may be tempted to set $d = 0$. In this case, however, the true spacing depends on the relationship between recorded depth and wavelength. In experiments with CrO_2, BaFeO, and MP tapes, an effective spacing of 0.05 μm (2 μin.) is the minimum spacing that will exist when the mechanical spacing is zero (Cramer and Slob 1991).

DOUBLE-COATED TAPES

There are obviously several good reasons for making tapes' and floppies' coating thickness δ smaller than the current minimum of 2 μm for data storage (5 μm for audio tapes is dictated by equalization standards). A thin coat will reduce demagnetization (higher $\Delta\phi$) and the length of the transition zone (smaller $\Delta x \rightarrow \Delta t$). The result is higher read voltage $\Delta\phi/\Delta t$. A thin-coat tape is now a reality, via double-coated tapes.

Double-coated tapes have fascinated recording researchers, starting with Otto Kornei's patent in 1949 (Ogawa 1991). A variety of products have been released since. Most designs were based on an evaluation of amplitude versus frequency characteristics, and improvements were verified by measurements. Not all satisfied the requirement that the two layers' coercivities must be alike; in such cases, severe transient time distortion occurs, as shown in Fig. 12.15.

Two coats on a tape have been difficult to manufacture. Slot die casting of the coats is necessary to ensure coating thickness uniformity. Casting first the bottom layer in one run and then coating the second layer on top gives poor results. Both layers need to be coated simultaneously, a process called "wet-on-wet" (Speliotis 1993). Figure 14.20 illustrates the principle (Inaba et al. 1993). The result is an MP tape with a nonmagnetic TiO-particle bottom layer and a 0.1 μm MP top layer (Fuji Photo Film 1995).

ERASEABILITY AND OVERWRITE

No single number can represent a media's eraseability. We may define a method of measurement, but the result would be applicable only to a particular set of circumstances, such as using a bulk degausser, or erasure by overwriting with new data or a separate erase head.

Magnetics becomes more difficult to erase after long-time storage. Elevated storage temperature makes erasure more difficult (Mountfield and Kryder 1989). A recording made with a high bias level is more difficult to erase than a tape recorded at a low bias level; it is also observed that high frequencies (short wavelengths) are more easily erased than low frequencies; see Fig. 8.27.

Investigations into the erasure process are surprisingly few. The problems of insufficient erasure with erase heads were covered in Chapter 8, and it is apparent that bulk degaussers do not completely erase tapes. The author found only a few papers on the topic. Manly (1976) examined the erasure process and the strange phenomenon that signals may recur after erasure and storage. Fayling (1977) continued this work, and found that there were directional differences in the bulk erasure process, which requires very large fields to accomplish the erasure.

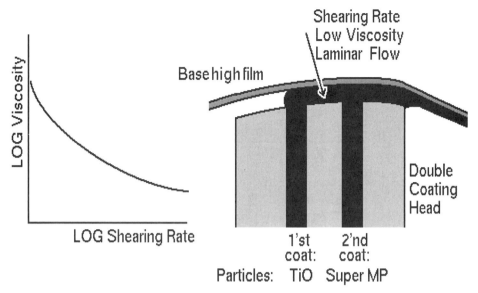

FIGURE 14.20   The fabrication of a double-coated tape requires a wet-on-wet process, where the extrusion of the first layer is followed immediately by extrusion of the top layer (*after Ogawa, 1991*).

The problems are aggravated for users of high-energy tapes (Burke and Sanders 1985). BaFeO is the hardest to erase and CrO_2 the easiest, and particle interaction play a major role in the erasure process (Lekawat et al. 1993 (2)).

Overwriting old data with new is used in digital systems, and presents problems of its own. During overwrite there is a re-recording phenomenon. The author explains this by considering the recording of flux that is generated by sensing incoming, not-yet erased data (see Fig. 8.28). Another contribution can be explained with contact printing model (Sugaya et al. 1993). Finally, there may be residual magnetization from the original recording that plays a role.

A recent investigation (Christensen and Finkelstein 1985) ranks the methods of erasure as follows (from poor to good): dc-erasure, two-gap dc-erasure (bringing M_{rs} to zero prior to rewriting data), ac-ferrite head erasure, and finally bulk erasure. Only small differences were found between CrO_2 and Co-treated γFe_2O_3 tapes.

Measurement of erasure in audio and instrumentation tape is made by recording a 1000 Hz signal at the standard recording level, plus 5 dB with standard bias current. Then a portion of the recording is erased. The signal-to-erase-noise ratio is defined as the difference in dB between the playback output level of the recorded portion and that of the erased portion.

STRESS DEMAGNETIZATION

When a tape loops through a recorder it is deformed several times as it passes guides and heads. This induces stresses in the coating and consequently on the particles. These forces will cause demagnetization of particles that are magnetostrictive.

Magnetostriction is a material property whereby mechanical forces cause an alignment of the domains in magnetostrictive material. A magnetization will, conversely, cause an elongation or contraction of the material.

The cobalt-doped iron oxide particles possess a rather large magnetostriction (magnetostriction constant λ). Their stress demagnetization will appear as a signal loss that increases with the number of passes. Measured data confirms this; see Fig. 14.21 (Daniel and Naumann 1971).

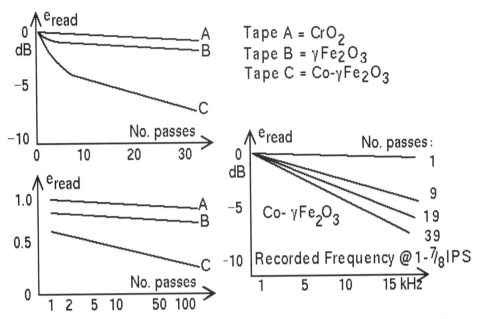

FIGURE 14.21 Demagnetization and magnetostrictive losses after multiple-pass tests (*after Daniel and Naumann, 1971*).

These measurements were accelerated by running a prerecorded tape loop through a transport while observing the playback signal during each pass. The losses were found to be proportional to the number of passes and the recorded frequency. Such losses are significant in applications where many repeated passes of a tape are anticipated, as in the analysis of signals in instrumentation recording, and these should be accounted for. Stress demagnetization has been studied by Hoshi et al. (1984) and Izawa (1984). Measurements on $CoFe_2O_3$ are reported by Flanders (1979), on thin-film disks by Terada et al. (1983) and Flanders (1983), and on videotapes by Woodward (1982).

When a flying head impacts on a rigid disk, enough energy may affect the disk magnetics so that it loses magnetization at the point of impact. This corresponds to the well-known effect of reducing the strength of a permanent magnet by hitting it with a hammer.

Hard-disk impact may occur during dynamic loading, contact-start/stop, and track access. A theoretical analysis has shown that these impact stresses may be insufficient to cause loss (Jeong and Bogy 1993). The author has witnessed data losses during track accessing where the head would travel over regions of the disk where vibration amplitudes were large enough to cause impacts between disk and head. This could demagnetize recorded data via magnetostriction, like reducing magnetization in a magnet by hammering on it.

DROPOUTS

Dropouts are momentary signal losses in the write/read process, leading to errors in digital systems and degraded performance of audio, and in particular video, playback signals.

The definition of a dropout varies according to the application. A signal drop exceeding 50 percent of the normal value is a dropout for low-packing-density digital recording (< 25,000 bpi), while a drop of only 35 percent may classify as a dropout in high-density recording (> 25,000 bpi). The length, and hence duration, of a dropout is significant in selection and construction of error-detecting and error-correcting codes.

A dust particle or a nodule in the coated surface will lift the tape away from the heads. During recording, this will move the tape away from the otherwise properly adjusted write field and reduce the recorded flux. During playback, it will reduce the signal by the normal spacing loss equal to 54.6 d/λ dB, where d is the spacing between the tape and head surfaces and λ is the recorded wavelength, also equal to 2 bit lengths.

Dropouts are primarily caused by foreign particles that raise the tape surface in a tent-like fashion; see Fig. 14.22, top. In the early days of tape recording, dropouts were often cause by imperfections of all sorts in the coated surface (Radocy 1955; Kramer 1955; and van Keuren 1970). Nowadays media can be manufactured virtually dropout-free, and high-grade computer tapes are certified to have no more than one error per each 100 feet, or even to be error-free. (Note: One error, or dropout, may comprise numerous bits.)

We can estimate the length of the dropout, if we know the height Y of the particle between the tape and the head, the latter made with a contact radius of R. This situation is illustrated in Fig. 14.22, top, and we find by comparing a couple of triangle ratios that the dropout length $2X$ is proportional to the square root of the height Y multiplied by the radius R (Baker 1977).

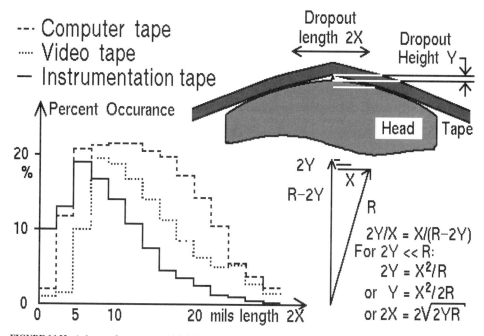

FIGURE 14.22 A dropout forms a tent with height Y. Historical data (probability distributions) for the duration of dropouts in three tape applications.

EFFECTS OF ENVIRONMENT UPON DROPOUT ACTIVITY

There is a remarkable correlation between much of the experimental data collected relating to dropouts. Errors in digital recording systems are characterized by burst errors; i.e., the signal may disappear for a duration lasting hundreds of bits, which is disastrous without the proper error-control coding.

Observations of the effective dropout length on a tape has resulted in a mean value of 0.15–0.25 mm (6–10 mils), which translates into asperity heights of 2–3 μm (80–120 μin). This mean value has been found by several independent observers; see Fig. 14.22 (Alstad and Haynes 1978; Meeks 1979).

They are, for the most part, not fixed errors in the surface of the media. Some investigators of dropouts would, when they detected an error, stop the transport and go back to the error location. In many cases there was nothing to be observed; the error was a dirt particle that moved about.

Another revealing experiment disclosed what is now known as the Wilmot effect: The dropout from a tape stored on a reel with solid flanges is an order of magnitude less than tape stored on a reel with the standard openings (Perry et al. (1982)). A detailed analysis of the dropout count from beginning to end of a tape is shown in Fig. 14.23: The tape portion that was exposed through the flange openings has the highest error count.

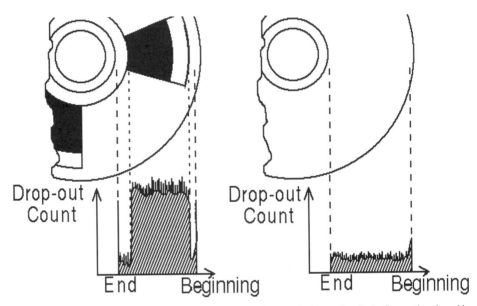

FIGURE 14.23   The dropout count along two tapes, the left wound on a reel with openings in the flanges, the other with closed flanges (glass).

It is easy to envision the air flow that must enter from the sides and along a tape when it is lifted away from the reel pack. The air contains dust particles, and many of them will settle down on the tape due to possible electrostatic attraction. Some may even become permanently embedded into the coating surface when wound with the tape onto the take-up reel.

The only cure for tape and diskette dropouts is, therefore, to have the best possible clean-room condition around the drive installations. Smoking is, of course, out of the question. The new IBM 3480 and 3490 tape cartridge and drive protects the tape from its environment, and should therefore be one order of magnitude better in performance than the comparable reel-to-reel tape drives.

The effect of dropout has been studied and described in several papers listed in the references: digital systems (Jack-Kee, and Middleton 1982; Nunnelley 1984; and Maedigger et al. 1984), video tapes (Lee and Papin 1982), audio (Pichler and Pavuza 1984) and instrumentation (Law 1991).

Dropouts along the length of a tape should occur along various tracks in a random fashion. An instrument has been designed to map the dropout activity over a tape's surface. Statistical data and a map of the tape surface will assist the tape user in appraising its quality and acceptability; see Fig. 14.24 (Waschura 1992 and 1994). Note the cyclic patterns that were traced to a once-around damage on a calender roller. This detailed information is useful in the design of error control codes, in system testing, and in tape certification.

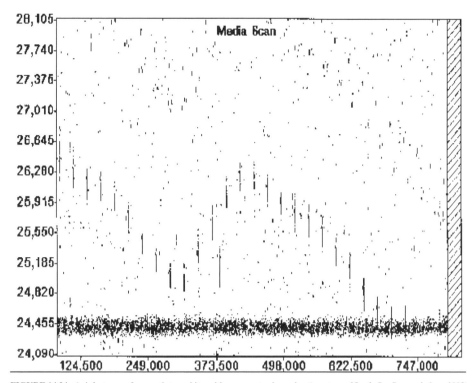

FIGURE 14.24   A defect map of magnetic tape, bit position versus track number (*courtesy of SyntheSys Research, Inc., 1992*).

STORAGE STABILITY

All magnetic particles exhibit a certain instability that, among other things, can lead to print-through, or transfer of signals recorded on one layer of tape onto the adjacent layers. This manifests itself as a pre-echo and a post-echo of the recording, and can be rather annoying in music recordings. Print-through will be discussed further on in this chapter.

The instability that causes print-through is a time- and temperature-dependent property. Another instability will cause permanent loss of signal, in various amounts, and it is a property of the coating particles' magnetostriction. Time effects on the magnetization were covered in Chapter 12 (see Fig. 12.12).

Storage stability of tapes and corrosion resistance of thin-film disks are very important issues in archival data storage. Tapes have been under scrutiny for many years now, and recommendations can be made for tape selection and storage conditions. Archieval stability of recordings are covered in Chapter 31, and a chart in Fig. 31.5 shows the general results of storage stability for the various particulate media.

This chart appeared around 1992, and was quite critical of metal particles (Iron). Improvements in MP stability have been achieved (Okazaki et al. 1992; Mathur et al. 1992). Each Fe particle now has a 10-20 Å thick coat of $Al_2O_2 \times SiO_x$, and tapes made with such particles are very stable.

Magnetic tape is typically used in pack form on metal or plastic reels, and is generally packaged in a cassette. When a tape is brought from cold to hot and/or from dry to wet environments, then the tape will gradually respond. This may very well affect the tape runability, and a recent study has shown that the change in temperature and moisture content in magnetic tape packs can be described by the heat diffusion equation for a hollow cylinder (Vos et al. 1994).

The difference between the tape middle and tape surface was only slight. A 40°C change in temperature surrounding a 1-inch-wide tape pack would cause the tape temperature to reach within a few degrees of its final temperature in about 3 hours, if it was in a cassette. When the tape was exposed directly to calm air, the time was 2 hours, versus 1 hour in circulated air. Moisture diffusion took considerably longer to penetrate and stabilize, with times measured in hundreds of hours.

AUDIO AND INSTRUMENTATION RESPONSE USING AC BIAS

In order for the end user to have some kind of a yardstick, there are test and alignment tapes on the market. They are designed for the purpose of bringing the reproduce head, amplifier, and equalization within some defined tolerances. But the record side (bias, levels, equalization) must be set up using the selected tape. (Some test tapes have a blank reference section at the beginning or the end for this purpose, but it is not likely to match the tape the user plans to operate on the recorder).

NORMAL RECORD LEVEL

An example will shed some light on this subject: the definition of a normal recording level. This level will first of all depend on the amount of bias used in the recording (which in general is defined by a standard or operating practice). The recording level is next adjusted to produce a one-percent or a three-percent third harmonic distortion. Let us choose the last, at a signal frequency of 1 kHz.

When we follow this procedure for a group of modern audio tapes, for example, we will find that the absolute flux levels on the tapes may range from 200 to 1000 nWb/m. This has been of some concern in the standardization of audio recording levels. Practical reasons dictate that interchangeable tapes within the broadcast industry should all have identical program levels. This will eliminate the need for gain adjustments from program to program.

Deutscher Rundfunk in Germany has in the past used peak levels of 320 nWb/m for full-track tapes, and 510 nWb/m for stereo tapes. Peak levels are measured with peak-reading instruments and displayed by LED devices or oscilloscopes. New standards define a normal reference level as 320 nWb/m at tape speeds of 19 cm/sec and higher, and 250 nWb/m at 9.5 cm/sec and lower.

The normal reference level is set by first playing a prerecorded standard reference tape while adjusting the VU-meter gain control for a reading of 0 VU. Next, a recording is made on the selected tape with the bias properly adjusted. The record gain is now adjusted so a normal input voltage (for instance, 0 dBm) results in a level on the tape that indicates 0 VU on the playback VU-meter.

The recorded level depends on the amount of bias current. Figure 14.25 shows the relationship between output level and bias current; each frequency has an optimum bias current, and a compromise must be made. A high bias current produces a clean, in-depth recording with low harmonic distortion, while a small bias current results in excessive high-frequency response with high distortion levels.

A useful guide on distortion is that the one-percent distortion level (at a long wavelength) is about 10 dB below tape saturation, and the five-percent distortion level is 5 dB below tape saturation. This is universal for all tapes, γFe_2O_3, Co-γFe_2O_3, CrO_2, and Fe, provided the record signal corresponds to a recorded wavelength of 5 mils (125 μm) and that bias is adjusted for maximum signal output at that wavelength.

The standards for the various systems will be covered by a future IEC publication, No. 94 (interested readers can write IEC and/or ITA for further information; see the end of Chapter 26 for the address). Current practice is spelled out below, following the procedures spelled out on manufacturers' tape specification sheets.

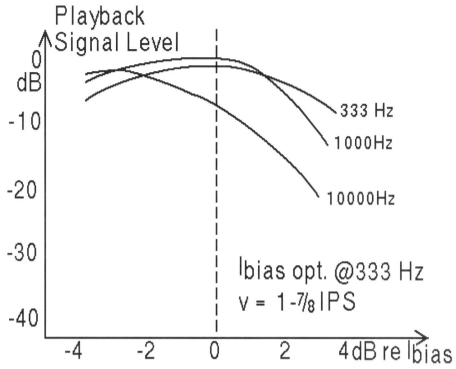

FIGURE 14.25 Amplitudes of three test frequencies, plotted as a function of the ac-bias current level.

AUDIO RECORDING

1. Adjust the bias level so a maximum playback signal results when recording a low-level tone of 1000 Hz at 19 cm/sec and higher speeds, 333 Hz at 9.5 cm/sec and lower speeds (set the level at –20 dB relative to the normal level). It is quite difficult to repeat this adjustment due to the flat maximum of the low-frequency tone, as shown in Fig. 14.25. It is better to use a higher frequency (shorter wavelength) and establish how much overbias is required to match the maximum output at the low frequency. One manufacturer recommends 0.5 dB signal drop when using a frequency of 4000 Hz, another 4 dB drop at 10,000 Hz.

2. Now adjust the signal record level to produce a 0 VU reading on the playback meter. This is the normal reference level (as we described a few paragraphs earlier in this text).

3. Measure h_3, the third harmonic distortion level; it should be below 1 percent.

4. Appraise the tape by measuring the maximum output levels it can produce (MOL). This is gener- ally done at two frequencies:

   MOL_{333} is the maximum level output when the record level has been adjusted to produce three percent harmonic distortion on the tape (some manufacturers use five percent). MOL_{333} versus i_{bias} is shown in Fig. 14.26 for a typical tape.

   MOL_{10} is the difference in output between the normal reference level and the tape level at satura- tion, recorded at 10 kHz (some use 12.5 kHz). A typical curve is shown in Fig. 14.26.

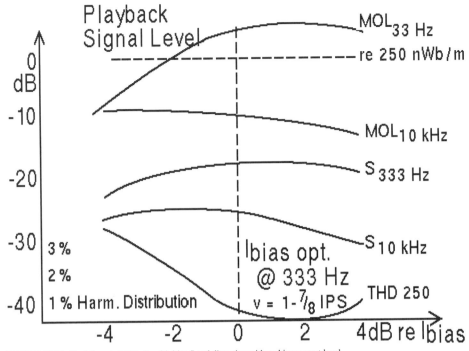

FIGURE 14.26 Variations in MOL, Sensitivities S and distortion with ac-bias current level.

A different procedure is used in instrumentation recording, where the record level is adjusted to reach a tape magnetization level that has either a one percent or three percent third-harmonic distortion level. Either could be used as the normal reference level in a manner similar to that used in a studio recorder, but the level will not necessarily produce the standard flux level of 320 nWb/m (a thick coating may result in 1000 nWb/m, while a thin coating will produce only 200 nWb/m). For further details consult with IRIG Document 118-79, Vol. III.

The frequency response of ac-bias recordings depends on several factors:

- Switching field distribution *SFD*
- Recorded thickness δ_{rec}
- Surface smoothness R_a
- Bias setting

The recorded thickness δ_{rec} enters into the formula for coating thickness loss:

$$A = 20 \times \log \left(\left(1 - e^{-2\pi\delta_{rec}/\lambda} \right)/2\pi\delta_{rec}/\lambda \right) \text{ dB}$$

This loss results in a voltage-versus-frequency response curve that increases at a rate of 6 dB/octave at low frequencies, and becomes level above a frequency f_c that corresponds to a recorded wavelength $\lambda = 2\pi\delta_{rec}$; see Fig. 14.27a. We will therefore find different crossover frequencies for different coating thicknesses, assuming they all are normally biased; i.e., $\delta_{rec} = \delta$ (Fig. 14.27b).

The voltage-versus-frequency response is equalized in the reproduce amplifier by means of an integrator with a shelf having a response that corresponds to one of the curves in Fig. 28.1. The question is: which one should be standardized? If the tapes are recorded with a constant current, then the equalizer should be set to one of the crossover frequencies f_δ listed in Table 14.2. This will assure a flat voltage-versus-frequency response during playback.

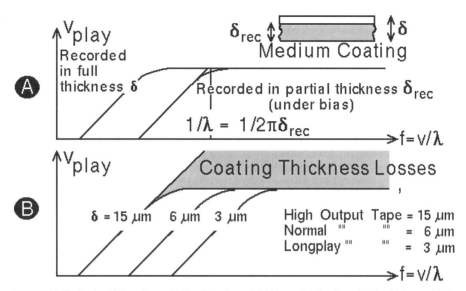

FIGURE 14.27   Coating thickness losses: (a) The effect of recorded thickness. (b) The effect of coating thickness, with the record level optimized for maximum low-frequency (long-wavelength) output for each tape thickness.

TABLE 14.2   Tape Speed, Coating Thicknesses and Time Constants for Audiotapes

Application:	Speed IPS	Coating δ μm	f_δ Hz	$\tau_\delta =$ $\frac{1}{2}\pi f_\delta$ μS	IEC/DIN μS	NAB μS
Studio	30	10	11,395	13	35	10
Professional	15	5	11,395	13	35	50
Home A	7½	5	5,970	27	50–70	50
Home B	3¾	5	2,985	54	90	90
Cassette	1⅞	2.5	2,985	54	120	90

Matters are complicated by the fact that the studio recorder uses overbias, the home recorder normal bias, and the cassette recorder slight underbias. Furthermore, it is common practice to boost the record current to achieve a uniform record level-versus-frequency curve. This is permissible for speech and music programs that have little energy at high frequencies, but it is not a recommended practice in instrumentation recorders.

The amount of boost has been standardized by NAB in the USA and DIN, CCIR, and IEC in Europe. The standards specify that the record current is boosted above a certain frequency in order to provide a recorded flux that is constant down to a wavelength corresponding to a defined τ_δ.

The selected standards for τ_δ are listed in Table 14.2. A proper wavelength response (or frequency response) of a tape requires that it is biased properly and that the record current equalization is adjusted to provide a flat response through a playback system that has been equalized to the chosen standard.

TEST OF RESPONSE IN AUDIO AND INSTRUMENTATION TAPES

A-B comparisons of tapes are often carried out by readjusting only the bias level and not changing the record boost. The relative changes in playback output are therefore a measure for the tapes' ranking.

Tapes can be categorized into four types:

1. Normal γFe_2O_3 for normal bias; $\tau = 120 \, \mu S$
2. CrO_2, Co-treated γFe_2O_3; $\tau = 70 \, \mu S$
3. Ferrichrome, double coat; $\tau = 70 \, \mu S$
4. Metal (Fe) (has more head room); $\tau = 70 \, \mu S$

A tape can be tested for amplitude versus frequency response by adjusting the recorder using a standard alignment tape to adjust first the playback equalizers, then the bias and record equalizers for a flat response using a known tape; call it the reference. Then load the unknown tape to be compared, leaving all control settings alone. If the new response falls within a given envelope (say plus/minus 3 dB), then the new tape is compatible with the reference.

Next, adjust the bias in accord with the rules for setting up the reference tape, and remeasure the response; if the record current needs additional boost, then the new tape's response is poorer than the reference; and vice versa. The amount of boost or reduction in record equalization is a direct measure of the difference between the tapes' frequency responses.

The recommended bias setting is to first adjust the audio signal level at 1 kHz to 10 dB below MOL. Increase the bias level until the signal level upon playback reaches a maximum, and then decreases again. This procedure results in an overbiased condition; it is recommended to adjust the bias so the signal level drops the following amounts:

$$0.25 \text{ dB at } 19.6 \text{ cm/sec (7.5 IPS)}$$
$$0.75 \text{ dB at } 39 \text{ cm/sec (15 IPS)}$$
$$1.25 \text{ dB at } 78 \text{ cm/sec (30 IPS)}$$

It is important to measure amplitude versus frequency response at a level that is well below the normal reference level, or the one percent 3rd harmonic distortion level. 20 or 30 dB below the reference level will provide the correct response without tape overload at short wavelengths. When this overload happens, erronous response curves result that show poor high-frequency response (Fig. 14.28). This overload is partly inherent in the write process, and partly due to the preequalization employed.

It is not always necessary to measure the entire response in order to evaluate and compare tapes. Standard measures have been defined by the sensitivities $S_{333 \text{ Hz}}$ and $S_{10 \text{ kHz}}$. Both are measured as the dB difference in output between the standard reference level less 20 dB (i.e., 25 nWb/m), and the corresponding levels measured at 333 Hz and 10 kHz, recorded 20 dB below normal. Relative frequency response is defined as the difference in dB between the S-values, against those of the reference tape.

A figure of merit in evaluating a tape's short-wavelength response was introduced in Chapter 5; it is the *switching field distribution*, abbreviated *SFD*. The switching field distribution is proportional to the steepness of the slope of the *MH*-loop, and a high value is synonymous with good short-wavelength characteristics.

It should in closing be mentioned that the record head field characteristics will influence the resulting response, as well as the reproduce head losses: AB-tests are valid only when carried out on the same recorder and all bias and equalization settings are specified.

LONG-WAVELENGTH UNIFORMITY AND DROPINS

Amplitude variations are common in the analog reproduce voltage and they are, at long wavelengths, directly related to variations in coating thickness. This applies to all methods of recording and reading, including digital.

Diskettes may have once-around amplitude variations which are the result of incomplete randomizing of the particles' orientation in the plane of the coating. Variation in output is called modulation, and should be within ±10 percent.

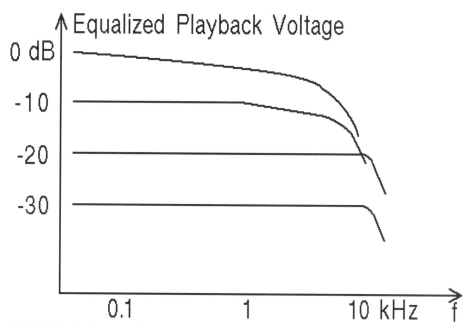

FIGURE 14.28 Amplitude-versus-frequency-response curves must be made at low levels (−20 to −30 dB below normal reference level) in order to avoid overload.

Amplitude variations in tapes may vary in a cyclic fashion for every few feet of tape, in which case they originate from an eccentric roller in the coater during tape manufacturing. These variations are small, typically ±0.5 dB, down to ±0.1 dB over a few feet of tape.

Slow changes can occur from the beginning to the end of a reel of tape, and are traceable to minute changes in the coating slurry during a coating run. These variations are in the order of 0.5 to 1.5 dB. Level changes from one reel to another are in the same range (0.5 to 1.5 dB).

Changes of very short duration may occur due to particle agglomeration or voids. The first are named drop-ins, the latter dropouts. Both are detected quite easily by recording the tape with a dc current (to saturation). A flawless tape has perfectly constant dc flux, and therefore zero read voltage. Defects will appear as voltage pulses $d\phi/dt$ because ϕ changes.

The long-wavelength level variations can be annoying in instrumentation recording applications, but are of seemingly little consequence in audio work due to the insensitivity of the ear to level changes below ±1dB.

Measurements are made simply by connecting a strip-chart recorder to the tape recorder output. Sensitivity uniformity is defined as the difference in dB between the maximum and minimum playback levels of the sample tape, upon which a recording has been made over the entire length under the same circumstances as sensitivity measurements.

SHORT-WAVELENGTH UNIFORMITY

Changes in short-wavelength response occur quite frequently in tape production due to variations in particle dispersion, coating viscosity, and drying rates. All contribute to a surface that, after calendering and other surface treatment techniques, results in a few microinches of waviness.

The uniformity of the short-wavelength output is easily measured by connecting a strip chart recorder to the tape recorder output. Variations are ±1dB for an exceptionally good tape recorded with a wavelength of 2.5 μm.

The surface regularity can be tested by the measurement of dc noise after the tape has been recorded with a dc record current that was adjusted for maximum noise output. The noise voltage output represents surface noise rather than coating dispersion noise.

DEAD LAYER AND VELOUR EFFECT

Many users have observed that a medium may exhibit a difference in short-wavelength output when operated in the forward motion versus reverse motion. This is attributed to the orientation of particles in the coating because they, during the coating and following magnetic alignment process, end up laying at a very small angle to the coating surface.

The remanence that exists in the very surface of a recorded media is perpendicular (Bate and Dunn 1974). Some of the particles also appear at almost a right angle to the nearly perpendicular field lines at the interface between the coating and a recording head. Therefore, they may not be influenced strongly enough by the recording field; taking an average, one finds less remanence in the coating surface. This is in general referred to as the dead layer, and is related to the velour effect.

Figure 14.29 illustrates the differences in the interrelation between the head field and the particles in the surface of the coating. A weak recording results in (A) where the particles are perpendicular to the field, whereas in (B) a stronger recording results; more particles agree with the field.

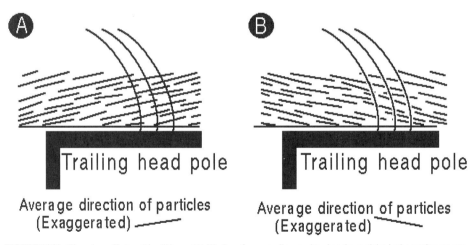

FIGURE 14.29 The velour effect, or "dead" layer: (A) Weak surface recording results when the particles in the coating are perpendicular to the field lines on the gap's trailing edge. (B) Strong surface recording results when the particles are not perpendicular to the field lines.

PRINT-THROUGH

Some magnetic tapes suffer from a flaw called print-through, which is found in thin base tapes, and in particular those with small particle coatings. When the tape is wound onto a reel, adjacent coating layers are separated by a distance equal to the thickness of the base material. Flux lines from one layer will reach adjacent layers and, under the influence of time, temperature, and external fields, cause a weak "recording" on the adjacent layers. This printed signal can be very annoying in audio recording, while it is seldom observed in video and instrumentation applications. A quiet pause before a loud orchestra opening may, by print-through, contain a faint prelude of the opening. This calls for caution in using thin-based tapes.

A distribution of particle sizes in an early audio tape coating has included some very small particles. The γFe_2O_3 particles have small coercivities below a length of about 0.3 μm, and eventually become super-paramagnetic below a length of 0.2 μm. If we assume a length-to-width ratio of 5, then the corresponding volumes are 0.001 μm^3 and 0.0003 μm^3. One early sample contained 4 percent particles that were smaller than 0.001 μm^3, and less than 0.5 percent smaller than 0.0003 μm^3.

The direction of magnetization in these particles can be reversed by the thermal energy kT, where k is Boltzmann's constant and T the absolute temperature. The probability of this occurring within a given time period can be expressed by a time constant where:

$$1/\tau = e^{-uM_sh_{ci}}/2kT \tag{14.4}$$

where u is the particle volume, M_s is the magnetization, and h_{ci} is the coercivity of the particle.

A particle characterized by a time constant τ will be susceptible to a magnetization reversal during the time τ. This time is relatively short (hours, days) for very small particles and is further shortened by an increase in T (τ for normal-sized particles is well in excess of 50 years).

In the absence of an applied field, the probabilities of the particle ending up magnetized in one direction or the other are equal. In the presence of a field, such as that ascending from an adjacent layer of magnetized tape, the particle will tend to end up magnetized in the direction of the field, even when the field strength is very small. The total printed magnetization would then be obtained by integrating over all particles' τ within the range 0 to t(from time of recording to time of playback).

If this time t is comparable to τ for the smallest particles (below 0.0003 μm), of which there were 0.5 percent, the print-through could be as much as 20 log(0.5/100) = –46 dB relative to the normal record level.

Longer storage times combined with elevated temperatures may then include the larger particles, and the print-through would be worse, equal to 20 log(4/100) = –28 dB.

Print-throughs will be present on both of the tape layers adjacent to the recording. As the tape is played back, the first signal heard would be a preprint, then the original signal, and finally a postprint (there may be additional, weaker pre- and postprints).

Pre- and postprints are different in magnitudes. This phenomenon arises because the printed magnetization has both longitudinal and vertical components and these, on playback, are additive for one set of prints and subtractive for the other set. The situation is shown in Fig. 14.30.

Daniel (1951) has expressed the ratio of the magnitudes of the prints as shown in the illustration. The maximum print-through occurs when the wavelength of the recorded signal is approximately equal to the total tape thickness $\delta + b$, where b is the base film thickness. Under these conditions equation (14.4) reduces to:

$$P_{pre}/P_{post} \approx 1.3 \times (\chi_x + \chi_y)/(\chi_x - \chi_y) \tag{14.5}$$

The multiplier will be 1 instead of 1.3 at shorter wavelengths.

When the tape is isotropic, $\chi_x = \chi_y$ and the preprints will vanish, while the post-prints will double. If the anisotropy is predominantly longitudinal, then $\chi_y = 0$, and the prints are equal.

In modern tapes $\chi_y = 0.25 \times \chi_x$, resulting in:

$$P_{pre}/P_{post} = 2.2, \text{ or } 6.8 \text{ dB.}$$

A practical consequence is that European studios and broadcast radio stations wind their tapes with the oxide out. This makes the pre-print (the more annoying print) weaker than the post-print. The same practice is followed in audio cassettes.

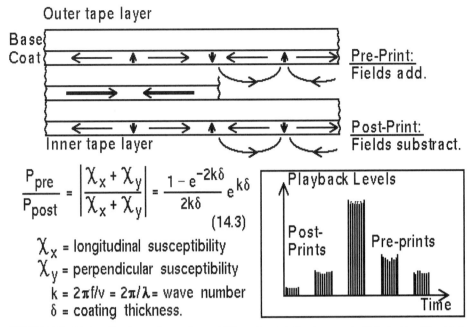

FIGURE 14.30 Printing effects in stored magnetic tape: (a) Pre- and postprint in a tape wound with the backing side out (normal). (b) Playback levels of printed signals (*after Daniel, 1951*).

Maximum print-through occurs when a frequency of 1200 Hz is recorded at 15 IPS on 2-mil-thick tape. This is, unfortunately, in the region where the ear is most sensitive. The situation is better for audio cassette tapes, where the speed is only 1⅞ IPS and the total thickness about 1 mil, resulting in a maximum print-through at a frequency of only 300 Hz (where the ear is less sensitive). More recent work on print-through has been done by Tochihara et al. (1970), Stafford (1976), and Corradi et al. (1984).

REDUCTION OF PRINT-THROUGH

Print reduction is possible by moving the tape past a very weak permanent magnet (Radocy 1959), or by feeding a very small amount of dc or ac current through the record head during playback. This is sufficient to erase the unstable particles that were responsible for the print effect.

One method of adjusting this print-erase current is to measure the output of a prerecorded 15-kHz pilot tone well before the actual recording on the tape. Adjust the erase current for about 1 dB erasure of the pilot tone; this will ensure print erasure while leaving the recording intact.

MEASUREMENT OF PRINT-THROUGH

A tape's print-through characteristics can be measured in the following way: record a very short section (a few cm) with a 1000 Hz tone, or a worst-case signal (square wave at $f = v/2\pi(\delta + b)$) at maximum recording level (6 dB below saturation). Rewind immediately and store the tape for 24 hours at 30° C (some standards call for 20° C. That will produce less print-through).

Now play back the tape and record the output signal on a strip chart recorder. This will provide a record of the levels of pre-prints, original signal, and post-prints. Such records will assist in classifying tapes for storage and archival use, because the amount of printing is in direct relation to the tape's instability over prolonged periods of time and possibly elevated temperature.

CONTACT DUPLICATION OF TAPES

Duplication of tapes is a simple process whereby a master playback machine is connected to a number of slave recorders. A prerecorded master tape and blank tapes are placed in their units, and all machines started at the same time. It is possible to duplicate at speeds up to 180 IPS, with equalizer settings properly scaled.

This method is costly in manpower and equipment. Today two methods are available for a much faster and cheaper way of making copies. Both methods use a high-coercivity, stable master tape that has a recording that is a mirror image of the final tape's recorded track pattern. The copies are made by a printing method, one by heat and the other with ac bias.

The copy method by heat uses the low Curie temperature of CrO_2 particles. The master tape and the slave tape (CrO_2) move together through a special transport so the coated surfaces touch each other over a distance of several centimeters. A ray of infrared light will locally heat the CrO_2 particles above their Curie temperature ($110°–120°$ C) whereby they become nonmagnetic. They will, upon cooling, remagnetize with a polarity that is opposite the field sensed from the master pressed against it. In this way, a faithful copy is made. The duplication can be done at very high tape speeds, as long as there is enough time for the CrO_2 magnetization process and no heat damage is done to the binder (Hiller 1987).

The ac-bias method was initially somewhat similar. A high-frequency ac-bias field would be applied in the zone where the master and slave made contact, and an anhysteretic "recording" was made from master to slave.

Both methods involve rather precise transports in order to prevent tape slippage during the duplication. A third method prevents such possible errors: The reel with several lengths of the copy-master is wound onto another reel together with a blank slave tape, so the tapes' surfaces contact. This winding is done carefully and possibly with a pack wheel. This master/slave reel is next exposed to an increasing and then decaying ac field that acts as ac bias for a recording of the master magnetization onto the slave. Voila! All that remains is only unwinding and loading sections of the slave into cartridges. Tapes for master and slaves are optimized for duplication. The new BaFeO particles show promising characteristics for duplication (Noda et al. 1992; Kawamaki et al. 1992).

REFERENCES

MECHANICAL PROPERTIES OF SUBSTRATES

Behr, M.I., and J.K. Osborn. Nov. 1981. Technique for Measuring Dynamically the Dimensional Stability of A Flexible Disk. *IEEE Trans. Magn.* MAG-17 (6): 2748–2750.

Bhushan, B. 1992. *Mechanics and Reliability of Flexible Magnetic Media.* Springer Verlag. 565 pages.

Izraelev, V. Sept. 1983. On Determination of Thermal and Hygroscopic Expansion Coefficients of Pet Floppy Disk Substrates by the Recording Method. *IEEE Trans. Magn.* MAG-19 (5): 2253–2256.

Mann, D. Aug. 1994. *Head RVA vs Substrate RVA Research*. Head/Media Research Journal, 1 (3): 29 pages.

Schaake, R.C.F., H.J.M. Pigmans, J.A.M.v.d. Hejikant, and H.F. Huisman. Jan. 1987. Tensile Testing of Magnetic Tapes. *IEEE Trans. Magn.* MAG-23 (1): 109–111.33.

Thompson, J.A., and P.B. Mee. Sept. 1984. A Statistically Designed Investigation of Substrate Effects on CoCr Thin Films for Perpendicular Magnetic Recording. *IEEE Trans. Magn.* MAG-20 (5): 785–787.

MECHANICAL PROPERTIES OF TAPE COATINGS

Akiyama, S., S. Nakagawa, and M. Naoe. Nov. 1991. Electrically conductive layer of wear-resistant Fe-Mo-B alloy for protecting magnetic recording tape. *IEEE Trans. Magn.* MAG-27 (6): 5094–96.

Aoyama, S., and M. Kishimoto. March 1991. The Behavior of Lauric Acid Lubricant in a Magnetic Coating Layer and Its Effect on Mechanical Properties of Magnetic Media. *IEEE Trans. Magn.* MAG-27 (2): 791–94.

Au-Yeung, V. Sept. 1983. FTIR Determination of Fluorocarbon Lubricant Film Thicknesses on Magnetic Disk Media. *IEEE Trans. Magn.* MAG-19 (5): 1662–1664.

Lee, T-H. D. Jan. 1990. Selection of Lubricants for Metal Evaporated Tape. *IEEE Trans. Magn.* MAG-26 (1): 171–73.

Nakamura, K., K. Momono, Y. Ota, A. Itoh, and C. Hayashi. Sept. 1984. Organic Lubricant Evaporation Method, A New Lubricated Surface Treatment for Thin Film Magnetic Recording Media. *IEEE Trans. Magn.* MAG-20 (5): 833–835.

Perry, D.M., P.J. Moran, and G.M. Robinson. April 1984. Three Dimensional Surface Metrology of Magnetic Recording Materials Through Direct Phase Detecting Microscopic Interferometry. *Intl. Conf. Video and Data 84*, IERE Publ. No. 59, pp. 57–62

Robinson, G.M., C.D. Englund, and R.D. Cambronne. Sept. 1985. Relationship of Surface Roughness of Video Tape to its Magnetic Performance. *IEEE Trans. Magn.* MAG-21 (5): 1386–1388.

Robinson, G.M., P.J. Moran, R.W. Peterson, and C.D. Englund. Sept. 1984. Applications of Interferometric Measurements of Surface Topography of Moving Magnetic Recording Materials. *IEEE Trans. Magn.* MAG-20 (5): 915–917.

Sumiya, K., Y. Yamamoto, K. Kaneno, and A. Suda. 1989. The Role of Polymer Binder in Magnetic Recording Media. In *Polymers in Information Storage Technology*, Plenum Press, pp. 291–298.

Wahl, F., S. Ho, and K. Wong. July 1983. A Hybrid Optical-Digital Analysis Processing for Surface Inspection. *IBM Jour. Res. Devel.* 27 (4): 376–385.

Wierenga, P.E., J.A.v. Winsum, and J.H.M.v.d. Linden. Sept. 1985. Roughness and Recording Properties of Particulate Tapes: A Quantitative Study. *IEEE Trans. Magn.* MAG-21 (5): 1383–1385.

Wright, C.D., and H.G. Tobin. August 1975. Surface Lubrication of Magnetic Tape. *ITC Conf. Proc.*, pp. 336–344.

MECHANICAL PROPERTIES OF RIGID DISK COATINGS

Agarwal, S., E. Li, and N. Heiman. Jan. 1993. Structure and Tribological Performance of Carbon Overlayer Films. *IEEE Trans. Magn.* MAG-29 (1 - Pt. 1): 264–269.

Ager III, J.W. Jan. 1993. Optical Characterization of Sputtered Carbon Films. *IEEE Trans. Magn.* MAG-29 (1 - Pt. 1): 259–263.

Babcock, K., V. Elings, M. Dugas, and S. Loper. Nov. 1994. Optimization of Thin Film Tips for Magnetic Force Microscopy. IEEE Trans. Magn. MAG-30 (6): 4503–05.

Burke, E.R., R.D. Gomez, and I.D. Mayergoyz. Nov. 1994. Analysis of 3-D Magnetic Fields Measured Using a Magnetic Force Scanning Tunneling Microscope. *IEEE Trans. Magn.* MAG-30 (6): 4488–90.

Gomez, R.D., A.A. Adly, I.D. Mayergoyz and E.R. Burke. Nov. 1993. Magnetic Force Scanning Tunneling Microscopy: Theory and Experiment. *IEEE Trans. Magn.* MAG-29 (6): 2494–96.

Hansma, P.L., V.B. Eiling, O. Marti, and C.E. Bracker. Oct. 1988. Scanning Tunneling Microscopy and Atomic Force Microscopy: Application to Biology and Technology. *Science*. 242 pp. 209–242.

Marchon, B., M.R. Khan, and D.B. Bogy. Nov. 1991. Scanning Tunneling Microscopy Studies of Carbon Overcoats of Thin Film Media. *IEEE Trans. Magn.* MAG-27 (6): 2067–69.

Merchant, K., P. Mee, M. Smallen, and S. Smith. Sept. 1990. Lubricant Bonding and Orientation on Carbon Coated Media. *IEEE Trans. Magn.* MAG-26 (5): 2688–90.

Streator, J.L., B. Bhushan, and D.B. Bogy. Jan. 1991. Lubricant Performance in Magnetic Thin Film Disks With Carbon Overcoat - Part I: Dynamic and Static Friction. *Jour. Tribology* 113: 22–31.

Streator, J.L., B. Bhushan, and D.B. Bogy. Jan. 1991. Lubricant Performance in Magnetic Thin Film Disks With Carbon Overcoat - Part II: Durability. *Jour. Tribology* 113: 32–37.

Wickramasinghe, H.K., Oct. 1989. Scanned-Probe Microscopes. *Scientific American*. 261 (4): 98–105.

Yang, M., S.K. Ganapathi, R.D. Balanson, and F.E. Talke. Nov. 1991. The Frictional Behavior of Thin Film Magnetic Disks. *IEEE Trans. Magn.* MAG-27 (6): 5157–59.

ABRASIVITY, DURABILITY OF FLEXIBLE COATINGS

Osaki, H., E. Oyanagi, H. Aonuma, T. Kanou, and J. Kurihara. Jan. 1992. Wear Mechanism of Particulate Magnetic Tapes in Helical Scan Video Tape Recorders. *IEEE Trans. Magn.* MAG-28 (1): 76–83.

Osaki, H., K. Fukushi, and K. Ozawa. Nov. 1990. Wear Mechanism of Metal-Evaporated Magnetic Tapes in Helical Scan Video Tape Recorders. *IEEE Trans. Magn.* MAG-26 (6): 3180–85.

Patton, S.T., and B. Bhushan. 1995. Effect of Interchanging Tapes and Head Contour on the Durability of Metal Evaporated, Metal Particle and Barium Ferrite Magnetic Tapes. To be published in *Tribology Trans*. (ASME).

Terada, A., O. Ishii, and S. Ohta. Sept. 1985. Wear Resistance and Signal-to-Noise Ratio in the gamma-Fe_2O_3 Thin Film Disks. *IEEE Trans. Magn.* MAG-21 (5): 1520–1523.

Tomago, A., T. Suzuki, and T. Kunieda. Sept. 1985. Effects of the Surface Configuration of Magnetic Layers on Durability Against Mechanical Stress. *IEEE Trans. Magn.* MAG-21 (5): 1524–1526.

SHORT-WAVELENGTH RESPONSE

Cramer, H.A.J., and C. Slob. Nov. 1991. Determination of Spacing Loss and Intrinsic Frequency Response of Recording Media by Scaling. *IEEE Trans. Magn.* MAG-27 (6): 4966–68.

Inaba, H., K. Ejiri, N. Abe, K. Masaki, and H. Araki. Nov. 1993. The Advantages of Thin Magnetic Layer on a Metal Particulate Tape. *IEEE Trans. Magn.* MAG-29 (6): 3607–12.

Kornei, O. 1949. *US Patent* no. 2643130.

Ogawa, N. Jan. 1991. Double-Coating Technology for Video Tapes. Fuji Photo Film, private communication.

Schildberg, H.P., G. Fischer, A. Hagemeyer, and H. Hibst. Jan. 1993. Spacing Loss of ME tape. *Jour. Magnetism and Magn. Matls*. 120 (1-3): 292–295.

Smith, D.P. June 1989. Reliability Issues in Data Tape Cartridges. *Proc. 5th Intl. Congress Tribology* Vol. 3, pp. 216–221.

Speliotis, D.E. Nov. 1993. Double-Layer Particulate Magnetic Recording Tapes. *IEEE Trans. Magn.* MAG-29 (6): 3613–15.

ERASEABILITY AND OVERWRITE

Burke, E.R., and D.R. Sanders. Sept. 1985. The Erasure of High Energy Tapes. *IEEE Trans. Magn.* MAG-21 (5): 1374–1376.

Christensen, E.R., and B.I. Finkelstein. Sept. 1985. Erasure Methods for High-Density Recording. *IEEE Trans. Magn.* MAG-21 (5): 1377–1379.

Fayling, R.E. Sept. 1977. Anisotropic Erasure and Demagnetization Characteristics of Recording Tapes Comprising Particles with Uniaxial Magnetocrystalline Anisotropy. *IEEE Trans. Magn.* MAG-13 (5): 1391–1393.

Lekawat, et al., 1993. Erasure and Noise Study in BaFeO Tape Media. *Jour. Appl. Physics 73* : 6719–21.

Lekawat, L., G.W.D. Spratt, and M.H. Kryder. Jan. 1993. Erasure of high-energy magnetic tape recording media. *Jour. Magnetism and Magn. Matls.* 120 (1-3): 103–105.

Manly, W.A. Nov. 1976. Erasure of Signals on Magnetic Recording Media. *IEEE Trans. Magn.* MAG-12 (6): 758–760.

Mountfield, K.R., and M.H. Kryder. Sept. 1989. The Effect of Aging on Erasure in Particulate Disk Media. *IEEE Trans. Magn.* MAG-25 (5): 3638–40.

Sugaya, H..T. Arai, and Y. Ueda. Jan. 1993. Overwrite recording on different recording media. *Jour. Magnetism and Magn. Matls.* 120 (1-3): 106–111.

STRESS DEMAGNETIZATION

Daniel, E.D., and K.E. Naumann. Nov. 1971. Audio Cassette Chromium Dioxide Tape. *Jour. AES*, 19 (10): 822–828.

Flanders, P.J. May 1979. Magnetostriction and Stress-Induced Playback Loss in Magnetic Tapes. *IEEE Trans. Magn.* MAG-15 (3): 1065–1067.

Flanders, P.J. Sept. 1983. Elastic Stress-Induced Coercive Field Changes in Ni-Co-P Films Used in a Rotating Disk. *IEEE Trans. Magn.* MAG-19 (5): 1680–1682.

Hoshi, Y., M. Matsuoka, M. Naoe, and S. Yamanaka. Sept. 1984. Demagnetization of Co-Cr Films Induced by Stress and Heat. *IEEE Trans. Magn.* MAG-20 (5): 797–799.

Izawa, F. July 1984. Theoretical Study on Stress-Induced Demagnetization in Magnetic Recording Media. *IEEE Trans. Magn.* MAG-20 (4): 523–528.

Jeong, T.G., and D.G. Bogy. Nov. 1993. Dynamic Loading Impact- Induced Demagnetization in Thin-Film Media. *IEEE Trans. Magn.* MAG-29 (6): 3903–05.

Terada, A., O. Ishii, and K. Kobayashi. Jan. 1983. Pressure-Induced Signal Loss in Fe_3O_4 and γFe_2O_3 Thin-Film Disks. *IEEE Trans. Magn.* MAG-19 (1): 12–20.

Woodward, J.G. Nov. 1982. Stress Demagnetization in Videotapes. *IEEE Trans. Magn.* MAG-18 (6): 1812–1818.

DROPOUTS

Alstad, J.K., and M.K. Haynes. Sept. 1978. Asperity Heights on Magnetic Tape Derived from Measured Signal Dropout Lengths. *IEEE Trans. Magn.* MAG-14 (5): 749–751.

Baker. B.R. July 1977. A Dropout Model for a Digital Tape Recorder. *IEEE Trans. Magn.* MAG-13 (5): 1196–1199.

Bate, G., and L.P. Dunn. Oct. 1974. The Remanent State of Recorded Tapes. *IBM Jour. Res. Devel.* vol. 18, pp. 563–569.

Jack-Kee, T., and B.K. Middleton. 1982. Dropouts and Their Effects on Error Rates in a Digital Magnetic Tape Recording System. *Intl. Conf. Video and Data 82*, IERE Publ. No. 54, pp. 43–49.

Kramer, A. March 1955. Locating Defects in Magnetic Recording Tape. *Jour. AES*, 3 (3): 143.

Law, E.L. Sept. 1991. Correlation Between Tape Dropouts and Data Quality. *THIC Meeting*, 6 pages.

Lee, T.D., and P.A. Papin. Nov. 1982. Analysis of Dropouts In Video Tapes. *IEEE Trans. Magn.* MAG-18 (6): 1092–1094.

Maediger, C., H. Voelz, H.K. Wiollaschek. Aug. 1984. Analysis of Signal Statistics and Dropout Behaviour of Magnetic Tapes. *IEEE Trans. Magn.* MAG-20 (5): 765–767.

Meeks, L.A. July 1979. Characterization of Instrumentation Tape Signal Dropouts for Appropriate Error Correction Strategies for High Density Digital Recording Systems. *Intl. Conf. Video and Data 79*, IERE Publ. No. 43, pp. 199–215.

Nunnelley, L.L. Jan. 1984. Determination of Defect Length in Disk Coatings by Autocorrelation. *IEEE Trans. Magn.* MAG-20 (1): 93–95.

Perry, M.A., F. Blackwell, and R. Harris. 1982. The Importance of Dropout Measurements in Ensuring Good Short Wavelength Recording on Magnetic Tape. *Intl. Conf. Video and Data 82*, IERE Publ. No. 54, pp. 23–41.

Pichler, H., and F. Pavuza. Oct. 1984. Criteria for the Selection of Audio Tapes for Analog and Digital Recording According to Their Dropout Characteristic. *Audio Eng. Soc. Preprint.*

Radocy, F. Jan. 1955. Defects in Magnetic Recording Tape: Their Cause and Cure. *Jour. AES*, 3 (1): 31.

Van Keuren, W. Jan. 1970. An Examination of Dropouts Occuring in the Magnetic Recording and Reproducing Process. *Jour. AES*, 18 (1): 2.

Waschura, T.E. March 1992. Beyond Conventional Bit Error Rate Testing. *THIC Meeting*, 4 pages. Avail. from SyntheSys Res. Inc., 415-364-1853.

Waschura, T.E. March 1994. Techniques for Analyzing Error Performance on ID-1 Tape Recorders. *THIC Meeting*, Orlando, Fl. 19 pages. Avail. from SyntheSys Res. Inc., 415-364-1853.

STORAGE STABILITY

Mathur, M.C.A., G.F. Hudson, and L.D. Hackett. Sept. 1992. A Detailed Study of the Environmental Stability of Metal Particle Tapes. *IEEE Trans. Magn.* MAG-28 (5): 2362–64.

Okazaki, Y., K. Hara, T. Kawashima, A. Sato, and T. Hirano. Sept. 1992. Estimating the Archival Life of Metal Particulate Tape. *IEEE Trans. Magn.* MAG-28 (5): 2365–67.

Vos, M., G. Ashton, J. Van Bogart, and R. Ensminger. March 1994. Heat and Moisture Diffusion in Magnetic Tape Packs. *IEEE Trans. Magn.* MAG-30 (2): 237–242.

PRINT-THROUGH AND DUPLICATION

Corradi, A.R., S.J. Andress, C.A. Dinitto, D. Bottoni, G. Candolfo, A. Cecchetti, and F. Masoli. Sept. 1984. Print-Through, Erasability, Playback Losses: Different Phenomena from the Same Roots. *IEEE Trans. Magn.* MAG-20 (5): 760–762.

Daniel, E.D. Oct. 1951. Accidental Printing in Magnetic Recording. *BBC Quarterly*, Vol. 5, pp. 241–256.

Hiller, D.M. Sept. 1987. Copy Tape Parameters for Thermomagnetic Duplication. *IEEE Trans. Magn.* MAG-23 (5): 3184–86.

Kawakami, S., K. Kusumoto, and M. Kishimoto. Sept. 1990. Contact Duplication Characteristics of High Density Ba-Ferrite Tapes. *IEEE Trans. Magn.* MAG-26 (5): 2104–06.

Noda, M., Y. Okazaki, K. Hara, and K. Ogisu. Jan. 1990. Characteristics of Barium Ferrite Tape for Tape Magnetic Contact Duplication. *IEEE Trans. Magn.* MAG-26 (1): 81–86.

Radocy, F. March 1959. A New Device for the Reduction of Print-Through. *Jour. AES* 7 (3): 129.

Stafford, M.K. Sept. 1976. A Print-Through Constant for Magnetic Particles. *IEEE Trans. Magn.* MAG-12 (5): 583–584.

Tochihara, S., Y. Imaoka, and M. Namikawa., Dec. 1970. Accidental Printing Effect of Magnetic Recording Tapes Using Ultrafine Particles of Acicular Gamma-Fe2O3. *IEEE Trans. Magn.* MAG-6 (4): 808–811.

BIBLIOGRAPHY

GENERAL INTEREST

Babcock, K., M. Dugas, S. Manalis, and V. Eilings. 1995. Magnetic Force Microscopy: Recent Advances and Applications. To appear in *Proceedings of Materials Research Society* meeting (Boston). 12 pages.

Barndt, R.D., A.J. Armstrong, and J.K. Wolf. Jan. 1993. Media Selection for High Density Recording Channel. *IEEE Trans. Magn.* MAG-29 (1): 183–188.

Bhushan, B. 1990. *Tribology and Mechanics of Magnetic Storage Devices.* Springer Verlag. 1018 pages.

Murdock, E.S., R.F. Simmons, and R. Davidson. Sept. 1992. Roadmap for 10 GBit/in^2 Media: Challenges. *IEEE Trans. Magn.* MAG-28 (5): 2271–76.

NML Staff. July 1992. Futures in magnetic recording: A materials view. *NML Bits,* 3 pages. Avail. from NML 612-733-0468.

Palmer, D.C. and J.D. Coker. Jan. 1993. Media Design Considerations for a PRML Channel. *IEEE Trans. Magn.* MAG-29 (1): 189–194.

Simmons, R., and R. Davidson. Jan. 1993. Media Design for User Density of up to 3 Bits per Pulse Width. *IEEE Trans. Magn.* MAG-29 (1): 169–176.

RECORDING CHARACTERISTICS

Kawanabe, T., J.G. Park, and M. Naoe. Nov. 1991. Magnetic and Recording Characteristics of Co-Cr-Ta/Cr Thin Film Prepared on Si substrate. *IEEE Trans. Magn.* MAG-27 (6): 5031–33.

Mapps, D.J., G. Pan, M.A. Akhter, S. Onodera, and A. Okabe. Jan. 1993. In-contact magnetic recording performance on Pt/CoCrTa thin films on glass computer disk. *Jour. Magnetism and Magn. Matls.* 120 (1-3): 305–309.

Mian, G., J.A. Potter, and T.D. Howell. Nov. 1993. Effects of Media Orientation and Intergranular Exchange Coupling on Nonlinear Transition Shift. *IEEE Trans. Magn.* MAG-29 (6): 3700–02.

O'Grady, K., R.W. Chantrell, and I.L. Sanders. Jan. 1993. Magnetic Characterization of Thin Film Recording Media. *IEEE Trans. Magn.* MAG-29 (1 - Pt. 1): 286–91.

Pressesky, J.L., S.Y. Lee, N. Neiman, D. Williams, T. Coughlin, and D.E. Speliotis. Sept. 1990. Effect of Chromium Underlayer Thickness on Recording Characteristics of CoCrTa Longitudinal Recording Media. *IEEE Trans. Magn.* MAG-26 (5): 1596–98.

Richter, H.J., H.P. Schildberg, and H. Hibst. Sept. 1992. Angular Dependent Recording Behaviour of ME Tape. *IEEE Trans. Magn.* MAG-28 (5): 3288–90.

Speliotis, D., W. Lynch, J. Burbage, and R. Keirstead. Jan. 1991. High Density Recording on Barium Ferrite and Very High Coercivity Thin Film Rigid Disks. *Jour. Magn. Soc. Japan.* 15 suppl. (S2): 51–56.

Takano, H., T.T. Lam, J-G. Zhu, and J.H. Judy. Nov. 1993. Effect of Orientation Ratio on the Recording Characteristics of Longitudinal Thin Film Media. *IEEE Trans. Magn.* MAG-29 (6): 3709–11.

Velu, E.M.T., and D.N. Lambeth. Sept. 1992. High Density Recording on SmCo/Cr Thin Film Media. *IEEE Trans. Magn.* MAG-28 (5): 2349–54.

MEDIA NOISE

Arnoldussen, T.C. 1992. Theoretical Considerations of Media Noise. *In Noise in Digital Magnetic Recording,* World Scientific, pp. 101–139.

Hoinville, J.R., R.S. Indeck, and M.W. Muller. Nov. 1992. Spatial Noise Phenomena of Longitudinal Magnetic Recording Media. *IEEE Trans. Magn.* MAG-28 (6): 3398–3406.

Murdock, E.S. 1992. Measured Noise in Thin Film Media. In *Noise in Digital Magnetic Recording,* World Scientific, pp. 65–100.

APPLICATIONS OF TAPE

Eiling, A. Sept. 1990. The Importance of Magnetostatics in Video Recording: A Theoretical Analysis. *IEEE Trans. Magn.* MAG-26 (6): 3173–79.

Goetz, K. Jan. 1993. CrO_2 particles for DCC tape. *Jour. Magnetism and Magn. Matls.* 120 (1-3): 19–24.

Iseaka, K.Y. Fujimaki, T. Nakamura, S. Takahashi, K. Kobyashi, and S. Leader. March 1989. The Application of High-Coercivity Cobalt Iron Oxide Tape for Digital Video Recording. *SMPTE Journal* 98 (3): 168–72.

Ito, J., Y. Sato, and I. Nakamura. Jan. 1990. Characteristics of S-VHS tape. *IEEE Trans. Magn.* MAG-26 (1): 87–90.

Speliotis, D.E., and K. Peter. Jan. 1993. High density digital recording on 4 mm metal particle and Ba-ferrite tapes. *Jour. Magnetism and Magn. Matls.* 120 (1-3): 28–32.

Suzuki, T., T. Oguchi, Y. Kishimoto, and N. Saito. Jan. 1993. Ba ferrite 8 mm data tapes. *Jour. Magnetism and Magn. Matls.* 120 (1-3): 25–27.

Yokoyama, H., T. Ito, M. Isshiki, K. Kurata, and T. Fukaya. Sept. 1992. Barium Ferrite Particulate Tapes for High-Band 8mm VCR. *IEEE Trans. Magn.* MAG-28 (5): 2391–93.

CHAPTER 15
TAPE TRANSPORTS

The drive apparati required to move the magnetic media over the write and read heads are conceptually simple mechanisms, tailored to move the medium at a constant speed. There are only four basic configurations: A long ribbon (tape) wound onto a spool, a short, wide ribbon (card), and a circular, rotating sheet that is thin and flexible (diskette) or thick and rigid (disk).

The tape or disk drives are precision mechanisms that move the medium over the head assembly at a perfectly constant speed. The data may be distorted due to time variations (constancy of medium speed, flutter, or TDE), or amplitude and phase variations (head-medium spacing, drop-outs, tracking, or coating thickness uniformity).

This chapter will cover the topics that pertain to tape drives:

- Tape transports with fixed heads
- Capstan and pinch roller
- Reeling and handling of tapes
- Cassettes and cartridges
- Tape transports with rotating head assemblies

Tape drives are vital components in data measurement and computer systems. Their demise has been predicted numerous times, and they have each time survived—lately in the form of numerous back-up drives, and in the new IBM 3480 and 3490 tape drives. Several file storage and high-rate instrumentation data storage applications are discussed in Chapter 27.

Tape drives have been designed and manufactured for cassettes, cartridges, or reels with tapes of different widths and lengths. Tape widths and their applications are:

4 mm	R-DAT (audio), data
6.25 mm	Quarter-inch cartridge
	Reel-to-reel audio
8 mm	Video (Hi-8), data
½-inch	Studio audio
	VHS and BETA video
	Computer data (reel-to-reel)
	Computer data (3480, 3490)
	Instrumentation
¾-inch	Video, digital data
1 inch	Audio, video, instrumentation
2 inch	Audio, early video, instrumentation

LAYOUT OF A TAPE TRANSPORT

The physical layout of a magnetic tape transport depends on its application. Figure 15.1 shows four different layouts used for analog transports. Figure 15.1a illustrates the reel-to-reel concept that is so commonly found in home-type recorders (including cassette) and in many instrumentation recorders. Figure 15.1b shows a compact version where one reel is concentric with the other, a feature that saves space. A cartridge design (Fig. 15.1c), used in many automobile installations and broadcast stations, is shown together with an endless loop contained in a bin (Fig. 15.1d). The latter design is often used where constant monitoring of data is required without the need for prolonged storage; one such application is in flight recorders, where only the last few minutes of cockpit conversation must be stored.

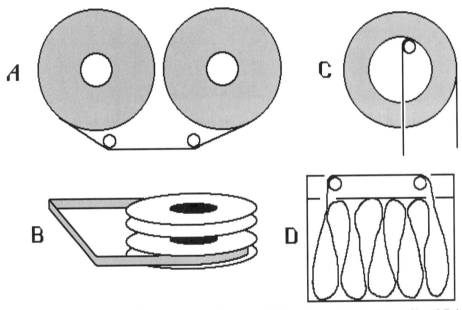

FIGURE 15.1   Four arrangements for handling tape: (a) Side-by-side. (b) Stacked reels. (c) Endless-loop cartridge. (d) Endless-loop bin storage.

REEL-TO-REEL TRANSPORTS (DIGITAL DATA, INTERMITTENT)

The tape transports must satisfy several requirements. Very fast starts and stops are necessary, and movement of the tape must be in a perfectly straight path. The last requirement is particularly critical in parallel-track applications with high bit densities; here the bits from the tracks need to be read simultaneously. Tape skew will often necessitate the use of deskew buffers to correct for time-displacement errors between tracks.

The fast start-stop cycles place heavy demands on the tape drive. The tape may have to move from standstill to a speed of 200 IPS over a very short span of tape. Calculations show that a linear acceleration of 9×10^5 in/sec^2 is required during less than 1.5 milliseconds.

Such high accelerations were not possible in early digital tape drives. It was common to bring the tape into motion by pressing it against a rotating shaft called a capstan by means of a puck. A forward and a reverse capstan are used; see Fig. 15.2. Stopping tape motion is done by releasing the puck and simultaneously clamping the tape between the puck and a fixed post.

This system was improved by introducing a perforated capstan, wherein a vacuum would attract the tape and move it with the capstan; see Fig. 15.3. To stop the tape, air pressure would force the tape against a fixed surface near the capstan.

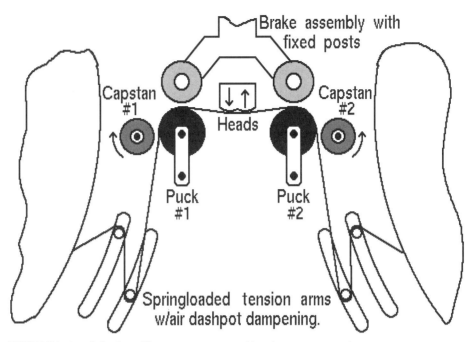

FIGURE 15.2   An early fast-forward/fast-reverse computer tape drive using two counter-rotating capstans.

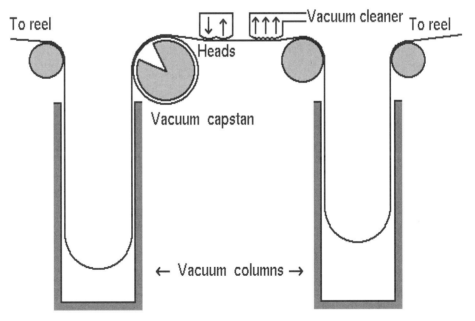

FIGURE 15.3   A fast-forward/fast-reverse computer tape drive using one low-inertia vacuum capstan.

In the 1960s, very-low-inertia motors with printed circuit armatures were developed. These motors could meet the tape acceleration requirements and, with a programmed servo control of the speed, the designer could also achieve good tape speed stability within the allotted acceleration time. The tape is typically wrapped around the capstan over 120 to 180 degrees, and follows its motions precisely.

Another improvement in tape handling was the introduction of vacuum columns instead of tension arms to absorb the tape slack. Reel speed is servo controlled, which allows for control of the torque applied to the reel. When a reel of tape is subjected to excessive torque, it may happen that the tape pack shifts, causing cinching and subsequent damage to the tape. The tension with which the tape is wound on the reel contributes to the forces that prevent layers from slipping. These forces will decrease during storage because tape is a visco-elastic material.

REEL-TO-REEL TRANSPORT (AUDIO OR INSTRUMENTATION)

Figure 15.4 shows the traditional layout of a reel-to-reel tape transport. Thousands of recorder designs use this format for recording audio or data. There are recorders for all the tape widths listed above, and for tape lengths from a few hundred feet to well over 10,000 feet. The tape transports all have several common design denominators, such as capstan drive, reeling system, and tape guidance.

A constant tape speed is paramount. Speed variations called flutter are easily detected by the human ear as a vibrato in music. They are likewise sensed by detection circuits in instrumentation and digital equipment. Flutter implies that recorded events are not played back with the original timings between them, and a measure for this is defined in the time displacement error (TDE).

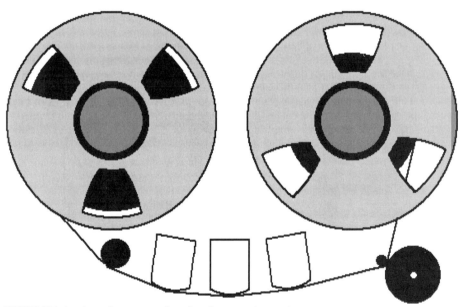

FIGURE 15.4 A reel-to-reel tape transport for audio or instrumentation recorders.

CASSETTE TRANSPORT (AUDIO)

Thirty-five years ago the audiocassette was introduced by Philips, and years later Sony provided the Walkman for it. Its success has been formidable. The basic cassette unit is shown in Fig. 15.5. A very similar design was introduced in 1993, also by Philips. The new cassette is called DCC, digital compact cassette. Details about this digital sound system are found in Chapter 28.

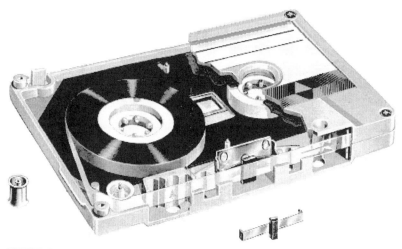

FIGURE 15.5 A Philips audiocassette.

3480/90 CARTRIDGE TRANSPORT

In mid-1984, IBM introduced the model 3480 ½-inch cartridge drive; see Fig. 15.6. The cartridge measured $1 \times 4 \times 5$ inches, and housed a single reel; the drive employs an 18-channel thin film head for 18-track recording on chromium-dioxide tape, storing about 200 Mb at 38 kbpi density. The write function is done by inductive heads, while the read is by MR elements. The drive was recently upgraded to the model 3490, which records 36 channels at double to quadruple the linear density.

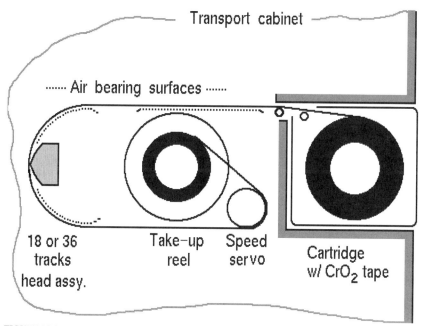

FIGURE 15.6 The IBM 3480/90 ½-inch tape cartridge drive.

QUARTER-INCH-CARTRIDGE TRANSPORT

The amount of data that many microcomputers need to store in a backup archive, disks, or tapes, will often require less tape than normally found on a ½" tape reel. A smaller tape unit has thus emerged, the quarter-inch tape cartridge shown in Fig. 15.7. Two cartridge sizes are available: The DC-600, and the DC-2000.

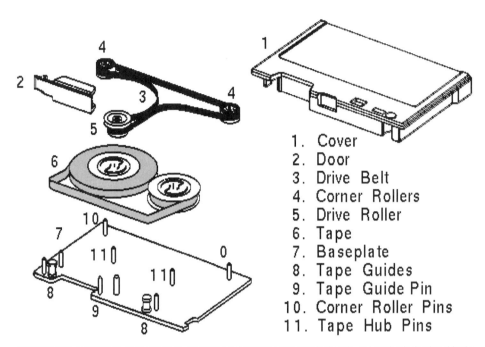

1. Cover
2. Door
3. Drive Belt
4. Corner Rollers
5. Drive Roller
6. Tape
7. Baseplate
8. Tape Guides
9. Tape Guide Pin
10. Corner Roller Pins
11. Tape Hub Pins

FIGURE 15.7   QIC (quarter-inch-cartridge). An endless polyurethane belt pulls both reels, and tension is developed in the head/tape area.

The housing consists of an aluminum base plate where a number of rollers and guides are installed. The aluminum provides the rigidity and precision; the rest of the housing is molded clear plastic.

The tape is located on two hubs without flanges, and is driven in one or the other direction by an endless polyurethane belt, which in turn is powered from a rubber-coated drive roller on the drive motor shaft. The head assembly is pushed into contact with the tape through an opening in the plastic housing; the opening is covered by a hinged door when the cartridge is not inserted into a drive.

One disadvantage is a varying tension profile from beginning to end of tapes. The heads must have a special profile with additional contact points (outriggers) to assure a minimum change in wrap angle around the heads' contact points.

HELICAL SCAN TAPE TRANSPORTS

Helical scan recorders have heads mounted on a rotating drum; these scan the tape in a helical path as it moves past the head assembly. From a discarded invention in the mid-1950s, it has become the overwhelming choice for high-density transports. The volumetric packing density surpasses any other media. Figure 15.8 shows the very popular VHS format; others are covered in Chapter 29. Transport mechanisms are covered in this chapter.

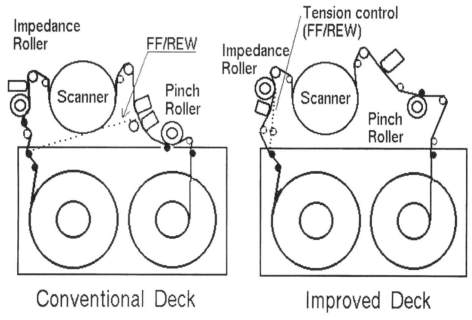

FIGURE 15.8   The original and the improved tape path in VHS video transports, using helical-scan heads.

DIRECT DRIVE CAPSTAN CONFIGURATIONS

The most important element in a magnetic tape transport is the capstan—a precision shaft against which the tape is pressed, often by a pinch roller. The speed of the capstan determines the tape speed. The diameter of the capstan shaft varies from about 1.5 mm (¹⁄₁₆ of an inch) for slow-speed cassette recorders to about 75 mm (3 inches) for high-speed, precision instrumentation recorders.

There are essentially three basic capstan drive systems (Fig. 15.9). The oldest, and still a much-used concept (Fig. 15.9a), is the open-loop drive, where the capstan pulls the tape over the head with tension provided by holdback torque or a felt pad arrangement. These transports often have an idler that provides isolation from any tape vibrations from the supply reel. In this regard, the closed-loop drive (Fig. 15.9b) is superior in isolating any speed variations from the reeling system. This idea is used in several instrumentation recorders.

Figure 15.9c shows the best drive system using two capstans with associated pinch rollers. One capstan rotates at a slightly higher speed than the other, thereby providing tension on the heads. Both capstans could rotate at the same rpm. If the diameter of the outgoing capstan is slightly larger than the other, then tension will be generated. This differential drive is in wide use in instrumentation recorders, and recently has been in studio and home recorders, in reel-to-reel as well as cassette configurations.

Figure 15.9d illustrates yet another drive layout, where the tape's contact with the large-diameter capstan is provided by a vacuum. This method is called a zero-loop drive, and has the advantage that the tape is supported all the way past the heads, which reduces scrape flutter.

CAPSTAN MOTORS

The capstan shaft is driven by either an ac or a dc motor. In inexpensive home entertainment re-corders, the drive motor is quite commonly a squirrel-cage configuration that drives the capstan

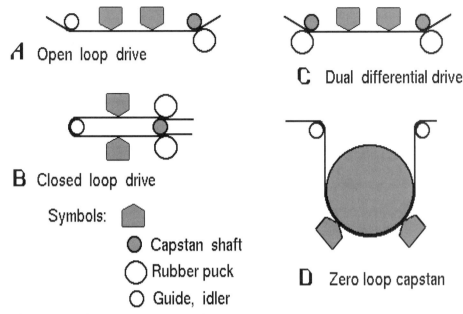

FIGURE 15.9   The four basic capstan arrangements in tape transports.

through a belt system. With such an indirectly driven system, the capstan shaft usually has a flywheel on it to smooth out any motor cocking or other speed variation. More expensive units use a hysteresis-synchronous motor in which the speed corresponds to the power-line frequency. Again, a flywheel is used on the capstan shaft, or with higher-grade synchronous motors the capstan shaft may be the motor shaft itself.

Construction details of each motor are shown in Fig. 15.10. The stator, contained within the motor housing, is the same for either motor. A rotating magnetic field is generated when the stator winding is connected to a power-line source. The rotor on the squirrel-cage motor is constructed from stacked, insulated laminations with short-circuiting connections between the two end stators. The rotating magnetic field from the stator generates currents through the short-circuiting rods, and these currents in turn generate a magnetic field with a polarity opposite to that of the stator field. The stator currents therefore magnetize the rotor, which tries to follow the rotating magnetic field from the stator. When the rotor approaches the speed of the field from the stator, the induced currents weaken, and if we assume that the rotor followed the stator field exactly, there would be no current induced and consequently no magnets would be formed in the rotor. Therefore, it would slow down. The squirrel-cage motor does, therefore, have a certain amount of slip. That is, it does not have a perfectly constant speed; the speed depends on the load. This slip can vary from two to ten percent. That's why a squirrel-cage motor is used only in inexpensive recorders.

The hysteresis-synchronous motor has a composite rotor with an outer shell of a hard magnetic material and an inner core of soft magnetic material. The starting torque is higher, because of the rotor's permanent magnetization due to the remanence in the hard magnetic material. The speed of the hysteresis-synchronous motor follows the power line frequency exactly, because any slip would mean a constant remagnetization of the rotor. Figure 15.11 shows the difference in the speed-versus-torque curves for the two types of motors. It is evident that the hysteresis-synchronous motor is superior regarding constant speed as well as higher starting torque.

Most recorders are provided with two and sometimes three speed ranges. This is achieved in two ways: one, by changing the pulley ratio between the drive motor and the capstan shaft flywheel, which usually means mechanically shifting the drive belt (often a seamless cloth belt). Such a change can obviously only be made while the capstan motor is running. The second, and more practical solution, is ob-

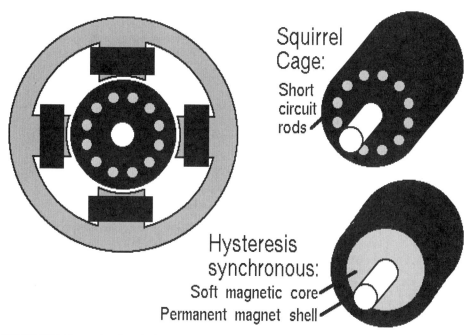

FIGURE 15.10 Construction of a squirrel-cage motor, left and top-right. The armature in a hysteresis-synchronous motor shown in the right-hand lower corner.

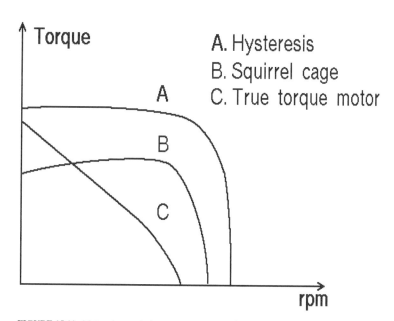

FIGURE 15.11 Motor characteristics, torque-versus-speed: (a) Hysteresis-synchronous. (b) squirrel-cage. (c) true torque motor.

tained by using a two- or three-speed motor, with two or three times the number of poles on the stator. The speed selection is made by merely coupling all or only some of the poles to the power source.

In the instrumentation recorders, the speeds commonly used range from 120 IPS down through 60, 30, 15, 7½, 3¾ and even down to 1⅞ IPS. To make speed changes by means of a pulley and belt in this wide speed range is quite impractical. The choice of a drive system for such a demanding system, therefore, falls on a servo-controlled system, where the capstan drive motor is a dc type with a suitable tachometer mounted on the drive-motor shaft. The output frequency of the tachometer is compared with a control reference frequency, and any frequency deviation results in an error signal that corrects the motor speed. Here again, the drive motor can be coupled to the capstan shaft by means of a belt drive or the capstan shaft can be the motor shaft itself.

SERVO-CONTROLLED SPEED

In earlier instrumentation recorders, speed control was attained by recording a 17-kHz control frequency that was amplitude-modulated at a rate of 60 Hz. During playback, this signal was demodulated and fed into a phase-comparison circuit, which fed a precision 60-Hz motor-control oscillator whose output after amplification drove the synchronous capstan motor. This technique is referred to as speed control, because the servo system operated through a high-inertia drive system with flywheels and belt drive and the maximum error rate that could be allowed was around 1 Hz.

Instrumentation recordings are often made under adverse conditions (vibrations, etc.). The inherent wow and flutter both from the recorder and the external vibrations will change the frequency of the recorded tones. This will affect any analysis work that is performed on data from the tape. Any change of speed causes a change of signal frequency, which is translated to a change in timing. Therefore, it is vital that this error, also called time-base error, be reduced. This requirement led to a family of instrumentation recorders that have a drive system as illustrated in Fig. 15.12.

The capstan drive motor is a low inertia, high-peak-torque motor, driven by a power amplifier that is controlled by a servo amplifier. When the tapes are recorded originally, one track is

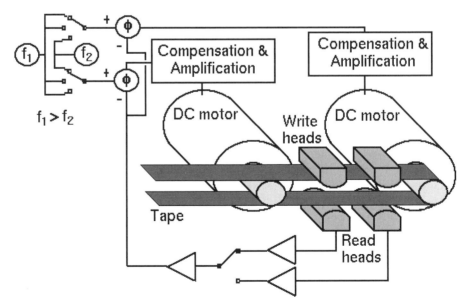

FIGURE 15.12 A dual motor-speed servo control, shown in the playback mode.

recorded with a control tone frequency ranging from 200 kHz down to 6.25 kHz, all dependent on the speed selected. Upon playback, this signal is amplified and fed to a phase detector and compared with a crystal-generated control frequency. Any speed deviation results in a phase-error signal, which in turn controls the servo amplifier and causes the tape drive to speed up or slow down. The inertia of the capstan motor, and possibly the coupling elements, restrict the time for making speed corrections.

A measure of merit is saturation acceleration (equal to the peak torque of the drive motor divided by the inertia). A low-inertia drive with high saturation acceleration is far more capable of correcting speed variations than any of the more conventional transports with their heavier motors and flywheels.

The degree to which a servo can suppress a flutter component is a function of the open-loop gain of the servo at the frequency of the particular flutter term. If the velocity of the capstan is considered to be the same as the velocity of the tape (a valid assumption when the distance between head and capstan is short) the transfer function relating the tape velocity ω_o (in rads/sec) to the reference frequency ω_{ref} is (see Fig. 15.13, top):

$$\omega_o/\omega_{ref} = G(s)/(1 + G(s)H(s)) \qquad (15.1)$$

where $G(s)$ = transfer function for the forward path and $H(s)$ = transfer function for the feedback path.

It can be shown that the relationship between output F_o and input flutter F_i is:

$$F_o/F_i = 1/(1 + G(s)H(s)) \qquad (15.2)$$

This equation shows that the output flutter is attenuated by the factor $(1 + G(s)H(s))$. Consequently, to obtain the maximum attenuation, it is necessary to make $G(s)H(s)$ as large as possible and to provide this correction over as wide a bandwidth as can be achieved.

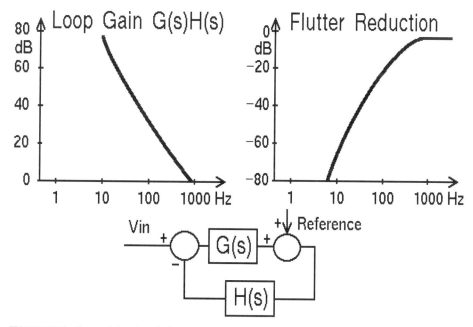

FIGURE 15.13 Characteristics of a typical capstan servo system.

The complete transfer function becomes unwieldy when the individual transfer functions are substituted into our equation. A number of meaningful conclusions can, however, be reached by considering the general form of the expressions. Then graphical techniques can be employed to evaluate servo performance of the actual transfer function.

Our previous equation exhibits unity in the frequency range where $G(s)H(s) \gg 1$ (Fig. 15.13). Because $G(s)H(s)$ has two poles at zero frequency (the compensation and feedback each introduce $1/s$ terms) the dc gain is theoretically infinite and $\omega_o/\omega_f = 1$. Consequently, there can be no static error between the shaft velocities and the reference.

Equation (15.2) also shows that the useful bandwidth of the servo equals the frequency range for which $G(s)H(s) \gg 1$. Consequently, the servo will provide correction only for flutter that occurs within this bandwidth.

The degree of correction is practically limited by several terms in $G(s)$ arising from mechanical resonances in the system. The tape itself is elastic, with a spring constant K that enters into the servo loop. Speed corrections propagate with the speed of sound in PET film (2000 meters/second), and cause a minor delay. The capstan shaft and the motor itself have rotational resonances (Diamond 1965; Alexander and Ling 1967; Hu 1974). The head stacks may vibrate as cantilevers if they are not rigidly mounted. Most troublesome is still the tape, because all flutter components are inside the servo loop and a closer examination of tape flutter is worthwhile (Law 1976).

Example 15.1: This example covers a few calculations for a typical example of the numbers pertaining to flutter reduction and the dynamics involved in correcting tape speeds.

Assume a disturbing flutter (as it exists on the recorded tape) of 1 percent at 15 Hz, at the speed of 120 IPS. This corresponds to a *TBE* of ±53 µsec:

$$TBE = 0.01 \times Flutter/2\pi f$$
$$= \pm 0.005/2\pi 15$$
$$= \pm 53 \ \mu sec$$

and the actual tape displacement error at the reproduce head is ±0.000636 inches. If we assume an ideal dampened idler with infinite inertia (Wente and Mueller 1941), then the capstan must correct these relatively instantaneous displacement errors by increasing or relaxing the tension in the tape. The slope of the stress/strain curve for the tape is 750,000 psi around $\sigma = 1250$ psi (corresponding to 10 ounces of tension in the ½", 1-mil tape).

When the calculations are carried through, we find that the tension variations range from 19.5 oz to 0.5 oz. Had the flutter been larger, a tape slack condition would have occurred during each flutter cycle (negative tension).

This capstan configuration (open loop) would obviously not provide a good servo system. The dual capstan system provides an advantage: it will assist in pushing (or holding back) the tape, which moves as a platform under constant tension.

An instantaneous time displacement error of 53 µsec will still require a displacement correction of 6 mils at the reproduce head. But the speed of sound in tape is around 2000 m/sec (78,800 inches/sec.), and if each capstan corrects for one-half of the length of the tape platform, the theoretical tape limit for error correction in a 4" long tape is ½ × 4 × 1/18,800 = 25 × 10⁻⁶ seconds, or 25 µsec; this corresponds to an upper frequency limit of 20 kHz.

CAPSTANS AND PINCH ROLLERS

The objective of the capstan drive is to meter the tape motion at a constant speed, and to generate or maintain a constant tape tension over the magnetic heads. The very basic capstan drive is a rotating shaft moving the tape at a constant speed.

The capstan shaft may be driven from a two-speed hysteresis-synchronous motor by way of a belt, or the motor shaft itself could serve as the capstan. This requires a good motor, one free from hunting (Clurman 1971). Modern recorders use low-inertia dc motors with a tachometer and a servo system for speed control, or brushless dc motors.

The tape is pulled by friction. To maintain a reasonable tension over the heads (approximately 20 ounces per inch of tape width), it is necessary to clamp the tape to the capstan. An early practice still in widespread use today is the method of pressing the tape against the capstan with a pliable pinch roller. This forms a nip where the tape is clamped. The tape speed is now determined by the pinch roller. (The area on the pinch roller and the coefficient of friction between the roller and the tape is greater than between the capstan and the tape.)

The true tape speed may not be equal to the surface speed of the capstan (Fig. 15.14). The pressure causes an indentation in the flexible pinch roller, and the effective radius of the roller is reduced. The tape enters and follows the pinch roller at angle β_1 with a certain speed, and leaves the roller at angle β_2. An answer to the complexity of the capstan/pinch roller problem may come someday from detailed analysis (Durieu 1984).

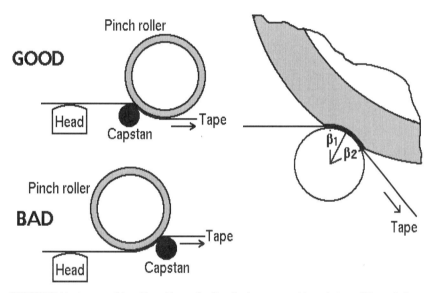

FIGURE 15.14 A capstan drive with a rubber puck roller. Absolute tape speed is nearly impossible to calculate.

The surface speed of the pinch roller in the nip area is somewhere between v_2 and v_3, and so is the tape speed. The tape leaves at the angle β_2, and there the surface speed of the pinch roller is greater than the tape speed v. Therefore the tape will slip, and the relative speed difference will be so small that stick-slip friction occurs. This would also be true if the tape contacted the pinch roller before the capstan (see Fig. 15.14, lower left)—a bad situation that generates high flutter in the head area. The pinch roller is omitted in the friction capstan shown in Fig. 15.15. The precision-ground capstan is made from steel with a high-friction rubber/plastic overcoat.

The tape and the pinch roller are both elastic materials, and slight variations in the take-up and holdback tension from the reeling system will cause deviations in the tape speed. The tendency is for the tape to slow down toward the end of a reel. The implication may be a noticeable change in pitch if the recorded material is edited.

A few manufacturers specify the maximum speed deviation that may occur as slip, and it is normally within one percent. The effect of slip upon editing can be minimized by using large tape reels (10½" rather than 7" diameter) and recording on less than half of the tape length. The change in pitch from sections spliced together may then not be noticeable.

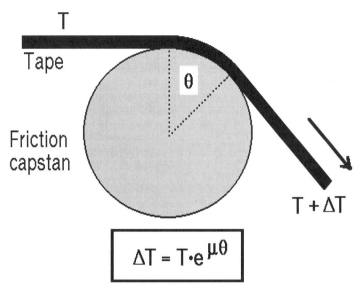

FIGURE 15.15   A capstan friction drive, which needs much attention for cleanliness.

Much can be said against the application of pressure felt-pads that press the tape against a tape guide post for tension control. But they will, when properly maintained and adjusted, provide constant tension and hence assure constant tape speed throughout a reel.

THE QIC DATA CARTRIDGE

The *quarter-inch cartridge* (QIC) was invented in 1971 by von Behren at 3M Company. The cartridge contains two spools of tape, and requires only one rotating capstan to move the tape. There are no reeling motors for tension control and winding. Motion of the reels, and hence the tape, is provided for by an endless polyurethane belt pulled by a drive roller that is rotated by an outboard capstan shaft; see Fig. 15.16.

THE BELT DRIVE MECHANISM

The tape transport is shown in Fig. 15.16, where an endless belt in contact with the two tape reels moves the tapes from one hub to the other. How is tension over the head developed? And how is a "normal" tape wrap angle obtained when a cartridge is recorded on one machine and played back on another? And what tolerances may be required?

The generation of tension in the tape is the key element in the patent. The drive force from the outboard capstan shaft will turn the drive roller for the belt, causing it to move in the directions shown. A pulling stress occurs at (a), while a push occurs at (b). The forces are augmented by the fact that rollers (c) in two corners contain a grease. Therefore: The belt is stressed and elongated at (a), while it is pushed and shortened at (b).

The belt moves continuously, and it must therefore move locally faster at (a) and slower at (b); otherwise, it would break up at (a) and/or pile up at (b). This leads to the conclusion that the surface velocity at the periphery of roll A is faster than of roll B. The tape is itself elastic, and will therefore stretch between points (d) and (c), and thus a tension developed at the head contact (von Behren 1989).

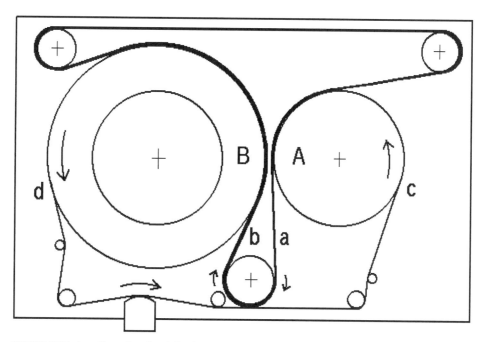

FIGURE 15.16 An endless polyurethane belt pulls reels A and B, and tension is developed in the head/tape area.

To maintain a steady tape speed, high friction conditions must exist at the drive capstan/drive roller and drive roller/belt interfaces. Similarly, high friction must exist at the belt/tape interface, in particular at high tape speeds (120 IPS). It was early found that air would entrap at the belt/tape interface. This could be eliminated by texturing the belt surface (Smith 1989). The tension developed in the 5.25" cartridge is now 2 to 2.3 ounces in the speed range of 40 to 180 IPS, and remains quite constant from beginning to end of a pass (Smith 1989; Gills 1989).

Photomicrographs of the textured belt surface are shown in Fig. 15.17. Typical surface textures are in the 1500- to 3000-nm range. A longitudinally textured pattern allows the air molecules to escape from the belt/tape interface during fast tape motion. This air can become trapped under the tape when it comes to rest against the tape pack on one of the hubs. This prevents the tape from settling down, and it may even skew out to one or the other side of the intended tape pack.

The distance between tape and tape pack may initially be 2–3 μm, settling down to 1.5 μm after one full revolution; that is, that tape is still floating on air, and may not assume its final position until several revolutions have passed. Here is where the belt helps by squeezing the air out from under the tape; a distance of only 0.5 μm is achieved after one revolution. This is essentially contact, and a good tape pack with smooth sides is achieved. This is enhanced by making the tapes with a nonslip, antistatic backcoat of thickness 1 μm.

The precision tape guides have a coefficient of friction of 0.25–0.3, with a surface texture of 400 nm rms with 950 nm summit-to-valley. Tape tracking has been improved by relieving the tightly screwed connection between the aluminum base plate and the plastic cover. They would warp each other during temperature changes, due to different coefficients of thermal expansion. Currently, the cover is attached to a couple of posts at two noncritical points of the base plate.

The mean tracking accuracy is ±0.5 mils (±12.5 μm) (Topham 1989). Part of this tracking error is associated with the direction the tape moves in, so it could be +0.5 mils in the forward direction and –0.5 mils in the reverse direction. There appear to be two causes for this error: One is from the slitting of tape that is made on a PET film that early in its life was stretched to achieve high strength. The built-in stresses of the web are highly nonuniform (see Fig. 11.13), and their release when the web is slit causes some curl (see Fig. 14.1).

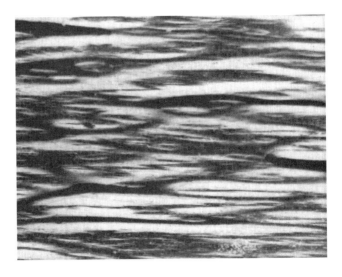

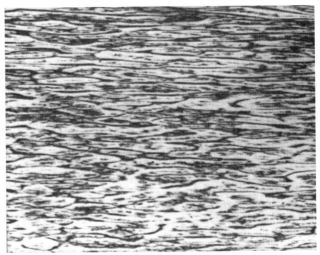

FIGURE 15.17a. Textured drive belt for QIC drive. The top's magnification is four times that of the bottom (*courtesy of 3MCompany*).

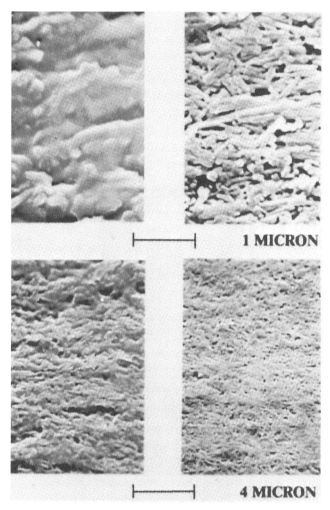

FIGURE 15.17b. Surfaces of a standard and a high density QIC tape. The top pictures' magnification is four times that of the bottom. (*courtesy 3MCompany*).

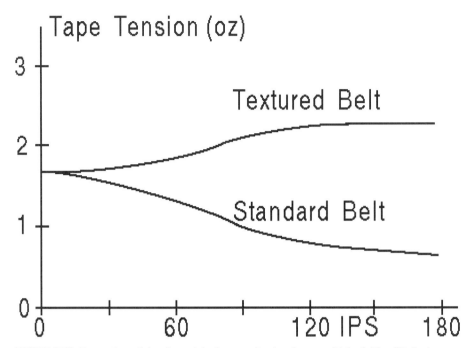

FIGURE 15.18   Tape tension variations for two belts shown as a function of tape speed (*after Smith and Topham*).

The polyurethane belt may also be causing the directional tracking error. The belt was originally a small flat disk with a hole in the center. It was forcibly wrapped onto two parallel mandrels, and under heat was turned while the mandrels were removed from each other. This would stretch the rubber while simultaneously forming it into a belt.

It appears that all QIC drives have this differential tracking error. Extra allowance must be made for the land between tracks, and the limited number of tracks recorded on a tape; see Fig. 15.19 (left). The track format shown follows a straightforward serpentine pattern, where the next track is recorded adjacent to the previous. Using a stepping device with equal distances between steps would result in a reduction in land between some tracks, an increase between others. Early recognition of this tracking problem provided a simple solution: All odd tracks are recorded on the top half of the tape, while the even tracks are recorded on the bottom, as shown in Fig. 15.19, (right).

One last item of interest is the occurrence of vibrations in the free spans of tape between points (c) and (d). That distance measured approximately 180 mm; it results in a fundamental allowed resonance frequency near 5 kHz. This is the longitudinal oscillation. Other, potentially more damaging vibrations are in the transverse mode; they convert to longitudinal speed errors by way of Poisson's ratio (see Chapter 17.)

TRACK-FOLLOWING SERVO MECHANISM

The current layout of a head for use in a QIC drive is shown in Fig. 15.20 (left). It is intended for a single data channel and has a write head (TFH) in the center, followed on either side by an inductive or MR read head. The newer three-channel head is shown in Fig. 15.20, center, and a multichannel head is shown to the right.

The three-channel head is shown in operation in Fig. 15.21 (Schwartz 1992), in six situations: In case A, the bottom head scans on the top of the two prerecorded servo bands, and will place the two upper

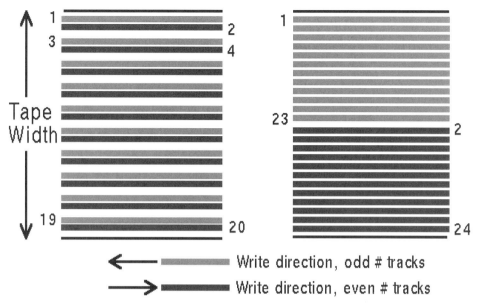

FIGURE 15.19 Shown left are tracking problems in a purely serpentine pattern, using a stepper motor for head positioning. Even and odd tracks should be grouped on the top and bottom halves of the tape, as shown at right.

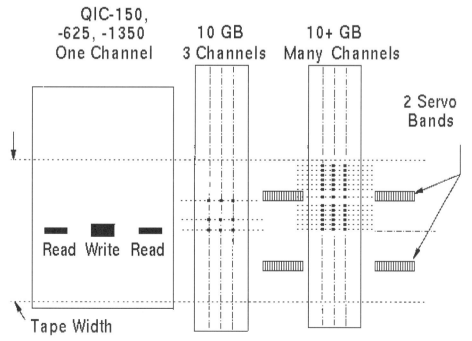

FIGURE 15.20 Single-channel, three-channel, and multichannel heads for present and future QIC formats.

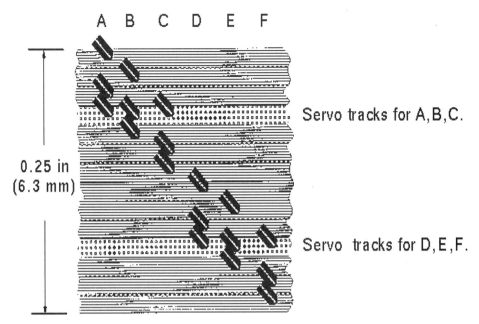

A B C D E F

0.25 in
(6.3 mm)

Servo tracks for A,B,C.

Servo tracks for D,E,F.

FIGURE 17.21 A three-channel head shown in six different positions, utilizing a two-band set of servo tracks.

channels on tracks within bands 1 and 3. In case B, the center head scans the top servo band and places the top and bottom heads on tracks within bands 2 and 4. In case C, the top track scans the top servo band and the two bottom heads scan tracks 5 and 6. A similar arrangement holds for cases D, E, and F.

This technique works very well, and will permit track densities to go beyond the traditional limit of 100 tpi, up toward 1000 tpi. The improved tracking is evident from Fig. 15.22 (Molstad and Jewett 1991).

TAPE GUIDANCE AND SKEW

When bytes are recorded onto tapes, they are sometimes written in parallel over eight tracks, with a ninth track allocated for parity check bits. When later read, they should occur simultaneously; that is, in time synchronization. This situation never exists in practice because of tolerances in the manufacture of multitrack heads and in the slitting of tapes. The nominal width of a computer tape is 0.498 +0.0 –0.002 inches, and it may have a longitudinal curvature of up to ⅛" over 36 inches. The curvature may, among other factors, cause the tape to "weave" through the drive and it could, over a distance of say 3 inches, depart as much as ±0.002 inches from the ideal, straight path. This translates into an angle of:

$$\beta \approx \tan\beta$$
$$= \pm 0.002/3$$
$$= \pm 0.00067.$$

This will result in an edge-to edge displacement of $0.5" \times \tan\beta = 335$ μin. To this error from dynamic skew should be added the gap scatter and head tilt (from one drive to another), and the total maximum error could be as high as 1 mil between edge tracks.

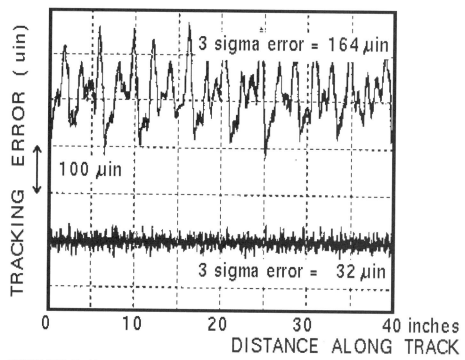

FIGURE 15.22 Tracking error without servo head control (top) and with servo control (*after Molstad*).

An error of 1 mil is almost equal to one bit length at the conservative packing density of 800 bpi, and the simultaneous read-out of a byte would be impossible. The problem is solved by storing the bit stream from each track in individual buffers, which are clocked in synchronism. The skew changes quite slowly, and the timing of the deskew buffers is initiated by a known bit pattern stored in front of a block of records. Only the edge tracks need to be read for the pattern to be recognized by the logic in the controller. The buffers are normally kept half-full (on the average), and time displacement errors between tracks are completely removed (Fig. 15.23).

REELING AND WINDING OF TAPES

While the capstan drive system plays a major role in maintaining speed accuracy, equal attention must be given to the reeling system. The supply reel must feed the tape to the head area in a perfectly smooth motion, while the takeup reel winds the tape onto the reel under constant tension and with a resulting smooth tape pack.

During winding or rewinding, when the capstan puck is disengaged, the tape should move with the highest practical speed from one reel to the other. When the winding or rewinding mode ends, both reels should stop in such a fashion that the tape will neither be overstretched nor throw a loop. This again puts a demand on the design of the braking mechanism for the reel system.

The best reeling is obtained by the use of two separate reel motors, which should be motors that provide for an almost constant tape tension (Holcomb 1952). Such motors have an inverse speed-torque curve, as shown in Fig. 15.11, curve C. When the reel is empty and the speed is high, the torque is low and a given tape tension is provided. Next, when the reel is filling up the speed slows down and the torque increases, but the radius of tape accumulating on the takeup reel is increasing

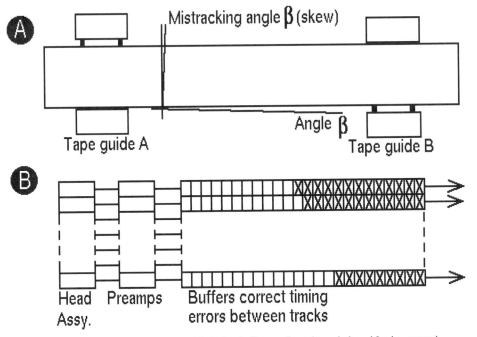

FIGURE 15.23 Tape skew between guides A and B. Deskew buffers are allocated to each channel for skew removal.

and the tension remains constant from the empty to the full reel (Hartmann 1958). The almost-constant tension is desirable, because tapes stored for a long period under excessive tension, or a highly varying tension through the reel, may become permanently damaged.

Winding of a tape at high speeds does not always happen with a smoothly flowing motion. Minute fluctuations in the reel-motor torque (cocking) may cause the tape to flap back and forth, and the pack on the take-up reel will thus be bad. If the transport designer is faced with these problems, he may find help by studying a similar phenomenon that occurs in band saws and other line structures (Mote 1965; Thurman and Mote 1969; Mote and Wu 1985). The designer will also benefit from a paper on the handling of wide webs (Pfeiffer 1977).

Constant tension during record or play modes is important for the absolute tape speed, in particular with open-loop transport drives. The speed is affected by the supply reel tension and, in poorly designed recorders, the absolute speed of the tape can change several percent from the beginning to the end of a recording. If played back on the same recorder, the same tension variation exists and the playback speed is equal to the record speed. But if the tape is edited, and especially if a portion is moved from the beginning to the end or vice versa, there will be a noticeable change in the frequency or pitch of the recorded material.

The braking system, which (as earlier mentioned) works in conjunction with the reeling motors, must be carefully designed to avoid tape damage. A reeling motor brake may consist of a flat steel band with a felt coating that faces a flat drum fixed on the motor shaft. The braking action is differential. In other words, if the tape is traveling from left to right when the brakes are engaged, the left wheel will receive a slightly greater braking action than the right wheel, and vice versa. This braking system is disengaged during playing or winding either by means of a lever or by solenoids. The use of solenoids for engagement of the brakes is an attractive solution, because any power failure to the recorder will cause the brakes to automatically engage.

Spring loading for the brakes must also be given consideration during the design of a recorder. Most home recorders are designed for the use of 7-inch reels, but quite often such recorders are used with, say, a 7-inch reel for takeup and a 3-inch reel for supply (in the exchange of tape letters, for example). Here the dynamics of the braking action are different from that originally designed for two reels of the same size, and the operator should pay great care in starting and stopping a recorder with different reel sizes. The same problem exists in instrumentation recorders where the two commonly used reel sizes are 10½-inch-and 14-inch-diameter reels. The larger reel has far more inertia than the smaller reel. Therefore, some of these recorders are equipped with a switch that selects the proper braking action for the various reel sizes.

Dynamic braking is commonplace in instrumentation recorders and in better audio recorders. Proper current levels are fed to the reel motors and possibly servo-controlled by sensing arms in the tape path, bringing the tape motion to a gentle stop. It is important that the tape path from the supply reel to the take-up reel be in one perfectly flat plane. This requires a rigid transport deck (which often is a casting) and carefully mounted and adjusted tape guides. The tape guides must provide perfect edge guidance of the tape. Normally, two types of fixed guide are used, as shown in Fig. 15.23.

Rotating guides, with or without flanges, are also employed. They may be flat or crowned rollers, mounted on precision roller bearings; the reader is referred to the papers listed for further details (Eshelman et al. 1973; Paroby et al. 1975; Clurman 1981).

Flanged guides in the head area are supposed to guide the tape without skew, or deviation, from the straight path. The clearance through the guide is manufactured to the exact specified maximum width of the tape. The actual tape width may be less by as much as 2 mils, because the tolerance of state-of-the-art tape slitting is ±1 mil. This may cause considerable "wandering" of the tape as it runs through the transport, with a maximum deviation of 2 mils over the length of the path between two guides. When the tape is recorded and played on the same transport, chances are that the tape will wander in the same pattern. But if it is recorded on one machine and played back on another, a full deviation of 4 mils may occur over a span of, say, 4 inches.

This is equal to 0.001" per inch, or a misalignment of:

$$\alpha = \arctan(0.001)$$
$$= 0.001 \text{ rads}$$
$$= 0.0057 \text{ degrees}$$

The signal loss can then be calculated from the alignment loss formula in Chapter 6. A 2-MHz signal at 120 IPS could suffer a worst-case attenuation of 14.4 dB if the tape clears the guides in opposite ways during record and playback, while a realistic worst case would be only about 7 dB. A more annoying outcome is the amplitude variation associated with an uneven tape passing through the guide.

A similar skew problem occurs when the tape path is changed from one plane to another. This happens in recorders with coaxial reels. Sufficient lengths must be allowed for the tape to make it twist, so in no place is it subjected to a stress exceeding 3000 psi. Higher stresses may cause permanent tape damage, often in the form of stressed edges.

Some useful features may, in conclusion, be incorporated with the guides on a transport. These are: tape-break and end-of-tape sensors, footage counters, and tape cleaners (vacuum, possibly with ionization).

HEAD MOUNTING

In magnetic recorders with low packing densities, it is quite common to find the heads mounted on a fixed base plate, which in turn is mounted on the capstan drive precision plate. Where the packing densities (i.e., wavelengths per length of tape) exceed 5000 Hz per linear centimeter of tape, it becomes necessary to mount the reproduce head on an alignment plate so its azimuth can be adjusted to be exactly equal to that of the record head. The record heads are normally fixed-mounted, with a certain tolerance allowed for the azimuth that is obtained by careful grinding of the recording head mounting surface in the final step of its production.

Reproduce heads must be carefully shielded with a mu-metal housing or another suitable metal in order to minimize hum pickup from the drive motors, transformers, and any other external interference source. In wideband recorders that operate in the range of up to 4 MHz, additional shielding becomes necessary in order to avoid pickup of signals in the midwave radio-frequency band. Powerful broadcast stations in this range may be picked up by a sensitive reproduce head, and appear as noise in the recorder output. In the design of the head mounting, attention should also be paid to the fact that signals from the record head may radiate into the reproduce head during recording, which makes alignment of the recorder's electronics difficult.

DESIGN FOR LOW FLUTTER

In order to keep flutter components low, the magnetic heads should be placed at a point where the tape speed is smoothest, and this is normally as close to the capstan as possible. Attention should be paid here to the fact that the magnetic tape itself is an elastic medium, and that longitudinal oscillations (similar to those of a violin string) occur. Whenever the tape runs over a fixed guide or magnetic head, it rubs against these fixed elements, resulting in oscillations (scrape flutter; see Chapter 17). Guide and head surfaces must be as smooth as possible, to minimize flutter, and the wrap angle around the heads should be a minimum; normally in the order of 3 to 5 degrees. If the heads are located between the idler and the capstan, scrape flutter will be reduced. A rotating high-inertia guide will likewise serve as a grounding point for the longitudinal oscillations, and such guides are frequently incorporated into precision tape drives.

Flutter and time displacement errors must be minimized in a tape drive, both for the sake of listening pleasure and for data integrity. It is true that modern digital recorders will remove flutter by data buffers on the read side, with sizes proportional to the largest speed variations that must be removed. This fact should not cause the transport designer to relax his design efforts.

In all longitudinal tape oscillations there is a natural resonance frequency, which by a rule of thumb is 100,000 Hz per centimeter. So, if we have a free tape span from the in-going idler to the capstan of say, 12 cm, the resonant frequency will be at 8300 Hz. If, on the other hand, the tape span is 20 cm, the resonant frequency will be at 5000 Hz. Thus, a short tape span is important. In the design of a precision tape recorder, it is quite useful to apply an equivalent electrical diagram of the electromechanical components in the recorder, as shown in Fig. 15.24.

A valuable tool is available to the mechanical designer of tape drive mechanisms: The electromechanical analog. This approach is advantageous because the electrical engineers have developed a set of convenient symbols for circuit elements that can be used to represent a mechanical system with masses, springs, forces, etc. The circuit diagram can be examined and its behavior treated; the results are then applied to predict the mechanical systems behavior, and design corrections applied.

Excellent tutorial examples are found in two now classic papers by Wolf (1966) and by Jacob (1963), supported by a classic textbook on the topic (Olson 1966). A simplified layout of the mechanical components is easily transformed into an electrical admittance diagram, which then is converted into an impedance diagram. Now all the tools of modern electrical system engineering can be used: LaPlace transform, impulse response, etc. The system's performance can be traced back to the individual circuit components or their mechanical counterparts, and needed design changes can be made.

HELICAL SCAN TAPE TRANSPORTS

The tape transport moves the tape past the rotating scanner from supply reel to takeup reel. The tape is stored in a cassette in all consumer products, and in the D-1, D-2, and D-3 digital video recorders.

Tape guides are placed at precise angles before and after the scanner. This places the tape at an angle to the head wheel so a slanted track is written. Early recorders had the reels at different planes for the slant track recording. Compact instrumentation recorders with stacked reels also have a change of reference plane for the tape (see Fig. 15.1.)

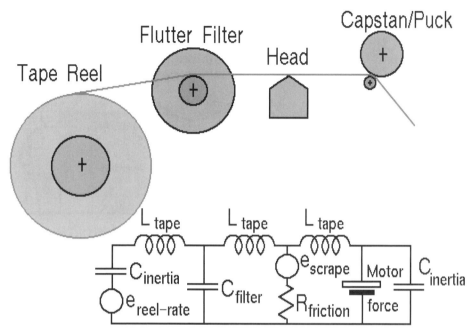

FIGURE 15.24 Electromechanical analogies permit circuit analysis of tape drive performance and design.

A tape cassette inserted into a helical recorder is first moved into its proper position by an elevator mechanism. Then a gate closes the entry opening to prevent interference with the cassette and/or the tape. The tape is pulled out from the cartridge and led over several guides, and finally wrapped around the scanner. It is advantageous to keep the count of moving parts to a minimum, and rigidity is paramount.

The wrap angle around the scanner varies from system to system. The larger the wrap, the greater the frictional forces around the scanner. This leads to tape motion instability, a topic we will address in the following sections on the scanner and the magnetic tape surface.

The Beta format has the most complicated path, called full-load, or U-threading. The tape is pulled around the scanner and stays there during record or play and during winding. This allows for the time to enter record or play modes, and provides faster, more accurate editing relative to VHS; but winding time suffers.

Most other recorders, including VHS, use half-load, or M-threading. The tape is wrapped around the scanner for record or play, and after that retracted into the cassette. This was shown in Fig. 15.8 (left), and results in slow start/stop times, but fast winding. It makes editing difficult, because winding in the cassette introduces slip due to imperfect contact with the head.

Figure 15.8 (right) shows an improved version of the M-threading. The tape is always in good contact with the scanner. Another improvement is the pinch roller mechanism; in the new arrangement, a geared elevator moves it into place to contact the tape's backing side, not the more vulnerable magnetic coating.

Figure 15.25 shows the tape path in a DAT system. The wrap angle is 90° only, and the tape stays in contact for record, play, and winding. This allows for fast search with the tape in contact with the scanner.

The C-format recorder, a reel-to-reel transport, has an intricate tape path; see Fig. 15.26. The tension around the drum varies considerably due to frictional forces, and the slanted pattern makes the resulting stresses nonuniform.

The D-1 and D-2 transports both have somewhat simple M-threading (Fig. 15.27), again with a pinch roller that moves up and contacts the tape on the backing side. Three different sizes of cassettes can be accommodated.

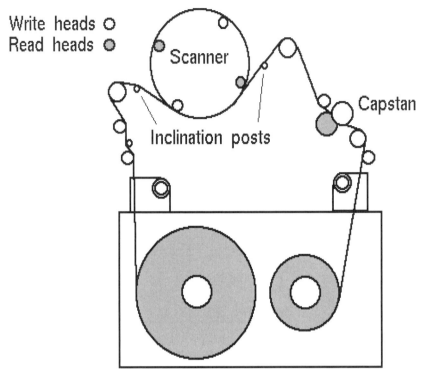

FIGURE 15.25 The tape path in a DAT transport.

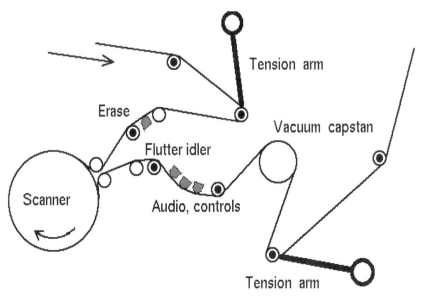

FIGURE 15.26 The tape path in a C-format transport.

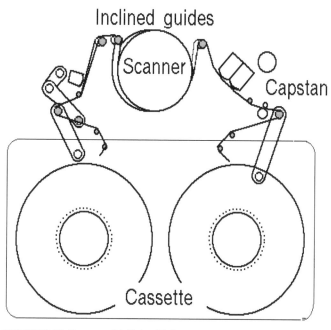

FIGURE 15.27   The tape path in D-1 and D-2 transports.

SCANNERS

Figure 15.28 shows a helical-scan recorder from the 1950s. It employs a fixed upper and lower drum, and the long tape path causes many flutter and tracking problems.

Figure 15.29 shows a head wheel with four sets of record and playback heads. Three rotating transformers for signal couplings are shown in Fig. 15.30. The three units are identical, using flat annular coils and using the divided ferrite cores to form a closed path for the flux. This construction also minimizes crosstalk between the transformers, and reduces RFI (radio frequency interference).

The head wheel with heads and amplifiers is shown in Fig. 15.29. Normally a head wheel for video recording with a 180° wrap would have two opposite-mounted heads, both serving as either record or playback heads. The head wheel shown is for a 180° wrap system, and it will actually record and playback four channels simultaneously, because there are eight set heads.

Higher data rates can be handled by using four channels instead of one. The incoming digital signal is divided up into four channels (interleaved), and then expanded four times on the time scale; see Fig. 15.31. Each channel is now within the passband of the recorder. They are, after playback, time-compressed four times and then reassembled to the original data stream.

We will now address the fundamental problem that exists when a flexible magnetic tape moves around the scanner containing the rotating head assembly. Figure 15.32 shows the tape path. Two slanted guides are used in order to make the tape conform to the tilted drum assembly; You can see these guides in many camcorders if you look inside before inserting the tape cassette. They are also shown in the tape path in Fig. 15.27. Edge guidance around the drum is provided for by pulling the tape down toward the milled bottom edge or reference edge, also called the fence.

The surface along the drum generates a fair amount of friction when the tape moves over it. This is partly eliminated in consumer recorders by using an upper drum that rotates with the heads. The speed difference between the tape and the upper drum ranges from 3 to 7 m/s. An air bearing is formed between the tape and the drum surface, and the friction is nearly zero. The C- and D-2 recorders also have a rotating upper drum, against the longitudinal tape speed. Only the D-1 and the ID-1 recorders have a stationary upper and lower drum. Figure 15.33 shows the different configurations.

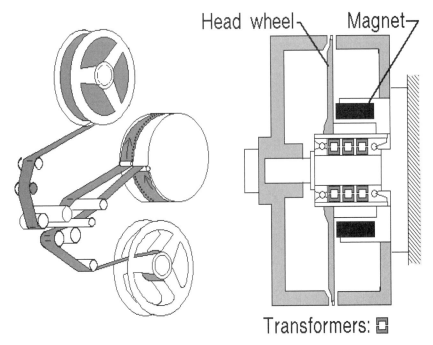

FIGURE 15.28 An early helical scanner (*after Philips*).

FIGURE 15.29 A head wheel with eight write heads with drivers, and eight read heads with preamplifiers (*courtesy of Datatape*).

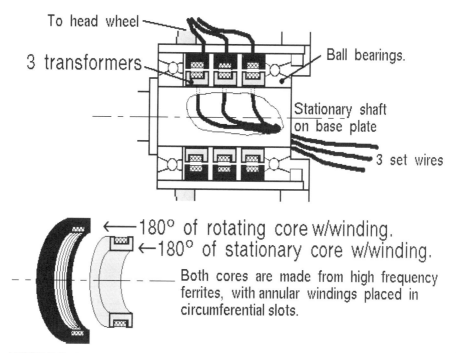

To head wheel

3 transformers

Ball bearings.

Stationary shaft on base plate

3 set wires

←180° of rotating core w/winding.
←180° of stationary core w/winding.

Both cores are made from high frequency ferrites, with annular windings placed in circumferential slots.

FIGURE 15.30   A rotating transformer assembly. Windings are placed in the donut-shaped grooves in the ferrite rings.

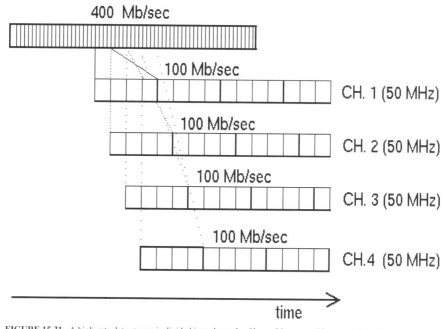

400 Mb/sec

100 Mb/sec   CH. 1 (50 MHz)

100 Mb/sec   CH. 2 (50 MHz)

100 Mb/sec   CH. 3 (50 MHz)

100 Mb/sec   CH.4 (50 MHz)

time

FIGURE 15.31   A high-rate data stream is divided into channels of lesser bit rates and is recorded simultaneously on four tracks, i.e., 400 MBps becomes 4 × 100 MBps.

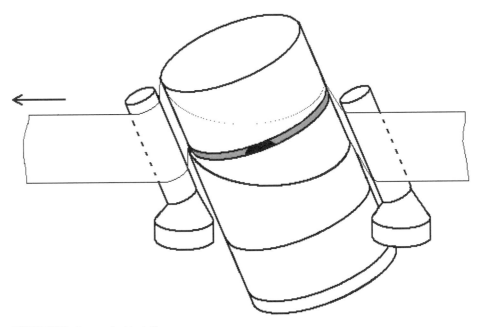

FIGURE 15.32 A tape path with a half-wrap scanner.

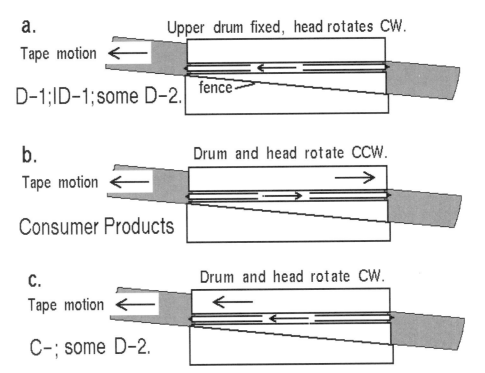

a.

Upper drum fixed, head rotates CW.

Tape motion ←

D-1;ID-1;some D-2.   fence

b.

Drum and head rotate CCW.

Tape motion ←

Consumer Products

c.

Drum and head rotate CW.

Tape motion ←

C-; some D-2.

FIGURE 15.33 Scanner configurations: (a) Fixed drums, heads protruding; (b) rotating head wheel with rotating upper drum, following tape (VHS, 8 mm); (c) rotating upper drum, against tape (C-format).

The frictional forces along the tape are not uniform. There is clearly appreciable friction when the tape initially hits the lower, stationary drum. Further along, the fixed surface area reduces, and the tape is essentially supported by the air bearing formed by the rotating upper drum. Toward the end of its travel around the drum, the air bearing becomes thinner, and sporadic frictional forces appear. The problem is complex and will be analyzed in the next section.

TRACKING

The head needs to stay on track; that always requires a servo mechanism. The VHS format has a track width of 58 microns and a length of 97,400 microns, supported only by the thin flexible tape. This is a very long path for the head to follow, in particular when that path changes due to variations in friction, tension, temperature, and humidity. Before we cover these changes, let us look at ways to tell if the head is on or moving away from a track.

First, the recorded control track along one edge of the tape acts as an electronic perforation analogous to motion picture film. The control playback signal is compared with a reference. Any deviation will produce an error signal to control the tape transport speed in relation to the head wheel rotations so the head will be on-track (see Figs. 15.45).

A couple of methods exist to obtain information about the tracking of a head: Recorded pilot signals, and the AFT. 8 mm recorders use the first method, where four different frequencies are recorded along with the video information. The frequencies change from track to track, and are located below the color-under information.

If the head strays off track, it will pick up the neighboring track frequency, and a difference between the two frequencies is detected. Figure 15.34 lists the four possible combinations that result in two different signals: 16.5 or 46.0 kHz. The detected difference frequency will cause the head to be moved up or down (and/or the transport to speed up or slow down.)

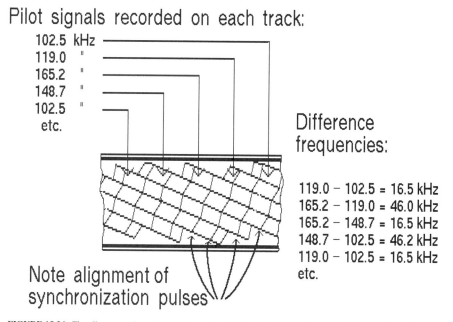

FIGURE 15.34 The alignment of synchronization pulses. Pilot signals are recorded simultaneously with the video information, and are located below the color-under carrier.

R-DAT uses a technique called AFT, area-divided track following, in which separate parts of the track are set aside for track-following. This is shown in Fig. 15.35. A set of four frequencies is used, like in the pilot-track servo method. Amplitude and difference frequencies will inform the head position servo of needed correction.

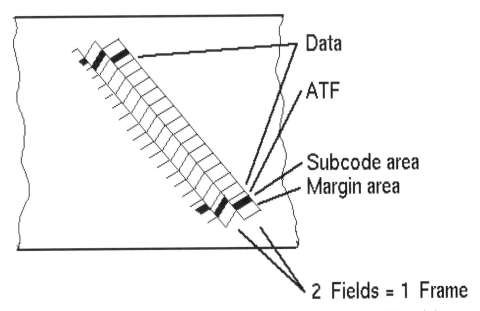

FIGURE 15.35 Each individual slanted track contains information at beginning and end for synchronization and other matters, such as tracking; here they are shown using AFT.

The required motion of the head is made by inserting a bimorph member between the head wheel and the head block, shown in Fig. 15.36. Application of a dc voltage across the barium titanate will cause it to bend, and therefore to deflect the head from its rest position.

Clever application of a dithering motion superimposed on the head during playback will in itself deliver servo information for the head motion. The head will vibrate up and down when an ac voltage is applied to the barium titanate. A 600 Hz signal will provide ten full cycles per track (one field, 60 per second). This will amplitude modulate the playback signal. The phase of the detected signal will have a 180° phase difference between the head being above or below the track. The detected amplitude also will show the amount of correction needed. The amplitude modulation is removed by clipping prior to demodulation of the video signal.

Tracking problems exist during tape speeds different from the nominal speed. Figure 15.37 shows the alternating tracks recorded in an azimuth recording as clear and grey-colored slanted bars. Shown below is the sequencing of track scanning. We assume that the head has correct azimuth no matter what track portion it follows. Starting with the normal speed, the head follows along the heavy black line right through the center of both tracks. Then it goes on to the next set. The picture is perfect.

During standstill, we see that the head crosses from one track set to the next, and the signal is therefore marginal halfway through its travel. The result is a noise bar in the center of the picture. The same thing happens during double speed, while triple speed results in two noise bars, quadruple speed in three noise bars, etc.

MISTRACKING IN HELICAL SCAN RECORDERS

An unusual behavior of tape motion occurs at the low speed employed in consumer products. The frictional forces combined with the elastic properties of tape will cause the tape to move in a jerking

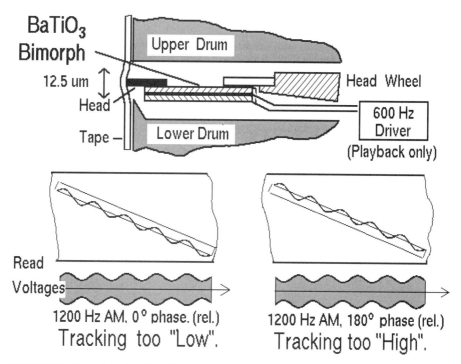

FIGURE 15.36 Dynamic head tracking with a bimorph piezoelectric bender.

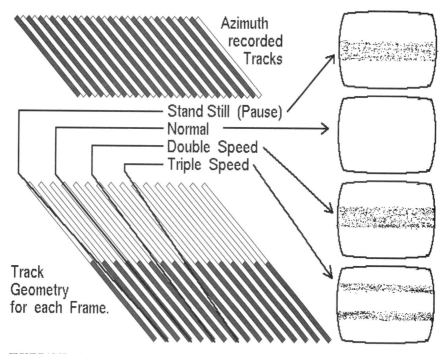

FIGURE 15.37 Video track loci at different speeds. These patterns are seen during fast search on a VCR.

motion, called stick-slip. The critical speed at which this occurs can be figured from the formula shown in Fig. 15.38. It is typically between a few mm/s and a few dozen mm/s of tape speed.

This is the range of linear tape speed in consumer video recorders. We have probably all experienced poor home video picture quality (with timing errors that no time-base corrector can fix), and wow and howling sound quality that no ear can stand.

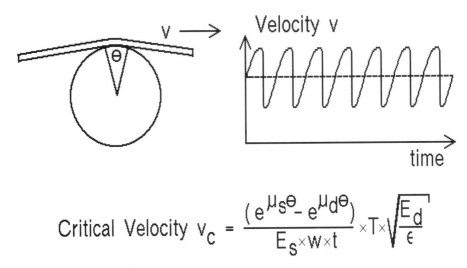

$$\text{Critical Velocity } v_c = \frac{(e^{\mu_s \theta} - e^{\mu_d \theta})}{E_s \times w \times t} \times T \times \sqrt{\frac{E_d}{\epsilon}}$$

FIGURE 15.38 Stick-slip motion of tape occurs at low tape speeds.

Mistracking will also occur due to the differential frictional forces around the head drum. The frictional forces vary in an exponential fashion when a tape moves over and around a cylinder. The formula for the frictional force developed is shown in Fig. 15.39. The constants and variables are listed below the formula. An expression for the tape's elongation is also listed.

A numerical example will illustrate the magnitudes of frictional forces and resulting track displacement. Our example will apply to a VHS recorder, where the track angle is 6°. The tape wrap is 180° around a drum of diameter 62 mm, and the length of the track L'' is therefore:

$$L'' = \pi \times 62/2 = 97.389 \text{ mm}$$

and the corresponding length of the tape stretch L' is:

$$L' = L'' \times \cos 6° = 96.957 \text{ mm}.$$

The width of the tape covered by video tracks is:

$$\begin{aligned} w &= \text{Trackwidth} + L' \times \tan 6° \\ &= 0.058 + 9.165 \\ &= 9.223 \text{ mm} \end{aligned}$$

Assuming an ingoing tension of 3 ounces and a coefficient of friction equal to 0.1, then the increase in tension on the outgoing side is:

$$\begin{aligned} T &= 3 \times \exp(0.1 \times \pi) \\ &= 4.1 \text{ oz force} \\ &= 1.14 \text{ N} \end{aligned}$$

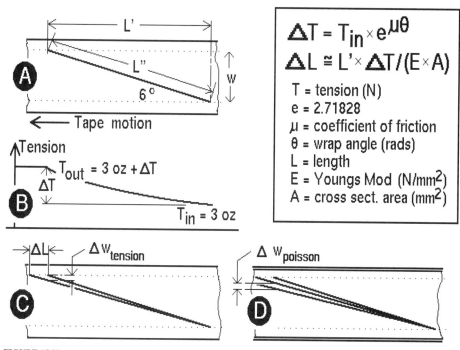

FIGURE 15.39 Distortion of track dimensions, due to friction or temperature/humidity.

(The conversion factors are: 1 oz force = 0.278 N and 1 N = 3.60 oz force.)
This increased tension will, on the average, cause an elongation of the tape equal to

$$
\begin{aligned}
L &= L' \times T/(E \times A) \\
&= 96.957 \times 1.14/(5000 \times 12.7 \times 0.025) \\
&= 110.709/1588 \\
&= 0.0697 \text{ mm}
\end{aligned}
$$

Figure 15.39c shows how this elongation will cause a lowering of the track by w, which is found by comparing two triangles:

$$
w/L = w/L', \text{ or}
$$

$$
\begin{aligned}
w &= Lw/L' \\
&= 69.7 \times 9.223/96.957 \\
&= 0.0066 \text{ mm}
\end{aligned}
$$

Another, smaller mistracking occurs simultaneously due to the Poisson ratio (0.3) of the tape's base film; see Fig. 15.39d. By geometry considerations we find:

$$
\begin{aligned}
w' &= 0.3 \times w \times T/(5000 \times 12.7 \times 0.025) \\
&= 0.0020 \text{ mm}
\end{aligned}
$$

The total track misplacement is the sum of w and w', which is equal to 0.0086 mm, or 15% of the track width. These changes are difficult to compensate, because the coefficients of friction between tapes and the metal surfaces of drums, guides, and heads vary a great deal. The coefficient of friction increases as a tape is worn, for the static (start-stop) and dynamic value as well.

Humidity changes cause a dramatic change in the coefficient of friction, as shown in Fig. 15.40 center. There is no method to compensate for these changes, as one can implement for expansions and contractions due to temperature. Finally, it should be mentioned that lubricants play a major role in the mistracking problems in recorders.

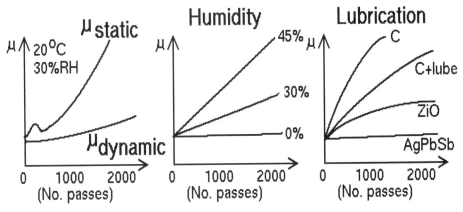

FIGURE 15.40   Changes in coefficient of friction due to tape wear, humidity, or temperature.

A detailed analysis of the tracking problems in a C-format transport is outlined in Fig. 15.41, where a finite element analysis includes all stresses resulting from the frictional forces around the scanner. Variations in the scanner drum friction, or in tape tension, will cause variations in tape displacements around the scanner. Figure 15.42 shows the computed resulting mistracking or track displacement (Cole 1989, see bibliography).

Another recent analysis, expanding on the sample calculation presented in Fig. 15.39, shows the variation in tension around the scanner, Fig. 15.43, which is similar to B in Fig. 15.39. Displacements caused by temperature or humidity changes are shown in Fig. 15.44 (Zahn 1989, see bibliography).

SERVO SYSTEMS

A simplified servo system for a helical-scan recorder is shown in Fig. 15.45. The separate tape transport and scanner motors are individually synchronized to an external source during recording. This source can be the incoming video signal or the studio genlock signal. A genlock signal makes certain that all pieces of video equipment in a studio are in phase together.

Both motors generate tachometer signals that are compared with a derived sync signal so they are running at the proper speed. The sync signal is recorded on a longitudinal track for use during playback.

The servo will, during playback, make certain that the head wheel motor and capstan motor operate at the proper speeds. The sync signal is played back and compared with a reference. This can be the genlock signal or an internal reference in the recorder. Fine adjustment via a tracking control is possible. Any speed deviations will be detected in the phase detectors, and will generate error signals to the motor amplifiers.

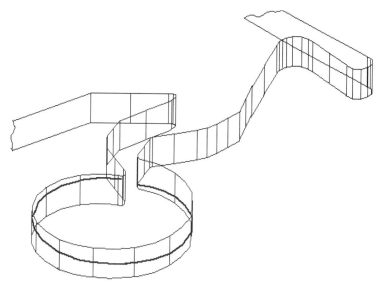

FIGURE 15.41 Computed stresses in tape around a scanner (*after Kevin Cole, Eastman Kodak Co.*).

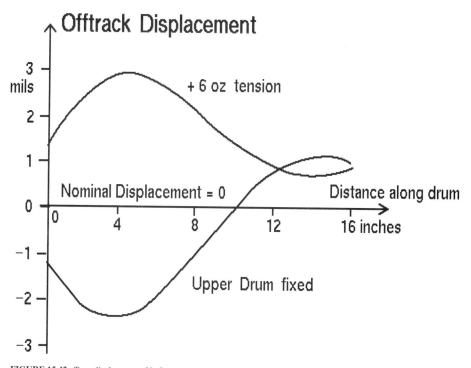

FIGURE 15.42 Tape displacement: Up for an additional 6 oz. of tension, down for an increase in upper-drum friction.

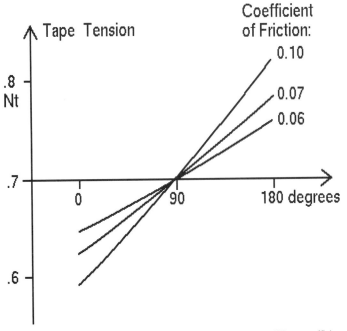

FIGURE 15.43 Tape tension around a drum at midspan tension .7 N, at different coefficients of friction (*after Zahn*).

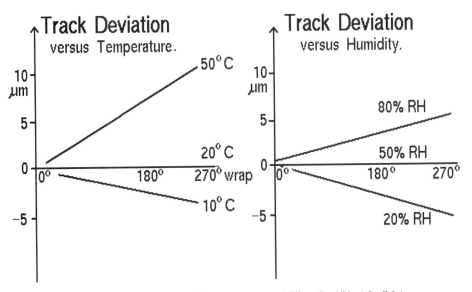

FIGURE 15.44 Track deviation as a function of different temperatures and different humidities (*after Zahn*).

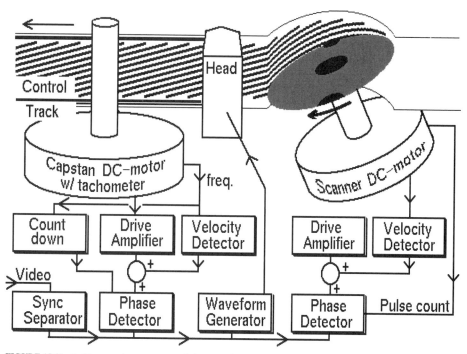

FIGURE 15.45 A video recorder servo system during recording.

REFERENCES

GENERAL INTEREST

Becker, J.C., D.A. DiTommasok and S.J. Mortensen. March 1985. Low Cost, Highly Reliable Tape Backup for Winchester Disc Drives. *HP Journal* 36 (3): 34–35.

Curtis, H.L., and R.L. Turley. March 1985. Development of a High Performance, Half-Inch Tape Drive. *HP Journal* 36 (3): 11–16.

Cutler, D.S. April 1982. Design Considerations for Streaming Tape Transports. *Intl. Conf. Video and Data*, IERE Publ. no. 54, pp. 113–125.

Dong, J.W., D.J. Van Maren, and R.D. Emmerich. March 1985. Streaming Tape Drive Hardware Design. *HP Journal* 36 (3): 25–29.

Ruska, D.W., V.K. Russon, B.W. Culp, A.J. Richards, and J.A. Ruf. March 1985. Firmware for a Streaming Tape Drive. *HP Journal* 36 (3): 29–31.

SERVO-CONTROLLED TAPE SPEED

Alexander, A.G., and C.C. Ling. Feb. 1967. Calculating the Natural Frequencies of Rotating Machine Elements. *Machine Design* pp. 137–140.

Diamond, A. July 1965. Inertially Damped Servomotors: Performance Analysis. *Electro-Technology* pp. 28–32.

Hu, P.Y. May 1974. Vibration Characteristics of the Printed-Circuit Motor. *Jour. Engr. for Industry* pp. 541–46.

Law, E. 1976. Frequency Response of Tape Transport Servo System. *ITC Conf. Proc.* 12: 196–201.

Wente, E.C., and A.H. Mueller. Oct. 1941. Internally Damped Rollers. *Jour. SMPTE* 50 (10): 406–17.

CAPSTAN AND PINCH ROLLER

Clurman, S. Sept. 1971. On Hunting in Hysteresis Motors and New Damping Techniques. *IEEE Trans. Magn.* MAG-7 (3): 512–17.

Durieu, J., and M. Petit. April 1984. A 2-D Solution of the Contact Problem in the Capstan/Tape/Roller Mechanism of Magnetic Recorders. *Computer Methods in Appl. Mech. and Engr.* 43 (1): 21–36.

QIC DRIVES

Gills, D. Aug. 1989. Reliability Assessment of a Quarter-Inch Cartridge Drive. *HP Journal* 40 (4): pp. 74–78.

Schwartz, T. Sept. 1992. QIC Migration Path. *1st Intl. Symp. on BaFeO & Adv. Recording*, Kalamata, Greece. 24 viewgraphs.

Smith, D., and R. von Behren. Oct. 1986. Squeeze-Film Analysis of Tape Winding Effects in Data Cartridge. *ASLE Spec. Publ. on Tribology and Mechanics of Magn. Storage* Vol. 6, pp. 88–92.

Smith, D.P. June 1989. Reliability Issues in Data Tape Cartridges. *Proc. of the 5th Intl. Congr. on Tribology* Vol. 3, pp. 216–221.

Topham, A.D. Aug. 1989. Mechanical Design of a New Quarter-Inch Cartridge Tape Drive. *HP Journal* 40 (4): 67–73.

von Behren, R.A. July 1989. Data Cartridge Theory of Operation. *Tech. Services Prod. Bull. 3MCo*, 2 pages.

von Behren, R.A., "Belt-Driven Tape Cartridge", *U.S. Patent* 3,692,255, 1971.

REELING AND WINDING OF TAPES

Clurman, S., Nov. 1981. A Simple Tape Wrap Around a Guide: Some Complexities. *IEEE Trans. Magn.* MAG-17 (6): 2754–56.

Eshleman, R.L., A.P. Meyers, W.A. Davidson, R.C. Gortowski, and M.E. Anderson. 1973. Feasability Model of High Reliability Five-Year Tape Transport, *IITRI Proc. E6225*, NASA contract NAS5-21692, 3 volumes.

Hartmann, G. Feb. 1958. Zur Bremsvorgang bei Magnettongeraeten. *Elek. Rundschau* 2 (2): 45–49.

Holcomb, A.L. Jan. 1952. Film-Spool Drive with Torque Motors. *Jour. SMPTE* 58 (1): 28–35.

Mote, C.D. Dec. 1965. A Study of Band Saw Vibrations. *Jour. Franklin Inst.* 279 (12): 430–44.

Mote, C.D. Jr., and W.Z. Wu. Sept. 1985. Vibration Coupling in Continuous Belt and Band Systems. *Jour. Sound and Vibr.* 102 (1): 1–10.

Paroby, W., and R. DiSilvestre. 1975. Tape Tracking and Handling for Magnetic Tape Recorders. *Intl. Telemetry Conf. Proc.* Vol. 11, pp. 345–57.

Pfeiffer, J.D. Dec. 1977. Web Guidance Concepts and Applications. *Tappi* 60 (12): 53–60.

Thurman, A.L., and C.D. Mote, Jr. March 1969. Free, Periodic, Nonlinear Oscillation of an Axially Moving Strip. *Jour. Appl. Mech.* pp. 83–91.

DESIGN FOR LOW FLUTTER

Jacoby, G.V. Jan 1963. The Design of a Magnetic-Recording-Tape Transport for Very-High Timing Accuracy. *IEEE Trans. Comm. and Ctrl* pp. 491–99.

Olson, H.F. 1966. *Solutions of Engineering Problems by Dynamical Analogies.* Van Nostrand 277 pages.

Wolf, W. June 1966. Electromechanical Analogs of the Filter System Used in Sound Recording Transports. *IEEE Trans. Audio and Electroacoustics* AU-14 (2): 66–85.

BIBLIOGRAPHY

GENERAL INTEREST

Adams, R.J. Nov. 1992. The Influence of Rolls and Reels on Flutter and Windage. *Tappi* 75 (11): 215–22.

Dong, J.W., K.A. Proehl, R.L. Abrahamson, L.G. Christie, Jr., and D.R.Dome. June 1988. A Reliable, Autoloading, Streaming Half-Inch Tape Drive. *HP Journal* 39 (3): 36–54.

Harris, J.P., Phillips, W.B., Wells, J.F., and W.D. Winger. Sept. 1981. Innovations in the Design of Magnetic Tape Subsystems. *IBM Jour. Res. & Dev.* 25 (5): 691–99.

Jorgensen, F. Oct. 1984. Mechanical Design Considerations in Magnetic Tape and Disk Drives—A Tutorial Review. *ASLE Spec. Publ. on Tribology,* (1): 1–8.

Totzek, U., D. Preis, and J.F. Boehme. April 1987. A Spectral Model for Time-Base Distortions in Magnetic Recording. *Archiv fuer Elek. Uebertragung* 41 (4): 223–231.

Wildman, M. Sept. 1974. Mechanical Limitations in Magnetic Recording. *IEEE Trans. Magn.* MAG-10 (3): 509–14.

Winarski, D.J. et al. Mechanical Design of the Cartridge and Transport for the IBM 3480 Magnetic Tape Subsystem. *IBM Jour. Res. & Dev.* 30 (6): 635–43.

HELICAL-SCAN RECORDERS

Broese van Groenou, A., J.G. Fijnvandraat, S.E. Kadijk and J.P.M. Verbunt. Oct. 1989. The Shape of Heads for Tape Recording: Experiments and Calculations for Various Tapes. *ASLE Spec. Publ. on Tribology and Mechanics of Magn. Storage* Vol. 6, pp. 120–27.

Cole, K., Jan. 1989. Two-Dimensional Rotary Recorder Transport Analysis Using Finite Elements. *THIC Meeting* 18 pages viewgraphs.

Coleman, C.H. March 1971. A New Technique for Time-Base Stability of Video Recorders. *IEEE Trans. Broadcast* BC-17.

Desserre, J.R., and Y. Toyzier. April 1990. Non-destructive Procedure for Viewing the Formats on Tape. *Intl. Conf. on Video, Audio and Data Rec.*, IEE Publ. No. 319, pp. 26–31.

Doyama, Y., K. Kubo, K. Yamada. Aug. 1986. Head Actuator Servo for VTR Using 4 Frequency Pilot Signals. *IEEE Trans. Consumer Electronics* CE-32 (3): 388–97.

Eguchi, T. Feb. 1987. The SMPTE D-1 Format and Possible Scanner Configurations. *SMPTE Journal* 96 (2): 166–70.

Eshel, A. Oct. 1987. On Head-to-Tape Interface Phenomena in Helical Scan Video Recording. *ASLE Spec. Publ. on Tribology and Mechanics of Magn. Storage Vol.* 4, pp. 68ff.

Eshel, A. Oct. 1990. Simulation of Head Contour Effects in Helical Scan Video Recording. *ASLE Spec. Publ. on Tribology and Mechanics of Magn. Storage* Vol. 7, pp. 21–24.

Fujiwara, Y., T. Eguchi, and K. Ike. Sept. 1974. Tape Selection and Mechanical Considerations for the 4:2:2 DVTR. *SMPTE Journal* 83 (9): 818–29.

Gregory, S. 1988. *Introduction to the 4:2:2 Digital Video Tape Recorder.* Pentech Press, London 200 pages.

Jacobi, H., and U. Nowak. Oct. 1990. The Influence of Al_2O_3-Particles on the Tribological Behaviour of Gamma-Fe_2O_3-Particulate Video Tapes. *ASLE Spec. Publ. on Tribology and Mechanics of Magn. Storage* Vol. 7, pp. 114–22.

Kaku, N., S. Ozaki, S. Yokoo, T. Ozawa, Y. Niguchi, H. Ono, K. Ogiro, and H. Yokota. Aug. 1989. Mechanical Considerations in the Design of a Composite Digital VTR. *SMPTE Journal* 98 (8): 568–574.

Kocherscheidt, G., and H. Thiemer. May 1985. Capstan Drive for the Smallest Professional Video-Tape Cameras. *F&M Feinwerktechnik & Messtechnik* 93 (3): 169–71.

Kuga, T., T. Hattori, Y. Yuzawa, M. Kataoka, Y. Sato, M. Katsuki, Y. Sawamoto, and K. Ezure. March 1992. Development of Mechanism for VHS Camcorder. *Sharp Tech. Jour.*, No. 52, pp. 43–46.

Lacey, C.A., and F.E. Talke. Sept. 1990. Tape Dynamics in a High-Speed Helical Recorder. *IEEE Trans. Magn.* MAG-26 (5): 2208–10.

Liu, Y., H. Chen, J. Yang, and J. Yi. Sept. 1992. An Accurate Measuring System for VTR Magnetic Track. *IEEE Trans. Magn.* MAG-28 (5): 2473–75.

Ogiro, K., N. Kaku, K. Masuda, and K. Nagai. Aug. 1986. A Precision Mechanism for High Density Magnetic Recording of 8 MM VCR. *IEEE Trans. Consumer Electronics* CE-32 (3): 379–87.

Osake, H., K. Fukushi, and K. Ozawa. Nov. 1990. Wear Mechanisms of Metal-Evaporated Magnetic Tapes in Helical Scan Videotape Recorders. *IEEE Trans. Magn.* MAG-26 (6): 3180–85.

Ryan, D.M. 1978. Mechanical Design Considerations for Helical-Scan Video Tape Recorders. *SMPTE Journal* 87 (?): 767ff.

Sakai, K., T. Terayama, N. Okamoto, and H. Tokota. Oct. 1987. Clearance Characteristics Between Magnetic Tapes and Heads of Video Cassette Recorders. *ASLE Spec. Publ. on Tribology and Mechanics of Magn. Storage* Vol. 4, pp. 61ff.

Sievers, L., M.J. Balas, and A. von Flotow. June 1988. Modeling of Web Conveyance Systems for Multivariable Control. *IEEE Trans. Autom. Control* 33 (6): 524–31.

Sundaram, R., and R.C. Benson. Oct. 1989. A Green's Function with Improved Convergence for Cylindrically Wrapped Tapes. *ASLE Spec. Publ. on Tribology and Mechanics of Magn. Storage* Vol. 6, pp. 98–110.

Sundaram, R., and R.C. Benson. Sept. 1990. Tape Dynamics Following an Impact. *IEEE Trans. Magn.* MAG-26 (5): 2211–13.

Talke, F.E., and R.C. Tseng. Nov. 1976. Submicron Transducer Spacing in Rotating Head/Tape Interfaces. *IEEE Trans. Magn.* MAG-12 (6): 725–27.

Uchiyama, S., T. Suezawa, T. Eto, T. Osaki, N. Sugihara, and M. Katsuki. June 1991. Development of Small R-DAT Mechanism. *Sharp Tech. Jour.* No. 49, pp. 37–40.

van Groenau, A.B., and H.J.J. Meulenbroeks. Oct. 1988. On The Laws of Head Wear in Video Tape Recording: The Influence of Pressure and Time. *ASLE Spec. Publ. on Tribology and Mechanics of Magn. Storage* Vol. 5, pp. ?.

van Groenou, A.B., and H.J.J. Meulenbroeks. Oct. 1987. Tests on Materials for Video Heads by CrO_2 Tapes in the Sphere-on-Tape Apparatus. *ASLE Spec. Publ. on Tribology and Mechanics of Magn. Storage* Vol. 4, pp. 152ff.

van Groenou, A.B., H.J.J. Meulenbroeks, and R.C.F. Schaake. Oct. 1987. Wear by Magnetic Tapes in Sphere-on-Tape Apparatus *ASLE Spec. Publ. on Tribology and Mechanics of Magn. Storage* Vol. 4, pp. 143ff.

Wada, Y. 1973. Track Straightness in Helical Scan VTR. *Intl. Conf. on Video, Audio and Data Rec.* IERE Publ. No. ?, pp. 51–60.

Wada, Y. Dec. 1975. Track Straightness in Helical Scan Video Tape Recorders. *SMPTE Journal* 84 (12): 954–58.

Watkinson, J. 1991. *RDAT.* Focal Press 244 pages.

Williams, C.H. April 1990. The Measurement and Classification of the Impairment for DVTR Transports. *Intl. Conf. on Video, Audio and Data Rec.* IEE Publ. No. 319, pp. 67–78.

Yamaguchi, T. Aug. 1985. A New Tape Path Analysis for all VCR Tape Transport Systems. *IEEE Trans. Consumer Electronics* CE-31 (3): 398–404.

Yoneda, K., and T. Sawada. Nov. 1988. Simulation of Tape Flying Characteristics Above VTR Drum Considering In-plane Stress. *IEEE Trans. Magn.* MAG-24 (6): 2766–68.

Zahn, H.L. June 1988. Friction and Tolerance Relations between Tape and Drum in a Rotary Recorder. *THIC Meeting* 30 pages.

Zahn, H.L. July 1989. Friction—Its Influence in Rotary Magnetic Tape. *SMPTE Journal* 98 (7): 520–524.

ASSORTED PAPERS

Bayer, R.G., and N.G. Payne. Feb. 1984. Wear Evaluation of Molded Plastics. *Lubrication Engr.* 41 (5): 290–93.

Bogy, D.B., and F.E. Talke. 1986. Mechanics-Related Problems of Magnetic Recording Technology and Ink-Jet Printing. *Appl. Mech. Rev.* 39 (11): 1665–77.

Goosey, M.T. (Ed.). 1985. *Plastics for Electronics.* Elsevier Appl. Sci. Publ. Ltd. (Essex) 380 pages.

Mathew, N.M. Jan. 1991. Ozone Damage of Rubber and Its Prevention. *Fractography of Rubbery Materials.* pp. 217–246.

Swavely, D.S. 1991. Finishing and Machining Plastics. *Plastics Engineering Handbook.* pp. 657–92.

Uchiyama, Y.L. Aug. 1986. The Effect of Environment on the Friction and Wear of Rubber. *Wear* 110 (3–4): 369–78.

Yu, J.S., A.H. Wagner, and D.M. Kalyon. Jan. 1992. Simulation of Microstructure Development in Injection Molding of Engineering Plastics. *Jour. Appl. Poly. Sci.* 44 (3): 477–89.

Zilling, F. April 1987. Measuring the Roughness, Straightness and Parallelism in Small Boreholes. *FM Feinwerktechnik & Messtechnik* 95 (3): 199ff.

CHAPTER 16
DISK DRIVES

Disk drives (rotating memories) provide fast access to a very large number of data records; this is in contrast to tape drives, where the tape first is wound to the location of the data.

Rotating memories were born in 1957. This was the IBM 350 disk drive, consisting of 100 fixed recording surfaces of 24 inches in diameter. There were 20 tracks in parallel per inch radius (20 tpi), and the bit density was 100 bpi, so the areal packing density was 2000 bits per square inch. The data rate was 8.8 kbps (Harker et al.).

In the 1960s came the *disk pack*, which contained a number of gamma ferric oxide-coated aluminum disks upon which data were stored. These packs were removable from the drive and head mechanisms. Limited data capacity made the users revert to drives with permanently mounted disks. The magnetic oxide coatings were replaced with the much thinner magnetic thin films. These drives have evolved into today's marvels. One example was announced in August 1994 by IBM: a 17 mm-high drive with three 2½-inch platters, with a capacity of approximately 800 MB. It is expected to be up to 1.6 GB by spring 1995.

One can make a curious observation: Densities in 1980 were around 10 MB/in^2. At the end of the 1990s we can anticipate 1000 MB/in^2, a hundredfold increase as depicted in Fig. 2.4 (areal density). The decreased size of disk drive is accompanied by smaller components. The head assembly that holds the heads (TFH, MR) and provide gliders for the flying function has decreased in length, from 4 to 1 mm. The width and height have also shrunk by a factor of four, and the net volume reduction is $4 \times 4 \times 4 = 96$ times! So in disk drives, the rule is: "small is good."

The speed of development is dazzling at times, and is the result of progress in materials, processes, servo designs, and precision manufacturing rather than any novel write/read techniques.

The drives are very advanced electromechanical assemblies. A typical small disk drive is shown in Fig. 16.1, and its main components are displayed in Fig. 16.2. Figure 16.3 shows how the 3½-inch drives have gained in popularity, while the 2½-inch drives are exhibiting slower growth, reflecting conditions in the laptop market.

Data on a disk are stored in circular tracks (see Fig. 16.4). The disks range in diameter from 35.6 cm (14 inches) down to 3.3 cm (1.3 inches), rotating at speeds from 7200 to 300 rpm. The recorded information of the disk surface is divided into sectors. Servo information is recorded in the radially continuous narrow wedges between sectors. The information in the servo sections are:

- Track number
- Sector number
- Tracking information

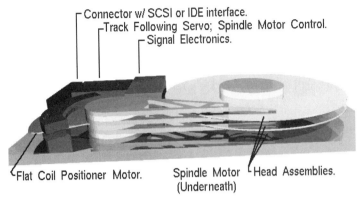

Connector w/ SCSI or IDE interface.
Track Following Servo; Spindle Motor Control.
Signal Electronics.

Flat Coil Positioner Motor.   Spindle Motor   Head Assemblies.
                              (Underneath)

FIGURE 16.1   A rigid disk drive with two platters and four heads.

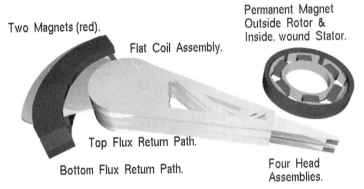

Two Magnets (red).

Permanent Magnet
Outside Rotor &
Inside. wound Stator.

Flat Coil Assembly.

Top Flux Return Path.

Bottom Flux Return Path.

Four Head
Assemblies.

FIGURE 16.2   Magnetics in a disk drive: disk(s), head(s), spindle motor, and actuator motor.

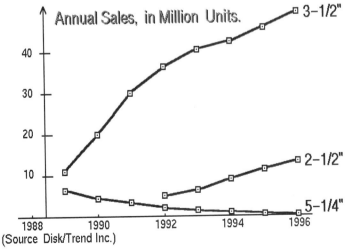

Annual Sales, in Million Units.   3-1/2"

40

30

20

10

                                   2-1/2"

                                   5-1/4"

1988   1990   1992   1994   1996
(Source Disk/Trend Inc.)

FIGURE 16.3   Changes in production of three popular disk drive sizes.

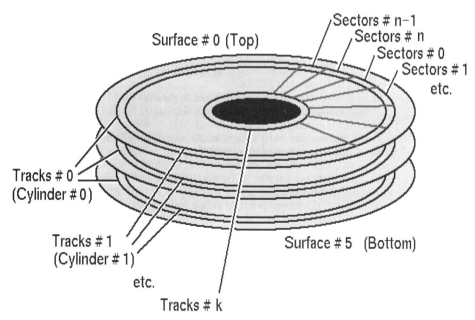

FIGURE 16.4 Data are recorded in sectors along concentric tracks. All tracks with the same number make a "cylinder."

A 3½-inch disk will have between 64 and 128 sectors. The outer track length is approximately 10 inches. Seven to ten percent thereof is used for servo information, leaving 9.3 to 9 inches total track length for data.

The inside track length may be as small as four inches. Zone recording is used to fully utilize the disk area. The data rate is 2.5 times larger at track 0 than at the inside track. There may be 10 to 16 zones across the disk. The data rate increases toward the outside tracks to assure fairly constant, maximum packing density.

We will first treat the flexible- and rigid-media drives, and then describe linear and rotary actuators for moving the heads. At low track densities it is relatively easy to position the head; at higher densities, it is necessary to use a track-following technique by using a closed-loop servo.

Overall performance of a disk drive is measured by its seek time, which is the average time it takes for a head to move from one track to another (including the time to settle over the track) plus the latency, which is the time for one revolution (some may quote the time for one-half revolution as the average latency). One should also consider the bit rates for the data that may be correlated with the interface standards, SCSI or IDE.

Space limitations will at times make this chapter a tutorial. The interested reader can gain further insight by studying earlier papers like one by Harker et al. on early disk drive development, or by Workman on digital servo design. Sierra's book has much useful information, while many details are described in two chapters in Mee and Daniels' *Magnetic Recording*. IDEMA (International Disk Drive Equipment and Materials Association) continues to develop and publish standards.

FLEXIBLE DISK DRIVES AND REMOVABLE RIGID DISKS

The 8-inch-diameter flexible disk was introduced in 1973 by IBM as a device for loading control microprograms and diagnostics. Since then the applications have broadened to distribution and exchange of data, programs, microcode, diagnostic procedures, and other digital information, and as a removable disk in system files (Engh).

The 5¼-inch diskette followed in 1976. The 3½-inch format made its strong and lasting entry in 1982. Their current capacities are 0.72, 1.44, and 2.88 MB, with corresponding transfer rates of 0.25, 0.5 and 1 MHz. Track densities are 48 or 135 tpi. Extended performance is achieved in the Bernoulli drive by Iomega Corp., and in the Floptical drive by InSite Peripherals.

Many design details for flexible disk drives are similar to those of a tape drive, such as signal losses, friction, and wear. Different are the vibration modes, which now correspond to those of a circular membrane, clamped at the center and free at the edges.

A high areal packing density on flexible disks can be achieved by applying a large number of concentric tracks, measured in tracks-per-inch (tpi). This design effort is hampered by the mechanical tolerance buildup. The exact center location of a diskette varies from drive to drive. Add to this the changes in disk dimensions with changes in temperature and in humidity. These matters were discussed in the beginning of Chapter 10, about selection of track widths.

Flexible disks are circular PET disks, typically of thickness 75 mm (3 mils), and packaged as shown in Figs. 16.5 and 16.6. The 8- and 5¼-inch formats are housed inside a PVC jacket, while the 3½-inch format is enclosed in a rigid plastic housing. The 5½-inch formats for high packing densities are further enclosed in a rigid structure, or inside a cartridge.

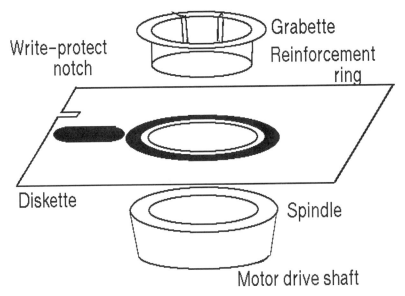

FIGURE 16.5 Clamping of a 5¼-inch flexible diskette to the driveshaft.

The rotary motion of the disk at 300–360 rpm is achieved by clamping the disk at the location of the center hole. This may be done as shown in Fig. 16.5, where the disk ends up seated against a spindle, centered and clamped by a cone-shaped clamp. This clamping is done in conjunction with closing the door to the disk drive. The centering of the diskette may not always be perfect upon the first clamping, due to friction; it is recommended to center the disk manually, as best can be done, and then to double clamp; that is, close then open and close the drive gate one extra time after the disk has been inserted.

The clamping of the 3½-inch diskettes is done by a magnet that is brought up into contact with the center metal piece, and a pin on the drive spindle interlocks with the positioning hole in the center piece (Fig. 16.6). This ensures repeatable positioning of the track's center, drive to drive.

Clamping and centering of disks in the disk cartridges require both precision and strong clamping because they operate at 1500–1800 rpm, and dynamic balancing is important.

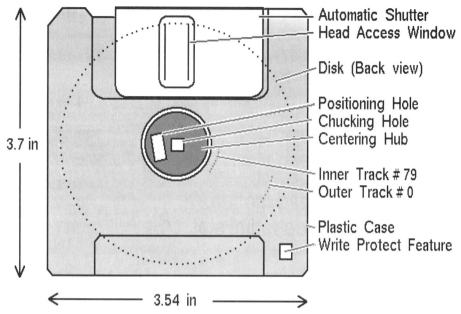

FIGURE 16.6 Dimensions of the 3½-inch diskette.

A hybrid floppy/rigid drive exists in the Bernoulli drive from Iomega Corp. A cookie, which is the double-sided, coated floppy disk without its protective jacket, is housed inside a rigid plastic shell. A cover is slid away when this shell is inserted in the drive mechanism. The cookie is grabbed at its center and placed into contact with a polished plane, then brought to rotation at 2000 rpm. Air is brought in at the center and a Bernoulli bearing formed by the outward airflow (M. Carpino). The result is a cookie that now has very stiff characteristics and a head can be placed against it to perform write/read operations; Fig. 16.7 (K.C. Tripathi).

A cookie rotating by itself will undulate and change geometries in a most undesirable way, as was learned during early studies of optical recording systems (R.T. Pearson). The Bernoulli drives employ servo methods from the rigid-drive systems. A capacity of 50 to 150 MB per 5½-inch disk is possible.

A high-density (20 MB) floppy system was introduced in 1989, named the Floptical (J. Godwin). 1530 concentric grooves are inscribed with a high-power laser on one of the surfaces of a 3½-inch magnetic medium (BaFeO cookie). The grooves are spaced at 20 μm (0.8 mil) intervals corresponding to a track density of 1250 tpi; see Fig. 16.11, left.

A main head carriage can be positioned to any of 135 tracks, and a secondary carriage is positioned by a VCM (voice coil motor). An infrared LED illuminates the grooved disk surface in a floodlight configuration through an elongated hole in the center of the dual head assembly. The reflected light beam provides information to the head positioning servo. A dual-head assembly is used to ensure backward compatibility with standard 3½-inch disks.

The removable storage media can also be a rigid disk housed in a shell, as shown in Fig. 16.8 (SyQuest Corp). The shell is front loaded, a magnetic hub holds the disk, and the heads are loaded onto track 000. Thereafter the disk spins up to a couple of thousand rpm (A. Roman). It can be noted that the SyQuest drive is the sole survivor of eighteen makes offered since 1983. Capacities range from 44 to 270 MB.

The 3½-inch floppy systems may be further improved through better media (BaFeO, according to Imamura et al. 1994). This will increase the density along the track toward 35 kbpi. Servo mechanisms may also be explored to increase the track density from 135 to 406 tpi (T. Tsujisawa et al.) or even 542 tpi in the perpendicular mode (T. Sugaya et al.).

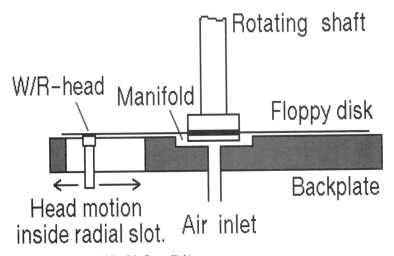

FIGURE 16.7 The principle of the Bernoulli drive.

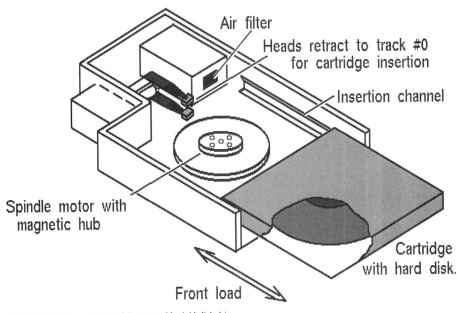

FIGURE 16.8 The principle of the removable rigid-disk drive.

HEAD POSITIONING IN FLEXIBLE DRIVES

The magnetic head must be positioned in contact with the correct track before writing or reading. This is accomplished by mounting the head on an arm that moves radially to the section of the disk allocated to the write/read operation. There are two simple methods for moving the arm, or actuator: The capstan-band and the lead screw, both operated by a stepper motor.

The capstan-band system is shown in Fig. 16.9. The metal band is actually a spring that is fastened to the capstan on the stepper motor shaft. The pliance of the band preloads the positioner arm such that backlash is minimized, and repeat positioning of the head improved. The stepper motors are inexpensive, and operate with steps of typically 1.8 degrees with a 3-percent accuracy (Kenjo). The lead screw was the earliest method used; it is shown in Fig. 16.10. The screw is powered by a stepper motor, which rotates the screw in small equal steps. These steps are translated into a linear movement of the head assembly. Tracking and maximum track widths are discussed by Resnik, while the first 3½-inch disk drive is described by Katoh et al.

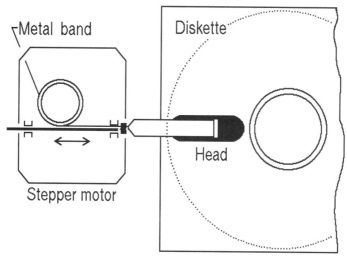

FIGURE 16.9 Head-positioning mechanism in a floppy disk drive, using a stepper motor with flexible-band coupling.

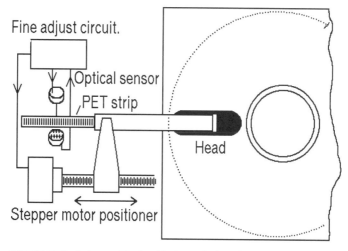

FIGURE 16.10 An improved head-positioning mechanism using coarse and fine positioning stepper motors, plus compensation for humidity changes.

RIGID DISK DRIVES

Rigid disk drives operate at much higher speeds (2400 to 7200 rpm) and track densities (1000 to 5000) than do flexible diskettes. This places higher demands on mechanical rigidity and tight tolerances; the disk, or disks, are permanently installed on a spindle inside the drive.

A rigid disk drive in the office environment is in a constant run mode, in contrast to laptop drives that only rotate during write and read cycles. The heads in a rigid disk drive are flying over the surface, except for start and stop, in which they take off or land on designated areas on the disk. Some designs do not have landing areas, but rather the heads are lifted away from the disks when the disks are not rotating.

Disk surface finish, flatness, and dynamic stability are usually considered to be the most critical mechanical parameters in a disk drive. However, there are also the aerodynamic problems of mechanical as well as thermodynamic nature, generated by the airflow around the spinning disk(s).

Any kind of large-scale unsteady flow or pressure fluctuation is a random excitation source for vibration in the disk and positioner-arm assembly. These relative motions between disk and head introduce amplitude variations and off-track errors. Severe disk flutter causes data losses that are due to slider-disk interference (magnetostrictive impact losses) or even head crashes.

A nonuniform flow will also generate heat, and thermal expansion results. The heat energy generated in an eight-platter 14-inch disk drive can be as high as 50 watts.

The amount of disk flutter and actuator arm vibration depends upon several factors (Lennemann):

- The flutter may increase by a factor of five when two disks are used instead of one. The increase may be up to 20 for more disks.

- The form and dimensions of the disk drive housing will heavily influence the disk flutter.

- Steady airflow reduces disk flutter by an order of magnitude, and aids in cooling the disk(s).

- Proper form of the actuator arm and head assembly is important: A spoiler between disks prevents the development of unsteady airflow.

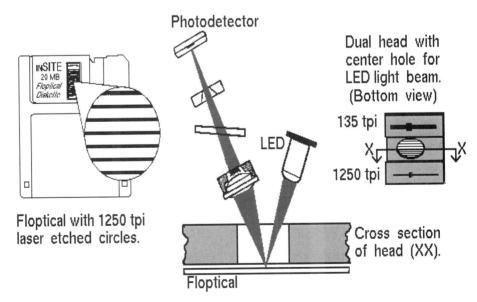

FIGURE 16.11   A floptical disk (a pre-embossed flexible magnetic disk that provides optical servo control for locating the magnetic track).

Many drives use 8-inch disks that produce far less air turbulence than the 14-inch disks do. The result is less disk flutter, arm vibration, and generated heat. Also, 10-inch disks are advantageous in this respect (Mizoshita and Matsuo).

Fairly large motors are used to drive the precision spindle holding the disk(s). A high starting torque is needed to overcome the static friction from the often high number of heads resting on the disks (in landing zones). Hence brushless dc-motors are used, often mounted inside the spindle (Fig. 16.12), or underneath the base plate casting (Fig. 16.13). Preloaded and sealed ball bearings are used to support the spindle shaft in both ends; some drives actually have spindle shafts that are supported fully in both ends to avoid any cantilever effects.

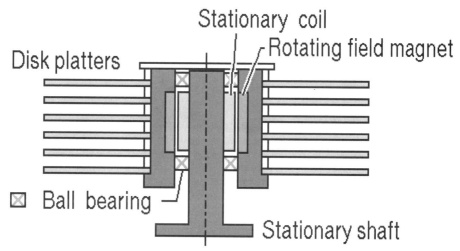

FIGURE 16.12 An in-hub motor mount.

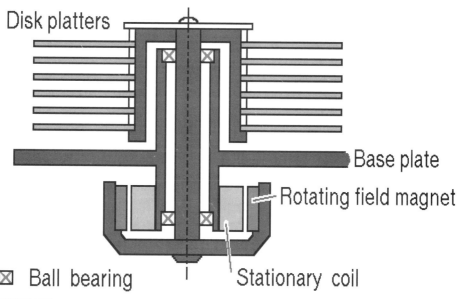

FIGURE 16.13 An underslung motor mount with a rotating magnet on the outside.

An enlightening design outline for the classic IBM 3340 disk drive is found in Mulvany's paper (1974), and a discussion of disk file manufacturing in a paper by Mulvany and Thompson (1981). A paper by Kaneko and Koshimoto (1982) describes an early 3200 MB, 8-platter, 8-inch disk drive with a density of 15.5 MB/in^2.

Design of a rigid disk drive includes considerations for shock and vibration. Already mentioned are the landing zones for heads, or a mechanism that moves the heads away from the disk when it is not rotating at normal speed. Another measure to be taken is the inclusion of a clamp that prevents the disk(s) from rotating during transport of the disk drive.

The choice of disk diameter is crucial in designing highly reliable HDA mechanisms. The decrease in overall size was not done to just satisfy the available room (computer "tower" cases, laptops) but was also done to decrease vibration magnitudes and power requirements (Takanami et al.; Wagner & Colah):

- Vibration of a disk decreases with the 4th power of the disk diameter.
- Rotational windage loss decreases with the 4.6th power of the disk diameter.

Further vibration decrease can be achieved by reinforcing base plates and covers with ribs. Sound-deadening materials can successfully be applied to these and other surfaces (Lee and Singh; Nashif et al.). Windage loss and disk flutter may also be reduced by using a helium atmosphere.

Shock absorbers will, if properly designed, isolate the drive from bumps and vibrations. It is important to use the proper compliance with consideration for allowable sway space. It is worthwhile to consult suppliers of shock mounts plus applicable literature (Chen et al.; Thomson, Nashif et al.). A disk drive should be able to withstand accelerations of 40 Gs when non-operating (such as in shipment), and 10 Gs while operating (Rahimi).

For the ultimate design assurance, one does well in following the guidelines and spirit of the military specification MIL-E-5400. (Caution: strict adherence causes mental breakdown in designers.)

GENERAL SPINDLE MOTOR CONSIDERATIONS

The rotation of disk(s) can be provided by ac as well as by dc motors, through a belt or direct drive where the motor shaft is the spindle shaft. ac motors are simple and reliable, and run at speeds controlled by the line frequency. The latter is a disadvantage, because four models must be provided for to accommodate any mix of 110 or 220 V, and 50- or 60-Hz line frequencies.

The dc motors used today are brushless permanent-magnet motors, and require a dc power supply plus speed-control circuits. These motors can easily be made in a flat package, well suited for the shallow disk drives in use today. They are also much simpler to turn on and off, as may be required for each write or read cycle (laptops).

The principle of a brushless dc motor is shown in Fig. 16.14. The stator windings W1 and W2 are controlled by a Hall element, which in turn is switched by the rotating permanent-magnet rotor. This provides for the commutation required in a dc motor. Modern spindle motors have eliminated the Hall elements, and make use of the back-emf voltage for commutation and speed control. The back-emf is a voltage induced in the winding when the permanent magnet rotates. A modern spindle motor is shown in Fig. 16.15; the individual components are shown in Fig 16.16.

The characteristic of a permanent-magnet dc motor is ideal for disk drives: The starting torque is very high, and it decreases uniformly as the speed increases. The force on each conductor in the winding is:

$$f = BLI \text{ Newtons}$$

where:

B = flux density in Wb/m^2
L = length of wire in m
I = current in A

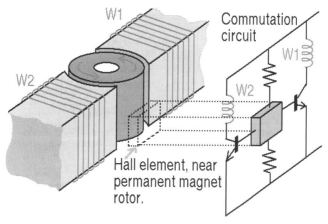

FIGURE 16.14   The use of Hall elements for commutation of coil currents in a brushless dc-motor.

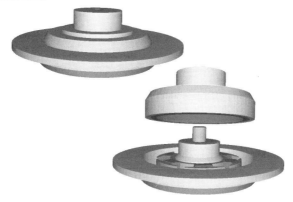

FIGURE 16.15   A flat-spindle motor design.

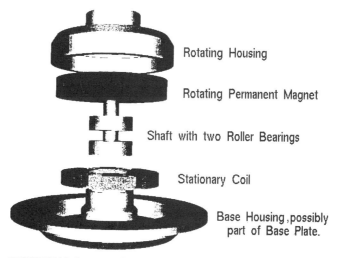

Rotating Housing

Rotating Permanent Magnet

Shaft with two Roller Bearings

Stationary Coil

Base Housing, possibly part of Base Plate.

FIGURE 16.16   Components in a spindle motor.

If the average radius to a wire is R, then the torque T is:

$$T = RBLI \text{ Nm}$$

The torque is proportional to the current I because R, B, and L are fixed. Further development will lead to a torque constant K_T for the motor (unit: Nm/A). The torque is then:

$$T = K_T \times I$$

The induced back emf in each winding is:

$$e = vBL$$

where v is the velocity of the conductor. We will arrive at a back-emf constant K_E that happens to equal K_T (through reciprocity). K_T and K_E are both proportional to the flux density B in the motor field, the winding effective radius R, the number of conductors n, and their length L exposed to the field. The back emf is next determined by:

$$E = K_E\Omega$$

where Ω is the rotational speed (rpm). K_E has unit of V/krpm, (Kenjo and Nagamori).

A motor that uses permanent magnets to supply the field (Fig. 16.17) is represented by the equivalent circuit in Fig. 16.18. Here R_a is the armature winding resistance. We will ignore the voltage drop in the commutating circuitry, and find an equation for V:

$$V = R_aI + K_E\Omega$$

The resulting armature current is:

$$I = (V - K_E\Omega)/R_a$$

and the torque is:

$$\begin{aligned} T &= K_TI \\ &= (K_T/R_a) \times (V - K_E\Omega) \end{aligned}$$

The starting torque T_S is:

$$T_S = K_T \, V/R_a$$

where V is the applied voltage and R_a is the winding resistance. Note the high torque at standstill, exactly what is needed in a disk drive where the frictional force between the head (in the landing zone) and disk is high.

The no-load speed is:

$$\Omega = V/K_E$$

Low battery voltage requires motors with low R_a. Reduction of number of turns will reduce R_a but also K_T. A large cross-section conductor is required. Speed control and switching (commutating) of windings is by Hall sensors placed near the rotating magnet, or by sensing the back emf and extracting the pulsating signal portion that reflects the rpm.

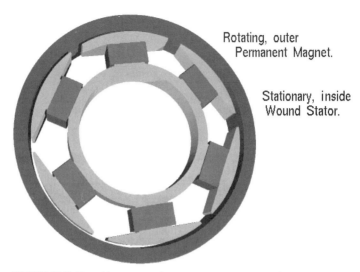

FIGURE 16.17 Essential components in a permanent-magnet motor: The round, stationary armature, and the round, rotating magnet.

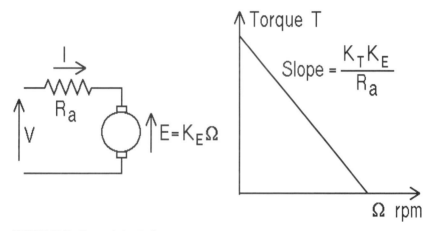

FIGURE 16.18 Characteristics of a dc permanent-magnet motor.

Next-generation spindle motors may feature a printed-circuit stator that will provide further reduction in motor height (Bursky; Jabbar; Cho and Fussell). The general characteristics sought are:

- Very flat design
- Low power consumption
- High torque constant
- High efficiency
- Low run-current
- Low torque ripple
- Low flux leakage
- Low temperature rise
- Low run-out

The motor may be integrated with the base plate (Packard).

A reduced motor height will reduce the precision in tracking, due to the more closely placed bearings for the spindle shaft. Wobble and NRR (nonrepeatable runout) will increase. Current spindle motors with ball bearings have NRR $\approx 0.12 - 0.25$ µm (5 – 10 µin.) at 3600 to 7200 rpm (Huber). Air bearings can provide an order of magnitude of improvement, i.e., NRR ≈ 0.012 µm (0.5 µin.) (Fluid Film Devices).

A proposed spindle motor is shown in Fig. 16.19, where the bearing is a $0.5 - 1$ µm (20 – 40 µin.) air film between two polished ceramic surfaces (Fluid Film Devices, Chihara). This motor has a very high resistance to shock damage, because the stiffness of the bearing increases with the inverse spacing (Gross), and the absence of surface contact while operating avoids wear and wear debris.

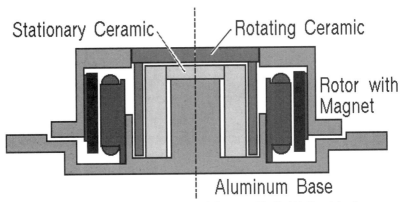

FIGURE 16.19 A proposed high-speed, low-runout spindle motor with a liquid (oil) or air bearing.

HEAD POSITIONING IN RIGID DISK DRIVES

The magnetic head must be positioned in contact with the correct track before writing or reading. This is accomplished by mounting the head on an arm that moves radially, or in an arc, to the section of the disk allocated to the write/read operation. The mechanics, magnetics, and servo system required are described in the following three sections.

In rigid disk drives, many head assemblies and support arms are mounted on one common carriage that is moved as one unit. This is shown in Fig. 16.20, where a four-platter disk drive is shown in the top portion, and the head carriage is moved back and forth by a linear voice-coil motor identical to the "motor" in a loudspeaker.

A disk drive with four heads per platter is shown in Fig. 16.21. Here the heads only have to travel half the distance, which means shorter access time, and write/read operations can be optimized for the inner (low speed) and outer (high speed) heads individually.

The linear actuator, which is very much like a loudspeaker coil/magnet mechanism, was developed in the early 1970s (Brown and Ma; Inoue et al.). The coil inductance is proportional to the square of the number of turns, and it must be low to allow for a rapid current rise so the actuator quickly shifts the heads to new track positions. On the other hand, a certain number of coil turns are necessary to achieve the desired force $f = BLI$, where L is the total wire length in the field B, and I is the drive current. The force in a large (14-inch) disk drive positioner used to be in the range of 20-200 N (Schinköthe). The inductance value can be reduced by a clever design using a shorted turn inside the coil, mounted on the center magnet pole (Fig. 16.22).

The shorted turn will act like a secondary winding on a transformer, with the motor coil being the primary winding. Its action is easiest to understand if we consider the value of the shorted turn as an overcoupled impedance from the secondary. This impedance can be expressed as:

$$(\omega M)^2/(R_2 + j\omega L_2)$$

where $M = k \times \sqrt{(L_1 L_2)}$ is the mutual coupling.

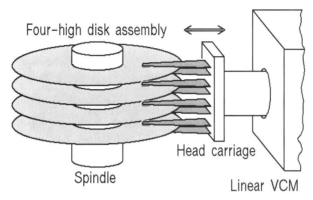

FIGURE 16.20   A linear motor for head assembly positioning.

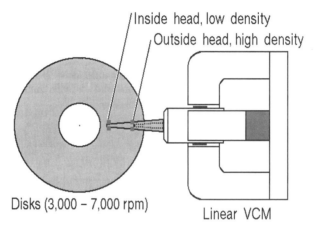

FIGURE 16.21   A linear voice-coil motor with two-zone recording heads.

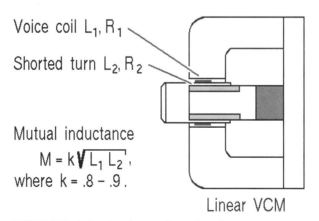

FIGURE 16.22   A shorted turn improves linear voice-coil motor response.

k is the coefficient of coupling, L_1 is the coil inductance and L_2 is the inductance of the shorted turn; R_2 is the resistance of the shorted turn. The total input impedance is:

$$Z = R_1 + j\omega L_1 + (\omega M)^2/(R_2 + j\omega L_2)$$
$$= R_1 + j\omega L_1 + [(\omega M)^2/(R_2{}^2 + \omega^2 L_2{}^2)] \times (R_2 - j\omega L_2)$$
$$= (R_1 + k_2 R_2) + j\omega(L_1 - k_2 L_2)$$

The net result is an increase of resistance and a decrease of the primary inductance, and hence improvement in the rise time of the current. The shorted turn will absorb a fair amount of power, and must have good thermal conductance to the motor structure, serving as a heat sink. Improvements in the response time range from 15 to 40 ms down to 10 to 15 ms (Wagner 1982 and 1983). A detailed analysis of the primary impedance will include the overcoupled impedance and eddy current losses in the magnet structure (Vanderkooy, Zmood et al.).

The performance of an actuator system is limited by mechanical dynamics and the heat generated in the coil. Excessive heat can destroy coil bonding materials, and may cause thermal warp of the overall disk drive. A recent analysis deals with the power dissipation problem (Cooper, Hallamsek and Horowitz). The reader may also benefit from the interdisciplinary findings in two audio engineering papers' treatment of heat and its dissipation in a loudspeaker (Zuccatti, Button). The VCM drive may be constant-current or constant-voltage; tradeoffs are discussed in a paper by Sohn and Chainer.

The linear motor can be shaped in various ways. One approach results in a dual-path actuator that has more evenly distributed flux lines and lesser stray flux (Dong). A parallelogram actuator design is shown in Fig. 16.23 (Chainer et al.). Two parallel leaf springs (stainless steel) are fixed to the base plate and carry the head suspension back and forth over the disk. The lowest lateral resonance frequency in these springs is 1500 Hz, with lower frequencies for the head pitch and roll. Rotary head positioners allow for more compact designs, and are found in most disk drives today; see Fig. 16.24 (Heath; Hearn; Winfrey et al.) and Fig. 16.25.

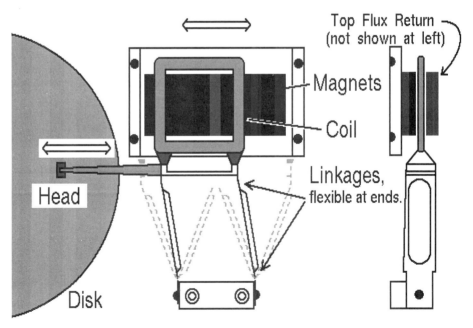

FIGURE 16.23 A flexural in-line actuator.

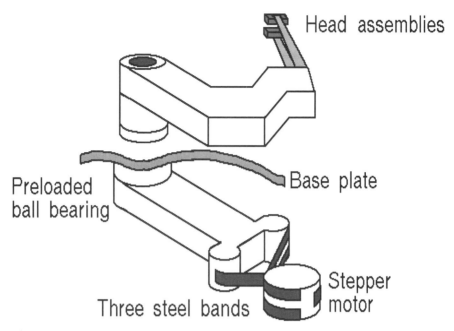

FIGURE 16.24   A rotating head stack moved by a stepper motor with a no-backlash steel belt arrangement.

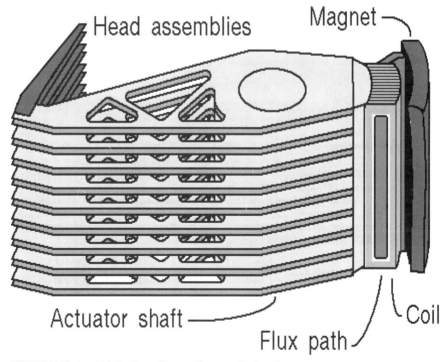

FIGURE 16.25   A multiple-head assembly moved by a curved voice-coil motor.

Modern rotary actuators all use voice coil motors as shown in Fig. 16.25 and 16.26. Some experimental disk drives have track densities of well over 10,000 tpi. It may then be advantageous to use a two-stage actuator where the suspension is connected to the main positioner arm through a piezoelectric ceramic. Coarse tracking, to within a few tracks, is done with the positioner arm while the final positioning is done with the piezoelectric element (Mori et al.).

The stray fields from the permanent magnet structures may present problems for the drive designer: It has been established that ferrite heads are susceptible to a field of 400 A/m (5 Oe) and thin-film heads to 240 A/m (3 Oe). Write/read operations are impaired at fields equal to or higher then these. At 2400 A/m (30 Oe), the drive is rendered useless.

The distance between heads and the air gap in the VCM magnet is about a few inches, and the gap fields are approximately 320 kA/m (4000 Oe) from an Alnico magnet, and 800 kA/m (10,000 Oe) from samarium cobalt.

An increase in VCM force with a high-flux-density magnet is possible if the flux carrying capability of the yoke material is increased (Wagner).

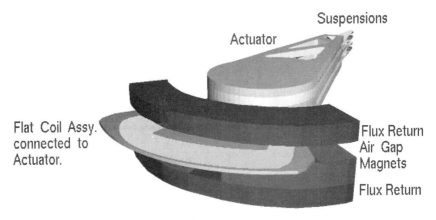

FIGURE 16.26   A flat voice-coil motor with stationary magnets.

HEAD SUSPENSIONS

The arm(s) that reach from the actuator point of rotation to near the head assemblies are very rigid, lightweight structures. At the end, a flexible suspension carries the head(s); see Fig. 16.27.

The heads will float on an air bearing when the relative speed between the head and the disk exceeds a couple of hundred IPS. A controlled flying height is established by applying pressure by means of the suspension.

The load varies between 0.15 N (15 grams) for the standard 3370 slider (4 mm long) to 0.035 N (3.5 grams) for the pico slider (1 mm long). The load remains when the disk slows down and the head contacts its surface. This is a delicate process that should not result in damage to the head or the disk surface. It is known as CSS (contact start/stop), and will be discussed in Chapter 18. CSS should only occur when the head is located in the loading zone near the disk center. The suspension holds the head by a gimbal or a flexure that is laser-welded to the head. This allows the head to pivot on a precisely located dimple.

The suspension must hold the head at the correct angles (pitch and roll). It must also accommodate motions of the head due to undulations in the disk and irregularities in disk rotations. Figure 16.28 shows changes in roll and skew; flying characteristics are described in Chapter 18.

The suspensions are made from stainless steel or beryllium copper (Bagheri and Miller). The flanges along the sides of the suspension serve to stiffen it and raise the resonance frequencies to 1.7–2.3 kHz for its first torsion mode, and 4.0–9.8 kHz for the sway (yaw) mode (Hutchinson).

Dampening of vibration modes can be achieved by application of dampening coating material (Nashif et al.).

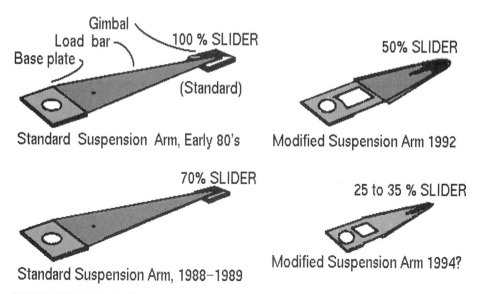

FIGURE 16.27 Evolution in slider and suspension arms (*courtesy Hutchinson Inc.*).

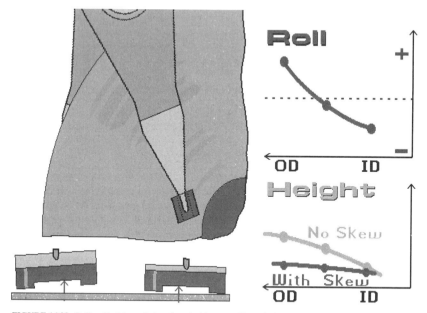

FIGURE 16.28 Roll and height variations from inside to outside tracks in a disk drive with a standard head slider.

TRACK ACCESSING IN RIGID DISK DRIVES

Rigid disk drives demand faster and more precise positioning of the head. The access time should be in the 10- to 30-ms range. To position a head accurately, full forward power of the actuator is applied for half the time required. Then reverse power is applied until the actuator comes to zero velocity. When the process is done, if the control system functions properly, it will position the head precisely on target.

This is what servo engineers call a "bang bang" control; see Fig. 16.29 (Ananthanarayanan). The servo system is needed to keep track of where the head is, and this is done with a difference counter that contains the number of tracks to the target, and which is updated as tracks are crossed (Oswald 1974, 1980).

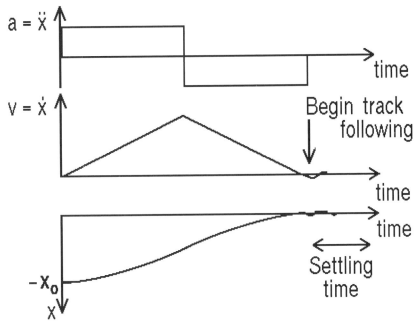

FIGURE 16.29   A bang-bang servo shortens response time.

TRACK-FOLLOWING IN RIGID DISK DRIVES

When the head has been positioned over the right track, the next task for the disk controller's servo is to keep it on track irrespective of mechanical vibrations and aerodynamic disturbances, not to mention changes due to varying temperature. At high track densities, a direct-position feedback from the head itself is necessary to correct for any mistracking.

Track misregistration (TMR) is caused by several items:

• Spindle runout

• Resonances and disk flutter

• Thermal track shift

• Head settling

• Actuator interactions

• Improper servo writing

A disk drive's TMR is therefore a summation of all the head position errors summed up over a period of time. This results in a probability density function (PDF) as shown in Fig. 16.30.

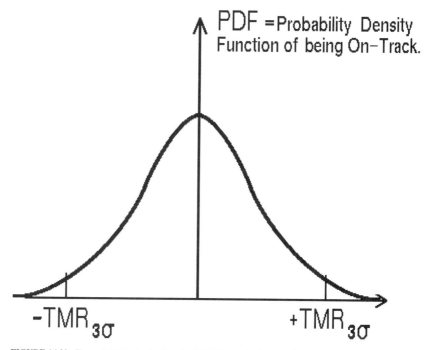

FIGURE 16.30 The probability density function (PDF) for track misregistration (*TMR*).

An analysis will show that the allowable track density D_t is related to the effective 3σ value of the write-to-read track misregistration TMR by

$$D_t = 1/(8TMR_{3\sigma})$$

(Comstock and Workman; Mee and Daniel)

High-density disk drive design must therefore start with the best possible mechanical design. This will result in the smallest TMR. Then a servo design is implemented to reduce the tracking errors.

The runout of spindles can be measured by placing a capacitance transducer in close proximity to the spindle, and observing the time-variant displacement signal between the stationary probe and the rotating spindle.

The mean value of runouts of ball bearings is in the range of 0.02–0.03 μm (0.8–1.2 μin.) (Bouchard et al.). The values for ferrofluid or air bearings are at least one order of magnitude smaller.

Resonances in base plate, actuator, arm assemblies, and disk(s) must be carefully monitored during design. The allowed frequencies should in all cases be made very high, which require low masses, high stiffness, and small dimensions.

Disk drives operate over a wide range of temperatures. Deformation caused by differences in thermal expansion ratios is often the dominant cause of off-track errors. This difference exists in the thermal expansion ratios of the aluminum spindle (or actuator) and the steel bearings.

As much as 3.5 μm (140 μin.) deformation of the bearing fitting in the spindle occurred when the temperature changed 50° C in an eight-platter 5¼-inch disk drive (Aruga et al. 1991). A structural change that applied a separate buffer sleeve for the bearing mounting reduced the deformation to 0.5 μm (20 μin.). Thermal off-tracking is minimized by improving heat transport through fins, and by unifying the temperature distribution inside the HDA (Takanami et al.).

POSITION ERROR SIGNAL (PES)

In any event, a servo signal must be provided on the disk to direct the actuator motor to find and then follow a track. In early disk drives with many platters, it was common practice to dedicate one recording surface to servo tracks. Because all heads are mounted on a common carriage, they will all be positioned over their proper tracks (all these associated tracks making up one so-called cylinder). In modern high-tpi systems, the mechanical tolerances for doing this become unbearable.

Some early disk drives with thick magnetic coating used buried servo signals. These are initially written onto a disk with a long-gap recording head, so the entire thickness of the coating has a continuous servo track for each concentric data track; a short-gap then erases the surface portion to stabilize the servo amplitude.

Short-gap read-write heads then write and read high-frequency data signals without interference. Using frequency-separation filters and an ac-biased pulse write method, an inductive head reads servos while simultaneously writing data (Haynes, Hansen).

Modern disk drives with high tpi all use embedded servo information. Figure 16.4 illustrates how the data are divided up into sectors. A few percent of the total disk surface is allocated to track servo information that is stored in the wedges between data sectors; see Fig. 16.31. The wedges are radially continuous, and store information about the sector number, the track number, and how to center the head on a track. The servo information is sampled at a rate equal to (number of servo wedges) × rpm/60 Hz. For example: $64 \times 3600/60 = 3840$ Hz.

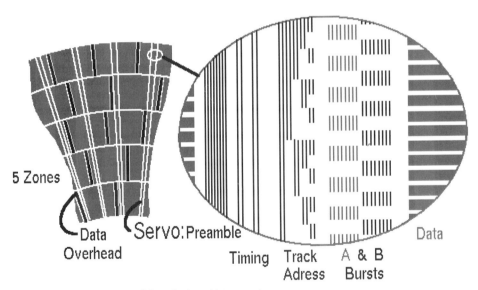

FIGURE 16.31 Prerecorded servo information located between sectors.

The data are stored in concentric zones in which the packing density along the track is kept fairly constant. The data rate is highest at the outer track, and changes from zone to zone. This is easy to do in software, and the benefit is a near doubling in data capacity (Sierra).

The servo signals are not affected by the division into wedges. They are fixed-frequency signals, and contain information about the track number (grey code) and the track center (A and B signals). The A and B servo tracks are very short full-track sections, while the data track width usually is 70 to 80 percent of the track pitch.

The servo information A and B are written between data sectors as a 1F signal, or as a null pattern as shown in Fig. 16.32. In this pattern, magnetic transitions of opposite polarity appear in sequence

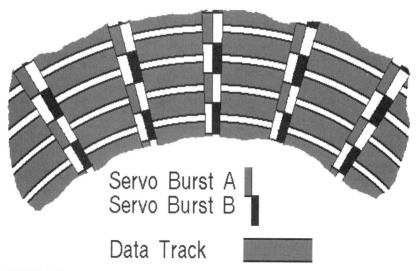

FIGURE 16.32 Embedded servo bursts A and B straddle the land between tracks.

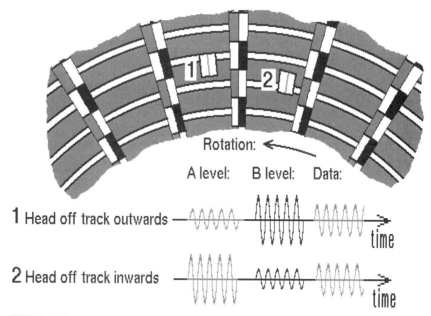

FIGURE 16.33 Position information is contained in the relative magnitudes of A and B servo signals. After processing, a position-error signal (PES) is generated and the head actuator moves the head(s) onto the track.

on either side of the track center line. When the head is on track, these two transitions result in equal pulses of opposite polarity, indicating that the head is on track. In an off-track position, a large or small read voltage of A will first occur, followed by small or large voltage for B. This information will enable the servo to move the head back on track, because the difference voltage amplitude equals the amount of off-track position and the direction to move the head in.

SERVO WRITER

The A and B servo burst must be recorded with great precision. The wedges must be precise radially, and the recorded bursts must straddle the land between future tracks.

The servo information is recorded onto the disk after the HDA has been assembled. The drive is mounted and clamped onto a precision servo writer mechanism that positions a single head to write a reference signal at a track located at the disk's perimeter. The precision of this clock signal will directly affect the recording of the servo signal, which must be radially continuous. The clock signal is written over several times until all timings between pulses are of equal length. This is particularly difficult for the gap between the first and last pulse.

The recording of track and servo information is done by the heads on the drive. Their track position must be very precise, because each A and B burst is recorded twice. The drive's heads are 20 to 30 percent narrower than the track pitch, and the servo recording is therefore recorded once and the head moved a half track width to record the same information to yield a full-width servo track.

The precise actuator control is achieved by employing a laser interferometer measurement system such as shown in Fig. 16.34. The interferometer diverts half the laser source energy to a reference beam and half to the retroreflector mounted on the HDA actuator arm. When recombined in the interferometer receiver, the phase difference between the beams generates a position control signal to drive the servowriter heads. At the retroreflector, the position accuracy may be controlled within 1 minute of angle (Lee, Freedland; See also papers by Brown and Srijayantha; Ono).

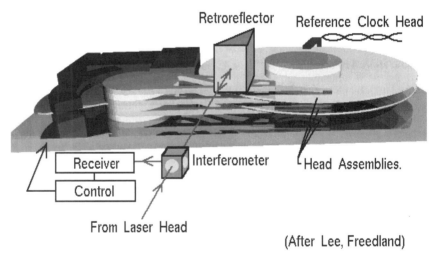

FIGURE 16.34 The principle of servo writing (see text).

SERVO DESIGN

The design of servo mechanisms and associated electronics requires an analysis of the open-loop response of the actuator-plus-motor system (Samuels, Edwards), and careful design of filters (Sidman). The servo signal itself should be analyzed and optimized (Siegel and Marcus; Cooper).

The open-loop system response will exhibit several electro-mechanical resonances, typically starting at a few hundred Hz and continuing up to several kHz. The phase response is normally unpredictable at and above the lowest resonance, and it behooves the designer to identify and possibly increase that resonance frequency to make the servo as effective as possible (increase its bandwidth).

Chapter 17 will discuss some of the encountered resonances in detail. Vibration resonances can be due to disk-to-slider mounts, and can be measured by laser Doppler techniques (Miu et al.). Also, airflow from the rotating disks can cause variations in the head flying height in large-diameter disk systems (Tokuyama et al.). The frequencies observed are the natural frequencies of the suspension rather than the dominant frequencies of the airflow forces.

Outside vibrations will upset the head positioning function, and are best fought with a disturbance compensation called "acceleration feedforward." It compensates for the positioning error caused by disk movement in the seek direction by attaching an accelerometer to the HDA base (Aruga et al.).

Another glitch in the servo operation can also occur when switching from track accessing to track following. An initial value compensation method can assure a smooth transition (Yamaguchi et al.).

Refined techniques are necessary in future drives with very high tpi. One approach follows InSite's idea; see Fig. 16.11. A laser beam is used to follow a prerecorded optical track at 17 ktpi (Akagi et al.; Futamoto et al.).

The data signal quality is best when the read head is positioned correctly over the track. This is shown in Fig. 16.35, which shows the variation in signal amplitude as the read head crosses a very narrow recorded track, i.e., 0.5 μm wide (Jensen et al.). The read profile corresponds to the head track width of 3.6 μm.

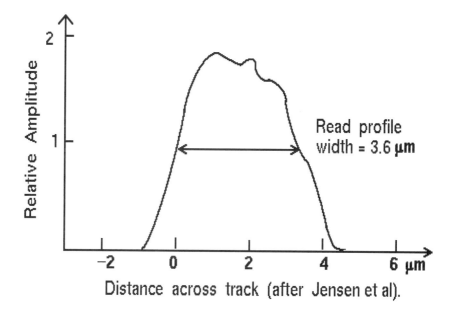

(From measurements with a 3.6 μm wide read element and a track width of 0.5 μm.)

FIGURE 16.35 The read signal is attenuated when the read head moves away from the track centerline.

A good measure of the off-track performance deterioration is measured with the bathtub curve (so named because of its shape). Shown in Fig. 16.36, it illustrates the increase in BER when the head slides off center.

We earlier determined (measured) the TMR probability density function PDF(TMR); see Fig. 16.30. The expected error rates of off-track write-read signals can be determined by the product curve of the PDF and the bathtub curve, as shown in Fig. 16.37. This is called the weighted error rate curve (Arnett and McCown). The integral of the PDF × bathtub curve provides the expected BER.

The inclusion of an MR element next to the inductive TFH presents unusual problems. Figure 16.38 shows the two heads and their separation (drawn out-of-scale for illustrative purpose). If the read head

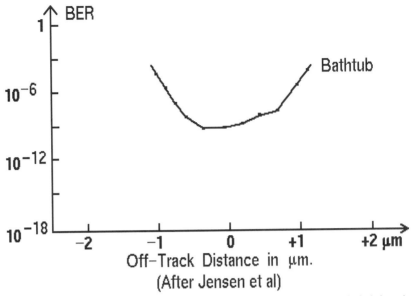

FIGURE 16.36 BER increases when the read head moves off-track. The curve is called the bathtub due to its shape.

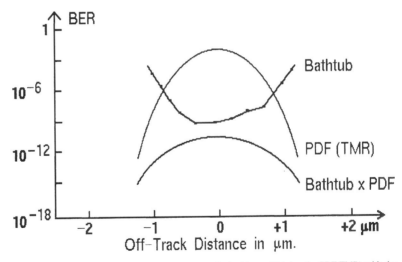

FIGURE 16.37 The weighted error-rate curves are obtained by multiplying the PDF(TMR) with the bathtub curve.

is on-track (as shown), then the write head is off-center and overlaps the next track. The $TMR_{3\sigma}$ for the write-read process must be very small to achieve a high track density (Cahalan and Chopra).

Another conclusion is that the read width must be smaller and smaller for high tpi. Such considerations depend entirely on the servo scheme. It would be ideal to read servo information with the inductive head when writing, and to read servo with the MR element when reading data.

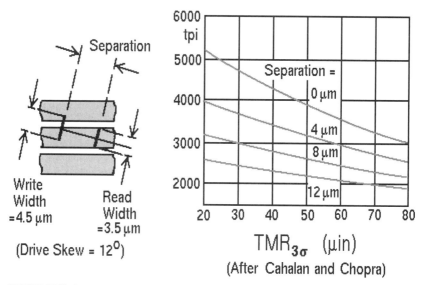

FIGURE 16.38 Separation between the inductive TFH and the MR element must be small to achieve high tpi.

The prime task of the disk drive's tracking servo is to reduce the TMR by controlling the actuator motion according to the PES. The design of today's high performance disk drives is outlined in this chapter's references. The designs are gradually changing to become digital (Hasegawa et al.), which offers the advantages of intelligent functions and more flexible control. The use of integrated circuits further downsizes the disk drive package.

TESTING DISK DRIVES

The data written on disks are in serial format, and the write and read processes must be synchronized. There will be speed variations between the input and output data due to four timing functions: The speed at which the device generating the write signal operates, the rotational speed of the disk at the time of writing, the rotational speed at the time of reading, and finally the speed of operation of the device that is trying to interpret the recorded information.

The best way to synchronize the data is to carry the timing information along with the data stream in the form of clock pulses, that are separate from the data pulses. In single density encoding all data bits are stored in defined bit cells. Every bit cell contains a timing bit that is called the clock bit; the leading edge of the clock bit determines the start of each bit cell. If there is a data bit in the bit cell, it will occur at the center of the bit cell. A detection circuit will normally identify this pulse as a "1," while the absence of a pulse is a "0." For example, 2-microsecond and 4-microsecond time intervals correspond to 500,000 and 250,000 bits per second respectively. Typical clock and data pulses would be approximately 200 nanoseconds wide at these bit rates. A bit stream that contains only clock pulses would represent the 1f (lower) frequency of operation of this particular system. If all data were 1s, then the bit stream would represent the 2f (higher) frequency of operation. There can be only 1f or 2f frequency signals, and the encoding scheme is therefore called binary frequency modulation.

The existence of a clock bit at the start of each bit cell makes data recovery straightforward. The recovery electronics lock onto the leading edge of each clock pulse, and a window is shortly thereafter opened to look for the presence (a 1) or absence (a 2) of a transition. The window length for data recovery is typically 50 percent of the bit cell, or 2 microseconds.

Single-density, or FM, encoding is easy to implement, and reliable because there is a clock bit in every bit cell. The price is that half the storage space has been allocated to timing bits. That situation is improved in MFM, or double-density encoding, where all clock timing transitions are eliminated except between zeros (see Chapter 24).

The detection circuit's window must now be timed from a free-running oscillator that is adjusted to be operating at the correct rate, which in this case is 250,000 windows per second. Whenever a timing pulse occurs the oscillator is retimed; the circuit for this type oscillator is named a phase-locked oscillator. It can also be an oscillator that operates at twice the speed, and therefore can use all transitions (ones or clock pulses) to adjust its phase; the window then operates at half the oscillator rate.

Drift in disk speed (typically specified within a couple of percent) and other timing variations (peak-shifts; see Chapters 19 and 21) will shift the transitions to be detected back and forth within the window, and may even fall completely outside. In that case an error occurs, and their number must be limited to something like only one in 10 billion. These events would be rare, and would not provide much information about such things as the drive speed stability or the entire disk system's stability. A time interval test has evolved to examine the shifts of transitions within a window, the so-called phase margin test. It basically consists of dividing the detection window up into smaller time windows, and during reading of data examining the timing of the leading edges and logging them into the appropriate timing windows. This will eventually provide a histogram of the system's timing accuracy, and allows characterization by maximum, minimum, mean, and variance figures.

Instruments are available for testing disk drives (Lerma, Rosenblatt), and preparing test disks (Mackintosh 1982), and there are extant discussions of test results (Ruoff). Additional testing will be discussed in Chapter 23 (Detection) and in Chapter 26 (Data Storage), looking at storage devices as a whole. The author has listed some assorted papers that may be of interest to the reader. They are found in the reference section titled Testing, Materials, and Processes.

REFERENCES

Comstock, L.R., and M.L. Workman. 1988. Data Storage on Rigid Disks, Chapter 2 in *Magnetic Recording*, Vol. II. Ed. by Mee and Daniel, McGraw-Hill, pp. 19–129.

Franklin, G.F., J.D. Powell and M.L. Workman. 1990. Case Design: Disk-Drive Servo., pp. 703–47, in Digital *Control of Dynamic Systems*. Addison-Wesley, 837 pages.

Harker, J.M., D.W. Brede, R.E. Pattison, G.R. Santana, and L.G. Taft. Sept. 1981. A Quarter Century of Disk File Innovation. *IBM Jour. Res. & Dev.* 25 (5): 677–89.

Sierra, H.M. 1990. *An Introduction to Direct Access Storage Devices*. Academic Press 261 pages.

Standards, *IDEMA* (International Disk Drive Equipment and Materials Association), Ringbinder, updated regularly (Phone in US: 408-720-9352.)

Yaskawa, S., and J. Heath. 1988. Data Storage on Flexible Disks, Chapter 3 in *Magnetic Recording*, Vol. II. Ed. by Mee and Daniel, McGraw-Hill, pp. 130–169.

FLEXIBLE DISK DRIVES

Carpino, M. April 1991. The Effect of Initial Curvature in a Flexible Disk Rotating Near a Flat Plate. *Jour. Tribology* 113 (2): 355–360.

Engh, J.T. Sept. 1981. The IBM Diskette and Diskette Drive. *IBM Jour. Res. & Dev.*, Sept. 1981, Vol. 25, No. 5, pp. 701–10.

Godwin, J. Jan. 1989. An Introduction to the Insite 325 Floptical Disk Drive. *Proc. SPIE* No. 1078, pp. 71–79.

Imamura, M., Y. Ito, M. Fujiki, T. Hasegawa, H. Kubota, and T. Fujiwara. Sept. 1986. Barium Ferrite Perpendicular Recording Flexible Disk Drive. *IEEE Trans. Magn.* MAG-22 (5): 1185–87.

Pearson, R.T. Jan. 1961. The Development of the Flexible-Disk Magnetic Recorder. *Proc. IRE* 49 (1): 164–74.

Roman, A. June 1982. 3.9-in. Winchester Features Removable Media. *Mini-Micro Systems.* 15 (6): 239–44.

Sugaya, T., C. Kawakami, and K. Yamamori. Oct. 1991. Perpendicular Recording and Track Following Servo for High Density Flexible Disk Drives. *INTERMAG* Poster Session. 9 pages.

Tripathi, K.C. April 1990. Development of Removable Mass Information Storage Bernoulli Principle. *Intl. Conf. on Video, Audio and Data Rec.* IEE Publ. No. 319, pp. 79–83.

Tsujisawa, T., H. Murayama, and H. Inada. April 1988. Modern Control Theory Application in High Track Density FDD. *NEC Res. & Dev.* No. 85, pp. 41–46.

HEAD POSITIONING IN FLOPPY DISK DRIVES

Katoh, Y., M. Nakayama, Y. Tanaka, and K. Takahashi. Nov. 1981. Development of A New Compact Floppy Disk System. *IEEE Trans. Magn.* MAG-17 (6): 2742–44.

Kenjo, T. 1984. *Stepping motors and their microprocessor control*s. Clarendon Press, 244 pages.

Resnik, D. March 1980. Drive Mechanism Design Reduces Errors in Mini-Floppies. *Computer Design* 19 (3): 160–68.

RIGID DISK DRIVES

Chen, Y.S., T.J. Hsu, and S.I. Chen. March 1991. Vibration Damping Characteristics of Laminated Steel Sheet. *Metallurgical Trans.* 22 (3): 653–656.

Kaneko, R. and Y. Koshimoto. Nov. 1982. Technology in Compact and High Recording Density Disk Storage. *IEEE Trans. Magn.* MAG-18 (6): 1221–26.

Lee, M.R., and R. Singh. Sept. 1992. Identification of Pure Tones Radiated by Brushless DC Motors Used in Computer Disk Drives. *Noise Ctr. Engr.* 39 (2): 67–75.

Lennemann, E. Nov. 1974. Aerodynamic Aspects of Disk Files. *IBM Jour. Res. & Dev.* 18 (?): 480–87.

Mizoshita, Y. and N. Matsuo. July 1981. Mechanical and Servo Design of a 10 Inch Disk Drive. *IEEE Trans. Magn.* MAG-17 (4): 1387–91.

Mulvany, R.B. Nov. 1974. Engineering Design of a Disk Storage Facility with Data Modules. *IBM Jour. Res. & Dev.* 18 (?): 489–505.

Mulvany, R.B. and L.H. Thompson. Sept. 1981. Innovations in Disk File Manufacturing. *IBM Jour. Res. & Dev.* 25 (5): 711–23.

Nashif, A.D., D.I.G. Jones, and J.P. Henderson. 1985. *Vibration Damping*, John Wiley & Sons, 453 pages.

Rahimi, A. Oct. 1984. Designing Hard Disk Drives to Take Abuse. *Computer Design* 23 (10): 141–48.

Takanami, S., M. Mizukami, and T. Kakizaki. Jan. 1988. Large-Capacity Fast-Access Magnetic Disk Storage. *NEC Res. & Dev.* 36 (1): 91–95.

Thomson, W.T. 1987. *Theory of Vibration with Applications*, Prentice-Hall, 493 pages.

Wagner, J.A., and M.S. Colah. July 1990. Power into Spindle Motors. *DISKCON Tech. Conf.* 7 Viewgraphs.

SPINDLE MOTORS

"Self Acting Bearing Spindle Performance", Fluid Film Devices (Phone: US 619-452-1768; UK +44(0)-794-514551.)

Bursky, D. April 1989. Flat Coils Compression Spindle Motor Size. *Elec. Design* 36 (8): 40–44.

Cho, C.P., and B.K. Fussell. Nov. 1993. Detent Torque and Axial Force Effects in a Dual Air-Gap Axial-Field Brushless Motor. *IEEE Trans. Magn.* MAG-29 (6): 2416–18.

Gross, W.A. Jan. 1963, Gas Lubrication. *International Science and Technology*, p. 32 ff.

Huber, A. May 1990. Assessing the Bearings of Precision Engineering. *F&M Feinwerktechnik & Messtechnik.* 98 (5): 209–214.

Ichihara, J. Sept. 1991. Applications of Herringbone Grooved Gas-Journal Bearing for Computer Apparatus. *Jour. Japan Soc. Tribo.* 36 (9): 665–671.

Jabbar, M.A. March 1993. Advanced Motor Technology/ Small Designs", *Components Technology Review*, Peripheral Research (Santa Barbara), Singapore, pp. 9a–96aa.

Kenjo, T., and S. Nagamori 1985. *Permanent-Magnet and Brushless DC Motors*, Clarendon Press, 194 pages.

Packard, Ed. March 1993. Spindle Motor Design: Integration, Key to the Small Form Factor Drive. *Component Technology Review*, Peripheral Research (Santa Barbara), Singapore, pp. 97–108.

HEAD POSITIONING IN RIGID DISK DRIVES

Brown, C.J. and J.T. Ma. 1968. Time-Optimal Control of a Moving-Coil Linear Actuator. *IBM Jour. Res. & Dev.* 12 (?): 372 ff.

Button, D.J. Jan. 1992. Heat Dissipation and Power Compression in Loudspeakers. *Jour. AES* 40 (1/2): 32–41.

Chainer, T.J., W.J. Sohn, M. Sri-Jayantha, D.H. Brown, and N.C. Apuzzo. Nov. 1991. A Flexural In-Line Actuator for Magnetic Recording Disk Drives. *IEEE Trans. Magn.* MAG-27 (6): 5295–97.

Cooper, E.S. May 1988. Minimizing Power Dissipation in a Disk File Actuator. *IEEE Trans. Magn.* MAG-24 (3): 2081–91.

Dong, C. Sept. 1983. Dual-Path Electromagnetic Actuator for a High Performance Magnetic Disk Drive. *IEEE Trans. Magn.* MAG-19 (5): 1689–91.

Hallamsek, K.T., and R. Horowitz. Sept. 1989. Comments on Minimizing Dissipation in a Disk File Actuator. *IEEE Trans. Magn.* MAG-25 (5): 4353–4354.

Hearn, A.R. Feb. 1980. Actuator for an Eight-Inch Disk File", *Disk Storage Technology*, IBM, pp. 83–88.

Heath, J.S. June 1976. Design of a Swinging Arm Actuator for a Disk File. *IBM Jour. Res. & Dev.* 20 (?): 389–97.

Inoue, Y., Y. Sata, and K. Hashizume. March 1974. New Linear Motion Actuator for Head Positioning. *Fujitsu Sci. & Tech. Jour.* 10 (?): 95–118.

Mori, K., T. Munemoto, H. Otsuki, Y. Yamaguchi, and K. Akagi. Nov. 1991. A Dual-Stage Magnetic Disk Drive Actuator using a Piezoelectric Device for a High Track Density. *IEEE Trans. Magn.* MAG-27 (6): 5298–5300.

Oswald, R.K., J.A. Wagner, and K. Wasson. 1993. *The disk file moving coil actuator: An introduction to the magnetics and control*, Short course, Santa Clara University, Maple Press, San Jose, CA, 1993.

Schinkïthe, W. May 1986. Gleichstromlinearmotoren für die Gerätetechnik. *Feingerätetechnik* 35 (5): 207–11

Sohn, W.J., and T.J. Chainer. May 1990. Constant-Current Versus Constant-Voltage VCM Drive Analysis. *IEEE Trans. Magn.* MAG-26 (3): 1217–24.

Vanderkooy, J. March 1988. A Model of Loudspeaker Driver Impedance Incorporated Eddy Current in the Pole Structure. *Jour.* AES 27 (3): 119–128.

Wagner, J.A. Nov. 1982. The Shorted Turn in the Linear Actuator of a High Performance Disk Drive. *IEEE Trans. Magn.* MAG-18 (6): 1770–73.

Wagner, J.A. Sept. 1983. The Actuator in High Performance Disk Drives: Design Rules for Minimum Access Time. *IEEE Trans. Magn.* MAG-19 (5): 1686–88.

Wagner, J.A. Nov. 1993. The limit of magnet remanence Br in moving coil actuators for disk drives. *IEEE Trans. Magn.* MAG-29 (6): 2920–22.

Winfrey, R., C.M. Riggle, F. Bernett, J. Read, and P. Svendsen. July 1981. Design of a High Performance Rotary Positioner for a Magnetic Disk Memory. *IEEE Trans. Magn.* MAG-17 (4): 1392–95.

Zmood, R.B., D.K. Anand, and J.A. Kirk. Feb. 1987. The Influence of Eddy Currents on Magnetic Actuator Performance. *Proc. IEEE* 75 (2): 259–60.

Zuccatti, C. Jan. 1990. Thermal Parameters and Power Ratings of Loudspeakers. *Jour. AES* 38 (1/2): 34–40.

HEAD SUSPENSIONS

Bagheri, R., G.A. Miller. March 1993. Fatigue and Corrosion Fatigue of Beryllium-Copper Spring Materials. *Journal of Testing and Evaluation* 21 (2): 101–106.

Nashif, A.D., D.I.G. Jones, and J.P. Henderson. 1985. *Vibration Damping*, John Wiley & Sons, 453 pages.

TRACK FINDING AND FOLLOWING SERVOS

Akagi, K., T. Nakao and Y. Miyamura. Nov. 1991. High-Density Magnetic Recording Tracking/ Method Using a Laser Diode. *IEEE Trans. Magn.* MAG-27 (6): 5301–03.

Ananthanarayanan, K.S. May 1982. Third-Order Theory and Bang-Bang Control of Voice Coil Actuators. *IEEE Trans. Magn.* MAG-18 (3): 888–93.

Arnett, P.C., and D. McCown. July 1992. TMR and Squeeze at Gigabit Areal Densities. *IEEE Trans. Magn.* MAG-28 (4): 1984–87.

Aruga, K., Y. Mizoshita, and M. Sekino. Jan. 1991. Structural Design for High-Performance Magnetic Disk Drives. *Fujitsu Sci. & Tech. Jour.* 26 (4): 365–77.

Aruga, K., Y. Mizoshita, M. Iwatsubo, and T. Hatagami. March 1990. Acceleration Feedforward Control for Head Positioning in Magnetic Disk Drives. *JSME III* 33 (1): 35–41.

Bouchard, G., L. Lau, and F.E. Talke. Sept. 1987. An Investigation of Non-Repeatable Spindle Run-Out. *IEEE Trans. Magn.* MAG-23 (5): 3687–89.

Brown, D.H., and M. Sri-Jayantha. April 1990. Development of a Servowriter for Magnetic Disk Storage Applications. *IEEE Trans. Instr. and Meas.* 39 (2): 409–415.

Cahalan, D., and K. Chopra, "Effects of Magneto-Resistive Head Track Profile characteristics on Servo Performance. *IEEE Trans. Magn.* MAG-30 (6): 4203–05.

Cooper, E.S. Nov. 1985. A Disk File Servomechanism Immunized to Media Defects. *IEEE Trans. Magn.* MAG-21 (6): 2592–94.

Edwards, S.A. Jan. 1984. High-Capacity Disc Drive Servomechanism Design. *HP Journal* 35 (1): 23–27.

Freedland, R. July 1991. Disk Drive Evaluation: The Key Is Servowriting. *Evaluation Engineering*, 3 pages.

Futamoto, M., F. Kugiya, M. Suzuki, H. Takano, Y. Matsuda, N. Inaba, Y. Miyamura, K. Akagi, and T. Nakao. Nov. 1991. Investigation of 2 Gb/in^{2}FD Magnetic Recording at a Track Density of 17 kTPI. *IEEE Trans. Magn.* MAG-27 (6): 5280–85.

Hansen, N.H. Nov. 1981. A Head-Positioning System Using Buried Servos. *IEEE Trans. Magn.* MAG-17 (6): 2735–38.

Hasegawa, S., K. Takaishi, and Y. Mizoshita. Jan. 1991. Digital Servo Control for Head-Positioning of Disk Drives. *Fujitsu Sci. & Tech. Jour.* 26 (4): 378–90.

Haynes, M.K. Nov. 1981. Magnetic Recording Techniques For Buried Servos", *IEEE Trans. Magn.* MAG-17 (6): 2730–34.

Jensen, R.A., J. Mortelmans, and R. Hauswitzer. Sept. 1990. Demonstration of 500 Megabits per Square Inch with Digital Magnetic Recording. *IEEE Trans. Magn.* MAG-26 (5): 2169–71.

Lee, C. May 1991. Servowriters—A Critical Tool in Hard Disk Manufacturing. *Solid State Tech.* 34 (5): 207–211.

Miu, D.K., G. Bouchard, D.G. Bogy, and F.E. Talke. Sept. 1984. Dynamic Response of a Winchester-type Slider Measured by Laser Doppler Interferometry. *IEEE Trans. Magn.* MAG-20 (5): 927–29.

Ono, H. Jan 1993. Architecture and Performance of the ESPER-2 Hard-Disk Drive Servo Writer. *IBM Jour. Res. & Dev.* 37 (1): 3–11.

Oswald, R.K. Feb. 1980. The IBM 3370 Head-Positioning Control System. *Disk Storage Technology*, IBM, pp. 41–44.

Oswald, R.K. Nov. 1974. Design of a Disk File Head-Positioning Servo. *IBM Jour. Res. & Dev.* 18 (?): 506–12.

Samuels, F.A. July 1978. Beyond the Second Order System in Track Following Servos. *IEEE Trans. Magn.* MAG-14 (4): 178–81.

Sidman, M.D. May 1978. An Adaptive Filter for the Processing of Position Error Signal in Disk Drives. *IEEE Trans. Magn.* MAG-14 (4): 185–87.

Siegel, P. and B. Marcus. Sept. 1984. Worst Case Code Patterns for Magnetic Buried Servos. *IEEE Trans. Magn.* MAG-20 (5): 906–08.

Takanami, S., M. Mizukami, and T. Kakizaki. Jan. 1988. Large-Capacity Fast-Access Magnetic Disk Storage. *NEC Res. & Dev.* 36 (1): 91–95.

Tokuyama, M., Y. Yamaguchi, S. Miyata, and C. Kato. Nov. 1991. Numerical Analysis of Flying-Height Fluctuation and Positioning Error of Magnetic Head Due to Flow Induced by Disk Rotation. *IEEE Trans. Magn.* MAG-27 (6): 5139–41.

Yamaguchi, T., K. Shishida, H. Hirai, K. Tsuneta, and M. Sato. Sept. 1992. Improvement of Servo Robustness for Digital Sector Servo System. *IEEE Trans. Magn.* MAG-28 (5): 2910–12.

TESTING; MATERIALS; PROCESSES

Adank, G. Sept. 1990. The Suitability of Elastomers for Disk Drives", *DISKCON Tech.* Conf. 5 pages and 9 viewgraphs.

Bryzek, J., K. Petersen, and W. McCulley. May 1994. Micromachines on the march. *IEEE Spectrum* 31 (5): 20–31.

Chainer, T.J., and E.J. Yarmchuk. Nov. 1991. A Technique for the Measurement of Track Misregistration in Disk Files. *IEEE Trans. Magn.* MAG-27 (6): 5304–06.

Denton, N.L., R.J. Bernhard, and P.K. Baade. Jan. 1990. A Comparison of Disk Drive Fan Noise Emissions with Standard Fan Data. *Noise Ctr. Engr.* 34 (1): 25–33.

Dirzus, J., D. Fischer, and B. Salle. May 1984. Application of Aluminium in Disk Storage Drives for Computer Systems. F&M *Feinwerktechnik & Messtechnik* 92 (4): 208–11.

Hiratsuka, Y. Jan. 1992. Present State of Surface Cleaning Technology and Its Future Problems. *Jour. Japan Soc. Tribo.* 37 (3): 218–224.

Kapala, M. Sept. 1990. Effects of Chemical Adsorbers Against Striction in Disk Drives. *DISKCON Tech. Conf.*, 10 Viewgraphs.

Kaymaram, F., N. Saka, and J.P. Sharma. Jan. 1992. Runout Effects in the Failure of Thin Film Rigid Magnetic Recording Disks", *Tribology Trans.* 35 (1): 21–28.

Lerma, J. April 1985. Analyzer Integrates Disk Drives with Controllers Systems. *Digital Design* 15 (4): 132–34.

Mackintosh, N.D., and Miyata, J.J., "A Standard Disk for Calibrating Head-Disk Interference Measuring Equipment", *IEEE Trans. Magn.*, Sept. 1982, Vol. MAG-8 (6): 1230–1232.

Rosenblatt, A. Dec. 1985. Analyzer for Checking Floppy-Disk Drives Draws on Video-Game Techniques. *Elec. Prod. Mag.* 28 (13): 42–46.

Ruoff, J. May 1984. How to Test Winchester Disk Drives. *Computer Design* 23 (5): 81–86.

Smith, J.T. Oct. 1990. Materials Analysis for the Magnetic Disk Drive Industry. *Solid State Tech.* 33 (10): 103–106.

Tsai, C.J., D.Y.H. Pui, and B.Y.H. Liu. March 1992. "Transport and Deposition of Wear Particles in Computer Disk Drives", *Particle & Particle Sys. Char.* 9 (1): 31–39.

Whyte, W. Jan. 1991. An Introduction to the Design of Clean and Containment Areas. *Cleanroom Design* pp. 1–22.

CHAPTER 17
VIBRATION IN TAPES, DISKS, AND SUSPENSIONS

A linearly moving tape, a rotating disk and the head positioner arm may vibrate at a number of frequencies; this will manifest itself as flutter and timing errors in data streams (window margins and peak shifts). It can also cause excessive wear of heads and media. The vibration modes for tapes, disks, and actuators will be treated separately.

There are many items that can excite vibrations:

- Friction between heads and medium.
- Internal drive vibrations (motors, ball bearings etc.)
- Airborne noise that the media picks up just as a microphone membrane would.
- External vibrations affecting the drive unit.

The latter has been verified when listening to the detected output of flutter meters; the signal contains airborne sound from nearby conversation. Consider the fact that early ribbon microphones were constructed by mounting a flat conductive ribbon between two magnet poles, as shown in Fig. 17.1. A voltage will be induced in the ribbon when it moves back and forth in the flux lines (Beranek 1976; "Pressure Gradient Microphones").

The effects of vibrations are annoyance while listening to music from an audio recorder. This is illustrated in Fig. 17.2 where the human annoyance threshold for frequency modulation of tones is shown, together with the threshold for amplitude variations (Stott and Axon 1957; Sakai 1970; McKnight 1972). It is also possible to identify a recorder from measurements of its flutter (McKnight and Weiss 1976). Data from instrumentation recorders will suffer in accuracy, as several papers in the references describe (Prager 1959; Kashin et al. 1968; Minukhin 1969; Slepov 1970; McKnight 1962; and Ratz 1964).

The reader may thrust all this aside by referring to the flutter elimination that a digital system easily provides by reading and dumping the digitized bits into a buffer, and then clocking the bits out in a perfect time sequence; it is only necessary to keep the buffer about half-full of bits most of the time, and to regulate the playback speed to assure this level of operation. The transport designer should still strive for the best possible tape motion; that leads to a buffer size smaller than a poorly designed transport would require.

This buffering is not required in a self-clocking digital read/write system for data and databases, where the speed accuracy of the read mechanism easily can be a couple of percents off. Now speed

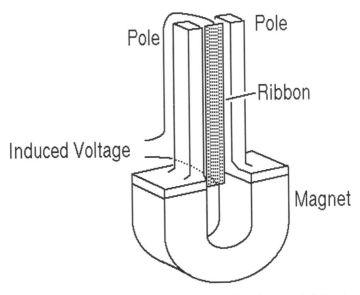

FIGURE 17.1 A ribbon microphone. A flat metal ribbon vibrates in a magnetic field, and a voltage is induced.

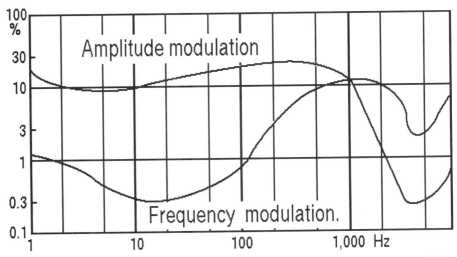

FIGURE 17.2 The human ear's sensitivity to amplitude and frequency modulation of tones as a function of the modulation frequency (*after Stott and Axon*).

variations show up in deterioration of the window margins, as discussed in Chapter 16, and again we find it beneficial to design a mechanism that will move the media at a perfectly constant and smooth speed; vibrations will affect the latter.

This chapter will therefore focus on vibrations that can take place in the media, in the head actuator, and between the media and the head. Most are determined by the dimensions and material properties of the media. Vibrations at large are already well treated in several textbooks (den Hartog 1984; Thomson 1981; Lalanne et al. 1983); the latter has several interesting computer programs written in BASIC.

VIBRATIONS GENERATED BY FRICTION

The frictional behavior of two surfaces sliding against each other is a complicated science in which a large amount of knowledge comes from observations and measurements. A good first approximation for the vibration generating characteristics of friction comes from equating it with a white noise generator that produces vibration components at all frequencies.

Friction cannot occur without some abrasion and wear (see Chapter 18). The force in a dry frictional process follows the general law that the force F is proportional to the coefficient of friction μ multiplied with the force P perpendicular to the surface, and independent of speed:

$$F = \mu P \qquad (17.1)$$

This relationship is shown in Fig. 17.3, top left. When a lubricant is added, the law no longer holds true. The drag force is now proportional to the perpendicular pressure as well as to the velocity (Fig. 17.3, top middle). When the velocity is very low, there may be isolated contact points where the lubricant film is broken and the frictional forces increase drastically (Fig. 17.3, top right).

Tape and disk coatings contain minute amounts of lubricants. This results in a composite picture of the frictional forces versus tape velocity as shown in Fig. 17.3, bottom. The dynamic coefficient of friction is lower than the static coefficient.

This results in a small region at very low speeds where the slope of the curve for the frictional force versus tape velocity is negative, causing a "stick-slip" motion. An interesting study of the irregular tape motion is found in a paper by Ristow (1985), while Kalfayan et al. (1972) have many practical measurements and observations to add to our knowledge.

When the speed approaches 100 IPS (or 9.1 km/hour; 6 MPH), enough air is drawn along with the media surface to reduce the coefficient of friction to zero. At higher speeds, an airfoil bearing is formed, causing separation between the tape and the head.

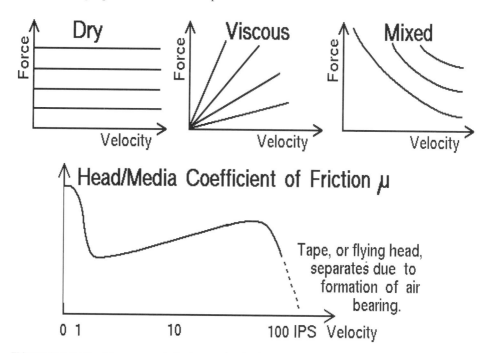

FIGURE 17.3 Frictional force versus velocity does not obey the simple law presented in formula (17.1). Dry and/or wet conditions change the outcome dramatically. For a typical tape/head system, the bottom curve applies.

Mean values of the coefficient of friction μ in an ambient environment are listed in Chapter 18. They will vary with binder formulation, and they all show a generally upward trend after many passes, as illustrated in Fig. 17.4 (left). This increase is understood by realizing that high points on the coating are worn down, providing for a larger contact area.

The increase due to rising temperature and humidity is not well understood, although it correlates with the fact that abrasion increases with humidity, which in turn requires higher pulling forces.

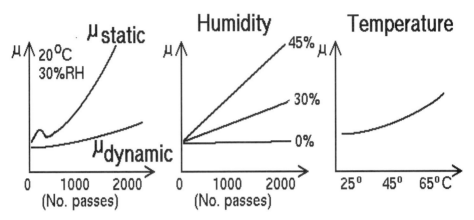

FIGURE 17.4   Coefficients of friction are not constants. They depend on wear conditions, humidity, and temperature.

STICK-SLIP FRICTION

Tapes have been found to move with a jerky motion at very low speeds due to a static coefficient of friction that is larger than the dynamic coefficient of friction. This is normally (and justly) disregarded in high-speed transports, but it has occasionally been mentioned in discussions regarding possible tape slippage over capstans (such slips may occur at the low capstan-to-tape speeds (ideally a relative stand-still), and any stick-slip activity would sort of "pluck" the resonance frequencies of the tape and thus cause high flutter).

If a tape moves along at a speed v_1, it would sense less friction if the speed moved up to v_2. And if there is sufficient tension and compliance it will do just that for a short moment, then advance to get too far ahead, and then stop—and then go through v_1 and v_2 again. This jerky motion is stick-slip, and can be expected to occur at tape speeds less than the critical speed v_c, shown in Fig. 17.5.

The critical speed where stick-slip occurs can be calculated using formula (17.2) in Fig. 17.5. Steinhorst (1966) introduced the dynamic value of Young's modulus of elasticity, which uses Poisson's ratio for its calculation. Poisson's ratio is defined as the ratio of lateral unit deformation to the unit longitudinal deformation when a material is strained; pulling or pushing a rubber eraser illustrates the Poisson's ratio.

The critical velocity is typically in the range of a few cm/sec; it is not exact because the coefficient of friction changes with several parameters, as we saw in Fig. 17.4.

VIBRATIONS IN TAPES

Flutter is quite often introduced into a recording by either eccentric reels, an improper hold-down mechanism for the reels, or by cogging in the reel motors. These flutter components are particularly

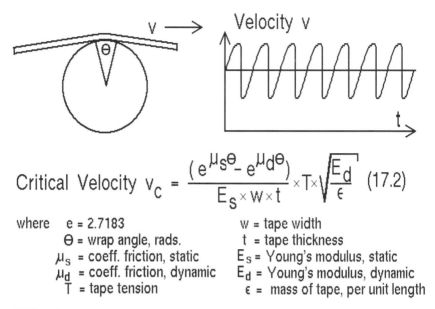

$$\text{Critical Velocity } v_c = \frac{(e^{\mu_s\theta} - e^{\mu_d\theta})}{E_s \times w \times t} \times T \times \sqrt{\frac{E_d}{\epsilon}} \quad (17.2)$$

where $e = 2.7183$ w = tape width
        θ = wrap angle, rads. t = tape thickness
        μ_s = coeff. friction, static E_s = Young's modulus, static
        μ_d = coeff. friction, dynamic E_d = Young's modulus, dynamic
        T = tape tension ϵ = mass of tape, per unit length

FIGURE 17.5 Stick-slip friction causes uneven motion and subsequent timing errors (*after Steinhorst*).

noticeable when the reel is almost empty, and they can be avoided only by a careful mechanical design. Several tape recorder manufacturers incorporate sensing arms in the tape path. The arms have transducers that detect any tension variations, and can therefore compensate for reel tension variations or reel motor cogging by servo-controlling the torque on the reel motors.

Flutter from the capstan area is more often caused by a soiled pinch-roller or worn bearings in the pinch roller or capstan assembly. Another (although rare) cause of flutter may be an eccentric capstan. These driveshafts are ground to a high degree of precision, but may inadvertently have been bent or otherwise damaged.

LONGITUDINAL RESONANCES

One contributor to flutter is the resonance frequency for the longitudinal oscillations in a free tape span. This has long been recognized as one cause of flutter and is minimized by shortening all free tape sections, such as capstan to head and head to head (Hadady and Bentley 1967).

The length of the tape is determined by the length between any two adjacent points of support, i.e., capstan(s), rollers, guides, or heads. Fixed heads and guides are generators of frictional forces, but are also grounding points insofar as the mechanical vibrations are concerned. And in many cases one or more high-inertia rollers are placed in the tape path to break it up into shorter spans in order to increase the resonance frequency and obtain higher losses. (The resonance frequency is inversely proportional to the length; see formula (17.3) in Fig. 17.6.)

Note that the frequency f_n is independent of the tape tension. The amplitude of oscillation is limited by the Q of the mass-compliance system. The example shows that the resonance frequency for a 5-cm piece of 1-mil-thick tape is 23.6 kHz.

A rule of thumb is, therefore, that the fundamental frequency of the longitudinal resonance is about 20 kHz per 5 centimeters of tape, or 100 kHz per cm length of free tape; for any given length L simply divide 100 kHz by L (in cm) to find the resonance frequency (the harmonics are rarely of any practical value).

Design of servo systems for speed control of tape drives will encounter this frequency as the ultimate limitation for the servo bandwidth. Any corrections produced by the capstan servo will reach

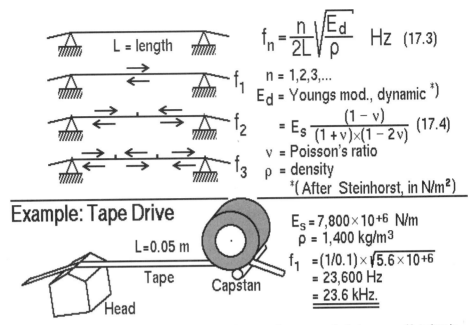

FIGURE 17.6 Longitudinal vibrations in a free span of tape have several resonance modes that are comparable to those in a string.

the read head with the speed of sound in the tape. That speed is related to the fundamental resonance frequency f_1 by $v = 2L \times f_1$.

Support of the tape with a soft, endless rubber belt has been found to reduce the vibration amplitude (Steinhorst 1966). This is illustrated in Fig. 17.7, which shows a ¼-inch tape moving past a magnetic head. Recording of a 2.7-kHz sine wave resulted in the playback signals shown in Fig. 17.8. One curve shows the playback noise floor, measured with a variable-bandpass filter.

Modulation noise occurs due to flutter during playback, as shown by the other curves. The use of a co-belt clearly reduces the noise, or flutter. The very pliant rubber belt adds to the tape mass per unit length, and will therefore reduce the flutter frequencies; its presence does also significantly lower the Q of the mechanical system.

TRANSVERSE RESONANCES

Another set of vibrations, seemingly more troublesome than the longitudinal mode, are the low-frequency transverse resonances.

The formula for the transverse vibration resonances for a string is shown in Fig. 17.9, top (Equation 17.5), and for a membrane, bottom (Equation 17.6). There exists a host of resonance frequencies, and the short spans of tapes in tape recorders should be regarded as membranes—not as strings.

A membrane is a plane sheet whose stiffness is negligible compared to the restoring forces due to tension. Analysis of the motions of a membrane is more complicated than that of the motions of the corresponding one-dimensional system, the flexible string. The membrane has more freedom in the way it vibrates than the string has.

It is useful to treat the tape span between two "ground" surfaces (guide, head, capstan) as a membrane. The allowed transverse oscillations of a square membrane can be calculated (Morse 1936), and verified by measurements. The author believes that these vibrations are the cause of so-called scrape-flutter.

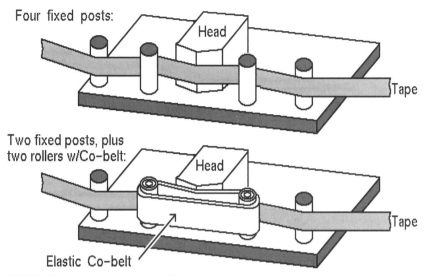

FIGURE 17.7 The reduction of vibration (flutter) by using a compliant co-belt (*after Steinhorst*).

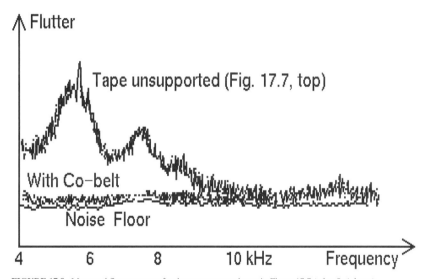

FIGURE 17.8 Measured flutter spectra for the two systems shown in Figure 17.7 (*after Steinhorst*).

The resonance frequencies can be calculated from formula (17.6) in Fig. 17.9, and there are actually a double infinity of solutions (m,n); a few of the vibration modes are shown in Fig. 17.10, and the computed resonance frequency spectra are shown in Fig. 17.11.

A given tape span need to be fitted to formula (17.6); a is the length of the tape (being clamped or passing over a head), but b is not equal to the width of the tape because the edges are free. A value of b can, however, be fitted from appropriate resonance measurements, and has been found to vary from several times the width of the tape for very light tension to slightly larger than the width for high tensions.

The spectra offer an explanation for the broad spectrum for scrape flutter. We see that the band of frequencies shift toward higher values for higher tape tension. And the individual frequencies will simultaneously move closer together (as a result of a smaller effective value of b).

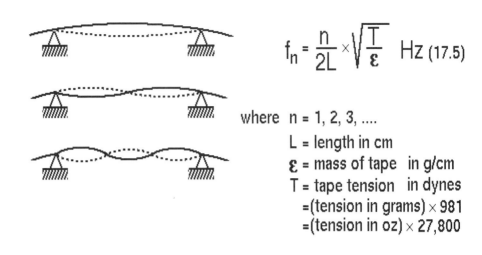

$$f_n = \frac{n}{2L} \times \sqrt{\frac{T}{\varepsilon}} \text{ Hz (17.5)}$$

where n = 1, 2, 3, ...

L = length in cm

ε = mass of tape in g/cm

T = tape tension in dynes

=(tension in grams) × 981

=(tension in oz) × 27,800

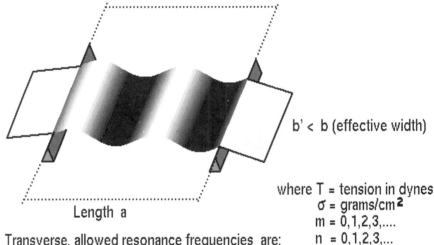

b' < b (effective width)

Length a

Transverse, allowed resonance frequencies are:

$$f_{mn} = \frac{1}{2} \times \sqrt{\frac{T}{\sigma}} \times \sqrt{\left(\frac{m}{a}\right)^2 + \left(\frac{n}{b}\right)^2} \text{ Hz (17.6)}$$

where T = tension in dynes

σ = grams/cm^2

m = 0,1,2,3,....

n = 0,1,2,3,...

(m=0, n=4 shown)

Dimensions in cm.

FIGURE 17.9 Transverse vibrations in a string also have an infinite number (*n*) of allowed modes, while those in a membrane (tape) have a double infinity (*n,m*) of allowed modes.

Note that these frequencies are tension-dependent (*T*). The tape tension in any recorder varies from one instant to the next due to any or all of these causes:

• Mechanical run-outs of rotating parts

• Changes in coefficients of friction

• Changes in the tape cross section

• Changes in the tape elastic modulus

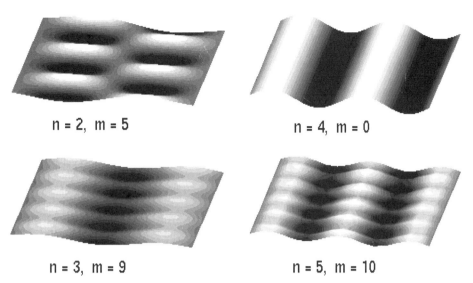

FIGURE 17.10 Computer simulation of some vibration modes in a section of tape between heads or between heads and a capstan.

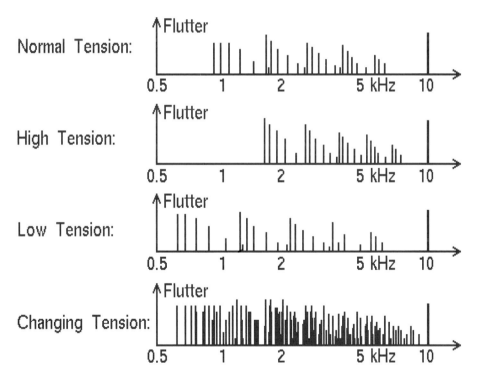

FIGURE 17.11 The calculated value of allowed resonance frequencies for a 1.5-inch-long tape, 1-inch wide (*like between two IRIG heads*).

Further variation in tape tension may be generated by a servo-controlled capstan. Large tension changes will, in particular, occur in a servo drive when it corrects for a large amount of flutter that was caused by vibrations of a recorder during recording. Tension variations from 0 to many ounces would result in a resonance spectrum extending almost down to dc, and this would display a pattern looking like grass on an oscilloscope connected to the flutter meter.

Reduction of the oscillation amplitudes is not possible by the conventional means of adding a dampening layer; it would have to be prohibitively thick (Schwarzl 1958; Dietzel 1967). A possible solution may be the installation of a low-friction pebbled surface parallel to, and immediately next to, the tape; this would deter transverse vibrations (author).

VIBRATIONS IN DISKS

Disk drives employ disks that are clamped at their centers and free on the outside; the flexible disks are best treated as circular membranes, while the rigid disk are similar to rigid, circular plates. Early investigations of the flexible disk drive were made by Pearson; fluttering of the flexible disk could be severe (Fig. 17.12) unless supported by a back plate air bearing (Fig. 17.13). The latter principle is used in the Iomega cartridge drive.

Resonance frequencies of the thicker, rigid-disk platters can be computed as shown in Fig. 17.14. It is an approximation because the formula is for a disk simply supported at its center, not an annular support. The reader is referred to the work by Leissa (1969), or Priola and Sitjia (1984) for details on annular clamped disks. The allowed resonance frequencies for a 3.5-inch aluminum disk are computed, and are: 490 Hz, 825 Hz, 2,355 Hz, 4,213 Hz,—etc.

The vibration flexures are shown in the bottom of Fig. 17.14. Note that the existence of an allowed frequency does not mean that the disk will start vibrating at that, or at any of the other allowed frequencies. But it may, and will, if excited.

The allowed frequencies in plate and membrane vibrations do not follow the common pattern of harmonics. They are the solutions to the wave equation in solids, and their values often are determined by nulls of Bessel functions; these functions are encountered later in the book in connection with frequency modulation; see Fig. 27.7.

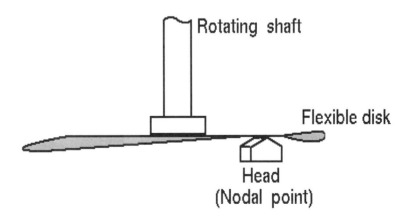

FIGURE 17.12   A flexible PET disk will shape and conform to a fixed object (*after Pearson*).

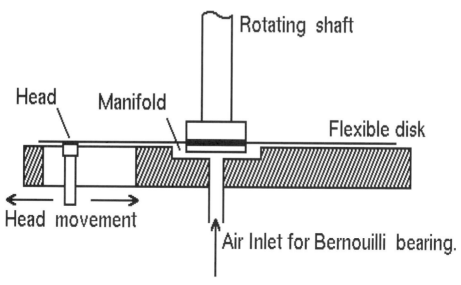

FIGURE 17.13 A flexible disk supported via thin air bearing to a polished, rigid back plate (Bernouilli bearing) (*after Pearson*).

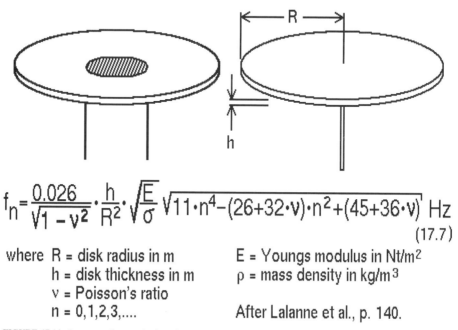

$$f_n = \frac{0.026}{\sqrt{1-\nu^2}} \cdot \frac{h}{R^2} \cdot \sqrt{\frac{E}{\sigma}} \sqrt{11 \cdot n^4 - (26+32\cdot\nu)\cdot n^2 + (45+36\cdot\nu)} \text{ Hz}$$

(17.7)

where R = disk radius in m   E = Youngs modulus in Nt/m²
          h = disk thickness in m   ρ = mass density in kg/m³
          ν = Poisson's ratio
          n = 0,1,2,3,....   After Lalanne et al., p. 140.

FIGURE 17.14 Resonance frequencies in a circular membrane (disk), clamped at the center and free at the outer, round edge. Pictures shown are computer simulations of some allowed modes.

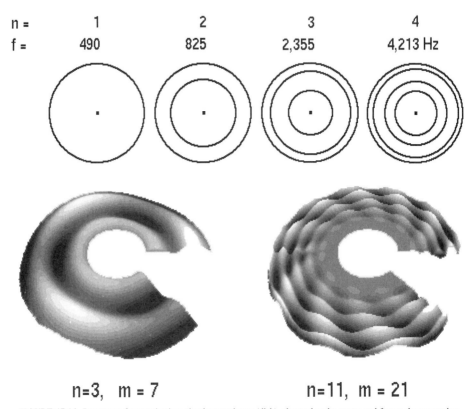

Example: $R = 3.5''/2 = 0.0445$ m
$h = 50$ mils $= 0.00127$ m
$E = 1E+7$ psi $= 68,950E+6$ Nt/m^2 (Aluminium)
$\rho = 2,710$ kg/m^3
$v = 0.3$

n =	1	2	3	4
f =	490	825	2,355	4,213 Hz

n=3, m = 7   n=11, m = 21

FIGURE 17.14 Resonance frequencies in a circular membrane (disk), clamped at the center and free at the outer edge. Pictures shown are computer simulations of some allowed modes (*Continued*).

The use of soap films to demonstrate allowed frequencies in thin membranes are shown on pages 186–187 in a book by French (1971). A simpler experiment is shown in Fig. 17.15, where a 5¼-inch disk (jacket removed) is mounted (by press fit) onto a small paper cylinder glued to the center of a small loudspeaker. Excitation of the disk is done by connecting the loudspeaker to the output terminal on a sinewave oscillator. Pre-sprinkle the disk with an even, thin layer of ordinary salt (or any granular substance), and vary the signal frequencies. At certain frequencies patterns will develop. There exist m nodal circles and n nodal diameter vibration modes. When a rotating head without R/W head is excited in the axial direction, only the m_0 mode ($m = 0,1,2,\ldots$) appears on the disk. However, if a head system is attached to the disk, all modes are excited on the disk (Jiang et al. 1990). Also, aerodynamics play a role (Lennemann 1984).

Modern investigations use finite element analysis to study vibration modes (Good and Lowery 1985; Benson 1983; and Benson and Takahashi 1991); see Fig. 17.16. Further information about vibration in membranes, rectangular as well as circular, can be found in the books listed (Morse 1936;

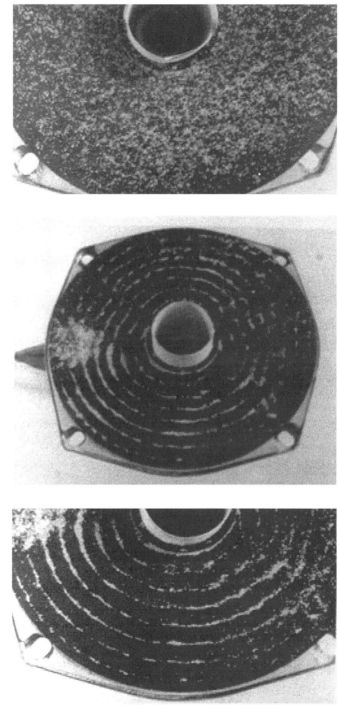

FIGURE 17.15 Oberservation via sand patterns of resonance frequencies in a 5¼-inch floppy diskette.

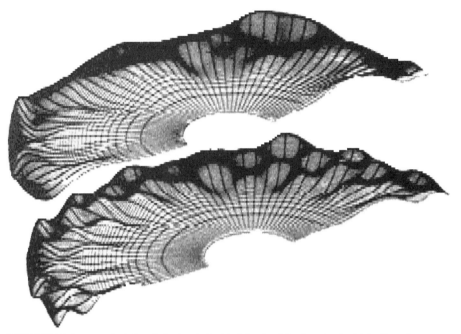

FIGURE 17.16   Computer simulation of vibrations in flexible disk, using finite element method (*after Benson*).

Baldock and Bridgeman 1981; and French 1971), while a good introduction to Bessel function is a book by Bowman (1958). Other aspect of vibrations in disk drives are covered by Richards (1978), Ohta et al. (1985), and Yamaguchi et al. (1986).

VIBRATIONS IN ACTUATOR ARMS

The actuator arm that holds the magnetic R/W head and positions it to the desired track on a disk is a spring-mass component that has many vibration modes. A typical 3370 suspension arm is drawn in Fig. 17.17, going from the mounting plate to left through a load beam that is flexible over a short distance and thereafter rigid due to the bent-up edges providing stiffeners, ending in the attachment to the head slider.

The suspension shown is the in-line configuration used in the rotary actuators in modern, compact disk drives. Earlier disk drives used a linear actuator motor, and the head was turned 90° from the position shown in Fig. 17.17.

The short, flexible portion is needed so the head can move up and down during load/unload operation. The flexure will also absorb any disk warp. Three modes of flexure are possible: Cantilever (head moves up/down), pitch, and roll. The resonance frequencies are 80, 345, and 320 Hz, respectively, for a 3370-type slider (Miu et al. 1990). For the newer and smaller assemblies, the frequencies are somewhat higher.

They play a role during load/unload operations. The air bearing pressure under the slider is very low before loading into its final flying position a few microinches above the disk surface. The slider is just held by the suspension, and is easily perturbed by the air flow due to the disk runout. This will excite the roll mode, plus another mode of slightly higher frequency (Jeong and Bogy 1992). This will increase the danger of slider/disk damage during the loading process. The cantilever mode is also excited by airflow, with resulting fluctuations in slider flying-height (Yamaguchi et al. 1990).

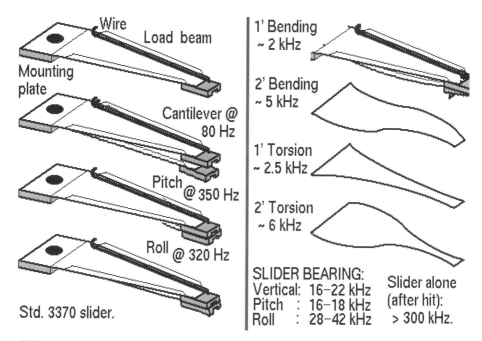

FIGURE 17.17   Vibrations in slider suspension arms.

The slider itself acts as the mass in a resonance system where the compliance (spring) is the air cushion between slider surfaces and the disk surface. The resonance frequencies are in the 15- to 40-kHz range; see Fig. 17.17 (Briggs and Talke, 1990).

A pivot between the suspension arm and the slider transmits the load force to the slider and gives the slider rotational freedom. To reduce spacing fluctuation at low frequencies, the moment arm can be reduced by moving the pivot from the slider to the arm, as shown in Fig. 17.17, top right (Ohwe et al. 1990). The net result is a reduction in spacing fluctuations during load/unload (access).

The slider itself can vibrate. This may be a bending mode in either or both directions, or a twist. The amplitudes are small, but may affect operations in near-contact applications. The resonance frequencies have been determined to be in the 300- to 600-kHz range (Briggs et al. 1992), and may be activated by collisions between slider and disk asperities. An analysis of R/W-head suspension dynamics in a high-performance floppy drive system is found in a paper by Frees and Miu (1990).

The last type of vibrations is related to the suspension arm itself, from mounting plate to slider pivot point. They are also the vibrations most likely to receive attention, because they are in line with the correction motions sent from the tracking servo. Any correction motion of a frequency above a suspension arm resonance frequency will not operate properly.

Figure 17.17 shows the four lowest fundamental modes pertaining to bending and torsion, starting at 2 kHz and going up in frequency. The need is clearly to increase these, and small disk drive designs automatically do this. A comprehensive review is found in a paper by Miu and Karam (1991). Design methology is discussed in papers by Yumura et al. (1993) and Ono and Teramoto (1992).

It is natural to seek a method of dampening the arm vibrations, once a design that provides reasonable stiffness for the load pressure and compactness has been achieved. The addition of a layer of dampening material (Scotchdamp #SJ2015SX Type 112, according to Henze et al. 1990.) did reduce the amplitude of the offtrack twisting motion from 11.5 to 0.8 μm, but did not reduce the amplitude of a sway mode that adds a sideways motion of the head slider to the motions shown for bending/torsion modes in Fig. 17.17.)

Dampening proves difficult to do, because the amount of dampening material must be quite thick in relation to the arm metal. The problem is analogous to the dampening of airborne sound though the surface of an aircraft, to the interior. Only a honeycomb structure has been successful. An arm consisting of two thin metal layers with dampening material between will offer a solution. One such construction is the *constrained layer damping* (CLD) patch where a viscoelastic layer ($t = 125$ μm) is placed so it is between the arm and a thin elastic layer (usually steel or aluminum; $t = 25$ μm) (Harrison and Talke 1993).

VIBRATIONS IN DISK DRIVES: ANALYSIS AND MEASUREMENTS

Head disk assemblies (HDA) are normally isolated from vibrations by means of conventional vibration mounts, which can be rubber or metallic spring systems. Unfortunately, HDA shock mounting leads to wind-up, a self-induced excitation of HDA/shock mount modes. It appears during high-acceleration seeks, where the actuator applies a torque to the rotary actuator and a reaction torque on the HDA that supports it. The result is rotational vibration of the HDA, which also supports the disk spindle. Because the angular displacement of the HDA and the radial position of the data tracks correlate, positioning errors between the data heads and data tracks arise. Mispositioning due to external vibrations and wind-up can be reduced by feedback into the servo of sensed drive angular accelerations (Davies and Sidman 1993).

Several instruments are available for the measurement of flutter. A flutter meter is a constant-frequency generator whose output is coupled to the record amplifier, while a frequency discriminator demodulates the output from the playback amplifier. The output of the discriminator is fed to a meter, which can be reading either RMS or peak-to-peak. The output is available for observation on an oscilloscope.

Flutter during later playback of a tape may give a higher flutter reading than measured while recording. It also may be found that some low flutter-frequency components will cancel each other so the playback flutter oscillates between a low value and a high value.

It is important to assure a high signal-to-noise ratio in flutter measurements. It is advisable to insert a band-pass filter during playback to eliminate low-frequency and high-frequency noise. Any noise will affect the discriminator output and result in a reading that appears as flutter. Also, adjust the record level for saturation of the tape during recording.

The performance of disk drives is measured with an instrument that measures the timing accuracy in the data stream. The technique is referred to as time interval analysis (TIA), and is discussed in Chapter 23.

Direct observation of tape or disk vibrations is possible in several ways. The illustrations in this chapter generally show exaggerated amplitudes of vibrations that seldom are found in real drives. The simplest instrument that has been used to observed tape or disk vibrations is simply a stereo phono cartridge of a type that has very low equivalent mass (for example Ortofon, with 0.6 to 0.9 mg effective needle mass). Transverse or longitudinal vibrations are measured by connecting the coils in series or in parallel; this method was used to verify the computations shown in Fig. 17.10.

A secondary indication of disk vibrations is hidden in the motions of heads and heads plus arm rest; method of observation are high-speed motion picture film (Millman et al. 1986), and a two-beam interferometer (Best et al. 1986). A precise evaluation of a head assembly motion has been done using a piezoelectric tripod shaker (Mori et al. 1993). The shaker had a displacement of 4 μm, a frequency range of 10 kHz, and a mode isolation of over 20 dB. Vibration of the head assembly has been observed with a multichannel laser doppler vibrometer system (Yoshida and Yanagiisawa 1992).

Direct observation of disk vibrations have been made with a laser doppler vibrometer (Bouchard and Talke 1986; Wiezin et al. 1984), monomode fibre optic interferometer (Lewin et al. 1985), and electrostatic capacitance probes (ADE Corp., Newton, Massachusetts).

BIBLIOGRAPHY

Beranek, L.L. 1976. *Acoustics*. McGraw-Hill, New York. 481 pages.

Den Hartog, J.P. 1984. *Mechanical Vibrations*. Dover Publ. 436 pages.

Lalanne, M., P. Berthier, and J. Der Hagopian. 1983. *Mechanical Vibrations for Engineers*. John Wiley and Sons. 266 pages.

McKnight, J.G. March 1972. Development of a Standard Measurement to Predict Subjective Flutter. *IEEE Trans. Audio & Electroacoustics* AU-20 pp. 78–80.

McKnight, J.G., and M.R. Weiss. Nov. 1976. Flutter Analysis for Identifying Tape Recorders. *Jour. AES* 24 (11): 728–734.

Sakai, H. March 1970. Perceptibility of Wow and Flutter. *Jour. AES* 18 (3): 290.

Stott, A. and Axon, P.E. May 1957. The Subjective Discrimination of Pitch and Amplitude Fluctuations in Recording Systems. *Jour. AES*. 5 (3): 142–155.

Thomson, W.T. 1981. *Theory of Vibration with Applications*. Prentice-Hall. 493 pages.

TIMING ERRORS

Minukhin, V.B. Sept. 1969. Signal Distortions in a Magnetic Recording Channel in the Presence of Oscillations of a Magnetic Tape Speed. *Telecom. and Radio Eng.* 25/26 (9): 101–107.

Kashin, F.A., T.M. Nemeni, and N.N. Slepov. Oct. 1968. Errors in Magnetic Recording Caused by Tape-Speed Instability. *Radio Engineering* 23 (10): 98–101.

McKnight, J.G. Jan. 1962. Time Base Distortion in Continuous Recording Systems: Its Terminology, Measurement, Causes and Effects. *Jour. AES*, Vol 10, pp.44–48.

Prager, R.H. June 1959. Time Errors in Magnetic Tape Recording. *Jour. AES* 7 (2): 81–88

Ratz, A.G. Dec. 1964. The Effect of Tape Transport Flutter on Spectrum and Correlation Analysis. *IEEE Trans. Space Electr. and Telemetry*. SET-10, pp.129–134.

Slepov, N.N. Dec. 1970 and Jan. 1971. Effect of Tape-Velocity Fluctuations on the Spectra of Single-Sided Pulse-Width Modulation of the Second Kind. Part 1 and Part 2, *Radio Engineering* 25 (12): 116–121, and 26 (1): 128–133.

VIBRATIONS IN TAPES

Dietzel, R. 1967. Vergleichende Untersuchung uber den Verlustfaktor einfacher und eingezwängter Dämpfungsbelage auf dünnen Blechen. *Hochfreq. und Elektro Akustik*. Vol. 76, pp.189–197.

Dietzel, R. 1967. Zur Bestimmung des Verlustfaktors von eingezwängten dämpfungsbelagen auf dünnen Blechen. *Hochfreq. und Elektro Akustik*. Vol. 76, pp.151–162.

Hadady, R. and Bentley, R. 1967. Low Flutter High Environment Recorders. ITC Conf. Vol. 3, pp.637–659.

Kalfayan, S.H., R.H. Sil, and J.K. Hoffman. April 1972. Study of the Frictional and Stick-Slip Behavior of Magnetic Recording Tapes. *Jet Prop. Lab. Tech. Report* No. 32-1548, 19 pages.

Morse, P.M. 1936. *Vibration and Sound*. McGraw-Hill, 351 pages.

Ristow, J. Dec. 1985. Die elektrische Nachbildung des Stick-Slip-Vorgänges, I and II, *Hochfrequenz und Elektroakustik*, 74 (12): 191–198 and 75 (1): 1–9.

Schwarzl, F. March 1958. Forced Bending and Extensional Vibrations of a Two-Layer Compound Viscoelastic Beam. *Acoustica*. 8 (3): 164–172.

Steinhorst, W. April 1966. Elastische Longitudinalschwingungen in Tonbändern speziell bei Stick-Slip-Anregung. *F&S Feinwerktechik & Messtechnik*. Heft 4, pp.172–184.

Steinhorst, W. March 1966. Theoretische Betrachtungen zum elastische Verhalten von Tonbändern. *F&S Feinwerktechnik & Messtecknik*. Heft 3, pp.114–119.

VIBRATIONS IN DISKS

Baldock, G.R., and T. Bridgeman. 1981. *The Mathematical Theory of Wave Motion*. Halsted Press, Div. of John Wiley and Sons. 261 pages.

Benson, R.C. Sept. 1983. Observations on Steady-State Solution of an Extremely Flexible Spinning Disk with a Transverse Load. *Jour. Appl. Mech.* 50 (3): 525–530.

Benson, R.C., and K.A. Cole. Oct. 1990. Tranverse "Runout" of a Nonflat Spinning Disk. *ASLE Spec. Publ. on Tribology and Mechanics of Magn. Storage.* Vol. 7, pp. 1–8.

Benson, R.C., and T.T. Takahashi. 1991. Mechanics of Flexible Disks in Magnetic Recording. *Adv. in Info. Storage Systems*, Vol. 1, pp. 15–35.

Bowman, Frank. 1958. *Introduction to Bessel Functions*. Dover Publ. 135 pages.

French, A.P. 1971. *Vibrations and Waves*, M.I.T. Physics Series, W.W. Norton & Co. 316 pages.

Good, J.K., and R.L. Lowery. May 1985. The Finite Element Modelling of the Free Vibration of a Read/Write Head Floppy Disk System. *Jour. Vibr., Acoustics, Stress and Reliab. in Design (ASME)* 107 (3): 329–333.

Jiang, Z.W., S. Chonan, and H. Abé. Jan. 1990. Dynamic Response of a Read/Write Head Floppy Disk System Subjected to Axial Excitation. *Jour. Vibr. Acoustics.* 112 (1): 53–58.

Leissa, A.W. 1969. *Vibration of Plates*. NASA SP-160, NTIS No. N7018461. 353 pages.

Lennemann, E. Nov. 1984. Aerodynamic Aspects of Disk Files. *IBM Jour. Res. Devel.* Vol. 18, pp. 480–488.

Ohta, N., J. Naruse, and T. Hirata. July 1985. Vibration Reduction of Magnetic in Disk Drive Mechanism, *Bull. of the JSME* 28 (241): 1489–1496.

Pearson, R.T. Jan. 1961, The Development of the Flexible-Disk Magnetic Recorder. *Proc. IRE* 49 (1): 167–174.

Priolo, P. and C. Sitjia. Jan. 1984. Efficiency of Annular Finite Elements for Flexural Vibrations of Thick Disks. *Jour. Sound and Vibr.* 92 (1): 21–32.

Richards, D. July 1978. The Relationship between Disk Surface Acceleration and Head-To-Disk Interaction, *IEEE Trans. Magn.* MAG-14 (4): 194–196.

Yamaguchi, Y, K. Takahashi, H. Fujita, and K. Kuwahara. Sept. 1986. Flow Induced Vibration of Magnetic Head Suspension in Hard Disk Drive. *IEEE Trans. Magn.* MAG-22 (5): 1022–1024.

VIBRATIONS IN ACTUATOR ARMS

Briggs, C.A., and F.E. Talke. Sept. 1990. The Dynamics of "Micro" Sliders Using Laser Doppler Vibrometry. *IEEE Trans. Magn.* MAG-26 (5): 2442–44.

Briggs, J.C., M-K. Chang, and M-K. Tse. 1992. High Frequency Slider Vibrations During Asperity Impacts in Rigid Magnetic Disk Systems. *Adv. in Info. Storage Systems*, Vol. 4, pp. 181–194.

Frees, G.M., and D.K. Miu. Jan. 1990. Experimental and Numerical Analysis of Read/Write Head Suspension Dynamics for High-Performance Floppy Drive Systems. *Jour. Vibr. Acoustics.* 112 (1): 26–32.

Harrison, J.C., and F.E. Talke. Nov. 1993. Combined Tuned & Constrained Layer Damping of a Type 13 Magnetic Recording Head Suspension. *IEEE Trans. Magn.* MAG-29 (6): 4098–4100.

Henze, D., R. Karam, and A. Jeans. Sept. 1990, "Effects of Constrained-Layer Damping on the Dynamics of a Type 4 In-Line Head Suspension", *IEEE Trans. Magn.* MAG-26 (5): 2439–41.

Jeong, T.G., and D.B. Bogy. Sept. 1992. Unloaded Slider Vibrations Excited by Air Flow Between Slider and Rotating Disk. *IEEE Trans. Magn.* MAG-28 (5): 2539–41.

Miu, D.K., and R.M. Karam. 1991. Dynamics and Design of Read/Write Head Suspensions for High-Performance Rigid Disk Drives. *Adv. in Info. Storage Systems*, Vol. 1, pp. 145–153.

Miu, D.K., G.M. Frees, and R.S. Gompertz. Jan. 1990. Tracking Dynamics of Read/Write Head Suspensions in High-Performance Small Form Factor Rigid Disk Drives. *Jour. Vibr. Acoustics.* 112 (1): 33–39.

Ohwe, T., S. Yoneoka, K. Aruga, T. Yamada, and Y. Mizoshita. Sept. 1990. A Design of High Performance Inline Head Assembly for High-Speed Access. *IEEE Trans. Magn.* MAG-26 (5): 2445–47.

Ono, K., and T. Teramoto. 1992. Design Methodology to Stabilize the Natural Modes of Vibration of a Swing-Arm Positioning Mechanism. *Adv. in Info. Storage Systems*, Vol. 4, pp. 343–359.

Yamaguchi, Y., A.A. Talukder, T. Shibuya, and M. Tokoyama. Sept. 1990. Air Flow around a Magnetic-Head-Slider Suspension and its Effects on Slider Flying-Height Fluctuations. *IEEE Trans. Magn.* MAG-26 (5): 2430–32.

Yumura, T., K. Funai, and T. Yamamoto. 1993. A New Design Method of Inline Suspensions Based on Dynamic Characteristics. *Adv. in Info. Storage Systems*, Vol. 5, pp. 87–99.

VIBRATIONS IN DISK DRIVES: ANALYSIS AND MEASUREMENTS

Best, G.L., D.E. Horne, A. Chiou, and H. Sussner. Sept. 1986. Precise Optical Measurement of Slider Dynamics. *IEEE Trans. Magn.* MAG-22 (5): 1017–1018.

Bouchard, G., and F.E. Talke. Sept. 1986. Non-Repeatable Flutter of Magnetic Recording Disks. *IEEE Trans. Magn.* MAG-22 (5): 1019–1021.

Davies, D.B., and M.D. Sidman. 1993. Active Compensation of Shock, Vibration, and Wind-up in Disk Drives. *Adv. in Info. Storage Systems*, Vol. 5, pp. 5–20.

Lewin, A.C., A.D. Kersey, and D.A. Jackson. July 1985. Non-contact Surface Vibration Analysis Using a Monomod Fibre Optic Interferometer Incorporating an Open Air Path. *Jour. Phys., E (Scient. Instr.)* 18 (7): 604–609.

Millman, S.E., R.F. Hoyt, D.E. Horne, and B. Beye. Sept. 1986. Motion Pictures of In-Situ Air Bearing Dynamics. *IEEE Trans. Magn.* MAG-22 (5): 1031–1033.

Mori, K., J. Shimizu, T. Masukawa, T. Takahashi, Y. Takeuchi, and S. Imai. Nov. 1993. Experimental evaluation of a head assembly by using a piezoelectric tripod shaker. *IEEE Trans. Magn.* MAG-29 (6): 3921–23.

Wiezien, R.W., D.K. Miu, and V. Kibens. July/Aug 1984. Characterization of rotating flexible disks using a laser Doppler vibrometer. *Optical Engr.* 23 (4): 436–442.

Yoshida, S., and M. Yanagisawa. 1992. Vibration Analysis of Flying Head Sliders in Seek Motion with a Multi-Channel Laser Doppler Vibrometer System. *Adv. in Info. Storage Systems*, Vol. 4, pp. 361–370.

CHAPTER 18
TRIBOLOGY AND HEAD/MEDIA INTERFACE

This chapter will describe and discuss the following topics:

- Friction
- Lubrication
- Wear of tape and heads
- Contouring of heads
- Head/disk interface in disk drives
- Disk Surfaces, sliders, and air bearings
- Contamination

In modern science these are part of Tribology, which comes from the greek word *tribos*, meaning "rubbing." It is therefore a science of rubbing, which entails the topics listed above. It includes description of techniques to observe, classify and describe surfaces. The head/medium interface is the kernel in a tape or disk system: Here the signal is laid down as a magnetic pattern, and later picked up during reproduction.

High storage capacity in a recording system is achieved by recording many bits per linear inch of track length (named BPI, or bits per inch), and many tracks side-by-side (named TPI, or tracks per inch). A good, short bit-length performance requires a minimum of spacing between the head and the medium surface; zero spacing would be preferable in order to minimize signal losses.

SIGNAL LOSSES

Several signal losses occur during the magnetic write and read processes. The read voltage can be analyzed, and its mathematical expression will disclose three such losses (see Chapter 6):

Coating thickness loss: $A = (1 - e^{2\pi\delta_{rec}/\lambda})/(2\pi\delta_{rec}/\lambda)$

Spacing loss: $B = e^{-2\pi d/\lambda}$

Read gap loss: $C = \sin(X)/X$ where $X = \pi g/\lambda$

The reader will note that nowhere in these expressions do we find the variable f, used for frequency. The losses are dependent upon dimensions only—i.e., the recorder wavelength λ, the recorded thickness δ_{rec}, the spacing d between the head and the medium surface, and the gap length g. The recorded wavelength λ is equal to two bit lengths, and relates to a recorded frequency f as $\lambda = v/f$, where v is the head-to-track speed.

Only a portion of a recorded coating produces a flux. This is evident from the loss term called coating thickness loss. The formula for the loss can be reduced to as a simple rule: 75 percent of the flux that contributes to the read head voltage comes from an effective coating thickness δ_{eff}, which equals 0.44 times the bit length.

$$\delta_{eff} = .44 \times BL \tag{18.1}$$

If the bit length $BL = 3\ \mu m$ (120 $\mu in.$), then the coating need not be thicker than $0.44 \times 3 = 1.3\ \mu m$ (53 $\mu in.$). It would be detrimental to make it thicker, because this may lead to another type of loss called demagnetization loss (see Chapter 4).

A distance between the head and the medium will cause a spacing loss that can be expressed as:

$$\text{Loss} = 55 \times d/\lambda \ \text{dB} \tag{18.2}$$

where d is the effective head/medium spacing. This formula was found by taking 20 times the $\log_{10}$ of the spacing loss factor, a convention often used by electrical engineers to keep numbers reasonable, and in accord with the way many things in nature change (For example, subjective hearing levels in relation to sound pressure levels are logarithmic).

The waveforms of recorded bits are basically rectangular patterns, which can be decomposed into a fundamental signal with a number of odd harmonics. The latter provide for the sharp transition times and the square corners in the waveform. They are of shorter wavelengths than the fundamental, and therefore have greater spacing losses. The net result of the spacing loss is a signal attenuation combined with decreased resolution.

In-contact operation is therefore desirable in tape and diskette (floppy) drives. It requires a medium coating that has a perfectly flat surface. This can cause problems; binder debris may build up on the head and encourage stick-slip motion between head and medium. The result is time-base errors and potential rapid head and/or medium wear.

Build-up of debris on the head surface will aggravate stick-slip. Epoxies in the head interface are often the starting point for debris buildup, or it may be polymer deposits from the coating itself. Once a cake of debris is formed, it may grow and eventually form a thick layer, causing spacing loss. The layer may be clear, or may be discolored by frictional heat from tape running over it; hence its name, "brown stain." It is difficult to remove by conventional means, and it may be necessary to run green abrasive tape over the heads at slow speed, cooled by pure alcohol. This debris formation is pronounced from temperature-damaged tapes.

High data rates require a high velocity between head and medium, and in-contact operation is not always feasible. Rigid disk drives have heads hovering over the rotating disk surfaces at distances of 0.025 to 0.25 μm (1 to 10 $\mu in.$).

Cleanliness and freedom from airborne dust particles are important in both systems. Many particles are large enough to make scratches in head or medium surfaces, and to cause head crashes. The latter is a catastrophic failure; disk drives are therefore sealed units to keep dust particles out. Large foreign particles between a head and a passing tape has also been known to cause gap smear.

The gap loss leads to a simple relationship between the required gap length and the bit length: A long gap is desired for high head efficiency, but the signal output is zero when the gap length equals two bit lengths. The gap length is therefore made equal to or less than one bit length:

$$\text{Gap length} <= \text{Bit length} \tag{18.3}$$

TRIBOLOGY IN TAPE AND FLOPPY SYSTEMS

The head/media interface presents an environment of its own. There are frictional forces and wear processes. The head may be in contact with the medium or spaced away by an air film.

FRICTION

The engineer or technician that seeks solutions to problems regarding head wear or tape/disk abrasivity will benefit from studying the interdisciplinary topic of tribology. This term covers all sorts of mechanical and chemical interface problems between two contacting surfaces, at stand-still or sliding past one another. The books by Rabinowitz (1974), by Halling (1978), and a chapter by Talke and Bogy (1987) in Mee and Daniel's handbook are recommended for an introduction to tribology. Two books by Bhushan (1990, 1992) cover tribology in magnetic recording. And two journals cover this special topic.

It is common for tape and flexible diskette coatings to contain minute amounts of lubricant. This tends to make the drag force proportional to the contact pressure as well as to the velocity. When the velocity is very low, there may be isolated contact points where the lubricant film is broken and the frictional forces increase drastically. This results in a coefficient of friction versus head/medium speed that starts at a high value at zero speed, decreases to a minimum around 2.5 cm/s (1 IPS), rises slowly to a maximum in the vicinity of 100 cm/s (40 IPS), and then decreases to zero at speeds near 200 cm/s (80 IPS). At higher speeds, a hydrodynamic air bearing is formed between the tape and the head. A characteristic curve for the coefficient of friction versus speed in tape recording systems is shown in Fig. 18.1. The negative slope at velocities less than 1 IPS gives rise to a "stick-slip" motion, see in Chapter 17 (Mukhopadhyay 1978).

Friction is needed in the capstan nip area to grip a tape and move it through a tape transport. Typical values for the coefficient of friction of an oxide surface are, at a speed of 25 cm/s (10 IPS) at 18° C, 40% RH:

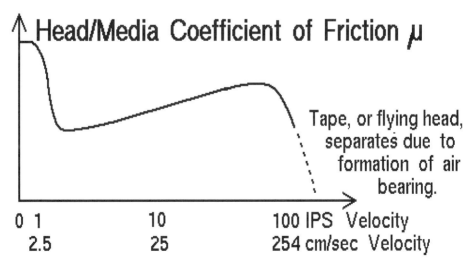

FIGURE 18.1 Friction versus head/meadia speed in a typical tape recorder.

$$\mu_{static} \text{ against rubber} \approx 0.8$$

$$\mu_{dynamic} \text{ against rubber} \approx 0.35$$

$$\mu_{static} \text{ against aluminum} \approx 0.1\text{--}0.2$$

$$\mu_{dynamic} \text{ against aluminum} \approx 0.05\text{--}0.1$$

The values vary with binder formulation, and they all show an upward trend after many passes and with increasing humidity. Graphite is often used as a solid lubricant because it offers the additional benefit of lower electrical resistance. A tape may act like a Van de Graaff generator that discharges the accumulated electrostatic charge at the head. This can result in noise in the read signal, or erratic frictional behavior (Schnurmann and Warlow-Davies 1942). The critical nature of a tape or floppy coating surface was also discussed in Chapter 14. Friction of videotapes against the head scanner present tracking problems; see Chapter 15 under Mistracking in helical-scan recorders.

Sliding friction normally generates heat. The temperature rise at the head/tape interface has been examined and reported on in Bhushan's book (1990). The tape coating's behavior is purely elastic, and the average temperature rise is on the order of 5–7° C. Only when an asperity, such as a protruding particle, goes past the head can a much higher flash temperature develop, on the order of 1000° C.

LUBRICATION

A lubricant is a normal ingredient in modern tapes. It is part of the coating, and resides in small pockets throughout the coating from which it oozes to the surface, where it provides lubrication (Lorenz et al. 1991; Aoyama and Kishimoto 1991).

The application of a lubricant to a solid metal film is a different and difficult matter. This problem was discussed in Chapter 14, under Surface Properties of Tapes and Flexible Disks.

A deposition process places a lubricant layer on top of a particulate medium, such as on a floppy (Hilden et al. 1990). A thin amorphous carbon film was sputtered onto the surface of the disk. The functional properties were found to be exceptionally good. Greater than 20,000 start/stop cycles were achieved for various types of media formulations.

WEAR OF HEADS AND WEAR OF TAPES

Media surfaces contain hard particles that are highly abrasive: γFe_2O_3, CrO_2, Fe, BaFeO, etc. The resulting wear of the magnetic heads presents a costly maintenance problem, and the selection of low-wear tapes is a difficult task. Experience is still the master in this matter, although attempts constantly are made in order to find a quick and reliable test for the abrasion of a medium (see later).

It is possible to compute and predict the contour of a head after wear, and to optimize a head's contour. The wear of heads presents more of a maintenance problem than the mere replacement of worn heads. The electrical characteristics of the head change with wear: the reduced gap depth increases the efficiency. This means that less write current is required, and the read voltage from a worn head is higher than that from a new head. The increase in write efficiency will cause a changed magnetization pattern in the coating, and hence a changed read voltage waveform. The electronics must either be tolerant to these changes, or the system must be calibrated from time to time.

Foreign dust particles and media surface imperfections can cause scratches in heads and/or dropouts (A dropout is the temporary loss of signal). Examinations of media defects have in the past disclosed embedded dirt particles, resulting in further demands upon cleanliness in the "wet end" of media coating facilities. As a result, modern tapes and disks are essentially dropout-free; dropouts are introduced by the users!

Measurements of dropout activities in many recorder installations under varying operating conditions have revealed that loose dust particles are serious generators of dropouts. The use of reels with solid flanges, instead of flanges with spoke openings, netted an order of magnitude reduction in dropout count (the "Wilmot" effect; see Chapter 14).

Video recorders using rotating-head assemblies operate with very high head-to-media speeds, in the range of 3 to 25 m/s (120–1000 IPS), or 11 to 90 km/hour! It is generally recognized that an air film forms between a head and a medium surface when the speed is greater than 2–3 m/s. This means that a spacing loss will occur in most video recorders if measures are not taken to force the head approximately 50 μm (2 mils) into the tape on the units with highest speed, less at lower speed (tip penetration). That explains why still-framing causes severe wear and deterioration of the tape. The associated temperature rise can be high, and head materials must be selected that will remain magnetic at that temperature (i.e., have a sufficiently high Curie temperature).

MECHANISM OF WEAR

Excessive wear of magnetic heads and/or disks is a problem that often occurs when a new recording medium or a novel medium drive is introduced, or when operating conditions are changed. Experience in the field of abrasion and wear does not always allow accurate prediction of performance, and each system must be evaluated separately.

Any vibrations in the head/medium interface area will aggravate wear, as will high humidity and high temperature. The chemistry of the coating binder appears crucial, and it is now common practice to add minute amounts of an abrasive powder (alumina) to the binder in order to abrade the head free from any debris formation.

This section will briefly review the fundamental wear theories, and analyze how well they relate to the wear of magnetic heads and media. Various test methods are described, and results from these should be useful in future developments of media and/or drives.

THE NATURE OF WEAR

The technology of wear is part of tribology, the science and technology of interacting surfaces in relative motion. It is without the exact laws of other sciences, and the guidelines set forth in this chapter are just that: Suggestions for an optimum design, based on past data and experiences.

There are four generally accepted modes of wear:

Adhesive Wear. This process involves the joining or welding of small interfering projections in mating surfaces, where material then is torn away from the weaker material by continuing motion.

The contacting surfaces touch each other only at high points, where plastic deformations take place as shown in Fig. 18.2. The real area of contact between a polymeric magnetic medium and a rigid surface has recently been investigated and appear to be elastic in nature (Bhushan 1991).

Adhesion will clearly occur in cases where head and medium have been in prolonged stationary contact. Spacecraft recorders have this problem if left unattended; the tape drive must be exercised regularly.

Adhesive wear is also present in normal recorder operation, where continuing sliding causes the junctions to be sheared and new junctions formed. This leads to an oversimplified theory of friction, where the coefficient of friction is equal to the shear strength divided by the yield pressure. With further rubbing, some of the transferred material is detached to form loose wear particles.

A. Light load

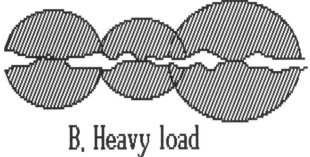

B. Heavy load

FIGURE 18.2 The formation of plastic zones under the high-pressure contact points between two surfaces.

The wear volume per unit distance of sliding is given by (Archard 1953):

$$Q = kW/3p_o \tag{18.4}$$

where:
k = probability of an asperity contact producing a wear particle
W = total load
p_o = yield pressure

This law of wear is found to be true for a wide range of conditions, and expresses that the volume of wear material is:

- proportional to the distance of travel
- proportional to the load
- inversely proportional to the yield stress, or hardness, of the softer material

At high loads we will find plastic zones below the contacting surfaces, as shown in Fig. 18.2. This leads to large-scale welding and seizure; the wear rate increases dramatically, as indicated in Fig. 18.3. This occurs when the load pressure increases beyond one third of the hardness of the softer material. A couple of tape abrasion tests operated in this region, raising a question about their validity.

Abrasive Wear. Basically, abrasive wear involves various cutting actions in which a hard material sliding under pressure across a softer surface tends to plough into the softer surface. There are many variations of this basic action, depending on surface size, shape, velocity, pressure, number of passes, and the properties of the materials.

Many abrasive situations are covered by an equation of the form

$$Q = k_a W/H \tag{18.5}$$

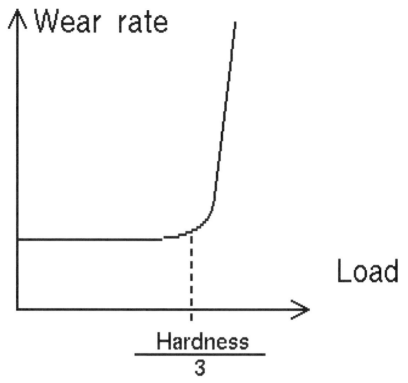

FIGURE 18.3   The wear rate for many materials increases dramatically above a load that equals ⅓ of its hardness.

where:

k_a = an abrasive constant
W = total load
H = hardness of the softer material

This mode alone is not responsible for wear of magnetic heads. Figure 18.4 illustrates the results of a test carried out at the U.S. government's Jet Propulsion Laboratories in the late 1960s, showing the wear pattern left after two different tapes were run over the head. The scalloped patterns show that there was no correlation between wear of the cores (wide laminations) and the shields (narrow laminations) and their hardnesses!

Further: The operating environment for many tape and disk drives are not exactly clean-room conditions, and dust from the atmosphere may be a major cause of abrasive wear. This fact may contribute to the spread in wear data measured for a certain tape type, but under different conditions and by different methods.

Corrosive Wear.   Corrosive wear involves corrosion and abrasive wear, either combined or alternating. Both involve removal of surface material: Corrosion by chemical action (such as oxidation), and wear by mechanical action.

Contact Stress Fatigue.   This concerns the mechanism of fatigue in removal of comparatively large particles from surfaces under high contact pressure, combined with a degree of sliding and rolling.

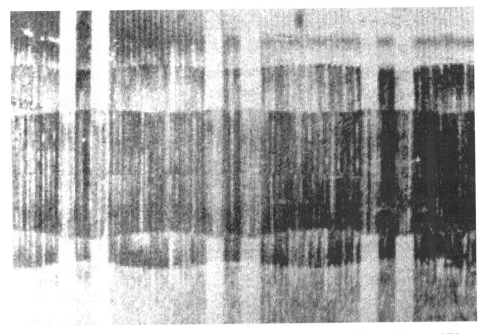

FIGURE 18.4   A tape head surface showing reversal of wear intensity by changing from tape X to tape Y (*courtesy of JPL*).

WEAR OF MAGNETIC HEADS

With so many ways to remove surface material by mechanical action, it is not surprising that the analysis of wear failures is difficult. Knowledge and study of the surface conditions causing wear are obviously necessary. A proper analysis will also depend on careful visual examination on both the macro and micro scales, as well as upon consideration of a number of factors such as hardness examination and possible chemical interactions. Figures 18.5 and 18.6 illustrate a tape surface before and after surface treatment.

Abrasive wear and some amount of corrosive wear are largely responsible for head wear. The adhesive mode is observed in the stick-slip motion of a tape at slow tape speeds (½ IPS and slower), and may in part be the cause of changes in frictional forces (drag), which vary greatly for different coatings, different head materials, and different environments. The coefficient of friction also increases with the number of passes, whereby heads and coatings develop wear patterns that are replicates of one another, with resulting larger contact area and reduced wear.

Abrasive wear is also observed as severe scratches on head and coating surfaces, caused by debris formation or foreign particles. This mode can lead to a destruction of the otherwise perfectly straight gap, and occasionally drags material across the gap (gap smear), and causes a magnetic short of the head. This failure mode can only be remedied by resurfacing the head by passing a highly abrasive tape across it.

Several methods have been devised since the mid-1950s to measure the amount of wear. They were described in the author's earlier editions of this handbook. Other reviews are found in papers by Williams and Lewis (1982) and Jorgensen (1972). They all measured, in one way or another, the amount of material removed from the dummy head of the test element used.

A recent proposal (Rogers and Hinteregger 1993) greatly simplifies and expedites measurement of wear (see the end of the section on head contouring). For a detailed study of head wear, the reader is referred to the third (or second) edition of this handbook; only key results are reported in the following.

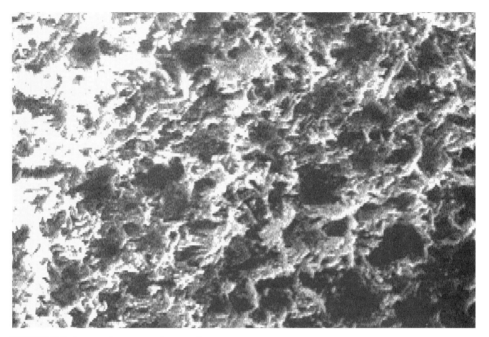

FIGURE 18.5 A tape surface, as coated, untreated.

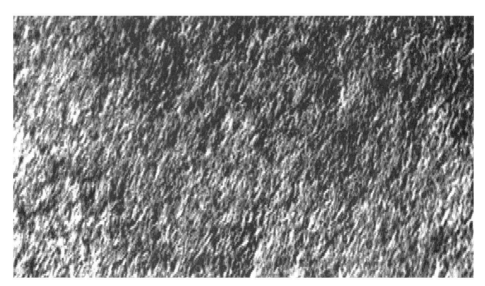

FIGURE 18.6 A tape surface after calendering and polishing.

Effect Of Hardness: Wear predictions based upon head core material hardness are unreliable. It does remain a fact that head cores fabricated from Al-Fe alloys wear about ten times slower than Ni-Fe cores (Mu-Metal, Permalloy, etc). This should hold also for the amorphous MetGlass cores, which have a very high hardness. The wear rate data for ferrites are scattered. NiZn remains much in use as shield and tape bearing surface in MR heads due to its high wear resistance.

Effect of Load Pressure: The wear rate is proportional to the load pressure. This was learned in some of the earliest tests. Clurman's analysis of head contours after the lapping effect (wear) of tapes correlates very well with actual observations (See under Head Contouring). He assumed that the wear rate is proportional to the local pressure. He also found that the pressure contour is practically constant over the contact area.

Effect of velocity: The wear rate is constant for velocities less than 10–15 IPS. The wear rate decreases in inverse proportion to the speed above 15 IPS, and becomes zero at 60 IPS at low tape tensions, ranging up to 200 IPS at very high tensions.

These observations have no theoretical counterpart. The reduced wear is due to the formation of an air bearing. Measured wear rates are shown in Fig. 18.7, using 3–4 degrees of wrap angle and a tape tension of 8 oz per ½ inch of tape width.

Effect of Multiple Passes: The wear rate decreases after multiple passes, possibly by an order of magnitude. The general data are shown in Fig. 18.7, and are the results of loop testing (author's notes).

Multiple passes of tapes are not a common occurrence, with the exception of multiple replays of sections of music, video, or data. Only still-frame operation of a videotape can cause excessive damage to a tape. It is a general requirement that still-framing should take place without tape damage for at least five minutes. This is equivalent to $5 \times 60 \times 30 = 9000$ passes of the video scanner on the same segment of tape.

Effect of Humidity: The wear rate increases dramatically with relative humidity of the operating environment. The wear rate is affected only slightly by the humidity condition of the medium (tape or disk). The results are shown in Fig. 18.7.

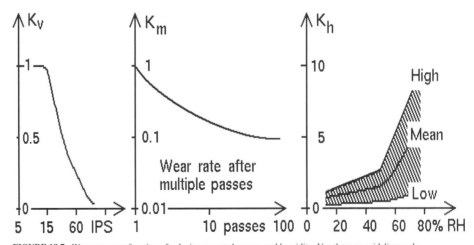

FIGURE 18.7 Wear rates as a function of velocity, repeated passes, and humidity. Use these as guidelines only.

GUIDELINES

A wear rate unit of APF (Angstroms per foot) has been selected to provide the reader with a convenient number for wear rates.

The chart in Fig. 18.8 summarizes the limited data found scattered in the literature, representing actual head wear measurements. Many sets of data from simulated tests were not applicable for this chart because they were in arbitrary units that either were nonconvertible, or one or more test conditions were ill-defined or lacking. No contradictory results were found, however.

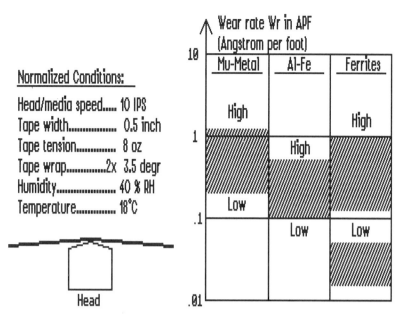

FIGURE 18.8 Measured wear rates; normalized values and the range of values observed to date. Use formula (18.6) to calculate a wear rate for conditions other than the normalized case.

Figure 18.8 will, in conjunction with the K-multipliers from Fig. 18.7, allow for a budgetary estimate of head wear from a formula:

$$\text{Wear rate} = W_r K_t K_v K_m K_h \tag{18.6}$$

where:

W_r = Wear rate from Fig. 18.8.
K_t = tension constant
    = (Tension in oz.) $\times$ (tan α) $\times$ (0.5/tape width)
    = head-media pressure in oz.
K_v = velocity constant from Fig. 18.7.
K_m = repeated pass constant from Fig. 18.7.
K_h = humidity constant from Fig. 18.7.

Having decided that a simple model is better than no model, the reader must use caution in applying formula (18.6). It does not include the severe wear problems that occur when a diskette drive starts to "sing," i.e., exhibit vibration problems that very likely are to be found at the head/media interface. (Remedy: Use a cleaning diskette, or repair or replace the drive.) This vibration will aggravate wear just as a vibrating support for sandpaper accelerates the abrading action. The author has found the formula quite good for estimating the wear of heads; and experience can only improve the model, as expressed by the four graphs and formula (18.6).

Example 18.1: What is the estimated head wear produced by ten passes of a 30-minute audio-cassette, of good quality (low abrasivity)? A mu-metal head is used; tape tension is 2 oz, the width .15", wrap angle 5°, and humidity conditions are 50% RH.

Answer: The tape tension constant is:

$$K_t = 2(\tan 5°)(0.5/0.15) = 0.58.$$

The tape speed is 1⅞ IPS, so $K_v = 1.0$.

After ten passes, K_m has fallen to 0.2; the mean value for the ten passes is 0.7.

At 50% RH we read $K_h = 1.3$.

The wear rate from Fig. 18.8 is judged at approximately 0.1 APF, a low-abrasivity tape; hence the estimated wear rate is $0.10 \times 0.58 \times 1 \times 0.7 \times 1.3 = 0.053$ APF. The length of tape passed over the head for ten single passes is $10 \times 2 \times 30 \times 60/12 = 3000$ feet, and the total wear equals

$$3,000 \times 0.053$$

$$= 159 \text{ Angstrom}$$
$$= .016 \text{ } \mu m$$
$$= 0.64 \text{ } \mu in.$$

If the gap depth is 2 mils, then 10 cassette playings can be performed $2000/0.64. = 3144$ times—for a total of 31,400 playings, or 654 days of continuous playing. A poor grade of tape (i.e., a no-brand-name type) may, on the other hand, wear out the head 10 to 100 times as fast.

CHANGES IN RECORDER PERFORMANCE WITH HEAD WEAR

Changes in the sensitivities and impedances of magnetic heads will occur when the front gap depth is reduced due to wear from magnetic tapes. The changes are, in general, so small that they only develop after many passes of reels of tape.

The trend is always an increase in sensitivity and a decrease in impedance. They should therefore be considered in programs that assess tape interchangeability, and certainly in any operation pertaining to standards work.

When the extent of changes versus wear have been established for a particular family of heads, then there may be additional steps to be included in a recorder maintenance schedule. The 1980 edition of this handbook showed several graphs that illustrated what happens to a head's read efficiency, write current requirements, and impedance levels as a function of the gap depth (representing wear). The improved models of magnetic heads that have been developed since will show different behaviors for different heads.

All heads will still exhibit improved read efficiency as the head wears. The write efficiency should include the stray fields around the gap and the coil, and each head must be analyzed separately before a prediction can be made of its changes in write efficiency and impedance during wear.

OPTIMUM HEAD CONTOURS IN LONGITUDINAL TAPE DRIVES

An example of tape wrap and general contouring is illustrated in the head assembly in a QIC drive. The head is mounted on a support arm that first will bring it into the opening of the QIC cartridge when inserted into the transport. The head is then in contact with the tape at a certain, specified wrap angle. The head will later move up and down on the tape to find a given track. This motion is most simply generated through a linkage to a stepper motor, or more intricately by a servo-controlled voice coil motor. The latter will be used in the QIC drives with track densities above 50 tracks per quarter inch (200 tpi).

The QIC drives are seeking bit densities along the recorded track that equals or exceeds that of precision instrumentation recorders or helical-scan recorders. The tape tension in the QIC cartridge is less than half the norm for instrumentation recorders (8 ounces per ½" of tape width), and the consistency of the value of the tape wrap angle becomes critical.

The front of the next-generation multichannel head has, like its predecessors, three "bumps" for the thin-film head structures as shown in Fig. 18.9. There are two additional bumps, or outriggers, at the edges of the head. It is the function of these outriggers to remove wrap angle errors of the tape around the head. Slight variations from drive to drive will result in differing wrap angles, and the potential of insufficient tape/head contact.

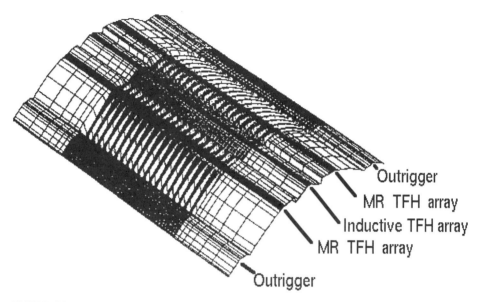

FIGURE 18.9   The contour of the front of a 36-track QIC head. The inductive TFH in the center with MR heads on each side is there so reading after writing is possible from either tape motion direction.

A wear pattern of the head front may look like that shown in Fig. 18.10, left. If another cartridge is inserted, then the wrap angle may exceed the nominal value (due to unfavorable tolerance buildup.) Now the tape no longer makes perfect contact with the head at the gap area, with subsequent spacing losses (write as well as read). This error is minimized by making the heads with outriggers to control the wrap angles within the head contact area. Now a deeper or lesser head penetration will result in much smaller wrap angle errors; see Fig. 18.10, right.

It is useful to design the heads with a contour that truly will reflect the one that will exist after some tape wear. It is then properly seated against the tape. Figure 18.11 shows how the contour curve *y* versus distance *x* from the gap line can be calculated, using a model of a simple deflected beam. Plotting additional curves by using the formula for *y* in Fig. 18.12 has resulted in curves that were completely similar to Clurmans (1978). Other head contours were discussed and evaluated by Van Groenou et al. (1989).

It is a curiosity that the detrimental effect of head wrap (without outriggers) recently has been employed in an elegant and accurate way of measuring head wear. The method eliminates the usual concern about how realistic the given method of abrasion is: A head is worn in to a certain (but arbitrary) wear pattern. The head is next moved farther "into" the tape path, and a situation as shown in Fig. 18.10, left and bottom, exists. The replay voltage V_1 (dB) from a prerecorded short wavelength signal is now attenuated due to the spacing loss. When the tape runs over the head it will gradually wear, the spacing loss will reduce, and a new voltage V_2 (dB) measured. The wear depth can now simply be calculated from (Rogers and Hinteregger 1993):

$$Worn\ depth = (V_2 - V_1) \times wavelength/55$$

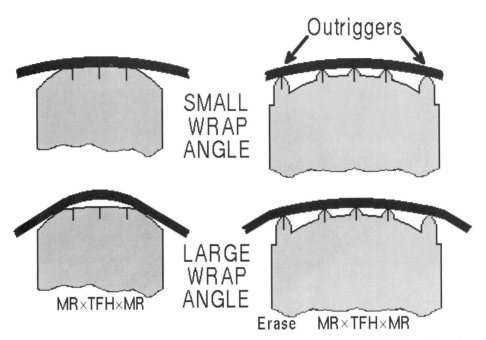

FIGURE 18.10 Wear patterns will prevent good tape/head contact if the wrap angle is increased (left). This problem can be eliminated by using outriggers as shown to the right.

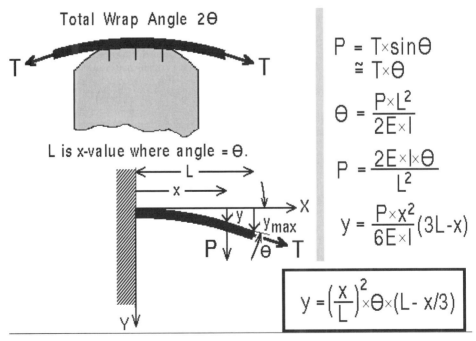

FIGURE 18.11 Calculation of tape deflection y versus distance x from gap, using a deflected-beam model.

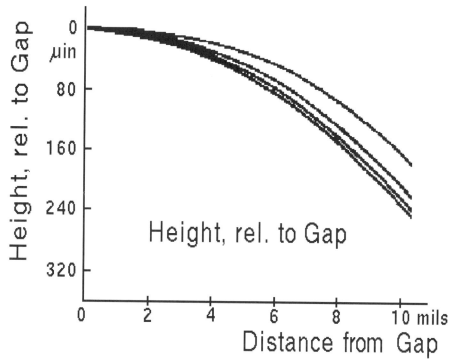

FIGURE 18.12 Head contours after prolonged wear (*after Clurman*).

The head wear is less than normally experienced in tape transports. One set of wear data is shown in Fig. 18.13 (Gills 1989). The wear rate is initially only 0.4×10^{-3} angstrom/foot and decreases thereafter. This lower wear is in part due to the lesser tape tension, and in part due to carefully balanced tapes that have a minute amount of abrasive particles embedded in the coating to cause a small amount of abrasion of primarily debris formations on the head. The balance has evolved to a point where the head wear is only slightly above what it normally would be without the abrasives.

Proper tape wrapped around the front of a general tape head must ensure good contact for small spacing losses and high resolution. Tape tension must also be a minimum in order to prevent excessive head wear and/or tape stretching damage. The air film drawn along with the moving tape will at some point create an air bearing that separates the tape from the head. This aerodynamic bearing has most recently been studied and measured by Lacey and Talke (1992).

Their approach permits tape and head asperities to contact, and they define spacing as the difference between the local mean height of the tape and the local mean height of the head, neglecting surface roughness. Because the roughness of tape is typically much larger than head surface roughness, contact pressure vs. spacing measurements were made with a magnetic tape sample on glass. A sensitive interferometer was used to measure tape-to-head spacing as function of pressure. An example of their findings is shown in Fig. 18.14.

Recently a wear-resistant overcoat has been developed for a small Philips tape cassette system (Zieren et al. 1994). The heads are manufactured by Seagate Corp., and contains, like the QIC heads, arrays of TF inductive and MR heads. The throat heights of all the individual heads are important and normal head wear is unacceptable. Overcoats of 65 nm CrN improved wear resistance in a normal home environment. The severe automotive environment (from a cold winter day to a hot and humid summer day) required a better coat and a proprietary SPL (super-protective layer) was developed.

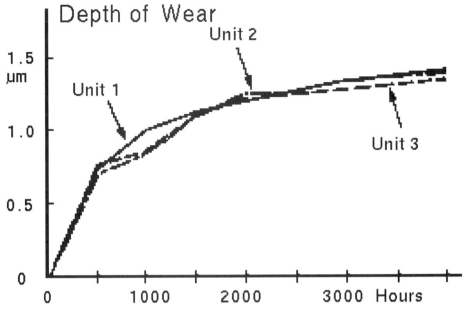

FIGURE 18.13   Head wear as function of test hours (*after Gills*).

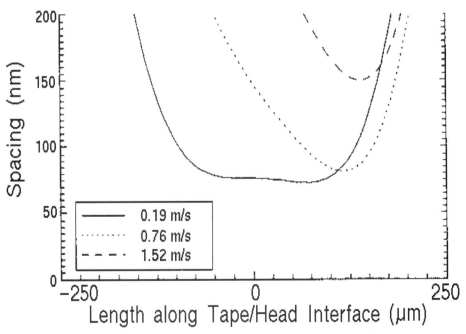

FIGURE 18.14   Spacing between tape (moving toward the right) and a head surface (*after Lacey and Talke*).

HELICAL-SCAN RECORDERS

The rotary scanner in helical-scan recorders moves past the nearly stationary tape at a velocity that ranges from 4 m/sec (for 8 mm) to 35 m/sec (for D-1 digital video recorders). At these speeds, an air bearing will normally form and defeat efforts to gain high resolution. The heads are consequently mounted to protrude 25–50 μm (1–2 mils) beyond the head drum perimeter.

The head geometry has a profound effect on the wear rate and the contact with the tape. An experimental project has led to the head geometry shown in Fig. 18.15, left (Maegawa et al. 1992). Head-to-tape contact was observed by using transparent heads and an optical interference fringe method. It is recommended to use heads having small r/R values and a smaller head area at the trailing edge, where R is a radius of head along its longitudinal direction and r is a radius along its transverse direction.

Head-tape spacing measurements are important during recorder development, and spacing is a parameter that can be measured by interferometry. Experience from an R-DAT unit revealed that the region where the head/tape spacing is less than 100 nm changes significantly as a function of tape tension in the range of 0.05 to 0.15 N. The small spacing region was not affected much by changes in scanner speed (Lacey and Talke 1991).

Tape dynamics in a helical recorder have revealed the presence of a bow wave in the tape. Figure 18.15 show these chevron-shaped waves, quasi-stationary with respect to the head. They were observed at high head speeds and low tape tensions (Lacey and Talke 1990). The nature of these waves need to be understood in order to develop very-high-data-rate helical-scan recorders.

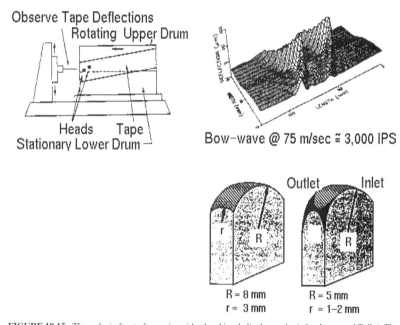

FIGURE 18.15 The wake in front of scanning video head in a helical recorder (*after Lacey and Talke*). The optimum shape of video heads (*after Maegawa et al.*).

FLEXIBLE DISKS

The heads for diskette drives are single-channel structures embedded in a housing that mounts on a movable arm, together forming a slider assembly. The head surface is rounded and the heads are named button heads.

The single-sided disk application is straightforward, as shown in Fig. 18.16, top. The double-sided disks require a head on both sides. This is best done as shown in Fig. 18.16 center, where the top head is mounted on a lever that is part of the arm system. In early designs, both heads were mounted on springs. That caused numerous failures due to vibrations in the head/diskette mechanical system.

The vibration problem is made worse by the dimple in the disk formed by the head. This contact problem has been analyzed in the past (Charbonnier 1976; Greenberg 1978; Adams 1980; and Licari and King 1981).

It is unfortunate that all the work was done by spinning a disk in close proximity to a rigid back plate, with the head protruding through a hole in the plate. The plate served as a stabilizing platform for the disk. This method eliminated the real disk vibrations that otherwise occur when they are rotating in their jackets only (see Chapter 17).

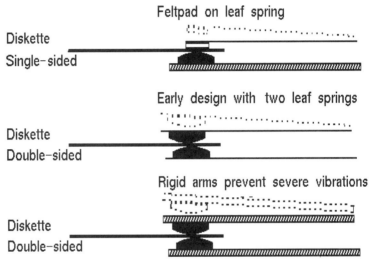

FIGURE 18.16   The heads for single-sided and double-sided diskettes.

TRIBOLOGY IN RIGID DISK DRIVES

Modern computers are insatiable with regard to the speed with which they wish to write and read data. This in turn requires higher head-to-media speeds $v = \lambda f = BL \times BR$, where BL is the bit length, and BR the bit rate. The required speeds are above the 2–3 m/s threshold for the formation of an air film, and spacing losses will occur.

The flying head in a disk drive is associated with three identifiable tribology topics:

- Potential adhesion to the disk during long periods of nonoperation
- Sporadic contact with the disk during start-stop operations (CSS)
- Flying attitude

THE NATURE OF DISK SURFACES AND TEXTURING

The disk is made from a flat material, which traditionally is an AlMg alloy. Recently glass and ceramic sheet materials have been used. Glass disks appear better than the best AlMg disk by having less glide noise and amplitude modulation of the read signal.

The disk surface is made as smooth as possible, on the order of a few angstroms rms. This will, unfortunately, invite stiction of the head to the surface, inhibit startup, and possibly cause permanent damage to slider/disk. Texturing was introduced, whereby a large contact area is broken into many smaller areas. Even though the unit area pressure now increases, the net starting torque (and potential pull-out damage to disk and/or head) is reduced (Ito et al. 1990).

The texturing does not appear to degrade the signal write/read operations. This is, of course, the strength of using digital signals. We saw earlier how an amplitude-modulated video color carrier signal would degrade when recorded/played on a rough surface (see Figs. 14.8 and 14.9). There is likewise little effect from texturing upon flying height, while landing speeds for a smoother disk were slower (Suk et al. 1990).

The effects on both signal and flying attitude will increase as the flying height goes to 0.05 μm (2 μin.) and below. Disks will be made with the smoothest possible surface for low flying heights and quiet operation. A narrow band near the disk center will be textured and assigned to start/stop landings.

The making of disk texturing may occur during a lathe operation where the disk surface is turned absolutely flat with a diamond tool; see Fig. 18.17, left. Or it can be crosshatched by, for instance, interfacing the disk with another eccentric disk, with abrasive powder between the two (Fig. 18.17, right).

Disks with crosshatched texture have lower TOV (takeoff velocity) and better tribological performance than circumferentially textured disks (Lee et al. 1992).

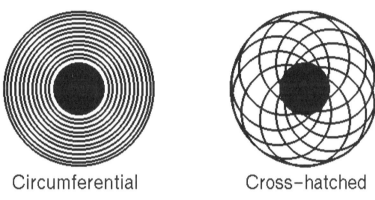

Circumferential   Cross–hatched

FIGURE 18.17   The texturing of rigid-disk AlMg substrates will reduce static friction and CSS time.

OVERCOATS, LUBRICATION, AND FRICTION

Overcoats of the disk must be provided to protect the magnetics against the environment, including slider contacts. The overcoat structure should also provide lubrication. Traditionally a carbon layer is sputtered, providing a diamond-like hard film of thickness 2–300 Å. Another overcoat material is Zirconia (ZrO_2) (Yamashita et al. 1988).

A lubricant layer should have a thickness of 20–40 angstroms (Marchon et al. 1990), sometimes thicker for rigid disks. Typical lubricants are perfluoropolyethers. Another paste-like overcoat is AgPbSb (Cohen 1992). It is claimed to be more durable than other lubricants, and the coefficient of friction remains unchanged after several thousand CSS's, where it normally increases dramatically.

EXAMINATION OF DISKS

A number of precision tools, mostly optical, have been developed for the examination of disks as well as sliders with heads. Optical interferometry (Wickramasinghe 1989) and microscopy is used for many such tasks, and well-known instrument makers are THôT, WYKO, and ZYGO (Tools and techniques are updated annually in a test equipment report; see bibliography section). Recently the atomic force microscope has gained inroads for surface analyses both mechanical and magnetic.

The flatness of a magnetic disk should be verified to assure proper disk drive operation. Flatness can be inherent in the disk itself, but might change when clamped onto the spindle motor shaft (Freeman and Eckerman 1994).

Microwaviness is a disk surface distortion that has a frequency higher than two cycles per revolution and less than three cycles per head length. This defect can generate violent oscillations of the head flexure. An example is shown in Fig. 18.18.

Asperities on a disk surface may cause damaging crashes, and should be below a certain maximum height. The largest may be removed during the disk fabrication by a burnishing head. This is a slider with a surface that looks like a waffle iron. An optical glide test (A trademark of THôT) is a noncontact method of measuring small (4 μm or 0.1 μin.) asperities protruding above the disk surface in the presence of large amounts of axial runouts. A laser vibrometer (Nakamura and Freeman 1994) is used for measuring, and collected data can be displayed as shown in Fig. 18.19.

Rather sophisticated levels of material control, processing, and cleanliness are found in disk drive manufacturing. There are many measurement and analysis tools available, and a reader will find a review from Surface Science Laboratories helpful (Smith 1990).

FIGURE 18.18 Microwaviness of rigid disk (*courtesy THôT, 1995*).

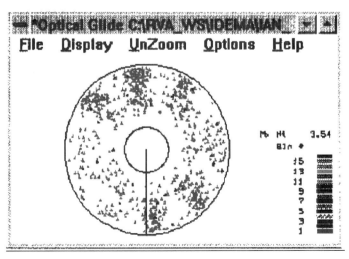

FIGURE 18.19 The optical glide test (*courtesy THôT, 1995*).

THE FLYING HEAD ASSEMBLIES

The head assembly is designed to fly over the disk surface, originally at a distance of 3 μm (120 μin.). The rpm of the disk was chosen to be high enough to allow for bit lengths that are long enough so that spacing losses are small. In this fashion, the rigid disks were born, at first using disks of diameter of 35.6 cm (14").

The flying attitude of the air-lubricated head platform with sliders has been analyzed by several researchers. The structure known as the IBM 3370 slider is shown in Fig. 18.20, with the air bearing pressure contours shown in Fig. 18.21. Definitions of key parameters that define the slider attitudes are shown in Fig. 18.22.

The sliders' behavior during the first couple of milliseconds after start-up is also shown in Fig. 18.22. The vibrations do settle rather quickly, and the flying height is thereafter determined by the speed, the load force, and the width of the slider bearing surfaces.

Any flutter of the disk will excite head vibrations. An analysis that relies on linear superposition is questionable due to a very complex behavior of the air (gas) film between the sliders and the disk surface. The usual modelling of the gas film is not valid for spacings less than 0.5 μm; in this operating region, the head clearance dimension equals only a few molecular collisions worth of gas volume.

Experimental observations must, therefore, provide us with most of our knowledge. A typical test set-up is shown in Fig. 18.23, where a head is loaded against a rotating glass disk. By shining light onto the head surface, a set of fringe patterns develop, allowing for the determination of flight height and flying attitudes. The technique has been extended to laser Doppler interferometry, which allows for the use of an ordinary coated disk (Miu et al. 1984).

These observations gain further value when information about the slider arm vibrations are added. Finite element analysis are reported in several papers (Tagawa and Mashimoto 1985).

Additional information is available from a miniature accelerometer mounted on the slider arm (Yeack-Scranton 1986). It measures the noise generated by acoustic emission (AE), which are high-frequency elastic waves generated by the deformation or destruction of solids. AE is generated by the slider-to-disk contact (Kita et al. 1980).

The AE rms voltage and slider flying height are shown in Fig. 18.24. There are three distinct ranges. In range I, the whole flat part of the slider touches the disk surface continuously. In range II, the leading edge of the slider begins to fly, and in range III, light measurements indicates that the slider flies without contact; obviously there are occasional collisions between the slider and projections of the disk surface.

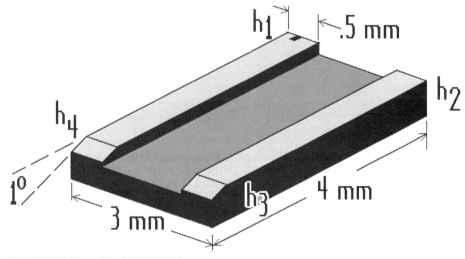

FIGURE 18.20 The traditional IBM 3370 slider.

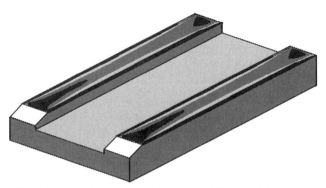

FIGURE 18.21 The formation of an air bearing under 3370 sliders (*after White*).

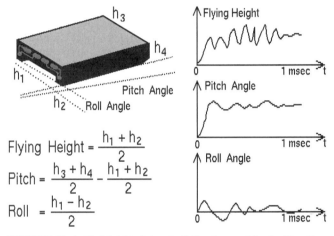

$$\text{Flying Height} = \frac{h_1 + h_2}{2}$$

$$\text{Pitch} = \frac{h_3 + h_4}{2} - \frac{h_1 + h_2}{2}$$

$$\text{Roll} = \frac{h_1 - h_2}{2}$$

FIGURE 18.22 A slider's height, pitch, and roll. How they are defined and how they behave after the start-up of disk rotation.

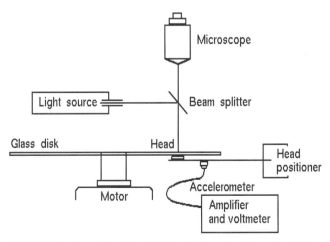

FIGURE 18.23 A test of flying height with laser Döppler setup. Suspension vibration can be monitored with a small piezoelectric transducer.

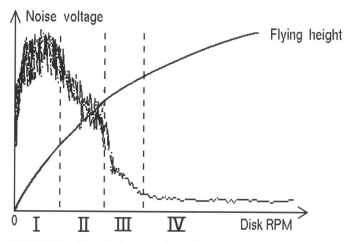

FIGURE 18.24 Glide noise decreases as flying height increases.

The fearsome head crash problems have been analyzed in two papers (Kawakubo et al. 1984; Kumaran and Chou 1985). There is no easy cure: ensure utter cleanliness, air filtration, perfect disk surfaces without large projections, and good parking zones for the head, and avoid pressure from the marketing department.

AIR BEARINGS AND LINEAR ACTUATOR SYSTEMS

Air bearings have been studied and described in several papers in the *IBM Journal of Research and Development*, November 1974 issue. A foil-bearing theory for the tape-head interface is described (Stahl et al. 1974), together with measurements (Vogel and Groom 1974). A like treatment is made for the disk-head interface (Tseng and Talke 1974), along with a technique for the measurement of flying height (Fleischer and Lin 1974). A frequency domain analysis for the dynamics of a slider bearing is important for the design and appraisal of the entire disk/head/arm system (Ono 1975.)

The air bearing theory is cast in the Reynolds equation, a single differential equation relating pressure, density, surface velocities, and film thickness for a lubricating film. An exact solution is impossible, and numerical methods are employed for a modified Reynolds equation. This modification is necessary because the film thickness between head and disk no longer is large compared to the mean free path of air molecules (64 nm, or 2.6 µin.). Several papers describe solutions, and are also found on page 675 in Bhushan's book (1990).

Figure 18.20 showed the traditional IBM 3370 head that is found in smaller sizes in current small disk drives (70-, 50-, and 30-percent scalings). The 3370 head assembly is made from a ceramic slider that is held against the disk with a pressure of 9–15 grams, which is countered by the air bearing pressure.

The air flows under the taper and pressure builds up under the rails, of largest magnitude near the taper and at the end where the head is located. Between these points the pressure changes gradually, as shown in Fig. 18.21. It decreases gradually to zero at the edges of the rails.

The suspension arm's pressure is applied through a gimbal at the center of the slider. This force will, together with the forces along the rails, cause the head to lift off from the moving disk surface and produce a pitch that brings the trailing end of the slider in closest position to the disk (i.e., flying height). Both pitch and flying height (at the head gap) increases as the head moves toward the outside tracks, due to the higher velocity there.

AIR BEARINGS AND ROTARY ACTUATOR SYSTEMS

Figure 18.25 illustrates that the flying height of an in-line disk head on a rotary arm is proportional to the head-disk speed (in-line refers to the fact that the head sliders are parallel to the suspension arm rather than the dog-leg mounting found on drives with linear actuators, where the slider "direction" is perpendicular to the suspension arm.) It is also proportional to the width of the rails. A thorough examination of the attitudes of a head with its suspension arm is increasingly important as the flying height is reduced to 1–2 microinches. The inner and outer rails of the head are subject to slightly different aerodynamics, in particular when a rotary actuator is used. This will lead to roll in the head.

Figures 18.26 and 18.27 show variations in flying height and pitch as function of the head position near the center or the edge of the disk. The drawing also shows a source of roll imparted to the slider: Air gliding sideways in and under the slider. The amount of roll is characterized by the skew. Skew is defined as positive when the leading edge of the slider is nearest the disk center, and negative when the trailing edge is nearest; see Figs. 18.26 and 18.27.

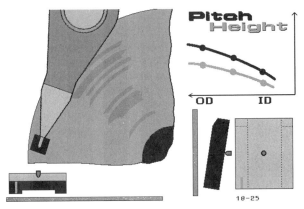

FIGURE 18.25 Pitch and height increase with the distance from the disk center due to the higher velocity.

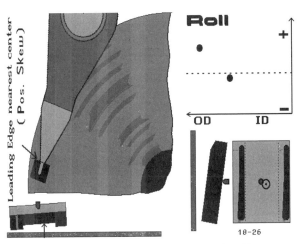

FIGURE 18.26 Roll increases at outer tracks, where skew is positive.

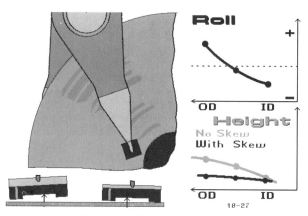

FIGURE 18.27 Roll becomes small and negative at inner tracks, where skew also is negative.

The lower right corner of both illustrations shows the slider as seen from above. Underneath, the pressure ridges are shown, and when the skew is positive we find these ridges pushed inwards by the air rushing in from the left outside. The center of the average air pressure is moved toward the disk center, producing a torque that results in a positive roll, where the inside rail is further away from the disk than the outside.

The negative skew produces a negative roll, as shown in Fig. 18.27. The head is always mounted at the outside rail, and the effect of the roll pushes the head closer to the disk, partly compensating the slider's larger height. This improvement toward constant flying height is a cause of roll that remains undesirable; it can potentially lead to a head crash.

MODERN LOW-PRESSURE SLIDERS

Different slider rails can be used to control the flying attitude, as shown in Fig. 18.28. The 3380K slider increases the pitch (Nishihara et al. 1988).

A zero-load design has a transverse rail that creates a vacuum on the middle of the slider. This can be used for self-loading. The stiffness is 5–10 times higher than for the 3370 head, and the slider is better in following disk wobble, etc. The variations in flying height from an inside to an outside track are also less than for other sliders.

Only a few grams of pressure from the suspension arm is required. This translates into smaller contact pressure during stops; danger of stiction is reduced, as is potential damage during CSS (contact-start-stop) (White 1983). Also, load/unload processes have special problems (Hashimoto and Tagawa 1989).

Insensitivity to skew is achieved by milling tapers on the sides of the rails, as shown in the TPC slider (Transverse Pressure Contour). Roll is essentially eliminated, and air-bearing dampening increases. Natural frequencies are above 30 kHz (White 1987).

The practical slider geometries were pretty well dictated by the capabilities of milling and grinding equipment. More exotic geometries had to wait until slider manufacturers gained experience with ion milling and similar tools. More elaborate designs have led to tapered rails, which provide fast takeoff and a flying height that is insensitive to disk velocity (Yoneka et al. 1991). The recess between the rails is a shallow 2 μm, sufficient to generate suction (negative pressure).

The characteristics listed above are found in today's modern sliders, Read-Rite's Tripad, and Seagate's AAB (advanced air bearing) (Hardie et al. 1994; Menon 1994). The slider geometries are shown in Fig. 18.29. Both sliders have a shallow recess area between the rails. The Tripad uses a neg-

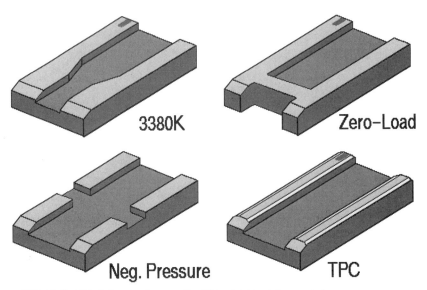

FIGURE 18.28   Slider designs that decrease roll and flying-height variations; see text.

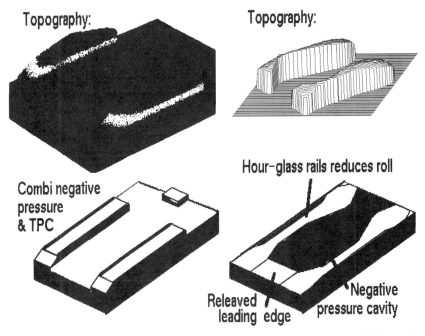

FIGURE 18.29   Read*Rite's TriPad (TM) and Seagate's AAB (TM) slider designs as seen by a WYKO or ZYGO instrument (*courtesy of Read*Rite and Seagate Technologies*).

ative pressure effect near the center of each rail, and an open stretch between side rails and center rail (with the TFH head structures). The AAB head relies on the hourglass shape of the rails. Additional papers of aspects of sliders and their flying attitudes are listed in this chapter's bibliography.

MEASUREMENTS OF FLYING HEIGHT

As dimensions shrink, new problems arise, some in the area of measurements. Optical instruments are traditionally used to measure distances, and we have already mentioned laser Doppler interferometry. Another proposed method uses two laser beams, one reflected of the slider back side, the other off the disk surface right next to the slider. Half the flying height is determined as the travel distance between the two beams, measured during stop and during operation (Veillard 1993, and Smith et al. 1993).

An interferometric system for measurement of spacing between a transparent disk and a slider was shown in Fig. 18.23. The effect of phase shift on reflection off the slider surface must be considered to obtain an accurate measurement. The intensity of light is determined by the path difference for the two interfering beams: It is zero when the travel distance (two flying heights) equals a multiple of wavelengths. The light is dark because the reflected light undergoes a 180° phase shift. Maximum light intensities are observed when the path difference is one half wavelength. When using three wavelengths, the light intensity varies as shown in Fig. 18.30. It is difficult to use a single light source (monochromatic), because readings for this method at short distances (a few microinches) are made on the initial slope on one of the curves in Fig. 18.30. An error is made because the reflected phase is assumed to be 180°. It is in reality dependent on the reflecting material.

A new technique uses an ellipsometric measurement of the slider's complex index of refraction, from which phase shift on reflection is calculated (Lacey et al. 1993). Measurements are further made by using the reflected yellow (580 nm), green (548 nm), and blue (436 nm) light beams from a mercury arc high-intensity light source. Then the flying height will correspond to three points on the different wavelength curves (Lacey et al. 1994). By including earlier ellipsometry data, a precise determination of flying height is possible with a high signal-to-noise ratio. Dynamic data are readily obtained by assistance from a computer. Additional papers on measurement techniques and topics are found in this chapter's bibliography.

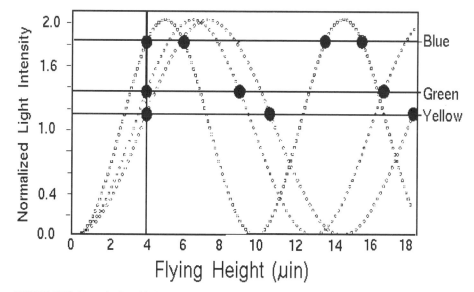

FIGURE 18.38 Determination of flying height by measuring with three wavelengths of light (*after Phase Metrics, Inc.*).

REFERENCES

Advanced in Information Storage Systems, 1991-present (Vol. 1 thr. 5), ASME Press Series, ASME, 345 Ea. 47th Street, New York, NY 10017, USA.

Bhushan, B. 1990. *Tribology and Mechanics of Magnetic Storage Devices*. Springer-Verlag, 1019 pages.

Bhushan, B. 1992. *Mechanics and Reliability of Flexible Magnetic Media*. Springer-Verlag, 564 pages.

Halling, J. (Editor). 1978. *Principles of Tribology*, The Macmillian Press Ltd., London.

Head/Media Research Journal, 1995-present, Don Mann Magnetics, Inc., 750 Camino Las Conchas, Thousand Oaks, Calif 91360, USA.

Rabinowicz, E. 1974. *Friction, Wear and Lubrication*. MIT, Cambridge, Mass.

Talke, F.E., and D.B. Bogy. 1987. Head-Medium Interface, in *Magnetic Recording*. Ed. by C.D. Mee and E.D. Daniel, McGraw-Hill, pp. 427–506.

TAPES AND FLOPPY SURFACES: FRICTION, LUBRICATION, AND WEAR

Aoyama, S., and M. Kishimoto. March 1991. The Behavior of Lauric Acid Lubricant in a Magnetic Coating Layer and Its Effect on Mechanical Properties of Magnetic Media. *IEEE Trans. Magn.* MAG-27 (2): 791–94.

Archard, A.R. Aug. 1953. Contact and Rubbing of Flat Surfaces. *Jour. Appl. Phys.* 24 (8): 981–988.

Hilden, M., J. Lee, V. Nayak, G. Ouano, and A. Wu. Jan. 1990. Sputtered Carbon on Particulate Media. *IEEE Trans. Magn.* MAG-26 (1): 174–178.

Jorgensen, F. Feb. 1972. Investigation and Measurment of Magnetic Head/Tape Interface Problems I-II. *Telemetry Journal* 7 (2): 13–16, 21–24.

Lorenz, M.R., V.J. Novotny, and V.R. Deline. Nov. 1991. Lubricant Migration in Particulate Magnetic Recording Media. *IEEE Trans. Magn.* MAG-27 (6): 5088–90.

Mukhopadhyay, A. Sept. 1978. Study of the Effect of Static/Dynamic Coulomb Friction Variation at Tape-head Interface of Spacecraft Tape Recorder by Nonlinear Time Response Simulation. *IEEE Trans. Magn.* MAG-14 (5): 333–335.

Schnurmann, R., and E. Warlow-Davies. Jan. 1942. The Electrostatic Component of the Force of Sliding Friction. *Physical Society* 54 (1): 14–27.

Williams, G.I., and H.M. Lewis. April 1982. Review of Methods for Measuring Abrasiveness of Magnetic Tape. *Intl. Conf.Video and Data Rec.* IERE Publ. No. 54, pp. 11–21.

HEADS FOR FLEXIBLE MEDIA: LONGITUDINAL TRANSPORT TAPE HEADS

Clurman, S. Sept. 1978. A Predictive Analysis for Optimum Contour-After-Wear for Magnetic Heads. *IEEE Trans. Magn.* MAG-14 (5): 339–41.

Gills, D. Aug. 1989. Reliability Assessment of Quarter-Inch-Cartridge. *HP Journal* 40 (4): 74–78.

Lacey, C.A., and F.E. Talke. Oct. 1992. Measurement and Simulation of Partial Contact at the Head Tape Interface. *Trans. Tribology* 114 (4): 646–652. Also presented at a THIC Meeting, Jan. 1992.

Rogers, A.E.E., and H.F. Hinteregger. Jan. 1993. Measurement of Magnetic Tape Abrasivity by Interchanging Tape Thickness. *Tribology Trans.* 36 b(1): 139–143.

Van Groenou, A.B., J.G. Fijnvandraat, S.E. Kadijk and J.P.M. Verbunt Oct. 1989. The Shape of Heads for Tape Recording: Experiments and Calculations for Various Tapes. *ASLE Spec. Publ. on Tribology and Mechanics of Magn. Storage*. Vol. 6, pp. 120–127.

Zieren, V., M. de Jongh, A. Broese van Groenou, J.B.A. van Zon, P. Lasinski and G.S.A.M. Theunissen. March 1994. Ultrathin Wear-Resistant Coatings for the Tape Bearing Surface of Thin Film Magnetic Heads for Digital Compact Cassette. *IEEE Trans. Magn.* MAG-30 (2): 340–345.

VTR HEADS

Lacey, C.A., and F.E. Talke. Nov. 1991. Interferrometric Measurement of Head/Tape Spacing in a High-Speed R-Dat Helical Recorder. *IEEE Trans. Magn.* MAG-27 (6): 5091–93.

Lacey, C.A., and F.E. Talke. Sept. 1990. Tape Dynamics in a High-Speed Helical Recorder. *IEEE Trans. Magn.* MAG-26 (5): 2208–10.

Maegawa, T., Y. Morioka, A. Kuroe, and M. Kobayashi. Sept. 1992. Optimum Magnetic-Head Shape Design for High-Density Recording. *IEEE Trans. Magn.* MAG-28 (5): 2557–59.

HEADS FOR FLOPPY DISKS

Adams, G.G. Jan. 1980. Analysis of the Flexible Disk/Head Interface. *ASME Jour. of Lubr. Tech.* 102 (1): 86–90.

Charbonnier, P.P. Nov. 1976. Flight of Flexible Disk over Recording Heads. *IEEE Trans. Magn.* MAG-12 (6): 728–730.

Greenberg, H.J. Sept. 1978. Flexible Disk-Read/Write Head Interface. *IEEE Trans. Magn.* MAG-14 (5): 336–338.

Licari, J. and King, F. Dec. 1981. Elastohydrodynamic Analysis of Head to Flexible Disk Interface Phenomena. *Journal of Applied Mechanics* 48 (?): 763–768.

RIGID DISK SURFACES

Cohen, U. Jan. 1992. A Stictionless Soft Overcoat for Non-textured Media. *ECS Proc. 2nd Int. Symp. on Magn. Matl. Proc. & Dev.*, 6 pages, 4 ill.

Freeman, I., and J. Eckerman. 1994. A Study of Disk Flatness and Clamping Distortion using Doppler Laser Vibrometry. A white paper from, *THôT Technologies*, *Inc.* 271 E. Hacienda Ave., Campbell, CA 95008.

Ito M., T. Shinohara, K. Yanagi, and K. Tanaka. Sept. 1990. Influence of Surface Texture of a Thin-Film Rigid Disk. *IEEE Trans. Magn.* MAG-26 (5): 2502–04.

Lee, J.K., A. Chao, J. Enguero, M. Smallen, H.J. Lee, and P. Dion. Sept. 1992. Effect of Disk Cross Hatch Texture on Tribological Performance. *IEEE Trans. Magn.* MAG-28 (5): 2880–82.

Marchon, B., M.R. Khan, N. Heiman, P. Pereira, and A. Lautie. Sept. 1990. Tribochemical Wear on Amorphous Carbon Thin Films. *IEEE Trans. Magn.* MAG-26 (5): 2670–75.

Nakamura, H., and I. Freeman. 1994. The History and Development of Ultra Low Glide Capability, A New Approach Using Doppler Lasers. A white paper from , *THôT Technologies*, *Inc.* 271 E. Hacienda Ave., Campbell, CA 95008.

Smith, J.T. Oct. 1990. Materials Analysis for the Magnetic Disk Drive Industry. *Solid State Technology*, 4 pages.

Suk, M., B. Bhushan, and D.B. Bogy. Sept. 1990. Role of Disk Surface Roughness on Slider Flying Height and Air-Bearing Frequency. *IEEE Trans. Magn.* MAG-26 (5): 2493–95.

Wickramasinghe, H.K. Oct. 1989. Scanned-Probe Microscopes. *Sci. American.* 261 (4): 98–105.

Yamashita, T., G.L. Chen, J. Shir, and T. Chen. Nov. 1988. Sputtered ZrO2 Overcoat with Superior Corrosion Protection and Mechanical Performance in Thin Film Rigid Disk Application. *IEEE Trans. Magn.* MAG-24 (6): 2629–34.

FLYING ATTITUDE OF SLIDERS

Fleischer, J.M., and Lin, C. Nov. 1974. Infrared Laser Interferometer for Measuring Air-bearing Separation. *IBM. Jour Res. Devel.* ? (18): 529–533.

Hardie, C., A. Menon, P. Crane and D. Egbert. March 1994. Analysis and Performance Characteristics of the Seagate Advanced Air Bearing Slider. *IEEE Trans. Magn*. MAG-30 (2): 424–432.

Hashimoto, M., and N. Tagawa. Sept. 1989. Self-Load/Unload Slider Dynamics for Non-Contact Start Stop Operations with Negative Pressure Flying Head Mechanisms. *IEEE Trans. Magn*. MAG-25 (5): 3719–3721.

Kawakubo, Y., H. Ishihara, Y. Seo, and Y. Hirano. Sept. 1984. Head Crash Process of Magnetic Coated Disk during Start/Stop Operations. *IEEE Trans. Magn*. MAG-20 (5): 933–935.

Kita, T., K. Kogure, Y. Mitsuya, and T. Nakamishi. Sept. 1980. New Method of Detecting Contact between Floating-Head and Disk. *IEEE Trans. Magn*. MAG-16 (5): 873–75.

Kumaran, A.R., and Y.S. Chou. Sept. 1985. Effect of Head/Disc Imperfections on Gas Lubricated Slider Performance. *IEEE Trans. Magn*. MAG-21 (5): 1515–1517.

Lacey, C., J.A. Adams, E.W. Ross, and A. Cormier. 1994. A New Method for Measuring Flying Heights Dynamically. A *white paper from Phase Metrics*, 3978 Sorrento Valley Blvd., San Diego, CA 92121.

Lacey, C., R. Shelor, A.J. Cormier, and F.E. Talke. Nov. 1993. Interferometric Measurement of Disk/Slider Spacing: the Effect of Phase Shift on Reflection. *IEEE Trans. Magn*. MAG-29 (6): 3906–08.

Menon, Aric. March 1994. Challenges of Sub-2 Microinch Flying Height. *IDEMA's Diskcon Singapore '94 Intl. Tech. Conf.* pp. 123–147.

Miu, D.K., G. Bouchard, D.G.Bogy, and F.E. Talke. Sept. 1984. Dynamic Response of a Winchester-type Slider Measured by Laser Doppler Interferometry. *IEEE Trans. Magn*. MAG-20 (5): 927–929.

Nishihara, H.S., L.K. Dorius, S.A. Bolasna, and G.L. Best. Oct. 1988. Performance Characteristics of IBM 3380 K Air Bearing Design. *Publ. on Tribology and Mechanics of Magn. Storage*. Vol 5, pp. 117–23.

Ono, K. April 1975. Dynamic Characteristics of Air-Lubricated Slider Bearing for Non-contact Magnetic Recording. *Jour. Lubrication Tech*. Vol. 97, pp. 250–260.

Smith, P.W., S.K. Ganapathi, and D.H. Veillard. Nov. 1993. Measurement of Head-Disk spacing using Laser Heterodyne Interferometry—Part II: Simulation and Experiments. *IEEE Trans. Magn*. MAG-29 (6): 3912–14.

Stahl, K.J., J.W. White, and K.L. Deckert. Nov. 1974. Dynamic Response of Self-Acting Foil Bearings. *IBM Jour. Res. Devel*. 18 (?): 513–520.

Tagawa, N., and M. Mashimoto. Sept. 1985. Submicron Spacing Dynamics for Flying Head Slider Mechanisms Using Building Block Approach. *IEEE Trans. Magn*. MAG-21 (5): 1506–1508.

Tseng, R.C., and Talke, F.E. Nov. 1974. Transition from Boundary Lubrication to Hydrodynamic Lubrication of Slider Bearings. *IBM Jour. Res. Devel*. ? (18): 534–540.

Veillard, D.H. Nov. 1993. Real-time tracking of the head-disk separation using Laser Heterodyne Interferometry. Part one: Instrumentation. *IEEE Trans. Magn*. MAG-29 (6): 3909–11.

Vogel, S.M., and J.L. Groom. Nov. 1974. White Light Interferometry of Elastohydrodynamic Lubrication of Foil Bearings. *IBM Jour. Res. Devel*. ? (18): 521–528.

White, J.W. Oct. 1987. Dynamic Response of the Transverse Pressure Contour Slider. *ASLE Spec. Publ. on Tribology and Mechanics of Magn. Storage*. Vol. 4, pp. 72–82.

White, J.W. 1983. Flying Characteristics of the zero-load slider bearing. *Jour. Lubr. Techn., Trans. ASME* Vol. 105 pp. 484–490.

Yeack-Scranton, C.E. Sept. 1986. Novel Piezoelectric Transducers to Monitor Head-Disk Interactions. *IEEE Trans. Magn*. MAG-22 (5): 1011–16.

Yoneoka, S., M. Katayama, T. Ohwe, Y. Mizoshita, and T. Yamada. Nov. 1991. A Negative Pressure Microhead Slider for Ultralow Spacing with Uniform Flying Height. *IEEE Trans. Magn*. MAG-27 (6): 5085–87.

BIBLIOGRAPHY

Bhushan, B., and B.K. Gupta. 1991. *Handbook of Tribology—Materials, Coatings, and Surface Treatment*. McGraw-Hill.

Gross, W.A., L. Matsch, V. Castelli, A. Eshel, T. Vohr, and M. Wilamann. 1980. *Fluid Film Lubrication*. Wiley.

IDEMA Sub 2-Microinch Fly Height Workshop. May 12, 1993. IDEMA, 109 pages.

Test Equipment Report. 1995. Peripheral Research Corp., 351 So. Hitchcock Way B-200, Santa Barbara, CA 93105, 125 pages.

TAPES AND FLOPPY SURFACES: FRICTION, LUBRICATION, AND WEAR

Bhushan, B. Feb. 1990. Magnetic Media Tribology—State of the Art and Future Challenges. *Wear* 136 (1): 169–197.

Harth, K., H. Hibst, H. Mannsperger, H.P. Schildberg, and A. Werner. Jan. 1990. Tribological Investigation of Metallic Layers on Flexible Substrates. *IEEE Trans. Magn.* MAG-26 (1): 156–158.

Kim, K.J., E.G. Kolycheck, F. Habbal, and B.J. Falabella. Jan. 1990. Magnetic Media Durability: A System's Approach. *IEEE Trans. Magn.* MAG-26 (1): 159–164.

Lacey, C., and F.E. Talke. Sept. 1992. Simulation of Wear of Tape Head Contours. *IEEE Trans. Magn.* MAG-28 (5): 2554–56.

Osaki, H., E. Oyanagi, H. Aonuma, T. Kanou and J. Kurihara. Jan. 1992. Wear Mechanism of Particulate Magnetic Tapes in Helical Scan Video Tape Recorders. *IEEE Trans. Magn.* MAG-28 (1): 76–83.

Osaki, H., H. Uchiyama, and N. Honda. Jan. 1993. Wear Mechanisms of Co-Cr Sputtered Deposited Magnetic Tapes in Helical Scan Recorders. *IEEE Trans. Magn.* MAG-29 (1): 41–58.

Osaki, H., K. Fukushi, and K. Ozawa. Nov. 1990. Wear Mechanisms of Metal-Evaporated Magnetic Tapes in Helical Scan Videotape Recorders. *IEEE Trans. Magn.* MAG-26 (6): 3180–85.

Texier, L.F., J.M. Coutellier, and N. Blanchard. Oct. 1992. Head to Tape Contact Characterization for a Fixed Multi-Track Magnetic Head. *Magnetic Materials, Processes, and Devices*. Electrochemical Society Proc. 92–10, pp. 171–176.

van Groenou, A.B. Jan. 1990. On the Interpretation of Tape Friction. *IEEE Trans. Magn.* MAG-26 (1): 144–146.

VTR HEADS

Eshel, A. Oct. 1991. Simulation of Head Contour Effects in Helical Scan Video Recording. *Tribology Trans.* 34 (4): 573–576.

Osaki, H., J. Kurihara, and T. Kanou. July 1994. Mechanisms of Head-Clogging by Particulate Magnetic Tapes in Helical Scan Video Tape Recorders. *IEEE Trans. Magn.* MAG-30 (4): 1491–1498.

Sundaram, R., and R.C. Benson. Sept. 1990. Tape Dynamics Following an Impact. *IEEE Trans. Magn.* MAG-26 (5): 2211–13.

DISK SURFACES

Ishida,S., and K. Seki. 1990. Thin Film Disk Technology. *Fujitsu Sci. & Tech. Jour.* 26 (4): 337–52.

Lee, J.K., M. Smallen, H.J. Lee, J. Enguero, and A. Chao. Jan. 1993. Effects of Chemical and Surface Properties of Hydrogenated Carbon Overcoats on the Tribological Performance of Rigid Magnetic Head/Disk Interface. *IEEE Trans. Magn.* MAG-29 (1): 276–281.

O'Connor, T.M., M.S. Jhon and C.L. Bauer. 1993. Diffusion and Degradation of Disk Lubricants. *Lecture notes from Data Storage System Center*, Carnegie Mellon University, Pittsburg, PA.

Trauner, D., Y. Li, and F.E. Talke. Jan. 1990. Frictional Behavior of Magnetic Recording Disks. *IEEE Trans. Magn.* MAG-26 (1): 150–152.

FLYING ATTITUDE OF SLIDERS: MEASUREMENTS

Bhushan, B. July 1992. Magnetic Head-Media Interface Temperatures—Part 3: Application to Rigid Disks. *Trans. ASME* Vol. 114 pp. 420–430.

Bogy, D.B., and X. Yun. March 1994. Enhancement of Head-Disk Interface Durability by Use of Diamond-Like Carbon Overcoats on the Sliders's Rails. *IEEE Trans. Magn.* MAG-30 (2): 369–374.

Bolasna, S.A. Nov. 1990. Air Bearing Parameter Effects on Take-Off Velocity. *IEEE Trans. Magn.* MAG-26 (6): 3033–38.

Chandrasekar, S., and B. Bhushan. April 1991. Friction and Wear of Ceramics for Magnetic Recording Applications .2. Friction Measurements. *Jour. Tribology.* 113 (2): 313–317.

Clark, B.K. Nov. 1991. An Experimental Correlation of Slider-Disk Contact Detection between Piezoelectric and Electrical Resistance Measurements. *IEEE Trans. Magn.* MAG-27 (6): 5151–53.

Clifford, G., and D. Henze. Sept. 1989. An Air Bearing Minimizing the Effect of Slider Skew Angle. *IEEE Trans. Magn.* MAG-25 (5): 3713–3715.

Henze, D., P. Mui, G. Clifford, and R. Davidson. Sept. 1989. Multi-Channel Interferometric Measurements of Slider Flying Height and Pitch. *IEEE Trans. Magn.* MAG-25 (5): 3710–3712.

Jhon, M.S., P.R. Peck, R.F. Simmons Jr., and Th.J. Janstrom. March 1994. Behavior of the Head-Disk Interface in Future Disk Drives. *IEEE Trans. Magn.* MAG-30 (2): 410–416.

Kumaran, A.R. Sept. 1989. Head/Disk Spacing Budget Analysis—An Optimization Tool in Drive Design for Reliability. *DISKCON Tech. Conf.,* 19 Viewgraphs.

Lee, H.J., R.D. Hempstead, and J. Weiss. Sept. 1989. Study of Head and Disk Interface in Contact Start Stop Test. *IEEE Trans. Magn.* MAG-25 (5): 3722–3724.

Leo, H-L., and G.B. Sinclair. Nov. 1991. So How Hard Does a Head Hit a Disk? *IEEE Trans. Magn.* MAG-27 (6): 5154–56.

Levi, P.G., and F.E. Talke. Sept. 1992. Load-Unload Investigations on a Rotary Actuator Disk Drive. *IEEE Trans. Magn.* MAG-28 (5): 2877–79.

Oshiki, M., and S. Hamasaki. 1990. Thin Film Head Technology. *Fujitsu Sci. & Tech. Jour.* 26 (4): 353–64.

Pan, X., and V.J. Novotny. March 1994. Head Material Effects on Interface Tribochemistry. *IEEE Trans. Magn.* MAG-30 (2): 433–439.

Sonnenfeld, R. Sept. 1992. Simultaneous Fly-Height, Pitch and Crown Measurements of Hard-disk Sliders by Capacitance Stripe. *IEEE Trans. Magn.* MAG-28 (5): 2545–47.

Suk, M., T. Ishii, and D.B. Bogy. Sept. 1991. Evaluation of Capacitance Displacement Sensors used for Slider-Disk Spacing Measurements in Magnetic Disk Drives. *IEEE Trans. Magn.* MAG-28 (5): 2542–44.Trans

Suzuki, S., and F.E. Kennedy, Jr. Sept. 1989. Measurement of Flash Temperature and Contact between Slider and Magnetic Recording Disk. *IEEE Trans. Magn.* MAG-25 (5): 3728–3730.

Tokuyama, M., Y. Yamaguchi, S. Miyata, and C. Kato. Nov. 1991. Numerical Analysis of Flying-Height Fluctuation and Positioning Error of Magnetic Head Due to Flow Induced by Disk Rotation. *IEEE Trans. Magn.* MAG-27 (6): 5139–41.

Yamamoto, T., M. Takahashi, and M. Shinohara. 1990. Head-Disk Interface. *Fujitsu Sci. & Tech. Jour.* 26 (4): 415–27.

Yoneoka, S., T. Ohwe, and Y. Mizoshita. 1990. Flying Head Assemblies. *Fujitsu Sci. & Tech. Jour.* 26 (4): 404–14.

ENVIRONMENTAL INFLUENCES

Analytical Tools Used to Characterize Surface Cleanliness. 1992. *Reflections*, from Surface Science Laboratories (Phone: 415-962-8767.)

Chen, G-L., J. Xuan, T. Truong, and T. Sugiyama. Nov. 1993. In-Situ Measurement of Wet Stiction at HDI and Whole Surface Mapping for Glide Height Lower than 500 Angstroms. *IEEE Trans. Magn.* MAG-29 (6): 3963–65.

Denape, J., and J. Lamon, J. Aug. 1990. Sliding Friction of Ceramics—Mechanical Action of the Wear Debris. *Jour. Matl. Sci.* 25 (8): 3592–3604.

Li, Y, D. Trauner, and F.E. Talke. Sept. 1990. Effect of Humidity on Stiction and Friction of the Head/Disk Interface. *IEEE Trans. Magn.* MAG-26 (5): 2487–89.

Raman, V., and W.T. Tang. Environment Dependent Stiction and Durability in Hydrogenated Carbon Overcoated Thin Film Disks. *IEEE Trans. Magn.* MAG-29 (6): 3933–35.

Smallen, M., P. Mee, K. Merchant, and S. Smith. Sept. 1990. Contamination Induced Stiction in Drive Level Studies. *IEEE Trans. Magn.* MAG-26 (5): 2505–07.

Strom, B.D., D.B. Bogy, C.S. Bhatia, and B. Bhushan. Oct. 1991. Tribochemical Effects of Various Gases and Water Vapor on Thin Film Magnetic Disks with Carbon Overcoats. *Jour. Tribology*, Vol. 113, pp. 689–693.

Tian, H., and T. Matsudaira. Sept. 1992. Effect of Relative Humidity on Friction Behavior of the Head/Disk Interface. *IEEE Trans. Magn.* MAG-28 (5): 2530–32.

Tsai, C.J., D.Y.H. Pui, and B.Y.H. Liu. Jan. 1991. Wear Particle Generation in Thin Film Computer Disk Drives—Experimental Study. *Aerosol Sci. & Tech.*, 15 (1): 36–48.

Van Groenou, A.B., II.J.J.C. Meulenbrocks, and M. deJongh. Jan. 1990. Level Differences in Hybrid Heads after Contact with Various Tapes. *IEEE Trans. Magn.* MAG-26 (1): 153–155.

Xuan, J., G-L. Chen, and J. Chao. Nov. 1993. Organic Buildup on Slider Leading Edhe Tapers and Its Effect on Wet Stiction. *IEEE Trans. Magn*. MAG-29 (6): 3948–3950.

Yamamoto, T., M. Takahashi, and M. Shinohara. Oct. 1990. Influence of Component Outgassing on a Head-Disk Interface. *ASLE Spec. Publ. on Tribology and Mechanics of Magn. Storage*. Vol. 7, pp. 91–94.

CHAPTER 19
ADVANCED TOPICS IN RECORDING

Research efforts in the recording industry require a detailed understanding of the magnetic process whereby a transition is recorded and read back. The mechanism of pulse recording and the influence of head/media parameters has been studied extensively in the past. The recording or write process has not been open to a rigorous and accurate treatment due to the nonlinear behavior of the magnetics, the interaction fields, and simultaneous satisfaction of several conditions (changing demagnetization during the record process). A fair agreement between measurements and some of the theories has been established for pulse recording at low densities, but modern recordings with a hundred thousand or more transitions per inch of track length have not been fully theoretically explained.

We will first review earlier models for the record process, and then consider the effects of the demagnetizing field from a just-written transition. This will cause a broadening and a forward shift of the transition. An immediately following transition will also be broadened, but shifted less. This will cause a time error between the two transitions, and it will be further affected if overwriting old data.

We have assumed that the medium's coercivity H_r is a constant parameter. This assumption allowed us to use the field strength contours from Figs. 5.15, 5.16, and 5.17 to find the outline of the transition (without demagnetization.) Recent data show that the coercivity varies with field angle, from along the media surface to perpendicular thereto. This will change the transition shape in the simple model, and will certainly also affect the results from more elaborate computations.

Perpendicular recording will be reviewed because it allows for very high densities. 600 kfci has been demonstrated. Narrow track recordings will require tighter servos for track following, because off-track performance degrades faster in perpendicular recording than in longitudinal recording.

A switching time limit of particulate media was mentioned at the end of Chapter 3. The phenomenon will be reviewed, showing a potential upper limit of a 200–300 MHz data rate for γFe_2O_3 and 100 MHz for BaFeO. The required record-field strength at high data rates will also be shown to be higher than at low rates.

Very-short-wavelength recording must be done with a short-gap method. A simple illustration will illustrate this.

EARLY MODELLING OF THE RECORDING PROCESS

Direct recording was introduced in Chapter 5, and was initially explained through a series of graphical projections. This recording (or write) model is extraordinarily simple; the recording zone is merely a thin sheet, placed perpendicular to the front plane of the write head (Fig. 19.1a). The coating is magnetized in its full thickness, in accord with the longitudinal field strength H_x, which is considered equal to the gap field strength H_g. The perpendicular field H_y is equal to zero everywhere, and the track width w is set equal to infinity.

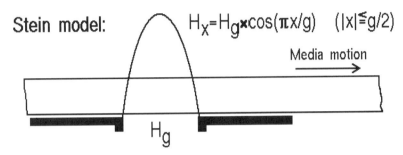

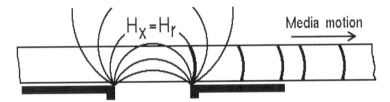

FIGURE 19.1   An early model of the recording process: Stein's remanence model and Bauer/Mee's bubble (or cylinder) model.

The remanent magnetization in this simple remanence model is calculated as the graphical projection onto the *M*-axis in the *MH* graph (Figs. 5.7 through 5.9). We will now review developments of models that represent the actual situation in better ways.

These models can be of great value for us in understanding the digital record process, and can explain some of the things we observe, like peak shifts; these are phenomena where the timing between the reproduced voltage peaks differ from the timing between the write current pulse transitions.

We should also like to obtain a correlation between write/read performance and the magnetic properties of the magnetic coating. If we can first design a model by using our knowledge and experience, and then verify the model by experiments, then we may use it both as a design guide and in calculations of performances.

Three recording models were proposed in publications in 1961: by Stein, by Bauer and Mee (see the reference section for Chapter 5), and by Schwantke (see the reference section for Chapter 20).

Stein modified the remanence model so the field strength across the gap varied as a cosine function. It was purely longitudinal, and was considered confined to the gap (Fig. 19.1, top):

$$H_x = H_g\cos(\pi x/g) \text{ for } |x| \le g/2$$

$$H_x = 0 \text{ for } |x| > g/2$$

$$H_y = 0 \text{ everywhere}$$

The initial magnetization curve was linear, and the *MH*-loops were simple parallelograms. The use of the model in calculations gave reasonable results when compared with experimental data for thin coatings and at long wavelengths.

Bauer and Mee recognized that the magnetization in a coating was not necessarily uniform in depth. Also, the remanence was obtained in a zone somewhere after the gap. They analyzed the recording field and postulated their bubble theory in 1961. Certain restrictions were necessary:

- All particles are identical, with the same coercive forces.
- There is no interaction between particles.
- The particles exhibit longitudinal anisotropy, and cannot support a perpendicular magnetization.
- The recorded wavelength $\lambda > g$
- The coating thickness $\delta_{rec} > g$

It is then permissible to consider only the horizontal component $H_x(x,y)$ of the total recording field:

$$H_x(x,y) = H_{total}\cos \beta$$

where:

$$H_{total} = (g/\pi)H_g/R \text{ and } R = \sqrt{(x^2 + y^2)}$$

or:

$$H_x(x,y) = (g/\pi)H_g \times y/R^2 \tag{19.1}$$

The field lines are circular, and the total field is formed from cylinders (bubbles) of varying field intensity: the small cylinders near the gap, having the largest field strength (Fig. 19.1), are fairly simple to predict.

The first three approximations listed above mean that the medium has a square hysteresis loop, and remagnetization occurs for fields exceeding the coercive force H_r. The particles inside a cylinder whose periphery corresponds to H_r are consequently magnetized to the polarity of the field. Outside the cylinder, the particles are not affected. It is easy to see that the magnetization pattern on the medium ends up as shown in Fig. 19.1b.

The Bauer-Mee model gave some insight into the distortion mechanism at long wavelengths, when recording with ac bias was used. It also explained the associated "self-erasure" effects at short wavelengths. Its drawback was its limitation to longitudinal magnetization and an "almost" ideal media coating; the method did not correlate the *MH*-loop and its magnetic properties to recording performance.

Schwantke's proposal offered understanding of the internal magnetization process through the concept of particle interaction. This was explored as early as in 1935 by the German physicist Preisach, and has been invaluable in explaining the recording process with ac bias (see Chapter 20).

Early studies of digital recording concentrated on investigations of the recording of a single transition (i.e., the leading or trailing edge of a bit). This is the position where the maximum flux change occurs and the maximum read voltage is sensed. It is also the position where a digital signal carries its information or clock signals.

A short transition zone is necessary in order to achieve a maximum output voltage and a maximum packing density. The latter is, as we shall see, partly diluted by the playback process, which involves a fairly broad sensing function as determined by the gap field.

The Bauer/Mee model did find use with magnetic coatings that had normal hysteresis loops. The magnetization on each side of a transition is longitudinal, and the transition zone is a magnetic pole, or a curved sheath of thickness Δx. We will now analyze the formation of this transition zone, and its final shape after removal of the recording field and subsequent removal of the highly permeable write head.

RECORDING OF A SINGLE TRANSITION

Let us see how the recording of a transition takes place. A presaturated coating of thickness δ is moving past a write head at a distance d (Fig. 19.2a). The record field is turned on in Fig. 19.2b and we shall, for now, consider only the longitudinal field components $H_x(x,y)$ of the head field. The lines of constant field strength are therefore equal to the circles (bubbles) in the Bauer-Mee model.

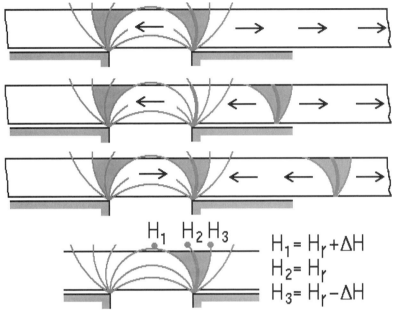

$$H_1 = H_r + \Delta H$$
$$H_2 = H_r$$
$$H_3 = H_r - \Delta H$$

FIGURE 19.2 The recording of transitions. Top: The write field is turned on. Middle: The write field is turned off (during switching). Bottom: The write field is turned on with opposite polarity. A sketch shows the outline of the recorded transition zone.

The coating's hysteresis loop is idealized, and the cylinder of strength H_1 will remagnetize 75 percent of the coating inside it. (Remember, the arrows show H_a, not the opposing H_d, which is the field direction after magnetization.) The magnetic particles (or domains) lying on line no. 2 of strength H_2 are brought to a nonmagnetic state, while only 25 percent of those on line 3 of strength H_3 are switched.

When the record field is removed (turned off as shown in c, or the coating moved away) then a transition zone of length Δx has been recorded. The magnetization changes by an amount of $2M_{sat}$ over the distance Δx, and from this we find the pole strength (by definition):

$$p = 2M_s/\Delta x \tag{19.2}$$

The voltage from the read head will be proportional to the pole strength, which will require that Δx is as short as possible. This will simultaneously ensure the highest packing density, and the recording will, in the limit, consist of a series of closely spaced transition zones.

It is therefore of great practical value to determine what the minimum transition length can be. Δx may be estimated as the distance between the two field lines of strength $H_r + \Delta H$ and $H_r - \Delta H$ as shown in Fig. 19.2, for the appropriately scaled values of δ and d. $\Delta H = H_r \times SFD/2$. The example chosen has a transition zone length Δx of 0.2 to 0.4 μm, which corresponds to maximum packing densities of 50,000 to 25,000 flux changes per cm. Such packing densities are made today with short-gap heads. The flux reversals can, of course, be made at a rate so high that a new transition zone is started before the preceding one is "completed," with loss in overall levels.

Calculation of the transition zone length Δx has for years been the focal point of much research. To do this, many simplifying assumptions were necessary. A detailed account is beyond the scope of this chapter, and the interested reader is referred to the bibliography. Most of the reported work is done for the longitudinal case only, and not until recently has the difficult task of calculation with vector magnetization begun. This is necessary for the high packing densities used today, and much theoretical as well as experimental work is required.

The earliest approach was to assume a pattern for the change of the magnetization M_s across Δx, and to calculate the field from the pole (the transition). This is the demagnetizing field we treated in Chapter 4, and its maximum value is always limited to the material's coercivity H_r. This permits a calculation of the minimum Δx; results were inconsistent with observations, and the magnetization pattern changed.

The best fit is achieved by assuming a gradual change in magnetization across the transition between two bits, following a complementary error function combined with the decreasing head field. It was found that an arctangent function served very well and is easier to use. Such simple transitions were computed and discussed in Chapter 5.

SELF-CONSISTENT, ITERATIVE COMPUTATION OF A TRANSITION

Each transition has a field strength due to its magnetization or pole strength p (formula (19.2)). This field strength equals the demagnetizing field, which can be as large as the material's coercivity H_r. This field is therefore not small compared to the recording field at that point, and it was in the late 1960s suggested that it should be added to the record field. This addition will cause the transitions to be recorded at slightly different positions.

This addition of the transition field to the head field is identical to the argument that the field acting upon a magnet includes its own demagnetizing field:

$$H_{core} = H_{applied} + H_{demag} \qquad (4.3)$$

The net effect is a weakening of the head field on the inside of the transition (near the gap), and a strengthening on the outside. The field patterns without and with a transition are shown in Figs. 19.3 and 19.4. The corresponding field magnitudes along the medium surface are shown below, and are useful for estimating the length of the transition zone Δx at that location. Note that this illustration reflects an assumed value of one for the coating's relative permeability; this is customary in most theories for the recording process.

It is evident that the inclusion of the transition's own field will broaden the transition and shift it outward (away from the gap). At a certain point, there will be a stable magnetization such that its demagnetizing field together with the head field results in that pattern. The computations are iterative and must be self-consistent, i.e., the final magnetization pattern must correspond to one that produces the field that (together with the head field) produces that pattern.

The details of these computations are beyond the scope of this chapter. One early computation is shown in Fig. 19.5, computed by Iwasaki and Suzuki in 1968. The short transition zone is the result of applying the head field alone to a saturated medium, while the longer zone results when both the head field and the demagnetization field H_d are present.

The result is an outward shifted and longer transition zone. When a following transition zone is recorded it is also shifted, but less so than the first zone was shifted if the two are close together. The reversed head field will assist the demagnetizing field from the prior recorded transition and will reduce its strength (partial erasure).

All data patterns have varying timings between their transitions, and the timing relationship between the recorded transitions will be destroyed by the peak shift introduced by the recording process.

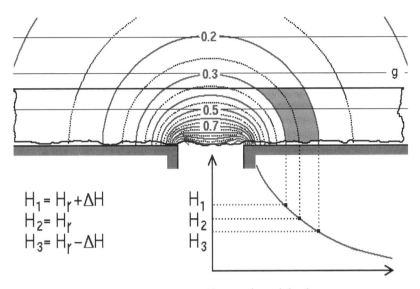

FIGURE 19.3 The head field alone produces a transition zone of a certain length.

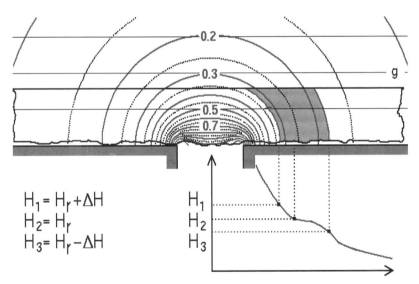

FIGURE 19.4 The total field (head field plus demagnetization from the transition) produces a longer transition.

A simple model for the recording process was introduced in Chapter 5. It showed the gradual change in magnetization across a transition, shown in Fig. 19.6 for a recording on a very thick ac-erased medium. It was quite easy to introduce a demagnetizing field in the model. This H_d field was placed at the transition center as it developed. Its value was set at 10 percent of H_r, and the result is shown in Fig. 19.7. The transition zone length is nearly doubled, and its position shifted a small distance "forward." Δx is increased as much as 2.5 times when $H_d \approx 0.25 \times H_r$.

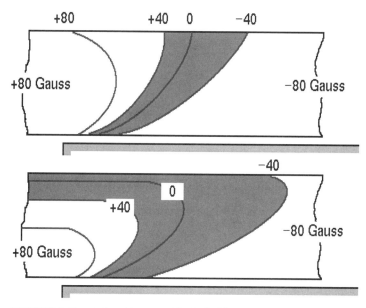

FIGURE 19.5 Computed transition zones. A longitudinal field alone produces a short transition zone, shown at top. The bottom figure shows a self-consistent, iterative computation including the transition's demagnetizing field as it develops. Hence, it shows a longer transition.

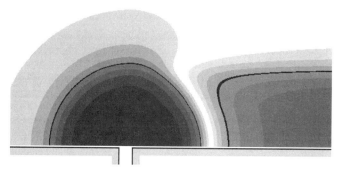

FIGURE 19.6 A computed transition resulting from a head field alone.

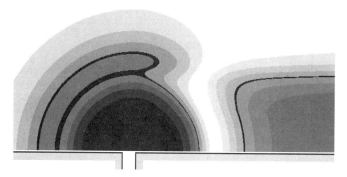

FIGURE 19.7 A computed transition including a demagnetizing field of one-tenth of the head field (*at the transition*).

CALCULATED TRANSITION ZONE LENGTH

The simple full-field model in Chapter 5 resulted in an expression for the length of the transition zone Δx:

$$\Delta x = SFD \times write\ level \qquad (5.3)$$

It does not include the demagnetization field H_d, and produces values that are somewhat smaller than other models.

Comstock and Williams, and later Maller and Middleton, formulated the conditions of recording with the added demagnetizing field and a maximum field gradient (smallest Δx). The result was:

$$a = [(M_r/\pi H_r) \times \delta(d+\delta/2)]^{1/2} \qquad (5.4)$$

Here a is called the transition parameter, which corresponds to half of Δx:

$$\Delta x = 2a$$

The influence of the demagnetizing field is evident from the appearance of M_r and H_r. But the SFD is absent from (5.4) for which the assumption that $S* = 1$ was made. Otherwise two additional terms must be accounted for (Bertram).

We can learn from the simple approach (5.3) and from the more elaborate (5.4):

1. Use material with small SFD, i.e., square loop.

2. Use material with low δM_r and high H_r.

3. Reduce dimensions (small δ and d equates to a small write level).

Note the absence of gap length g: Transitions are written after the gap, over the trailing pole. The field component that writes the transition (in contact or near contact) is perpendicular, even though the remanent magnetization in a thin film relaxes to a longitudinal remanence.

The Karlqvist field strength $H_y(x,y)$ from Fig. 4.24 can be expanded into a series:

$$H_y(x,y) = -(2H_g/\pi) \times (G + G^3/3 + G^5/5 + ...)$$

$$\approx -(ni\eta/\pi x) \times (1+G^2/4)$$

where $G = g/2x$. The variation in $H_y(x,y)$ with x, and therefore Δx, is essentially independent of g for $x > g$.

COMPUTED TRANSITION-ZONE LENGTH AND GEOMETRY

Computer methods are necessary when examining magnetizations in thick coatings (Potter and Schmullian 1971). Figure 19.5 shows one of the first 2D computations of a recorded transition (Iwasaki and Suzuki 1976). The computer models make possible the inclusion of not just demagnetization fields, but also those from the magnetic image of the transition inside the highly permeable head. The 2D models divide up the medium into a number of laminates to provide for pictures like Fig. 19.5, or charts of minute magnetizations and their magnitudes like Fig. 19.17. Also, 3D models have been computed (Beardsley 1986).

It appears that demagnetization fields play a minor role when recording on a thick medium, such as tapes (Bertram and Niedermeyer 1978 and 1982). Recent methods, computations, and results are described by Dinnis, Middleton and Miles, and by Wei, et al.

Micromagnetics plays an increasingly important role (Bertram and Zhu 1992) and we will see results thereof in the section on narrow-track recordings. The studious reader will enjoy the complete and detailed analysis of the write process in recent books by Ruigrok (1990) and Bertram (1994).

PERPENDICULAR RECORDING

Westmijze's in-depth analysis of magnetic recording did consider the recording of perpendicular magnetization in tapes (see the bibliography section in Chapter 5). This was contrary to the interest at that time, namely recording of audio at the highest possible signal level to achieve adequate SNR. This ruled out the perpendicular mode, which has very high demagnetization at long wavelengths.

A novel, circular magnetization mode was discovered in 1975 by Iwasaki. The following year, Suzuki revealed the importance of a perpendicular magnetization component, even in a thin medium. And the signal to be recorded was changing to a digital signal, with bit lengths as short as possible. Perpendicular recording was therefore proposed in 1977 (Iwasaki), and many theories and computations have been presented since (Potter and Beardsley 1980; Middleton and Wright 1982 and 1984).

Perpendicular recording looks good from all models and experiments. It is largely pursued by the Japanese recording industries, as evidenced by a recent book by Iwasaki and Hokkyo: 202 papers are of Japanese origin, 30 from US and Europe. It appears that the American/Dutch contributions culminated with two modelling works in 1986 (Zhu and Bertram; Beusekamp and Fluitman).

There are problems with off-track performance, but media and heads are nearly in place: We have for a long time had CoCr perpendicular disks with a thickness of 2500 – 5000 Å. A good tape is emerging with BaFeO particles. Even a particulate disk with BaFeO was once developed. Heads are still experimental, although a thin film head for in-contact is available, see Fig. 19.8 (Hamilton, et al. 1992). The perpetual question is "Does perpendicular recording have a future, and when will it emerge?"

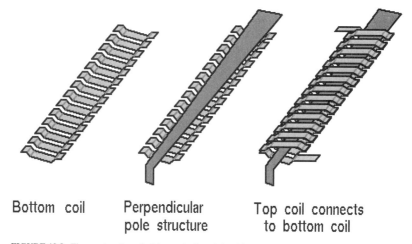

Bottom coil   Perpendicular   Top coil connects
                     pole structure   to bottom coil

FIGURE 19.8   The construction of a 14-turn single-pole head for perpendicular recording (*after Censtor*).

OVERWRITE

Recording of data in disk drives is done by writing at a level that simultaneously erases old data. This erasure is nowhere near perfect, and a residual signal from old data is common. Recording codes contain signals of a high and a low frequency, and it is practice to let the old data be a low-frequency signal (*1f*) because that is more difficult to erase (see Chapter 8).

The ratio of the new signal to the old signal is called overwrite. It is roughly proportional to the write level of the new signal.

An example will help to understand this. If the thin film disk has $S^* = 0.7$ (i.e., $SFD = 1 - S^* = 0.3$), then the write field in the back of the film needs to be 152% to assure 40 dB erasure (152% level equals $1.52 \times H_r$). This was outlined in Chapter 5. The levels listed in that section are actually less than achieved in conventional erasure with a magnetic head. There are several possibly simultaneous reasons therefore:

1. Recording a bubble of reverse magnetization into an otherwise saturated medium, i.e., hard direction (Bloomberg, et al. 1979; Tang and Tsang 1989).

2. Modulation of the new $2f$ signal by the old $1f$ signal (Fayling, et al. 1984).

3. The ingoing portion of $1f$ magnetization near the gap is not yet erased and has a fringing field across the gap. It may be strong enough to be recorded as a replica. The 2f signal serves as ac bias (Wachenschwanz and Jeffers 1985).

4. The ingoing $1f$ magnetization is sensed by the head sensing function and converted into flux. This flux is "data" for the $2f$ ac-bias signal. The linear head is able to read while writing. See also Chapter 8 under Erase Heads.

5. Old high-coercivity particles/grains in the coating back may carry enough old $1f$ data forward to interfere with the write process. Also, particle interaction fields may play a role.

6. Edges of tracks are not perfectly overwritten due to spindle runout and insufficient write-wide-read-narrow margin (Zhu, et al. 1994). A highly sensitive magnetic force microscope was used in their investigations.

A recent examination of overwrite's complexities can be found in the book by Bertram and in a paper by Lin, Zhao, and Bertram (1993).

NARROW-TRACK RECORDING

A high-density perpendicular recording system has demonstrated the feasibility of 680 kfrpi (26.8 kfr/mm) for near contact with a head of width 100 μm (4 mils) (Yamamoto, et al. 1987). Subsequent experiments investigated the possibility of a trackwidth of only 10 μm (400 μin) (Beaulieu, et al. 1989). The results were discouraging due to relatively poor offtrack capability (at 2000 tpi.)

This has been confirmed by micromagnetic modeling for a CoCr perpendicular film with a soft magnetic back layer (Zhu and Ye 1993). Their model is shown in Fig. 12.23. The simulation results are shown in Fig. 19.9 and show side-written information just outside the track edge. This will result in an off-track capability that degrades with increasing linear density.

Edge overwrite simulation (Fig. 19.9b) also shows that the new track and previously written track are directly connected at track edges without any separation. This is in contrast to overwriting in a planar isotropic longitudinal thin-film medium where edge overwrite generates a side-erase band that separates adjacent tracks (Ye, Zhu, and Arnoldussen 1993).

The simulation of a written bit (two transitions) is shown in Figs. 19.10 and 19.11. The isotropic film is clearly superior to the oriented film by producing far less sidewriting.

SHORT-GAP RECORDING

We saw in Chapter 5 how the combination of the head field gradient and the medium's switching field distribution SFD controlled the length of the transition recorded between bits.

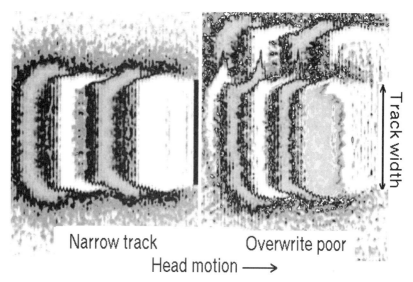

FIGURE 19.9 Results of computer modelling of the write process on a perpendicular medium (*after J.G. Zhu*).

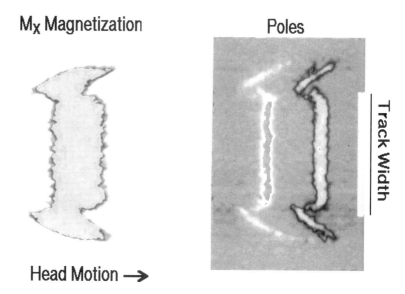

FIGURE 19.10 Computation of a bit (two transitions) written on a longitudinal, oriented medium (*after J.G. Zhu*).

Figure 19.12 show the situation for an in-contact recording. Clearly, the recording zone can be shortened by reducing the record field. That is, the length of the transition zone is proportional to the record field strength.

A very shallow recording is made with the field adjusted as shown in Fig. 19.12, middle. Now we do, however, have two recording zones, causing a very poor-quality recording—one zone on top of

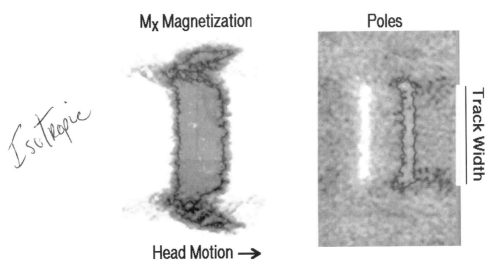

M$_X$ Magnetization　　　　　**Poles**

Track Width

Isotropic

Head Motion →

FIGURE 19.11 Computation of a bit (two transitions) written on a longitudinal, isotropic medium (*after J.G. Zhu*).

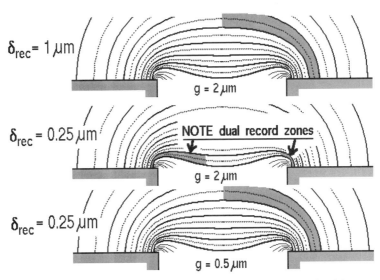

δ_{rec} = 1 μm

g = 2 μm

δ_{rec} = 0.25 μm

NOTE dual record zones

g = 2 μm

δ_{rec} = 0.25 μm

g = 0.5 μm

FIGURE 19.12 Recording zones in a short-wavelength recording. A short gap length is recommended.

another. This situation can only be remedied by reducing the gap length, as shown in Fig. 19.12, bottom. This will require an increase in the gap field strength, and such heads are therefore made from high-saturation materials such as Al-Fe or FeTaN alloy.

We learned in Chapter 5 that the length of the gap is of little influence on the length of the transition zone. That situation was different: When the recording depth remains fixed, then the gap length has little effect.

The length of the transition zone equals the length of the freezing zone, which is located by using field lines of constant strength in combination with the *SFD* to establish the borders of Δx.

EFFECTS OF DATA RATE (WRITE FREQUENCY)

The magnetic particles, or grains, have upper limits for the speed at which they can switch magnetization. This was introduced in Chapter 3 (Thornley 1975). Recent measurements have revised the upper frequency limits to at least 250 MHz for MP media, but less than 100 MHz for Co-γFe$_2$O$_3$ or BaFeO. Further work is in process, including frequency limits for thin-film media.

Another time-dependent feature is the frequency dependency of the coercivity H_r. Figure 12.11 shows the decay of coercivity with time referenced to the instant of magnetization change. Measurements of H_r occur at less than 10^{-2} s for the *MH*-meter (@ 60 Hz), while it is 1 sec or more for the VSM. Writing at a rate of 20 Mbps (10 MHz, or 2.5 MBps), the time frame is 10^{-8} sec.

The coercivities at that frequency are higher than the data sheet for the tape indicated. Typical values (Oe) are (after Sharrock 1990):

	Co-gamma	MP	BaFeO
H_r (60 Hz)	900	1530	1410
H_r (VSM)	800	1430	1180
Write (10 MHz)	996	1623	1660

Similar results are obtained by others (de Witte, et al. 1993). The experimental values or required write currents will therefore be higher than calculated values based upon H_r values from data sheets.

MAGNETIZATION PATTERNS IN RECORDED MEDIA

Changes in the coercivity of a coating have recently been reported to vary with field direction. Shown in Fig. 19.13 are graphs for the variation across 0 to 90 degrees incidence angle of the field. The coatings are S-VHS (γFe$_2$O$_3$), MP (Fe), BaFeO tapes (Suzuki), and CoNiPt and CoCrTa rigid-disk coatings (Speliotis.) The latter and the S-VHS shows decreasing H_r with angle. When this is factored into the model for drawing the transition zones, Figs. 19.14 through 19.16 result (no demagnetization was included).

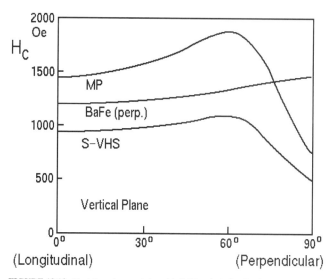

FIGURE 19.13 Variations in coercivity with field angle (*after Suzuki*).

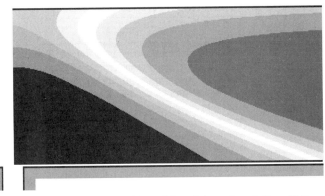

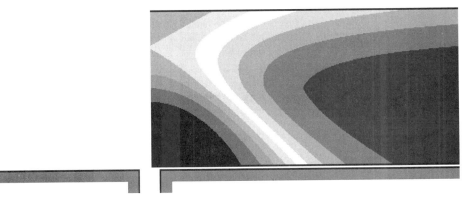

FIGURE 19.14   An estimate of a simple transition zone for S-VHS tape with coercivity variation as shown in Figure 19.13.

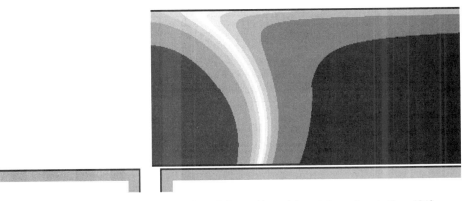

FIGURE 19.15   Estimate of a simple transition zone for MP tape with coercivity variation as shown in Figure 19.13.

FIGURE 19.16   Estimate of a simple transition zone for BaFeO tape with coercivity variation as shown in Figure 19.13.

For thin films, similar data are emerging (Huang and Judy 1992 and 1993). The data show that the coercivity approaches zero when the angle goes to 90°. This is obviously in conflict with the recording theory we have adopted, because the head field direction in future in-contact systems approached the perpendicular mode.

Between transitions, the magnetization varies in strength and direction. For ac-bias recording these magnetization vectors can be calculated with good approximation; see Chapter 20 on ac-bias recording.

For digital saturation recording we will further simplify (20.7) and (20.8) for the magnetization vector components (for $x \leqq R$ and $y \leqq R$):

$$\Delta m_x = S_x \times M_{rs} \times y/R$$

$$\Delta m_y = S_y \times M_{rs} \times x/R$$

where S_x and S_y are the longitudinal and perpendicular squareness and M_{rs} is the remanence after saturation. An example is shown in Figs. 19.17 and 19.18. Read voltages for these magnetization patterns will be discussed in Chapter 21, Advanced Topics in Playback (Read).

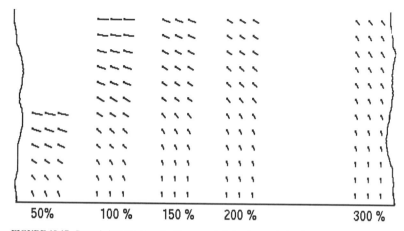

FIGURE 19.17 Recorded vector magnetizations recorded at various write levels.

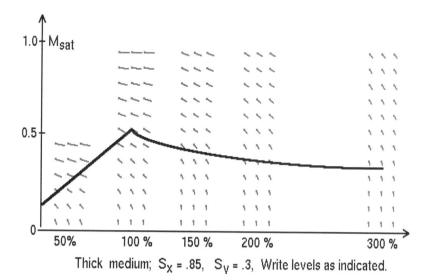

FIGURE 19.18 Net magnetization levels $((M_x^2 + M_y^2)^{1/2})$ from Figure 19.17.

BIBLIOGRAPHY

PROPERTIES OF TRANSITIONS AND THE RECORDING THEREOF

Beardsley, I.A. Sept. 1986. Three Dimensional Write Model for Magnetic Recording. *IEEE Trans. Magn.* MAG-22 (5): 361–363.

Bertram, H.N., and R. Niedermeyer,. Sept. 1978. The Effect of Demagnetization Fields on Recording Spectra. *IEEE Trans. Magn.* MAG-14 (5): 743–746.

Bertram, H.N., and R. Niedermeyer. Nov. 1982. The Effect of Spacing on Demagnetization in Magnetic Recording. *IEEE Trans. Magn.* MAG-18 (6): 1206–1209.

Bertram, H.N. 1994. *Theory of Magnetic Recording.* Cambridge University Press, UK. 356 pages.

Bertram, H.N. and I.A. Beardsley. Nov. 1988. The Recording Process in Longitudinal Particulate Media. *IEEE Trans. Magn.* MAG-24 (6): 3234–48.

Comstock, R.L., and M.L. Williams. 1971. An Analytical Model of the Write Process in Digital Magnetic Recording. *AIP Conf. Proc.*, Vol. 5 (1): 738–42.

Curland, N., and D.E. Speliotis. Sept. 1971. An Iterative Hysteresis Model for Digital Magnetic Recording. *IEEE Trans. Magn.* MAG-7 (3): 538–543.

Dinnis, A.K., B.K. Middleton, and J.J. Miles. 1993. Theory of longitudinal digital magnetic recording on thick media. *Jour. Magn. and Magn. Matls.* 120: 149–153.

Iwasaki, S., and T. Suzuki. Sept. 1968. Dynamic Interpretation of Magnetic Recording Process. *IEEE Trans. Magn.* MAG-4 (3): 269–76.

Maller, V.A.J., and B.K. Middleton. 1973. A Simplified Model of the Writing Process in Saturation Magnetic Recording. *IERE Conf. Proc.* Vol. 26: 137ff.

Middleton, B.K. 1987. Recording and Reproducing Processes, Ch. 2 in *Magnetic Recording, Vol. 1,* Ed. by Mee and Daniel, McGraw-Hill, pp. 22–97.

Potter, R.I., and R.J. Schmulian. Dec. 1971. Self-Consistently Computed Magnetization Patterns in Thin Magnetic Recording Media. *IEEE Trans. Magn.* MAG-7 (4): 873–880.

Ruigrok, Jaap J.M. 1990. *Short-Wavelength Magnetic Recording: New Methods and Analysis.* Elsevier Advanced Technology, Oxford, UK. 564 pages.

Stein, I. 1961. Analysis of the Recording of Sine Waves. *IEE Trans. Audio* AU-9: 146–155.

Wei, Dan, N.H. Bertram, and F. Jeffers. Sept. 1994. A Simplified Model of High Density Tape Recording. *IEEE Trans. Magn.* MAG-30 (5): 2739–49.

PERPENDICULAR RECORDING

Beusekamp, M.F., and J.H. Fluitman. Sept. 1986. Simulation of the Perpendicular Recording Process including Image Charge Effects. *IEEE Trans. Magn.* MAG-22 (5): 364–366.

Hamilton, H., R. Anderson, and K. Goodson. Nov. 1991. Contact Perpendicular Recording on Rigid Media. *IEEE Trans. Magn.* MAG-27 (6): 4921–26.

Iwasaki, S.I., and K. Takemura. Nov. 1975. An Analysis for the Circular Mode of Magnetization in Short Wavelength Recording. *IEEE Trans. Magn.* MAG-11 (5): 1173–1176.

Iwasaki, S. Jan. 1980. Perpendicular Magnetic Recording. *IEEE Trans. Magn.* MAG-16 (1): 71–76.

Iwasaki, S., and J. Hokkyo. 1991. *Perpendicular Magnetic Recording.* Ohmsha, Japan. 211 pages.

Iwasaki, S., and Y. Nakamura. Sept. 1977. An Analysis for the Magnetization Mode for High Density Magnetic Recording. *IEEE Trans. Magn.* MAG-13 (5):1272–77.

Lemke, J.U. Mar. 1982. An Isotropic particulate medium with additive Hilbert and Fourier field components. *Jour. Appl. Phys.* 53 (3): 2561–2566.

Middleton, B.K., and C.D. Wright. April 1982. Perpendicular Recording. *Intl. Conf. Video and Data 82* IERE Publ. No. 54, pp. 181–192.

Potter, R.I., and I.A. Beardsley. Sept. 1980. Self-Consistent Computer Calculations for Perpendicular Magnetic Recording. *IEEE Trans. Magn.* MAG-16 (5): 967–973.

Sharrock, M.P., and D.P. Subbs. Dec. 1984. Perpendicular Magnetic Recording Technology: A Review. *Jour. SMPTE* 93 (12): ?.

Suzuki, K. May 1976. Theoretical Study of Vector Magnetization Distribution Using Rotational Magnetization Model", *IEEE Trans. Magn.* MAG-12 (3): 224–230.

Wright, C.D., and B.K. Middleton. April 1984. The Perpendicular Record and Replay Processes. *Intl. Conf. Video and Data 84.* IERE Publ. No. 59, pp. 9–16.

Zhu, J., and H.N. Bertram. Sept. 1986. Computer Modeling of the Write Process in Perpendicular Recording. *IEEE Trans. Magn.* MAG-22 (5): 379–381.

OVERWRITE

Bloomberg, D.S., G.F. Hughes, and R.J. Hoffmann. Nov. 1979. Analytic Determination of Overwrite Capability in Magnetic Recording Systems. *IEEE Trans. Magn.* MAG-15 (6): 1450–1453.

Fayling, R.E., T.J. Szczech, and E. F. Wollack. Sept. 1984. A Model for Overwrite Modulation in Longitudinal Recording. *IEEE Trans. Magn.* MAG-20 (5): 718–720.

Lin, G.H., Y. Zhao, and H.N. Bertram. Nov. 1993. Overwrite in Thin Film Disk Recording Systems. *IEEE Trans. Magn.* MAG-29 (6): 4215–23.

Tang, Y-W, and C. Tsang. Jan. 1989. Theoretical Study of the Overwrite Spectra Due to Hard-Transition Effects. *IEEE Trans. Magn.* MAG-25 (1): 698–702.

Wachenschwanz, D., and F. Jeffers. Sept. 1985. Overwrite as a Function of Record Gap Length—A New Effect. *IEEE Trans. Magn.* MAG-21 (5): 1380–1382.

Zhu, J-G., Y. Luo, J. Ding, X-G. Ye, and E.A. Louis. Sept. 1994. MFM Study of Edge Overwrite in Perpendicular Thin Film Recording Media. *IEEE Trans. Magn.* MAG-30 (5): 2755–57.

NARROW TRACKS

Beaulieu, T.J., D.J. Seagle, M.A. Meininger, and C.J. Spector. Sept. 1989. Track density limitation for dual-layer perpendicular recording in a rigid disk environment. *IEEE Trans. Magn.* MAG-25 (5): 3369–71.

Yamamoto, S., Y. Nakamura and S. Iwasaki. Sept. 1987. Extremely high bit density recording with single-pole perpendicular head. *IEEE Trans. Magn.* MAG-23 (5): 2070–72.

Ye, X-G., and J-G. Zhu. Nov. 1993. Track Edge Overwrite and Easy Axis Orientation in Narrow Track Recording. *IEEE Trans. Magn.* MAG-29 (6): 3978–80.

Zhu, J-G., and X-G. Ye. Nov. 1993. Narrow Track Recording in Perpendicular Thin Film Media. *IEEE Trans. Magn.* MAG-29 (6): 3736–38.

Zhu, J-G., X-G. Ye, and T.C. Arnoldussen. Sept.. 1992. Side writing phenomena in narrow track recording. *IEEE Trans. Magn.* MAG-28 (5): 2716–18.

MAXIMUM WRITE FREQUENCY

de Witte, A.M., M. El-Hilo, K. O'Grady, and R.W. Chantrell. 1993. Sweep rate measurements of coercivity in particulate recording media. *Journal of Magnetism and Magn. Materials* 120 (1–3): 184–86

Doyle, W.D., and L. He. Nov. 1993. Measurement of the Switching Speed Limit in High Coercivity Magnetic Media. *IEEE Trans. Magn.* MAG-29 (6): 3634–36.

Sharrock, M.P. Jan. 1990. Time-Dependent Magnetic Phenomena and Particle-Size Effects in Recording Media. *IEEE Trans. Magn.* MAG-26 (1): 193–97.

Thornley, R.F.M. Sept. 1975. Pulse Response of Recording Media. *IEEE Trans. Magn.* MAG-11 (5): 1197–99.

MAGNETIZATION INSIDE COATING

Huang, M., and J.H. Judy. Nov. 1991. Effects of Demagnetization Fields on the Angular Dependence of Coercivity of Longitudinal Thin Film Media. *IEEE Trans. Magn.* MAG-27 (6): 5049–51.

Huang, M., and J.H. Judy. Nov. 1993. Calculations of the Angular Dependence of Apparent and True Coercivities of Thin Film Magnetic Recording Media. *IEEE Trans. Magn.* MAG-29 (6): 4083–85.

Speliotis, D.E., and J.P. Judge. Nov. 1991. Angular Dependence of the Remanence Coercivity in Magnetic Recording. *IEEE Trans. Magn.* MAG-27 (6): 4984–86.

Suzuki, T. Sept. 1992. Orientation and Angular Dependence of Magnetic Properties for Ba-Ferrite Tapes. *IEEE Trans. Magn.* MAG-28 (5): 2388–90.

CHAPTER 20
RECORDING WITH AC BIAS

The reader was introduced to the concept of ac-bias recording in Chapter 5, and it may be worthwhile to review it in these few pages. In this chapter we will develop a model for the ac-bias recording technique that will allow us to calculate the recording performance in terms of record current, head fields, and tape coating parameters.

What we know today about ac-bias recording is the sum total of many researchers' efforts to achieve an in-depth understanding of the anhysteretic record process, where two signals of different frequencies and amplitudes generate a remanent magnetization in the tape coating.

The graphic approach, using projections on the hysteresis loop, was tried early and did not provide any meaningful insight into the record process; also, it did at one point or another disagree with experiments. It became necessary to go back to the mechanism that causes hysteresis loops, namely the irreversible rotation of magnetization in single particles (Figs. 3.23 and 3.24).

THE PREISACH MODEL

The behavior of an assembly of small magnetic particles is best described by a model that ties together the individual coercivities and their collective interaction fields. Such a model dates back to 1935, when it was first described by Preisach in *Zeitschrift für Physik*. The application to anhysteretic recording was presented first by G. Schwantke. Independent and almost simultaneous presentations were also made by Woodward and Della Torre, and by Daniel and Levine.

The effect of interaction between the particles in a tape coating can be illustrated by placing the particles in a coordinate system where particles with zero interaction fields are placed on an H_c-axis in order of intrinsic coercivities. The ordinate axis is used to locate particles that are influenced by positive or negative interaction fields.

This is the Preisach diagram, and is shown in Fig. 20.1. It is a three-dimensional distribution $h(h_{ci}, H)$, as shown in Fig. 20.2. Illustration a is a conceptual drawing of such a distribution, while b is a histogram of a measured audio tape (Brock).

The remanent magnetizations are in Fig. 20.1 indicated by the dots, assuming that this assembly of particles was completely demagnetized, and therefore has a sum total of $M_r = 0$. The demagnetization can be performed by a slowly decaying ac field, and all particles with zero interaction field will end up magnetized at either $+M$ or $-M$; the chance for one or the other is fifty-fifty.

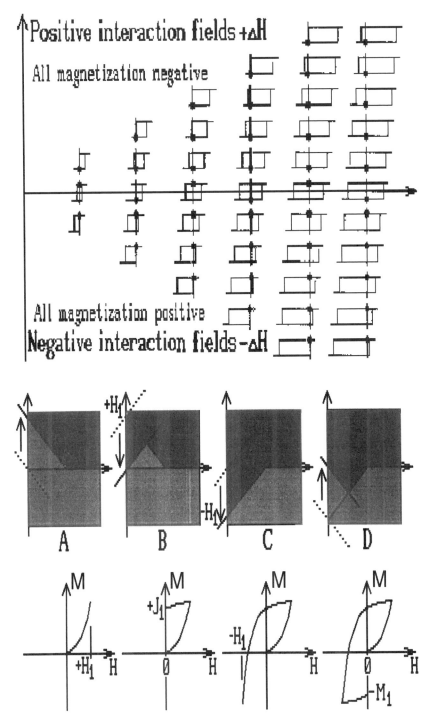

FIGURE 20.1 The placement of magnetic particles in the Preisach model coordinate system (top); the changing of magnetization distribution by the use of two simple rules (bottom; see text).

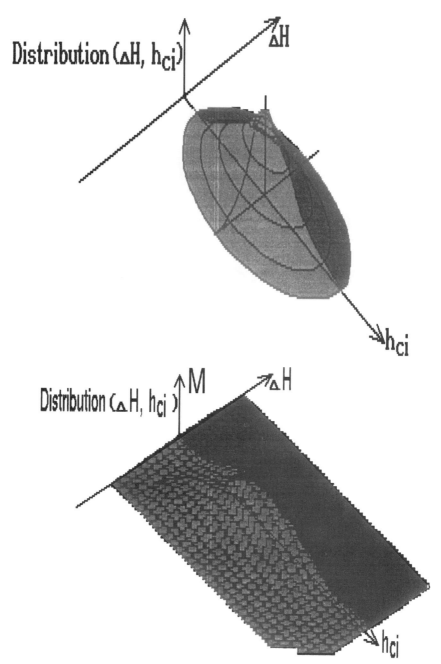

FIGURE 20.2 (a) Statistics of the Preisach diagram; (b) Measured density distribution (*after Brock, 1979*).

If the particle is under the influence of a positive interaction field (see upper right, Fig. 20.1), then the decaying field will at first be strong enough to alternate the particle's magnetization, but can no longer magnetize the particle to $+M$ after the field's peak value has decayed to below $H + H_c$. All particles in the upper half will therefore end up magnetized at $-M$.

Particles in the lower half will similarly be magnetized at +M. The total magnetization of the particle assembly will be zero, given that the distribution of the particles in the diagram is symmetrical about the H_c axis.

A field is now applied, *increasing* from zero to +H_1 (see Fig. 20.1, A). All particles below the H_c-axis are magnetized to +M_s, and those above will require that H_1 equals, and exceeds, $H_c + H$ before they switch to +M_s. This will first happen to those particles having small H_c and H (i.e., near the origin). As H_1 increases, the particles will change magnetization to +M_s as the line:

$$H = \Delta H + H_c \qquad (20.1)$$

sweeps up in the first quadrant. The line has an angle of 135° with the H_c-axis, and it crosses the H axis at $H = H_1$ ($H_c = 0$).

When H reaches H_1, all particles in the triangular area behind the line have been remagnetized and the net magnetization of the assembly is positive, as shown. If we now *decrease* H from H_1 to a negative value, another line will sweep into the fourth quadrant, at an angle of +45° with the H_c-axis. This line is determined by:

$$H = \Delta H - H_c \qquad (20.2)$$

where the proper sign (–) applies for ΔH when the line moves into the fourth quadrant due to a decreasing field strength.

When H reaches zero (b), we can readily see that a net magnetization remains, M_r; as H changes sign, the magnetization changes through zero to the negative value $-M_n$ (c).

We can finally reduce H back to zero, and $-M_r$ results (d). From this little exercise we obtain the following rules for the use of the Preisach diagram:

• An *increasing field strength H* moves at a 135° line upward; this line intersects with the H axis at the present value of H. All particles below this line become magnetized to +M_s, if not already at +M_s.

• A *decreasing field strength H* moves a 45° line downward; this line intersects with the H axis at the present value of H. All particles above this line become magnetized to –M_s, if not already at –M_s.

These rules are summarized in Fig. 20.3 and applied in Fig. 20.4 and in several examples this chapter. NOTE: The distinction is between increasing or decreasing field strength, NOT positive or negative field strength!

We will next apply these two rules and observe the outcome; first we will treat the problem of erasure, and then work into ac-bias recording, also called anhysteretic recording.

AC-ERASURE PROCESS

AC erasure of magnetic tapes, disks, or heads is done by applying an alternating field that at some time has a value that exceeds the material's coercivity by a factor of at least five. The field strength is next decreased slowly to zero, in a time span that corresponds to several hundred cycles duration; i.e., longer than 5 seconds for a demagnetizing field from a 60-Hz line.

The alternating field that increases in amplitude is initially applied as shown in Fig. 20.4. A maximum is reached at t_0 whereafter the field decays. This decay is accomplished by removing the degausser from tape, disk, or head, and in a recorder when a segment of a tape leaves the gap in an erase head.

The illustration shows that alternating and decreasing amounts of magnetization are left in the tape. Only a proper timing of the alternating cycles would leave the tape with a net magnetization of zero. A higher frequency must be used for complete erasure (see bottom, Fig. 20.4). This first example of the use of the Preisach diagram is in complete agreement with our practical experience.

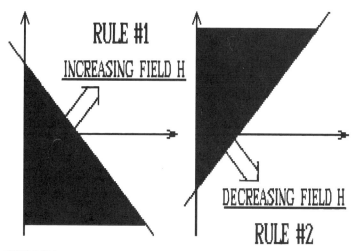

FIGURE 20.3 The two rules for changing magnetization in the Preisach diagram, following a change in field strength.

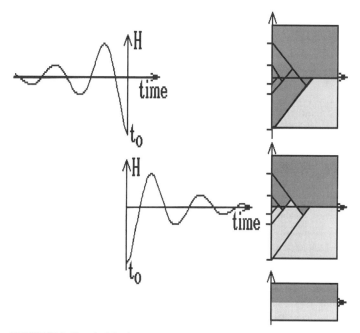

FIGURE 20.4 The principle of ac-erasure.

RECORDING WITH AC BIAS

We will in this section examine the record process by following a small coating segment dx by dy (by track width) on its way across the record head. It will experience an increasing and then decreasing field strength H, with a maximum in front of the gap in the head. This is illustrated in Fig. 20.5 for both a dc and an ac current through the head winding.

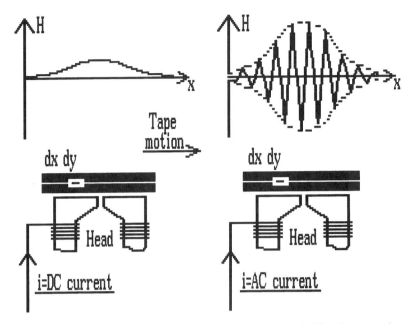

FIGURE 20.5 The fields *H* experienced by *dxdy* as it moves past a head energized by a dc-current and an ac-current.

In ac-bias recordings a high-frequency ac-bias signal is added to the data signal and both fed into the head winding (Fig. 20-6). Note the waveform of the added signals; the process of addition is linear.

The coating will experience a strong field in its front, in contact with the head while the strength will be less in the back. This will complicate an explanation of the ac-bias process a great deal, and we will therefore limit our initial discussion to the small element dxdy, as shown in Figs. 20.5 and 20.6; the third dimension of the element equals the track width.

The field variations experienced by the element as it moves across the head are shown in the variances in amplitude of the signal above the head.

MAGNETIZATION LEVEL VERSUS DATA CURRENT LEVEL

Figure 20.7 shows the remanences left after three levels of dc current only. They are nonlinearly related to the current levels, and illustrate the curvature of the initial magnetization curve. The magnetized regions of the Preisach diagram are shown by the clear triangles, and their areas increase as the square of the current value. This is in agreement with Fig. 20.1, and with the beginning of the initial magnetization curve from Fig. 3.44. It also illustrates the highly nonlinear nature of direct recording of signals; see Fig. 5.7.

Now we add an ac current to the data current, and the process will become linear. When at first a low-frequency ac-bias current is added to the dc current, then the field signal strength varies as shown in Fig. 20.8, leaving the element magnetized according to rules No. 1 and 2 from Fig. 20.3. Obviously, it is clear that the remanence is "cleaner" the higher the ac-bias frequency is.

This finding is the cause for a rule-of-thumb: *The bias signal frequency should be as high as feasible, at least five times the highest data frequency, and no less than three times this frequency.*

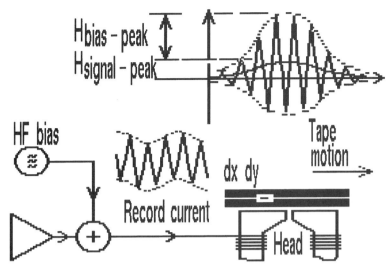

FIGURE 20.6 The addition of ac-bias to the data signal current produces an alternating field for tape element *dxdy* as it moves across the gap. Note the summation of currents, NOT modulation.

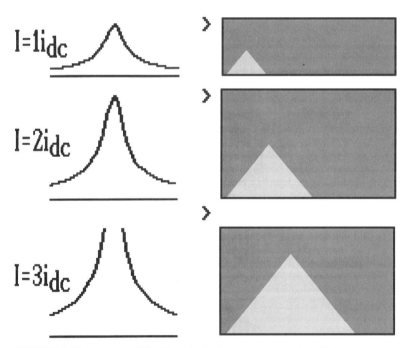

FIGURE 20.7 The remanence will ideally quadruple when the signal current is doubled.

A very high ac-bias frequency will result in an almost straight line, with very small "jaggies," and it will appear that the magnetized region in the Preisach diagram is a triangle that now is proportional to the data current, because the height of the triangle equals the corresponding data field level.

Figure 20.9 illustrates this relationship, summarized in Fig. 20.10. The levels' magnitudes relate as 1:2:3, and so do the heights of the triangles. This should lead to increases in magnetization as

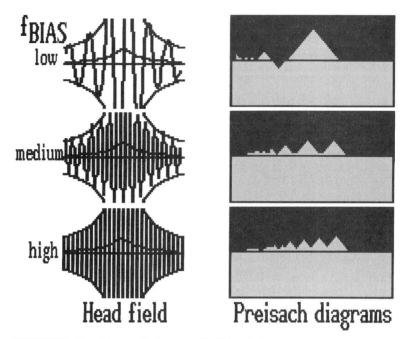

FIGURE 20.8   The ac-bias recording improves with a higher bias frequency.

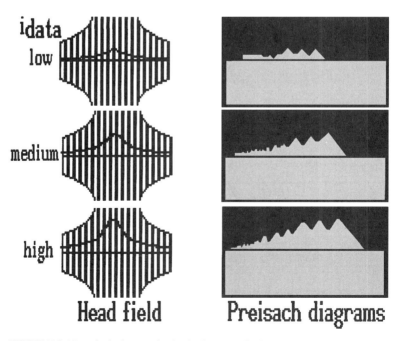

FIGURE 20.9   Magnetization is proportional to the data current level.

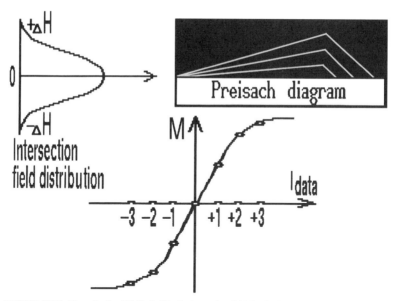

FIGURE 20.10 Magnetization *M* is limited by the interaction field distribution at high data signal currents.

1:2:3, but does not, as shown in the measured graph. At levels greater than 3, the curve levels off and we say that the material saturates.

It is more correct to say that there are no more particles to magnetize, as illustrated in the accompanying 3-D views of the number of particles magnetized (see Fig. 20.11). The height of the triangular region will, at saturation, reach into a region where only a few particles are located; i.e., particles under the influence of very large interaction fields. (NOTE: These fields can theoretically not be larger than the particle coercivities.) The measured distribution in Fig. 20.2B shows two humps at low field

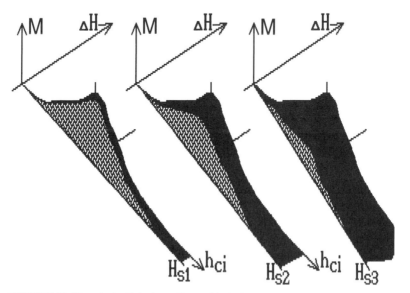

FIGURE 20.11 Magnetization limitation, as illustrated by the 3-D Preisach diagram.

values at regions where H is larger than H_c, and the blame placed on measurement errors. But the diagram merely indicates that some large particles (with low h_{ci}) are located in the vicinity of smaller particles with higher interaction fields.

The relationship between the magnetization M and the record current I_{data} is the recorder's transfer curve, and we observe its excellent linearity when using ac bias. It clearly reflects the distribution curve for the particle interaction fields (Fig. 20.12).

We will later return to the transfer curve for a discussion of linearity and distortion in ac-bias recordings.

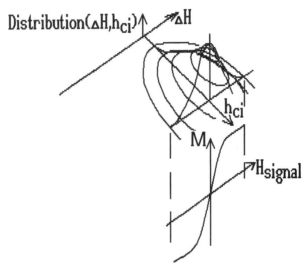

FIGURE 20.12 Correlation between the interaction field distribution and the overload curve.

MAGNETIZATION LEVEL VERSUS DATA FREQUENCY

When the frequency of the data signal is very high, then the magnetization process changes as the element $dxdy$ moves across the head. This has been discussed in connection with the length of the recording (or freezing) zone at the trailing edge of the gap in the record head, and the actual loss is computed in Chapter 19.

The Preisach diagram offers an additional view in Fig. 20.13, where the magnetized region now alternates around the zero line. Figure 20.14 offers a 3-D view of the number of particles magnetized; the + portion subtracts from the – portion, and the net magnetization is zero when the two are equal.

MAGNETIZATION VERSUS AC-BIAS CURRENT LEVEL

The ac-bias current level is now varied, while keeping the data current level constant at a small value (in the linear range). Successive applications of rules one and two result in the three graphs shown in Fig. 20.15, summarized in Fig. 20.16.

The magnetization level should ideally increase linearly, but the curve is S-shape (as shown in the actually measured curve). By placing the switching field distribution curve below and in proper relation to the Preisach diagram, we quickly see that there are no particles to magnetize at very low levels, and there are no further particles to magnetize at high levels; the result is the S-shape.

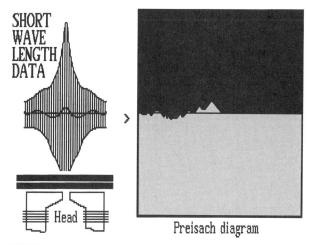

FIGURE 20.13 The recording head data field alternates too fast during the recording of very short wavelengths.

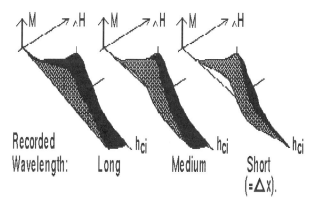

FIGURE 20.14 Remanent magnetizations at short wavelengths.

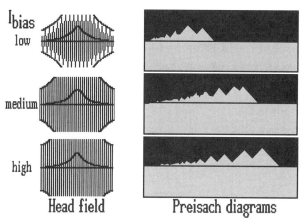

FIGURE 20.15 Magnetization is ideally proportional to the bias current level.

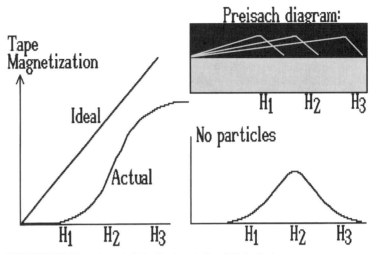

FIGURE 20.16 Magnetization is limited by the switching field distribution.

As we increase further the ac-bias level we will measure a decline in magnetization in a recorder (overbias; see Fig. 20.17). This phenomenon has elsewhere in this book been related to the steeper record field, and subsequent perpendicular remanence in the otherwise longitudinally oriented coating.

An additional factor now enters the picture: notice how the top side in the triangles in the Preisach diagram (Fig. 20.17) gently moved closer to the horizontal line at H_2 at very high ac-bias levels. The extension to the right of the triangles does not increase the magnetization because there are no particles there, and the lowering of the top side will, on the other hand, *reduce* the magnetized area where the particles are located.

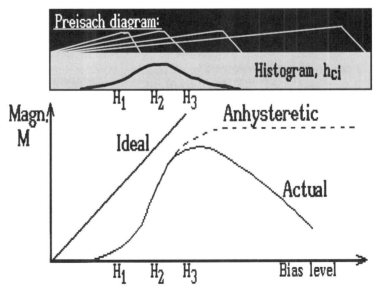

FIGURE 20.17 Increasing the ac-bias field will at first increase the remanence; thereafter it will decrease it.

This topic was investigated by Daniel and Levine. They made measurements on tape samples placed in the center of a long solenoid, and their instrumentation would decrease the bias field while keeping the data field constant. Their output-versus-bias level is shown with the broken line curve, called the *ideal anhysteretic* remanence, where the signal level is the same everywhere during the decay of the bias field. The ac-bias process when both fields decay simultaneously is named *modified anhysteretic*.

In early investigations, a fairly large discrepancy was found between the measured and the calculated output results; in some cases as much as 8 dB. This mystery was resolved by Bertram, who included the reduced coating magnetization due to the perpendicular magnetization that is pronounced at high bias levels; his analysis reduced the difference between measured and calculated remanences to about 1 dB.

The difference between the ideal anhysteretic curve (broken line) and the actual magnetization (solid line) in Fig. 20.17 is therefore made up of a lowering of the number particles to become magnetized (area under triangles, between H_1 and H_3), and the transition toward perpendicular magnetization (dead layer, velour effect).

An ac-bias tape specification can be expressed by its susceptibility, which is its low-level signal sensitivity:

$$M_r = \chi \times H_s \text{ gauss (Oe)} \tag{20.3}$$

Susceptibility has its electronic counterparts in the amplification factor μ of a vacuum tube, or β of a transistor.

Most modern tapes are manufactured to provide a coating with longitudinally oriented particles. This provides for a several-dB-higher output across the frequency spectrum. We may measure a value χ_x for the longitudinal sensitivity of susceptibility; we will also find that the coating is sensitive to perpendicular magnetization. The *perpendicular susceptibility* χ is quite a bit less than y for oriented tapes:

$$\chi_y \cong \chi_x/4 \tag{20.4}$$

It has also been found that the perpendicular remance is significant in many short wavelength recordings. This is the area where all fields of application are looking for improvements.

REMANENT MAGNETIZATION

We will first examine the remance after a recording is made with normal bias setting (Fig. 20.18). We will assume a coating permeability μ_r of 1, so the field lines are semiarcs; the latter approximation is acceptable so long as $R = \sqrt{x^2 + y^2} > L_g$:

$$M_x = \chi_x \times H_x \tag{20.5}$$

$$M_y = \chi_y \times H_y \tag{20.6}$$

The fields H_x and H_y can be expressed as:

$$H_x = H_s \times \sin\alpha$$
$$= H_s \times y/R$$

and

$$H_y = H_s \times \cos\alpha$$
$$= H_s \times x/R$$
$$= H_s \times \sqrt{(R^2 - y^2)}/R$$

The remanent magnetization components are now:

$$M_x = \chi_x \times H_s \times y/R \tag{20.7}$$

$$M_y = \chi_y \times H_s \times x/R \tag{20.8}$$

from which we can compute the composite magnetization vector.

The remanent magnetization M is therefore proportional to the total field strength H_s multiplied with susceptibility χ.

Our previous formula will also apply for the underbias condition, provided that gap length is smaller than the effective coating thickness δ_{eff} of the magnetized layer, and that the expression in the formula is *multiplied by* (δ_{eff}/δ) for underbias (Fig. 20.18). The overbias condition broadens the recording zone, and the remanence becomes more perpendicular.

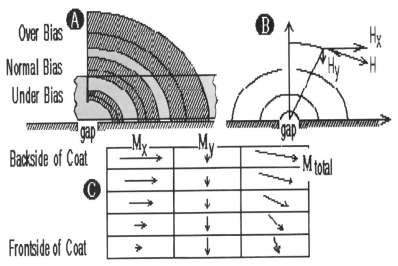

FIGURE 20.18 Vector magnetization after recording with ac-bias: (a) Recording zones, shown cross-hatched; (b) Components of the recording field; (c) Remanent magnetization in the coating.

A vectorial picture of the calculated remanent state in the recorded tape is shown in Fig. 20.18. Comparisons with 5.5 and 21.7 show good agreement. Demagnetization and interaction between the magnetized areas will modify the final remanent state.

Two observations are important:

• The remanent magnetization is stronger at the back side of the coating than in the surface.

• The surface is predominantly perpendicular.

The first statement has been verified by other workers in the field. Bertram found a linear increase in H_x from zero at the coating surface to a maximum at the backside, and he also includes the perpendicular component. He proved, in the latter paper, that the y-component must be included into a total vector field in order to achieve agreement between theory and measurement of the effective anhysteretic susceptibility.

Hersener conducted experiments with a scale model, and his results verify both statements. Bate and Dunn found a perpendicular magnetization that ranged from 2 to 15 percent of the in-plane component. Further proof of the perpendicular component is evident from recent years' work, which goes further to make media that will enhance perpendicular magnetization.

Figure 20.19 shows three recorded patterns, using our previous formula for calculation of the magnetization in segments of the coating, or packets, each consisting of a very large number of particles. The similarity of Fig. 20.19 with the results of experiments of Tjaden and Leyten, and Hersener and Iwasaki is evident.

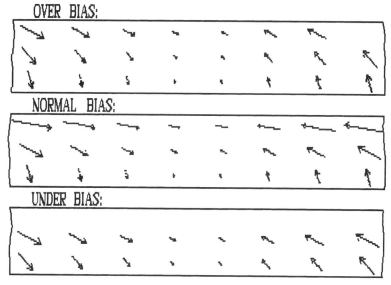

FIGURE 20.19   Magnetization patterns in ac-bias recorded coating.

LINEARITY AND DISTORTION

Figure 20.20 shows the transfer curve for ac-bias recording; the low-level crossover distortion from direct recording without bias is nonexistent. This transfer curve was produced by Koester, who made measurements on a large number of tapes. Some 6 to 10 specimens of different tape materials (Fe_2O_3, Co-Fe_2O_3, CrO_2, Fe) in thicknesses from 2 to 6 μm were measured and their transfer curves all normalized to a number m_{ar} for the anhysteretic remanence, $m_{ar} = M_r/M_{sat}$. All curves were found to be *equal to a universal transfer curve* to within ±2.5 percent!

This is the curve shown in Fig. 20.20 and it can be expressed:

$$m_{ar} = (2/\pi) \arctan (h) \tag{20.9}$$

where h is the normalized record current i_{data} divided by the record current $i_{50percent}$ required for a remanence that is 50 percent of the saturation value. Its universal applicability makes it a quite useful curve, and it can expressed as:

$$m_{ar} = a_1 h + a_3 h^3 + a_5 h^5 \tag{20.10}$$

where:
    $a_1 = 0.57$
    $a_3 = -0.076$
    $a_5 = 0.0065$

By substituting $h = h_o \sin \omega t$, the *third harmonic distortion* can be calculated, and the result is shown below the universal transfer curve.

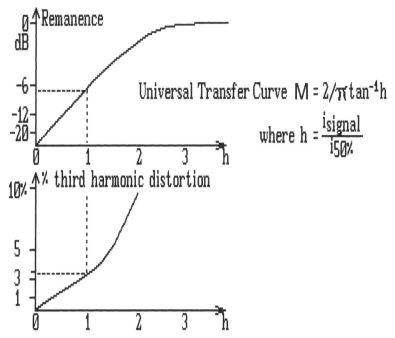

FIGURE 20.20 The overload curve and distortion in ac-bias recordings.

Note that the levels for one percent distortion are 10 dB below saturation, and it is only 5 dB below for 5 percent distortion. This margin, or *headroom*, is essentially the same for all tapes by nature of their close fit to the universal curve, which is valid for bias set to optimum output as noted in the figure.

It is common practice during adjustment of bias to record a low-frequency signal and observe the playback signal level. The bias is often adjusted for peak signal (maximum susceptibility), or until the signal level has dropped one or two dB. The latter provides for better recording with less noise and distortion. If the operator also monitors the third harmonic distortion he will see a pronounced *dip in distortion* somewhere before the signal level reaches its peak. This is illustrated in Fig. 20.21.

The reason for this dip can be understood if we examine the transfer curves for three different bias levels (Fig. 20.21, right). Three bias levels are shown: normal (optimum bias) "o," underbias "a," and overbias "b." Curves "a" and "b" have opposite initial curvatures, and the third harmonic distortion does therefore change phase in relation to the fundamental when the bias level changes from "a" through "b." The transfer curve for "o" is the straightest line and at the bias level there is minimum distortion.

One may object by referring back to Fig. 20.18 and pointing out that the recording conditions vary through the tape coating thickness due to anisotropy. And indeed, the curves in Fig. 20.21 are valid only for a very *thin* coating. The *distortion analysis* was verified in the first published analytical treatment of distortion by Fujiwara in 1979. His treatment of the subject is a testimony to the Preisach statistical approach to recording theory, as described in this chapter. Fujiwara calculated the magnetization curves based on the distribution of particles switching fields, measured the material constants for these, and next calculated the sum total of distortion in the magnetization in several layers of the coating. A computed playback signal, using reciprocity, gave results that correlated well with measurements.

The remanence from an ac-biased recording can be determined from the two-dimensional density function $h(H_c, H)$ in the Preisach diagram. The net magnetization can be expressed as the difference between the particle magnetizations in the upper $(+H)$ and lower $(-H)$ portions of the diagram.

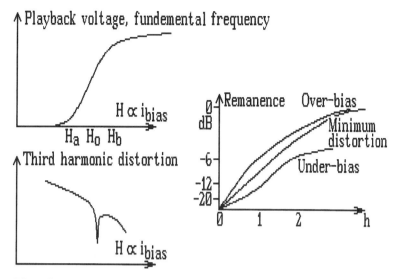

FIGURE 20.21 Output and distortion as functions of the bias current level (left); Remanence versus signal-level at three bias levels.

This difference is shown graphically in Figs. 20.10 and 20.11 for three signal levels $H_{s1} < H_{s2} < H_{s3}$. Clearly, as H_s increases, the net remanence increases. The relationship is linear for small values of H_s. The magnetization M versus H_s is shown in the middle of Fig. 20.12 and can be calculated, if we assume that the density function is gaussian with respect to the coercivity distribution as well as to the interaction fields. The *net magnetization M* can then be expressed as:

$$M = .5 \; \mathrm{erf} \; \alpha((H_s - H_c + H_b)/H_c) + \mathrm{erf}(\alpha((H_s + H_c - H_b)/H_c)) \tag{20.11}$$

where:

    erf = error function (see tables in mathematical handbooks).
    α = material constant = $H_c/\sqrt{2} \times \sigma_c$.
    H_c = mean value of coercivity.
    σ_c = standard deviation of switching field distribution.
    H_b = bias field.
    H_s = data signal field.

When the bias field equals H_c, we find:

$$M = .5 \times (\mathrm{erf}(\alpha H_s/H_c) + \mathrm{erf}(\alpha H_s/H_c))$$
$$= \mathrm{erf}(H_s/(\sqrt{2}) \times \sigma_c))$$

This expression can also be expanded into a series (Lagrange polynomial):

$$M = AH_s + BH_s^3 + CH_s^5$$

where:

    A = 0.398
    B = −0.062
    C = 0.002

This transfer curve compares well with the curve found by Koester.

INTERMODULATION: EVEN HARMONIC DISTORTION

Two other distortion figures need to be mentioned: *intermodulation*, which occurs when two or more signals mix in a nonlinear transfer curve, and *second harmonic distortion*. Intermodulation is annoying in music recording and a source of errors in instrumentation recording. Its calculation and prediction are difficult to the point of not being worth the effort, and its occurrence should be avoided by using fast peak-level indicators (oscilloscopes, L.E.D. devices, or similar tools, rather than VU-meters).

The second harmonic distortion does not really exist in tape recording because the transfer curve is symmetrical. So when it shows up, something is wrong in the form of a dc component:

- Magnetized heads
- Magnetized tape guides
- Leakage of dc current into record head (Faulty coupling capacitor)
- Amplifier distortion
- Asymmetrical ac-bias waveform

The reader is referred to Chapter 30 on maintenance, and of course the manual for the recording equipment, in order to correct any of these malfunctions.

BIAS OSCILLATOR

The high-frequency bias current added to the record-head current improves the linearity of the magnetic recording process. Bias is generated by an *oscillator*, and in audio and home-type recorders the same oscillator is used to produce current for the erase head. The bias current, as mentioned earlier, may cause intermodulation in the output stage of the record amplifier (beats). The bias frequency in audio or instrumentation recorder is normally five times the highest frequency to be recorded. For home recorders this means that a 60- to 75-kHz bias oscillator frequency is required. This is also what the erase current for the erase head requires. In instrumentation recorders with extended bandwidth to 4 MHz or higher, the bias frequency has to be 14 to 20 MHz, which in turn imposes a design limitation on the record head, extending its self-resonance beyond 25 MHz.

The bias current level is normally 8 to 10 times that of the data current. In order to *avoid* beats between bias oscillators in multichannel recording, it is a common design practice to use one master oscillator with buffer amplifiers in each record-amplifier section. This applies in particular to instrumentation recorders where as many as 14 channels may be recorded simultaneously.

While high-frequency bias added to the record current improves the linearity of analog magnetic recordings, its waveform must be purely sinusoidal in order to avoid excessive noise and distortion.

PLAYBACK AMPLITUDE VERSUS FREQUENCY RESPONSE

We can now, finally, determine the recorded flux levels. We will consider three bias conditions: underbias, normal bias, and overbias.

Normal bias is here defined as that bias field which records the tape in its exact coating thickness (Fig. 20.19, b). We then have, at the shortest wavelength, a demagnetization loss, and a recording zone loss. To this we must add the normal coating thickness loss, and the net available flux is shown as the bottom curve in Fig. 20.22.

An *underbias* condition (see c in Fig. 20.19) only records the surface portion of the coating. Both demagnetization loss and coating thickness losses are less than for a normal bias setting, and so are the recording losses due to the shallower recording zone.

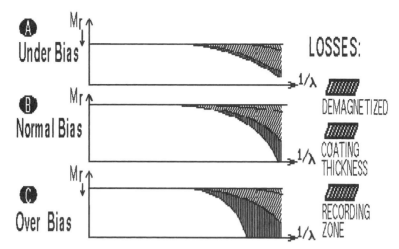

FIGURE 20.22   Available flux levels in ac-bias recordings, as a function of frequency (1/wavelength): (a) Under bias. (b) Normal bias. (c) Over bias.

This appears deceitfully promising for high frequencies, but the quality is poor. Distortion is higher, the long wavelength output lower, and the sound in audio recording is unsteady and "gurgled."

Overbias, on the other hand, does not increase the level at long wavelengths beyond the levels for normal bias; on the contrary, the recording zone now becomes so wide that the record losses extend into the long-wavelength area. See Fig. 20.19a. Coating thickness and demagnetization losses are, on the other hand, unchanged from normal bias.

CONCLUDING COMMENTS

Overbias is used in top-quality audio recording because of its better sound fidelity; the excessive short-wavelength loss is traded for a higher tape speed. It is beneficial to control the bias level by the record signal to achieve less distortion and better high-frequency response (Selmer Jensen).

The information in this chapter pertains to the recording of analog as well as digital signals, although the latter rarely occurs (digital data are recorded directly). When recording digital data it often becomes necessary to synchronize the bias oscillator to the data clock rate, and assure that the bias frequency has an integer relationship to the clock frequency.

The ratio between data and bias signal amplitudes is much less in digital recording than in audio/instrumentation. The maximum ratio of bias-to-data levels is generally between 10:1 and 30:1 in audio (at high signal levels), while it can be as low as 6:4 in digital. The author has found very little documentation on the latter issue.

REFERENCES

Bate, G., Statistical Stability of the Preisach Diagram for Particles of Gamma-Ferric-Oxide. *Jour. Appl. Phys.*, July 1962, Vol. 33, No. 7, pp.263–269.

Bertram, H.N., and Niedermeyer, R., The Effect of Demagnetization Fields on Recording Spectra. *IEEE Trans. Magn.*, Sept. 1978, Vol. MAG-14, No. 5, pp. 743–745.

Bertram, H.N., Long Wavelength AC Bias Recording Theory. *IEEE Trans. Magn.*, Dec. 1974, Vol. MAG-10, No. 4, pp. 1039–1048.

Bertram, H.N., Wavelength Response in AC Biased Recording. *IEEE Trans. Magn.*, Nov. 1975, Vol. MAG-11, No. 5, pp. 1176–1179.

Brock, M., Preisach Distribution in Magnetic Tapes. *Thesis for M.Sc.*, Tech. Univ. of Cph., 1979.

Daniel, E.D., and Levine, I., Determination of the Recording Performance of a Tape from Its Magnetic Properties. *JASA*, Feb. 1960, Vol. 32, No. 2, pp. 258–267.

Daniel, E.D., and Levine, I., Experimental and Theoretical Investigation of the Magnetic Properties of Iron Oxide Recording Tape. *JASA*, Jan. 1960, Vol. 32, No. 1, pp. 1–15.

Fujiwara, T., Even-Order Harmonic Distortion in AC Bias Recording. *IEEE Trans. Magn.*, Sept. 1979, Vol. MAG-15, No. 5, pp. 1336–1339.

Fujiwara, T., Nonlinear Distortion in Long Wavelength AC Bias Recording." *IEEE Trans. Magn.*, Jan. 1979, Vol. MAG-15, No. 1, pp. 894–899.

Fujiwara, T., Wavelength Response of Harmonic Distortion in AC-Bias Recording. *IEEE Trans. Magn.*, May 1980, Vol. MAG-16, No. 3, pp. 501–507.

Hersener, J., Modelluntersuchungen zur Magnetisierungsverteilung im Magnetband bei sinus-formigen Signalen. *Wiss. Ber. AEG-Telefunken*, 1973, Vol. 46, No. 1, pp. 15–24.

Koester, E., A Contribution to Anhysteretic Remanence and AC Bias Recording. *IEEE Trans. Magn.*, Nov. 1975, Vol. MAG-11, No. 5, pp. 1185–1188.

Koester, E., The Reversible Susceptibility of Fine-Particle Assemblies and the Magnetic Anisotropy of Gamma-Fe_2O_3 and $CrO2$ Particles. *IEEE Trans. Magn.*, Sept. 1969, Vol. MAG-5, No. 3, p. 263.

Schwantke, G., The Magnetic Recording Process in Terms of the Preisach Representation. *Jour. AES*, Jan. 1961, Vol. AU-9, pp. 139–145.

Selmer, Jensen J., and Pramanik, S.K., Dynamic Bias Control with HX Professional. *Audio*, Aug. 1984, Vol. 68, No. 8, pp. 34–41.

Woodward, J.G., and Della Torre, E., Magnetic Characterics of Recording Tapes and the Mechanism of the Recording Pro. *Jour. AES*, Oct. 1959, Vol. 7, No. 4, pp. 189–195.

BIBLIOGRAPHY

Daniel, E.D., Axon, P.E., and Frost, W.T., A Survey of Factors Limiting the Performance of Magnetic Recording Systems. *Proceedings I.E.E.*, March 1957, Vol. 104, No. 14, part B, pp. 158–168.

Dolivo, F., and Closs, F., A Simulator for the AC-Biased Magnetic-Recording Channel. *IEEE Trans. Magn.*, Sept. 1978, Vol. MAG-14, No. 5, pp. 737–740.

Holmes, L.C., and Clark, D.L., Supersonic Bias for Magnetic Recording. *Electronics*, July 1945, Vol. 18, pp. 126–136.

Iwasaki, A., An Analysis on the State of AC-Biased Recording." *Ann. N.Y. Acad. Sci.*, March 1972, Vol. 189, pp. 3–20.

Stewart, W.E., *Magnetic Recording Techniques*, McGraw-Hill, New York, January 1958, 272 pages.

Straubel, R., Der Aufzeichnungsvorgang der Magnetspeichertechnik mit Wechselfeldvormagnetisierung in phaenomenologischer Sicht. *Hochfrequenztechnik und Elektroakustik*, May 1966, Vol. 75, pp. 153–162 and 2160–225.

Thiele, H., On the Origin of High-Frequency Biasing for Magnetic Audio Recording. *Jour. SMPTE*, July 1983, Vol. 92, No. 7, pp. 752–754.

Thurlings, L., and Kipzen, W., On the Mechanism of Particle Interaction in Magnetic Recording Media. *IEEE Trans. Magn.*, Sept. 1980. Vol. MAG-16, No. 5, pp. 1120–1122.

von Braumuhl, H.J., and Weber, H., An Improved Magnetron Process. *Z. Ver. deuts. Ing.*, July 1941, Vol. 85, p. 628.

Woodward, J.G., and Pradervand, M., A Study of Interference Effects in Magnetic Recording. *Jour. AES*, April 1961, Vol. 9, No. 4, p. 254.

PRINCIPLE OF AC BIAS

Albach, W., The Action of High-Frequency Bias in Magnetic Tape Recording. *Funk und Ton*, Dec. 1953, Vol. 7, pp. 628–630.

Arndt, W., and Carraro, U., Elektromekanische Simulation des Preisach-Modells. *Hochfreq. und Elektroakustik*, 1968, Vol. 77, pp. 7–11.

Axon, P.E., An Investigation into the Mechanism of Magnetic Tape Recording. *Proc. Inst. Elec. Engrs.* (London), May 1952, Vol. 99, pt. III, pp. 109–126.

Bertram, H.N., Monte Carlo Calculation of Magnetic Anhysteresis. *J. Phys.* (Paris), March 1971, Vol. 32, pp. C1 684–685.

Camras, M., Graphical Analysis of Linear Magnetic Recording Using High-Frequency Excitation. *Proc. I.R.E.*, May 1949, Vol. 37, pp. 569–573.

Daniel, E.D., The Influence of Some Head and Tape Constants on the Signal Recorded on Magnetic Tape. *Proc. IEE (London)*, May 1953, Vol. 100, Pt. III, pp. 168–175.

Dolivo, F., and Closs, F., A Simulator for the AC-Biased Magnetic-Recording Channel. *IEEE Trans. Magn.*, Sept. 1978, Vol. MAG-14, No. 5, pp. 737–739.

Dunlop, D.J., Grain Size Dependence of Anhysteresis in Iron Oxide Micropowders. *IEEE Trans. Magn.*, June 1972, Vol. MAG-8, No. 2, pp. 211–213.

Fritsch, K., and Scholz, Chr., Zur Verwendung des Preisach-Modells in der magnetslpeichertechnik. *Hochfreq. und Elektroakustik*, 1965, Vol. 74, pp. 25–30.

Huisman, H.F., Particle Interactions and Hc: Experimental Approach. *IEEE Trans. Magn.*, Nov. 1982, Vol. MAG-18, No. 6, pp. 1095–1098.

Kneller, E., Friedlander, F.J., and Pushert, W., AC Field "Freezing" and "Melting" of Magnetization in Fine-Particle Assemblies. *Jour. of Appl. Phys.*, March 1966, Vol. 37, No. 3, pp. 1162–1163.

Kneller, E., Puschert, W., Pair Interaction Models for Fine Particle Assemblies. *IEEE Trans. Magn.*, Sept. 1966, Vol. MAG-2, No. 3, p. 250.

Kneller, E., Magnetic-Interaction Effects in Fine-Particles Assemblies and in Thin Films. *Jour. of Appl. Phys.*, Feb. 1970, Vol. 39, No. 2.

Kneller, E., Relation Between Anhysteretic and Static Magnetic Tape Parameters. *IEEE Trans. Magn.*, Sept. 1977, Vol. MAG-13, No. 5, pp. 1388–1390.

Kneller, E., Static and Anhysteretic Magnetic Properties of Tapes. *IEEE Trans. Magn.*, Jan. 1980, Vol. MAG-16, No. 1, pp. 36–41.

Korolev, Towards Analysis of Magnetic Recording on a Relatively Thin Carrier by Preisach's Method. *Elektrichestvo*, September 1980, No. 9, pp. 66–69.

Mallinson, J.C., and Bertram, H.N., Write Processes in High Density Recording. *IEEE Trans. Magn.*, Sept. 1973, Vol. MAG-9, No. 3, pp. 329–331.

McCown, D.P., Barbosa, L.C., and Howell, T.D., Comparative Aspects of AC Bias Recording. *IEEE Trans. Magn.*, Nov. 1981, Vol. MAG-17, No. 6, pp. 3343–3346.

Minnaja, N., Magnetization Reversal of a Pair of Interacting Linear Dipole Distributions. *IEEE Trans. Magn.*, Sept. 1970, Vol. MAG-6, No. 3, pp. 649–662.

Sawamura, S., and Iwasaki, S.I., Application of Internal Reaction Field on the Analysis of Anhysteretic Magnetization Process. *IEEE Trans. Magn.*, Sept. 1970, Vol. MAG-6, No. 3, pp. 646–649.

Spindler, S., Eine Analyse des Preisachmodells. *Arch. elektr. Inform. und Energietechnik*, Jan. 1973, Vol. 3, No. 1, pp. 31–35.

Straubel, R., and Nietzsch, J., Analoge Simulation des Aufzeichnungsvorganges der Dynamischen Magnetspeichertechnik. *Hochfreq. und Elektroakustik*, Jan. 1967, Vol. 76, pp. 182–187.

Straubel, R., A Method for Calculating Hysteresis Loops of Interacting Single-Domain Particle Systems in a Nonmagnetic Binder. *IEEE Trans. Magn.*, Sept. 1969, Vol. MAG-5, No. 3, pp. 263–264.

Volz, H., Gerat zur halbautomatischen Messung von Preisachbelegungen bei Magnetbandern. *Hochfreq. und Elektroakustik*, Jan. 1969, Vol. 78, pp. 101–105.

Wohlfarth, E.P., Review of the Problems of Fine Particle Interactions with Special Reference to Magnetic Recording. *Jour. Appl. Phys.*, March 1964, Vol. 35, pp. 783–790.

CHAPTER 21
PLAYBACK WAVEFORMS (READ)

Faraday's law provides the simple expression for the playback voltage from a magnetic recording:

$$e = -n \, d\phi/dt \qquad (21.1)$$

where ϕ is the flux from the recording. We have used this formula for a sinusoidal flux pattern, and found:

$$e = -n \, \phi_m \, \omega \cos\omega t \qquad (21.2)$$

The amplitude of the flux ϕ_m can be determined by the principle of reciprocity, where for each element $w\Delta x\Delta y$ in the coating we have:

$$\Delta(x,y) = H_s(x,y) \times m(x,y) \times \cos \beta \qquad (21.3)$$

where $m(x,y)$ is the magnetization (in Wbm) from element $w\Delta x\Delta y$ in the coating and $H_s(x,y)$ is the reproduce head's sensing function (in m^{-1}; see Chapter 6). β is the angle between the magnetization and the sensing function vectors. The total flux is found by substituting $m(x,y)$ with the recorded magnetization pattern and then carrying out a double integration, with x going from plus to minus infinity, and letting the y component go from d to $\delta_{rec} + d$ (d = spacing, or distance between head and tape; δ_{rec} = recorded thickness):

$$\phi(x) = w \int_{-\infty}^{\infty} \int_{d}^{\delta_{rec} + d} H(x - \tau, y) \, m(\tau, y) \, \cos\beta \, d\tau dy \qquad (21.4)$$

This integration is manageable when $m(x,y)$ is zero in the y-direction and varies sinusoidally in the x-direction. This is the classical longitudinal magnetization, and the result of the integration was presented in Fig. 6.9.

We can also integrate when the magnetization pattern is a step from negative to positive magnetization. It is common practice to use the arctangent model for the transition zone, and to calculate the resultant flux change and its derivative to find the voltage pulse.

PULSE WIDTH AND SENSING FUNCTION

The reproduced pulse is characterized by its length, as measured between the two points where the amplitude is 50 percent of the pulse peak voltage. This 50-percent pulse width, also named *PW50*, can be estimated from the calculations, and an approximation for a thin coating is:

$$PW50 = 2 \sqrt{d^2 + (g/2)^2} \qquad (21.5)$$

where *d* is the head-to-media spacing and *g* is the gap length (from (6.12)).

There have been considerable difficulties in finding a good agreement between the theories and measurements in pulse recording. The advances in high packing densities indicated several years ago that the perpendicular y-component of the magnetization should be included in the integration in formula (21.4).

The complexity of the playback process becomes obscured in the integration. A picture of the interplay between the head and the tape magnetization is biased by the awareness of a short gap, almost to the point where one envisions the flux being sensed in front of the gap only. This gives the playback head a sensitivity function much like the Stein model of the record process, a sheet in front of the gap, perpendicular to the head. The mathematical solutions (where differentiation of a sinusoidal flux gives a cosine voltage) emphasizes this simple picture, as shown in Fig. 21.1a.

The total flux is determined by the vector products of the magnetization vector and the head sense vector. This results in a considerably broader sensitivity function (the head field pattern). For sense a purely longitudinal magnetization, the playback of a transition gives a fairly broad voltage pulse (Fig. 21.1b).

The frequently observed asymmetry in the read pulses from recorded transitions has its origin in the vector magnetization of the magnetic coating. This pattern is the result of the media magnetization anisotropy and the write field shape at the trailing edge of the write gap. The transition zone it-

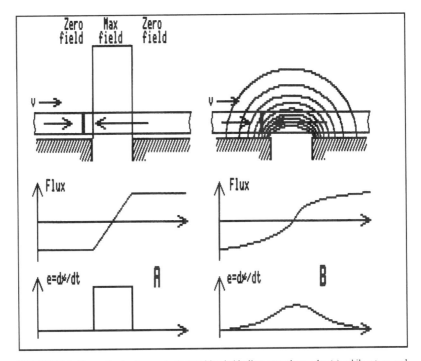

FIGURE 21.1 The read pulse from a single transition is ideally a very short pulse (a), while a true read pulse broadens out (b).

self, lengthened by demagnetization, has less influence due to the long sensing field of the read head, which extends many gap lengths on both sides of the gap: the read sensing function encompasses flux many gap lengths away from the gap.

It is relatively simple to compute the shape of a read pulse, when the magnetization pattern in the coating is known. Such information is useful in at least four areas:

1. The effects of various magnetization patterns on the shape of the read pulse can quickly be evaluated; this relates to the media magnetization anisotropy.

2. A predicted magnetization pattern is verifiable by comparing the computed and the measured read pulses from isolated transitions.

3. A Fourier transform of the response to a single transition will reveal the write/read system's transfer function (see Chapters 22 and 23).

4. The isolated pulse response is useful in a convolution with a write data pattern (corrected for write-peak shift) to determine the read voltage pattern.

COMPUTATION OF READ VOLTAGE

The output voltage from a reproduce head reflects the magnetization pattern in the tape or disk coating. The purpose of this chapter is to outline a method of voltage computation, and to highlight some general relations between magnetization patterns and voltage waveforms.

The approach is elementary; we know from Chapter 6 that the flux contribution from a magnetization is:

$$\Delta\phi = \Delta m \times \boldsymbol{H}_s \qquad (21.6)$$

where $\boldsymbol{H}_s$ is the reproduce head sensing function; this may be the Karlquist field for a longpole head (Fig. 4.24), or as given by Potter for a thin-film head (see reference in Chapter 8), or by Szczech and Fayling (see the bibliography of this chapter).

We can also express (21.6) as:

$$\Delta\phi = m_x \boldsymbol{H}_{sx} + m_y \boldsymbol{H}_{sy} \qquad (21.7)$$

where m_x and m_y are the magnetization components (see Fig. 21.2), and H_{sx} and H_{sy} are the sensing function components.

MAGNETIZATION PATTERNS

We will use a simple method for determination of the magnetization in the coating. It appears equally well-suited for biased or nonbiased direct recordings (Fig. 20.18).

Most coatings are anisotropic due to the use of oriented, elongated particles. A modem tape will typically have a longitudinal squareness S_x equal to 0.8 to 0.85, while the perpendicular squareness S_y is only 0.2. The magnetizations' x- and y-components are determined from equations (20.5) and (20.6), or from (20.7) and (20.8). For saturation recordings use (Fig. 21.2):

$$m_x = S_x M_{sat} \times y/R \qquad (21.8a)$$
$$m_y = S_y M_{sat} \times x/R \qquad (21.8b)$$

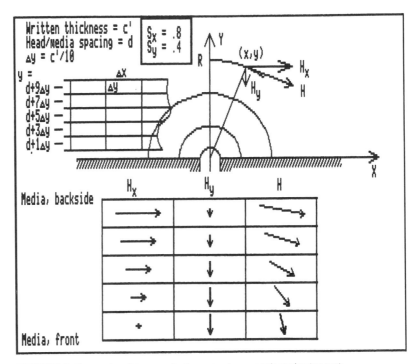

FIGURE 21.2   Computation of the magnetization vectors in a magnetic coating (see text).

RESPONSE TO A TRANSITION

Let us for a moment consider how we could determine the read system's response to magnetization changes. If we use a single impulse signal, then its resulting output is an impulse response. Such an impulse is obtained by differentiating a step function. The system is linear, and the differentiation can take place anywhere.

VERY SHORT TRANSITIONS

Let this be the action of the inductive read head: It differentiates a step change in magnetization into a voltage that ideally is an impulse. Thus a single magnetization transition at one $\Delta x\Delta y$ element produces a contribution to the impulse response. If it is done for all elements, we will obtain the read system's "impulse response," placed in quotation marks because it really is the response to a step change in magnetization.

We will therefore, with this short method, obtain the impulse response without moving a transition past the head. Merely sum up all for each column at a given x value, and use it for a point on the response curve at that x value. The resulting total curve is the response that an ideal, vertical transition of width Δx will produce Fig. 21.3.

If the transition zone is curved, then the vertical summation is changed to using the flux values that are displaced in accord with the curvature. One step in this summation is illustrated to the right of the gap in Fig. 21.4. The flux contribution is—at that point—higher than for a straight transition; and it is lower for summations to the left of the gap.

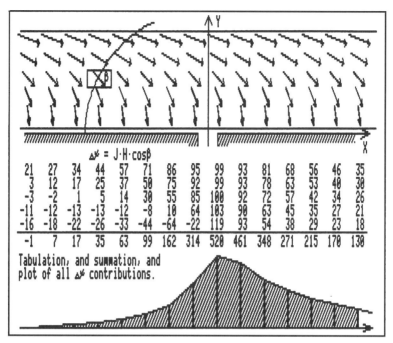

FIGURE 21.3 Computation of the read flux patterns and voltage from a magnetized coating (this is the ideal, short-transition response).

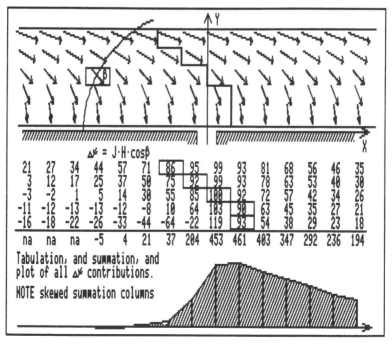

FIGURE 21.4 Computation of the read flux, including the effect of a curve recording zone, and hence transition zone.

LONG TRANSITIONS: THE COMPLETE COMPUTATION

We shall now determine the voltage response for a transition recorded on a coating, with an arbitrary (but known) magnetization pattern such as the one shown in Fig. 21.5.

The long way to compute the voltage is to sum all flux contributions, with the transition first at the leftmost value of x (say $x > 100$ times gap length g), the last at the rightmost value; and for all layers, from y_{front} to y_{back}. This represents the flux value at the first position on the flux curve.

Next, shift the transition one Δx to the right and recompute the flux sum, which is the value at the next point; then make another shift and recompute, etc. The waveform:

$$\Phi_n = \phi_n(n\Delta x) \tag{21.8}$$

is the flux waveform, and a differentiation will produce the induced voltage waveform.

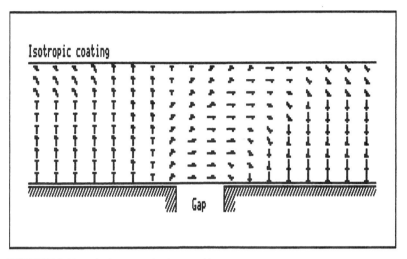

FIGURE 21.5 Magnetization patterns in a long transition.

LONG TRANSITIONS: THE QUICK COMPUTATION

For a broadened transition zone it becomes necessary to use a mixture of two computational methods. First, compute the response $\phi = f(x)$ for the uniform magnetization pattern that exists away from the transition zone, say in the middle of a bit (or in the coating after dc magnetization by the write head). Store the values for each x in array elements. This corresponds to the response for a very short transition (Fig. 21.3).

Now start stepping the transition zone along the x axis, starting with the transition just outside the leftmost value of x. Compute the total flux from all elements in the transition zone; add to this number the summation of all stored flux values to the left and to the right of the present position of the zone-block; see Fig. 21.6. This gives one point on the flux curve. Now step the zone one Δx to the right, and repeat the computation; repeat as many times as needed to make the transition zone move past the x value x_{end}.

This will give a curve that represents the flux level as a function of the position of the zone center line along the x-axis. Now differentiate this curve to get $d\phi/dx$, substitute $x = vt$, and we have now determined $d\phi/dt$.

Neither the first complete summation procedure nor the last shorter way are necessary for a 5- to 10-percent accurate determination of the output pulse shape. The short "once-through" computation

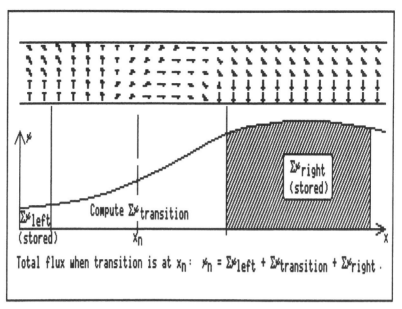

FIGURE 21.6   A method of fast computation of the response to a long transition (see text).

will frequently suffice because the vector magnetization pattern has a larger influence on the pulse shape than the zone broadening and/or curvature.

The saving in computation time is formidable: A reasonable resolution and accuracy can be obtained by using 10 Δy layers for the coating, and a couple of hundred Δx elements for the x scan (equal to a minimum of 20 gap lengths; verify that additional contributions from elements further out are not needed).

The number of computations required are:

The long method	$150 \times 150 \times 10 = 225{,}000$ computations
The short method	$1 \times 150 \times 10 = 1{,}500$ computations
The medium method	$1 \times 150 \times 10 + 20 \times 190 \times 10 = 39{,}500$ computations

The computation time ratios are 150:1:25.

INVESTIGATING MAGNETIZATION PATTERNS

The read waveform from a variety of magnetization patterns can now be evaluated, as a few examples will illustrate. The pattern from Fig. 21.5 results in curve shown in Fig. 21.7 (broken line). It compares well with the read pulse from Tjaden and Leyten's experiment (from Fig. 5.5). Both show a short rise and a long decay time.

This is characteristic for vector magnetization patterns. A purely longitudinal magnetization results in a perfectly symmetrical pulse, while a purely perpendicular magnetization results in a dipulse. Both are shown in Fig. 21.8, which also shows another response to a short transition with vector magnetization.

The usefulness of separate arrays for the computational result using formula (21.7) is evident from Fig. 21.9. Here the response to the transition in Fig. 21.5 is shown, made up from fluxes from the longitudinal and perpendicular components.

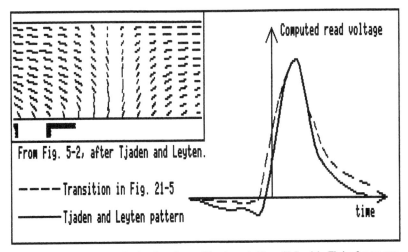

FIGURE 21.7   Computed read pulses from the transition shown in Figure 21.5 and the Tjaden-Leyten transition.

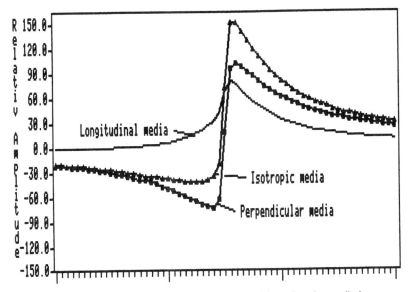

FIGURE 21.8   Read pulses from three different media: longitudinal, isotropic, and perpendicular.

Figure 21.10 shows three transitions: a short, strongly magnetized transition (top), a broad and strong transition (middle), and a broad and weak transition (bottom). The last should give a weak and broad read pulse, as compared to the top transition.

The computed responses are shown in Fig. 21.11, which verifies our predictions for the output levels. When the levels are normalized, we can also verify that the shortest response is obtained for the short transition. Notice the dipulse response; they are all predominantly perpendicular magnetized; see Fig. 21.12.

The effect of head-media spacing can easily be computed, and the result is shown in Fig. 21.13 for a 1.25 μm coating with a 0.7 μm gap head. Increased spacing causes reduced output and broadened pulses.

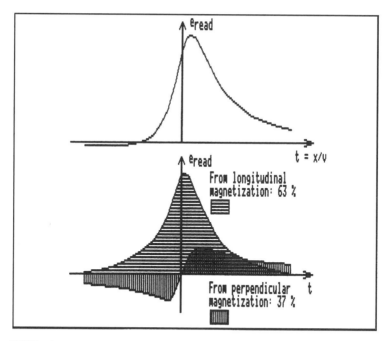

FIGURE 21.9 A read pulse for the transition shown in Figure 21.5 is composed of contribution from longitudinal and perpendicular magnetizations.

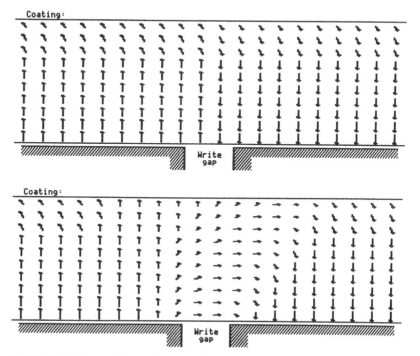

FIGURE 21.10 Three possible transition patterns from recording of a single transition on an isotropic medium: Short and strong, broad and strong, and broad and weak.

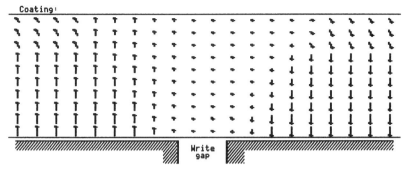

FIGURE 21.10 Three possible transition patterns from recording of a single transition on an isotropic medium: Short and strong, broad and strong, and broad and weak (*Continued*).

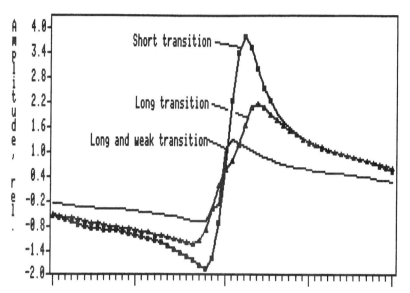

FIGURE 21.11 The computed responses from the three transitions in Figure 21.10.

The reader will find further discussions on the response to a single transition in Chapters 22 and 23, and will hopefully benefit from the knowledge of the various responses from various magnetization patterns in and around written transitions. This chapter's modelling is a valuable tool for all sorts of trend analysis.

The response to two transitions separated by T is shown in Fig. 21.14. Formula (6.13) for the Lorentzian pulse was used. Note the reduced and time shifted amplitudes. This read peak shift is shown for three and five transitions, Fig. 21.15 and Fig. 21.16.

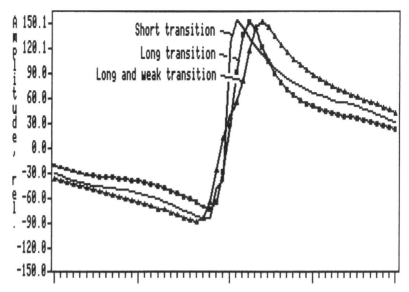

FIGURE 21.12   Normalized levels for responses in Figure 21.11.

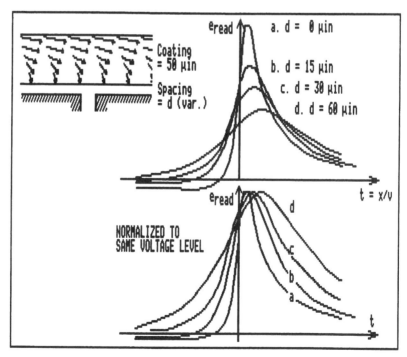

FIGURE 21.13   The effect of distance between a read head and the recording surface: Increased distance causes a reduced output level and a broadening of the pulse

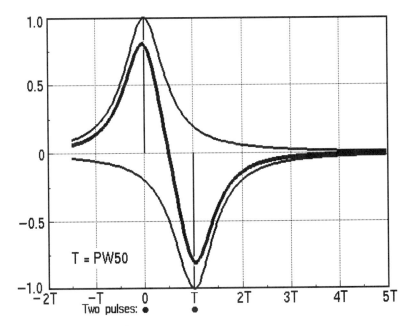

FIGURE 21.14 Di-bit interference.

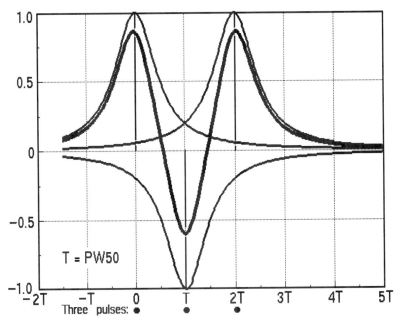

FIGURE 21.15 Tri-bit interference.

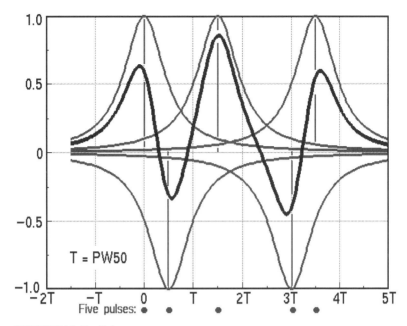

FIGURE 21.16   Five-bit interence.

BIBLIOGRAPHY

READ SENSITIVITY FUNCTION

Daniel, E.D., and P.E. Axon. May 1953. The Reproduction of Signal Recorded on Magnetic Tape. Proc. IEE (London), Paper 1499R, Vol. 100, Pt. III, pp. 157.

Gooch, B., R. Niedermeyer, R. Wood, and R. Pisharody. Nov. 1991. A High Resolution Flying Magnetic Disk Recording System with Zero Reproduce Spacing Loss. *IEEE Trans. Magn.* MAG-27 (6): 4549–54.

Heim, D. Sept. 1983. The Sensitivity Function for Shielded Magnetoresistive Heads by Conformal Mapping. *IEEE Trans. Magn.* MAG-19 (5): 1620–1622.

Lambert, S.E., F.D. Sargent, and M.L. Williams. Nov. 1991. Head Sensitivity Function Determined from Recording Measurements. *IEEE Trans. Magn.* MAG-27 (6): 2458–60.

Szczech, T. and R. Fayling. Nov. 1982. The Use of Perpendicular Head Field Equations for Calculating Isolated Pulse Output. *IEEE Trans. Magn.* MAG-18 (6): 1176–1177.

Vos, M.J., and J.H. Judy. Nov. 1988. Sensitivity Function and Spectral Response without Gap Nulls of an Inductive Perpendicular Head Geometry. *IEEE Trans. Magn.* MAG-24 (6): 2401–03.

COMPUTATIONS OF READ VOLTAGE

Baker, B.R. Nov. 1981. Response of Asymmetric Thin Film Heads to Vertical and Horizontal Magnetization. IEEE Trans. Magn. MAG-17 (6): 3123–3125.

Camras, M. Aug. 1946. Theoretical Response from a Magnetic Wire Record. *Proc. I.R.E.*, Vol. 34, (8): 593–602.

Laubsch, H., S. Kisch, and J. Sladek. Jan. 1970. Modelluntersuchungen zum Wiedergabevorgang. *Hochfrequenztechnik und Elektroakustik*. Vol. 79, 144–149.

Middleton, B.K. Sept. 1975. The Replay Signal from a Tape with Magnetization Components Parallel and Normal to Its Plane. *IEEE Trans. Magn*. MAG-11 (5): 1170–1173.

Minuhin, V.B. (pt. 2 with M. Steinback). Nov. 1985. Dependence of Readback Output on Medium Thickness in the Presence of a Permeable Underlayer—Part 1 and 2. *IEEE Trans. Magn*. MAG-21 (6): 2595–2619.

Minuhin, V.B. Jan. 1985. Theory of Playback Process with Soft Magnetic Underlayer. *IEEE Trans. Magn*. MAG-21 (1): 28-35.

Vos, M.J., S.B. Luitjens, R.W. de Bie, and J.C. Lodder. Sept. 1986. Magnetization Transitions Obtained by Deconvolution of Measured Replay Pulses in Perpendicular Recording. *IEEE Trans. Magn*. MAG-22 (5): 373–375.

MAGNETIZATION IN MEDIA

Bate, G. et al. Nov. 1974. The Remanent State of Recorded Tapes. *IBM Jour. Res. Develop.*, pp. 563–569.

Bate, G. et al. Sept. 1974. Magnetization in Recorded Tape. *IEEE Trans. Magn*. MAG-10 (3): 667–669.

Habbal, F., and W.T. Vetterling. June 1984. Angular Distribution of Magnetization in Coated Magnetic Films. *Jour. Appl. Phys*. 55 (6): 2291–93.

Hersener, Jurgen. 1973. Modelluntersuchungen zur Magnetisierungsverteilung im Magnetbandbei sinusformigen Signalen. *Wiss. Ber. AEG-Telefunken* 46 (1): 15–24.

Knowles, J.E. et al. Jan. 1980. The Angular Distribution of the Magnetization in Recording Tapes. *IEEE Trans. Magn*. MAG-16 (1): 42–44.

Minuhin, V.B. Sept. 1986. Theoretical Comparison of Readback Harmonic Responses for Longitudinal Recording and Perpendicular Recording with Probe Head over a Medium with Permeable Underlayer. *IEEE Trans. Magn*. MAG-22 (5): 388–390.

Valstyn, E.P., and J.E. Monson. Nov. 1979. Magnetization Distribution in an Isolated Transition. *IEEE Trans. Magn*. MAG-15 (6): 1453–1456.

Wells, R.B. Jan. 1985. Estimation of the Shape of Magnetic Transitions by a Deconvolution. *IEEE Trans. Magn*. MAG-21 (1): 14–19.

Williams, E.M. Nov. 1982. The Dorf Effect: Magnetization Ripple in Particulate Media. *IEEE Trans. Magn*. MAG-18 (6): 1086–1088.

FIELDS FROM MEDIA

Baird, A.W., W.F. Chaurette, and C.D. Lustig. Nov. 1981. Field Measurements Near High-Density Statically Recorded Transition in a Thin-Film Medium. *IEEE Trans. Magn*. MAG-17 (6): 2553–2555.

Fayling, R.E. Nov. 1979. Edge Profile Studies of Recorded Flux Transistions. *IEEE Trans. Magn*. MAG-15 (6): 1469–1470.

Indeck, R.S., and J.H. Judy. Sept. 1984. Measurements of Surface Magnetic Fields of Perpendicular and Longitudinal Magnetic Recorded Transitions Using a Magnetoresistive Transducer. *IEEE Trans. Magn*. MAG-20 (5): 730–732.

Iwasaki S., and T. Suzuki. January 1967. The Effect of Demagnetizing Field on 2-Bit Pattern in Digital Recording Process. *Rep. of Res. Inst. of El. Comm. Tohoku Univ*. 19 (3): 171–86.

Mallinson, J.C. Dec. 1973. One-Sided Fluxes—A Magnetic Curiosity? *IEEE Trans. Magn*. MAG-9 (4): 678–683.

McKnight, J.G. June 1970. Tape Flux Measurements Theory and Verification. *Jour. Audio Engr. Soc*. 18 (3): 250–259.

Monson, J.E. Nov. 1988. Fringing Fields from Step Transitions of Longitudinal and Perpendicular Magnetization of Finite Tracks. *IEEE Trans. Magn*. MAG-24 (6): 3108–10.

Schmidbauer, O. Oct. 1952. The Field of the Harmonically Magnetized Tape. *Frequenz.* Vol.6: 319-334.

Speliotis, D.E., and J.H. Judy. March 1971. Calculations of External Bit Fields. *IEEE Trans. Magn.* MAG-7 (1): 158–163.

Tagami, K., Y. Suganuma, and M. Nagao. July 1977. A New External Bit Field Observation Method. *IEEE Trans. Magn.* MAG-13 (5): 1689–1692.

Tagami, K., Y. Suganuma, and M. Nagao. May 1979. External Bit Field Analyses. *IEEE Trans. Magn.* MAG-15 (3): 1054–1060.

Westmijze, W.K. Aug. 1953. Studies on Magnetic Recording, IV: Calculation of the Field In and Around the Tape. *Philips Res. Rep.* 8 (4): 255–269.

Yeh, N.H. Sept. 1980. Ferrofluid Bitter Patterns on Tape. *IEEE Trans. Magn.* MAG-16 (5): 797–982.

PEAK SHIFTS

Bloomberg, D. Sept. 1983. Readback Bit Shift with Finite Pole-Length Heads on Perpendicular Media. *IEEE Trans. Magn.*, Vol. MAG-19, (5): 1617–1619.

Jorgensen, F. April 1982. Asymmetry & Peak Shift on the Read Pulse for a Single Recorded Transition. *IERE Conf. Video and Data*, IERE Publ. (54): 165–179.

Roscamp, Thomas, A. Nov. 1981. Origins of Playback Pulse Asymmetry in Recording with Thin Film Disk Heads. *IEEE Trans. Magn.*, Vol. MAG-17, (6): 2902–2904.

CHAPTER 22
SIGNAL RESPONSE AND EQUALIZATION

Faithful reproduction of magnetically recorded information requires compensation for a number of losses. Some of these occur during the recording (write) cycle, such as the loss in resolution due to the switching field distribution, the gradual decrease in head field strength, and due to head-medium spacing. There may also be signal attenuation at high frequencies due to head core losses. Further attenuation occurs in the media magnetization due to demagnetization, and during storage a small decay with time may occur. During the read process there are additional losses due to geometrical factors (recorded thickness, head-medium spacing, read gap length) and high frequency core losses.

When the losses are subtracted from the signal amplitude versus frequency curve, that has a slope of 6 dB/octave, then a response as shown in Fig. 6.2 results. Compensation is made by an amplifier circuit that boosts the low and the high frequencies, a process named *amplitude equalization*. The choice of equalizer circuits should not be made arbitrarily, since the majority of the losses have little or no associated phase shifts; ordinary RLC-circuits for compensation can equalize the amplitude-versus-frequency response very well, but may at the same time add considerable phase error into the signal.

This chapter will discuss the means of restoring the signal to as close a representation as possible of the original input signal to the record circuit. This implies restoration of the amplitude response to a flat response, and of the phase to a linear function of the frequency that intercepts the origin (frequency equal to zero) at zero phase (or $\pm n \times \pi$). The amplitude equalization is not necessarily carried out for low frequencies in digital recording; i.e., the read preamplifier is not followed by an integrator, but possibly by a differentiator to facilitate zero-crossing detections, which correspond to the induced voltage peaks occurring at transitions.

It is a mistake to believe that digital recording is a matter of digital processing only—the signal passes through the write amplifier, media magnetization, read amplifier, and filter/equalizer circuits, all of which are members of the analog world. These circuits are discussed in this chapter.

The phase response of a recorder must be correct in pulse (digital) recording equipment, which is not necessarily assured by a flat amplitude-versus-frequency response. An evaluation of the response to a single transition is important in optimizing a disk or tape recorder for digital data storage. The record/reproduce (write/read) channel can be modelled, optimized and verified by measurements. And powerful instruments are available, using the most up-to-date signal analysis methods (see Chapter 23).

RESPONSES OF AMPLITUDE AND PHASE VERSUS FREQUENCY

The amplitude response of a record/playback channel is well known with fairly precise characterizations of the various losses (see Fig. 6.2); a signal phase shift is not evident from the expression for the read voltage in Fig. 6.11 because it was derived for purely longitudinal magnetization in the medium. We would have found a similar formula for purely perpendicular magnetizations, but with a 90-degree phase shift.

Signal attenuation in ordinary electric networks made up of resistors, capacitors, and inductors have an associated change in the phase characteristic. Certain electrical networks are exceptions to this rule, such as the class of all-pass networks and certain lossy filters (Kallmann). A phase shift will also occur in magnetic recording, as we will now discuss.

The direction of the remanent magnetization changes along and through the coating thickness. Its pattern may look like Fig. 22.1. A simple experiment verifies such phase shifts: a row of dipole magnets are placed in groups, each corresponding to one bit length (or half a wavelength). This is shown in Fig. 22.2, where four bits are made up of three, three, two, one and one dipole, respectively.

The external flux lines are made visible with iron powder and they clearly show the 90-degree phase shift between the longitudinal (top) and perpendicular (middle) magnetization patterns. The bottom pattern for the 45-degree position of the dipoles (polarization) results in a phase shift that is somewhere between zero and 90 degrees. The bottom flux pattern also verifies the observed change in read waveforms that occur when a tape is played back in the forward and then in the reverse directions: Playback equalizers require a polarity change in the phase equalizer adjustment.

The magnetization in the coating is only purely longitudinal if the medium's susceptibility is zero in the perpendicular direction. The read voltage from a medium with both longitudinal and perpendicular magnetizations will be an asymmetrical pulse, as shown in Fig. 21.9, and is composed of signals from the longitudinal as well as the perpendicular magnetizations (see also Fig. 6.10).

The write/read process has thus split the input signal into two components. The first represents the original signal, while the other is a transform thereof called the Hilbert Transform. The latter is always in quadrature with the first (Bracewell).

The magnetization versus frequency (or, more correctly, versus 1/wavelength) is lagging at short wavelengths, as shown in Fig. 22.3. The magnetization poles are shifted and turned from the bit cell ends (transitions) toward the middle of the bit cells in such a way that they arrive later at the read head, and present a lag.

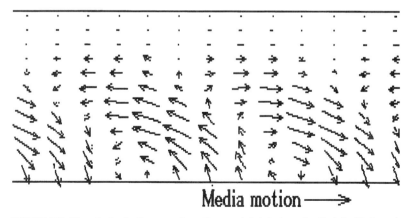

FIGURE 22.1   Magnetization inside a magnetic coating recorded at short wavelengths (*after Tjaden and Leyten*).

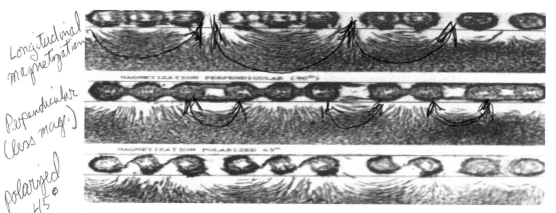

Longitudinal magnetization

Perpendicular (less mag.)

polarized 45°

FIGURE 22.2 The dipole model of a magnetized coating. Flux lines are made visible by application of iron powder. Top: Longitudinal magnetization. Middle: Perpendicular magnetization. Bottom: 45-degree magnetization (polarization).

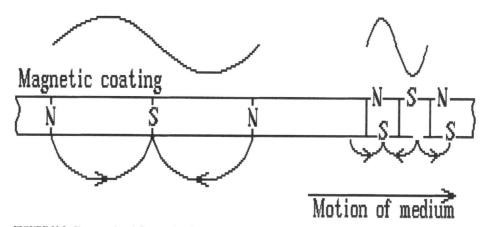

FIGURE 22.3 Short wavelength fluxes arrive "late" at the read head relative to the long wavelengths; i.e., the high frequencies are delayed.

This results in a phase versus wavelength response for the coating flux, as shown in Fig. 22.4, that agrees with the limited information available in the literature (Jacoby, Minukhin, Haynes). A different presentation is seen in Fig. 22.5, employing a linear frequency scale. The curves are quite linear over the recorder passband, but an extrapolation intersects the phase angle scale at a point different from zero degrees. This shift is carried along when the flux is differentiated in the read head winding, resulting in a +90 degree phase shift, and we obtain Fig. 22.6. The intersect different from zero (or $\pm n\pi$) is the source of the problem that requires phase equalization.

Phase distortion is illustrated in Fig. 22.7: A square wave is made up of a fundamental cosine waveform plus odd harmonics. The illustration uses only the third harmonic to recompose the waveform, and shows asymmetrical distortion of the waveform when the high frequencies arrive late. A correct waveform can be obtained by shifting the phase curve so it intersects the phase axis at zero degrees; see Fig. 22.8. This is most simply done with an all-pass network in the playback amplifier.

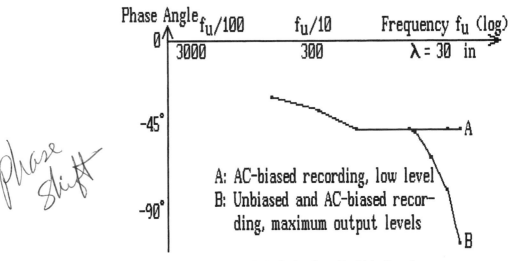

Phase shift

FIGURE 22.4 Phase shift curves for the flux into the read head (*after Haynes*).

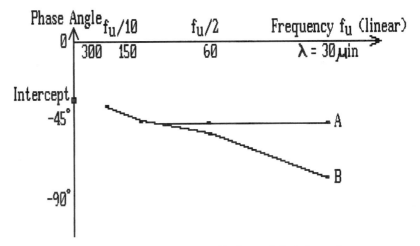

FIGURE 22.5 Phase shift curves for the flux, shown with a linear scale for the frequencies.

The photos in Fig. 22.9 illustrate how a square-wave input is distorted in an ordinary tape recorder. It can be restored to a symmetrical waveform by adding phase equalization to the electronics, and the result is the waveform shown in the bottom.

An estimate of the phase correction needed is possible from a knowledge of the ratio between the perpendicular and longitudinal magnetizations, M_{perp} and M_{long} (after Minukhin):

$$\text{Phase intersect at } (\omega=0) = \arctan (M_{perp}/M_{long}) \tag{22.1}$$

For purely longitudinal magnetization the angle is zero degrees, while it is 90 degrees for purely perpendicular magnetization.

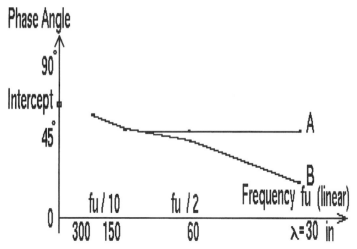

FIGURE 22.6 Phase shift curves for the read voltage (i.e., differentiated flux, at +90 degrees of phase shift).

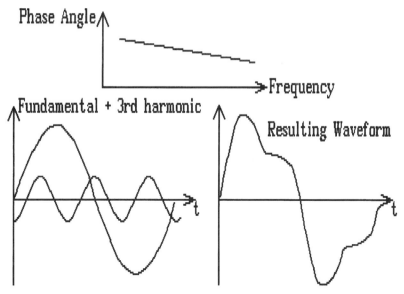

FIGURE 22.7 A linear phase-versus-frequency relation can result in phase distortion if the line intersects the phase axis at points different from 0 or ±n×180 degrees.

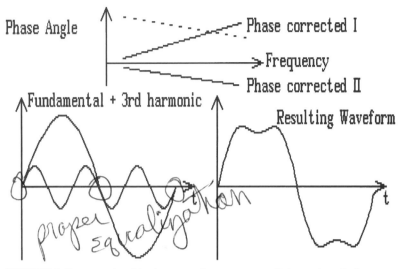

FIGURE 22.8   Phase correction of the phase-versus-frequency curve results in correct equalization.

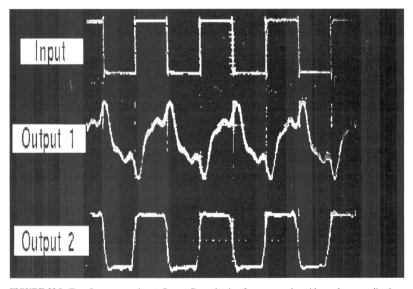

FIGURE 22.9   Top: Square wave input. Center: Reproduction from a recorder with no phase equalization. Bottom: A properly equalized playback waveform.

EQUALIZATION IN LINEAR (AC-BIASED) RECORDERS

A block diagram of an ac-biased recorder, either audio or instrumentation, is shown in Fig. 22.10. The input voltage V_{in} is converted into a current and passed through a preequalization network and into the head winding. The high-frequency bias signal is added through a capacitor.

The bias signal may cause intermodulation distortion at the output stage of the record amplifier; consequently, a tuned bias trap is inserted between the amplifier and the recording head. The record

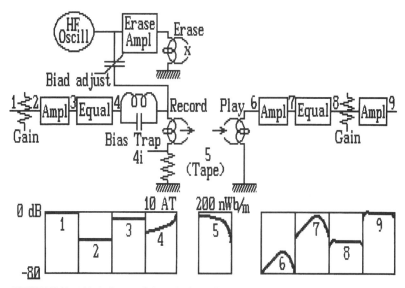

FIGURE 22.10   A block diagram of electronics in a typical ac-biased recorder, and a signal level diagram.

amplifier, bias oscillator and playback amplifier do not differ very much from conventional circuit designs, except that particular attention must be paid to such factors as low distortion (amplitude and phase) and low noise.

During playback, the flux from the recorded tape is differentiated in the reproduce head, amplified and then equalized. The quality of a recording can be monitored by a metering circuit, which can be switched to the playback position for an A–B comparison between recorder input and output. The level indicator can also be used for monitoring the record and bias current through a resistor (typically 10 ohms) in the ground leg of the record head—a useful feature during checkout or service of a recorder. The lower portion of Fig. 22.10 is a level diagram for the signal as it passes through the record amplifier, is recorded, played back, equalized, and amplified.

RECORD CIRCUITRY

The record amplifier is an amplifier with a flat amplitude-versus-frequency response that converts the input voltage into a record current. Its input impedance varies from 91 ohms for a wideband recorder (several MHz bandwidth) to 100 kilohms for a microphone audio amplifier. The corresponding sensitivities range from 1 volt rms to 1 millivolt rms, which in either case gives an input sensitivity on the order of 10 milliwatts. The record amplifier must be designed to provide a current that will saturate the tape prior to any amplifier distortion. If the record current is adjusted to produce one percent 3rd harmonic distortion on a tape (any type), then saturation will occur at a record current that is approximately 4 times higher (12 dB). The amplifier should have less than one percent distortion at that level.

The record head is an inductive load for the record amplifier, and its impedance, consequently, increases with frequency. This in turn requires that the record head be driven by a constant-current source which again dictates that the record amplifier has a high output impedance. A constant record current should, in reality, mean a constant record flux, but eddy-current and hysteresis losses make it necessary to increase the record current toward the higher frequencies in order to obtain a constant recording field strength versus frequency.

RECORD EQUALIZATION

Pre-equalization in the record amplifier compensates for eddy-current and hysteresis losses in the record head, and should also compensate for record zone losses (resolution) and demagnetization losses. The compensations for the latter must be done with a circuit that does not introduce phase shifts (see the delay line equalizer later in this chapter).

It is also desirable to compensate for the magnetization phase shift toward short wavelengths; this should be done with an all-pass network that preserves a flat amplitude response.

This nominal equalization will assure that the recorded magnetization level is uniform across the frequency, or better, wavelength range. Subsequent playback will require an additional boost of both low and high frequencies in order to compensate for thickness, spacing, and gap losses. This post equalization causes an increased amplification of any low-frequency noise (power supply hum, 1/f-noise) and high-frequency noise (tape, head, and amplifier noise), and can be minimized by applying the maximum permissible record current boost.

Most program materials do not have a flat energy spectrum, and pre-equalization may therefore be used to fully utilize the media magnetization; this will lessen the requirements for post-equalization, leading to an improvement in the overall signal-to-noise ratio.

Pre-equalization in audio recording was based originally upon the energy spectrum of music (i.e., level versus frequency). At first the classical work of Sivian, Dunn, and White was used in establishing pre-equalization standards. However, later findings and experience have shown that at times it is tolerable to record with a higher pre-emphasis than dictated by the energy spectra.

Here is an area where audio recordings differ from instrumentation recordings. In audio recordings, harmonic distortion is less noticeable and intermodulation distortion is often confused and intermixed with that of scrape flutter. In instrumentation recording, on the other hand, harmonic distortion and intermodulation in a multiplexed channel is readily observed as a foreign signal.

Record equalization in instrumentation recorders is therefore limited to that amount required to achieve a *constant flux* on the recorded tape. The pre-equalization is designed to correct for

- Head core losses (including phase shifts)
- Write resolution losses (recording zone, SFD; no phase shifts)
- Demagnetization losses (some phase shifts)

Adherence to this method of pre-equalization will produce tapes that are interchangeable from machine to machine, because they all will have a constant flux level independent of wavelength. The write resolution losses listed above corresponds to what generally is called bias self-erasure losses.

The head response, i.e., current in the winding versus frequency for a constant input current, must enter into the design considerations. A square-wave write current can produce undesirable ringing, as shown in Fig. 22.11. Pre-equalization should also include a phase-correcting network that will produce tapes that have identical phase response whether reproduced in forward or reverse motion (Johnson and Gregg).

Pre-equalization in audio recorders is tied in with the standards for playback equalization, and will be covered later in the chapter.

RECORD LEVEL INDICATOR

A recording level indicator is required to warn the operator of excessive recording levels. In its infancy, the magnetic recorder was provided with a VU meter (which stands for volume meter), a leftover from the early days in broadcasting, where it was used to monitor the signal level sent to the transmitter.

The VU meter is an averaging-type instrument, and quite inadequate for informing the recording engineer about the proper recording level. Distortion takes place at instantaneous peaks of the signal

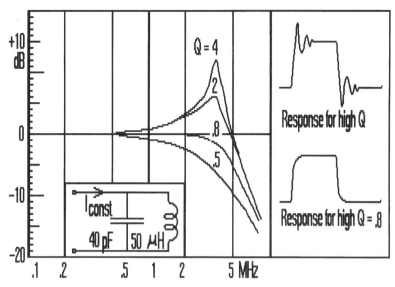

FIGURE 22.11 The response of current in a magnetic head for four values of Q.

to be recorded, causing both harmonic and intermodulation distortion. Peaks are much better detected by peak-reading indicators, which may be moving-coil instrument with a suitable peaking amplifier, or a row of light emitting diodes (LED).

A record-level indicator should be connected to the electronic circuitry after equalization, because constant-current recording (which produces a constant flux) provides a nominally constant distortion level. The disadvantage of connecting the level indicator at this point is that it will not show a true comparison between V_{in} and V_{out} (A–B test).

PLAYBACK AMPLIFIER

The playback amplifier's function is to amplify the weak reproduced signal to suitably high output level. The stages in the reproduce amplifier are normally of conventional design, with the exception that extreme care must be taken in the selection of the input circuit immediately following the reproduce head. The available voltage from the reproduce head at very low frequencies (20 Hz) is in the micro-volt region. This means that the input stage in the playback preamplifier must have an internal noise voltage that is only a fraction of a microvolt. Because the reproduce head is an inductive generator, it is further important that the input impedance be high, to avoid signal dropoff at high frequencies.

The output voltage from the reproduce head can be increased by increasing the number of turns on the reproduce head core. But this lowers the self-resonance of the head and limits the high-frequency response. Therefore, the head design criteria become a compromise between bandwidth and output level. The self-resonance of a magnetic head can readily be measured; see Fig. 9.40. Care must be taken to include the head cable capacitance and the preamplifier input impedance.

The reproduce head is an inductive component with a certain quality Q ($\omega L/R_s$. Values of Q are quite low for magnetic heads. The value can be around 5 to 15 at low frequencies, decreasing towards higher frequencies due to eddy-current losses, and will approach a value of one at very high frequencies. A Q of value greater than two at high frequencies may appear advantageous if the head resonance is "tuned" to the upper limit of the desired frequency response, and thus peaks the output voltage at those frequencies.

However, in reading pulses this may be a disadvantage, because the resonant frequency introduces an additional phase shift in the playback signal. It may be necessary to design the read amplifier input impedance to such a value that the head inductance plus input capacitance tuned circuit is critically dampened. The proper value of the preamplifier input impedance can easily be determined with the test setup shown in Fig. 22.12. A square-wave generator is connected to a 10-ohm resistor in the ground leg of the reproduce head. (Assuming that the normal load for the square-wave generator is 100 ohms, a 90-ohm resistor must be used in series with the connection.) With a one-millivolt peak-to-peak signal across the 10-ohm resistor, the variable input resistor (R) is adjusted until the amplified square wave shows little or no ringing; read the value of R to be used for the preamplifier input impedance or head termination.

Another factor to be considered is the large dynamic range the first stage in the playback amplifier must handle: At low frequencies, the induced voltage in the head is very low compared to the voltage from midrange signals. At the same time, the first amplifier stage must be able to handle a signal range of 50 to 60 dB. A high signal-to-noise ratio is therefore particularly difficult to achieve at the low frequencies, i.e., the long wavelengths.

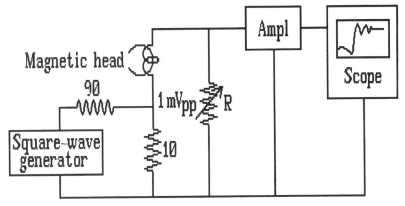

FIGURE 22.12   A test setup to determine the proper matching of a magnetic head.

PLAYBACK EQUALIZATION: AMPLITUDE AND PHASE RESPONSE

The flux level (in nWb/m) from a recorded tape is generally specified at a certain median wavelength, for instance 333 Hz at $v = 4.75$ cm/sec (1⅞ IPS), or at 1 kHz at higher tape speeds. That topic was discussed in Chapter 14 under Normal Record Level.

The coating thickness loss causes a reduction in flux level toward short wavelengths. This led to early definitions of the crossover point in equalization, i.e., where the low-frequency read signal gain of 6 dB/octave ceases. The associated time constants were listed in Table 14.2, and the standard flux level versus frequency curves are shown for four audio tape speeds in Fig. 28.1.

An amplitude-versus-frequency correcting network, with an inverse amplitude-versus-frequency response compared to the head output, may be inserted as a passive loss element between two amplifier stages or it may be incorporated in a feedback loop. It is important to notice that several of the losses in the playback process have no phase shift, while most passive networks made up of R, L, and C components will have a phase shift.

For example, from electrical network theory it is well-known that an RC network has a 6-dB-per-octave slope associated with a 90-degree phase shift. Therefore, such an equalization network will introduce a phase shift that is detrimental to a true reproduction of the recorded signal.

There are *post-equalization* networks that do not introduce phase shift. Let us first consider a simple RC network to produce an amplitude-versus-frequency response that is complementary to the playback head voltage response.

The circuit is shown in Fig. 22.13, and an analysis of this circuit and any others considered for equalization can readily be made by using Kirchhoff's laws and writing circuit equations for currents and voltages, and solving for the desired voltage, current, or impedance. The equations are invariably first- or second-order differential equations that must be solved together.

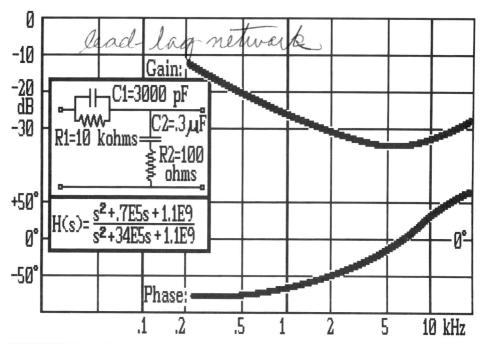

FIGURE 22.13   The amplitude and phase response of lead-lag network used for equalization.

A better, faster, and more informative way is to transform the equations by the *Laplace transformation*. Differentiation and integration transform, respectively, into multiplication and division; and signal waveforms transform into easy-to-handle algebraic functions. The Laplace transform is somewhat analogous to the logarithmic transformation that enables us to add numbers rather than multiplying them. The interested reader can study an elementary introduction (Bogart), or read books by Cheng, Thomas, and Rosa, or Maddock.

The bonus in using the Laplace transform to find the network response lies in its result: the *transfer function* $H(j\omega)$, which is easily applied to find either the amplitude-versus-frequency response, or the time response. Further, if $g_i(\omega)$ is the Fourier transform of the input, then we find the output spectrum function $g_o(\omega) = g_i(\omega)H(j\omega)$; and the response function $f_o(t)$ is found by taking the inverse Fourier transform of $g_o(\omega)$.

The analysis of the circuit in Fig. 22.13 shows that it has the desired amplitude-versus-frequency response, but is accompanied by a full 180-degree phase shift. This can be corrected by adding a single amplifier stage (shown in Fig. 22.14) where we take the output from the emitter at low frequencies, and from the collector at high frequencies; this provides a 180-degree phase correction, with a net overall response of –90 degrees (complementing the heads' differentiation at +90 degrees). The transfer functions for the two circuits are listed in Fig. 22.15.

A similar effect is achieved by the network shown in Fig. 22.16, where the essential element is a 1:1 transformer; the network operates as follows:

1. Low-frequency input voltages are integrated across capacitor C_1; this action is shelved by R_1, and a subsequent increase in amplitude provided for by L_1.

2. The signal phase is shifted 180 degrees, very much like in the circuit in Fig. 22.13. Here the phase is reversed, and hence corrected, by the opposite coupling of L_1 and L_2.

3. The additional boost at band edge is accomplished by tuning L_2 with C_2, damped by R_2. The variable value of R_1 makes it possible to shift midrange equalization; C_2 determines the band-edge frequency, and R_2 the amount of boost.

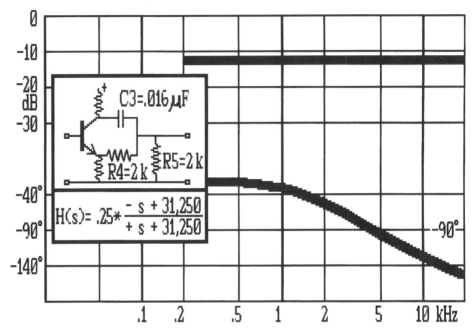

FIGURE 22.14 A phase-correction network for the network in Figure 22.13.

In addition to the equalizer phase shift, there is also the phase shift in the tape flux itself. This was discussed earlier in this chapter, and can be proved by recording a square wave on a tape, controlling the amplitude, and then equalizing the phase of the signal. When the tape is played in the reverse direction, the waveform becomes highly distorted. This explains the degradation occurring in tape copies made in sequences, where the transient response of copy number six, for example, has been severely distorted.

The phase shift from an LCR equalizer used to peak the response at the upper frequency band edge can be eliminated by replacing the LCR equalizer with the *Delay line* or *Aperture equalizer*. Its circuits and response curves are shown in Fig. 22.17 (Dennison, Jorgensen). The phase shift of the circuit is constant across the pass band (in the example shown in Fig. 22.17, the phase is +90 degrees up to 300 kHz), and does therefore not contribute to a phase shift as an ordinary tuned LC-circuit would.

The LC-network represents a delay line that in practice can be constructed from lumped elements, as shown. The 91-ohm resistor equals the line's characteristic impedance Z_0. The input voltage is transmitted along the delay line and received without attenuation at the amplifier. Because the termination here represents an open circuit, the signal is reflected and combines with the input voltage at the input. The termination at that point equals the characteristic impedance, and there is no further reflection. The voltage amplitude at the input will therefore vary cosinusoidally versus frequency because of the reflected signal.

The circuit gain can be expressed as:

$$Gain = (\mu741\ Gain) \times (1 - T \times \cos \omega\tau) \tag{22.2}$$

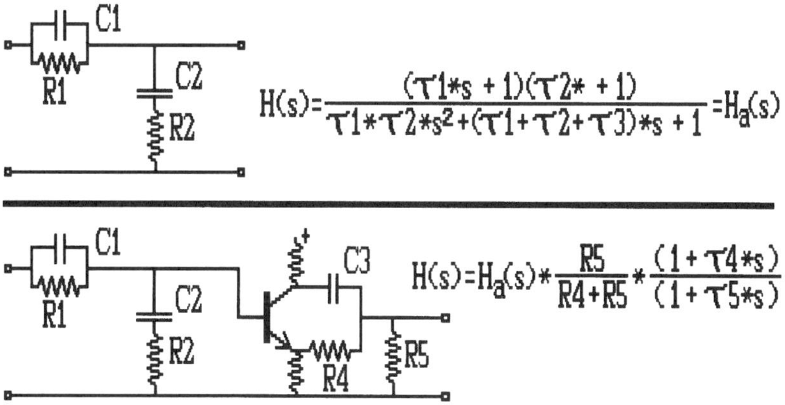

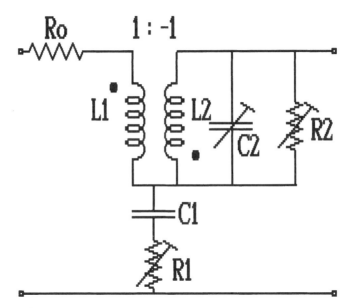

$$H(s)=\frac{(\tau 1*s + 1)(\tau 2* + 1)}{\tau 1*\tau 2*s^2+(\tau 1+\tau 2+\tau 3)*s + 1}=H_a(s)$$

$$H(s)=H_a(s)*\frac{R5}{R4+R5}*\frac{(1+\tau 4*s)}{(1+\tau 5*s)}$$

FIGURE 22.15   Transfer functions for the networks in Figures 22.13 and 22.14.

FIGURE 22.16   A transformer-coupled network for proper amplitude and phase equalization. The phase is constant (= +90 degrees) across the frequency range of the equalizer.

where $\omega = 2\pi f$, and T represents how far toward the high end the slider is positioned on the 91-ohm potentiometer. The gain rises cosinusoidally with frequency, attaining a maximum value at a frequency f_p at which the delay line is one-half wavelength long ($\omega \tau = \pi$). Thus the required line delay is $\tau = 1/f_p$.

The curves A, B, and C correspond to $T = 90$, 99, and 99.7 percent of the potentiometer range; it's always close to the top. This adjustment makes the delay-line equalizer well-suited for an all-around equalizer, with flat response obtained for curve A, and boost of the high frequencies only for curve C; the latter is advantageous in recording of digital signals, where the encoded signal has few or no low-frequency components.

It is assumed that all amplifier sections and couplings are free from phase errors, which can be ensured by proper design (Horwitz, Clark, and Hess). The residual phase error from the magnetic

Delay line equalizer, aka "pulse-slimming" equalizer

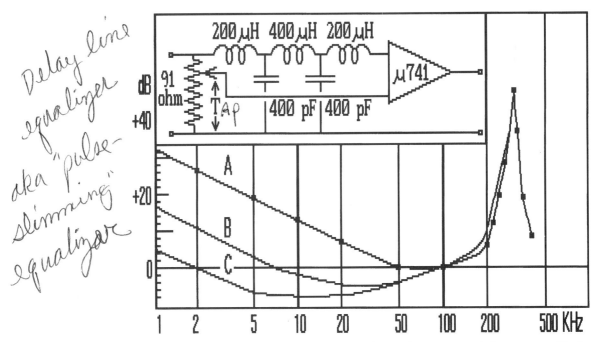

FIGURE 22.17   Gain of the delay line, or cosine, equalizer. The phase remains constant across the frequency range of the circuit. See the text for an explanation of a, b, and c.

write/read process as indicated by the intersection of the phase-versus-frequency curve in Fig. 22.5 needs correction, and this is generally done by inserting an all-pass network. The two generic types are shown in Fig. 22.18; the first (named type I) changes the phase from zero degrees at low frequencies to –180 degrees at high frequencies, while the second (Type II) starts at –180 degrees at low frequencies and changes to zero degrees (van Valkenburgh). The type II network is suitable.

Bridged-T networks are also candidates for phase equalization (Avins et al.; Cunningham), and can be designed to be adjustable (Pfitzenmmaier). Band-pass filters may also serve to correct phase (Herzog).

A final topic that belongs under signal restoration is the recovery of the original recording signal that has been nonlinearly distorted. This signal can be computed by an iterative process, given the distorted recording and the actual recording device or a model of the record-reproduce process (Preis and Polchloek).

EQUALIZATION OF DIGITAL SIGNALS

In Chapter 21 we learned that the vector magnetization pattern and the curved transition zone cause a peak shift, and a broadening of the pulse width *PW50*. The asymmetry of the pulse is also very evident. An equalization of this signal is made in anticipation of optimum detection of the earlier written bit pattern. The detection can be carried out by either detecting the level of the bit (voltage greater than zero for a "1" and less than zero for a "0"), or by detecting the transitions; either method will work.

Level detection requires a broadband equalization of the signal, very much as discussed in the earlier sections. This method is often used when wideband instrumentation recorders are used to record digital data (HDDR—high-density digital recorders).

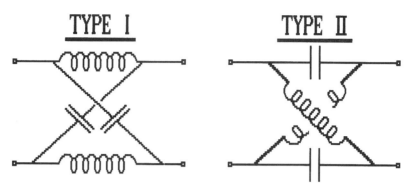

FIGURE 22.18 All-pass filters can correct phase while preserving a flat amplitude response.

Transition detection is more appealing to the digital engineers, because each transition produces a pulse, and its position is determined by detecting the peak of the signal. This is accomplished by differentiating and sampling for the signals zero-crossing, which corresponds to the peak of the original pulse.

It is still desirable to equalize the signal prior to any differentiation or other processing. Equalization means restoring the signal, which for the read pulse means to make it symmetrical and of shortest possible duration. Otherwise, adjacent pulses will overlap or interfere, and will result in a pattern that becomes more and more difficult to detect as the packing density increases.

Most work on equalizer designs has therefore concentrated on *pulse slimming* (Schneider). An early method used the aperture equalizer from Fig. 22.17, and offered a circuit description as shown in Fig. 22.19 (Kameyama et al.). The delay line is matched at the input and open at the other end. An incoming pulse will travel down the line and be reflected at the open termination, and its voltage doubles. Hence, for an input waveform of $f_i(t + \tau)$, the output of the equalizer can be expressed as:

$$f_o(t) = f_i(t) - (K) \times (f_i(t - \tau) + f_i(t + \tau)) \tag{22.3}$$

where K is the fraction of the signal across Z_o. The half-width of an isolated pulse is reduced to about 60 percent of the original pulse. It was found, however, that a higher SNR is required to obtain the improvement effectively. As an example, the recording density was increased by about 30 percent at $K = 0.7$ and $\tau = PW50$, when the SNR at the input is 35 dB (Kameyama et al.).

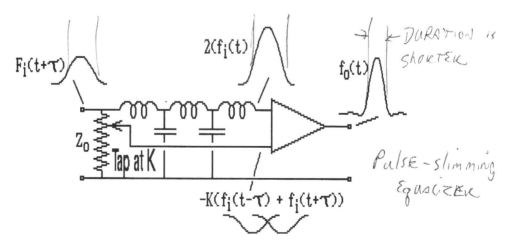

FIGURE 22.19 Time-domain analysis of a delay line equalizer (*after Kameyama et al.); see the text for details.*

Refinements to the equalization, and subsequent detection, have been made by adapting results from communication theory; in particular, from radar, where distinction between return echos (pulses) is important for high resolution. The analytical results (Huber, 1977) were employed in the design of preamplifier, a cos^4 equalizer, and AGC, and are shown in Fig. 22.20 (Geffon). The system was used in the first applications of 3PM coding (Jacoby; see also Table 24.1.) The topic of raised-cosine equalizers was subsequently examined (Chi), and the models for the read pulses examined (Mackintosh). But the generic block diagram of the read electronics remains pretty much as shown in Fig. 22.21.

The delay line equalizer is subject to improvements such as providing it with taps (Nishimura and Ishii), which of course is the original Kallman filter. Equalization of NRZ signals can also be accomplished, in spite of their heavy dc content (Wood and Donaldson; Huber 1981).

The latest member in the family of equalization techniques is the employment of decision feedback equalization (Bergman). An explanation of these equalizers is beyond the scope of this book, but the interested reader is referred to a recent book by A.P. Clark on the topic of equalizers for digital signals.

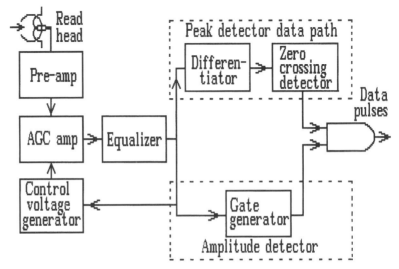

FIGURE 22.20 The inclusion of AGC in an equalizer (*after Geffon*).

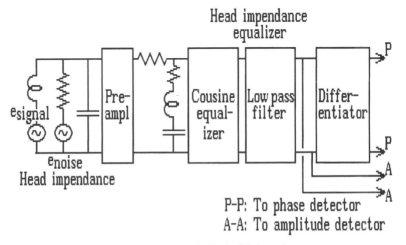

FIGURE 22.21 A general block diagram of read circuitry in digital recorders.

WRITE EQUALIZATION

The read waveform should, from the viewpoint of the communications engineer, look exactly like the input waveform. Equalization is a means to achieve this. If the record (write) current waveform is a square wave, we would also like to read a square wave after signal integration. Figure 22.22a shows the actual read voltage e' (integrated), and we note the asymmetry and lack of sharp corners.

The droop in the signal is caused in part by phase shift (the low frequencies arriving early; see Fig. 22.7), and in part by the lack of read-out response down to dc. The lack of well-defined corners is caused by the filtering action of the write-read channel.

The waveform can be analyzed by a mathematical method called a *Fourier Transformation*, which converts the amplitude-versus-time relationship for a periodic waveform into an amplitude-versus-frequency spectrum together with a phase-versus-frequency spectrum. The process of doing this used to be very time-consuming, but today it can be done in a very short time on a computer. Programmable engineering calculators can carry out a Fourier transform in a short time.

The result of such an analysis is shown in Fig. 22.22b, and we observe the following two important results:

First, the amplitude spectrum decreases faster with frequency than the sequence $1 - \frac{1}{3} - \frac{1}{5} - \frac{1}{7}$, etc.; that is the normal amplitude distribution for the components in a square wave. The relationship can be partly restored by amplitude equalization. Frequency components beyond the upper frequency limit of the recorder obviously cannot be restored.

Second, the phase relationship between the fundamental and the harmonics is not linear. This means that the components will make up a new waveform that can be expressed as (for the example shown):

$$V = E \sin\theta t$$

$$+ (E/6.9) \times \sin(3\omega t + \alpha_3)$$

$$+ (E/20.5) \times \sin(5\omega t + \alpha_5)$$

$$+ (E/51.0) \times \sin(7\omega t + \alpha_7)$$

$$+ \ldots$$

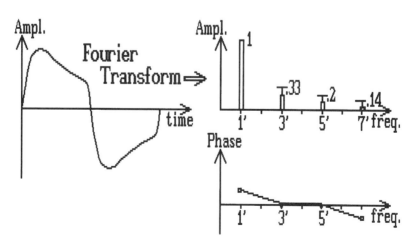

FIGURE 22.22   The integrated read waveform has a spectral amplitude response different from a square wave.

The angles α_n represent the phase distortion of the read signal. If they produce a phase shift-versus-frequency curve that deviates from a straight line that goes through zero at $\omega = 0$, then the phase distortion must be corrected. This is done by the techniques mentioned earlier.

The absolute amount of phase equalization is unimportant. The goal is to restore a linear relationship: if we include a phase angle β in the fundamental signal, then the requirement is:

$$\alpha_n = n \times \alpha_1 \tag{22.3}$$

For example: If $\alpha_1 = 20°$, then $\alpha_3 = 60°$ and $\alpha_5 = 100°$, etc. Phase distortion can also be evaluated by the *envelope delay*, defined as

$$\text{Envelope Delay } \tau = d\alpha/d\omega = d\alpha/2\pi df \tag{22.4}$$

where α = function(ω) defines the phase-versus-frequency characteristics. If the slope is constant, then the envelope delay τ is also constant, all harmonics arrive simultaneously at the network's output, and a square wave is restored to its original shape with the exception of missing high frequencies. Figure 22.23 illustrates one method of precompensation by using a suitable write current. Its waveform is determined by correcting for the deficiencies shown in Fig. 22.22; we wish to emphasize the amplitudes of the harmonics in connection with a phase shift, as shown in Fig. 22.23. After a Fourier transform from the frequency to the time domain, we arrive at a double-step waveform.

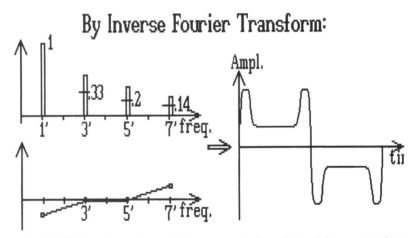

FIGURE 22.23   The write waveform required to compensate for the amplitude and phase responses shown in Figure 22.22.

A time-domain analysis of the double-step write waveform is shown in Fig. 22.24 (Jacoby). An isolated write transition (a) produces a read pulse as shown in (b). A reduction of its *PW50* can be obtained by superimposing a doublet on the transition, as shown in (c). Every transition in the write current is now preceded and followed by a minor transition of opposite polarity. The transition will, itself, still produce the read pulse from (b), but two additional responses are now added from the minor transitions pulses. This method affords a reduction of *PW50* to 60 percent of its original value (Jacoby).

This form of recording requires a linear recorder, which means that ac bias has to be applied. Its frequency must be synchronized to the clock frequency, and its level is adjusted so the bias current is approximately equal to the data current. This results in minimum phase distortion of the recording. ac bias further results in a quieter recording, with an overall improvement in the SNR.

Precompensation, including the transition time shifts discussed in Chapter 19, has received little attention in the literature (Jacoby; Schneider; Kato et al.). Kato illustrates how the fast rise time in a

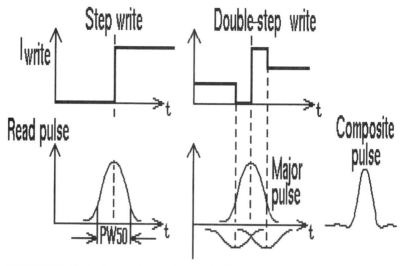

FIGURE 22.24   The double-step write waveform is derived from the time domain (*after Jacoby*).

high-Q (8 to 10) head can be used by superimposing a compensating current with ringing complementary to the head.

The fast rise time of the write current is beneficial to high packing density. It seems like another possibility exist in ac-bias recorders where a control of the bias current could be used to shift the write zone back or forth to compensate for transition time equalization.

A final consideration must be given to the fact that write current adjustment in disk drives is a compromise, because less current is required when the head is located at an inner track, where it is flying close to the disk surface, as compared to writing on an outer track.

ADJUSTING THE EQUALIZATION CIRCUITS (AC-BIAS RECORDERS)

The simplest and easiest way to adjust a recorder's playback equalization is to use prerecorded standard tapes. Such tapes are available from several manufacturers for the tape speeds used in audio work, and for instrumentation recorders as well. A typical alignment tape starts with a section that contains a short-wavelength recording for proper alignment of the reproduce head. A following section has a normal record level (one-percent distortion) for VU meter zero setting.

After these two tones, there follows a series of signals of varying frequency, and upon playback of these tones the controls in the equalization circuitry are adjusted for optimum flat frequency response. This procedure will assume that all prerecorded tapes for a given standard will be reproduced correctly. Prior to using a test tape, it is advisable to clean the magnetic heads and demagnetize them with a head demagnetizer. It is also advisable not to start and stop the recorder in the middle of a test tone, because this may produce weak transients on the tape and, consequently, shorten its useful life. A word of caution: When a full-track alignment tape is used for narrow tracks, there will be excessive output at long wavelengths due to fringing response.

If a test tape is not available, a different approach must be taken. This technique uses the figure-eight loop (see Fig. 9.39) or a straight wire in front of the reproduce gap, which is fed with current that has an amplitude-versus-frequency response as established below. (The input signal to the reproduce amplifier can also be provided by a voltage generated across a small resistor placed in the ground side of the magnetic head.) It is important that the signal be introduced to the playback amplifier through the reproduce head. If the signal is connected directly to the input stage, the head core losses and the self-resonance will not be included in the measurement and the reading will be incorrect.

Assuming, at first, that the current through the wire in front of the head is constant and independent of frequency, the flux through the head will be constant but limited by core losses. This corresponds, at low frequencies, entirely to the conditions existing during playback of a tape. But the external flux from a medium decreases due to coating thickness losses toward the shorter wavelengths. The flux through the head core will be further reduced toward the shorter wavelengths because of read gap losses, which can be determined as shown in Fig. 22.25.

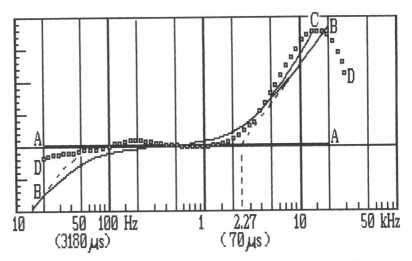

FIGURE 22.25 Curves to establish the proper post-equalization in a linear (ac-biased) recorder.

Assuming that the equalization has already been adjusted, the output voltage from the playback amplifier will increase toward the high frequencies when the wire in front of the head is fed with a constant current. It is relatively easy to predetermine how much this rise in output voltage should be for proper equalization. Referring to Fig. 22.25, the following steps show how to determine the rise in the output voltage:

1. On a sheet of linear-log graph paper, with the y-axis for amplitude and the x-axis for frequency, draw a straight line representing the desired final frequency characteristic (curve A in Fig. 22.25).

2. Add to this curve the difference between a constant-flux recording and a standardized tape flux. Figure 28.1 shows the current standards.

3. Measure the effective gap length of the reproduce head as shown in Fig. 9.38. Determine the curve for the gap loss from Fig. 6.17. Add this loss to curve B (Fig. 22.25), obtaining curve C.

4. Place the straight wire or the figure-eight loop in front of the reproduce head gap in such a position that the output voltage from the amplifier is maximum. If a constant current (versus frequency) is now fed through the wire, then the output voltage should closely match curve C in Fig. 22.25.

An actual measured curve may have a response as shown by curve D, where the low end will drop off due to amplifier limitations. It may have a peak at high frequencies due to self-resonance in the reproduce head. If the deviation between curve D and curve C is within 1 dB, the equalization is as good as can be, but deviations in the order of several decibels require readjustment (and possible redesign) of the playback equalizer.

Prior to measuring curve D it is advisable to connect an oscilloscope to the playback amplifier and monitor it so that the induced voltage in the playback head does not cause overload, distortion, and the accompanying erroneous reading of the output signal.

ADJUSTMENT OF RECORD PRE-EQUALIZER

A prerequisite for proper adjustment of the record amplifier's equalization network and record level setting is a properly functioning bias oscillator/amplifier. Even small amounts of distortion in the bias supply will generate both noise and distortion on the tape. The latter will affect the proper setting for normal record level (1-percent third-harmonic distortion on the tape).

The best way to check for a *proper bias waveform* is to connect the figure-eight loop to an oscilloscope, place it in front of the record head, and next energize the record and bias circuitry. The waveform of the bias signal (with data input short-circuited) must be a perfectly symmetrical sine wave.

Most recorders have a 10-ohm resistor in the ground leg of the record head winding, which provides a suitable test point to check the bias signal. Many bias oscillators, which also function as erase oscillators, have a balancing potentiometer. An easy way to check for a proper bias waveform is to listen to a tape via the playback amplifier and speaker, and to adjust the balancing control for minimum noise. The nature of poor bias or erase noise is a gurgling sound rather than a hiss.

Now connect a signal generator to the record amplifier input and observe the waveform across the test resistor (10 ohms) in the ground leg of the record head circuit (or from the figure-eight loop in front of the record head). As a first approximation in setting the correct record level, adjust the ratio between the bias amplitude and the data signal amplitude to approximately 20:1. The tape selected for use on the recorder is now threaded on the tape deck and tape motion started. In adjusting the proper bias level there are, unfortunately, no set standards, but the two following rules of thumb apply fairly well:

- For high-quality audio recordings, the bias current is adjusted for 2 dB overbias at 1 kHz. That is, the bias current is increased from a low value until the recorded signal level reaches a maximum; the bias current is now increased until the signal level has decreased 2 dB. This setting will ensure the largest dynamic range with a minimum of distortion.

- In instrumentation recorders with wide signal bandwidths, the bias setting is adjusted to achieve a high output at the shortest wavelength. For example, it is common practice in an instrumentation recorder operating at 120 IPS with a bandwidth of 4 MHz to adjust the bias while recording a 4-MHz tone. The bias level is increased from a low value until the signal on the tape reaches a maximum, and then it is increased slightly until the signal level drops 2 dB.

The recorder designer has a choice between these two extreme bias settings, and it is up to him/her to establish the compromise between bandwidth and dynamic range. The recorder user is referred to the manual for the adjustments on the particular recorder. It should be emphasized again that the bias level is dependent upon the type of tape used and the thickness of its oxide coating, particularly with audio recordings where the higher bias levels are used.

SETTING THE NORMAL RECORD LEVEL: MEASUREMENT OF DISTORTION

The "normal recording level" is that level that produces one percent harmonic distortion off the tape. With freedom from dc magnetization and a clean bias waveform, this distortion will be odd, from the third-harmonic component. The simplified formula for this distortion is:

$$\text{Distortion} = (V_3/V_1) \times 100 \text{ percent} \tag{22.5}$$

where V_1 is the level of the fundamental signal and V_3 is the level of the third-harmonic component.

Until recently, a frequency of 1 kHz was used in setting this level for instrumentation as well as for audio recorders. But it has been found that the distortion in wideband instrumentation recorders often increases with frequency. Therefore, it has become standard to set the record level in a wideband instrumentation recorder at a frequency that is one-tenth of the upper-frequency limit (for example, 150 kHz for a 1.5-MHz recorder).

Measurement of the third harmonic signal in a tape system differs from the measurement of distortion in an amplifier. A sharply-tuned selective wave analyzer will not accurately measure the harmonic components in the presence of wow and flutter from the tape (or disk) transport. The bandwidth of the signal analyzer must be quite wide in order to capture the third harmonic. The simplest way to do this is to measure V_1 with a voltmeter, insert a high-pass filter that attenuates V_1 by at least 50 dB, and then measure V_3 (plus any higher harmonics). The proper record level is now adjusted by trial and error to achieve a one-percent distortion (= –40 dB); the VU meter or other record level indicating device is then adjusted for a reading of 0 dB.

The recorder is now ready for adjustment of pre-equalization in the record amplifier. Because pre-equalization in audio recordings requires a boost of the high frequencies, which cause overload and distortion at the normal record level, the following adjustments should be made at a level that is as many dB below the normal record level as the amount of maximum equalization (boost) in dB. It is generally safe to carry out the measurements at a level of –20 dB re 0 dB VU.

A series of frequencies (in audio, for instance, they are 50, 100, 200, 500, 1000 Hz and 2, 5, 10, 12, 15 kHz) are recorded and played back, and the frequency response is plotted on an amplitude-versus-frequency graph. Without pre-equalization it will be found that the frequency response curve will drop off toward higher frequencies, and that a corrective network must be installed in a record amplifier or adjustments made to an existing network in order to achieve a flat frequency response. The amount of pre-equalization can be evaluated by measuring the record current across the 10-ohm test resistor in the ground leg of the record head, with the bias oscillator disabled.

REFERENCES

AMPLITUDE AND PHASE RESPONSE IN MAGNETIC RECORDING

Haynes, M.K. Sept. 1977. Experimental Determination of the Loss and Phase Transfer Functions of a Magnetic Recording Channel. *IEEE Trans. Magn.* MAG-13 (5): 1284–1286.

Jacoby, G.V. Sept. 1968. Signal Equalization in Digital Magnetic Recording. *IEEE Trans. Magn.* MAG-4 (3): 302–305.

Jorgensen, F. Oct. 1961. Phase Equalization Is Important. *Electronic Industries*. Vol. 20, No. 10.

Minukhin, V.B. 1975. Phase Distortion of Signals in Magnetic Recording Equipment. *Telecom. and Radio Engr*. Vol. 29/30, pp. 114–120.

Minukhin, V.B. Aug. 1973. Hilbert Transform and Phase Distortions of Signals. *Radio Engineering and El. Phys*. 18 (8): 1189–1193.

AMPLITUDE AND PHASE EQUALIZATION

Avins, J., B. Harris, and J. S. Horvath. Jan. 1954. Improving the Transient Response of Television Receivers. *Proc. IRE*, pp. 274–283.

Bogart, Th. F. 1983. *Laplace Transforms, Theory and Experiments*. John Wiley and Sons. 148 pages.

Bracewell, R.N. 1978. *The Fourier Transform and Its Applications*. McGraw-Hill, 444 pages.

Cheng, D.K. 1959. *Analysis of Linear Systems*. Addison-Wesley, 431 pages.

Clarke, K.K., and D.T. Hess. 1971. *Communication Circuits: Analysis and Design*. Addison-Wesley, 658 pages.

Cunningham, V.R. May 1966. Pick a Delay Equalizer, and Stop Worrying about Math. *Electronic Design*, pp. 62–66.

Dennison, R.C. Dec. 1953. Aperture Compensation for Television Cameras. *RCA Review*, Vol. 14, pp. 569–585.

Herzog, W. April 1983. Linearizing of the Phase at Band-passes (in German), *Archiv für Elektrotechnik* 66 (4): 187–194.

Horwitz, J.H. Aug. 1967. Reduce Delay Distortion at the Source. *Electronic Design*, pp. 116–120.

Johnson, K.O., and D.P. Gregg. Oct. 1965. Transient Response and Phase Equalization in Magnetic Recorders. *Jour. Audio Engr. Soc.* 13 (4): 323–330.

Kallmann, H.E. July 1940. Transversal Filters. *Proc. IRE*, Vol. 28, pp. 302–310.

Maddock, R.J. 1982. *Poles and Zeros in Electrical and Control Engineering*. Holt, Rinehart and Wilson. 216 pages.

Pfitzenmaier, G. March 1984. A Contribution to the Realization of Adjustable Delay Equalizer Swith Fixed Points (in German), *Frequenz* 38 (3): 54–62.

Preis, D., and H. Polchloek. Jan/Feb 1984. Restoration of Nonlinearly Distorted Magnetic Recordings. *Jour. AES* 32 (1/2): 26–30.

Sivian, L.J., H.K. Dunn, and S.D. White. May-June 1959. Absolute Amplitudes and Spectra of Cerytain Musical Instruments and Orchestras. *IRE Trans. Audio* AU-7 (3): 1–29.

Thomas, R.E., and A.J. Rosa, 1984. *Circuits and Signals: An Introduction to Linear and Interface Circuits*. John Wiley and Sons. 758 pages.

Van Valkenburg, M.E. 1982. *Analog Filter Design*. Holt, Rinehart and Wilson, 608 pages.

SIGNAL EQUALIZATION, DIGITAL

Bergmans, J.W.M. 1986. Decision Feedback Equalization for Digital Magnetic Recording Systems. *Intl. Video and Data Conf.* IERE Publ. No. 67, pp. 141–145.

Chi, C.S. Sept. 1979. Characterization and Spectra Equalization for High Density Disk Recording. *IEEE Trans. Magn.* MAG-15 (5): 1447–49.

Clark, A.P. 1985. *Equalizers for Digital Modems*, Halsted Press (John Wiley), 468 pages.

Geffon, A.P. Sept. 1977. A 6 KBPI Disk Storage System Using Mod-11 Interface. *IEEE Trans. Magn.* MAG-13 (5): 1205–1207.

Huber, W.D. Nov. 1981. Equalization of the D.C. Null in High Density Digital Magnetic Recording. *IEEE Trans. Magn.* MAG-17 (6): 3352–3354.

Huber, W.D. Sept. 1977. Maximization of Lineal Recording Density. *IEEE Trans. Magn.* MAG-13 (5): 1208–1210.

Jacoby, G.V. Sept. 1977. A New Look-Ahead Code for Increased Data Density. *IEEE Trans. Magn.* MAG-13 (5): 1202–1204.

Kameyama, T., S. Takanami, and R. Arai. Nov. 1976. Improvement of Recording Density by Means of Cosine Equalizers. *IEEE Trans. Magn.* MAG-12 (6): 746–748.

Mackintosh, N.D. July 1979. A Superposition-Based Analysis of Pulse-Slimming Techniques for Digital Recording. *Intl. Conf. on Video and Data 79*, IERE Publ. No. 43, pp. 121–147. *Also in The Radio and Elec. Engr.*, June 1980 50 (6): 307–314.

Nishimura, K., and K. Ishii. Sept. 1983. A Design Method for Optimum Equalization in Magnetic Recording with Partial Response Channel Coding. *IEEE Trans. Magn.* MAG-19 (5): 1719–1721.

Schneider, R.C. Nov. 1975. An Improved Pulse-Slimming Method for Magnetic Recording. *IEEE Trans. Magn.* MAG-11 (5): 1240–1242.

Wood, R.W., and R.W. Donaldson. July 1978. Decision Feedback Equalization of the DC Null in High Density Digital Magnetic Recording. *IEEE Trans. Magn.* MAG-14 (4): 218–222.

WRITE EQUALIZATION

Jacoby, G.V. July 1979. High Density Recording with Write Current Shaping. *IEEE Trans. Magn.* MAG-15 (3): 1124–1130.

Kato, T., R. Arai, and S. Takanami. Sept. 1986. Write-Current Equalization for High-Speed Digital Magnetic Recording. *IEEE Trans. Magn.* MAG-22 (5): 1212–1214.

Schneider, R.C. Nov. 1985. Write Equalization in High-Linear-Density Magnetic Recording. *IBM Jour. Res, and Devel.* 29 (6): 563–568.

BIBLIOGRAPHY

Barna, A. 1980. *High Speed Pulse and Digital Techniques*, John Wiley and Sons, 185 pages.

Baumgarten, W.H. 1985. *Pulse Fundamentals in Small Scale Digital Circuits*, Reston (Prentice Hall), 574 pages.

Chan, S.W.C., and B.K. Middleton. May 1992. Representation of a Digital Magnetic Recording Channel. *IEEE Trans. Magn.* MAG-28 (3): 1884–85.

Koren, N.L. Sept. 1990. Signal Processing in Recording Channel Utilizing Unshielded Magnetoresistive Head. *IEEE Trans. Magn.* MAG-26 (5): 2166–68.

Lin, Y., B. Buchan, and J.V. Howell. Nov. 1991. An Adjustable Write Equalization on QIC-1350 Tape Drives. *IEEE Trans. Magn.* MAG-27 (6): 4813–15.

Murayama, H. Oct. 1985. Read/Write Evaluation Considering Peakshift Characteristics for Perpendicular Recording. *NEC Res. & Devel.* No. 79, pp. 29–34.

Pan, T.W., and A.A. Abidi. June 1992. A Wide-Band CMOS Read Amplifier for Magnetic Data Storage Systems. *IEEE Jour. Solid-State Circuits* 27 (6): 863–73.

Rainal, A.J. June 1990. Reflections from Bends in a Printed Conductor. *IEEE Trans. Comp.* 13 (2): 407–413.

Schwetlick, H., and W. Kessel. Jan. 1987. Deconvolution of Band-Limited Signals Using a Priori Knowledge About the Impulse Response. *Arch. für Elek. Übertragung* 41 (1): 62–64.

Tachibana, M., and M. Ohara. July 1977. Optimum Waveform Design and Its Effect on the Peak Shift Compensation. *IEEE Trans. Magn.* MAG-13 (5): 1199–1202.

Thapar, H.K., P.A. Ziperovich, and R.W. Wood. April 1990. On The Performance of Symbol- and Fractionally-Spaced Equalization in Digital Magnetic Recording. *Intl. Conf. on Video, Audio and Data Rec.* IEE Publ. No. 319, pp. 168–175.

Wilhelm, W. March 1986. Propagation Delays of Interconnect Lines in Large-Scale Integrated Circuits. *Siemens Forsch. und Entwickl. Ber.* 15 (2): 60–63.

Wood, R. Sept. 1990. Enhanced Decision Feedback Equalization. *IEEE Trans. Magn.* MAG-26 (5): 2178–80.

Yeh., N.H., and R. Niedermeyer. Sept. 1990. Transition Asymmetry in High Density Digital Recording. *IEEE Trans. Magn.* MAG-26 (5): 2175–77.

Young, F.J., and A.T. Murphy. Dec. 1990. On Microstrip Capacitance and Impedance. *IEEE Trans. Compon. Hybrids Mfg. Tech.* 13 (4): 1121–1123.

CHAPTER 23
SIGNAL RECOVERY, NOISE, DETECTION MEASUREMENTS

The traditional illustration for the transmission of a signal from source to destination is shown in Fig. 23.1. It shows the application of analog-to-digital conversion, which today is used far more often than direct transmission of the signal. The transmission channel can be many things: A telephone wire, or antenna-to-antenna transmission of radio, TV, or other signals. It can be storage of data on magnetic or optical media for later playback.

The reasons for digitizing the signal are many. Small perturbations of digital signals caused by noise in the channel can be recognized and corrected, whereas analog signals are left at their perturbed values with no possibility of correction. However, the channel does not make any distinction whether the signal is analog or digital; they are treated alike. The superiority of digital signals is that they can be regenerated.

The noise that occurs in the write/read channel has many sources: It may be random noise from the magnetic medium, from the preamplifier, or from the power supplies. It may also be impulse noise of very short duration, or RFI (radio-frequency interference). The signal may also disappear for a very short duration due to dropouts.

SIGNAL WAVEFORMS THROUGH THE RECORDER CHANNEL

We will briefly review the origin of the train of bits, or "on-off" signals. Figure 23.2 shows an example of a pulse signal that represents the letters TAPE.

Each letter in the alphabet is represented by a byte that consists of 8 bits, where the last bit is an added 0 and the remaining seven are the ASCII code. (This 8-bit code is also called ASCII-8, and it is part of the commonly used code called EBCDIC, which has $2 \times 128 = 256$ different bytes.)

The sequence of the four bytes that represent TAPE forms a 32-bit signal (b), that in many systems would be recorded in NRZI (non-return-to-zero, inverted—see Chapter 24). The NRZI pattern is shown in line (c) and the resulting media magnetization in line (d).

The read head output is obtained by differentiating the flux from the magnetization (Faraday's law), line (e), and may further be equalized by integration as shown in line (f). The waveforms shown are ideal; the real-world signals are shown in line (g).

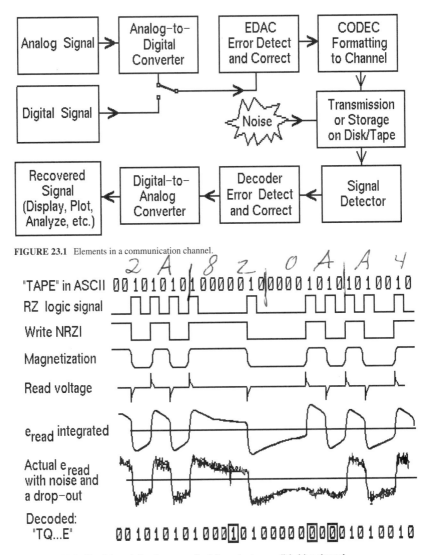

FIGURE 23.1 Elements in a communication channel.

FIGURE 23.2 Signal degradation due to transferal through a tape or disk drive channel.

The high-frequency limitation decreases the rise time of the leading and trailing edges of the individual pulses, while the lack of low-frequency response in the magnetic read process causes a droop in the flat portions of the pulse train. This lack of a dc component in the read signal may cause serious signal deterioration, and is avoided by use of modulation codes (see next chapter).

An examination of the RZ code versus the NRZI code for just the letters A and P shows a smaller dc component in the NRZI version. The excessive droops are caused by too may consecutive 0s (5 in the letter A).

The addition of noise can cause mistriggering of the detection circuits, in particular if noise spikes are present on a droop. One example is shown, causing a false reading of a zero-crossing, and hence a change of the word A's code 1000 0010 to 1000 1010 (which is the letter Q)!

An "open circuit" in the transmission through a tape recorder can be caused by a dropout in the head-media interface area. It may be a fault in the coating, but is much more likely to be caused by a foreign dust particle (smoke particles, floor dust) that lifts the coating away from the read gap (like a pole in a tent). A recorded change in flux polarity may not be sensed, as shown at the second bit in the letter P. It may further affect the correct detection of following flux changes, as for the second 1 in the example shown. The original letter "E" comes out as 0000 0000, which is "NUL" (Null).

This example has illustrated how the write/read channel's imperfections will degrade the digital signal. We shall now take a closer look at the write/read process for a single transition, and thereafter look at a larger view of the entire process for a sequence of transitions.

THE READ SIGNAL; RESPONSE TO A TRANSITION

The read (playback) process was analyzed in Chapter 21 and the result is shown in Fig. 23.3. The read-voltage waveform from a single, isolated transition has a shape that depends on, among other things, the pattern of the vector magnetization. A prediction of this magnetization presents something of a difficulty, because the saturation recording process is nonlinear and our knowledge of the recorded pattern is at best a rough estimate.

We may, however, modify the estimated pattern within a certain framework so the calculated response will closely match the measured one. We can, in other words, learn about the magnetization pattern through a reverse or backward analysis. This method must, however, be applied cautiously, because different magnetization patterns can produce the same response.

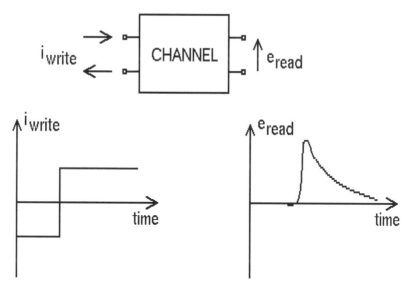

FIGURE 23.3 The recorded channel response to a step input current is a read voltage that resembles the response to an impulse, due to differentiation in the read head.

The response to an input step can now be applied to the write-read channel, because we have defined and "frozen" the nonlinear recording process. It is safe to assume that the read process itself is linear, and that we therefore can superimpose the response from several transitions to form a resulting read voltage waveform.

A comparison with other system tests are in place. An often-used test signal is the square-wave signal, which in a tape or disk system with inductive read heads will result in the step response shown in Fig. 23.3. Rounding of the corners is caused by the lack of high-frequency response, and the droop in signal level is caused by lack of response down to dc.

The response to just the leading (or trailing) edge of the square wave is called the step response. An examination of this response can reveal the network function of the channel; examples are shown in Fig. 23.4.

Estimating system response from observation of square wave response:

A. High frequencies attenuated; No phase error.
B. Fundamental attenuated; No phase error.
C. Phase distortion: Harmonics lagging (less than 5%).
D. Phase distortion: Harmonics leading (less than 5%).

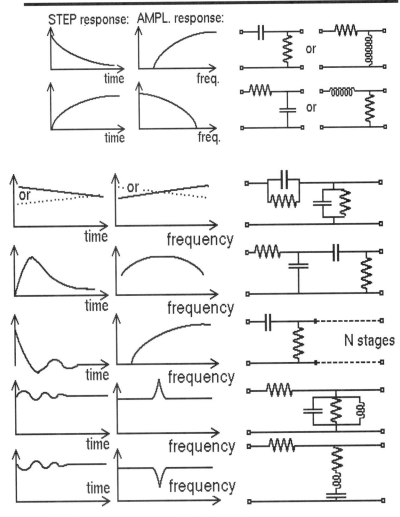

FIGURE 23.4 Responses to square-wave and step input functions.

Another popular test signal is the impulse, which gives a response called the impulse response. The impulse is a differentiation of the step, and the response is likewise the differentiated step response. Further, because the system (channel) is assumed linear, then a differentiation somewhere between input and output will result in the output being the impulse response when the input is a step.

In other words: A unit-step write current results in a unit-step response in the flux, which differentiates results in the impulse response.

This is very useful, because the read voltage now is determined by straightforward convolution of the write current with the impulse response, which is the recorder channel's response to a step write current. Any peak-shifts that occur during the record process must be determined and included in the write-current timings. Also, the method assumes inductive read heads. We can even go so far as naming the response an impulse response $h(\tau)$, similar to the notation used in systems theory.

DETECTION OF DIGITAL SIGNALS

A digital read circuit extracts the digital data by detecting either amplitudes (higher or lower than a certain + or – threshold), or by detecting the zero-voltage line crossovers. The first method is generally selected by the engineers with a background in signal transmission, while the other is selected by engineers with digital circuit experience.

The detection circuit needs timing information to interpret the pulse stream. The timing information is extracted from the coding of the bits and often from a "ready-phasing" signal just before the data bits. The timing signal is derived from amplitude detection as just described. It is used to control a clock that will produce pulses synchronous with the bit rate. The amplitude detection assumes an amplitude-equalized signal channel, while the zero-crossing detection assumes a differentiated signal (after the read-head preamplifier).

AMPLITUDE AND PEAK DETECTION

Amplitude detection circuits rely on the sampling of the levels of a pulse stream. The signal waveform is sampled at the center of each bit, and the sample value is compared with a preset discrimination level. If the level is above the discrimination level, a 1 will result, and if it is below, a 0 will result.

The strobing time should be much shorter than the bit length. The timing tolerances are reduced by peak shift, and the amplitude sensing mechanism is affected by bit amplitude changes caused by pulse crowding and by the general amplitude variations in a magnetic recording channel, including noise.

The effects of intersymbol interference and noise can readily be observed in the laboratory by using an oscilloscope to sweep the bit stream, with a horizontal sweep rate of $1/BR$ (where BR is the bit rate). The resulting display is widely known as an eye pattern from its resemblance to the human eye.

Two binary waveforms are shown in Fig. 23.5, one distorted, the other undistorted. When the waveform is undistorted, then all sampled values at middle of bit time are ± 1, and the eye pattern is wide open.

Intersymbol interference and noise will close the eye pattern due to both amplitude and timing shifts. The pattern will provide a great deal of information about the performance characteristics of a recording system. We can evaluate signal amplitude (A) and zero crossings (B).

Detection of signal levels is done with a threshold sensing of positive and negative levels; see Fig. 23.6. The exact time of detections that occur on the slope of the signal near peaks depends on the reference level as well as signal amplitude.

The differentiated voltage e' of the inductive read head voltage exhibits very sharp and well-defined crossings of the 0-level axis at the exact time of signal peaks. This makes it well-suited for the detection of bits.

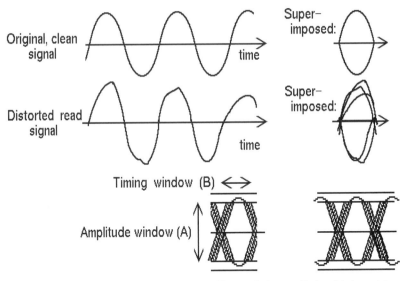

FIGURE 23.5 A repeat display of pulse signals on an oscilloscope discloses amplitude and timing variations in the so-called eye pattern. A clean signal produces the trace shown in the top row, a distorted signal produces the traces in middle.

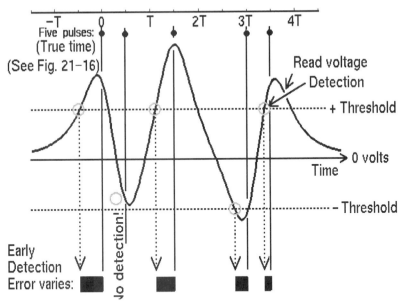

FIGURE 23.6 Amplitude and peak detection methods.

PARTIAL RESPONSE SIGNALLING

The ideal peak-detection channel has freedom from pulse interference. The price, therefore, is limited data rates at a given bandwidth. The pulse transitions will interfere with each other if the data rate is increased. Peak shifts result, the SNR is degraded, and the window margin reduced.

The maximum packing density along a track is largely determined by the transition parameter a (Formulas 5.3 and 5.4; $\Delta x = 2a$) and the fringing of the read pulse *PW50* (formula 6.12).

The following outline will attempt to explain a detection scheme that works on a three-level (ternary) read signal, and we will learn how such a signal can be created without an actual three-level record method that would have an SNR penalty.

Assume a recording system with return-to-zero (RZ) recording (see Fig. 24.1, top line). Each signal—1 or 0, mark or space—is recorded as a saturated pulse of the shortest possible duration. Readout of recorded information is, for instance, as shown in Fig. 23.7, where eight separate signals are recorded in four groups of two each.

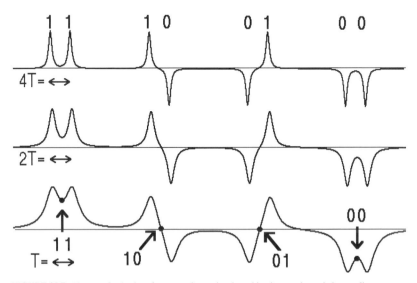

FIGURE 23.7   The use of pulse interference to detect signal combinations as shown in bottom line.

We now increase the bit density by increasing the data rate so the time between bits goes from $4T \rightarrow 2T \rightarrow T$. The result is severe pulse interference that in a normal peak-detection system would be unacceptable. If we level-detect at a time between pulse centers, we get a positive voltage from "11" and a negative voltage from "00," while obtaining zero results from either "10" or "01." Which one it should be depends on the just-prior detected digit: if a 1, then the zero is "10," and if a 0, then "01."

The result is related to the idea of multilevel recording, where 4 levels that occupy only one bit cell has a capacity of 2 ($2^2 = 4$) and 8 levels has a capacity of 3 ($2^3 = 8$). This multilevel scheme has been evaluated, and the improvements found to be cancelled by poorer SNR. For each increase of the exponent n in 2^n, a 6-dB loss in SNR is suffered. It can be recovered by reducing the system bandwidth to about half, and the capacity gain is cancelled (Mackintosh and Jorgensen 1981). And in the scheme just outlined, we started by losing 6 dB because of having to use RZ.

A different encoding of the binary signal can result in a binary recording (two levels) that produces a three-level read signal. The method was named duobinary, where the "duo" indicates doubling of the bit capacity of a simple binary system (Lender 1964 and 1966).

An example is based on the descriptions found in textbooks by Bennett and Davey, and by Lanski and Ingram. They encoded the incoming binary data stream (a) so a space is converted to a transition and a mark to no transition (b); see Fig. 23.8.

Readout of the recorded pattern in (b) with a system that acts as a comb filter $(1 + D)$, results in a waveform that is obtained by adding (b) to a one-bit delayed version of itself. The reader will recognize the original pattern by noting that zero levels corresponds to a space and positive or negative values to a mark.

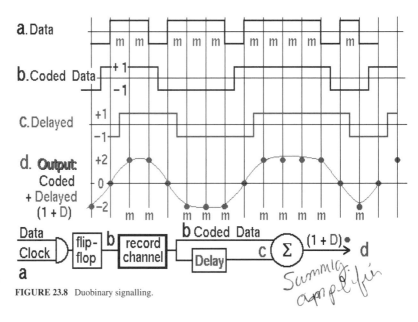

FIGURE 23.8   Duobinary signalling.

The code generated above is not ideal for recording, like the NRZ-M it is related to. It is neither dc-free nor self-clocking. Better codes have come along with self-clocking characteristics, such as partial-response maximum-likelihood codes. Partial-response relates to the spectral shape used in readout of the information, while maximum-likelihood refers to the detection process that uses the entire waveform in making decisions similar to the choice between "10" and "01" in Fig. 23.7. An advanced detector like the Viterbi decoder is required.

A PR-4 code is very well-suited for magnetic recording systems (Fig. 23.9); The incoming data stream (a) is encoded in the write circuitry to the form shown as (b). The read signal (b) is added to a two-bit-delayed and flipped version of itself (c) (A $1-D^2$ filter). The resulting waveform (d) is shown in Fig. 23.9. Again, we notice the three levels.

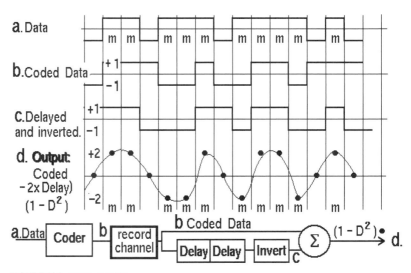

FIGURE 23.9   PRML signalling.

The frequency response of this 1-D^2 PRML code is a sinus curve with no response at dc (dc-null) nor at a frequency of one-half the clock rate. The net gain by using PRML over the peak-detection channel is a potential bandwidth gain of 25 percent. These levels are detected at mid-bit cell times, NOT at transitions as we have become used to. Signal detections are at amplitude levels.

We have essentially gone back toward the analog channel, but digitized it with PRML. There is no SNR penalty because the three levels (ternary) are created by clever signal processing from a binary saturation recording.

BIT ERRORS

The performance of any digital data channel is measured as its freedom from errors (missing or false bits). Ratings are typically 10^{-6} to 10^{-9} errors (raw).

A bit error rate (BER) of 10^{-6} means that, on the average, there is no more than one error per million bits. A poor BER figure in a data storage system can have many causes; today they primarily are caused by timing errors.

There are error-detecting and -correcting codes that can significantly improve a system's BER. This is invariably paid for by added electronics, which are often highly sophisticated circuits. Today LSI circuits are making fast advances, where ready-made chip for EDAC (error-detection and -correction) can be installed in the write and read circuits.

PROBABILITY OF ERRORS IN A MESSAGE

Figure 23.10 shows a waveform that resulted from detection in a recording channel. The zero crossings represents the data, and there are excursions due to timing variations. The data are detected by a gate circuit that opens for a short period, less than or equal to the bit period $BP = T$. If the crossing occurs at more than $T/2$ from the centerline, an error occurs.

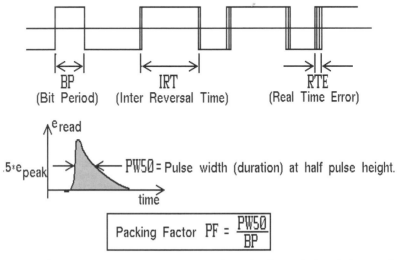

$$\text{Packing Factor } PF = \frac{PW50}{BP}$$

FIGURE 23.10 Time jitter in a detected signal (zero-crossings). Time jitter can be determined on an oscilloscope with delayed trigger.

This is similar to the traditional amplitude detection methods that most textbooks use for analysis of BER. Figure 23.11 shows how the amplitude detection is affected by noise superimposed on the signal levels.

We will assume that the signal detection is done by distinguishing between signals above or below zero volts, and will further assume a noise voltage with a normal (Gaussian) distribution and a mean noise power σ^2. This distribution or density is shown in Fig. 23.12. The crosshatched areas under the density curve represent the probability of a 0 sent and a 1 received, or a 0 received when a 1 was sent.

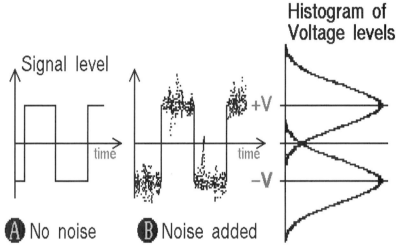

FIGURE 23.11   A bipolar signal with added noise. The histogram to the right shows the most likely occurrences of high and low voltages.

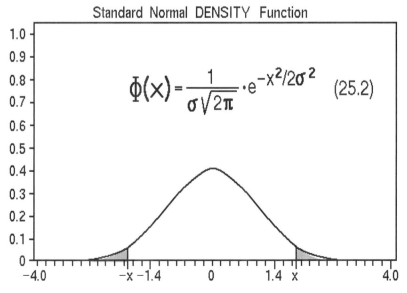

FIGURE 23.12   The standard normal density function.

An error may occur whenever the noise voltage exceeds a level equal to the peak signal voltage *V*. We wish to determine the probability thereof, based on the signal-to-noise ratio (SNR) of the system. This SNR is specifically the peak-signal (*V*) to rms-noise (σ) ratio, which we will call *K*:

$$K = V_{\text{peak}}/\sigma \qquad (23.1)$$

The standard normal density function is shown in Fig. 23.12, and the probability of a noise voltage exceeding the numerical value of *V* is found as the sum of the two crosshatched areas outside of −*x* and +*x*, or twice the integral of $\Phi(x)$ taken from minus infinity to $x = -k\sigma$.

The integral of $\Phi(x)$ is called the normal distribution function *F(x)*, and it changes in value from zero at $x = -\infty$ to one at $x = +\infty$, as shown in Fig. 23.13. The values of the integral are listed in several textbooks as erf(*x*), which has integration limits from zero to some positive value of *x*. This function is called the error function; we have erf(0) = 0.0 and erf(+∞) = 0.5000.

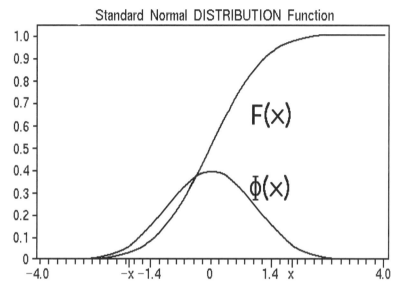

FIGURE 23.13 The standard normal distribution function.

For values of *x* beyond +*K*σ and −*K*σ we can therefore write the probability of measuring a noise voltage greater than *V*:

$$P\big\{(x > K\sigma)\text{ or }(x < -K\sigma)\big\} = 1 - \text{erf}\big(K/\sqrt{2}\big)$$

$$= \text{erfc}\big(K/\sqrt{2}\big) \qquad (23.2)$$

erfc(*x*) is named the complementary error function.

The result for the error probability for *x* > *K*σ or *x* < −*K*σ is:

$$P\big(x < -K\sigma\big) = P\big(x > K\sigma\big) = 0.5 \times \text{erfc}\big((K/\sqrt{2}\big) \qquad (23.3)$$

Notice that the error probability depends strictly on the SNR K. The error probability given in formula (23.3) is a reasonable first approximation for the BER. This BER versus SNR is plotted in Fig. 23.14. The BER-versus-SNR relationship for communication channels is determined by the method

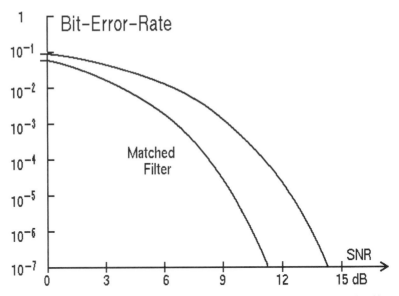

FIGURE 23.14 The bit-error-rate (BER) as a function of the peak-to-rms signal-to-noise ratio for a bipolar (magnetic recording) signal. A 3-dB improvement is possible by using matched filters prior to detection.

used to detect whether a 1 or a 0 was sent. A 3-dB improvement is possible for a well-designed data channel with matched filters (shown by the broken line).

BIT ERROR RATES

Figure 23.15 shows the distribution function for the Gaussian density in a linear plot to the left, logarithmic to the right. The Z-score values correspond to the σ-value. For instance, the probability of finding all scores between $x = -3\sigma$ and $x = +3\sigma$ is:

$$\Pr(-3\sigma \le x \le +3\sigma) = F(-3\sigma) - F(+3\sigma)$$
$$= 0.9987 - 0.0013$$
$$= 0.9974$$

The probability of finding values at x greater than 3σ is:

$$\Pr(x < -3\sigma \text{ or } x > +3\sigma) = 1 - 0.9974$$
$$= 0.0026$$
$$= 2 \times 0.0013.$$

The corresponding value of x for a Gaussian density is found from a table lookup of the x value that gives $F(x) = 0.0013$, which is $x = \pm 3.01\sigma$. These limits mean that 26 out of a total outcome of 10,000 would fall outside $x = \pm 3.01\sigma$. If these 26 extreme deviations represent timings in a detection circuit, they are not detected and therefore are classified as errors. We say that the bit error rate is 26 in 10,000, which we express as BER $= 2.6 \times 10^{-3}$.

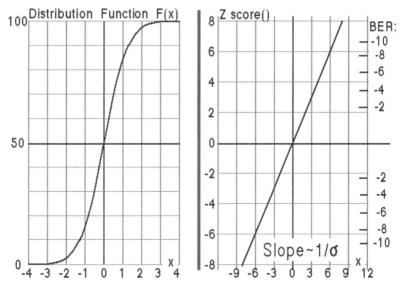

FIGURE 23.15   Linear and log plots of the standard normal distribution function (*courtesy of Tim Perkins.*)

To achieve a BER of 10^{-6} we need a detection probability of 0.999999, corresponding to $F(x) = 0.5 \times 10^{-6}$, or $2F(x) = 0.000001$, which requires $x = \pm 4.89\sigma$. And for BER $= 10^{-10}$, a probability of .9999999999 is needed, corresponding to $F(x) = 0.5 \times 10^{-10}$. This occurs for $x = \pm 6.47\sigma$.

The Z-score chart (or Weibull chart) in Fig. 23.15 is used to estimate the BER from a given probability, assuming the standard Gaussian distribution function (or vice versa). Do not confuse the term Z-score with the number Z often used to scale a distribution to become normal: $Z = (x_i - m)/s$.

The BER rating of a digital communication system, of which tape or disk drives are parts, determine the system's usefulness. Pictures (video) and audio information can have a fairly large number of errors; i.e., BER $= 10^{-5}$ is adequate because the errors can be concealed.

Scientific data demand much lower error rates, and accounting data must be error-free. The required bit error rates are very small, and only are achieved by applying error detecting and correcting codes. The demands to these codes will be less stringent the better the data channel is, and we will typically demand raw bit error rates in the order of 10^{-10}. The errors will then occur very seldom, every 1000 seconds on the average for a data stream with a bit rate of 10 Mbps. That roughly corresponds to one error every quarter of an hour, and days should be allocated to their measurement. We shall see that it actually may be done in a fraction of a second and then extrapolated to BER $= 10^{-10}$. This extrapolation can be done under the assumption of a Gaussian density.

The electronic circuits designed for signal detection will normally gate a trigger circuit for an amount of time called the window. It should capture one transition only, and its width (duration Δt) is therefore limited to the minimum inter transition time. This time interval is generally longer than the clock bit duration T, and depends on the data rate and the code. The minimum duration in a run-length limited code (1,7)RLL equals two bit-cell timings, $2T$.

The window is always of shorter duration. If we have BER $= 10^{-10}$, then the window duration may correspond to exactly $t = \pm 6.47\sigma$, where σ is the measured mean time of observed zero crossings. There is no margin for excursion beyond those values, and in that case we define the window margin to be zero. If the window's duration or opening $\Delta t = 2t$, then the window margin is said to be 50%.

The BER analysis presented above is ideal, and never found in a magnetic recording channel: the noise is not truly random, nor is the signal voltage constant. The amplitude in tape systems can be expected to vary ± 1 dB, with occasional large drops in level due to dropouts. These amplitude variations are named multiplicative noise, and must be included in the SNR (see Chapter 27, Amplitude

Fluctuations). The BER evaluation must also include the signal deterioration that occurs due to pulse interference and timing errors.

Multilevel signalling for recorders has been evaluated as a means of increasing the information stored. It can only be implemented successfully if automatic gain circuits are used to stabilize the signal amplitude variations in the read signal, in combination with sophisticated error detecting and correcting codes.

SOURCES OF TIMING ERRORS

Before we discuss measurements and simulation, let us review some of the mechanisms that contribute to timing errors.

NOISE IN GENERAL

Gaussian noise was early recognized and analyzed for its contribution to timing errors (Tahara et al. 1976; Katz and Campbell 1979).

The differentiated flux from the read head, and the subsequent differentiated read voltage are both shown in Fig. 23.16. The zero-crossing voltage is shown magnified, with superimposed noise. The latter will cause early or late triggering of the detector circuit with an effect that is proportional to its rms value divided by the slope of the differentiated head voltage (Howe 1986). A high value of the latter is provided for by sharp transitions in the recorded track (barium ferrite for particulate coatings) plus a head with a short sensing pattern (a thin film head).

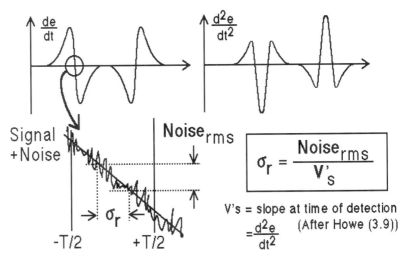

FIGURE 23.16   Noise superimposed on a data signal produces timing errors in a detector.

WRITE NOISE

The timing transitions that remain recorded along a track do not appear with bit lengths that correspond exactly to the time intervals between current reversals in the write head winding. One effect causing this is the proximity effect; see Fig. 23.17. There will be a certain amount of "forward" shift

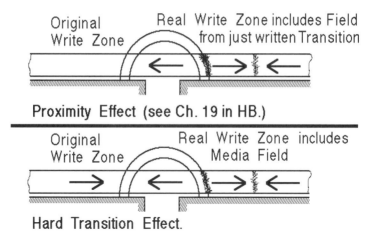

FIGURE 23.17 Transitions will shift during write due to 1. Demagnetization fields from itself plus fields from earlier written transitions (See chapter 19), and to 2. Fields from data being overwritten (hard transition effects).

of a transition due to field contribution from its demagnetizing field (see Chapter 19), and the immediately following transition will be shifted also, but less. This destroys the interval timing, and is called peak shift during write.

A similar effect occurs when a field is present from earlier recorded data, i.e., during overwrite. This is the hard transition effect. The field from old bits to be overwritten may work against the head field, and we refer to this as an unfavorable direction for the recording. It may also work with the head field, and we have a favored direction; see Fig. 23.17. Their effect on the BER is shown in Fig. 23.18 (Brittenham 1988). A technique for measurement of peak shifts during write has been used, using two adjacent transitions with time reference transitions placed far away (Tsang and Tang 1991).

The timing errors from both effects do not show up as random distributions, but as systematic discrete time events. These write errors (plus the later described read peak shifts) cause a flat top on the BER curve, as shown in Fig. 23.26, when using a log scale.

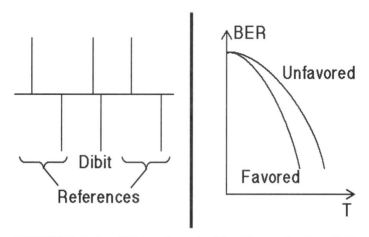

FIGURE 23.18 The time shift between the two signals in a dibit pattern (from Figure 21.14) will depend on the magnetic state of the media: Are fields from bits being overwritten with (favored) or against (unfavored) the head field? The early- and late-set digits are for timing references.

READ NOISE

Crosstalk and side reading are the information leaking from one track to another. When a head is positioned partly over a neighboring track, that track's data are read, as shown in Fig. 23.19. Side reading is most severe when all tracks are recorded with the same head, as in disk drives.

Helical-scan recorders have reduced sensitivity to side reading. The two heads that are displaced 180° on the head drum, are made to have, for instance, ±7° difference in alignment angles. This is shown in Fig. 23.19, bottom.

This affords excellent channel isolation at short wavelengths. Its application in a disk drive would require two head arms per disk surface, each with a head of opposite azimuth angle. This could eliminate the need for land between tracks and increase the area storage density.

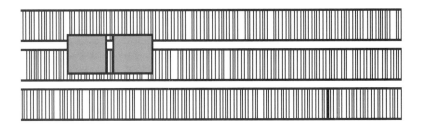

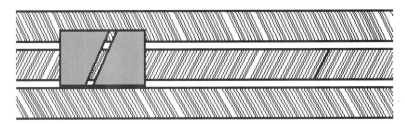

FIGURE 23.19   Crosstalk as it occurs in multichannel heads (see chapter 10). Sidereading is the sensing of flux from neighboring tracks. It is reduced when azimuth recording is used (see chapter 6).

MEDIA NOISE

This topic has been widely described in the literature, and many means to reduce noise have been tried out and implemented in media production. Two trends are obvious from recent work: first, to ensure that the metal films for rigid disks have a structure of very small grains. This is illustrated for a sputtered CoCr disk in Fig. 23.20, and the noise of this media approaches that of a small-size particulate coating. It is an analogy to low-noise photographic film, which requires very small grains. Second, noise reduction has recently been observed in multilayer coatings where the layers are isolated from each other, and yet coupled together magnetically (Murdock et al. 1990) The SNR appears to improve as the square root of the number of layers.

Other noises come from the zigzag transitions in metal-film media. Their lengths are inversely proportional to the film's coercivity. Irregularities in head fields at the gap ends can cause excessive zigzag patterns, as evidenced by the Lorentz microscopy picture sketched in Fig. 23.21. Reading the tracks with a wide read head results in higher timing errors than reading with a narrow track head placed in the center of the track does.

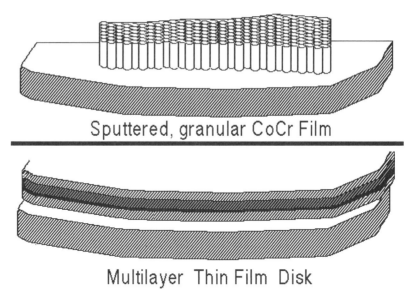

FIGURE 23.20   The sputtering of CoCr onto a rigid disk enhances the growth of grains oriented perpendicular to the disk. The division of the magnetic layer into several thinner layers improves the SNR (*Murdock*).

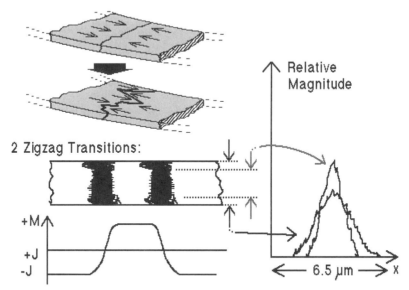

FIGURE 23.21   Zigzag transition zones produce read noise. It may be reduced by reading with a core narrower than that of the write heads.

HEAD NOISE

A read head produces random noise due to the resistive component of its impedance (Klaassen et al. 1992). An additional noise is often generated in thin film heads where only a few magnetic domains exist in the magnetic circuits that make up the head. Typical domain patterns are shown in Fig. 7.21, and the method of conducting alternating flux is by domain wall motions (left) or magnetization rotations (right). The latter works faster, as indicated by their respective resonance formulas (also Fig. 7.21).

The domain configuration shown to the right is enhanced during the film fabrication by the presence of a magnetic field, and by making the film from an iron-rich compound (with slightly negative magnetostriction). Some wall movements still take place during head operation, in particular during write cycles. The movements are not always reversible, i.e., a wall hangs up and will later snap back to its original location. This generates popcorn noise. Discontinuous wall motions in a head's magnetic circuit may cause Barkhausen noise. A head's characteristic concerning Barkhausen noise can be analyzed by a method where the head is placed between two pole shoes in an electromagnet, well-shielded with a copper housing to prevent electrostatic noise. An impulse field of 1 kOe will induce a transient voltage in the head, and it should decay in a smooth fashion. Any Barkhausen noise caused by irregular wall movements will cause signals that ride on top of the decay; one such Barkhausen noise spike can be seen in Fig. 23.22 (Klaassen 1990).

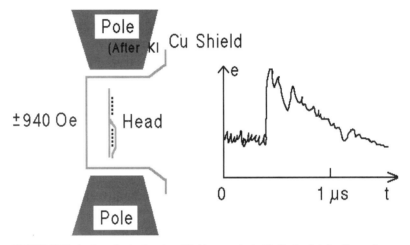

FIGURE 23.22   A scheme for the detection of Barkhausen noise in thin film heads (*after Klaassen*).

SYMBOL INTERFERENCE

Figure 23.23 shows a transition sequence of 2T1T2T3T3T1T. When these are recorded there will be predictable pulse interferences as shown in Fig. 23.24, and their cumulative effect is shown in Fig. 23.25. The timing errors are computed by knowing the pattern (2T1T2T3T3T1T, etc.) and the head response to a single, isolated transition. To these we now add all the previously discussed errors, and we have the total timing errors for the drive.

DRIVE ERRORS

BER due to drive errors is shown in Fig. 23.26. Timing errors are generated by the mechanics of moving the media. We are all familiar with the wow and flutter that affects audio recorders. Digital recorders have the same problem. It is seldom recognized by the user because of buffers that accept the data at whatever rate they arrive from the read head, and the buffer output is gated by a clock that follows the average data rate. This is why there is no detected wow and flutter from a digital tape recorder.

There are also changes due to change in temperature, and possibly humidity. An exposure of an operating drive, tape or disk, to vibrations should also be prevented.

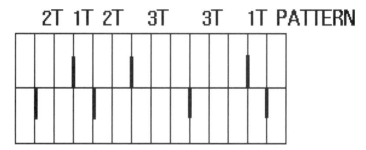

FIGURE 23.23 A pattern (transition locations) for a test signal.

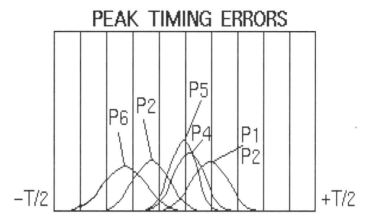

FIGURE 23.24 Timing errors in individual distributions.

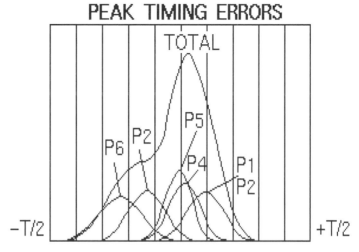

FIGURE 23.25 Timing errors in cumulative distribution.

We must finally remember that no disk or tape is perfect. A very low, but persistent, error rate may be measured below BER = 10^{-10}. This is shown in Fig. 23.26.

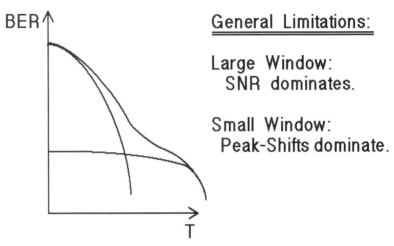

FIGURE 23.26   Possible BER in a disk drive, having two sources.

SEPARATION OF ERROR SOURCES

By performing a carefully scheduled sequence of measurements it becomes possible to separate some of the contributing errors that otherwise are hidden together in the BER plot; for an example, see Fig. 23.25. Most of the error sources are always active. If we can make one of the sources inactive for one set of measurements and active for another set, then the difference between some of the measured values will represent that source.

The planned experiment may initially be that 1000 sectors of a disk are written once and then read out many times each, maybe another 1000 times each. The read data do not include the write errors because all written transitions were laid down once. Let the measured, average variance be σ_{read}^2.

Now repeat the measurement, and this time write and read for each measurement. Let the new variance be $\sigma_{read+write}^2$. The read and write functions are independent, and we can therefore write:

$$\sigma_{write}^2 = \sigma_{read+write}^2 - \sigma_{read}^2$$

This technique is described in detail in the earlier mentioned HP application notes, and in a paper by Greenwood (1988) and Brittenham (1988). Further work toward separating the error sources has recently been described by Yuan and Bertram (1992).

RADIO FREQUENCY INTERFERENCE (RFI)

Man-made noise often limits the performance of a system. The sources of this noise range from noise spikes from dc-motor commutators, car engine's switching circuits, and so on, to the picking up of broadcast stations by the read head.

The rules are that such noises should be contained at their source. It is still necessary for the manufacturer of recording equipment to guard against RFI by using shielded enclosures and proper grounding techniques. Filtering of the ac power line is often necessary, as are low-pass filters on all ingoing and outgoing data lines.

SYSTEM NOISE

We will name the total of media noise, crosstalk, head noise, electronics noise, and RFI as system noise. Some of the noise components are randomly distributed (coating noise, head noise, and electronics noise); others are occasionally occurring noise spikes (dropouts and some RFI) and have a Poisson distribution. Finally there are cyclic components (RFI).

Each system has its individual system noise, and it is best represented by a spectrum showing band-pass-filtered noise versus frequency, plus statistical data for levels. The statistical information is needed to determine the number of times the noise level will be high enough to offset the detection of a bit or actually appear itself as a bit. These occurrences will determine the number for the system's bit error rate (BER).

SYSTEM MEASUREMENTS

Figure 23.27 shows a pseudorandom test signal where timings between transitions are T, $2T$, or $3T$. Such signals are generated with shift registers plus adders. An instrument to measure the time intervals was designed in the 1980s by the company Odetics. It has a circuit that triggers "on" for a transition and "off" for the next transition; i.e., it would measure time intervals (Mackintosh 1981). The instrument now rearms the triggering circuit, and during this time it is not possible to measure the immediately following interval. The time interval analyzer therefore samples some of the time intervals. Each measured time intervals adds to an array of storage registers where each register corresponds to a time interval length. The resulting accumulations are shown in Fig. 23.27 at bottom, with mean values of T, $2T$, and $3T$, and deviations around each.

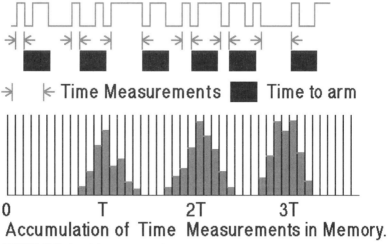

FIGURE 23.27 Sampled measurements of ΔT.

The instrument has been useful in establishing timing margins, which are the differences between the largest timing variations and the actual width of the timing window. But any attempt to analyze sampled time intervals will provide meaningless results, as illustrated in Fig. 23.28. Once the data have been added to the cumulative registers, valuable information is lost, and to extract their error causes would be impossible. Suppose the timing population has a disturbance that varies as shown in

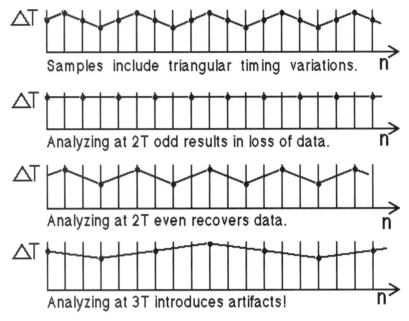

FIGURE 23.28   Sampling may cause filtering and/or aliasing when analyzing data.

the top row. If the sampling now occurs with an interval of $2T$, then the disturbances do not appear (2nd row) or they do (3rd row), all depending on the synchronization phase. On the other hand, if we sample with a frequency of $3T$, then artifacts (here a low-frequency component, 4th row) are generated. The sampling function acts as a digital filter, and the time observation should therefore be a continuous function with zero time for arming of the gate circuit.

The most recent generations of time interval analyzers have very short arming times so continuous measurements are possible, as shown in Fig. 23.29. Data are provided for the window-margin analysis, and they can additionally be stored sequentially in memory (Fig. 23.30). The stored data can later be analyzed, and will disclose information about the origin and nature of errors causing timing variations. The original data stream is no longer pseudorandom, but rather is specified, for example: $1T$ $2T$ $3T$ $3T$ $2T$ $1T$ $1T$ $2T$ $3T$, etc.

Instruments that are available include the HP5372A from Hewlett-Packard (Greenwood 1988; El-Fishaway 1991), and recently the HPE1725A TIA.

FREQUENCY RESPONSE

The measurement of a tape recorder's frequency response can be accomplished by the use of a standard alignment tape, which checks the frequency response of the playback amplifier at the same time. To check the overall frequency response, a signal generator is connected to the record amplifier input terminal, and the output is recorded at a number of frequencies and plotted on graph paper. When measuring the response of an analog recorder, it is necessary that the record level be 20 dB lower than the normal record level, because distortion and overload will occur at the higher frequencies. These result in a frequency response curve showing a dropoff toward high frequencies. This dropoff is unrealistic, and is caused by tape saturation due to pre-equalization.

A sweep generator is a very useful tool in setting up and aligning a tape recorder. The sweep generator output frequency is swept from low frequencies up to the highest frequency of interest, many times a second. This may be 50 or 60 times per second, providing an excellent display of the ampli-

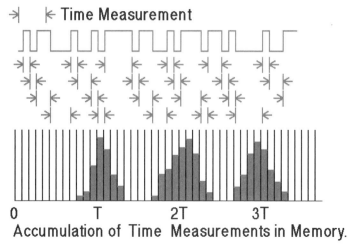

FIGURE 23.29 Continuous measurement of ΔT.

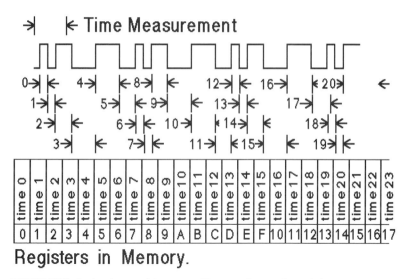

FIGURE 23.30 Continued measured data are stored in memory for exact data analysis.

tude-versus-frequency response on an oscilloscope. This greatly facilitates head alignment, bias, and equalizer adjustment. The single-frequency generator is more useful in the final plot of a recorder's frequency response.

SIGNAL-TO-NOISE RATIO (SNR)

A tape recorder's noise is always expressed in terms of a number that gives the ratio of the normal signal voltage to the noise voltage. This is referred to as the signal-to-noise ratio (SNR).

The SNR will depend strongly on the method used for its measurement. The normal level will in general correspond to a level associated with a one percent total harmonic distortion (although the

3rd harmonic is predominant). It may equally well correspond to three percent harmonic distortion, which gives a 4–5 dB higher SNR. The three percent distortion level is more reliable as a reference.

The noise level is more ambiguous: It can be measured with the tape transport stopped, or while moving a bulk-erased tape over the playback head.

It is important that the tape is in motion so that any electrical noise from the tape transport is included. A realistic measurement will further measure the noise from a tape recorded on the unit with the input terminal shorted. This will include bias noise, and will thus be a true SNR value for the recorder.

It may not be a realistic number for an audio recorder, where the ultimate judgment of an acceptable SNR is the human ear. It is fortunate that the ear's sensitivity versus frequency varies with the sound level. The response is fairly flat at normal listening levels, but is less sensitive at low and high frequencies at low sound levels. The latter means that the ear is less sensitive to the amplifier hum and the tape bias, which were both boosted in amplification along with the playback equalization. This results in a subjectively better judgment of the SNR than the bare number gives.

It is therefore common practice to measure an audio recorder's SNR as the ratio (in dB, decibels) between the normal level and the weighted noise level, resulting in, for example, 70 dB instead of 50 dB. This method is not applicable to instrumentation recorders, where at most a band-pass filter with cut-off frequencies set at lower and upper frequency limits is used.

Instrumentation recorders are often used to record several frequency multiplexed signals onto one channel. It is important that none of the individual signals cause harmonic distortion, because the harmonics then interfere with the reception and detection of the other channels.

It is even more important that the several signals do not interfere with each other via the tape's overload curve and thus cause intermodulation. A good test can be performed to assess a recorder's suitability for multiplex recording of several signals simultaneously on one channel. The record signal is simply white noise, which contains signals of uniform amplitude at all frequencies.

The components in the white noise will all cause intermodulation among each other, and the amount thereof is measured in the following way (Fig. 23.31, bottom). The write noise is recorded in full bandwidth or with part of its spectrum stopped by a band-stop filter, set at a frequency f_0. The filter bandwidth may be 10 kHz. The playback signal is measured through a narrower band-pass filter, also set at f_0.

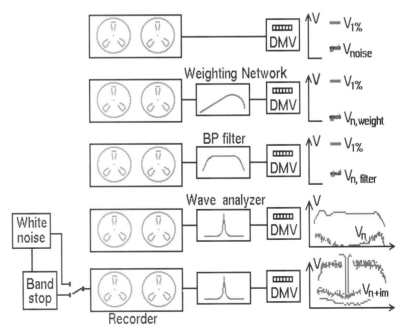

FIGURE 23.31 Test methods for measurements of noise and SNR.

When the band-stop filter on the record side is bypassed, a signal voltage V_s is measured. Next, the filter is inserted and a voltage V_n measured. The latter voltage is the sum of the reproducer noise plus any intermodulation products from the recording of the white noise outside the band-stop filter.

At low record levels, the intermodulation is insignificant, only reproducer plus tape noise is measured. The record level is below the normal level, and the $SNR = V_s/V_n$ is quite low (its value will differ from the slot-noise SNR, which is displayed as a noise-voltage-versus-frequency curve, measured with a wave analyzer. It is expressed in dB relative to a normal record level of a sine wave).

The signal-to-noise ratio in the white noise loading test is called the noise-power ratio (NPR) because two noise voltages are compared rather than a sine voltage and a noise voltage. The NPR range is therefore considerably smaller in dB than the normal SNR or slot-noise SNR. The first measured NPR can be improved by raising the record level. At some level the intermodulation will increase drastically, and the intermodulation products will spill into the slot around f_o and add to V_n. One may optimize the NPR by trial-and-error setting of the record level, or settle for a NPR number at the level where the noise voltage has increased 3 dB over the pure noise voltage (where the record input signal is equal to zero).

PHASE DISTORTION

Measurements of phase distortion or its companion, envelope or group delay, requires a specialized and difficult technique and is beyond the scope of this book.

Envelope delay is related to the delay that individual signals at different frequencies incur when passing through a system such as an amplifier, a cable, or a tape recorder. The definition of envelope delay is:

$$\text{Delay } \tau = d\phi/d\omega \qquad (23.6)$$

which is the slope of the phase versus frequency response curve. When the relationship between phase and frequency is linear, then τ is constant. All signals of various frequencies arrive at the same time and they will add up to a faithful reproduction of the original signal.

IMPROVEMENTS TO THE SNR

There are several ways in which the SNR of a recording system can be improved. They all involve some form of pre- and post-treatment of the signal, and they are not part of the magnetic record and/or playback processes.

LEVEL STRETCHING

The transfer curve for tape magnetization versus record current (field) departs from linearity at a level corresponding to 30 percent below saturation. Recordings at higher levels will produce harmonic and intermodulation signals that, on playback, show up as distortion.

It is possible to stretch the peak level of the record current to counterbalance the curvature of the overload curve, and thus record a higher level on the tape at a lower distortion.

A simple form of a record current stretcher is shown in Fig. 23.32. It is simply a resistive voltage divider with one resistor shunted by two identical diodes back-to-back. Its implementation is difficult, because the network must produce the exact inverse of the overload curve of the tape, and that overload changes with the recorded wavelength and it differs among the various tape types.

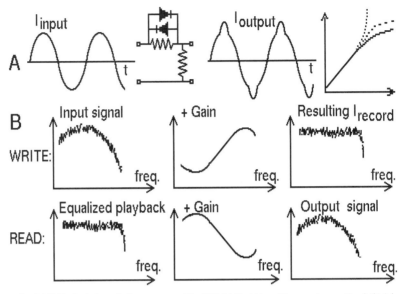

FIGURE 23.32   Techniques for improvement of the SNR: (a) Stretching of the record current level; (b) Additional pre- and post-equalization.

Add to these difficulties the fact that accidental overloads cause hard clipping, and in audio this results in a very harsh sound. Only a carefully designed circuit of fair complexity will provide a significant improvement in the apparent SNR (10 dB).

The stretcher circuit can be equally well employed on the playback side. It now restores occasional peaks that were "squashed" during recordings. Needless to say, these circuits must also be matched to the tapes overload curve.

ADDITIONAL EQUALIZATION

It follows from the discussions in Chapter 22 of equalization that it is desirable to record all signals in the record signal at the highest level possible. This reduces the requirements for post-equalization, with a resulting reduction in the equalized noise level.

Equalization standards need not be a common law in a closed organization such as a research laboratory or telemetry group. It may be very advantageous to examine the level-versus-frequency spectrum of the signal to be recorded, then design a simple pre-equalizer to provide a flat spectrum.

A matching post-equalizer will restore the data to its original spectrum. Neither amplitude nor phase distortion will be introduced if the pre- and post-equalizers are complementary.

BANDPASS FILTERING

SNR improvements are readily obtained by bandpass filtering of the reproduce signal. The desired fidelity determines the extent of filtering required (an amplifier's treble control is an example of a bandpass filter). Filtering of extraneous noise may also be advantageous during recording.

Simple RC-filtering will work, but is inferior to the use of low-Q filters. The rounding of pulses is shown in Fig. 23.33.

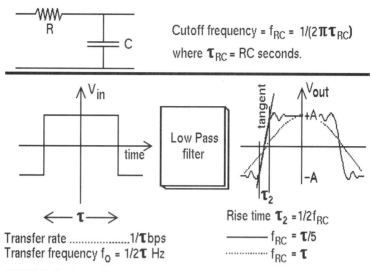

FIGURE 23.33   The rounding of pulse waveforms by low-pass filtering.

COMPRESSION/EXPANSION

Finally, there exists the method of compression of the signal prior to recording, and subsequent expansion after playback. This technique is useful when a signal with a large dynamic range is recorded, and it offers a subjective increase in the SNR.

The transfer curves for the expander and the compressor are shown in Fig. 23.34. The scheme is simple in theory, but complex in reality. It is not suitable for any instrumentation applications, and there are several requirements of the activation of the compression and expansion circuits concerning response time, hold time, and decay time.

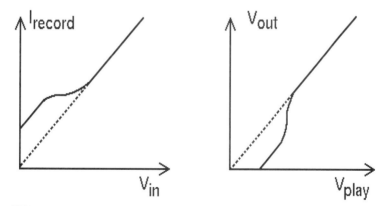

FIGURE 23.34   Signal compression during recording, and expansion after playback.

REFERENCES

Bennett, W.R., and J.R. Davey. 1965. *Data Transmission*. McGraw-Hill. 356 pages.
Brittenham, S. July 1988. A Review of Timing Effects in Magnetic Disk Drives. *hp Data Storage Test Solutions Symposium*. Paper no. 4, pp. 77–96.

Brittenham, S. Nov. 1988. Time Domain Characterization of Thin Film Head/Media Systems. *IEEE Trans. Magn*. MAG-24 (6): 2949–54.

Bylanski, P., and D.G.W. Ingram. 1980. *Digital Transmission Systems*. Peter Peregrinus Ltd. on behalf of the Inst. of El. Engrs., London. 431 pages.

El-Fishawy, K. April 1991. Improved Time Domain Characterization of Disk Drives. *hp 1991 Data Storage Industry Symposium*. Paper No. 3, 43 pages.

Greenwood, B. July 1988. Time Domain Characterization of Magnetic Disk Drives Using the HP 5371A Frequency and Time Interval Analyzer. *hp Data Storage Test Solutions Symposium*. Paper no. 5, pp. 97–124.

Howe, D.G.. Jan. 1986. Signal-to-Noise Ratio (SNR) for Reliable Data Recording. *Proc. SPIE* Vol. 695, pp. 255–61.

Katz, E.R. and T.G. Campbell. May 1979. Effect of Bitshift Distribution on Error Rate in Magnetic Recording. *IEEE Trans. Magn*. MAG-15 (3): 1050–54.

Klaassen, K.B., and J.C.L. van Peppen. Sept. 1990. Barkhausen Noise in Thin-Film Recording Heads. *IEEE Trans. Magn*. MAG-26 (5): 1697–99.

Klaassen, K.B., and J.C.L. van Peppen. Sept. 1992. Noise in Thin-Film Inductive Heads. *IEEE Trans. Magn*. MAG-28 (5): 2097–99.

Lender, A. Feb. 1966. Correlative level coding for binary-data transmission. *IEEE Spectrum*, vol. 6, pp. 104–115.

Mackintosh, N.D. Nov. 1981. A Margin Analyzer for Disk and Tape Drives. *IEEE Trans. Magn*. MAG-17 (6): 3349–51.

Mackintosh, N.D., and F. Jorgensen. Nov. 1981. An Analysis of Multi-level Encoding. *IEEE Trans. Magn*. MAG-17 (6): 3329–31.

Murdock, E.S., B.R. Natarajan, and R.G. Walmsley. Sept. 1990. Noise Properties of Multilayered Co-Alloy Magnetic Recording Media. *IEEE Trans. Magn*. MAG-26 (5): 2700–05.

Tahara, Y., H. Takagi, and Y. Ikeda. Nov. 1976. Optimum Design of Channel Filters for Digital Recording. *IEEE Trans. Magn*. MAG-12 (6): 749–750.

Tsang, C., and Y.S. Tang. March 1991. Time-Domain Study of Proximity-Effect Induced Transition Shifts. *IEEE Trans. Magn*. MAG-27 (2): 795–802.

Yuan, S.W., and H.N. Bertram. Jan. 1992. Statistical Data Analysis of Magnetic Recording Noise Mechanisms. *IEEE Trans. Magn*. MAG-28 (1): 84–92.

BIBLIOGRAPHY

SIGNAL AND COMMUNICATION THEORY

Bracewell, R.N. 1978. T*he Fouriers Transform and Its Application*. McGraw-Hill. 444 pages.

Brierley, H.B. 1986. *Telecommunications Engineering*. Edward Arnold, London. 278 pages.

Brigham, E.O. 1974. *The Fast Fourier Transform*, Prentice-Hall. 252 pages.

Carlson, A.B. 1986. *Communication Systems*, McGraw-Hill. 686 pages.

Chambers, W.G. 1985. *Basics of Communications and Coding*. Clarendon Press, Oxford. 240 pages.

Cheng, D.K., *Analysis of Linear Systems*, Addison-Wesley, 1961, 431 pages.

Cooper, G.R., and C.D. McGillem. 1971. *Probabilistic Methods of Signal and System Analysis*. Holt, Rinehart and Winston. 258 pages.

Cooper, G.R., and C.D. McGillem. 1974. *Continuous and Discrete Signal and System Analysis*. Holt, Rinehart and Winston. 395 pages.

Data sheets, application notes and design guides from several semiconductor manufacturers, plus information from journals listed in Bibliography for Chapter 26.

Lathi, B.P. 1983. *Modern Digital and Analog Communication Systems*. Holt, Rinehart and Winston. 708 pages.

Liu & Liu, 1975. *Linear Systems Analysis*. McGraw-Hill. 467 pages.

Papers from '94 TMRC Conference on Signal Processing. March 1995. *IEEE Trans. Magn*. MAG-31 (2): 1027–1213.

Schwartz, M. 1980. *Information Transmission, Modulation and Noise*. McGraw-Hill. 646 pages.

Senior, Th.B.A. 1986. *Mathematical methods in electrical engineering*. Cambridge Univ. Press. 272 pages.

Shannon, C.E., W. Weaver, *The Mathematical Theory of Communication*, 1948, reprinted by Dover, New York, Jan. 1975, 125 pages.

Whalen, A.D. 1971. *Detection of Signals in Noise*. Academic Press. 411 pages.

MODELLING OF THE SIGNAL CHANNEL

Abbot, W.L., J.M. Cioffi, and H.K. Thapar. Nov. 1988. Offtrack Interference and Equalization in Magnetic Recording. *IEEE Trans. Magn.* MAG-24 (6): 2964–66.

Aikawa, T., H. Mutoh, and T. Sugawara. Feb. 1991. Signal Processing for High Density Magnetic Recording. *Fujitsu Sci. Tech. Jour.* 26 (4): 391–403.

Klaassen, K.B. Nov. 1991. Magnetic Recording Channel Front-Ends. *IEEE Trans. Magn.* MAG-27 (6): 4503–08.

Lindholm, D.A. Dec. 1973. Fourier Synthesis of Digital Recording Waveforms. *IEEE Trans. Magn.* MAG-9 (4): 689–698.

Loze, M.K., B.K. Middleton, and A. Riley. March 1986. Simulation of Digital Magnetic Recording Systems. *Intl. Conf. Video and Data Recording*. IERE Publ. No. 67, pp. 185–193.

Mallinson, J.C. Dec. 1974. Correction of "Applications of Fourier Transforms in Digital Magnetic Recording Theory." *IEEE Trans. Magn.* MAG-10 (4): 1137.

Mallinson, J.C. March 1974. Applications of Fourier Transforms in Digital Magnetic Recording Theory. *IEEE Trans. Magn.* MAG-10 (1): 69–78.

Melbye, H.E., and C.S. Chi. Sept. 1978. Nonlinearities in High Density Digital Recording. *IEEE Trans. Magn.* MAG-14 (5): 746–748.

Price, R. Sept. 1978. An Experimental, Multilevel, High Density Disk Recording System. *IEEE Trans. Magn.* MAG-14 (5): 315–318.

Wood, R. and Donaldson, R. March 1979. The Helical-Scan Magnetic Tape Recorder as a Digital Communication Channel. *IEEE Trans. Magn.* MAG-15 (2): 935–943.

Wood, R.W., and R.W. Donaldson. Sept. 1980. Signal Processing and Error Control on an Experimental High-Density Digital Magnetic Tape Recording System. *IEEE Trans. Magn.* MAG-16 (5): 1255–1265.

EYE PATTERNS

Ainsworth, K. Nov. 1991. Monitoring the serial digital signal. *Broadcast Engr.* 33 (11): 32– (8 pages).

Dye, T.A., and E. Teose. Feb. 1992. Digital storage scopes advance. *IEEE Spectrum.* 29 (2): 38–41.

Feher, K. 1987. *Telecommunications Measurements, Analysis, and Instrumentation*. Prentice-Hall. 412 pages.

Kedem, B. Nov. 1986. Spectral Analysis and Discrimination by Zero-Crossings. *Proc. IEEE* 74 (11): 1477–93.

Mask Template Testing. Jan. 1990. *Tektronix Tech. Note*. Note 47W-7805, 6 pages.

Newcombe, E.A. and S. Passupathy. Aug. 1982. Error Rate Monitoring for Digital Communications. *Proc. IEEE.* 70 (8): ??.

DETECTION

Badger, D.M. August 1992. Stability of AGC Circuits Containing Peak-Detectors. *IEEE Trans. Consumer Elec.* 36 (3): 377–383.

Chopara, K., and D.D. Woods. Nov. 1991. A Maximum Likelihood Peak Detecting Channel. *IEEE Trans. Magn.* MAG-27 (6): 4819–21.

Curland, N., and R.J. Machelski. March 1994. Integrating an MR Head into a Peak Detection Channel. *IEEE Trans. Magn.* MAG-30 (2): 309–315.

Deeley, E.M., and A. Mitchell. Apr. 1984. Optimization of Nonlinear Clock Recovery Circuits for Digital Recording. *Intl. Conf. Video and Data Recording 84*. IEE Publ. No. 59, pp. 137–140.

Frederickson, L.J. Sept. 1990. Coding for Maximum Likelihood Detection on a Magnetic Recording Channel. *IEEE Trans. Magn.* MAG-26 (5): 2315–17.

Gottfried, N.H. Jan. 1993. Low Complexity Viterbi Detector for Magnetic Disc Drives. *IEE Proc.-E* 140 (1): 78–80.

Graham, I. H. July 1978. Data Detection Methods vs. Head Resolution in Digital Magnetic Recording. *IEEE Trans. Magn.* MAG-14 (4): 191–193.

Huber, W.D., J. Newman, and R.D. Fisher. Nov. 1981. Detection Window and Lineal Density in Digital Recording Systems. *IEEE Trans. Magn.* MAG-17 (6): 3355–3357.

Moon, J. Nov. 1991. Signal-to-Noise Ratio Degradation with Channel Mismatch. *IEEE Trans. Magn.* MAG-27 (6): 4837–39.

Moon, J., and L.R. Carley. 1992. Sequence Detection for High-Density Storage Channels. Kluwer Academic Publishers. 153 pages.

Moon, J., and L.R. Carley. Nov. 1990. Performance Comparison of Detection Methods in Magnetic Recording. *IEEE Trans. Magn.* MAG-28 (6): 3155–72.

Siegel, P.H. Nov. 1982. Applications of a Peak Detection Channel Model. *IEEE Trans. Magn.* MAG-18 (6): 1253–1255.

Spencer, R., and P.J. Hurst. Nov. 1991. Analog Implementations of Sampling Detectors. *IEEE Trans. Magn.* MAG-27 (6): 4516–21.

Thapar, H.K., N.P. Sanfs, W.L. Abbott, and J.M. Cioffi. Sept. 1990. Spectral Shaping for Peak Detection Equalization. *IEEE Trans. Magn.* MAG-26 (5): 2309–11.

Wood, R. Sept. 1990. A Theoretical Comparison of Detection Techniques on R-Dat. *IEEE Trans. Magn.* MAG-26 (5): 2157–59.

PRML

Coleman, C., D. Lindholm, D. Petersen, and R. Wood. April 1984. High Data Rate Magnetic Recording in a Single Channel. *Intl. Conf. Video & Data Rec. IERE Proc.* #59, pp. 151–157.

Kabal, P., and S. Patsupathy. Sept. 1975. Partial-Response Signaling. *IEEE Trans. Comm.* COM-23 (9): 921–934.

Kobayashi, H. Dec. 1971. A Survey of Coding Schemes for Transmission or Recording of Digital Data. *IEEE Trans. Comm.* Tech. COM-19 (6): 1087–1100.

Osawa, H., S. Tazaki, and S. Ando. July 1986. Performance Analysis of Partial Response Systems for Nonreturn-to-Zero Recording. *IEEE Trans. Magn.* MAG-22 (4): 247–52.

Shafiee, H., M. Melas, and P. Sutardja. Nov. 1993. Performance Comparison of EPRML and Peak Detection in High Density Digital Magnetic Recording. *IEEE Trans. Magn.* MAG-29 (6): 4015–17.

Wood, R.G. May 1990. Magnetic Megabits. *IEEE Spectrum* 27 (5): 32–38.

Wood, R.G., and D.A. Petersen. May 1986. Viterbi Detection of Class IV Partial Response on a Magnetic Recording Channel. *IEEE Trans. Comm.* COM-34 (5): 454–61.

BIT-ERROR RATES AND PROBABILITY

Brown, R. July 1991. All peaks aren't Gaussian. *Pers. Eng. & Instr. News* 8 (7): 51–54.

Cooper, G.R. and C.D. McGillem. 1986. *Probabilistic Methods of Signal and System Analysis*. Holt, Rinehart and Winston. 408 pages.

Howe, D.G. 1983. The nature of intrinsic error rates in high-density digital optical recording. *SPIE Vol. 421-Optical Disks Systems and Application*, pp. 31–42.

Howe, D.G. 1986. Signal-to-noise ratio (SNR) for reliable data recording. *SPIE Vol. 695 Optical Mass Data Storage II*, pp. 255–261.

Middleton, B.K., and T. Jack-Kee. Dec. 1983. Performance of digital magnetic recording channel subject to noise and drop-outs. *The Radio and Electr. Engr.* 53 (11/12): 393–402.

Pasian, F, and A. Crise. Jan. 1984. Restoration of Signals Degraded by Impulse Noise by Means of Low-Distortion Non-Linear Filter. *Signal Processing*. 6 (1): 67–76.

Pyzdek, T. 1989. *What Every Engineer Should Know About Quality Control*, Marcel Dekker, Inc. 251 pages.

SYMBOL INTERFERENCE

Bergmans, J.W.M. and Y.C. Wong. Jan. 1984. A Simulation Study of Intersymbol Interference Cancellation. *Archiv für Elek. Übertragung*. 38 (1): 9–14.

Damarowsky, M., and G.H. Guthhrlein. July 1988. Doublet resolving method using Fourier-Integral transformation (in German,) *Archiv für Elek. Übertragung*. 42 (4): 227–29.

Nunnelley, L.L., M.A. Burleson, L.L. Williams, and I.A. Beardsley. Sept. 1990. Analysis of Asymmetric Deterministic Bitshift Errors in a Hard Disk File. *IEEE Trans. Magn*. MAG-26 (5): 2306–08.

NOISE IN A MAGNETIC RECORDING CHANNEL

Arnoldussen, T.C., and H.C. Tong. Sept. 1986. Zigzag Transition Profiles, Noise, and Correlation Statistics. *IEEE Trans. Magn*. MAG-22 (5): 889–891.

Arnoldussen, T.C., and L.L. Nunnelley. 1992. Editors. *Noise in Digital Magnetic Recording*. Singapore, World Scientific. 280 pages.

Bennett, W.R. May 1956, Methods of Solving Noise Problems. *Proc. IRE*. Vol. 24, pp. 609–638.

Correira, G. April 1993. Shielding Products Against EMI and RFI. *Instr. & Control Systems* 66 (4): 67–69.

Darling, T.F. Jan-Feb 1987. Mathematical Noise Modelling and Analysis of Some Popular Preamplifier Circuit Topologies. *Jour. AES*. 35 (1/2): 15–23.

Finkelstein, B.I., and E.R. Christensen. Sept. 1986. Signal-to-Noise Ratio Models for High-Density Recording. *IEEE Trans. Magn*. MAG-22 (5): 898–900.

Fish, P.J. 1993. *Electronic Noise and Low Noise Design*. Macmillan, London. 278 pages.

Gravelle, L.B. and P.F. Wilson. May 1992. EMI/EMC in Printed Circuit Boards — A Literature Review. *IEEE Trans. Elmag. Comp*. 34 (2: 109–116.

Howell, T.D., P. Kasiraj, J.S.Best, F. Chu, and M.M. Yerry. Sept. 1986. A Study of Disk Noise Statistics. *IEEE Trans. Magn*. MAG-22 (5): 901–903.

Hughes, B.F., and R.K. Schmidt. Nov. 1976. On Noise in Digital Recording. *IEEE Trans. Magn*. MAG-12 (6): 752–754.

Keshner, M.S. March 1982. 1/f Noise. *Proc. IEEE*. 70 (3): 212–218.

Madrid, M., and R. Wood. Sept. 1986. Transition Noise in Thin-Film Media. *IEEE Trans. Magn*. MAG-22 (5): 892–894.

Mallinson, J. C. Sept. 1969. Maximum Signal-to-Noise Ratio of a Tape Recorder. *IEEE Trans. Magn*. MAG-5 (3): 182–186.

Netzer, Y. June 1981. The Design of Low-Noise Amplifiers. *Proc. IEEE*. 69 (6): 728–741.

Ott, H.W., *Noise Reduction Techniques in Electronic Systems,* John Wiley and Sons, 1976, 294 pages.

Palmer, D., P. Ziperovich, R. Wood, and T.D. Howell. Sept. 1987. Identification of Nonlinear Write Effects Using Pseudorandom Sequences. *IEEE Trans. Magn*. MAG-23 (5): 2377–79.

Roe, K., and R.F. Soohoo. Sept. 1986. Longitudinal and Perpendicular Magnetic Thin-Film Media Noise. *IEEE Trans. Magn*. MAG-22 (5): 886–888.

Sadashige, K. Dec. 1969. Study of Noise in Television Broadcast Equipment. *Jour. SMPTE.* Vol. 78, pp. 1069–1076.

Tavares, S.E. June 1966. A Comparison of Integration and Low-Pass Filtering. *IEEE Trans. Instr. and Meas.* IM-15 (1 and 2).

Thurlings, L. May 1980. Statistical Analysis of Signal and Noise in Magnetic Recording. *IEEE Trans. Magn.* MAG-16 (3): 507–514.

Van Gestel, W.J. March 1988. The Influence of Crosstalk on the Bit-Error-Rate in Magnetic Recording. *Intl. Conf. on Video, Audio and Data Rec.* IERE Publ. No. 79, pp. 35–44.

Wachenschwanz, D., and T. Carr. Nov. 1991. Modulation Noise Measurements of High Density Recording Channels Using MR Heads. *IEEE Trans. Magn.* MAG-27 (6): 5310–12.

Weinberg, L. April 1958. Exact Ladder Network Design Using Low-Q Coils. *Proc. IRE.* 46 (4): 739–750.

Wilmshurst, T.H. 1985. *Signal Recovery from noise in electronic instrumentation.* Adam Hilger Ltd., London. 193 pages.

Wood, R. Sept. 1989. Jitter vrs. Additive Noise in Magnetic Recording: Effects on Detection. *IEEE Trans. Magn.* MAG-23 (5): 2683–85.

MEASUREMENTS OF NOISE AND NPR

Blackman, R.B., and J.W. Tukey. 1958. *The Measurement of Power Spectra.* Dover Publ. 190 pages.

Buchholz, F.I., W. Kessel, and F. Melchert. Aug. 1992. Noise Power Measurements and Measurement Uncertainty. *IEEE Trans. Instr. and Meas.* 41 (4): 476–481.

Haynes, M.K. Sept. 1984. Density-Response and Modulation-Noise Testing of Digital Magnetic Recording Tapes. *IEEE Trans. Magn.* MAG-20 (5): 897–899.

Heidenreich, K. Heinz Jan. 1974. Der Rauschklirr-Belastungsversuch; Darstellung und Auswertung der Ergebnisse. *NTZ* No. 12, pp. 457–463.

Newcombe, E.A., and Pasupathy, S. Aug. 1982. Error Rate Monitoring for Digital Communications. *Proc. IEEE* 70 (8): 805–828.

Tang, Y.S. Sept. 1985. Noise Autocorrelation in Magnetic Recording Systems. *IEEE Trans. Magn.* MAG-21 (5): 1389–1391.

Tant, M.J. 1974. *Multichannel Communication Systems and White Noise Testing.* Marconi Instruments. 104 pages.

SYSTEMS, MEASUREMENTS, AND SIMULATION

Bergmans, J.W.M. March 1987. Performance Consquences of Timing Errors in Digital Magnetic Recording. *Philips Jour. Res.* 42 (3): 281–307.

Chang, T., D. Darrow, and S. Gupta. Sept. 1987. Statistical Characterization of Isolated Pulse Waveform of a Thin-Film Head. *IEEE Trans. Magn.* MAG-23 (5): 3155–57.

Characterizing Transient Timing Errors in Disk and Tape Drives. *hp Application Note* for 5372A, Jan 1990, Note 358-6, 22 pages.

Hardwick, J.O. March 1988. Track Misregistration and Its Effects Upon Bit Error Rate in Disc Drive Systems, Using Longitudinal Magnetic Recording. *Intl. Cont. on Video, Audio and Data Rec.* IERE Publ. no. 79, pp. 101–110.

Hardy, P. and D.J. Malone. Nov. 1991. Evolution of the Soft Error Rate Model. *IEEE Trans. Magn.* MAG-27 (6): 5313–15.

Howell, T.D, D.P. McCown, T.A. Diola, Y. Tang, K.R. Hense, and R.L. Gee. Sept. 1990. Error Rate Performance of Experimental Gigabit per Square Inch Recording Components. *IEEE Trans. Magn.* MAG-26 (5): 2298–2302.

Hudson, V.N., M.K. Loze, and B.K. Middleton. April 1986. Measurement of Error Rates in a Digital Recording System. *Intl. Conf. on Video, Audio and Data Rec.* IEE Proc. #67, pp. 177–183.

Kerwin, G.J. Sept. 1990. Insitu Measurement of Flyheight Variations in Magnetic Storage Devices Through Use of Sample Margin Signal Processing Techniques. *IEEE Trans. Magn.* MAG-26 (5): 2427–29.

Lutz, E., and K. Troendle. March 1980. Mean Value, RMS Value and Probability Density of Phase Jitter in Digital Transmission Systems (in German,) *Archiv für Elektr. Übertr.* 34 (3): 104–110.

Nunnelley, L., R.D. Harper, and M. Burleson. Sept. 1987. Time Domain Noise-Induced Jitter: Theory and Precise Measurement. *IEEE Trans. Magn.* MAG-23 (5): 2383–85.

Nunnelley, L.L., M.A. Burleson, L.L. Williams, and I.A. Beardsley. Sept. 1990. Analysis of Asymmetric Deterministic Bitshifts Errors in a Hard Disk File. *IEEE Trans. Magn.* MAG-26 (5): 2306–08.

Perkins, T., and R. Nakazawa. Feb. 1992. Paraphase Signal Analyzer V3.0. *A White paper from Applied Magnetic Corp.* 57 pages.

Petit, R.D.l. Feb. 1989. Margin Testing of Digital Videotape Recorders. *SMPTE Journal.* 98 (2): 128–31.

Schwarz, T.A. Sept. 1980. A Statistical Model for Determining the Error Rate of the Recording Channel. *IEEE Trans. Magn.* MAG-16 (5): 634–36.

Sousa, J.L. March 1991. Apply Square Waves to Simplify Measuring Frequency Response. *EDN* 36 (7): 151–157.

Using Histograms for Jitter and Noise Measurements. *Tektronix Tech. Note*, Jan. 1990, Note 47W-7804, 5 pages.

CHAPTER 24
MODULATION CODES

The Institute of Electric and Electronic Engineers defines a code as "a plan for representing each of a finite number of values or symbols as a particular arrangement or sequence of discrete conditions or events." Thus, coding is the process of transforming messages or signals in accordance with a definite set of rules.

The stream of 0s and 1s from a digital signal source represents data that has been coded into a binary format. The original data can represent digitized measurements, sound, video, data files, etc. A bit is a *bi*nary dig*it*.

The raw signal is not well-suited for recording and subsequent playback from a magnetic tape or disk; there is no bound on the dc content in a binary data stream, and the recording channel will not transmit (store and play back) dc levels.

The solution is modulation of the data into a more suitable code format that is dc-free. At the same time, modulation offers a higher packing density so more data can be recorded per cm of track length. This form of coding is a modulation process where one pattern of transitions is changed into another pattern more suitable for magnetic recording. It is assumed that any error-correction coding is done prior to modulation coding (see the next chapter for error-correction and detection coding).

Finding a digital code that maximizes the net linear density while maintaining satisfactory data reliability is a challenging objective. There exists a host of codes, and each is touted as being superior to the others. These claims may be true for certain applications, and for a given recorder and its setup.

Codes commonly used in data communication are RZ (return to zero), NRZ (no return to zero), and biphase. Further definitions are made by suffixes:

- -L (level) means that logical values are represented by different levels or transitions.

- -M (mark) means that logic 1 is defined as "mark" and is represented by a level transition, with the logical 0 defined by no transition.

- -S (space) means that logic 0 is defined as "space," and is represented by a transition, while logic 1 is represented by no transition.

We can appreciate the various codes if we follow the rationale for their developments.

DEVELOPMENT OF DC-FREE CODES

We have, in a couple of decades, come from a very simple RZ (return-to-zero) code to a sophisticated family of GCR (group-code recording) and RLL (run-length-limited) codes.

Foremost in our choice of a code lies the demand that it be essentially dc-free, because the magnetic recording channel does not reproduce dc. Next is the desire to stuff as much data as possible into a code for maximum use of the recorder's signal bandwidth. And in our attempts to satisfy these two demands, we must still have a code that is easy to manage in regard to clock extraction, synchronization, and signal detection.

RZ CODE

The earliest code is the RZ code (return-to-zero), which uses a positive pulse for a 1 and a negative pulse for a 0. Each pulse produces two output voltages on playback, where the first can be used to drive the clock for strobing to see if the pulse is positive (a 1) or negative (a 0). The code is, in other words, self-clocking. The RZ code is shown in the top line of Fig. 24.1.

The flux changes are from the zero level to saturation and back, and a low data density results because each bit requires two transitions. Another drawback is that you cannot record over old RZ data with new data; the tape must first be erased.

Digital logic circuits, including A/D converters, do not produce separate pulses for 1s and 0s. The signals are continuous trains of positive voltages for 1s and zero voltage for 0s (or vice-versa in some circuits), as suggested by the NRZ-L code.

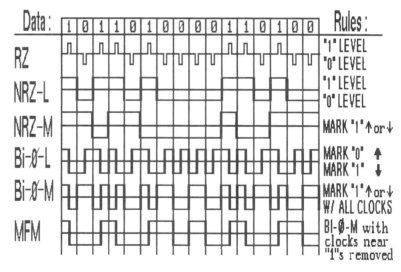

FIGURE 24.1 The initial development of codes.

NRZ-L CODE

We can record the digital signal as positive magnetization for all 1s and negative magnetization for all 0s. This method is known as no-return-to-zero, low, or NRZ-L, and is shown in the second line in Fig. 24.1.

The transition density of NRZ-L is less than half the density of RZ, but it is not self-clocking. The window for signal detection must be operated by an external oscillator that essentially is free-running and then rephased (or synchronized) as often as possible; the NRZ-L code has no built-in assurance that synchronization takes place often enough. The maximum time between transitions could be infinity.

The NRZ-L code is also prone to errors. If no flux change is detected at a clock timing, then the read logic assumes that the next bit is the same as the preceding bit. Now, if a string of 1s (or 0s) is recorded, then a spurious noise spike can make the logic think that a 0 (1) was present. The detection logic will continue calling every 1 a 0 and every 0 a 1 until another noise spike bumps it back on track. This is called error propagation, and must be prevented so that we can distinguish between 1s and 0s.

NRZ-M CODE (ALSO NAMED NRZ-I)

Error propagation is prevented in the NRZ-M code by generating a transition for every 1 and no transition for any 0; vice-versa for NRZ-S (space). NRZ-M was used in many early tape systems. The 1 transition is a flux reversal, without regard for direction; see Fig. 24.1, third line. When the read circuit detects a voltage from a flux change, it "knows" that a 1 is present. No flux change means that the bit is a 0.

This method will only be sensitive to single-bit errors, because subsequent bit detection is independent of previous polarity. But the NRZ-M code cannot, like NRZ-L, distinguish between a dropout and a 0, nor is it self-clocking. The latter problem is solved by having a transition at mid-bit time for each bit: A logical 1 starts out at a 1 level, but transitions to the 0 level at mid-bit time (vice versa for a logical 0). Additional transitions must therefore be inserted between like bits; see line 4 in Fig. 24.1 for this next code, Bi-Phase-L.

BI-PHASE-L CODE (PHASE-ENCODING (PE), SPLIT-PHASE, OR MANCHESTER II)

The inserted transitions change the NRZ-M code into a baseband form of phase-shift keying. It has essentially zero dc content, and is self-clocking. The latter is achieved by having the clock running at twice the bit rate so it can synchronize to all transitions, whether they represent data or are simply transitions inserted between like bits. This code is one of the recommended digital HDDR codes for the telemetry groups working in the aerospace industry (IRIG Doc. 106-86, section 6.11).

BI-PHASE-M CODE (BINARY FM, MANCHESTER, PE-M (PHASE-ENCODING MARK), OR DF (DOUBLE FREQUENCY))

A code similar to Bi-Phase-L was developed by merely adding transitions at the end of each bit cell to the NRZ-M code. This is the most commonly used code in magnetic tape recording and in a large number of communication systems, including fiber optics.

Its popular name FM (frequency modulation) came about by observing that two frequencies are present in the data pattern; the code is a form of frequency-shift keying. The FM code writes two transitions whenever a 1 is recorded. This is, of course, the double frequency and we can consider each 1 to be twice the clock frequency, while a 0 is equal to this frequency.

Appraisals of an FM-system are made on such items as resolution, which is the ratio between the signals at frequencies $2f$ and $1f$. Also, when considering the erasure of old data while new data is written (overwrite), reference is made to the frequencies $1f$ and $2f$.

Before we go on, let us pause and review the codes introduced so far. We paid a price for including a clock with a code: doubled bandwidth of the coded signal. It also affected the frequency spectrum of Bi-Phase-L and Bi-Phase-M. We see that Bi-Phase-M has little or no requirement for channel response to dc. The computer industry uses NRZ-M for tapes at low densities. Higher densities use phase encoding (Bi-Phase-L), while FM (Bi-Phase-M) is common for disks.

If we reexamine the Bi-Phase-M pattern in Fig. 24.1, we notice a couple of extra transitions we really do not need; namely, at the bit cell beginning and ending when writing a 1. The mid-bit 1 transition can serve as a clock synchronization signal. We will still want the clock transitions for a string of 0s.

MFM CODE (DOUBLE DENSITY, MILLER, DELAY MODULATION)

Let us now delete the excess clock transitions, leaving only those between successive 0s (again, a 1 is a positive- or negative-going midbit transition). We have now arrived at the modified Frequency Modulation (MFM) code from the FM code; see Figure 24.1, bottom.

This particular code was patented by Miller of Ampex in 1963, and is therefore often called the Miller code. The data information from FM has been preserved in MFM, and the potential problem of long strings of 0s has been eliminated, but the selfclocking feature for each bit has been lost. The major gain, though, is a reduction of about 2:1 in the required bandwidth and a large reduction in the requirement for dc. The signal spectrum is shown in Fig. 24.2. The signal's stability, in comparison with NRZ, is evident from the oscilloscope pictures in Fig. 24.3. Note the reduction in zero-level drift and baseline galloping when no dc is present.

The upper frequencies required to faithfully record and reproduce the three basic codes (Miller, NRZ-M, and Bi-Phase-M) are shown on the graphs in Fig. 24.2 as 2/3T, 1/T, and 3/2T.

We should note that the transitions in MFM signals have become critical; they are no longer spaced equally, but rather with 1, 1.5, or 2 times the bit cell length. We have deleted some clocking information, which previously reduced the transition spacing to ½ times the bit cell length. That is why we can increase the packing density to almost twice that of Bi-Phase-M for the same amount of intersymbol interference and bit crowding.

In the code developments we have so far achieved a complete removal of the dc content in the FM code, with a small residual amount in the MFM code. The effect of dc in a digital signal was shown in Fig. 24.2. It causes charge build-up on any capacitive element in the read amplifier chain; dc-restorer circuits are required.

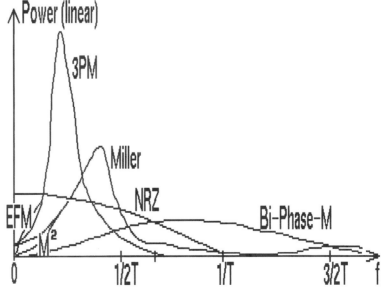

FIGURE 24.2 Signal power spectra for various codes.

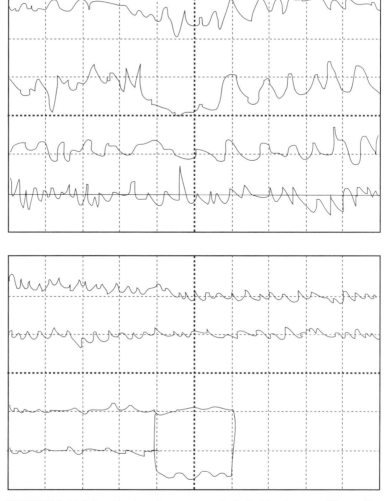

FIGURE 24.3 (a) Codes with a large dc-content cause "galloping" baselines at amplifier coupling stages. (b) Codes with small or no dc content (*courtesy of Ampex Corp.*).

PLACING DC CONSTRAINT ON MFM

The Bi-Phase-L and -M codes are dc-free; they effectively remove any dc component that may be in the NRZ-L data stream supplied by digital electronics.

The price is a doubling of the number of transitions in certain portions of the data stream. We can express this as a halving of the minimum time T_{min} between transitions in the code:

$$T_{min} = (\tfrac{1}{2}) \times T$$

where T is the original time between clock signals in the data.

The net effect is a necessary reduction in the length of the detection window T_d to half its original size:

$$T_d = (\tfrac{1}{2}) \times T$$

The MFM code was derived by deleting all the extra transitions around 1s. This brought T_{min} back to T, but remained the same data. T_d remains at $T/2$ in order to separate the 1 transitions from clock transitions between 0s.

The dc content can be evaluated by the *digital sum variation*, abbreviated DSV. This is the cumulative dc level, measured as the running integral of the area beneath the coded waveform. In computing the DSV the binary levels are assumed to be ±1. If the DSV of the code is bounded, then the code is dc-free. Figure 24.4 illustrates the DSV for MFM and Miller Squared.

The MFM code can be made dc-free by occasionally leaving out a transition: Whenever an isolated 0 is followed by an even number of 1s, then the final 1 will not be marked (i.e., no transition); this modified code is named Miller Squared, and ensures freedom from dc (Mallinson and Miller).

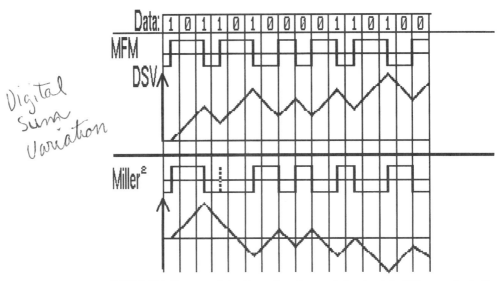

Digital Sum Variation

FIGURE 24.4 The average dc-level in a code can be controlled, such as in the Miller-squared code. The running integral of the dc voltage is measured as DSV (see text).

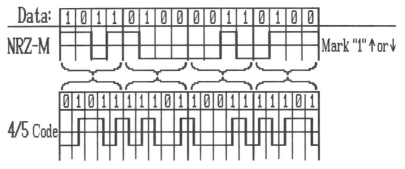

FIGURE 24.5 Generation of the 4/5 GCR code (*MFM with GCR error-detecting code*).

GROUP CODES

When coding is done by translation of blocks of m data bits each into another set of blocks with n bits each, where $m < n$, then the coding process is called group coding (recording), GCR. This technique actually belongs in the next chapter, which is about error detection and correction. However, the magnetic recording industry has historically favored the word "code" when developing techniques that borrowed from the sciences of both modulation and coding.

The rules for the recoding can be simple table look-up, or may be rather complex. The coding is in essence done by breaking the data code up in groups of m bits and mapping them into code words of n bits. The 2^m patterns of m-bit data words are stored in a library, with corresponding patterns for the n-bit code words. Such an arrangement is shown for the popular 4/5 GCR code in Table 24.1, left.

GCR basically uses the MFM format for 1s and 0s, but a restriction is added: there can be no more than two 0s in sequence ($k=2$—see the next section). This guarantees that flux changes occur at least once every three bit cells, and the variable-frequency clock need only be able to lock onto three pulses, corresponding to a succession of 1s, alternate 1s and 0s, or a 1 followed by two 0s.

The GCR-code is also advantageous when error-detection and correction is considered. Sixteen of the available 32 code words are not used; see Table 24.1. If a detected 5-bit word does not match any of the 16 allocated words, then an error has occurred. This is important information, although we do not have sufficient information to correct the error (see next chapter). The reader is referred to papers by Ringkjoeb and by Newton for details about the implementation of the GCR code in a 6250 BPI 9-track tape system.

TABLE 24.1 Look-Up Libraries for the 4/5 and the 3PM Codes

Number	4/5 GCR Data bits	4/5 GCR Code bits	3 PM Data bits	3 PM Code bits
0	0000	11001	000	000010
1	0001	11011	001	000100
2	0010	10010	010	010000
3	0011	10011	011	010010
4	0100	11101	100	001000
5	0101	10101	101	100000
6	0110	10110	110	100010
7	0111	10111	111	100100
8	1000	11010		
9	1001	01001		
10	1010	01010		
11	1011	01011		
12	1100	11110		
13	1101	01101		
14	1110	01110		
15	1111	01111		

A substitution of bits, plus the addition of a parity check bit, is found in a 7/8 code named E-NRZ, Enhanced NRZ. The encoding entails separating the NRZ-L data stream into seven-bit words. Bits 2, 3, 6, and 7 of each word are then inverted, and a single bit added as a parity (enhancement) bit to each of the words. This parity bit is added at the end of each word to make the total number of 1s in an eight-bit word an odd count (Severt). Figure 24.6 shows the implementation.

Zero Modulation (ZM-code) is a group code that modifies a Miller-code to provide dc-free sequences. The preceding bits in a coded block are entered into the evaluation of a block's dc content, and a single transition may be added or deleted to make the net dc in the bit stream equal to zero. For details, see Patel's paper.

A further advanced group code is the 3PM code, where three data bits are coded into six code bits. The encoding is made in accordance with the look-up library shown in Table 24.1, right. At the point

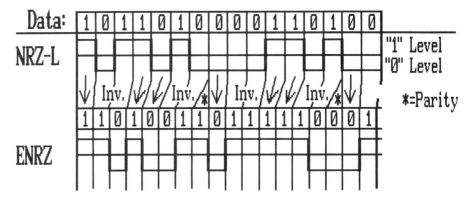

FIGURE 24.6 Generation of the E-NRZ code (NRZ-L with error detection) (*courtesy of Datatape*).

of joining code word bits, a pattern 101 may appear. This is a violation of the code, and it is replaced by 010. In order to make this possible at any time, the last code bit is always a 0.

The coding logic must therefore look ahead two data words (2×3 bits) so any merging resulting in 101 can be corrected prior to writing of the final code words. For details see Jacoby's papers.

RUN-LENGTH-LIMITED CODES AND FORMAL CODE DEVELOPMENT

It is constructive to examine the MFM code as a string of transitions (1s) and spaces (0s), at the bit rate of ($\frac{1}{2}$) $\times$ T. Note that for the MFM code there is a minimum of one 0 and a maximum of three 0s between two successive 1s.

There is no distinction between data and clock transitions, and the decoder must know what the encoder did. Formalizing this approach to describing codes uses the parameters d and k as the respective minimum and maximum number of 0s between 1s.

$$d = < z = < k \tag{24.1}$$

For the MFM code we have: $d = 1$ and $k = 3$. Codes with such (d,k) properties are called RLL (run-length-limited). The constraint d is used to control pulse-crowding effects, while k is used to ensure self-clocking ability (Franaszek; Davidson et al; Norris; Bloomberg).

To meet the (d,k) constraint, m bits of data are mapped into n bits of code, on the average, where $n>m$. We can use the numbers m and n plus the (d,k) to assist in evaluating codes, and to develop new coding schemes. Some of the descriptive numbers are defined in the following paragraphs. Good tutorial papers on the topic are by Watkinson and by Siegel. Early coding is described by Kiwimagi et al., and coding in general communication by Sanders.

Mapping determines a code rate R, which is:

$$R = m/n \tag{24.2}$$

Codes with a high R are less sensitive to timing jitter caused by noise and peak shift than those with small R. The minimum and maximum intervals between transitions are:

$$T_{min} = R \times (d + 1) \times T \text{ sec.} \tag{24.3}$$

$$T_{max} = R \times (k + 1) \times T \text{ sec.} \tag{24.4}$$

where T is the original bit cell time in seconds. The detection window T_d for the code is

$$T_d = R \times T \text{ sec.} \tag{24.5}$$

and the clock rate is $(1/R) \times (1/T)$. It is desirable to have a long detection window. The density ratio (or minimum transition interval) is:

$$DR = (m/n) \times (d + 1) = T_{min}/T \tag{24.6}$$

Codes with high DR may provide read signals with a high signal-to-noise ratio. Another ratio is:

$$P = T_{max}/T_{min} = (k + 1)/(d + 1) \tag{24.7}$$

Codes with a high P tend to have large peak-shifts, poor self-clocking ability, and a large spectral dc component.

RANKING OF CODES AND THE ERROR BUDGET

The matter of selecting a code for the recording and playback of digital data appears to involve more trade-offs and weighting of decisions than most engineering jobs require. The reader has been presented with a variety of codes and a fair amount of definitions of variables to describe them (m, n, k, d) and other definitions for comparing codes (R, T_{min}, T_{max}, T_d, DR, P, DSV). See Table 24.2 for a listing of codes.

Selection of a code must be based on the characteristics of the hardware (heads, media, resolution, and time error budget) such as RTE (real time errors) and IRT (interreversal time). The resulting maximum possible packing factor:

$$PF = PW50/(bit\ period) \tag{24.8}$$

has been analyzed for several codes by Mackintosh. All codes degrade under increasing time errors. Their maximum PF are within ±10 percent of each other for an RTE of approximately 16 percent $PW50$, so they perform quite similarly. The moral is: Select a code, make the best out of the system—and you will be near optimum performance (other papers on ranking of codes are by Huber, by Osawa et al., and by Gambe et al.).

The RTE is typically composed of many factors, such as skew, time displacement errors, incomplete erasure, crosstalk, print-through, medium noise, component tolerances and phase-locked loop errors.

A figure of merit has been suggested (Watkinson) as the product of the density ratio DR (from equation 24.6) and the detection window T_d (from equation 24.5):

$$Figure\ of\ Merit = FoM = DR \times T_d \tag{24.9}$$

Figure 24.7 displays the calculated FoM, compared with Mackintosh's findings.

APPLICATIONS

Computer tape drives and streamers use 4/5 GCR in the 6250 BPI drives, while computer disk drives use MFM in double-density floppies and in rigid disk drives. Many papers pertaining to these encoding methods have already been mentioned.

TABLE 24.2 A table of various codes with a listing of code parameters

Characteristics: Data: 1 0 1 1 0 1 0 0 0 0 1 1 0 1 0 0	d	k	m	n	$\frac{m}{n}$	T_{mi}	T_{ma}	Hi	Cl. R.	Max DSV	Max DC
RZ											
NRZ-L	0	∞	1	1	1	T	∞	T	$\frac{1}{T}$	$\pm\infty$	± 1
NRZ-M	0	∞	1	1	1	T	∞	T	$\frac{1}{T}$	$\pm\infty$	± 1
Bi-φ-L	0	1	1	2	$\frac{1}{2}$	$\frac{T}{2}$	T	$\frac{T}{2}$	$\frac{2}{T}$	$\pm T$	$\langle 0 \rangle$
Bi-φ-M	0	1	1	2	$\frac{1}{2}$	$\frac{T}{2}$	T	$\frac{T}{2}$	$\frac{2}{T}$	$\pm T$	$\langle 0 \rangle$
MFM	1	3	1	2	$\frac{1}{2}$	T	$2T$	$\frac{T}{2}$	$\frac{2}{T}$	$\pm\infty$	$\pm\frac{1}{3}$
E-NRZ	0	7	7	8	$\frac{7}{8}$	$\frac{7T}{8}$	$7T$	$\frac{7T}{8}$	$\frac{8}{7T}$	$\pm\infty$	$\pm\frac{3}{4}$
4/5 CODE	0	2	4	5	$\frac{4}{5}$	$\frac{4T}{5}$	$\frac{12T}{5}$	$\frac{4T}{5}$	$\frac{5}{4T}$	$\pm\infty$	$\pm\frac{2}{5}$
Miller²	1	5	1	2	$\frac{1}{2}$	T	$3T$	$\frac{T}{2}$	$\frac{2}{T}$	$\frac{\pm 5}{2T}$	0
ZM	1	3	1	2	$\frac{1}{2}$	T	$2T$	$\frac{T}{2}$	$\frac{2}{T}$	$\frac{\pm 3}{2T}$	0
3PM	2	11	3	6	$\frac{1}{2}$	$\frac{3T}{2}$	$6T$	$\frac{T}{2}$	$\frac{2}{T}$		

Multilevel encoding can increase the stored information by providing $2n$ signal levels within each data bit, where n is the number of levels. The application to a disk system have been examined and judged unsuitable in most practical systems, due to SNR limitations (Mackintosh and Jorgensen), while it nevertheless has been implemented with complex channel modulation and error-correction and detection-coding (Price et al.). The topic of maximum packing density on a disk system is discussed in a paper by Wood et al.

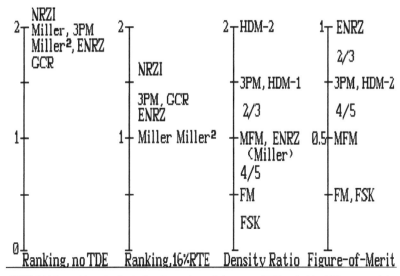

FIGURE 24.7 Ranking of codes: Left scales are for no timing errors (TDE) and with a 16% error rate (*after Mackintosh*); Right scales show ranking in accord with DR and FoM (*after Watkinson*).

HIGH-BIT-RATE DATA—HDDR

Several manufacturers of precision instrumentation recorders have developed codes for modulation of the data stream, and are using different methods:

Ampex (Kelly)	Miller Squared
Datatape (Staff)	ENRZ-L
EMI (Howard and Nottley)	3PM
Honeywell (Meeks)	NRZ-L
Sangamo (Stein and Kessler)	RNRZ-L

This makes interchange of recorded tapes impossible. The ANSI X3B6 and the ISO TC97-SC12 committees are still trying to make decisions on a standard, but the differences between the different systems are not all that subtle (Reynolds).

The Range Commanders' Council has recently issued IRIG Standard 106-86 with a choice between Bi-Phase-L for standard data gathering and RNRZ-L for extended-bandwidth data.

RNR-L is a randomized code related to convolutional codes. The randomizing is a form of polynomial division (modulo 2) of the data stream by a special number defined by a shift register circuit. It produces an output resembling, to a degree, a random sequence. This sequence, however, is an encoded form of the original sequence and can be decoded by the corresponding polynomial multiplication. The concept originated in scramblers for communication (Savage).

Figure 24.8 shows the standard 15-cell randomizer and derandomizer. The reader may easily test, on paper, how these work by applying a data stream containing all 0 values except for a single 1 (Stein and Kessler). The choice of Bi-Phase-L was made for the following reasons (IRIG 106-86, Appendix D, paragraph 3.6):

- Only a small portion of the total signal energy occurs near dc.

- The maximum time T_{max} is T, one bit period.

- The symbols for a 1 and a 0 are antipodal; that is, the symbols are exact opposites of each other (i.e., which is better than FM, where there is no symbol for 0s). Therefore, the bit-error probability-versus-SNR performance is optimum.

- Experiments further supported Bi-Phase-L because it is less sensitive to misadjustments of ac-bias and read equalizers than most other codes. Bi-Phase-L also performs well at low tape speeds and low bit rates.

RNRZ-L was chosen over the other codes for several reasons:

- RNRZ-L requires approximately one-half the bandwidth of Bi-Phase-L.
- The symbols for a 1 and a 0 are antipodal.
- The RNRZ-L decoder is self-synchronizing.
- The RNRZ-L data bits can easily be decoded in the reverse mode of tape playback direction.
- The RNRZ-L data are bit-detected and decoded using a clock at the bit rate. Therefore, the phase margin is much larger than that of codes that require a clock at twice the bit rate for bit detection.
- The RNRZ-L code does not require overhead bits.

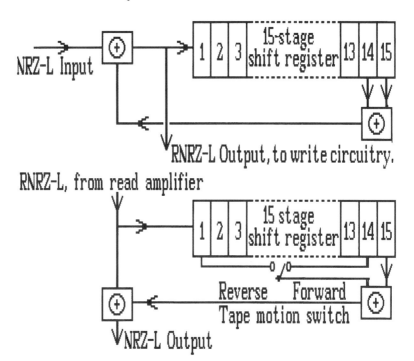

FIGURE 24.8 Randomizers for generation and decoding of RNRZ-L codes (*courtesy of Sangamo/Fairchild Weston*).

It is recognized that the RNRZ-L can have a large dc-content. Therefore, reproducing data at tape speeds that produce PCM bit rates less than 200 kB/s is not recommended unless a bit synchronizer with a specially designed dc and low-frequency restoration circuitry is used.

The codes listed above are used on multichannel high-speed recorders. High data rates can also be stored on rotating-head machines; suitable channel codes are discussed by Newby and Yen, and by Coleman et al.

The field of audio recording employed MFM in the first digital recordings. The compact disks use an 8/14 group code, while the masters for these disks are tapes made on rotary-head machines with transformer coupling to the heads, using FSK (Frequency Shift Keying). Other codes are constantly being conceived and evaluated, such as HDM-1 (Doi) and 8/10 coding (Fukuda et al.; Tazaki et al.). Papers from an AES convention in 1982, plus papers by Moriyama et al. and by Bellis, provide further insight into channel coding for digital audio.

In video recording binary FM is used for the SMPTE/BEU time code, and RNRZ-L in other instances; the latter code appears quite resistant to jitter. Four papers listed in the references section provide a good insight into the particular problems of digital video recording.

New modulation codes continue to be proposed. As of the writing of this 4th edition, the outline and principles in this chapter are still followed. Some of the latest code applications are 8/14, very much in use for the CDs, and for the D-3 and D-5 digital video tape recorders. The code is also named EFM (eight-to-fourteen modulation). Its *FoM* for Fig. 24.7 is 0.65, slightly above the 4/5 group code.

There are 256 combinations of input data bits (2^8), whereas there are 16,384 code bit combinations (2^{14}). The run-length-limited requirements to the 8/14 code is that there shall be no more than 5 identical bits in the first 6 bits of the code, and not more than 6 in the last 7 bits (Watkinson 1994). This leaves only 118 allowed codewords, and codes that are not dc-free will have to be accepted. They must be accounted for and the DSV from time to time brought back to zero if it is drifting. For details, see the book by Watkinson.

A 2,7 variable-length group code has been used in IBM disk drives (Patent: Franaszek 1972). With $d = 2$ there are always at least two zero channnel bits between ones, dividing the channel frequency by three. The density ratio is 1.5 and the *FoM* is 0.75. Also, a new book by Kees A. Schouhamer Immink has been added to the recording engineers' library. It is highly recommended for further studies into channel coding.

BIBLIOGRAPHY

Kiwimagi, R.G., J.A. McDowell, and H.H. Ottesen. Sept. 1974. Channel Coding for Digital Recording. *IEEE Trans. Magn.* Vol. MAG-10, No. 3, pp. 515–518.

Sanders, L.S. Jan. 1982. Pulse Codes in Serial Data Communications. *Computer Design*, Vol. 21. (1): pp. 203–210.

Schouhamer, Immink, K.A. 1990. Coding Techniques for Digital Recorders. Prentice Hall. 297 pages.

Siegel, P.H. Sept. 1985. Recording Codes for Digital Magnetic Storage, *IEEE Trans. Magn.*, Vol. MAG-21, (5): pp. 1344–1349.

Waggener, B. 1995. Pulse Code Modulation Technique. Van Nostrand Reinhold, 368 pages.

Watkinson, J.R. 1994. Chapter 4 Channel Coding, in book The Art of Data Recording. Focal Press. 518 pages.

Watkinson, J.R. Channel Codes for Digital Recording. Intl. Conf. on Video, Audio, and Data. 1986. IERE Publ. No. 67, pp. 129–140.

DEVELOPMENT OF DC-FREE CODES

Lindholm, D.A. Sept. 1978. Power Spectra of Channel Codes for Digital Magnetic Recording. *IEEE Trans.* Magn. Vol. MAG-14, No. 5. pp. 321–324.

Mallinson, J.C., and J.W. Miller. Apr. 1977. Optimum Codes for Digital Magnetic Recording. *Radio and Electronic Engr.* Vol. 47, No. 4. 4 pages.

Patel, A.M. July 1975. Zero-Modulation Encoding in Magnetic Recording. *IBM Jour. Res. and Dev.* Vol. 19, pp. 366–378.

Schouhamer Immink, K.A., and U. Gross. 1982 Optimization of Low Frequency Properties of EFM Modulation. *Intl. Conf. on Audio, Video, and Data.* IERE Publ. No. 54, pp. 375–383.

RUN-LENGTH-LIMITED CODES; FORMAL CODE DEVELOPMENT

Cohn, M., and G.V. Jacoby. Nov. 1982. Run-Length Reduction of 3PM Code via Look-Ahead Technique. *IEEE Trans. Magn.* Vol. MAG-18, No. 6. pp. 1253–1255.

Davidson, M., S.F. Haase, J.L. Machamer, and L.H. Wallman. Sept. 1976. High Density Magnetic Recording Using Digital Block Codes of Low Disparity. *IEEE Trans. Magn.* Vol. MAG-12, pp. 584–586.

Franaszek, P.A. July 1970. Sequence-State Methods for Run-Length-Limited Coding. *IBM Jour. Res. and Dev.* Vol. 14, pp. 376–383.

Jacoby, G.V. Sept. 1977. A New Look-Ahead Code for Increased Data Density. *IEEE Trans. Magn.* Vol. MAG-13, No. 3. pp. 1202–1204.

Jacoby, G.V., and R. Kost. Sept. 1984. Binary Two-Third Rate Code with Full Word Look-Ahead. *IEEE Trans. Magn.* Vol. MAG-20, No. 5. pp. 709–714.

Newton, M. July 1981. GCR Increases Data Recording Rates and Reliability. *Digital Design.* pp. 36–39.

Norris, K., and D.S. Bloomberg. Channel Capacity of Charge-Constrained Run-Length Limited Codes. *IEEE Trans. Magn.* Nov. 1981. Vol. MAG-17, No. 6. pp. 3452–3455.

Ringkjob, E.T. May 1975. Achieving a Fast Data Transfer Rate By Optimizing Existing Technology. *Electronics.* May 1975, pp. 86–91.

Severt, R.H. May 1980. Encoding Schemes Support High Density Digital Data Recording. *Computer Design.* Vol. 19, No. 5. pp. 181–190.

APPLICATIONS; THE CHOICE OF A RECORDING CODE

Gambe, HY., T. Matsumura, and T. Matsuda. Aug. 1984. Codec Evaluation Method Based on Measured DC Characteristics. *FUJITSU Scientific and Techncial Journ.* Vol. 20, No. 3. pp. 259–281.

Huber, W.D. Sept. 1980. Selection of Modulation Code Parameters for Maximum Lineal Density. *IEEE Trans. Magn.* Vol. MAG-16, No. 5. pp. 637–640.

Mackintosh, N.D. Apr. 1980. The Choice of a Recording Code. *The Radio and Electronic Engineer.* Vol. 50, No. 4. pp.177193; Also: IERE Conf. Proc. No. 43. Jan. 1979. pp. 77–119.

Osawa, H., S. Tazaki, and S. Andoh. 1982. Performance Comparison of Partial Response Systems for N.R.Z. Recording. *Intl. Conf. on Audio, Video, and Data.* Apr. 1982. No. 54.

COMPUTER DISK DRIVES

Mackintosh, N.D., and F. Jorgensen. Nov. 1981. An Analysis of Multi-Level Encoding. *IEEE Trans. Magn.* Vol. MAG-17, No. 6. pp. 3329–3332.

Price, R., J.W. Craig, H.E. Melbye, and A. Perahia. Sept. 1977. An Experimental, Multilevel, High Density Disk Recording System. *IEEE Trans. Magn.* Vol. MAG-14, No. 5. pp. 315–317.

Wood, R., S. Ahlgrim, K. Hallamasek, and R. Stevenson. Sept. 1984. An Experimental Eight-Inch Disc Drive with One-Hundred Megabytes per Surface. *IEEE Trans. Magn.* Vol. MAG-20, No. 5. pp. 698–702.

HDDR

Coleman, C., D. Lindholm, D. Peterson, and R. Wood. 1984. High Data Rate Magnetic Recording in a Single Channel. *Intl. Conf. on Audio, Video, and Data.* IERE Publ. No.9. pp. 151–157.

Howard, J.M., and G.C. Nottley. Sept. 1985. The Application of 3-Position Modulation Coding to Longitudinal Instrumentation Recording. *High-Density Digital Recording.* NASA Ref. Publ. No. 1111. pp. 195–215.

Kelly, J. Sept. 1985. Miller Squared Coding. *High-Density Digital Recording.* NASA Ref. Publ. No. 1111. pp. 127–142.

Meeks, L. Sept. 1985. The Honeywell HD-96 High-Density Digital Tape Record/Reproduce System. *High-Density Digital Recording.* NASA Ref. Publ. No. 1111. pp. 215–230.

Newby, P.S., and J.L. Yen. Sept. 1983. High Density Digital Recording Using Videocassette Recorders. *IEEE Trans. Magn.* Vol. MAG-19, No. 5. pp. 2245–2252.

Reynolds, S. Sept. 1985. High-Density Digital Recording (HDDR) Users Subcommittee Evaluation of Parallel HDDR Systems. *High-Density Digital Recording.* NASA Ref. Publ. No. 1111. pp. 281–292.

Savage, J.E. Feb. 1967. Some Simple Self-Synchronizing Digital Data Scramblers. *Bell Systems Tech. Jour.* Vol. 46, pp. 449–487.

Staff at Datatape. 1985. Parallel Mode High-Density Digital Recording: Technical Fundamentals. *High-Density Digital Recording.* NASA Ref. Publ. No. 1111. pp. 143–194. (Also available in a book of the same title from Datatape in Pasadena, Ca.)

Stein, J.A., and W.D. Kessler. Sept. 1985. The Development of a High-Performance Digital Recording Error Correction System. *High-Density Digital Recording.* NASA Ref. Publ. No. 1111. pp. 231–246.

AUDIO

Bellis, F.A. Oct. 1983. Introduction to Digital Audio Recording. *The Radio and Electronic Engr.* Vol.3, No. 10. pp. 361–368.

Blesser, B., B. Locanthi, and T.G. Stockman (Editors). 1982. Collected Papers from the AES Premiere Conference. *Digital Audio.* Rye, New York. 262 pages.

Doi, T.T. April 1983. Channel Coding for Digital Audio Recordings. *Jour. AES.* Vol. 31, No. 4. pp. 224–238.

Fukuda, S., Y. Kojima, Y. Shimpuku, and K. Odaka. Sept. 1986. 8/10 Modulation Codes for Digital Magnetic Recording. *IEEE Trans. Magn.* Vol. MAG-22, pp. 1194–1196.

Moriyama, Y., K. Yamagata, T. Suzuki, and T. Iwasawa. Nov. 1981. New Modulation Technique for High Density Recording on Digital Audio Discs. An AES Preprint No. 1827(I3). *AES Conv.* 7 pages.

Tazaki, S., T. Kaji, and H. Osawa. 1986. An Analysis of DC-Free Property on Run-Length Limited Code. *Intl. Conf. on Video, Audio, and Data* IERE Publ. No. 67. pp. 151–156.

VIDEO

Baldwin, J.L.E. Apr. 1984. Channel Codes for Digital Video Recording. *Intl. Conf. on Audio, Video, and Data* No.9.

Baldwin, J.L.E. July 1979. Codes for Digital Video Tape Recording at 10 M.bit/Sq. Inch. *Intl. Conf. on Audio, Video, and Data* No. 43. pp. 147–163.

Furukawa, T., M. Sept. 1984. Ozaki, and K. Tanaka. On a DC-Free Block Modulation Code. *IEEE Trans. Magn.* Vol. MAG-20, pp. 878–880.

Heitman, J.K.R. Feb. 1984. Digital Video Recording: New Results in Channel Coding and Error Protection. *Jour. SMPTE.* Vol. 93, No. 2. pp. 140–144.

CHAPTER 25
ERROR DETECTION AND CORRECTION CODING

Communication systems are subject to the affects of noise added to signals, or disruptions of a signal. The latter are in magnetic recording known as dropouts. A common measure of the system's freedom from these ill effects is a rating called bit-error-rate, abbreviated BER. The BER number states what the probability of errors in the data stream are. For instance, if BER = 10^{-6}, then an average of one bit out of one million may be in error, either missing or changed by a noise spike.

We will look at ways of adding extra bits to the signal and thereby making it possible to detect if an error occurred, and possibly correct it. This will lead us to the important concept of Hamming distance, and the design of block codes.

The technique of error correction will not eliminate errors, nor may it provide a drastic change of a BER rating, say 10^{-6} to 10^{-12} (from one error in a million bits to one error in one trillion bits). We will also examine bit error patterns, and find methods of predicting the BER before and after error detection and correction (EDAC) has been implemented into the system.

The chapter will provide the reader with insight into the concept of error-detecting and -correcting codes, using the basic block codes as examples. Longer codes can only be studied and appreciated if the reader is well-versed in abstract algebra (rings, fields, convolution, etc.). They are clearly beyond the scope of this book.

ERROR DETECTION AND CORRECTION

Error detection and correction are accomplished by adding information to the transmitted data. Take for instance a message you receive over the phone: if you are uncertain about its meaning, you may ask to have it repeated. And to make absolutely clear what was meant, a third repeat may be requested.

This very concept forms the basis for EDAC. Figure 25.1 shows a communication channel that transmits the status of a two-state condition; this can be the range of a temperature, high or low with regard to a reference, or the on or off condition of a relay controlled by a sensor circuit. We decide to transmit a 1 if the relay is closed, a 0 if it is open.

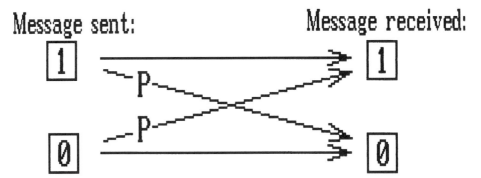

FIGURE 25.1   Signal transmission of an unprotected message.

The possibility now exists that a 1 is received as a 0, and vice versa. The *transition probability p* for this happening is equal to the transmission system's bit error rate, BER. There is no way of telling whether the received signal is correct or not.

We can improve this by always repeating the message, and will therefore transmit an 11 for the on condition, and 00 for the off condition; see Fig. 25.2. We have, in terms of coding language, added a *parity check bit*. Also, we have changed the *message words* 1 and 0 into *code words* 11 and 00, respectively.

A single error may occur with a probability of p. We will then receive a 10 or an 01, and we will know that an error occurred because neither belongs to the selected codewords of 11 and 00. But we cannot tell what the original message was; i.e., we cannot perform error correction. This is why we, in a telephone conversation, might ask what was said.

The probability of two errors, which will invert the transmitted messages without detection ($00 \rightarrow 11$, or $11 \rightarrow 00$), is equal to p^2 because the errors in two bits are independent of each other. The possibility of any double error is therefore extremely remote. (We will find, however, that the BER for a 2-bit word is different from the BER for a 1-bit word. See the example for a 7-bit versus a 4-bit word later in this chapter.)

Let us continue with our "intuitive" coding scheme—we will add another bit, so now three bits are transmitted for each single-bit message. The code words that belong to the 1 or the 0 will be 111 or 000. One, two, or three errors may occur, and the received code words may have any of the eight (2^3) patterns shown in Fig. 25.3.

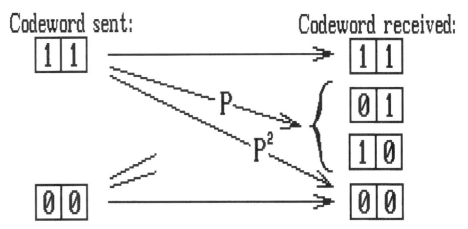

FIGURE 25.2   Signal transmission of a single-repeat code.

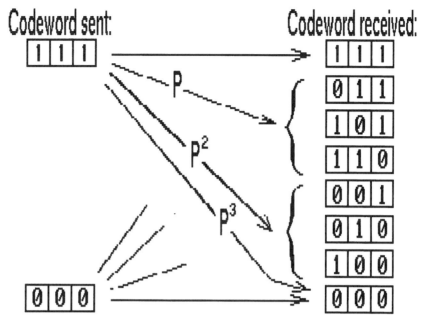

FIGURE 25.3 Signal transmission of a double-repeat code.

A single error in the 111 code word will result in the received words no. 2 through 4, and the decoder will assign 111 to any of these. It will do so because that is the *most likely* choice. The similar argument holds for the 000 code word. We have thus detected and corrected for one error.

Two errors will result in wrong decisions by the decoder, and an error in the decoded message results ($0 \rightarrow 1$, or $1 \rightarrow 0$).

More bits could be added to each of the example's two codewords, but we are already dealing with a rather inefficient way of providing error protection: Each bit is replaced with three bits, so the overhead is 200%, or the code is 33% efficient. It would be better to take a few bits at a time, and replace them with a longer set of bits where the extra bits will be used to detect for errors, and possibly correct them.

This changes the coding from the *repeat coding method* to the commonly used *block coding*, where *k* message or data bits are replaced by *n* code bits by adding *r check bits*. This method is similar to the 4/5 GCR and the 3/6 3PM coding schemes, except this time we must select additional bits so that they will assist in error detection and correction instead of reducing the code's DC content to zero (this is to say that the modulation encoding at times can merge with EDAC).

A POINT OF VANISHING RETURN

The incoming stream of data bits is divided up into message words, which in turn are altered by the addition of check bits. The codewords would quickly jam the encoder unless the data words are compressed in time so they (with the added check bits) occupy the same time frame as the original dataword.

This shortened time per bit corresponds to a widening of the signal bandwidth for the PCM signal, which results in a lowering of the SNR. Let us assume that SNR for the channel is 13.5 dB, which results in a BER of 10^{-6}. We now add one check bit to each message bit, and the time allotted for each bit must be cut in half. This means a reduction of 3 dB in the SNR to 5 dB, and the new BER (or p) is 0.5×10^{-3} (see Fig. 23.12). This is the new probability of a single error (which we can detect, though).

The probability of 2 errors, which we cannot detect, is $p \times p = 0.25 \times 10^{-6}$, so we have only improved this system by a factor of four! Much longer codewords will be needed, as we shall see later.

First, *a word of caution* before we proceed: There exists an unfortunate overlap in the use of symbols in coding theory, as shown in Fig. 25.4. The reader who plans to read further must be aware that the symbol n is always used for the number of bits coming out of an encoder. Then look for the use of n or d in either an (n,k) code (EDAC), or the (d,k) constraint in RLL codes, with corresponding rates k/n and m/n.

Let us now evaluate the (3,1) code (this designation is from the terminology (n,k) code, where k incoming bits are changed into n transmitted bits). The time allotted per bit in the codeword of the (1,3) code is reduced by a factor of 3, resulting in a 4.8 dB-less SNR. Continuing our previous example, we find the error probability corresponding to SNR = 8.8 dB equals $p = 2 \times 10^{-3}$. No final error results for one bit error in any of the three transmitted bits.

The probability of 2 bit errors equals $3 \times p^2 = 12 \times 10^{-6}$; the factor 3 accounts for the fact that there are 3 different patterns for two errors in three bits (this is a coincidence; b errors in a pattern of n bits results in a multiplier of $n!/(b!n-b)!$). This result is $12/(0.25) = 48$ times worse than the case where only one parity bit was added. The (3,1) code is therefore, at the SNR levels used, worse than no coding. We will later return to examine the value of longer codes.

The simple example with only two signal conditions helped us in getting a feel for the concept of error detection and correction. We will now expand the technique of adding check bits, and this time do it for a multiple bit data word.

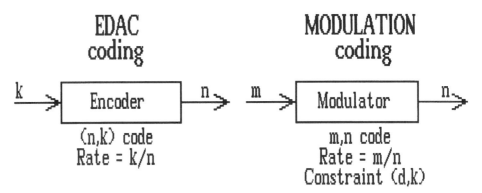

FIGURE 25.4 Conflicting use of symbols in two related disciplines.

BLOCK CODES

The incoming data stream is chopped up into *message words* **u** each of length k bits (following the notation used in Lin and Costello's book). An $r = (n-k)$ bits long *check word* **r** is generated, and added to the message word, forming a *code word* **v** of length n bits; Fig. 25.5 illustrates the process. (Lin and Costello place the check word ahead of the message, while earlier textbooks did the opposite. It doesn't really matter which way it is done; Lin and Costello's method does make the next step into cyclic codes smoother.)

We can write the code generation as:

$$\{n\text{-}k \text{ check bits}\} + \{k \text{ message bits}\} \rightarrow \{n \text{ code bits}\}$$

The task at hand is to find the n-$k = r$ check bits that will allow a decoder to determine and possibly correct errors. This is a formidable mathematical task for longer codes, and coding specialists must be apt mathematicians in advanced abstract algebra to delve into this challenge.

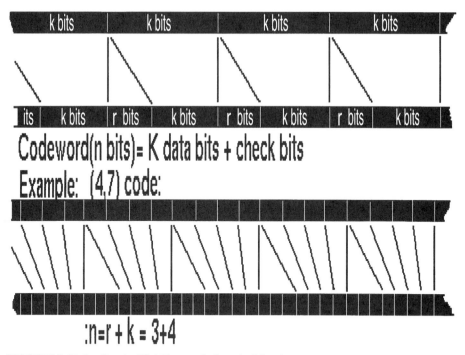

FIGURE 25.5   Block coding, simplified. The example shows the (4,7) code.

The term *abstract* reflects that the mathematicians that developed the systems divorced themselves from the physical world in order to develop logical, exact mathematical systems. Here the symbols like a, b, A, B, C, etc. do not stand for anything physical; they do not represent current, voltage, or time, for example.

The idea is to see whether or not a physical system, and in the present case, a coding system, falls within the definitions of one of the abstractly developed mathematical systems. If it does, then we can apply the rules of operations (addition, subtraction, multiplication, etc.) for that system, and remain assured that the outcome of the operation stays within the system, and therefore is correct.

The introduction to error detection leads us into the world of multi-dimensional vectors, and we will be required to carry out certain operations on these vectors such as addition and multiplication. The result of these operations must fall within the set of vectors that are allocated to the system.

A couple of examples will allow us to appreciate the processes without getting too deeply involved (there are plenty of references for those who wish to investigate further).

We saw earlier how addition of at first one bit, then another bit, took us from a one-dimensional to a three-dimensional solution set. This points toward treating the collection of bits as vectors; i.e., each n bits word is an n-dimensional vector. This sort of presentation is clearly beyond physical comprehension for humans in a 3-D world, but rather a delight for mathematicians. How can he/she add, multiply, or otherwise operate with these vectors and be sure that the results are correct?

Consider first our familiar decimal number system. We can add, subtract and multiply without much hesitation—the result is always right. But only a few hundred years ago, subtraction did not always exist because negative numbers were unheard of. Numbers were used to keep track of quantities of items, dimensions, and so forth. Negative numbers were invented by Italian bankers to provide loans to their customers.

We all apply certain restrictions though, subconsciously. If we add 4 dogs to 6 cats we know the result is 10 animals, and we would never dream of multiplying the two! Then how can we be sure of the operations we do on vectors that represent words? Our present concern is the application of algebra to codewords, which we block out in sets of n bits; how do we represent them in mathematical terms?

We are, first of all, only concerned about a number set consisting of a 0 or a 1. (There are multilevel codes but they are rarely employed in magnetic recording due to the poor signal amplitude stability of typically 10%.)

This small mathematical system works in accordance with the so-called *Modulo-2 arithmetic*, where the outcome of any operation is determined by dividing the result with 2—and the remainder used for the result. The symbol for this operation is $\oplus$.

For example:

$$0 \oplus 1 = 1/2 = 0 + \text{remainder } 1 = 1$$
$$3 \oplus 4 = 7/2 = 3 + \text{remainder } 1 = 1$$
$$3 \oplus 1 = 4/2 = 2 + \text{remainder } 0 = 0$$

The addition rules for the *finite set* (0 and 1) are:

$$0 \oplus 0 = 0$$
$$0 \oplus 1 = 1$$
$$1 \oplus 0 = 1$$
$$1 \oplus 1 = 0$$

For multiplication we find, similarly:

$$0 \times 0 = 0$$
$$0 \times 1 = 0$$
$$1 \times 0 = 0$$
$$1 \times 1 = 1$$

The results are easy to remember if we just keep in mind a clock with only two hours: 0 where 12 normally is located, and 1 where 6 is.

We then immediately understand why in modulo-2: $1 = -1$; add or subtract to your hearts delight, don't worry about signs. The remainder is always a 0 or 1.

CODEWORDS AND VECTOR SPACE

Let us now define a mathematical description for the block of 0 and 1 bits that make up a code word. We will define a *vector* in the sense it is used in algebra and coding theory, and will refer to it as a "vector $\mathbf{v}$ having n components over a field F."

We define a vector (in two dimensions) to be a pair of numbers $\{x, y\}$. The numbers, or scalars, x and y may be real or complex, or indeed elements of any field, and the set of vectors over this field consists of all possible ordered pairs (x, y). Two vectors are equal if and only if both components x and y are equal.

A three-dimensional vector is defined as an ordered triad of the numbers (x, y, z), where the components may be real, complex, or in any field.

The n components that make up the n-dimensional vector $\mathbf{v}$ are of course the n bits in the code word, and these components of the field F, the Galois field $(0, 1)$, form what we can call a n-tuple:

$$\mathbf{v} = \{v_1, v_2, v_3 ..., v_n\}$$

The vector given by all zero components except the ith, which is 1, is denoted by $\mathbf{w}$, and any vector may be expressed in the form:

$$\mathbf{v} = a_1\mathbf{w}_1 + a_2\mathbf{w}_2 +..., a_n\mathbf{w}_n$$

Set $\mathbf{w}_1$, $\mathbf{w}_2$..., $\mathbf{w}_n$ is called a *base* for the vectors. Normally we use binary values for the a_n terms. We are now defining the *rules of operation of vectors*:
Addition:

$$\mathbf{v}_1 + \mathbf{v}_2 = \{a_1 + b_1, a_2 + b_2..., a_n + b_n\}$$

Scalar multiplication:

$$c \times \mathbf{v}_1 = \{c \times a_1, c \times a_2..., c \times a_n\}$$

Product of two vectors:

$$\mathbf{v}_1 \bullet \mathbf{v}_2 = a_1 \times b_1 + a_2 \times b_2 +..., + a_n \times b_n$$

Our final definitions pertain to a *vector space over a field F*: It is a commutative group under addition (with components from $(0,1) = GF(2)$, the Galois field over 2, with modulo-2 arithmetic). It is closed under the operation of scalar multiplication, and the distributive and associative laws hold.

If we add a couple of 7-dimensional vectors we could write something like:

$$\{1,1,0,0,1,0,1\} + \{0,1,1,0,1,0,1\} = \{1,2,1,1,2,0,1\}$$

The two bilevel vectors have now become a trilevel vector, something our binary system is unable to handle. From abstract algebra we learn that we must apply *modulo-2 algebra* to force the solution into the binary system. A modulo-2 operation takes the outcome and divides it by two, and uses the remainder as its result. We did correctly operate on corresponding vector components in the example above, but should have used the modulo-2 operation on the end result:

$$\{1,1,0,0,1,0,1\} + \{0,1,1,0,1,0,1\} = \{1 \oplus 0, 1 \oplus 1, 0 \oplus 1, 0 \oplus 0, 1 \oplus 1, 0 \oplus 0, 1 \oplus 1\}$$
$$= \{1,0,1,0,0,0,0\}$$

HAMMING DISTANCE: WEIGHT

Let us review the on-off example before we go on to consideration of longer codes. The choice between a 1 or a 0 can be considered a one-dimensional matter—the selection of a point. With the addition of one check bit, the choice is two-dimensional—the selection of a vector. It must be either the zero vector $\{0,0\}$, or the vector to the point $\{1,1\}$. This is illustrated in Fig. 25.6, where also the erroneous vectors $\{1,0\}$ and $\{1,1\}$ are shown.

Going on to the three-bit code words, the selection is a three-dimensional choice. Vectors $\{0,0,0\}$ and $\{1,1,1\}$ correspond to the allowed code words; the other six vectors do not. Or better expressed: $\{0,0,0\}$ and $\{1,1,1\}$ is the *set* of transmitted code words, while all eight words belong to the set of possible received code words. We can show this by a three-dimensional illustration, or by a Venn diagram (Fig. 25.6).

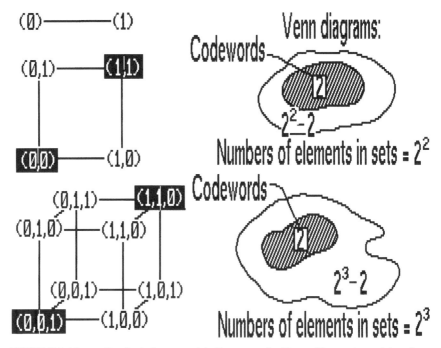

FIGURE 25.6 Message (0 or 1), single-repeat and double-repeat coding illustrated by one, two, and three-dimensional drawings, or by Venn diagrams.

Notice that the vectors {0,0} and {1,1} are different in two positions, and {0,0,0} and {1,1,1} in three positions. We refer to this difference as the *Hamming distance d*, introduced by professor Richard W. Hamming.

Another characterization of a code word is its weight, which is defined as the number of positions where it differs from zero. This number will be related to its distance from the zero vector.

The error correcting capabilities of a code can be evaluated (after Hamming):

The number t of errors that may be detected and corrected is:

$$t = (d{-}1)/2 \qquad (25.1)$$

where d is the minimum distance between codewords. It can be shown that d is equal to the smallest weight of any of the codewords.

The Hamming distance can be observed directly in the three-dimensional drawing in Fig. 25.6: Moving from {0,0,0} to {1,1,1}, from point to point, takes three steps. The two-dimensional drawing shows two steps in moving from {0,0} to {1,1}; hence $t = (2{-}1)/2 = .5 \to 0$ errors corrected, which is what we learned.

CODE GENERATION

The message, the check bits, and the code word are considered vectors of dimensions k, r, and n respectively. We will perform several mathematical operations on these "numbers," such as addition, subtraction, multiplication, and division.

Magnetic recording codes belong to the binary system, with symbols 0 and 1. This number system is also called a field, namely the Galois field GF(2), and is simpler than fields corresponding to multilevel or multisymbol codes. The rules for computations are simple, as shown.

Addition: Multiplication:

	0	1
0	0	1
1	1	0

	0	1
0	0	0
0	0	1

The results are always Mod(2) ($\oplus$), i.e., equal to the remainder after division with 2. Note also, that $1 \oplus 1 = 0$, or $+1 = -1$!

Example 25.1   Addition of two words:

$$\mathbf{a} = \{0,1,1,0,1,0,1,1,1\}$$
$$+\ \mathbf{b} = \{1,1,1,1,0,0,0,1,0\}$$
$$= \{0 \oplus 1,\ 1 \oplus 1,\ 1 \oplus 1,\ 0 \oplus 1,\ 1 \oplus 0,\ 0 \oplus 0,\ 1 \oplus 0,\ 1 \oplus 1,\ 1 \oplus 0\}$$
$$= \{1, 0, 0, 1, 1, 0, 1, 0, 1\}$$

Multiplication of two words:

$$\mathbf{a} = \{0,1,1,0,1,0,1,1,1\}$$
$$\bullet\ \mathbf{b} = \{1,1,1,1,0,0,0,1,0\}$$
$$= \{0 \times 1 \oplus 1 \times 1 \oplus 1 \times 1 \oplus 0 \times 1 \oplus 1 \times 0 \oplus 0 \times 0 \oplus 1 \times 0 \oplus 1 \times 1 \oplus 1 \times 0\}$$
$$= 1$$

Note that this vector multiplication is an inner product, or dot product, and the result is not a vector, but a scalar.

The incoming message bits are divided up into blocks of each k bits, and r parity check bits are appended to each block, making new code words of length $n = r + k$ bits each:

$$\mathbf{u} = u_1, u_2, \ldots , u_k$$
$$\rightarrow v_1, v_2, \ldots . v_r, \qquad v_{r+1}, v_{r+2}, \ldots . v_{r+k}.$$
$$< - r\ \text{parity bits} - > < \!\!\!-\!\!\!- k\ \text{message bits} \!\!\!-\!\!\!\!\longrightarrow >$$
$$= \mathbf{v}$$

The number of transmitted bits has increased from k to $r + k$. This reduction in efficiency is expressed by the *code rate*:

$$R = k/n = \text{information bits / transmitted bits}$$

The parity checks must be generated as linear combinations of the k-tuple message vectors by mod-2 summations ($\oplus$). Hence these codes are named *linear codes*, and are called *systematic* because the parity check bits are appended in a block of r bits at the beginning (or end) of the message bits.

The parity bits are calculated from:

$$v_1 = p_{11}u_1 \oplus p_{12}u_2 \oplus \ldots \ldots \oplus p_{1k}u_k$$
$$v_2 = p_{21}u_1 \oplus p_{22}u_2 \oplus \ldots \ldots \oplus p_{2k}u_k$$
$$\ldots \ldots \ldots$$
$$\ldots \ldots \ldots$$
$$v_r = p_{r1}u_1 \oplus p_{r2}u_2 \oplus \ldots \ldots \oplus p_{rk}u_k$$

which we can write in matrix form:

$$\{v_1, \ldots, v_r\} = \{u_1, u_2, \ldots, u_k\} \bullet \begin{Bmatrix} p_{11} & p_{21} & p_{r1} \\ p_{12} & p_{22} & p_{r2} \\ . & . & . \\ . & . & . \\ p_{1k} & p_{2k} & p_{rk} \end{Bmatrix}$$

$$= \mathbf{u} \cdot \mathbf{P}_{kr}.$$

P cannot be constructed in an arbitrary fashion. It can be shown that the rows must be uniquely defined, i.e., no rows can be alike, and they must not have all 0s, nor any single 1s in a row.

We can write the message bits in the code word as:

$$\{v_{r+1}, v_{r+2}, \ldots, v_{r+k}\} = \mathbf{u} \cdot \begin{Bmatrix} 1 & 0 & 0 & . & . & 0 \\ 0 & 1 & 0 & . & . & 0 \\ 0 & 0 & 1 & . & . & 0 \\ . & . & . & . & . & . \\ 0 & 0 & 0 & . & . & 1 \end{Bmatrix}$$

$$= \mathbf{u} \cdot \mathbf{I}_{kk}$$

The total code word is therefore:

$$v = \mathbf{u} \cdot \mathbf{P}_{kr} + \mathbf{u} \cdot \mathbf{I}_{kk}$$

or

$$v = \mathbf{u} \{\mathbf{P}_{kr} \ \mathbf{I}_{kk}\}_{kn}$$
$$= \mathbf{u} \bullet \mathbf{G}$$

where $\mathbf{G} = \{\mathbf{PI}\}$ is named the *code generator matrix*.

Example 25.2 Find the code word for messages $\mathbf{u}_1 = \{1\ 1\ 0\ 1\}$ and $\mathbf{u}_2 = \{0\ 1\ 0\ 1\}$, using the **P** matrix:

$$\mathbf{P} = \begin{Bmatrix} 1 & 0 & 1 \\ 1 & 1 & 1 \\ 1 & 1 & 0 \\ 0 & 1 & 1 \end{Bmatrix}$$

Answer:

$$\mathbf{u}_1 \cdot \mathbf{G} = \{1\ \ 1\ \ 0\ \ 1\} \cdot \begin{Bmatrix} 1 & 0 & 1 & 1 & 0 & 0 & 0 \\ 1 & 1 & 1 & 0 & 1 & 0 & 0 \\ 1 & 1 & 0 & 0 & 0 & 1 & 0 \\ 0 & 1 & 1 & 0 & 0 & 0 & 1 \end{Bmatrix} = \{0\ 0\ 1\ 1\ 1\ 0\ 1\}$$

$$\mathbf{u}_2 \cdot \mathbf{G} = \{0\ \ 1\ \ 0\ \ 1\} \cdot \begin{Bmatrix} 1 & 0 & 1 & 1 & 0 & 0 & 0 \\ 1 & 1 & 1 & 0 & 1 & 0 & 0 \\ 1 & 1 & 0 & 0 & 0 & 1 & 0 \\ 0 & 1 & 1 & 0 & 0 & 0 & 1 \end{Bmatrix} = \{1\ 0\ 0\ 0\ 1\ 0\ 1\}$$

For an additional exercise, find the remaining 14 code words.

ERROR PROBABILITY IN CODE WORDS

We evaluated, earlier in this chapter, the BER for a (3,1) code; we will now extend this to other codes. Whenever we transmit word **x** and receive word **y**, then a test for error-free transmission is that:

$$\mathbf{x} - \mathbf{y} = \mathbf{x} + \mathbf{y} = \{0\},$$

where $\{0\}$ is the zero vector:

$$\begin{aligned}
\mathbf{x} &= \{x_1, x_2, x_3, \ldots, x_n\} \\
-\mathbf{y} &= \{y_1, y_2, y_3, \ldots, y_n\} \\
\hline
&= \{0, 0, 0, \ldots, 0\}
\end{aligned}$$

A reasonable assumption is that any error will affect a bit, or symbol (or component), independent of errors in the others. An error vector adds to a signal vector, and it can be expressed as:

$$\mathbf{e} = \{e_1, e_2, e_3, \ldots, e_n\}$$

and the total error probability:

$$P(\mathbf{e}) = P(e_1) \cdot P(e_2) \cdot \ldots \cdot P(e_n).$$

The error probability, or BER, is also called *the transition probability*, where transition refers to a 1 changing into a 0, and vice versa. The probability of no errors is $1 - p$, and n independent symbols are therefore received error free with a probability of $(1 - p)^n$. We can summarize:

$$\begin{aligned}
P(\text{no errors}) &= (1 - p)^n \\
P(1\ \text{error}) &= p(1 - p)^{n-1} \\
P(2\ \text{errors}) &= p^2(1 - p)^{n-2}
\end{aligned}$$

$$P(b\ \text{errors}) = p^b(1 - p)^{n-b} \tag{25.2}$$

The actual probability will be greater, because errors can produce different *error patterns* (E = error), shown for a five-bit word:

$$\begin{aligned}
\text{ONE error: } &\text{E X X X X} \\
-\text{ or } &\text{X E X X X} \\
-\text{ or } &\text{X X E X X} \\
-\text{ or } &\text{X X X E X} \\
-\text{ or } &\text{X X X X E.}
\end{aligned}$$

For an n-dimensional word there are n different patterns for a single error, and the true error probability is therefore:

$$P(1\ \text{error}) = n \times p(1 - p)^{n-1}$$

Two errors will result in $n \times (n - 1)/2$ patterns:

EEXXX	XEXEX
EXEXX	XEXXE
EXXEX	XXEEX
EXXXE	XXEXE
XEEXX	XXXEE

A total of b errors will result in:

$$(n/1) \times ((n-1)/2) \times ((n-2)/3) \times ((n-3)/4) \times .. \times ((n-b-1)/b)$$

$$= n!/(b!(n-b)!) \text{ patterns,}$$

where

$$n! = n \times (n-1) \times (n-2) \times \ldots \times 3 \cdot 2 \cdot 1$$

The total error probability is therefore:

$$P(b \text{ errors}) = [\, n!/(b!(n-b)!) \,] \times p^b (1-p)^{n-b} \qquad (25.3)$$

Example 25.3: A (7,4) code. *Error probability, uncoded 4-bit message:* The probability of one error is $P(1 \text{ error}) = 4 \times p(1-p)^3 \approx 4p$ (for $p < 1$). We arrive at the following table, showing $P(1$ error) as a function of the system SNR:

SNR	p	P(1 error)
3.0 dB	10^{-1}	4×10^{-1}
7.2 dB	10^{-2}	4×10^{-2}
9.8 dB	10^{-3}	4×10^{-3}
12.5 dB	10^{-4}	4×10^{-4}
14.2 dB	10^{-7}	4×10^{-7}
16.0 dB	10^{-10}	4×10^{-10}

This table results from the use of formula (23.1).

Error probability, (7,4) code word: We have now added 3 check bits, and can for a reasonable selection of message words get a Hamming distance of $d = 3$. This will correct:

$$t = (d-1)/2 = (3-1)/2 = 1 \text{ error}$$

One error will therefore result when the number of errors in the code word is 2. Hence, using formula (25.3):

$$P(1 \text{ error}) =$$
$$P(2 \text{ code word errors}) = [\, 7!/(2!(7-2)!) \,] p^2 (1-p)^{7-2}$$
$$= 21 \times p^2$$

We must reestimate the values of p, in light of the reduced *SNR*. The message bit rate remains unaltered, and the code word rate must therefore be increased by a factor of $n/k = 7/4$; this corresponds to a bandwidth increase by 7/4, and will result in a 2.5 dB reduction of the *SNR*. A new table is:

SNR$_{7\text{BITS}}$	p	P(1 error)	(uncoded P(1))
0.5 dB	1.2×10^{-1}	3.0×10^{-1}	4×10^{-1}
4.7 dB	3×10^{-2}	1.9×10^{-2}	4×10^{-2}
7.3 dB	1×10^{-2}	2.1×10^{-3}	4×10^{-3}
8.9 dB	2×10^{-3}	8.4×10^{-5}	4×10^{-4}
11.7 dB	2×10^{-5}	8.4×10^{-9}	4×10^{-7}
13.5 dB	5×10^{-7}	5.3×10^{-12}	4×10^{-10}

The coding does provide improvement, about two orders of magnitude worth. This is characteristic for short codes; much longer code words are needed to demonstrate appreciable improvement.

PARITY-CHECK MATRIX

The received word $\mathbf{r}$ should now be checked for errors. We can take the last k bits of $\mathbf{r}$ (the data bits), multiply with $\{\mathbf{PI}\}$ and compare this generated word with the total $\mathbf{r}$ vector. If they are identical, we can conclude that there are no single errors; the message is then found by stripping its $n - k = r$ parity check bits. But if the comparison shows differences, then we know that there is at least one error, and the next question is how to find it (or them).

Note, during encoding:

$$\mathbf{v} = \mathbf{u} \{\mathbf{PI}\}$$
$$= \{\mathbf{uP\ uI}\}$$
$$= \mathbf{v}_r + \mathbf{u}$$

where $\mathbf{v}_r$ is the parity bit sequence:

$$\mathbf{v}_r = \mathbf{u}_k \cdot \mathbf{P}$$

The received code word $\mathbf{r}$ has the R check bits $\mathbf{v}_R$. An error free transmission would result in:

$$\mathbf{v}_R = \mathbf{v}_r$$
$$\mathbf{v}_R - \mathbf{v}_r = \mathbf{0}$$

where $\mathbf{0}$ is the null vector with all components equal to zero, or, (mod-2):

$$\mathbf{v}_r + \mathbf{v}_R = \mathbf{0}$$
$$\mathbf{u}_k \cdot \mathbf{P} + \mathbf{v}_R = \mathbf{0}.$$

Re-writing:

$$\mathbf{u}_k \cdot \mathbf{P} + \mathbf{v}_R = \left\{ \begin{matrix} \mathbf{u} \cdot \mathbf{P} \\ \mathbf{v}_R \cdot \mathbf{I} \end{matrix} \right\} = \left\{ \begin{matrix} \mathbf{uv}_r \\ \mathbf{I}_r \end{matrix} \right\} \mathbf{P} = \mathbf{0}$$

In other words, if the last vector-matrix product is zero, then the received parity checks are equal to the transmitted ones, and there are no detectable errors!

The last expression can be written:

$$\mathbf{vH}^{\mathrm{T}} = \mathbf{v} \left\{ \begin{matrix} \mathbf{P} \\ \mathbf{I}_r \end{matrix} \right\} = \mathbf{0}$$

where:

$$\mathbf{H} = \{\mathbf{P}^{\mathrm{T}}\mathbf{I}_r\}$$

is called the parity check matrix.

A code word must therefore satisfy the condition:

$$\mathbf{vH^T} = \mathbf{0} \qquad (25.4)$$

Example 25.4. Is $\{1\ 1\ 0\ 1\ 0\ 1\ 0\}$ a code word from the **G**-matrix used in example 25.2? The message word is $\{1\ 0\ 1\ 0\}$, assuming no errors. Its associated code word is:

$$\mathbf{v} = \{1010\} \bullet \begin{Bmatrix} 1\ 0\ 1\ 1\ 0\ 0\ 0 \\ 1\ 1\ 1\ 0\ 1\ 0\ 0 \\ 1\ 1\ 0\ 0\ 0\ 1\ 0 \\ 0\ 1\ 1\ 0\ 0\ 0\ 1 \end{Bmatrix} = \{0111010\}$$

This generated word has check bits $\{0\ 1\ 1\} \neq \{1\ 1\ 0\}$, and is thus *not* a codeword. Let us use the parity-check matrix:

$$H = \begin{Bmatrix} 1\ 1\ 1\ 0\ 1\ 0\ 0 \\ 0\ 1\ 1\ 1\ 0\ 1\ 0 \\ 1\ 1\ 0\ 1\ 0\ 0\ 1 \end{Bmatrix}$$

The word is only a proper code word if its product with the transposed H-matrix, $\mathbf{H^T}$, equals a zero word:

$$\{1101010\} \bullet \begin{Bmatrix} 1\ 0\ 1 \\ 1\ 1\ 1 \\ 1\ 1\ 0 \\ 0\ 1\ 1 \\ 1\ 0\ 0 \\ 0\ 1\ 0 \\ 0\ 0\ 1 \end{Bmatrix} = \{010\} \neq \mathbf{0}.$$

The condition is not satisfied—the word does not belong to the code words generated by the generator matrix **G**.

THE SYNDROME

It may seem trivial that the product of the received code word vector and the transposed parity check matrix should equal zero when the word is a proper code word.

But the interesting fact is that the result will tell us something about the possible error, if it is different from $\mathbf{0}$. Let us name the resulting vector the syndrome $\mathbf{S}$:

$$\mathbf{S} = \mathbf{r} \cdot \mathbf{H^T} \qquad (25.5)$$

where $\mathbf{r}$ is the received word.

If the channel is noisy then $\mathbf{r}$ equals the sum of the transmitted code word $\mathbf{u}$ and a noise vector $\mathbf{e}$:

$$\mathbf{r} = \mathbf{u} + \mathbf{e}$$

and **S** is therefore:

$$\mathbf{S} = \mathbf{r} \cdot \mathbf{H}^T$$
$$= \mathbf{u} \cdot \mathbf{H}^T + \mathbf{e} \cdot \mathbf{H}^T$$

or:

$$\mathbf{S} = \mathbf{e} \cdot \mathbf{H}^T$$

It turns out that we should compare **S** with the rows of **H**^{T}; *if it equals row no. i*, then we can correct the received word by changing bit number *i*!

Here's another exercise: An encoder uses a P-matrix given by:

$$P = \begin{matrix} 1\ 1\ 1 \\ 1\ 1\ 0 \\ 1\ 0\ 1 \\ 0\ 1\ 1 \end{matrix}$$

One received word is (1 0 1 0 0 0 1); it is not a code word (check). Correct the error.

EXAMPLES OF BLOCK CODES

Linear block codes are {*n,k*}-codes, where *k* message bits are changed into *n* transmitted (or stored) code words. There are several block codes in used in tape drives, such as the Hsiao or Patel and Hong coding schemes. Disk systems use Fire codes or Reed-Solomon codes.

A few simple codes are listed below:

Single-error correcting, $t = 1$ {$d = 3$}

n	k	Code	Eff = k/n
5	2	{5,2}	.4
6	3	{6,3}	.5
7	4	{7,4}	.57
15	11	{15,11}	.73

Double-error correcting, $t = 2$ {$d = 5$}

n	k	Code	Eff=k/n
10	4	{10,4}	.4
11	4	{11,4}	.36
15	5	{15,8}	.53

Cyclic block codes are extensions of the codes covered in this chapter. An example of an encoder of a (7,4) code is shown in Fig. 25.7. The theory behind these codes are founded upon the algebraic concepts of rings, fields, and polynomials, and would take us beyond the scope of this book.

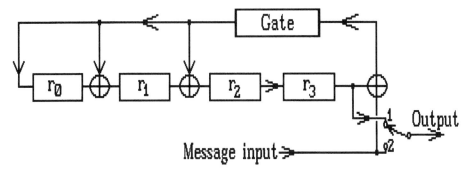

FIGURE 25.7   The generation of cyclic (7,4) block code.

Much longer codes are also needed to reduce the overhead, i.e., increase the efficiency. Members of these codes are the longer Fire codes, BCH (Bose-Ray-Chaudhuri-Hocquenghem) and RS (Reed-Solomon) codes. These codes use blocks consisting of many thousand bits. The interested reader is referred to books by Lin and Costello, and by Chambers.

Convolutional codes are mostly used in communication, and a simple encoder is shown in Fig. 25.8. It consists of a 3-stage shift register and two adders (modulo-2). At the start of any stage the next bits from the source is fed into the right of the shift-register, pushing the left-hand bit out. Then the sum mod-2 (modulo-2) of the first and the third bit is sent out, followed by the sum mod-2 of all three bits, providing an output consisting of bit-pairs.

A variation on the convolutional encoder was shown in Fig. 24.8 for encoding of NRZ-L data into randomized NRZ-L data. Applications of codes for magnetic tapes and disks are covered in Chapter 16 of Lin and Costello's book.

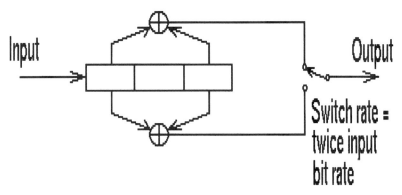

FIGURE 25.8   The generation of a ½ convolution code.

INTERLEAVING

A very powerful technique will ensure protection against errors, using even a simple error-correcting scheme. It consists of scrambling the data stream of encoded bits into a new sequence, which later is descrambled into the original sequence. This is illustrated in Fig. 25.9. Seven bit blocks of (7,4) codewords are stored in memory, and read out into a new memory as shown; at this point added protection can be made by generating check bits by taking mod 2 sums of the rows and columns.

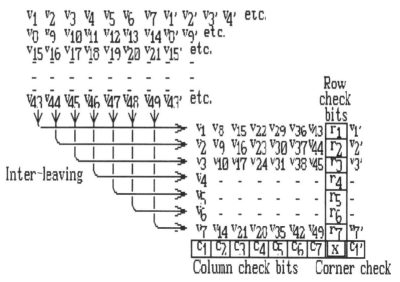

FIGURE 25.9 A simplified diagram showing interleaving, column and row check bits, and a corner check bit (X).

The data are now transmitted (stored/read) row by row, and the sequences have been scrambled. Now suppose that one entire row has been wiped out. After descrambling, the upper-left pattern will have one column in error, but each word (1-7, 8-14, etc.) will have only one bit in error, which can be detected and corrected.

This method offers protection against burst errors in single channel as well as multichannel applications. The latter is illustrated in Fig. 25.10, illustrating the signal path through a HDDR recorder. These recorders accept unusually high data rates, and the data stream is first encoded to protect against errors, and then divided into a large number of channels. Each of these are now modulated into one of the codes from Chapter 24, and recorded on the tape.

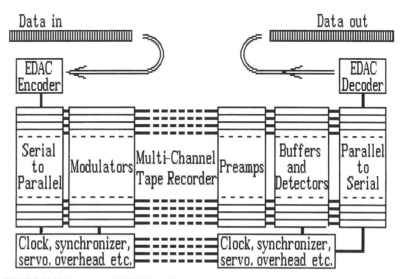

FIGURE 25.10 Interleaving in HDDR recordings.

The playback signals are amplified and temporarily stored in a buffer. They are clocked out in step with a reference clock (crystal controlled), thereby removing any time base errors. The buffers are monitored by a circuit that will signal the capstan servo to speed up or slow down. After detection (demodulation) a parallel-to-serial conversion follows. The resulting data stream is finally decoded and errors detected and corrected.

This scheme of EDAC plus interleaving is very powerful. Part of the overhead caused by the EDAC can even be recovered by not recording some of the tracks; the bit error rate will still be better than one in a trillion.

REFERENCES

Chambers, W.G. 1985. *Basics of Communications and Coding.* Clarendon Press. Oxford.. 240 pages.

Lin, S., and Costello, D.J. 1983. *Error Control Coding: Fundamentals and Applications.* Prentice-Hall. 603 pages.

ABSTRACT ALGEBRA

Childs, L. A. 1983. *Concrete Introduction to Higher Algebra.* Springer-Verlag. 340 pages.

Cohn, P.M. 1982. *Algebra Vol. 1.* John Wiley & Sons. 410 pages.

Fraleigh, J.B. 1976. *A First Course in Abstract Algebra.* Addison-Wesley. Reading, Mass. 478 pages.

Hall, F.M. 1980, 1981. *An Introduction to Abstract Algebra, Vol. 1 and 2.* Cambridge University Press. Cambridge, Mass. 388 and 300 pages.

McClellan, J.H. and Rader, C.M. 1979. *Number Theory in Digital Signal Processing.* Prentice-Hall, Inc. Englewood Cliffs, N.J. 276 pages.

Sawyer, W.W. 1978 (Original 1959). *Concrete Approach to Abstract Algebra.* Dover, New York. 234 pages.

Schroeder, M.R. 1984. *Number Theory in Science and Communication.* Springer-Verlag. 324 pages.

Singh, J. 1959. *Great Ideas of Modern Mathematics.* Dover, New York. 312 pages.

ERROR CORRECTION AND DETECTION CODING

Adi, W. Fast July 1984. Burst Error-Correction Scheme with Fire Code. *IEEE Trans. COMP.* Vol. 33, No. 7. pp. 619–625.

Berlekamp, E.R. May 1980. The Technology of Error-Correcting Codes. *Proc. of IEEE.* Vol 68, No. 5. pp. 564–593.

Bhargava, V.K. Jan. 1983. Forward Error Correction Schemes for Digital Communications. *IEEE Communications Magazine* Vol. 23, No. 1. pp. 11–19.

Blahut, R.E.B. May 1985. Algebraic Fields, Signal Processing, and Error Control. *Proc. of IEEE.* Vol. 73, No. 5. pp. 874–893.

Brown, D.T., and Sellers, F.F. Jr. July 1970. Error Correction for IBM 800-bit-per-inch Magnetic Tape. *IBM Jour. Res. and Dev.* Vol. 14, pp. 384–389.

Chi, C.S. Sept. 1984. Triplex Code for Quartenary High Density Recording. *IEEE Trans. Magn.* Vol. MAG-20, No. 5. pp. 888–890.

Couvreur, C., and Piret, P. May 1984. Codes Between BCH and RS Codes. *Philips Jour. of Research.* Vol. 39, No. 45.

Erdel, K. June 1985. Burst Error Correction for Digital 34 Mbit/s TV Signal Transmission with a Fire Code. *Frequenz.* Vol. 39, No. 6. pp. 165–169.

Ferreira, H.C. Sept. 1985. The Synthesis of Finite State Magnetic Recording Codes with Good Hamming Distance Properties. *IEEE Trans. Magn.* Vol. MAG-21, No. 5. pp. 1356–1358.

Franaszek, P.A. July 1970. Sequence-State Methods for Run-Length-Limited Coding. *IBM Jour. Res. and Dev.* Vol. 14, pp. 376–383.

Galen, P.M. Disc Drive Error Detection and Correction Using VLSI. Jan. 1984. *Hewlett-Packard Journal.* Vol. 35, No. 1. pp. 12–13.

Gillard, C.H. Error Correction Strategy for the New Generation of 4:2:2 Component Digital Video Tape Recorders. 1986. *Intl. Conf. on Video, Data and Audio* IERE Publ. No. 67. pp. 165–175.

Glover, N. 1982. *Practical Error Correction Design for Engineers.* Data Systems Technology Corp. 1801 Aspen Street, Broomfield, CO 80020. 379 pages.

Glover, N., May 1986. Error Correction and Detection (EDAC) Schemes. *Symposium on Mem. and Adv. Rec. Tech.* Paper No. WS-4-C. 17 pages.

Hodgart, M.S. July 1979. High Performance Low Redundancy Serial Error Correcting Codes for Audio Digital Recorders. *Intl. Conf. on Audio, Video, and Data* IERE Publ. No. 43, pp. 163–177.

Howell, T.D. March 1984. Analysis of Correctable Errors in the IBM 3380 Disk File. *IBM Jour. Res. and Dev.* Vol. 28, No. 2. pp. 206–211.

Hsu, I.S., Reed, Il S., Truong, T.K., Wand, K., Yeh, C.S., and Deutsch, L.J. Oct. 1984. The VLSI Implementation of a Reed-Solomon Encoder Using Berlekamp Bit-Serial Multiplier Algorithm. *IEEE Trans. on Computers.* Vol. 33, No. 10. pp. 906–912.

Isailovic, J. Jan. 1985. Codes for Optical Recording. *Proc. of SPIE.* Vol. 529, pp. 161–168.

Lee, R. July 1981. Cyclic Code Redundancy. *Digital Design.* Vol. 11, No. 7. pp. 77–85.

Meeks, L. 1982. A Dropout-Based Error Correction Method for Use on High Density Digital Recording Systems. *Intl. Conf. on Audio, Video, and Data,* 11 pages.

Melas, C.M. Jan. 1966. A New Group of Codes for Correction of Dependent Errors in Data Transmission. *IBM Jour. Res. and Dev.* Vol. 10, No. 1. pp. 58–65.

Nelson, B. Feb. 1982. Effortless Error Management. *Computer Design.* Vol. 21, No. 2. pp. 163–168.

Ohr, S. Dec. 1985. Error Checking and Correcting IC Slashes Optical Disk Defects. *Electronic Design.* Vol. 33, No. 29. p. 37.

Parker, M.A. April 1984. A Range of Combined Error Correction and Recording Channel Code Schemes. *Intl. Conf. on Audio, Video, and Data* No. 59.

Patel, A.M. and Hong, S.J. Nov. 1974. Optimal Rectangular Code for High Density Magnetic Tapes. *IBM Jour. Res. and Dev.* Vol. 18, pp. 579–588.

Patel, A.M. 1988. Signal and Error-Control Coding. chapter 5. *Magnetic Recording.* Vol. 2, Vol. 3. McGraw-Hill.

Peterson, W.W., and Welden, E.J. 1972. *Error Correcting Codes.* M.I.T. Press. Cambridge, Mass.

Rennie, L.J. June 1982. Forward-Looking Error Correction via Extended Golay. *Computer Design.* Vol. 21, No. 6. pp. 121–130.

Shenton, D., DeBenedictis, E., and Locanthi, B. Oct. 1982. Improved Reed Solomon Decoding Using Multiple Pass Decoding. Preprint No. 2035 (A-10). *AES Convention.* 9 pages.

Swanson, R. Aug. 1980. Matrix Technique Leads to Direct Error Code Implementation. *Computer Design.* Vol. 19, No. 8. pp. 101–108.

Varaiya, R. July 1978. On Ensuring Data Recoverability by the Use of Error-Correcting Codes. *IEEE Trans. Magn.* Vol. MAG-14, No. 4. pp. 207–210.

Watkinson, J.R. April 1984. Error Correction Techniques in Digital Audio. *Intl. Conf. on Audio, Video, and Data* No. 59.

CHAPTER 26
DIGITAL DATA STORAGE

The original purpose of this chapter was to outline PCM (pulse-code modulation) recording on tape and disk drives. This was a reasonable scope in the early 1980s when this chapter was created for the 3rd edition of this handbook.

Today, most information is stored digitized. To describe the methods and recorder applications would overflow the bounds of this chapter, which therefore is limited to comments on digital data storage on tapes and disks. The author's guilt has been lessened by the recent publications of a book and a bimonthly periodical: *The Art of Data Recording* by John Watkinson, and *Data Storage* (see references).

File sizes keep on growing. Ten years ago, a PC with a 20- or 40-MB hard drive was a good entry-level system. Today it is generally recommended to get a 1000-MB (1-GB) hard drive. Programs have swelled and must load just as fast as before; in other words, the transfer rate must go up.

FILES WITH PICTURE INFORMATION

The information highway will be full of pictures and video, both consuming large chunks of storage space. The illustrations in this book are mostly 640×480 bitmapped files, requiring enough information to plot 307,200 dots. The actual files will be smaller, because a first-level compression is done by storing information only when a dot changes during the video-like scan.

This may not bring much compression in a greyscale photo, where even uniformly changing surfaces have a lot of different pixels. To store a 256-level greyscale picture, we may therefore require 307,200 bytes. When adding color, the limit is 256 levels each of red, green and blue, i.e., $256 \times 256 \times 256 = 16,777,216$ colors. And four, not one, bytes are required per pixel, so the total maximum file could be $307,200 \times 4$ bytes (1.229 MB, $307,200 \times 32 = 9.8$ Mb).

Many applications do well with only a few thousand colors, but may need higher resolution, say by 4 times. These issues and associated compression techniques are regularly discussed in magazines as *Advanced Imaging* and *Imaging Magazine* (see references). The pixel resolution of 307,200 listed above compares with 20 to 30,000,000 for a high-quality 35 mm movie film image. If we were to apply such stringent demands for all our pictures, then the files would become totally unmanageable! Like 100 MB per picture, or 30 times more per second of motion picture, with a transfer rate of 3000 MB/sec = 24 Gb/sec = 12 GHz (for HDTV transmission, a bandwidth of 1 GB/sec is planned.)

Less than photorealistic quality is often sufficient, and the pictures in sequential motion pictures seldom change very much between frames. Motion picture compression utilizes this fact by first storing an entire frame, and thereafter storing only the changes from one frame to the next. The first picture and picture changes thereafter can themselves be compressed by clever algorithms. Well-known compression schemes are MPEG (Motion Picture Experts Group) and JPEG (Joint Photographic Experts Group). Fractal techniques can also be used.

In a couple of years the consumer may be able to select between a conventional videotape or one of the new double-sided optical video CDs, storing two movies in a 10- or 7-GB storage space (Toshiba/Time Warner and/or Philips/Sony; see Holyoke).

RECORDING SYSTEMS

The choice of recording system is not made easier for the end user with today's vast array of possibilities. The selection of the most reliable and least costly system must start with an assesment of the amount of storage space required and the maximum data rate necessary. Table 14.1 shows the many variations on tape systems. The user's selection from the many options depends on an appraisal of capacity, transfer rate, and cost (NASA Mass Storage Conference). Summaries of comparisons are regularly featured in the referenced magazines, and from a 1993 issue we have (after Balue):

	Capacity	**Cost**	**Data Rate**
	GB max.	\$/GB	MB/sec
3490	2.4	12.50	3.0
VHS	18	1.07	4.0
8 MM	5	7.00	0.5
4 MM	2	15.00	0.25
QIC	2	15.00	0.5
12" Optical	6.8	96.00	1.0

There are large price variations with each group of drives: A 2-GB QIC drive was listed at \$2000, while a 250-MB QIC drive was only \$250.

In 1994/95 two new ½-tape systems were introduced, one by Quantum using longitudinal, parallel tracks (*Linear Tape*) and another by StorageTek using helical-scan tracks (*Redwood (D-3) project*). Both will be used extensively in the huge libraries that governments, universities, and large businesses use. The QIC drives are now storing several GB by using narrower, servo-controlled tracks.

The choice of connection to tape or disk drive via SCSI (small computer system interface) or IDE (integrated drive electronics) interfaces is not discussed here, and the reader is referred to other sources for this information. There are several configurations of each interface.

Rigid disk drives come in a variety of sizes and storage capacities. They operate alone, or in a RAID (a redundant array of inexpensive disks). One of the disk drives in a RAID is redundant, and the information stored in a RAID is cleverly done so it can all be recovered if one drive fails. There is then adequate time to replace the faulty drive and rewrite the information back over all drives. This brings the reliability to a very high figure.

One subtle difference between tape and disk systems is the time it takes to locate a segment of the storage space. Tape has to be wound to a certain spot, and the medium is therefore called SAS—sequential access storage. Disk sectors can be reached within milliseconds, and disks represent Direct Access Storage. (Video on demand will utilize a combination of SAS and DAS.)

REFERENCES

Advanced Imaging—Solutions for the Electronic Imaging Professional. For subscription information, write to Advanced Imaging, 445 Broad Hollow Road, Melville, NY 11747- 4722. Att. Circulation.

Balue, T. May 1993. Half-Inch Tape Libraries Reach to The Terabyte and Beyond. *Computer Technology Review* 12 (6): 57–61.

Data Storage—Technology & Manufacture of Storage Devices. Bimonthly. For subscription information, write to DATA STORAGE, Circulation Department, Box 3004, Tulsa, OK 74101-3004. Or call: (918)-835-3161.

Fourth NASA Goddard Conference on Mass Storage Systems and Technologies. NASA Conf. Publ. 3295, March 1995, 396 pages.

Holyoke, L. Feb 20, 1995. Video Warfare: How Toshiba took the High Ground. *Business Week*, pp. 64–66.

Imaging Magazine—Document Solutions for Business. For subscription information, write to IMAGINE MAGAZINE, Subscription Dept., 1265 Industrial Highway, Southampton, PA 18966.

Watkinson, J. 1994. *The Art of Data Recording.* Focal Press, London. 518 pages.

BIBLIOGRAPHY

Ashton, G.R., editor, July 1994. *Storage Technology Assessment Final Report.* NML Technical Report RE-0016. Summarized in NML Bits 5 (1): 6–8. National Media Lab, Bldg. 235-1N-17, St. Paul, MN 55144-1000. Fax (612)733-4340.

Computer Data Storage, Vol. 2 of Magnetic Recording, edited by C.D. Mee and E.D. Daniel. McGraw-Hill, 1988. 408 pages.

Computer Technology Review. Published monthly in a newspaper format and quarterly with an added technology section. For subscription, write West World Publications, 924 Westwood Blvd. Ste. 630, Los Angeles, CA 90024-2927.

Giorgis, T.W., and J. McDonough. March 1995. 26 Safeguards against LAN Data Loss—Test report for backup tape drives. *BYTE* 20 (3): 144 ff.

Halfhill, T.R. How Safe is Data Compression? *BYTE* 19 (2): 56 ff.

IDEMA Insight on Drive Manufacturing Issues. Published bimonthly by IDEMA (International Disk Drive Equipment and Materials Association, 710 Lakeway, Suite 140, Sunnyvale, CA 94086. Phone (408)-720-9352.

Leathers, D. June 1994. Disk-based video storage. *Broadcast Engineering* 36 (6): 23–29.

Wallace, S. March 1994. Managing Mass Storage. *BYTE* 19 (2): 78ff.

Watkinson, J. 1995. *Compression in Video and Audio.* Focal Press, 256 pages.

CHAPTER 27
ANALOG INSTRUMENTATION RECORDERS

A great need for storage and playback of measurement data plus telemetry data occurred in the late 1950s. Missile and space exploration programs created this need, and the storage device became an extension of the German magnetophone: The high-speed multichannel instrumentation recorder.

Direct ac-biased recording was widely used on most tracks, while others carried recordings in FM (frequency modulation) or PCM (pulse code modulation). The latter was the forerunner of what we today call digital recording.

The ac bias serves to linearize the recording channel. Details are covered in Chapter 20. The tape or disk channel for the signal is not perfect, and the direct ac-bias recorded data may be contaminated by:

- Signal amplitude fluctuations
- Tape speed variations
- Errors in tape duplication
- Noise
- Distortion
- Dropouts
- Crosstalk.

All of these errors are predominant in direct analog recordings. They are of lesser affect in FM recordings, and in PCM recordings most of them can be reduced to a negligible level.

This chapter will describe fluctuations in signal amplitude and tape speed, plus possible duplication errors. Noise, distortion, dropouts, and crosstalk are described elsewhere in the book.

The reader may question the reasons for including this chapter's topics, because digital recording appears to be the cure for all those ills. The SNR can be made arbitrarily high by selecting enough sampling levels, and signal accuracy can be made high by frequently sampling the signal waveforms (at least twice per signal cycle). Flutter and timing errors are easily removed by the use of sufficiently large data buffers from where the data are clocked out in perfect step with an oscillator. So why spend time on a dissertation on the ills of analog recording?

One must remember that the digital signal is always recorded, stored, and read in an analog format. It is a nice, perfectly shaped rectangular waveform when it comes out of the TTL circuits; but after recording and read-out, it has been afflicted with all the problems of the analog signal process. The conscientious designer must therefore have in-depth knowledge about all facets of the analog signal channel. Chapter 20 offers insight into the ac-biased channel, while Chapters 22 and 23 explains signals through the direct channel (with or without bias). This chapter is devoted to the common weaknesses of tape or disk recording channels and how to overcome some of them.

AMPLITUDE FLUCTUATIONS AND SIGNAL ACCURACY

Direct analog magnetic recording is limited in performance by modulation noise and tape coating irregularities (nonuniform dispersion, a limited number of particles or grains "under" the read-head gap, nonuniform coating thickness, and magnetics variations). The modulation noise is proportional to the level of the recorded signal, and is also related to the coating irregularities. The amplitude of the read signal will fluctuate as if it was amplitude-modulated with a modulation index m that typically has a magnitude of ± 0.5 to ± 1.0 dB.

A figure of merit for a transmission system is the product of signal bandwidth B and the logarithm of the number of discrete levels n. This product is defined as the maximum rate of transmitting information and called the system capacity C:

$$C = B \times \log_2 n \tag{27.1}$$

The number of signal levels that can be distinguished from each other is proportional to the system's signal-to-noise ratio:

$$
\begin{aligned}
n &= (V_{\text{signal}} + V_{\text{noise}})/V_{\text{noise}} \\
&= 1 + (V_s/V_n) \\
&= 1 + SNR
\end{aligned}
$$

and C becomes:

$$C = B \times \log_2(1 + SNR)$$

or

$$C = 2B \times \log_2 \sqrt{1 + SNR} \text{ bits} \tag{27.2}$$

This classic law assumes a Gaussian distribution of the noise, and freedom from amplitude variations; the data transmission channel associated therewith is generally referred to as the AWGN channel, i.e., average white Gaussian noise. The magnetic tape or disk channel is not AWGN, and we must therefore modify (27.2) to include amplitude variations.

The fluctuations of the read level present a limitation on the number of amplitude levels that can be distinguished from one another. If the modulation index is $m = m \times V_{\text{signal}}/V_{\text{signal}}$, then the differences between two separate levels must be (see Fig. 27.1):

$$V_q(1 - m) - V_{\text{noise-peak}} = V_{q\text{-}1}(1 + m) + V_{\text{noise-peak}}$$

which results in a minimum amplitude difference of:

$$
\begin{aligned}
\Delta V &= V_q - V_{q-1} \\
&= m(V_q + V_{q-1}) + 2V_{\text{noise-peak}} \\
&= 2mV_q + 2V_{\text{noise-peak}}
\end{aligned}
$$

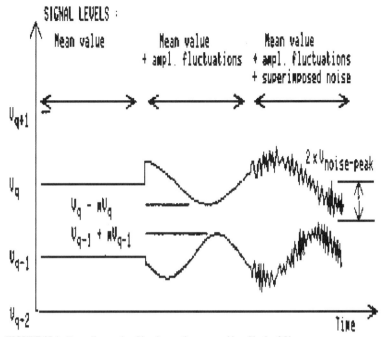

FIGURE 27.1 Two adjacent signal levels must be separated by μV_q plus 2 $V_{\text{noise-peak}}$.

for large signal-to-noise ratios. We may then also write:

$$\Delta V \approx 2mV_q$$

because $2mV_q > 2V_{\text{noise}}$. The maximum number of distinguishable amplitude levels is then:

$$n' = (1 + SNR)/2m$$

$$\approx SNR/2m \text{ (for } SNR > 1)$$

which equals a number of bits:

$$N' = \log_2(SNR/2m).$$

Volz has determined n' without approximations (called n''):

$$n'' = 1 + (\log_e(1 + 2m \times SNR) - \log_e(1 + 2m))/2m \qquad (27.3)$$

and the number of bits required:

$$N'' = \log_2(n'')$$

n'' is plotted in Fig. 27.2. Shannon's law (formula (27.2)) is also plotted, but applies only for low signal-to-noise ratios. The scale shows the number of levels and the equivalent instrument accuracy. A typical performance specification for a 2 MHz recorder/reproducer is 30 dB *SNR*, with

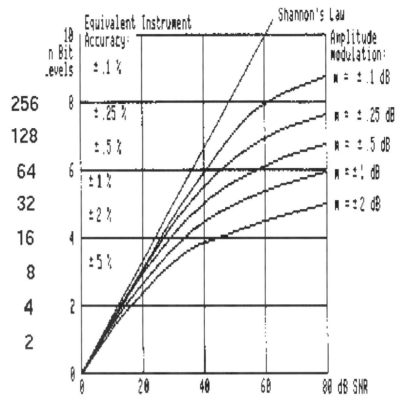

FIGURE 27.2 Maximum number n^n of distinguishable levels, and the equivalent instrument accuracy, in analog magnetic recording (*after Volz*).

$m = \pm 1$ dB. This corresponds to an equivalent instrument accuracy of about ± 4 percent. Improvements to the SNR alone will give diminishing returns unless means are found to also reduce amplitude variations.

The amplitude accuracy will suffer during repeated replays of a tape due to mechanical instability. The gradual decrease in signal amplitudes, which are wavelength-dependent, may quickly amount to errors greater than we have previously outlined. It is important to examine this effect for a given tape and recorder combination, and to apply corrections to the analyzed data (see Chapter 15).

The signal amplitude variations and the signal-to-noise ratio set a limit for the channel capacity of a magnetic tape recorder. It is important to recognize that the number of bit levels can be fairly large only if the mid-frequency range of a recorder is used (by band-pass filtering). Most recorders will, in this range, have a significantly higher *SNR* than the figure measured for the entire recorder bandwidth.

TAPE SPEED VARIATIONS

Magnetic tape transport for analog recording and playback are designed to provide a linear motion of the tape with as little speed variation as possible. This goal is limited by mechanical tolerances, and all tape transports therefore introduce errors into the data.

The speed variations are generally referred to as flutter. Their effect upon data in the form of noise in FM systems is well-known and discussed elsewhere in the literature; methods have been developed for electronic compensation thereof.

Factors other than mechanical tolerances contribute to variations in tape speed, because the data quite often are recorded on one transport and played back on another transport, or it may be a duplicate, adding a third transport into the chain. Because it is most unlikely that they all run at exactly the same speed, the final speed may be 0.5–1 percent off the speed at which the data were recorded. This results in an error if it is desired to reproduce "real time." Additional speed variations are often encountered because the recorder used in acquiring the data may have been subjected to heavy vibrations and gyrations, causing large variations and jumps in the tape speed (the latter from jerks from the reels) or the recorder may have been lightweight, with little control over moderate speed variations.

All these factors can add up to substantial errors in time and frequency domains. Figure 27.3 illustrates what takes place during recording (part a) with large speed variations and subsequent playback on three reproducers.

RECORDING ON A HIGH-INERTIA DRIVE

The speed variations are shown graphically: at time t_1 the tape is slowed down by, for instance, a gyration on the recorder, and at time t_2 the tape slips through the capstan drive with high acceleration due to a jerk or bump of the recorder. This in turn causes a loop to be formed at the supply reel, which slows down and, when the loop tightens at t_3, an instantaneous slowdown is caused before the tape motion stabilizes.

The data recorded during this interval is shown below as a three-frequency shift-keyed signal, but the recorded wavelengths will be modified by the speed variations because $\lambda = v/f$.

PLAYBACK ON A CONVENTIONAL HIGH-INERTIA TRANSPORT

In this case, we assume that reproduction is with a conventional recorder/reproducer with capstan flywheels and the synchronous motor drive in laboratory environment, and the tape speed is therefore constant with the exception of the inherent transport flutter. But the average speed may be off by as much as 0.5–1 percent, due to tolerance build-up between different transports. This is indicated by ΔV, and will result in an overall increase of all frequencies and faster display in the events, as shown by the deviation in the time base diagram.

With otherwise constant speed, we will now get the signal frequencies reproduced in correspondence with the recorded wavelength ($f = v/\lambda$). This is illustrated in the middle of the diagram of Fig. 27.3 under (b), where the following errors are observed:

- The frequencies are no longer a true replica of the original frequencies f_1, f_2, and f_3.

- The pulse lengths are no longer true replicas either.

- All the pulses are on the average shorter, and they all follow each other faster due to the overspeed ΔV of the reproducer. This time base error Δt is plotted in the lower diagram and illustrates the cumulative error in timing if the events recorded are observed from $t = 0$.

How much error of one or the other kind can be tolerated depends entirely on the type of data being recorded, its form, and the desired accuracy. Let us as an example assume that a frequency analysis of the recording is required.

The frequency deviations resulting from the initial speed variations will pose a limit on how selective the analysis filters can be made for a given accuracy. The average deviation ΔV results in a corresponding frequency shift Δf that can be used as a correction to the frequency scale. This is most easily obtained by recording a fixed-frequency control track on the tape. By comparing the frequency deviations during playback with a frequency equal to the one recorded, one can extract a speed error signal.

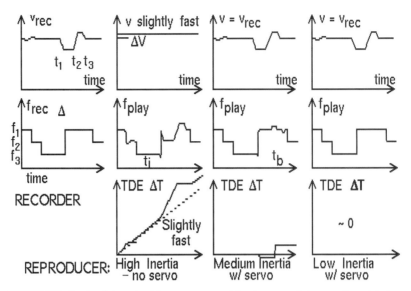

FIGURE 27.3 Speed variations, signal frequency errors, and time base errors during recording and during playback. Three playback units are shown: high-, medium-, and low-inertia drives.

The errors introduced by the transport flutter can be quite extensive. Very large variations (1 to 10 percent), as exemplified by t_1, t_2, and t_3, are virtually destructive and will in most cases be categorized as loss of data. The only means to recover the recordings is by using a servo-controlled tape drive during reproduction.

This was earliest done by recording a 17-kHz control frequency, amplitude-modulated at the rate of 60 Hz. During playback this signal was demodulated and fed into a phase comparison circuit, to which was also fed a precision 60 Hz voltage-controlled oscillator. The circuit's output, after amplification, drove the synchronous capstan motor. This technique is referred to as speed control. The maximum error rate that could be corrected was 1 Hz, and the method would therefore only handle the average speed variation ΔV, not the rapid speed changes.

PLAYBACK ON A LOW-INERTIA PLAYBACK TRANSPORT

A great amount of research and development in the industry has resulted in tape transports with medium- to low-inertia drives, employing elaborate servo circuitry and DC drive motors. This was illustrated in Fig. 15.12.

The increased accuracy during playback is shown in Fig. 27.3, where a distinction is made between a unit with medium inertia (c) and one with low inertia (d), assuming both are driven with identical forces. The inertia restricts the time for making speed corrections, and a useful figure of merit is the saturation acceleration (equal to the peak torque of the drive motor divided by the inertia) in connection with the cut-off frequency f_s for the servo-controlled system. This is exemplified by the acceleration dv/dt required at time t_2. The low-inertia drive is capable of handling this acceleration, while the medium-inertia drive cannot. It loses synchronization with the control frequency on the tape, and the "real" time is lost in addition to suffering severe frequency disturbances (as illustrated at the time corresponding to t_2).

The specification for speed accuracy is named TDE (time displacement error) or TBE (time base error), which gives the residual dynamic time difference from "real" time. In a low-TBE recorder-reproducer, the timing error is reduced by an order of two magnitudes, but the flutter is not reduced by the same amount. This is illustrated in Fig. 27.4, where the typical flutter spectrum is shown for a conventional transport, consisting of a number of discrete components and random distributed scrape flutter. When a servo mechanism is employed it removes the components below f_s, which is the upper frequency response of the electromechanical servo. When the cumulative flutter is measured (dc-10 kHz) there is a reduction of flutter, but it does not correspond to the reduction in TBE, except below f_s.

In spectrum analysis work, it has been shown that the criterion for full recovery of the power in a signal is:

$$BW \geq 3af_1 \tag{27.4}$$

where:

    BW = filter bandwidth
     a = peak flutter as a fraction of the average speed
    f_1 = frequency of flutter component

Below f_s the flutter components are removed, and the bandwidth BW should not be narrower than that stated above, where f_1 corresponds to the analysis frequency and can be taken as the cumulative peak flutter up to that frequency. The design of a low-inertia TBE recorder-reproducer should therefore not trade off flutter for low TBE, which easily can occur if the servo system has resonances in the scrape flutter range. Figure 27.4 also lists representative tape transport specifications to serve as a guideline in establishing accurate requirements for recorders/reproducers.

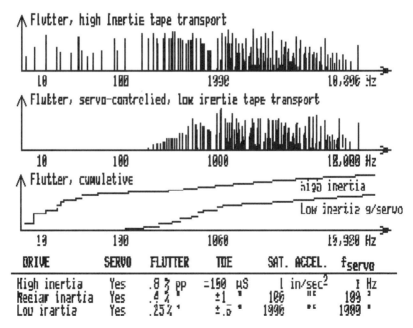

FIGURE 27.4 Flutter spectra for a high- and a low-inertia tape drive. Flutter reduction is only effective below the mechanical servo's high frequency.

TIMING ERRORS BETWEEN TRACKS (SKEW)

A multichannel recorder-reproducer exhibits some predictable time differences between its various data channels. These might or might not cause problems in correlating time events between the tracks. There are several causes for the interchannel timing errors, generally called skew:

- Gap-scatter
- Head-stack spacing
- Head tilt

    Dimensional changes with temperature, head wear

    Tape motion irregularities, slitting tolerances.

A worst-case analysis provides the following numbers to assist in an assessment of skew error relating to a 1.5-inch staggering head set:

- Gap scatter   5 µm
- Stack spacing   50 µm
- Head tilt   14 µm
- Tape tension change 1 ounce   6 µm
  50° F temperature change   19 µm
  100% humidity change   41 µm
  Dynamic tape skew   6 µm
                         Total: 141 µm

which is equivalent to a *TDE* of 47 µsec at 120 IPS.

The errors from the 1.5-inch tape span between heads in a staggered head assembly can be eliminated by using only one head stack:

- Gap scatter   5 µm
- Head tilt   14 µm
  Dynamic tape skew   6 µm
              Total: 25 µm

or *TDE* = 8.5 µsec at 120 IPS.

The items marked with • can, upon playback, be corrected by means of fixed delay lines. The dynamic skew from the tape motion irregularities can be reduced by means of electrically variable delay lines. The extensive use of digitized data for today's recordings makes skew elimination easy. Each data channel has a storage buffer for the read signals, which are recombined in perfect synchronization by the clock (see later under HDDR).

TAPE DUPLICATION

A certain amount of signal degradation occurs during duplication of tapes. Frequency response and signal-to-noise ratio are affected, along with an increase in envelope distortion (which will result in distortion of complex waveforms).

Envelope distortion is a measure for the delay times for the various frequencies in a transmission system. A typical figure is 500 nanoseconds peak-to-peak, for a bandwidth of 100 Hz to 1.2 MHz in an instrumentation recorder. This distortion occurs partly in the write/read process, and partly in the magnetization pattern in the coating. This distortion can become quite severe when several generations of duplicates are made. The last duplicate may have transients entirely different from the original signal.

The cumulative phase errors in magnetic recording are one of the reasons why the "direct-to-disc" recordings have been successful in the audio industry.

The phase error can be reversed by duplicating backwards. First, for critical recordings, it is always advisable to begin each reel of tape with a recording of not only selected frequencies for response setting, but also square waves. This will allow for proper setup of the equalizers in the system where tapes are copied (or reproduced).

Copies will have better fidelity if duplicated backwards. The procedure is as follows: Leave the recorded tape on the take-up reel, and transfer this reel to the supply reel hub. Now duplicate by running this tape through the recorder (and the duplicate will be correct, with the beginning of the recording at tape start). For additional copies, rewind the master tape each time and duplicate backwards again.

We close this section with a useful graph that should be applied to correct meter readings when the signal-to-noise ratio of a signal is less than 10 dB (Fig. 27.5).

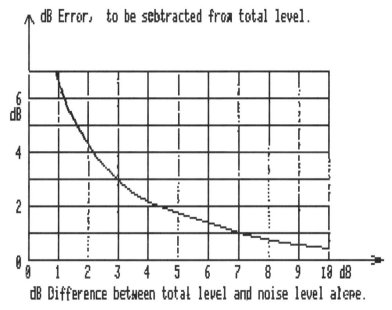

FIGURE 27.5 Correction of signals measured with poor SNR.

FM AND PCM RECORD TECHNIQUES

Signals that contain dc components cannot be played back by conventional inductive read heads. Flux-sensitive heads have never been successful in practical application for readout to DC. Some form of modulation is required.

Two basic methods for recording and playback that extend to DC are available: frequency modulation and pulse modulation. This chapter will explain the principles of FM and how it is employed in recording equipment. The prime area for FM is in instrumentation and television recording, while PCM is the prevailing technology in recording of audio, high speed instrumentation (HDDR, or high-density digital recording), studio television, and desktop video (computer-generated pixel graphics). Add to this that all computer programs and files are digital streams of bits, i.e., a form of PCM.

The material in this section is limited to the fundamental aspects of FM recording. FM provides response to dc, and it is immune to the signal amplitude variations in direct recording (with or without ac bias). Amplitude noise is reduced in proportion to the modulation index β, which we will soon characterize. Disturbances in tape motion (flutter) will act as a modulation of the FM signal, and they will consequently be detected as noise. A nonlinear phase-versus-frequency response will likewise introduce distortion.

The bandwidth of the recorder channel depends upon the selected value of β. It will in general be several times wider than the bandwidth of the data. (The signal-to-noise ratio is proportional to β. The channel bandwidth is also proportional to β.)

FM-recording consists of recording a signal that varies in frequency with the data input. An oscillator in the record amplifier provides an output signal of a frequency that depends on the signal input voltage. This is called a voltage-controlled oscillator, or FM modulator (Fig. 27.6). For zero voltage input, the frequency of the signal is constant and we are recording what we call the carrier. A plus or minus voltage to the FM modulator will change the frequency up or down. When these frequencies are played back, we can detect and demodulate them, which results in an output voltage that is identical to the input signal transmitted to the FM modulator.

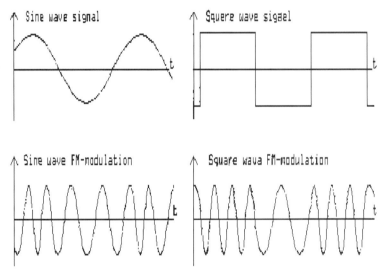

FIGURE 27.6 The principle of frequency modulation.

The expression for a periodic signal may be:

$$f(t) = A \times \cos(2\pi f_c t)$$

where A is the signal amplitude and f_c the signal frequency, also called the carrier frequency. When f_c is varied by a data or video signal we obtain frequency modulation. In general, this can be expressed as:

$$f(t) = A \times \cos(2\pi f_c t + \beta \times \sin(2\pi f_s t)) \tag{27.5}$$

where:

f_c = carrier frequency
f_s = signal frequency
β = modulation index

The modulation index β is a factor that results in a maximum deviation of $2\pi\Delta f_c t$ of β (when $\sin(2\pi f_s t) = 1$), where Δf_c is the frequency deviation away from f_c.

We can normalize f_s to Δf_c by:

$$\beta = \Delta f_c / f_s$$

or

$$\Delta f_c = \beta \times f_s.$$

We can similarly normalize f_c to Δf_c by introducing the deviation ratio D:

$$D = \Delta f_c / f_c$$

β expresses how "hard" we modulate the carrier frequency f_c, and is proportional to the amplitude of the data signal; just how fast f_c is varied is of course determined by the term $\sin(2\pi f_s t)$. We could also write $\beta = B \times \beta'$, where B is the data signal bandwidth and β' is the sensitivity of the FM modulator. β' is then equal to:

$$\Delta f_c / f_c$$

THE FM SIGNAL SPECTRUM

Our formula for frequency modulation can be expressed as a carrier signal of frequency f_c, surrounded by an often large number of sidebands of frequencies that are equal for f_c plus and minus f_s, $2f_s$, $3f_s$, etc. Their amplitudes vary in a very complex fashion according to a series of Bessel functions, and (27.5) becomes:

$$f(t) = A\{J_o(\beta)\cos\omega_c t$$

$$+J_1(\beta)\cos(\omega_c + \omega_s)t - J_1(\beta)\cos(\omega_c - \omega_s)t$$

$$+J_2(\beta)\cos(\omega_c + 2\omega_s)t - J_2(\beta)\cos(\omega_c - 2\omega_s)t$$

$$+J_3(\beta)\cos(\omega_c + 3\omega_s)t - J_3(\beta)\cos(\omega_c - 3\omega_s)t$$

$$+J_4(\beta)\cos(\omega_c + 4\omega_s)t - J_4(\beta)\cos(\omega_c - 4\omega_s)t....\} \qquad (27.6)$$

where $J_n(\beta)$ is the Bessel function of the first kind with argument β and order n, n being an integer. The values of $J_n(\beta)$ are tabulated in all books on FM and in mathematics handbooks. Figure 27.8 shows the variation of $J_o(\beta)$ = carrier signal, $J_1(\beta)$ = the first set of side bands, $J_2(\beta)$ = the second set of side bands, $J_3(\beta)$ = the third set of side bands, etc.

The distance between the individual sidebands and the carrier is equal to ω_s, the signal frequency. The number of sidebands is proportional to the modulation index. Typical sideband spectra are shown in Fig. 27.7.

It is apparent that there is an infinite number of sidebands around the carrier. We do not need all of them for demodulation and we can limit the transmission bandwidth (of the recorder) to include only those sidebands that are greater than one percent of the unmodulated carrier amplitude.

We can then determine the bandwidth (BW) of a sinusoidally modulated FM carrier for any modulation index β by determining the number of Bessel functions $J_n(\beta)$ that exceed 0.01, and then multiplying by $2\Delta f_s$. This bandwidth is plotted in a universal form in Fig. 27.8.

An example is the FM broadcast signal, and we are all familiar with its excellent quality. The maximum modulation frequency is 15 kHz, and the broadcast stations might use a maximum deviation of

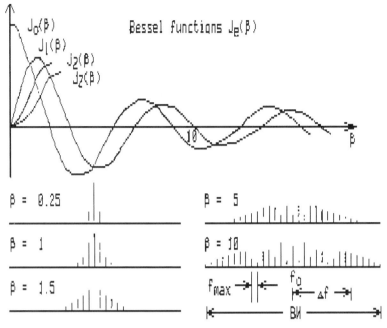

FIGURE 27.7 Bessel functions and spectra of a frequency-modulated signal carrier.

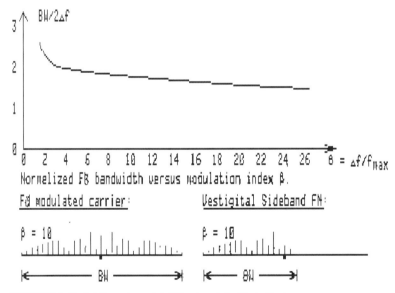

FIGURE 27.8 Bandwidth requirements in FM systems (*after Clarke-Hess*).

75 kHz for this signal. The value of β is therefore five. From the graph we determine $BW/2\Delta f = 1.6$. Consequently a bandwidth of $2\Delta f \times 1.6 = 150 \times 1.6 = 240$ kHz is required for transmission of the 15 kHz signal.

In order to record this signal, a recorder bandwidth of 250 kHz minimum would be required. This alone makes FM a poor candidate for audio recordings. Add the problem of flutter, which would not only be heard as a wobble of the pure tones but would also produce signals at any distinct flutter frequencies.

Modern VHS, S-VHS, and Hi-8 video recorders do record and play FM stereo (and four channels) with great fidelity. The recording is not along the longitudinal track, but on the slanted tracks through frequency multiplexing with the video and color signal. The flutter is nonexistent due to the high speed and inertia of the head wheel, while the tape essentially is standing still. The quality is great.

Instrumentation recorders are often used to record FM/FM telemetry, and they use a deviation ratio of about five. In wideband recordings the FM deviation ratio ranges from one to two.

Recordings of high accuracy, high *SNR*, and reasonable bandwidth are accomplished through a combination of digital coding (A/D) and recording technology from instrumentation, i.e., an ac-biased linear channel.

STANDARDS

Standards on instrumentation recording covers input and output characteristics, impedances, frequency response, tapes, reels, track configurations, etc. They are issued by ANSI, ISO, and IRIG. Write to ANSI for theirs and the ISO standards. IRIG's standard 106 (telemetry) and 118 (text methods) can be obtained from:

National Technical Information Service
U.S. Department of Commerce
5285 Port Royal Road
Springfield, VA 22161

BIBLIOGRAPHY

Kalil, F., and A. Buschman (editors). 1986. *High-Density Digital Recording*. NASA Ref. Publ. No. 1111, 313 pages. (Order from Superintendent of Documents, U.S. Government Printing Office, Washington, D.C. 20402.)

Lemke, J.U., 1988. Instrumentation Recording. Chapter in *Magnetic Recording*, ed. by C.D. Mee and E.D. Daniel, McGraw Hill 1995.

Mallinson, J.C. June 1990. Achievements in Rotary Head Magnetic Recording. *Proc. IEEE* 78 (6): 1004–16.

Moulin, Pierre, 1975. L'Enregistrement Magnetique d'Instrumentation, *Editions Radio*. 416 pages.

Shcherbitskii, V.G. July 1991. Recorder of Pulse Arrival Times. *Inst. and Exp. Techn.* 34 (4): 823–26.

Shibaya, H., K. Yokoyama, T. Kido, M. Matsumoto, I. Obata, and R. Tsunoi. March 1991. A Hi-Vision VTR for Industrial Applications Using a ½-in. Videocassette. *SMPTE Journal* 100 (3): 173–77.

Wood, R. May 1990. Denser Magnetic Memory—Magnetic Megabits. *IEEE Spectrum* 27 (5): 32ff.

Wright, A. Feb. 1992. Analogue Data Storage—Speaking of the Future. *Electr. World & Wireless World*. 98 (1671): 110–113.

Yan, C., and J.C. Diels. June 1991. Amplitude and Phase Recording of Ultrashort Pulses. *Jour. Optical Soc. America* -B 8 (6): 1259–63.

AMPLITUDE VARIATIONS

Voelz, H..1967. Spurhoehe, Spurzahl und Kanalkapacitaet bei der magnetischen Speicherung. *Hochfreq. und Elektroakustik:* 76 172–175.

TAPE SPEED VARIATIONS; TDE

Chao, S.C. March 1966. Flutter and Time Errors in Instrumentation Magnetic Recorders. *IEEE Trans. Aerospace and Elecr. Syst.* AES-2 (2): 214–223.

Mullin, J.T. March 1953. Flutter Compensation for FM/FM Telemetering Recorder. *Conv. Rec. I.R.E.* Pt.1—Radar & Telemetry, pp. 57–65.

Ratz, A.G. Dec. 1964. The Effect of Tape Transport Flutter on Spectrum and Correlation Analysis. *IEEE Trans. Space Electr. and Telemetry* pp. 129–134.

FREQUENCY MODULATION

Broch, J.T. 1966. FM Magnetic Tape Recording. *Bruel and Kjaer Tech. Rev.* 15 pages.

Jorgensen, F., "Distortion in Wideband FM Magnetic Tape Recording Systems," *Telemetry Jour.*, May 1968, Vol. 3, No. 3, pp. 53–56.

Robinson, J.F. 1981. *Videotape Recording*, Chapter 6 (FM Theory). Focal Press 362 pages.

Selsted, W.T. April 1953. A Low-Noise FM Recording System. *Jour. AES* 1 (?): 213–215.

Shaper, H.B. Nov. 1945. Frequency-Modulated Magnetic Tape Transient Recorder. *Conv. Proc. I.R.E.* 22 (11): 118–121.

CHAPTER 28
AUDIO RECORDING
ANALOG AND DIGITAL

Analog recording of audio was once the leading technology in magnetic recording. It was used extensively for recording high quality FM broadcasts, often the only way to capture select music sections.

Today the digital CD (Compact Disk) has replaced not just the vinyl LP record, but also the need for making recordings off the air (e.g., FM broadcast stereo programs). The music companies offer an abundance of both new recordings and restored earlier recorded masterpieces.

The next step would have been digital audio recorders for home use. Eight-millimeter PCM (pulse code modulation) and DAT (digital audio tape) were available, but expensive. Two new products have brought digital audio recording within reach of the consumer: The MO (magneto-optical) disk system from Sony (see Chapter 30) and the DCC (digital compact cassette) from Philips. Listener tests have rated the DCC system comparable to the audio CD, but the initial high cost of the hardware is delaying market penetration.

This chapter will deal with some aspects peculiar to the recording and playback of sound; music and speech in particular. Sound phenomena in acoustical research are recorded on special instrumentation recorders.

THE AUDIO SIGNAL AND EQUALIZATION

The allowable pre-equalization in audio recordings depends on the type of music to be recorded, because different instrument groupings and musical compositions have different energy spectra (Sivian et al. 1959). In order to have interchangeable tape recordings, the need for standards is obvious. There is a general agreement that any standard must pertain to the remanent flux on the recorded tape, and it is common among the standards that there is adherence to a flux level- versus-frequency response that essentially follows the coating thickness loss. This is not frequency but wavelength-dependent, which in turn means that equalization networks (or their component values) must change with different tape speeds. The variance in standards is found in the selected crossover frequency values for tape flux, or the equivalent time constant (70 μs, 120 μs, etc.).

The table in Fig. 28.1 illustrates the standards in USA (NAB, EIA) and Europe (CCIR, IEC) at four tape speeds. At the speed of 1⅞ IPS (4¾ CM/S) there is a spread in the crossover frequency of nearly 2:1, which gives a 6-dB level difference at high frequencies. A few recorders offer a selector switch between (for example) CCIR (Europe) and NAB (U.S.A.) standards, although this may be rather academic; most recordings are tailored to the individual recording engineer's tastes, which includes his control of bass and treble.

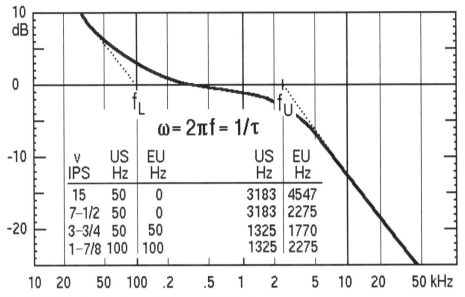

$$\omega = 2\pi f = 1/\tau$$

v IPS	US Hz	EU Hz	US Hz	EU Hz
15	50	0	3183	4547
7–1/2	50	0	3183	2275
3–3/4	50	50	1325	1770
1–7/8	100	100	1325	2275

FIGURE 28.1 Audio equalization standards for recorded flux levels.

The choice of standardizing organizations is also evident in the fact that NAB in the U.S.A. recommends a low-frequency boost during recording with the argument that less post-equalization is required with less amplifier hum and low-frequency noise, and therefore a lower general noise level (McKnight 1962).

On the other hand, CCIR in Europe argues that this very easily results in over-recording and intermodulation in program materials rich in low-frequency energy, such as organ music. European broadcast stations have not to this date recognized the NAB standard (Borwick 1980).

Improved tapes may result in different flux curves, but the different standards are nevertheless inconsistent. Different concepts are used and the same terms have conflicting uses, resulting in confusion when standards are compared. Standards should at times be updated, because tapes and tape recorders are improved through the ongoing evolution of about +1 dB/year. Certainly, current systems' SNR can be improved by as much as 10 dB with proper use of external pre- and post-equalizers (Fielder 1985).

Post-equalization in a tape recorder is most expediently adjusted by the use of a standard alignment tape, which contains first a high-frequency tone for playback head azimuth adjustment, then a normal record level tone (0 dB), and finally a series of different frequency tones for equalizer adjustment. The next step is setting the bias level for the selected tape and finally the trimming of the pre-equalizer network to obtain an overall flat level-versus-frequency response. These steps were covered in detail in Chapter 23.

The design of equalizer networks must be done so overall phase distortion in the record-tape-playback channel is minimized (Preis 1982). Variations in tape characteristics requires the use of one selected type only, unless the user is prepared to readjust the bias and possibly the record equalizer for each tape type; this operation may also be a built-in feature of the recorder (Sakamoto et al. 1982).

NOISE AND PRINT-THROUGH

Several noise sources exist within the magnetic record/playback process. Noise can originate in the heads, the electronics, and the tape itself (Skov 1964). Ideally, one would like the recorder to be absolutely quiet, with the tape as the only noise source.

Electronics noise is quite trivial, originating from power supply hum, low-frequency noise in the input stage, and resistor noise (hiss). These noise generators can be fairly well controlled by careful design and component selection, such as the use of low-noise resistors and high-grade transistors. If, for example, an insertion-type equalizer is used after the preamplifier, great care must be taken in optimizing the signal level and equalizer impedance to avoid added noise from the resistive components.

High-frequency noise in a wideband instrumentation recorder (1 to 2 MHz) is often generated in the reproduce head itself. It is caused by eddy-current losses, and the equivalent loss resistance of a playback head may be higher than the equivalent noise resistance of the preamplifier (50–100 ohms). That situation should be corrected by selecting a thinner lamination for the head core, or by using a ferrite core, possibly with hard tips.

Electronic noise sources are aggravated by the required post-equalization boost of low and high frequencies. A word of caution when evaluating a recorder's signal-to-noise ratio: it is normally measured without filters or a weighting network. The measured noise voltage is representative for the low- and high-frequency noise and does not show that the mid-band noise is much lower, about 20 to 40 dB for an audio recorder, and as much as 50 dB for a constant-flux instrumentation recorder. It would be entirely proper to use a weighting network when evaluating an audio recorder, because the shape of the noise spectrum happens to be close to Fletcher-Munson's curves for the ear's sensitivity (McKnight 1960). The subjective signal-to-noise ratio, therefore, is closer to 70 to 80 dB for a recorder that measured 50 dB on a voltmeter, without filtering. It is likewise more useful to evaluate an instrumentation recorder from a spectrum of its noise, rather than by a single broadband measurement.

Another major noise source in a tape recorder is the tape itself, and it shows up in various ways that are characteristic for magnetic recording. If a virgin tape is played in an audio recorder, one will notice a hissing sound. This noise is caused by the limited and varying number of particles that the playback head "sees." Each particle is a minute permanent magnet and if a sufficiently large and constant number were "seen" by the playback head, their fields would cancel out and no noise voltage would be generated.

If the same tape is again played on the recorder, this time with the record button pressed, the noise level will be several dB higher. This bias-induced noise remains a limitation of the signal-to-noise ratio in modern recorders, and can at best be reduced to a level of 3 dB higher than virgin tape (or bulk-degaussed tape). Investigations of this noise indicate two categories of bias noise.

When the bias frequency is higher than that resulting in a wavelength of 2 µm (80 µin), noise appears to arise from a combination of the bias field and the interaction field between particles. A lower bias frequency does not only increase the noise level, but it is now generated as modulation noise. This is illustrated in Fig. 28.2a.

AM modulation noise (or noise behind the signal) is best understood if for a moment we consider the dc magnetization noise of a magnetic tape. A dc magnetized tape can have a noise level 20 to 30 dB higher than a virgin tape. This is due to the nonuniformity of the coating dispersion, the backing surface, and the coating surface. An ac signal likewise generates noise, and can be considered a slowly varying dc signal. The noise is formed as sidebands around the signal frequency (Fig. 28.2b). This noise is not very noticeable, because it is masked by the signal. Only when pure tones are recorded and played back can the ear detect AM noise.

Returning to the bias-induced noise, it is likely that a low-frequency bias generates AM noise with sidebands reaching into the pass band of the recorder. These findings were made with a distortion-free bias oscillator (Ragle and Smaller 1965). Additional noise will be generated if the bias signal contains even harmonic components. These in essence constitute a dc signal, for which we have already explained the noise mechanism. The intermittent, gurgling sound of this noise is quite different from hiss. It is best avoided by using a push-pull type oscillator, possibly with filtering. Some

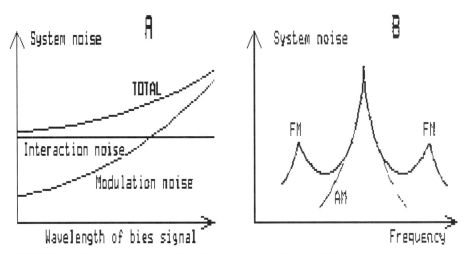

FIGURE 28.2 (a) Bias noise shown as function of bias wavelength. (b) AM versus FM modulation noise.

recorders use a bugging dc current through the record head winding, but that is only a cure, not a solution. AM modulation is often mistaken for FM modulation noise, which is caused by scrape flutter in the tape transport mechanism. This generates sideband, as shown in Fig. 28.2b, and can be very disturbing in changing the sound of a pure tone into a harsh-sounding note.

There are several sources for noise in recording equipment. Some have already been discussed, such as the ones above, plus the section on print-through in Chapter 14. Another noise-like effect originates in surface imperfections, called dropouts.

NOISE FROM DROPOUTS

The noise from dropouts may be quite annoying. The nature of this noise is really a momentary reduction (attenuation) of the playback signal and its value can be rated in terms of the duration and the amount of the attenuation (Bauer et al., 1967). The dropouts vary in duration. The average value is typically represented by an equivalent tape length of 10 mils, which in time corresponds to 5 milliseconds at a speed of 4.75 CM/S (1⅞ IPS).

The momentary loss of signal is an amplitude modulation, with associated sidebands in the frequency domain. When a sinusoidal tone drops out and recovers quickly enough, clicks can be heard, which are loudest during the steepest transitions.

Measurements have established that dropouts shorter than a few ms are in practice inaudible (Trendell 1969; Admiraal et al. 1977). Admiraal established an annoyance scale for use in listening tests:

$$h = 0 \quad \text{no annoyance}$$
$$= 1 \quad \text{audible, but not annoying}$$
$$= 2 \quad \text{annoying}$$
$$= 3 \quad \text{very annoying}$$

dropouts longer than 20–50 ms were rated $h = 0$ for a 3-dB attenuation, $h = 1$ for a 5-dB attenuation, and $h = 2$ for a 8- to 10-dB attenuation. They were very annoying only for attenuations near 20 dB. Repeat dropouts were more annoying than single dropouts.

A 25 ms dropout at 4.75 cm/s corresponds to a tape defect that is about 50 mils long. This can easily be seen without a magnifying glass, and such tape defects should not exist in even the cheapest tapes manufactured today. They are much more likely to occur because of leaving tapes lying around without a protective case.

It appears that the earlier described modulation noise (see Chapter 12) is more annoying in recordings, lending support to the success of "direct-to-disc" recordings and (recently) to the optical disks. Admiraal noted that flute music is the most sensitive of all to modulation noise. For flute music, the modulation noise is perceptible as soon as the signal-to-modulation noise ratio goes below about 40 dB. With orchestral music, however, the modulation noise does not become audible until the signal-to-modulation noise ratio is less than about 20 dB.

CROSSTALK

The leakage of signals from recordings on adjacent tracks should be nonexistent. Some crosstalk can be tolerated in stereo recordings, where the difference between channels is small.

HEAD NOISE

The noise voltage from the losses in audio heads may be limiting for the SNR.

AMPLIFIER NOISE

A number of factors contribute to the noise in amplifiers modified by equalization.

SYSTEM NOISE

The sum of the noise voltage from sources listed above will limit the signal-to-noise ratio of any high-fidelity sound system. The effects depend heavily on the specified record and playback equalizations and the allowed maximum distortion level.

The designer and user of analog recorders must bear in mind that the concepts of system grounding, shielding, cable type (often two-conductor, overall shield) and potential ground loops are very applicable.

NOISE REDUCTION

Noise reduction methods are today equated with a Dolby system. There are several versions of this method, which essentially raises faint record levels during recording to bring the magnetization well above the noise levels. Upon playback, a complementary reduction in playback gain is made. This brings about a higher *SNR*. Something must tell the playback amplifier to ride gain, and this is what the Dolby system does (Dolby 1967). Because the details are beyond the scope of this handbook, the reader is referred to the special literature.

The ac-bias level in audio recorders is generally fixed at a level that corresponds to the application. This will be a high level in high-speed studio recorders and a lower level in home type recorders.

The high bias level is desirable for low distortion and the associated short-wavelength signal loss is eliminated by using a high tape speed in studio recorders (15 IPS). The loss would be excessive in home cassette recorders at the speed of 1⅞ IPS, and the bias setting is therefore lower than in the studio recorder.

Dynamic control of the bias setting has been implemented in some home recorders by automatic reduction of bias for low-level, high-frequency signals and increased bias for high-level, mid-frequency signals; the accompanying short-wavelength loss is masked by the higher signal level (Selmer Jensen, and Pramanik 1984).

MAKING RECORDINGS, EDITING, AND COPYING

The operation of any particular recorder is described in its instruction booklet or operating manual, which should be studied carefully. Familiarity with the manufacturer's instructions will prevent improper operation, and will make you aware of the recorder's capabilities and limitations. However, there are certain fundamental rules that apply to most recorders, and one concerns the proper recording level. This applies in particular to music recording.

It is essential to a good recording that it be free of distortion, and that it has a maximum signal-to-noise ratio. All recorders are limited by amplifier and tape noise, and the optimum signal-to-noise ratio is, therefore, attained by recording with as strong a signal level on the tape as possible. Recording at too high a level, however, will cause both harmonic and intermodulation distortion. In music and speech the natural sound is distorted, and in instrumentation recording the playback data will be erroneous.

The proper recording level can be attained only with a certain knowledge of the program material and some means of monitoring the recording level, for instance with a VU meter. A magnetic tape's overload sensitivity is essentially the same at all frequencies. In almost all cases, by contrast, the program material has an uneven amplitude-versus-frequency distribution, and the highest level that can be tolerated is represented by the peak level in the amplitude distribution. For example, a male voice is rich in low frequencies, while a female voice is rich in high frequencies. Figure 28.3 illustrates both the average amplitudes and the peak amplitudes for an orchestra, a pipe organ, and a piano. It is clearly seen that overload during recording of a pipe organ will be caused by the low frequencies, while for a piano it will be the middle frequencies. These examples also show that the sound level drops off rather fast toward high frequencies, and audio recorders incorporate pre-equalization that boosts the record current at high frequencies to take better advantage of the capabilities of the tape.

The curves in Fig. 28.3 are not universal, because they will change with orchestral composition, the type of music being played, and with the conductor. They will be further enhanced by the studio acoustics and the applied microphone technique. To provide interchangeable tapes, standards have been established for the amount of pre-equalization that can be applied at the various speeds.

THE RIGHT RECORDING LEVEL

Referring again to Fig. 28.3, the reader may question why two sets of graphs are shown for music amplitudes. The left set of curves shows the average amplitude as it would be measured with a VU meter, while the right set of curves shows the peak amplitude that could be monitored with a peak-reading instrument. The recording level is monitored by various devices, such as an inexpensive neon lamp, a VU meter, a peak-reading meter, or the earlier type "magic eye." In some instrumentation recorders, a series of small cathode ray oscilloscopes are installed for monitoring the input signal to each channel.

The most annoying type of distortion in tape recordings, both music and instrumentation, is intermodulation distortion. It occurs at the instant the peak level exceeds a certain amplitude. The peak-reading devices are clearly advantageous over the average type indicators by showing the operator when a recording level is too high. The VU meter is useful only when the operator or recording engineer has previous experience with the program material to be recorded. This experience can be obtained by rehearsing the record level setting before the actual recording. This is relatively easy to do

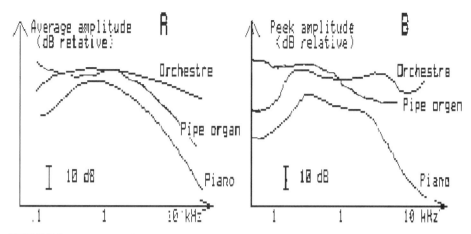

FIGURE 28.3   (a) Average amplitudes and (b) peak amplitudes for music programs (*after Sivian, Dunn, and White*).

on a recorder that has separate record and reproduce electronics; it is done by alternatively listening to the input signal and the playback signal, and comparing the sound quality. The gain control for the recording input is adjusted upwards to a level that does not result in intolerable distortion. If the recorder does not have separate record and reproduce electronics, it is necessary to record a small portion, rewind the tape and listen to it, and by memory compare the quality of the input signal with the playback signal. The next step is to repeat this process and so, by trial and error, obtain the proper record level.

Orchestral music often has a dynamic range that exceeds that of the recorder, and in this case it is common practice to "ride gain." The recording engineer will be familiar with the program, and can anticipate loud passages and very slowly reduce the gain before such a passage. This requires great skill and experience, and should not be attempted by a novice. The inexperienced operator would be better off with a recorder that has an automatic gain control. Such a control unit can be bought as an accessory for the recorder; but the true value of an automatic gain control is attained only if, upon playback, the AGC action is reversed, which in turn requires an additional control track on the tape. Recorded on the control track is a signal that represents the AGC action, and upon playback is fed to the control of another AGC unit. This method is employed in the Dolby systems.

The rapid rise of distortion versus input level is illustrated in Fig. 28.4. There is only a 10-dB margin between the one percent harmonic distortion level and full saturation of the tape. Intermodulation distortion, as a rule of thumb, is three times higher than the harmonic distortion level.

As a gauge, audio recording engineers use a level referred to as MOL, which is the maximum operating level. It is normally defined as being the magnetization in nWB/m that corresponds to three percent harmonic distortion of a 1-kHz fundamental tone. Figure 28.5 shows how the MOL of a typical modern tape (EMI 816) varies with bias current (Borwick 1980).

DIGITAL TECHNIQUES

Digital recording of sound was first made available to the consumer in the form of the Philips optically recorded disk (CD), while in the recording studios it was implemented in the form of digital multitrack tape recorders, much as the HDDR equipment described under digital applications of instrumentation recorders. These audio digital recorders use the same design philosophy concerning selection of interleaving, error correcting codes, and modulation codes as do HDDR instrumentation recorders.

There are some special considerations; selection of the sampling frequency has been of concern to ensure compatibility with the many interfacing systems, such as radio, TV, film, other studios, etc.

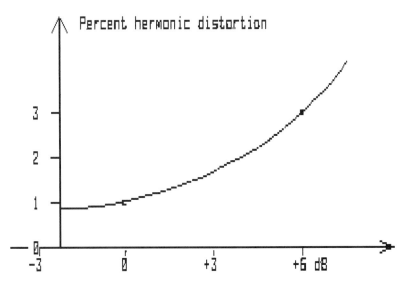

FIGURE 28.4   A typical curve for harmonic distortion versus record level in an ac-biased magnetic tape recorder.

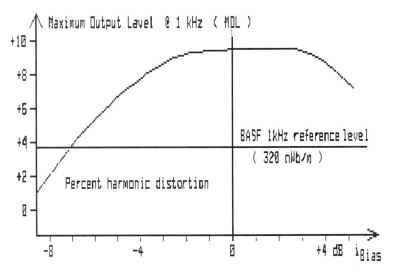

FIGURE 28.5   MOL (maximum overload level) as a function of ac bias level in a modern audio tape.

(Mruaoka et al. 1978). The particular application of digital techniques to audio recording is covered in two books (Watkinson 1991; Thomsen 1983), and three papers (Willcocks 1978; Blesser 1978; Bellis 1983).

Editing of digital recordings is not as straightforward as the splicing method for ordinary audio tapes. The methods are electronic, and they use delay lines and sophisticated controls (Youngquist 1982).

The aspect of distortion and quantization noise has been treated and shown to become minimized when the signal to become quantized contains a wide-band noise dither with an amplitude of approximately the step size (Venderkooy and Lipshitz 1984).

High-quality audio can now be enjoyed not only by the means of the optical digital disk (CD), but in the form of VCR tapes on newer machines with facilities for the recording and playback of FM audio in stereo (see standards for VHS and S-VHS in Chapter 29).

The next step in digital sound recording and playback came in the form of the small DAT (digital audiotape) cassette with 3.81-mm tape (Sakamoto 1984), the 8 mm VCR format (Watani et al. 1984) and today's DCC.

8 mm PCM (PULSE CODE MODULATION)

The audio signal in the 8 mm recorders is first converted into a digital signal, and then time-compressed by a factor of 6. Thus 6 channels can be recorded by dividing the 221 degrees of tape wrap around the capstan into 6 sectors of each 36 degrees, as shown in Fig. 28.6 (Itoh et al. 1985).

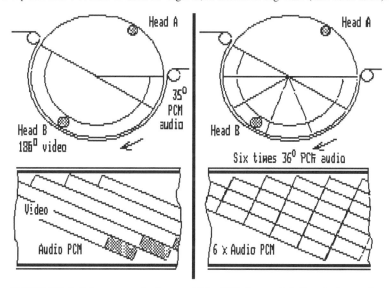

FIGURE 28.6 Left: 8 mm camcorders have an additional 36° of wrap, which allows for the recording of PCM sound. Six sectors would allow for 6-channel recording of PCM data.

R-DAT (ROTATING-HEAD DIGITAL AUDIO TAPE)

The new recorders using the 3.81 mm tape format are named R-DAT. The proposed cartridge measures only $73 \times 54 \times 10$ mm, and the tape is wrapped 90 degrees around the capstan drum (see Fig. 28.7). The relative velocity between the head gap and the track is approximately 5 m/sec (Watkinson 1991).

The audio signal is sampled at a rate of 44.1 kHz, and must be time-compressed due to the use of only two heads and the wrap angle of 90 degrees. The signal goes into a buffer memory, and is released to the head in blocks when either of the two heads are in contact with the tape. During playback a similar buffer memory stretches the data back into a continuous digital bit stream.

Synchronization between the rotating head and the tape is ordinarily done by a longitudinally recorded control track. The R-DAT recorders use a method developed by Philips called ATF, for automatic track finding. A short section of each track is recorded with a relatively low-frequency signal that has a frequency that alternates among four values for each half-rotation of the head assembly. During playback, the heads pick up more than the one track at a time (side reading at low frequencies), and proper alignment is obtained when the levels of the pilot tones from each track next to the

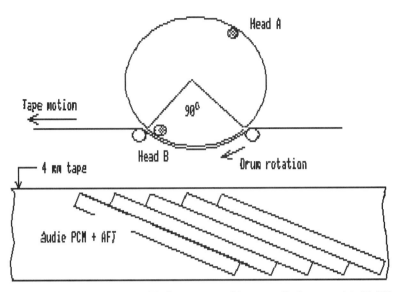

FIGURE 28.7   R-DAT recorders use 90° of tape wrap around the scanner. Tracks are recorded with AFT for track-following control.

one being scanned are equal. This system makes a tracking-control track unnecessary (Feldman 1986; Ranada 1987).

These new recorder developments have caused great concern in the audio industry, which fears widespread use of R-DAT in the unlawful duplication of CDs, with reduced revenues for the recording industry. In 1987 this led to an import restriction of the R-DAT into the United States.

DCC (DIGITAL COMPACT CASSETTE)

Philips has cleverly combined the latest signal-coding techniques, thin-film heads, and MR heads in a compact recorder design. The cassette is of the same form as the conventional consumer cassette. The new transports for the DCC will accept the conventional stereo cassette, thus offering backward compatibility.

The DCC recorders apply *precision adaptive subband coding* (PASC) to the digital audio data stream before recording. The 384 kb/s audio data output by the PASC encoder is doubled by error correction to 768 kb/s, which in turn is spread across eight parallel audio data tracks. Therefore each track operates at 96 kb/s, which at a tape speed of 1⅞ IPS translates into a packing density of 50 kb/in. The tape is chromium dioxide or cobalt-doped ferroxide.

The result is a data encoding method that requires only a quarter of the bit rate used in the CD (16-bit linear PCM) format, without any loss of fidelity. Typical audio performance is a frequency response from 5 Hz through 22 kHz, with a dynamic range of 105 dB (Piehl 1993).

The digital sound signal is recorded on eight parallel tracks (see Fig. 28.8), four tracks for the left stereo channel and four tracks for the right channel. A ninth track is for control search information, display, and subcode information. All nine tracks occupy half of the tape width.

The DCC head must have the ability to write and read the nine tracks on the DCC tapes plus the conventional stereo cassette tapes. The write heads are inductive thin-film heads, and the read heads are magnetoresistive elements. The written track width is 185 μm and the read width is 70 μm. The write-wide-and-read-narrow technique is used to facilitate interchange ability, and to reduce sensitivity to azimuth errors (Wood and Bonyhard 1992).

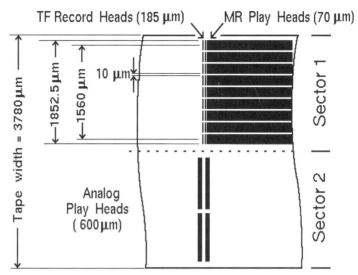

FIGURE 28.8   Arrangements of 9 parallel tracks in Philips' DCC recorder/player.

NT (NON-TRACKING)

Sony has developed a miniature tape format for broadcast-quality voice and noncritical music recordings in the field. The tape cassette measures $1.23 \times 0.82 \times 0.19$ inches ($30 \times 22 \times 5$ mm) and holds enough tape to record up to two hours (Pizzi 1993). The format's name refers to a radical departure from the tracking approach used in traditional rotary-head systems. There are two heads on the head drum. They are switched so that during recording a single write scan is made per revolution, while two read scans are made per revolution during playback. This means that playback scanning will overlap multiple recorded scans, as shown in Fig. 28.9. The NT head rotates at 3000 rpm while the tape speed is 0.25 IPS (6.35 mm/s).

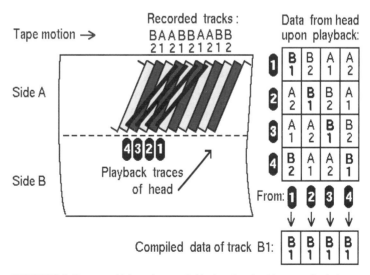

FIGURE 28.9   Recovery of information recorded in slanted tracks without a mechanical servo mechanism (see text for explanation).

After several consecutive playback passes, all the data in a recorded track of matching azimuth will have been read and sent to a buffer. It will be in scrambled order, but can be recompiled in proper order based on address data embedded in the recorded signals, as shown in Fig. 28.9.

STANDARDS

Several standards for the recorded flux level on magnetic tapes have been established in Europe and in the United States, standards that impose design restrictions on both recording and playback amplifiers. In transforming a record current to a remanent flux on the tape, a certain amount of pre-emphasis is necessary to overcome record-head losses. Likewise, in transforming the flux from the tape to a voltage at the playback amplifier input, reproduce-head core losses and gap losses affect the overall frequency response. It is therefore important that the designer not only know the head impedances, but also has the data that tells how the head losses vary with frequency.

The audio equalization standards were shown in Fig. 28.1. Details pertaining to these and other standards (reel, cassette, and cartridge sizes, track configurations, and dimensions) can be obtained from:

ITA—International Tape/Disk Association
10 Columbus Circle, Suite 2270
New York, N.Y. 10019

ANSI—American National Standards Institute
1430 Broadway
New York, N.Y. 10018

NAB—National Association of Broadcasters
1771 N Street N.W.
Washington, D.C. 20036

EIA—Electronic Industries Association
Engineering Department
2001 Eye Street N. W.
Washington, D.C. 20066

CCIR—International Radio Consultative Committee
International Telecommunication Union
Place des Nations
Geneva, Switzerland

Other standard organizations are:

- SMPTE—Society of Motion Picture and Television Engineers
- DIN—Deutscher Normenausschlus (DNA)
- BSI—British Standards Institute
- JIS—Japanese Standards Association
- IEC—International Electrotechnical Commission

REFERENCES

THE AUDIO SIGNAL AND EQUALIZATION

Borwick, J. (Editor), 1980. Sound Recording Practice, A Handbook compiled by the Association of Professional Recording Studios, Oxford University Press, London, 503 pages.

Fielder, L.D. Nov. 1985. Pre- and Postemphasis Techniques as Applied to Audio Recording Systems. *Jour. AES* 33 (9): 649–658.

McKnight, J.G. April 1962. The Case Against Low-Frequency Pre-Emphasis in Magnetic Recording. *Jour. AES* 10 (4): 106,107.

McKnight, J.G. Jan 1959. Signal-to-Noise Problems and a New Equalization for Magnetic Recording of Music. *Jour. AES* 7 (1): 5–12.

McKnight, J.G. July 1960. The Frequency Response of Magnetic Recorders for Audio. *Jour. AES* 8 (7): 146–153.

Preis, D. Nov. 1982. Phase Distortion and Phase Equalization in Audio Signal Processing—A Tutorial Review. *Jour. AES* 30 (11): 774–794.

Sakamoto, N., T. Kogure, M. Ogino, and H. Kitagawa. Sept. 1982. A New Magnetic Tape Recorder with Automatic Adjusting Functions for Bias and Recording Conditions. *Jour. AES*, 30 (9): 596–606.

Sivian, L.J., H.K. Dunn, and S.D. Write. July 1959. Absolute Amplitudes and Spectra of Certain Musical Instruments and Orchestras. *IRE Trans. Audio* AU-7 (3): 1–29.

NOISE, DROPOUTS

Admiraal, D.J.D., B.L. Cardozo, G. Domburg, and J.J.M. Neelen. Jan. 1977. Annoyance Due to Modulation Noise and Dropouts in Magnetic Sound Recording. Philips Tech. Review 37 (2/3): 29–37.

Bauer, B.B., and F.A. Comerci, E.J. Foster, and A.J. Rosenheck. Feb. 1967. Audibility of Tape Dropouts. *Jour. AES* 15 (2): 147ff.

NOISE

Dolby, R.M. Oct. 1967. An Audio Noise Reduction System. *Jour. AES* 15 (4): 383–388.

McKnight, J.G, Mar. 1960. A Comparison of Several Methods of Measuring Noise in Magnetic Recorders for Audio Applications. *IRE Trans. Audio* AU-8 (?): 39–42.

Ragle, H.U., and P. Smaller June 1965. An investigation of high-frequency bias-induced tape noise. *IEEE Trans. Magn.* MAG-1 (2): 105–110.

Selmer Jensen, J. and S.K. Pramanik. Aug. 1984. Dynamic Bias Control with HX Professional. *Audio Engineering* 68 (8): 31–41.

Skov, E. P. Apr. 1964. Noise Limitations in Tape Recorders. *Jour. AES* 12 (4): 280ff.

Trendell, E.G. June 1969. The Measurement and Subjective Assessment of Modulation Noise in Magnetic Recording. *Jour. AES* 17 (6): 644–653.

DIGITAL TECHNIQUES

Bellis, F.A. Oct. 1983. Introduction to Digital Audio Recording. *The Radio and El. Engr.* 53 (10): 361–368.

Blesser, B. A. Oct. 1978. Digitization of Audio: A Comprehensive Examination of Theory, Implementation, and Current Practice. *Jour. AES* 26 pp. 739–771.

Feldman, L. July 1986. Digital Audio Tape Decks. *Modern Electronics* pp. 18–22.

Itoh, S., Y. Watatani, N. Azuma, K. Kaniwa, H. Masui, T. Nakama, A. Shibata, K. Watanabe, and Y. Mogi Aug. 1985. Multi-Track PCM Audio utilizing 8 MM Video System. *IEEE Trans. Cons. Elec.* 31 (3): 438–446.

Mruaoka, T., Y. Yamada, and M. Yamazaki, Sampling-Frequency Considerations in Digital Audio. *Jour. AES*, Apr. 1978, Vol. 26, pp. 252–256.

Piehl, M. Feb. 1993. The Philips Digital Compact Cassette (DCC). *Broadcast Engineering* 35 (2): 52–55.

Pizzi, S. Feb. 1993. New audio recording formats (NT). *Broadcast Engineering* 35 (2): 60–63.

Ranada, D. Feb. 1987. Why 8 mm? *High Fidelity* pp. 47–52.

Sakamoto, N., T. Kogure, H. Kitagawa, and T. Shimada. Sept. 1984. On High-Density Recording of the Compact-Cassette Digital Recorder. *Jour. AES* 32 (9): 640–645.

Thomsen, D. 1983. *Digitale Audiotechnik*. Franzis' Verlag, Muenich, 192 pages.

Venderkooy, J., and S.P. Lipshitz. Mar. 1984. Resolution Below the Least Significant Bit in Digital Systems with Dither. *Jour. AES* 32 (3): 106–112.

Watatani, Y., S. Itoh, A. Shibata, and K. Mohri. Nov. 1984. The FM Audio Signal Recording System for 8 mm Video. *IEEE Trans. Cons. Elect.* 30 (4): ?.

Willcocks, M. Feb. 1978. A Review of Digital Audio Techniques. *Jour. AES* 26 (?): 56–64.

Wood, W.F., and P.I. Bonyhard. Dec. 1992. Magneto-Resistive Heads To Spur DCC. *Computer Technology Review* pp. 16–21.

Youngquist, R.J. Dec. 1982. Editing Digital Audio Signals in a Digital Audio/Video System. *SMPTE Journal* 91 (12): 1158–1160.

BIBLIOGRAPHY

Ballou, G.M. 1991. 2nd edition. *Handbook for Sound Engineers: The New Audio Encyclopedia,* Howard Sams 1506 pages.

Benson, B.K. (Editor), 1988. *Audio Engineering Handbook*, McGraw-Hill 1040 pages.

Dare, P.A., and R. Katsumi. Oct. 1989. Rotating Digital Audio Tape (R-DAT): A Format. *SMPTE Journal* 96 (10): 943–48.

Henriques, J.A., T.E. Riemer, and R.E. Trahan, Jr. Sept. 1990. A Phase-Linear Audio Equalizer: Design and Implementation. *Jour. AES* 38 (9): 653–66.

Huber, D.M. 1988. *Microphone Manual, Design and Application*. Howard Sams 297 pages.

Iwaki, T., T. Okuda, K. Koyanagi, Y. Yokomachi, C. Yamawaki, and T. Sasada. Aug. 1990. Signal Processing of a 20-Bit 8-Channel Digital Audio Recorder. *IEEE Trans. Consumer Electronics* 36 (3): 647–654.

Koenig, B.E. Jan. 1990. Authentication of Forensic Audio Recordings. *Jour. AES* 38 (1/2): 3–33.

Miyaoka, S. March 1984. Digital Audio is Compact and Rugged. *IEEE Spectrum* 21 (3): 35–39.

Monforte, J. Dec. 1984. The Digital Reproduction of Sound. *Scientific American* 251 (6): 78–84.

Olson H.F. 1967. *Music, Physics and Engineering*, Dover, New York, 460 pages.

Rundstein, R.E., and D. Miles Huber, 1986. *Modern Recording Techniques*, Howard W. Sams and Co., 384 pages.

Schetina, E.S. 1993. *Complete Guide to Digital Audio Tape Recorders: Including Troubleshooting Tips*. Prentice Hall 208 pages.

Sinclair, W.L., and L.I. Haworth. Aug. 1991. Digital Recording in the Professional Industry. 2. Studio Techniques. *Elec. & Comm. Engr. Journal* 3 (4): 177–184.

Thiele, H. Aug. 1988. Magnetic Sound Recording in Europe up to 1945. *Jour. AES* 36 (8): 396–408.

Transmission Format for Linearly Represented Digital Audio Data. Dec. 1985. *Jour. AES* 33 (12): 975–984.

Watkinson, J. 1991. *RDAT*, England, Focal Press, 244 pages.

Watkinson, J. 1993, 2nd edition. *Art of Digital Audio*. Focal Press 500 pages.

Zaza, A.J. 1991. *Mechanics of Sound Recording*. Prentice Hall 400 pages.

CHAPTER 29
VIDEO RECORDING

The recording and playback of television pictures place unique requirements on the tape transports and the electronics. The signal channel must cover a range from very low frequencies (essentially dc) to several MHz, the *SNR* must be 40 dB (or better), and the time base errors must be within fractions of a microsecond.

The field of video recording is vast, with numerous tape and machine formats in operation. This chapter describes what the an~'og and digital video signals are. The reader is referred to the book edited by Bens~~ ~out television signals, systems, and recorders. The SMPTE Journal i~ ew developments (see the Bibliography at the end of this chapter). ion Picture and Television Engineers.

ht intensity (luminance) and color (chrominance), and is continu-
~r requires that pictures are displayed in rapid sequence so the
on without staccato. Complete pictures are in the USA shown at
·ope, the standard is 25 per second.

FINN JORGENSEN, M. Sc.

MAGNETIC DATA STORAGE

Seminars - Consulting

DANVIK   1201 Bel Air Drive, Santa Barbara, CA 93105, USA. (805)-682-2102 FAX 682-6463

SIGNAL; RESOLUTION

~nal depends on the display of each frame. Each frame consists
~~, called fields. A close look of the picture tube shows horizontal lines where
...~s (1-3-5-...–262½) belong to the odd field, and the even lines (½-2-4-262) to the even field, making a total of 525 horizontal lines in the NTSC system (625 in PAL). NTSC stands for National Television System Committee, and PAL is Phase Alternation Line-rate.

The picture's width is larger than its height by the aspect ratio 4:3 (which is slated to increase with HDTV). We can define a measure for the picture resolution: All even lines are white and all odd are black, allowing for 525 alternating black-and-white pixels on a vertical line. The maximum number of pixels along a horizontal line corresponds to $(4/3) \times 525 = 700$ pixels. The total number of pixels per frame is $525 \times 700 = 367,500$. Some lines at top and bottom (and some pixels at both sides) are not visible (overscan); but they do have to be recorded and played back.

These pixels are shown 30 times a second, resulting in a throughput of $30 \times 367,500$ pixels, which corresponds to a signal frequency of $11,025,000/2 = 5,512,500$ Hz = 5.5 MHz.

This represents the ultimate in picture quality from an NTSC system. In practice, fewer picture elements (exclusive of overscan) are shown per frame:

NTSC	210,000
PAL	300,000
HDTV	1,000,000
35 mm color slide	30,000,000

It can be seen that even HDTV will not have photographic quality. The practical frequency response in current NTSC systems is from dc to 3.5 MHz, and the signal energy is near maximum at low frequencies (and dc), falling gently off to zero at 3–3.5 MHz.

The dc content has made magnetic recording of video very difficult, because playback is based on Faraday's principle ($e = n \times d\phi/dt$, formula 6.1); no voltage is induced for dc levels, although they are easily recorded.

This problem was solved by using FM modulation of the video signal, which will be covered shortly. Let's first look at the video signal we wish to record, including color information.

THE ANALOG COLOR VIDEO SIGNALS

A color video signal consists of three separate signals, generated in the camera: Red, Green and Blue. This set is called the RGB-signal, and to this must be added a synchronization signal.

Figure 29.1 shows the sensitivities of a typical RGB signal, which contains red, green, and blue signals. Two new signals are generated from the RGB signal, namely the luminance and chrominance signals.

The luminance signal, also called the Y signal, is a monochrome signal that describes the distribution of luminance levels in the picture. The chrominance signal, also called the C signal, is a combination luminance/chromaticity data stream, because it describes the distribution levels and chromaticity values in the picture (R,G,B).

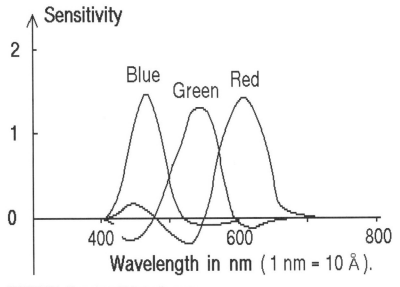

FIGURE 29.1 The eye's sensitivity to color ranges.

The luminance Y is generated from the RGB signals in a simple network, as shown in Fig. 29.2. Two additional signals are generated in similar networks, namely an I and a Q signal from which the chrominance signal C is generated. Both I and Q signals are modulated, but with signals that are 90 degrees apart (a sine and a cosine term), and then added to form the C signal as shown in Fig. 29.3.

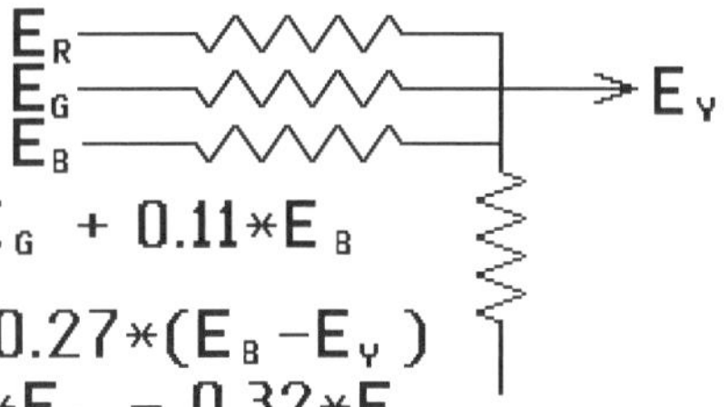

$$E_Y = 0.30 * E_R + 0.59 * E_G + 0.11 * E_B$$

$$E_I = 0.74 * (E_R - E_Y) - 0.27 * (E_B - E_Y)$$
$$= 0.60 * E_R - 0.28 * E_G - 0.32 * E_B$$

$$E_Q = 0.48 * (E_R - E_Y) + 0.41 * (E_B - E_Y)$$
$$= 0.21 * E_R - 0.52 * E_G + 0.31 * E_B$$

FIGURE 29.2 Resistance networks add the proper amounts of red, green, and blue to form the luminance (Y = monochrome information), and an I and a Q signal.

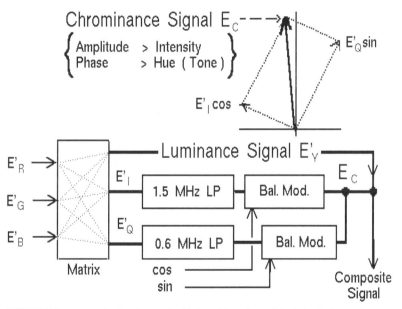

FIGURE 29.3 A composite chrominance signal is formed from the I and the Q signals. Composite NTSC is added to the Y signal.

The NTSC composite video signal is therefore an AM luminance signal with bandwidth from dc to 4.2 MHz plus a 3.58 MHz subcarrier, AM modulated by two color signals having 90° phase displacement. The C signal's amplitude represents the saturation level (intensity), while its phase contains the hue (tone). Any phase error (a few degrees only) will change the image color, and NTSC has therefore been nicknamed "newer the same color."

The PAL composite signal is similar to NTSC, but the color difference signals U and V (similar to I and Q) reverse in phase on alternating lines, and phase errors tend to cancel.

Demodulation of the composite signal is accomplished by a network, as shown in Fig. 29.4. The bandwidth of the composite NTSC video signal, and its typical spectrum, is shown in Fig. 29.5.

Commercial TV calls for a black-and-white video information signal bandwidth of 3 MHz. Figure 29.6 shows the signal levels for 1.5 lines. The video signal level varies between 0 percent (black) and 85 percent (white), with the continuous full range of grayscale levels in between. A horizontal synchronization signal follows each line so they all start in synchronism.

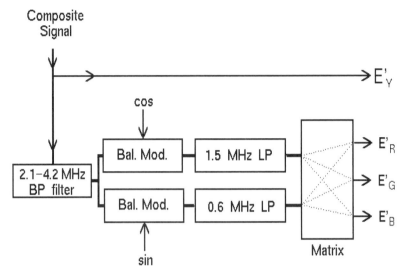

FIGURE 29.4 The extraction of Y and demodulation of the composite signal into R, G, and B information.

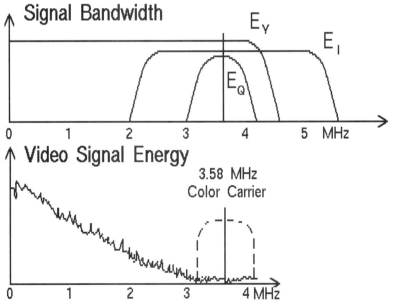

FIGURE 29.5 Bandwidth requirements for quality video information via NTSC.

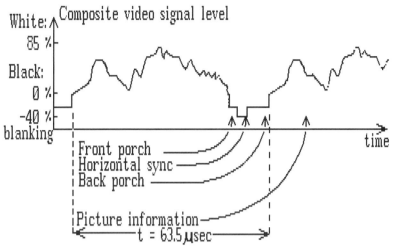

FIGURE 29.6 Waveforms and levels of 1½ video line.

Figure 29.7 shows the video signal required to produce four bars (white-black-white-black) on the screen. This illustration should help us appreciate that the average picture level produces a dc signal (here at the 42.5 percent level) without which the picture would be shown with whiter-than-white or blacker-than-black levels. Hence the signal channel must provide for response to dc. The high frequency components are necessary to produce the sharp transitions between bars.

For the sake of completeness, we should add that synchronization pulses are generated at the end of each set of 262½ lines in order to synchronize the picture sequence by signalling when to start a new scan.

The signals and frequencies used for present color television were created with the restriction that existing black-and-white receivers could still use the signal, and that color receivers could also use a

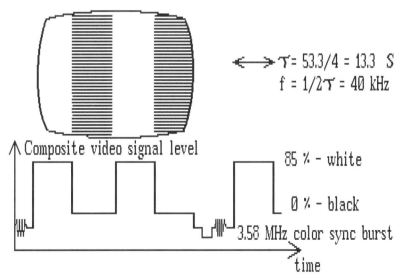

FIGURE 29.7 Four bars (2 white and 2 black) are displayed by the video line voltage waveform shown.

black-and-white signal. Thus, color pictures, which are created by adding red, green, and blue video signals, must use the same bandwidth of only one video signal. This is done with a clever combination of the three signals into a luminance signal and two chrominance signals, which fit in a single video-signal bandwidth.

The luminance signal by itself gives a black-and-white picture. The chrominance signals (I and Q) modulate a 3.58 MHz subcarrier, and the luminance signal are used to recreate the red, green, and blue video signals for color televisions. In order to correctly extract the chrominance signals, a phase reference is provided by placing 8 to 11 cycles of the 3.58 MHz subcarrier on the back porch of the color video signal. This is known as the color burst.

The bandwidth of the chrominance or color signal is generally only about 0.5 MHz, which corresponds to a resolution of 8-dot-wide patterns. This appears satisfactory for most color pictures transmitted to the home user, and is therefore also the color resolution used in home VCRs (videocassette recorders). There are VCRs with wider color bandwidth in the broadcast and studio fields.

The total signal consisting of video (luminance signal), synchronization pulses, color subcarrier, and color burst is called the composite video signal. The frequency spectrum of this signal is shown in Fig. 29.5.

This signal cannot be recorded and played back directly on a magnetic recorder, because it contains a large dc level. A solution was found by engineers at Ampex by using FM, frequency modulation. This had already been employed by RCA and Bing Crosby Enterprises for the low-frequency to dc range.

The normal FM techniques require a fairly large bandwidth, often 10–20 times the signal bandwidth. This is out of the question for the 4-MHz video signal, and instead a low modulation index was used so few sidebands were generated. General FM theory defies such a system, but it worked, and a paper by Felix (1965) explains its characteristics.

The first recorders (1958) used a frequency deviation of 1.2 MHz total, as shown in Fig. 29.8. There would occasionally be sidebands that on the low-frequency side would fold over (negative frequencies) and generate Moire patterns. Higher carrier frequencies and wider deviations were introduced to improve the video quality, first the High Band recorders in 1964 and later the Super High Band in 1971. Their spectra are shown in Fig. 29.8.

The color synchronization is refreshed for each line by the 3.58 MHz color burst, or each 63.5 µs. A noticeable change in hue (color) occurs for just a few degrees deviation from the 3.58 MHz sinu-

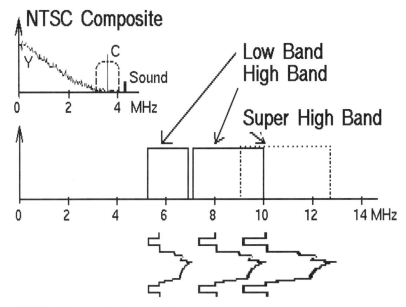

FIGURE 29.8 Bandwidth allocation for three quadruplex video recorders.

soid, and a reasonably good reproduction requires a phase accuracy of plus or minus 4 degrees, corresponding to $(1/3.58) \times (4/360) = 0.0039$ μs = 3.9 nS time differential during each 63,500 nS interval. There are no problems in building a reference oscillator with that stability—the problem will lie in providing a time-stable tape transport for recording and reproducing the signal.

Let us also assume that we have recorded an alternating white-black dot pattern, similar to the one we used to analyze the picture resolution. If two consecutive scans are synchronized at their start, what is the maximum time difference we can tolerate? This is a subjective question, and let us say that a one-bit shift at the end of the lines is unacceptable (the vertical line pattern would become a checkerboard on the right-hand side of the screen). Such a shift corresponds to the duration of one bit, or $53.3/450 = 0.118$ μs.

From this we can predict a reasonable limit for the time base error allowed for the tape transport:

$$\text{Maximum short term } TBE = \pm50 \text{ ns}$$

This corresponds well with the maximum error output of TV recorder/reproducers with time base error correction. The long-term error in the transport speed can be quite large, provided the TV monitor will stay in synchronization: Maximum long-term $TBE = \pm2000$ ns.

The required stability of the reproduced color signal presented an even more serious problem. The time base error of a transport must be less than 4 ns in order to ensure that the phase of color subcarrier is changed less than five percent, which is the maximum allowable. The 4 ns applies when the color carrier is at 3.58 MHz; it can be higher if the color carrier is shifted down in frequency, which is done in the Beta and VHS machines.

This shifting down in frequency, called the "color-under" principle, extracts the 3.58-MHz color information from the composite video signal and places it at 562.5 kHz by a modulation process; the color carrier plus sidebands occupy about 1 MHz (Fig. 29.9).

The video, also called the luminance signal, FM modulates a carrier located at 4.05 (VHS) or 5.17 (Beta) MHz, with maximum excursions of $\pm3\frac{1}{4}$ MHz. Spectra for S-VHS and Super-Beta are shown in Fig. 29.10. It is important to place the FM band as high as possible and to further adjust the carrier level so it serves as a bias for the color signal, because this improves the linearity of the color levels. The two signal bands may overlap slightly to allow for a luminance signal bandwidth of up to 2.7 MHz, corresponding to a 235-line resolution in the reproduced picture.

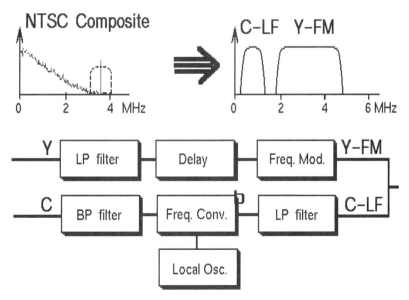

FIGURE 29.9 The color-under principle (*after W.R. Johnson*).

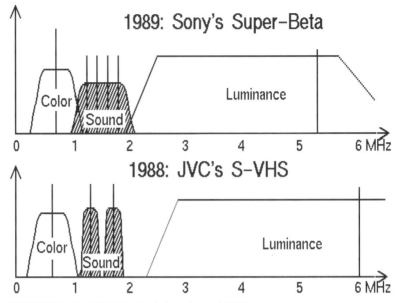

FIGURE 29.10 Bandwidth allocations in Super-Beta and S-VHS.

The signal-to-noise ratio (*SNR*) depends on a great number of factors such as trackwidth, system resolution, and tape quality. VTRs are always pushed to provide as long a playing time as possible, which means narrow tracks and slow head-to-tape speeds.

The *SNR* for video transmission is normally expressed as:

$$SNR = (\text{peak signal})/(\text{rms noise})$$

Noise in a television picture displays as excessive graininess or a pattern like snow. TASO (television allocation study organization) has provided a picture rating table for pictures of a 4 MHz bandwidth:

Excellent (no perceptible snow in picture)	45 dB
Fine (snow just perceptible)	35 dB
Passable (snow definitely perceptible)	29 dB
Marginal (snow somewhat objectionable)	25 dB

Amplitude uniformity must also be considered and the low frequencies (1–300 kHz) should be within a 0.2 dB limit, while a 0.8-dB envelope applies for higher frequencies.

The latter is achieved by using FM modulation for the video signal. An adequate signal-to-noise ratio requires special equalization, much like in audio recordings. When we look at Fig. 29.5 we observe that the average video picture has very little signal energy at high frequencies, and a direct recording thereof would fail to use the tape magnetization at the corresponding short wavelengths.

It is therefore common practice to boost the high frequencies during recording, and deemphasize them during playback. The pre-equalizers used in the broadcast field are shown in Fig. 29.11, and a complimentary deemphasis circuit is used during playback.

Additional equalization is needed during playback due to the read losses (see Chapter 6). They are: coating thickness, spacing, and gap losses. The reader will recall that none of these losses have

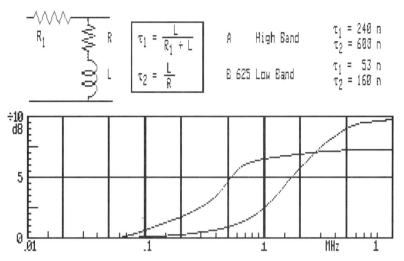

FIGURE 29.11 Preemphasis curves for broadcast video recorders (*after Robinson*).

associated phase shifts. They must therefore be compensated with the cosine equalizer described in Chapter 23. This equalizer was first used in sound film and in television, and was called an aperture equalizer (from sound film tracks scanned by a slit, or an aperture). The reader is referred to Robinson's book for a detailed analysis of the circuit function.

THE DIGITAL VIDEO SIGNAL

The composite video signal is subject to degradation when recorded and played back through an analog channel. Time displacement errors were large in early recorders, and amplitude and phase variations can occur anywhere across the frequency range.

These deficiencies are eliminated when the signal (or signals) is converted to the digital domain. Currently references are made to the D-1 and D-2 recorder systems, which both exhibit much larger bandwidth that the analog video recorders.

The D-1 Component Digital Recorder has three digitized signals:

Signal:	Y	I	Q
Max. frequency:	4	1.5	0.5 MHz
Sampling frequency:	13.5	6.75	6.75 MHz
Resolution:	8	8	8 bits

Total BW: 27 MHz sampling 8 bits = 216 Mbits/sec.

The sampling rates relate like 2:1:1, and because the sampling of the Y-signal is four times the bandwidth of that signal, the digitizing method for D-1 is also called 4:2:2.

The D-2 composite NTSC color video digitizes the composite color video signal, which is sampled at 4 times the color carrier at 3.58 MHz (i.e., 14.32 MHz sampling 8 bits = 115 Mbits/sec).

The D-2 has an advantage over the D-1 format in requiring only about half the bandwidth, and the playing time can be doubled. This is augmented by using azimuth recording in the D-2 recorder; the D-1 format uses slanted tracks separated by guard bands.

The most recent formats are the D-3 composite format and the Digital Beta-Cam. D-3 uses the same signal format as D-2, but uses a ½" wide tape. An advanced 8–14 modulation code is also used; see Chapter 24.

THE EVOLUTION OF HELICAL SCAN RECORDERS

The very first attempts at recording video (television pictures) and sound were made on transports that were extensions of the Magnetophone-type audio recorders. The shortest possible recorded wavelength was about 10 μm (400 μin), and the tape speed had to be v = (highest frequency × shortest wavelength) = $4,000,000 \times 0.0004 = 1600$ IPS. This presented a formidable problem, as it is well known that tape separates from heads at speeds greater than 120 IPS unless very high tensions are applied, and that is possible only at up to 180 IPS.

The solution was to divide the signal up into several bandpass sections and modulate them all to be low-frequency signals, say 20 channels of each 200 kHz bandwidth. Such schemes were employed first by RCA and later by Bing Crosby Enterprises, later Mincom Division of 3MCo. Figure 29.12 shows the longitudinal recorder used by Mincom to record television signals at a tape speed of 320 IPS.

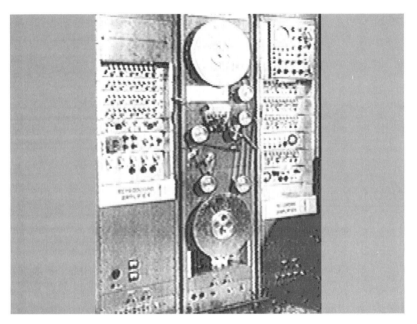

FIGURE 29.12   The first reel-to-reel video recorder (RCA).

Improved heads and tapes (the first chromium dioxide samples) made it possible to construct video recorders in the early 1960s operating at 120 IPS with a single channel, with a bandwidth of 3 MHz (Winston/Fairchild).

None of the above recorders had the required time base stability for reproduction of video signals. That made recombination of the separate channel nearly impossible, and caused severe geometry distortion of the reproduced pictures. The duration of a single pixel is 53.3 μs/700 = 76 ns based on the earlier ideal assumption of 700 pixels per line (the duration allocated per line is 63.5 μs; the additional 10.2 μs is used for the front and back porches, plus the horizontal sync signal (see Figs. 29.6 and 29.7).

The time displacement errors in ordinary longitudinal tape transports are about ±1000ns peak, which could lead to zig-zag errors in a vertical line of ±1000/76 = ±13 pixels! A very unpleasant picture. The same time displacement error plus intertrack skew made the proper recombination of multichannel video recording impossible, and the picture quality was therefore mediocre.

The only solution was to make the tape stand virtually still and move a magnetic head past it at high velocity, with high inertia to reduce the time errors.

The first approach was a transport where a wheel with magnetic heads was placed in plane contact with the tape so arcuate tracks were written; see Fig. 29.13. This principle did not work very well, and was soon abandoned and efforts concentrated on the transverse-scan recorder as shown in Fig. 29.14. These recorders were very successful. A few are still in use today, but they are fast being replaced with the helical-scan recorders.

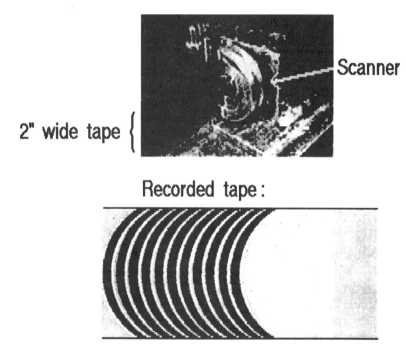

FIGURE 29.13   The accurate scan principle, first tried by Ampex in the early 1950s.

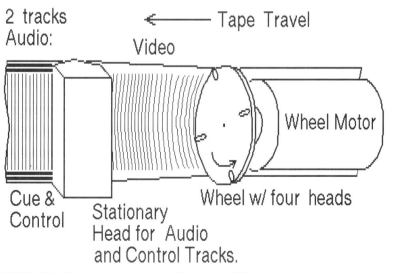

FIGURE 29.14   The transverse scan introduced by Ampex in 1950s.

The 1956 transverse recorder was invented and designed by a team at Ampex Corp., headed up by Charlie Ginsburg. A 2-inch-wide tape is scanned by a head drum with four heads (quadraplex) rotating at 14,400 rpm, or 240 rps, and 30 frames (60 fields) per second leaves 4 complete rotations per field. With four heads, this means that each track contains $\frac{1}{16}$ of a field.

This segmentation showed up in the picture as a venetian blind effect, and additional electronics were necessary to time-correct this phenomenon. Extra care was taken to make sure that the four heads were separated by exactly 90.0°. The head wheel was forced into the wrapped tape by 1 to 2 mils to penetrate the air bearing and ensure good head-tape contact; the head-to-tape speed was 240 × 4 × 1.56 = 1500 IPS, because only 1.56 inches of the possible 2-inch-long track was recorded with video. The edges on the tape were recorded longitudinally with audio, servo, and timing information.

A longer track could be recorded if the head wheel scanned sections of the tape at a small angle with the tape motion direction. This inspired Alex Maxi at Ampex to invent the helical-scan recorder with the Omega-wrap (see Fig. 29.15). A rather large drum was used (eight inches in diameter) and the aim was to record one field per track. Several prototypes were developed in the late 1950s, and in 1960 the Ampex VR-8000 was ready. The large head wheel (Fig. 29.16) provided a 24-inch-long track, and the tape motion suffered from instabilities when sliding along such a long surface. Later a smaller wheel, or drum, was tried, and segmentation of a field into 5 tracks was used.

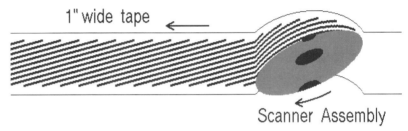

FIGURE 29.15 The transverse scan principle (*after Alex Maxi*).

FIGURE 29.16 A prototype helical-scan recorder (*courtesy Sid Damron*).

FIGURE 29.17 Compact, stacked reels in a helical-scan recorder/preproducer for airborne use (*courtesy Datatape, Inc.*).

Ampex decided not to sell the VR-8000 recorder because it would compete against their transverse recorders. But the companies Matsushita, JVC, and Sony continued development of helical-scan recorders, and introduced the U-Video, U-Tape, and U-Matic (SMPTE E-format) in 1966. These industrial-type recorders used ¾-inch tape and the tracks were laid down at an 11° angle. The head-to-tape speed had been reduced to 410 IPS thanks to improvements in tape and heads. They also introduced the color-under scheme invented by W.R. Johnson in 1958.

In 1970 the A-format broadcast helical-scan recorders were introduced by Ampex, followed by the B-format in 1976, and the C-format in 1978. Helical-scan recorders have fairly high time base errors, and full broadcast use of these units was not possible until 1973, when digital time-base correction was introduced. The reader may experience the degradation of pictures due to time base errors: the compact VHS cassettes in some camcorders suffer from this!

Other broadcast formats for ENG (electronic news gathering) have followed, just to mention a few: M-format (1982—1" tape), L-format (Betacam 1984—½" tape), M-11 (1986—½" tape), and Betacam-SP (1987—½" tape). And today disk drives are used for ENG and other short topics.

Home video recorders are all helical-scan recorders, and they started in 1968 with Matsushita's B&W system, using ±7° azimuth record techniques. Sony introduced Beta in 1975, JVC followed with VHS in 1976, and Philips came along with their V2000 system in 1980. Better tapes and heads again made improvements possible, so in the next go-around JVC came first with their S-VHS in 1988, followed by Sony in 1989 with Super-Beta.

One new format appeared in 1984: Sony's 8 mm format. It quickly was seized on as a candidate for data storage, thanks to its compact cassette design and high packing density. And the last application of helical-scan techniques is in audio, where digitized sound now can be recorded on the DAT (Digital Audio Tape) recorders, introduced in 1985.

The reader interested in backups and libraries in computer memories may view the 8 mm and DAT systems as something revolutionary. They are both the products of evolution, though, and are not the only helical-scan devices used for data storage. As early as 1961 the first transverse-scan recorder was used for storing instrumentation and telemetry data. The first helical-scan recorder dedicated to data recording was introduced by Echo-Science in 1973, with several other companies following suit.

Today data storage is being pursued by employing the analog VHS, DAT, 8 mm, C-format, and the digital D-1, D-2, D-3, D-5, and BetaCam recorders. And recently the NT recorder (see Fig. 28.9).

The number of heads on the head wheel in a helical machine ranges from one to four, and the corresponding wrap angle of the tape ranges from almost 360 degrees to 90 degrees. One scan of a head along the tape will generally correspond to one frame, or one picture of video information. This makes it possible to still frame, i.e., show just one 525-line picture on the screen at a time.

Most machines are two-headed, such as the Beta, VHS, and the new 8 mm formats. The count of heads in advertisements includes stationary erase, sound, and synchronization tracks. There are new schemes underway where four heads are mounted on the head wheel: One set with ±15° azimuth offset for recording/playback of modulated sound (FM or digital), and the other set with ±6° or 7° azimuth offset for recording video. The sound system uses a record level that penetrates the coating, while the video information is recorded in the surface; then the two systems do not interfere with each other.

DIGITAL RECORDERS, TAPE OR DISK

The first digital VTR (video tape recorder) used the D-1 format, without azimuth recording. This was really an extension of the analog C-format used in broadcast video (see Figs. 29.21 and 29.22).

The D-1 data rate is 216 Mb/s. Four digital audio channels are recorded in time multiplex. The audio tracks are laid down in the center of the tape in order to protect them from tape edge debris and damage. This was the result of 2 years of debate in standard committees. When the D-2 came along, the audio segments were positioned near the edges. They remain there in the D-3 format.

Broadcast work involving the editing and operation of commercials and so on often require very short record/play times, in the range of seconds or minutes. It is today's practice to perform these operations using a disk drive.

The highest data rates for single disk drives are 4–7 MB/s (Seagate's Barracuda 2 GB 1994). The D-2 digitized composite NTSC signal requires 115/8 = 14.4 MB/s for full studio quality, more than twice the disk drive's capability. There exist two solutions: Compress prior to recording, or record to a RAID (array of inexpensive disk drives).

Less quality than studio standard will do for certain applications and picture subjects. The video signal can then be compressed by a method called MPEG-2 (Motion Picture Experts Group). Only every 15th frame is recorded, and the changes in between are saved, processed, and recorded. This brings the data rate down to .5 to 5 MB/s for full-screen pictures. Desktop computers often operate with the lesser MPEG-1, which provides a ¼-screen size, 352×240 pixels.

The degradation due to compression can be prevented by recording the video to a RAID. These are available with rates of 12–17 MB/s, i.e., full studio quality. Capacities exceed several GB, so full-length videos maybe stored and played.

Rigid disks have also found application in cameras. Many years of evolution have produced today's 2½" removable hard disk for cameras. The picture through the lens is captured by a 2000×2000 CD array, digitized and recorded onto the disk. The file will be in the MB range for high-quality pictures. These electronic photos are readily transmitted over phone line or satellite to anywhere in the world, a great bonus for journalists.

TAPE RECORDING FORMATS

The past decades of developments in magnetic storage show a steady increase in packing density of bits per area: A doubling every four years, or a tenfold increase every decade. This has been made possible by increased coercivity, smoother tapes, and better heads. This increase is reflected in the decrease in the diameter of the scanner drums, as shown in Fig. 29.18.

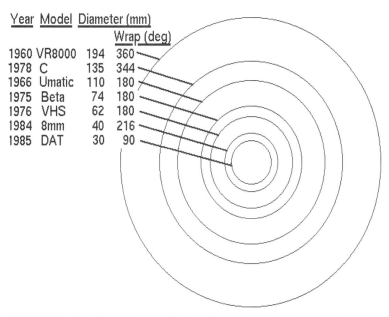

FIGURE 29.18 The reduction in scanner diameters reflect the decreased wavelength requirements for the same (or even better) frequency response.

We will consider the mechanical track formats, which result from using different tape widths, linear speeds, drum diameters, track lengths, and track angles. Four consumer and three broadcast formats are shown in Figs. 29.18 through 29.23. All formats use azimuth recording except the C and D-1 formats.

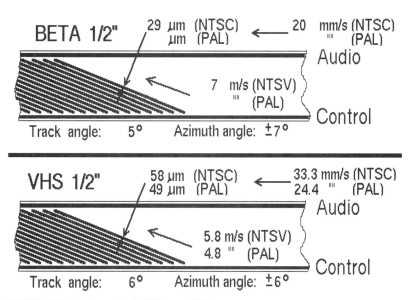

FIGURE 29.19 Track standards for ½" Beta and VHS formats.

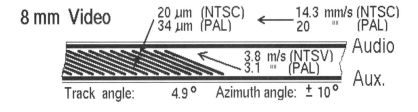

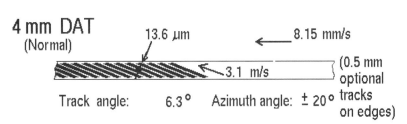

FIGURE 29.20   Track standards for 8 mm and 4 mm DAT formats.

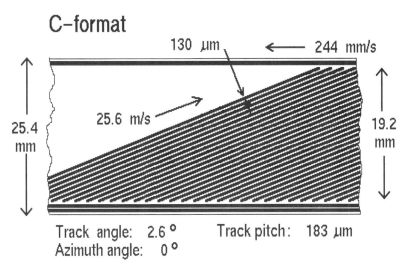

FIGURE 29.21   Track format, C-broadcast format.

Details of the figures need no comments, but do notice that all illustrations are looking at the coated side of the tape, that the tape moves from right to left, and that the head wheel follows the tape motion in all consumer formats while rotating opposite in all professional recorders. The argument for the latter method is a slight increase in the relative head-to-tape speed.

Azimuth recording does not offer perfect isolation from track to track: When the head strays into an adjacent track, the signal attenuation due to the gap misalignment is only effective at short wavelengths. The signal loss due to the azimuth loss was shown in Fig. 6.19. The true loss is shown by the thin line, while the heavy line shows the minimum isolation. To this loss must be added the registration loss when the head strays in over an adjacent track, and crosstalk when the head is off-track. The crosstalk loss is very close to the well known $55 \times d/$wavelength loss (in dB).

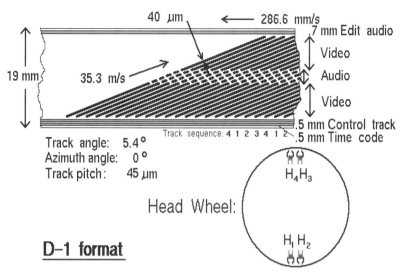

FIGURE 29.22 Track format, Digital, Component D-1 video recorder.

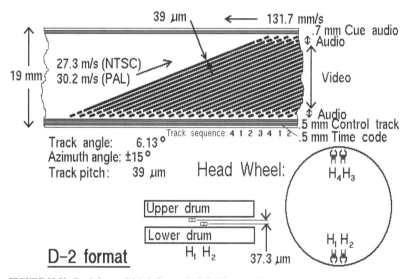

FIGURE 29.23 Track format, Digital, Composite D-2 video recorder.

Azimuth recording saves tape by eliminating the need for land between tracks. It cannot be used indiscriminately; knowledge of the signal spectrum is necessary, and it may have to be tailored to use the dips of high attenuation as shown in Fig. 6.19; this can be accomplished for digital signals by the choice of a proper recording code.

One troublesome source of crosstalk in recording video signals are the sync pulses. It is necessary to line up the sequences of each field recorded so the sync pulses line up, as shown in Fig. 29.24. This places restrictions on the choice of track geometry, tape speed, and head-versus-tape speed.

In designing a helical-scan VTR, first of all consider that the minimum recording wavelength is determined by the latest magnetic recording technology. The head wheel diameter is determined by the maximum video frequency that corresponds to the picture quality. After the tape width is determined,

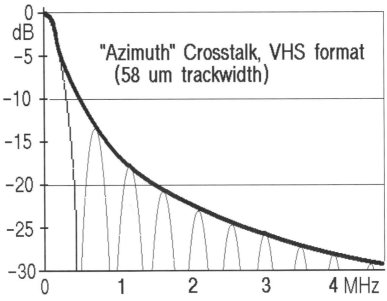

FIGURE 29.24 Some crosstalk exists in helical recorders due to incomplete attenuation of azimuth loss.

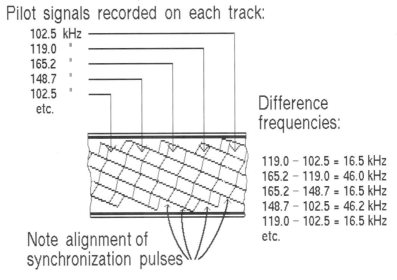

FIGURE 29.25 The track and servo format with servo signal alignments on DAT tape.

the video track-width and the tape speed are determined in order to obtain continuity of video signals from one track to another. It is usually necessary to align the horizontal sync pulses for the PAL and SE-CAM systems (SECAM stands for sequential color with memory). See Fig. 29.26.

The VHS format, using "no-guard-band recording," can record and playback the three basic television systems by switching only the electronic circuitry. In this way one basic mechanical transport assembly can be used all over the world. Caution: Recorded tapes are NOT interchangeable.

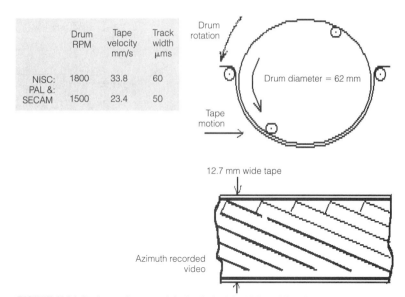

	Drum RPM	Tape velocity mm/s	Track width μms
NISC:	1800	33.8	60
PAL & SECAM	1500	23.4	50

Drum rotation

Drum diameter = 62 mm

Tape motion

12.7 mm wide tape

Azimuth recorded video

FIGURE 29.26 Basic tape formats and the head wheel for VHS NTSC and PAL/SECAM recorders (*after Sygaya*).

The new 8 mm format signal spectrum is shown in Fig. 29.26. The tape is contained in a cassette measuring 95 by 62.5 by 15 mm. The drum (head wheel) diameter is only 40 mm, so each track is $40\pi/2 = 62.8$ mm long. From our earlier considerations we find that the shortest wavelength is about $62.8/(4,000,000/30) = .47$ μm = 18.8 μin. Adequate performance can only be achieved by using a short-gap-length head in combination with a high-coercivity tape, such as a metal or Ba-ferrite.

The sound quality of home VCRs is often poor because the tape speed is one-half to one-quarter of the normal audiocassette speed, and because of the narrow track. The slow speed also falls in the danger zone of erratic tape motion due to stick-slip friction against the heads.

Recent developments have greatly improved the sound signal by not recording it in baseband or direct analog form along an edge track, but by FM modulating a signal carrier and multiplexing this signal in with the video as indicated in Fig. 29.10. And the latest is to digitize the audio signal and time-multiplex it in with the video; an example is the 8 mm system, where the tape wrap has been increased by 30 degrees to allow for the recording of digitized sound; see Fig. 29.26.

DUPLICATION

Duplication of prerecorded tapes is a time- and equipment-consuming process for videotapes. Audio tapes can be duplicated at high speeds, which is difficult with videotapes due to the rotating head assemblies and the much higher signal frequencies. The method is still one of loading a master tape onto a duplicator system with a large number of slave recorders. It is possible to operate the duplicating machines at higher than normal operating speed. The duplication center is faced with high labor and maintenance cost.

The difference in Curie temperature of tapes has long been recognized as a potential solution for fast in-contact duplication (Morrison). The process is fundamentally based on using a master tape made from high-Curie-temperature material (any iron oxide material) that is brought into contact with a low-Curie-temperature slave tape, such as one made from CrO_2 particles. They are heated together to a temperature between the two Curie temperatures, at which point the CrO_2 becomes nonmagnetic.

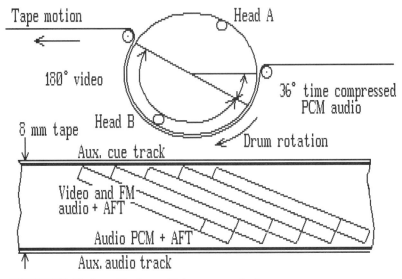

FIGURE 29.27 The 8 mm video format with time-compressed audio.

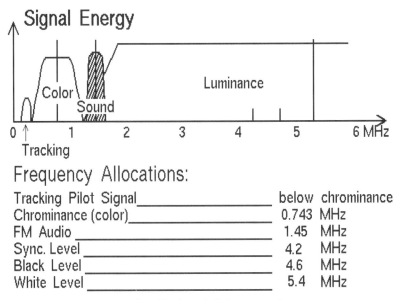

FIGURE 29.28 Chrominance, audio, and luminance in the 8 mm camcorders.

When cooled, the CrO_2 becomes magnetic, and each particle now becomes magnetic with a polarity that opposes any external field present; specifically, the one on the master tape. This will result in a mirror-image magnetization of the higher temperature tape's recording. If that recording is made in the proper format the result will be a duplicated tape. The format must correspond to the convention used in pressing records, where the pressing matrix has ridges instead of grooves.

The problems have in the past been mechanical misalignments, and the difficulties were in heating a tape to a temperature above the CrO_2 Curie temperature of about 110°C without permanent damage to the PET base film. Expansion and contractions are unavoidable, which in turn makes short-wavelength duplication difficult. Currently duplicators can operate at 165 times real-time duplication for NTSC SP mode (Standard Play). The heating is provided for by a 300 watt Nd-YAG single crystal laser that uses a Krypton gas pump (Otari Inc.).

The heat must be applied for only a very short time so slip problems between master and slave are minimized (Cole, Odagri). This has today been accomplished by selective radiation heating of the CrO_2 particles, without affecting the special binder.

EDITING OF TAPES

Editing of videotapes presents another formidable problem. Film editing is so simple: Cut and splice at the frame line. This method was applied in the transverse scan machines, and required great skill and care.

In helical machines, a cut and splice across the tape width will lead to a transition where the next sequence of pictures rolls up or down in a wiping motion. Only electronic editing will give a satisfactory result. This may be accomplished on-the-fly by operating two machines, or better yet by using a disk drive or a RAID for temporary storage of frame sequences to be spliced. This all points to costly equipment that is not for the amateur recording person; he/she will do best with the two-machine on-the-fly approach, or should rent time at a local editing facility.

REFERENCES

Benson, K. Blair, and J. Whitaker. 1992. *Television Engineering Handbook*. McGraw-Hill.

Felix M.O., and H. Walsh. Sept. 1965. F.M. Systems of Exceptional Bandwidth. *Proc. IEE* 112 (9): 1659–68.

Robinson, Joseph F. 1981. *Videotape Recording*. Focal Press. 362 pages.

BIBLIOGRAPHY

Baron, S. Nov. 1988. Television Tape Recording Nomenclature. *SMPTE Journal*. 97 (11): 928–936.

Cover Story: Video storage. August 1995. *Broadcast Engineering*. 37 (9): 28–75.

Gregory, Stephen, 1988. *Introduction to the 4:2:2 Digital Video Tape Recorder*. London, Pentech Press. 200 pages.

Inglis, A.F. 1992. *Video Engineering: NTSC, EDTV & HDTV Systems*. McGraw-Hill, 338 pages.

Kolb Jr., F.J. et al. Jan. 1991. Annotated Glossary of Essential Terms. *SMPTE Journal*. 100 (2): 111–26.

Mee, C.D., and E.D. Daniel. 1996. Magnetic Recording, McGraw-Hill.

Proposed Standard for television digital recording—19-mm type D-2 composite format—tape record. July 1990. SMPTE Journal. 99 (7): 588–607.

Schouhamer-Immink, Kees. 1991. *Coding Techniques for Digital Recorders*. Prentice Hall. 360 pages.

Watkinson, John. 1991. *The D-2 Digital Video Recorder*. Focal Press., England. 272 pages.

Watkinson, John. 1992. *D-3 Digital Video Recorder*. Focal Press. 288 pages.

Watkinson, John. 1994. 2nd edition. *The Art of Digital Video*. Focal Press. 608 pages.

VIDEO RECORDING

Abramgon, A. 1973. A Short History of Television Recording: Part II. *SMPTE Journal*. 82 (?): 188ff.

Anderson, C.E. 1957. The Modulation System of the Ampex Video Tape Recorder. *SMPTE Journal*. 66 (?): 132ff.

Ginsburg. G.P. April 1957. Video Tape Recorder Design: Comprehensive Description of the Ampex Video Tape Recorder. *SMPTE Journal*. 66 (4): 177–82.

Mullin, J.T. May 1954. Video Magnetic Tape Recorder. *Tele-Tech*. 13 (5): 127–129.

Oku, M., I. Aizawa, N. Azuma, S. Okada, K. Hirose, and M. Ozawa. Sept. 1989. High Picture Quality Technologies for an S-VHS Portable VCR. *SMPTE Journal*. 98 (9): 636–639.

Olson, H.F., W.D. Houghton, A.R. Morgan, J. Zenel, M. Artzt, J.G. Woodward and J.T. Fisher. 1954. A System for Recording and Reproducing TV Signals. *RCA Review*. 15 (?) 3ff.

Sadashige, K. 1980. An Overview of Longitudinal Video Recording Technology. *SMPTE Journal*. 89 (?) 501ff.

Sadashige, K. Dec. 1984. Developmental Trends for Future Consumer VCR's. *SMPTE Journal*. 93 (12): 1138 ff.

Sadashige, K. May 1985. An Introduction to Analog Component Recording. *SMPTE Journal*. 94 (5): 477–85.

Shiraishi, Y. Dec. 1985. History of Home Videotape Recorder Development. *SMPTE Journal*. 94 (12): 1257–63.

Stanton, J.A. and M.J. Stanton. March 1987. Video Recording: A History. *SMPTE Journal*. 96 (3): 253–63.

Sugaya, H. Sept. 1978. Recent Advances in Video Tape Recording. *IEEE Trans. Magn.* MAG-14 (5): 632–37.

DIGITAL RECORDING OF VIDEO

Bentz, C., and R. Lehtinen. Aug. 1991. Applied Technology: The D-1 and D-2 Formats. *Broadcast Engineering*. 33 (8): 76 ff.

Brush, R. March 1988. Design Considerations for the D-2 NTSC Composite DVTR. *SMPTE Journal*. 97 (3): 182–93.

Chan, C. Jan. 1994. Studio videotape recorders. *Broadcast Engineering*. 36 (1): 56–60.

Diermann, J. June 1978. Digital Videotape Recording: An Analysis of Choices. *SMPTE Journal*. 87 (6): 375–78.

Engberg, E. et al. Oct. 1987. The Composite Digital Format and Its Applications. *SMPTE Journal*. 96 (10): 934–42.

Fibush, D.K., and B. Elkind. Sept. 1992. Test and Measurement of Serial Digital Television Signals. *SMPTE Journal*. 101 (9): 622–631.

Gallo, L. Oct. 1977. Signal System Design for a Digital Video Recording System. *SMPTE Journal*. 86 (10): 749–956.

Hashimoto, Y. Sept. 1987. Experimental HDTV Digital VTR with a Bit Rate of 1 Gbps. *IEEE Trans. Magn.* MAG-23 (5): 3167–72.

Hedtke, R. Sept. 1986. Measurements Methods and Diagnostics Techniques for the Digital Television Tape Recorder (DTTR). *SMPTE Journal*. 95 (9): 878–88.

Kawamura, T., S. Kasai, T. Tominaga, H. Sato, and M. Inatsu. May 1987. A New Small-Format VTR Using an 8 mm Cassette. *SMPTE Journal*. 96 (5): 466–72.

Livingston, P. Aug. 1991. The ½-inch digital format. *Broadcast Engineering*. 33 (8): 68–72.

Livingston, P., and J. Safar. Sept. 1992. The D-3 Composite Digital VTR Format. *SMPTE Journal*. 101 (9): 602–605.

Luitjens, S.B., G.J. van den Enden, and H.A.J. Cramer. April 1990. A Way to Assess the Performance of Heads and Tapes for Digital Video Recording. *Intl. Conf. on Video, Audio and Data Rec.* IEE Publ. No. 319, pp. 43–49.

Morrison. Sept. 1982. Videotape Recording: Digital Component Versus Digital Composite Recording. *SMPTE Journal*. 91 (9): 789 ff.

Mullen, S. Jan. 1995. AudioVideo Hard Drives. *Digital Video Magazine*. 3 (1): 60–66.

Proposed SMPTE Standard for Television Digital Component Recording 19-mm Type D-1. March 1992. *SMPTE Journal*. 101 (3): 203–224.

Sadashige, K. Jan. 1989. Video Recording Formats in Transition. *SMPTE Journal*. 98 (1): 25–31

Sasaki, S., M. Ohtsu, M. Chiba, T. Bannai, H. Taniguchi, R. Tsunoi, and I. Obata April 1990. ½-Inch Video Cassette Recorder for Baseband HDTV. *Intl. Conf. on Video, Audio and Data Rec.* IEE Publ. No. 319, pp. 93–100.

Umemoto, M. Jan. 1992. Professional HDTV Digital Recorder. *IEEE Jour. Select Area Communication.* 10 (1): 80–85.

Umemoto, M., Y. Eto, H. Katayama, and N. Ohwada. April 1990. Record and Playback Systems for 1.2 Gb/s HDTV Digital VTR. *Intl. Conf. on Video, Audio and Data Rec.* IEE Publ. No. 319, pp. 54–59.

Wilkinson, J.H. April 1987. A Review of the Signal Format Specification for the 4:2:2 Component Digital VTR. *SMPTE Journal*. 96 (4): 1166–72.

COMPRESSION, ELECTRONICS

Akagi, H., H. Katata, M. Hyodo, Y. Fujiwara, Y. Noguchi, Y. Maezawa, H. Sakaguchi, T. Hirota, Y. Yamana, and N. Kako. March 1993. Development of Image Compression System for Digital VTR. *Sharp Tech. Jour*. No. 55, pp. 21–26.

Beaumont, J.M. Oct. 1991. Image Data Compression Using Fractal Techniques. *British Telecom Tech. Jour*. 9 (4): 93–109.

Dewith, P.H.N., M. Breeuwer, and P.A.M. Vangrinsven. Jan. 1992. Data Compression Systems or Home-Use Digital Video Recording. *IEEE Jour. Select Area Communication*. 10 (1): 97–121.

Matsumoto, T., K. Fujii, F. Koga, H. Ohta, O. Hosoi, S. Suzuki, and K. Yamamoto. Aug. 1990. All Digital Video Signal Processing System for S-VHS VCR. *IEEE Trans. Consumer Electronics*. 36 (3): 560–566.

Mokry, R., D. Anastassiou, and D. Teichner. Nov. 1990. Bandwidth Reduction for HDTV Transmission—Alternatives and Subjective Results. *IEEE Trans. Consumer Electronics*. 36 (4): 837–846.

Wallace, G.K. Feb. 1992. The JPEG Still Picture Compression Standard. *IEEE Trans. Consumer Electronics*. 38 (1): R18–R34.

DUPLICATION

Cole, G.R., L.C. Bancroft, M.P. Chouinard, and J.W. McCloud. Jan. 1984. Thermomagnetic Duplication of Chromium Dioxide Video Tape. *IEEE Trans. Magn*. MAG-20 (1): 19–23.

Odagri, Y., and T. Sato. Aug. 1984. High-Speed Video Tape Duplication Using Contact Printing. *IEEE Trans. Consumer Electronics*. 30 (3): 397–401.

Powell, S.J., and G.L. Kennel. Jan. 1987. Noise in Film-to-Video Transfers. *SMPTE Journal*. 96 (1): 16–27.

CHAPTER 30
MAGNETO-OPTICAL RECORDING

Magneto-optical (MO) recording uses temperature to alter the magnetization along a track on a rotating disk. The disk's magnetic layer resists changes in its magnetization, because its coercivity is about 2000 Oe.

A rise in the temperature of a magnetic material causes a decrease in its magnetization, and magnetization becomes zero at the material's Curie temperature. When the material cools the magnetization recurs, but it has forgotten its prior polarity. In media terms, it has been erased.

If some field H_a is present during the cooling period, then the magnetization will polarize to provide a field H_d that opposes H_a (very much as described in Chapter 4 about demagnetization; see formula 4.3). Hence, an alternating plus or minus magnetization M_d along the track is controlled by the temperature and the presence or absence of the field H_a.

The MO medium is a disk of diameter 64 mm (2½") or higher. The tracks are concentric, with a density of 20,000 tpi. The substrate is glass or high-quality plastic. The thin magnetic coating is a deposited material with perpendicular anisotropy.

WRITE/READ OPERATIONS

The principle in recording is shown in Fig. 30.1. A laser beam of around 5 mW in power is used to heat a spot in the recording layer. The magnetization inside the spot becomes zero after heating, and will become downward-magnetized during cooling when no external field H_a is present. This is simply a reaction to the surrounding upward-directed magnetization. We could define the recorded spot a mark, or 1.

If we wish a space or 0, then the laser power is turned off. The magnetization of the "spot" now becomes that of its surroundings. The matter of erasure is simple: For the duration of erasure, the laser power and field H_a are turned on. Everything along that track will be magnetized in the "up" direction.

Read-out of recorded information is done by Kerr effect (introduced in Fig. 3.51): The power of the laser beam is greatly reduced so that no heating beyond the Curie temperature takes place. The reflected light undergoes a slight phase change, less than one degree, and of polarity according to up or down magnetization. This phase change is detected and read accordingly as a 0 or a 1.

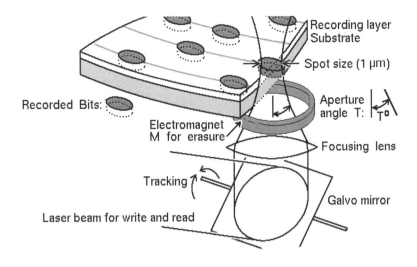

Recording layer
Substrate

Spot size (1 µm)

Aperture
angle T:

Focusing lens

Galvo mirror

Recorded Bits:

Electromagnet
M for erasure

Tracking

Laser beam for write and read

FIGURE 30.1 The principle in MO (magneto-optical) recording and playback.

SIZE OF RECORDED SPOT

The storage density of MO recording is greater than for any conventional tape or disk system. The spot size is roughly 1 µm (40 µin) diameter in current systems, which use a near-infrared laser of wavelength $\lambda = 780$–830 nm.

The spot size d is diffraction limited ($d = \lambda/2\sin T$) and will be reduced in next-generation MO drives by using a red laser ($\lambda = 680$ nm, $d \to 0.8$ µm. Tsutsime et al. 1993). The final target is use of the blue laser ($\lambda = 430$ nm, $d \to 0.5$ µm). The area density will roughly increase by $1/(0.52)^2 = 4$ times. This will bring the current capacity of a 5¼ MO disk of 1.3 GB to 5 GB. This is predicted to occur at the end of this century (Asthana IBM 1994).

A vintage 1995 picture shows the recorded spots between the track-following patterns; see Fig. 30.2. It was produced on an Atomic Force Microscope (courtesy M. Dugas, Advanced Research Lab, and V. Elings, Digital Instruments).

While MO technology excels in density, its performance lags in terms of access time and data rate. Access time is particularly delayed in the write cycle, where a complete revolution is required for erasure of old data. Direct overwrite is not possible with current technologies.

TRACK ARRANGEMENTS

Data are recorded on concentric tracks, in sectors. The disk rotates at a constant number of revolutions per minute (2400 to 3600 rpm), a method called CAV, constant angular velocity. This method was standardized in the late 1980s.

Spacing between data blocks varies, while the constant rpm assures that the same amount of data passes under the optical head at any point of the disk. The drawback is a lowering of the density at the outer tracks, and the inner track radius is half of the disk radius for maximum density (see Chapter 16).

In 1989, the ZCAV (zoned constant angular velocity) format was introduced. This format calls for increased data rates at the outer tracks where velocity is large. The zones with a fixed number of KB

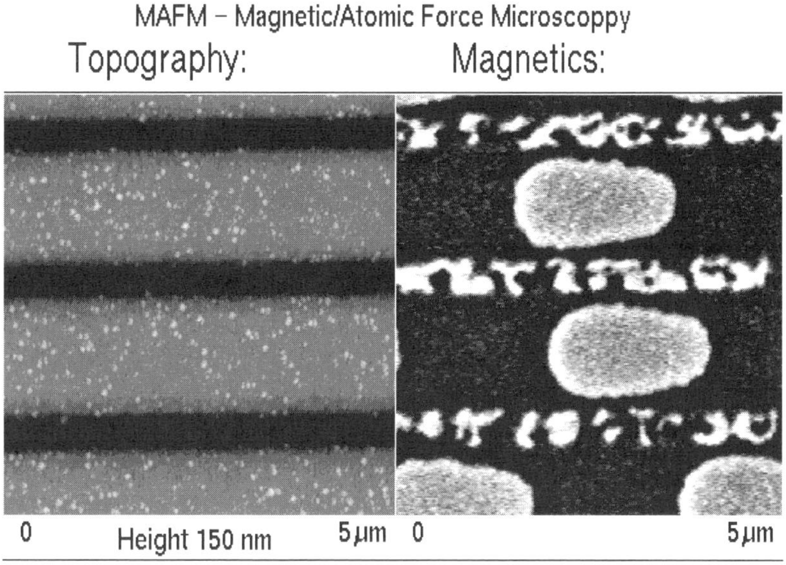

MAFM – Magnetic/Atomic Force Microscoppy

Topography: Magnetics:

0   Height 150 nm   5 μm 0   5 μm

FIGURE 30.2   An ATM picture of an MO disk's surface and recorded magnetic bits (*courtesy Digital Instruments, Inc.*).

now achieve the same length along the track. The entire disk surface can be used, and the total capacity was increased from 650 MB to 1 GB. Simultaneous reduction of the track pitch to 1.4 μm from 1.6 μm created the 1.2 to 1.3 GB capacities, also called the 2X format. The next generation, termed 3X, will increase density further by incorporating pulse-width modulation (Stevens 1993).

THE MO-DISK AND MEDIA MAGNETICS

Early experiments with MO recording used an MnBi alloy with a Curie temperature of 200°. The Kerr effect was large, 0.7 degrees, but the material was polycrystalline and had a grainy optical appearance. The magnetic domains were also irregular in size, following the crystallite dimensions. All the polycrystalline film had very poor SNR.

Researchers at IBM discovered in the late 1970s a class of MO alloys that were amorphous and had adequate Kerr effect (0.4 degrees). These materials were alloys made from the well-known magnetic (transition) materials (TM) like iron (Fe) or Cobalt (Co), and one or more of the rare earth (RE) elements like gadolinium (Gd), terbium (Tb), platinum (Pt), bismuth (Bi), etc.

The research into these materials was hampered by their easy corrosion when exposed to air. Protective overcoats had to be developed and applied. This problem is today a minor issue that essentially has been solved.

The RE/TM alloys are ferrimagnetic in such a fashion that the two components form opposing magnetizations. In ferromagnetic materials, all magnetons cooperate to form magnetization in one direction. In ferrimagnetic materials some magnetons are opposite the balance. This generally lowers the maximum magnetization and causes, for instance, ferrites to become less magnetic than iron or nickel. This appears as a lowering of the B_{sat} value.

When we redraw an earlier sketch (Fig. 3.40), we will show the magneton from, say, RE as pointing upward and TM as pointing downward; see Fig. 30.3. Their sums have different values that change with temperature—more so for the rare earth material than for the transition metal, as shown in Fig. 30.4 (Marchant 1990). Their values equal and cancel at a temperature of T_{comp} that can be

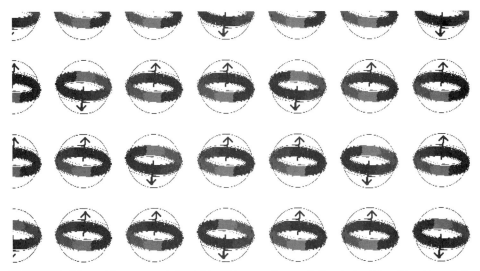

FIGURE 30.3 MO materials are composed of materials with two opposite magnetizations, having different Curie temperatures.

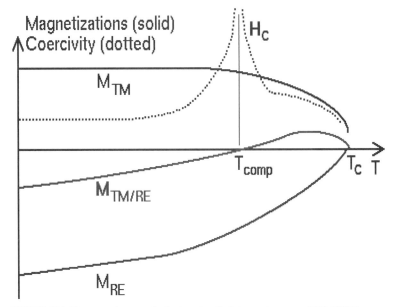

FIGURE 30.4 The composite magnetization has a low Curie temperature, around 100–200° C.

greater than or less than the ambient temperature. These magnetization curves can be tailored to have Curie temperature around 200° C (Fe-based RE/TM films) or 300° C (Co-based films). Practical materials have T_{comp} between the ambient temperature and T_{Curie}.

It is also desirable to keep T_{Curie} low and thereby limit the laser power required. Whenever writing takes place the spot on the film is heated to well over T_{Curie}. The temperature in the center of the spot is quite high, and is less at the edge of the recorded mark. This causes varying degrees of heat expansion and annealing of the material. An excessive number of write cycles could partially anneal the film and gradually reduce its performance.

Substrates for MO disks can be glass or plastic. Glass will partially oxidize an RE/TM coating, and plastics are permeable to oxygen and moisture (Marchant). A duplex overcoat/undercoat can provide corrosion protection (Frankenthal et al.). Aluminum acts as a self-passivating barrier, separated from the RE/TM film by a thin barrier layer of niobium. This layer is needed because the aluminum film could interdiffuse with the MO film.

THE MINIDISC (MD)

The MD is a new disk-based digital audio recording format used to achieve 74 minutes of record or play time on a 64 mm (2½") MO disk. Light reflected off physically molded grooves in circular patterns are used for tracking and spindle servo control. The MO recording is done with a direct overwrite process that uses a new magnetic-field modulation (MMO) system. This system modulates the magnetic field at high speed, creating specific magnetic orientations to represent the input signal (Kawakami 1993).

The audio playing time is comparable to the optical CD in spite of the smaller size. This is possible by severe audio signal compression called ATRAC (Adaptive Transform Acoustic Coding). The bit rate reduction is 5:1, while it is only 4:1 in Philips digital cassette cartridges (DCC). It is claimed that only critical music listeners can detect a slight degradation in the sound from the MD, while the DCC is on par with conventional CD.

The MiniDisc stores around 120 MB and may someday find its way into data storage for PCs. It will have to contend with competition from the announced 3½" floppy with a very thin magnetic coating (Fuji Film) and a density of 100 MB (Iomega).

REFERENCES

Asthana, P. Oct. 1994. A long road to overnight success. *IEEE Spectrum* 31 (10): 60–66.

Frankenthal, R., R. vanDover, and D. Siconolfi. 1987. Duples coatings for the protection of magneto-optic alloys against oxidation. *Appl. Phys. Letter*. 51 p. 452.

Kawakami, K. Feb. 1993. The Sony MiniDisc (MD). *Broadcast Engineering* 35 (2): 56–58.

Marchant, A.B. 1990. *Optical Recording*, A Technical Overview. Addison Wesley 408 pages.

Stevens, J. May 1993. Rewriteable Optical Manufacturers Get Ready for Next Generation. *Computer Technology Review* 12 (6): 91–95.

Tsutsumi, K., Y. Nakaki, T. Tokunaga, T. Fukami, and Y. Fujii. Nov. 1993. High Density Recording with A Visible Laser Diode in Direct Overwrite MO Disk. *IEEE Trans. Magn*. MAG-29 (6): 3760–65.

BIBLIOGRAPHY

Dauner, D.R., R.C. Sherman, M.L. Christensen, J.L. Methlie, and L.G. Christie, Jr., Dec. 1990. Mechanical Design of an Optical Disk Autochanger. *HP Journal* 41 (6): 14–23.

Fan, G.J. Sept. 1971. Magneto-optic Storage. *IEEE Trans. Magn*. MAG-7 (3): 590–94.

Greidanus, F.J.A.M. Jan. 1990. Status and Future of Magneto-Optical Disk Drive Technologies. *Philips Tech. Res*. 45 (1): 19–34.

Haeb, R., and R.T. Lynch. Jan. 1992. Trellis Codes for Partial-Response Magneto-optical Direct Overwrite Recording. *IEEE Jour. Select Area Commun*. 10 (1): 182–190.

Hatwar, T.K. Nov. 1991. High Reliability Magneto-Optic Media. *Jour. Appl. Phys.* 70 (10, Pt.II): 6335–37.

Johann, D., and A. Burroughs. Nov. 1991. Servo and Data Channels for High Performance Magneto-Optical Disk Drive. *IEEE Trans. Magn.* MAG-27 (6): 4496–4502.

Klahn, S., P. Hansen, and J.F.A.M. Greidanus. April 1990. Recent Advances in Thin Films for Magneto-Optic Recording. *Vacuum* 41 (4–6): 1160–65.

Levenson, M.D., R.T. Lynch, and S.M. Tan. Jan. 1991. Edge Detection for Magneto-optical Data Storage. *Applied Optics* 30 (3): 232–252.

Meada, M., Y. Hashimoto, K. Nakashima, H. Inoue, and S. Ogawa. Sept. 1989. Study on Readout Stability of TbFeCo Magneto-Optical Disks. *IEEE Trans. Magn.* MAG-25 (5): 3539–3571.

Patel, H.C., R.H. Noyau, and E.W. Williams. April 1990. A Review of Noise in Magneto-Optic Recording. *Intl. Conf. on Video, Audio and Data Rec.* IEE Publ. #319, pp. 1–6.

Saldanha, K.S., and C.T. Howe. Dec. 1990. Qualification of an Optical Disk Drive for Autochanger Use. *HP Journal* 41 (6): 35–37.

Sasaki, Y., T. Sekiguchi, H. Inada, and M. Okada. April 1990. A Video Disk Recorder Using Magneto-Optical Disks. *NEC Res. & Devel.* 33 (2): 26–16.

Takahashi, M., H. Shoji, Y. Hozumi, T. Wakiyama, Y. Takeda, and Y. Itakura. Nov. 1994. Giant Magnetic Kerr Rotation for MnSbPt Films with NiAs Structure. *IEEE Trans. Magn.* MAG-30 (6): 4449–54.

Wright, C.D., and N.A.E. Heyes. Nov. 1991. A Comparison of Various Readout Techniques in Magneto-Optic Recording. *IEEE Trans. Magn.* MAG-27 (6): 5127–29.

Yamanaka, T., D. Yamane, T. Takemoto, N. Horie, S. Taniguchi, and N. Sakamoto. Sept. 1990. Magneto-Optical Disk Drive. *Sharp Tech. Jour.* No. 46, pp. 66–69.

CHAPTER 31
CARE AND MAINTENANCE

A tape recorder or disk drive, like other instruments (for example, a motion picture camera), is subject to malfunctions that are best prevented by proper maintenance. In addition, a reel of magnetic tape or a disk (like a roll of photographic film) is sensitive to handling, storage, and shipment hazards. Routine maintenance of the equipment and the observance of a few fundamental rules about tape and disk handling will ensure the user of better performance and longer life for both equipment and media.

A media can be damaged in several ways, either mechanical (by handling) or chemical (environment). These failure mechanisms are analyzed, and will dictate a few guidelines for handling and storage of a media.

Equipment can (and will) malfunction, causing not only signal degradation but also possibly permanent media damage. Certain maintenance procedures are common to all tape and disk drives, while others are dependent upon the particular drive. An outline of general maintenance procedures is given, while the manufacturer's procedures are found in their respective manuals.

The chapter concludes with a listing of potential failures, causes, and remedies.

CAUSES OF FAILURE IN MEDIA

The following are some common causes of media failure.

IMPROPER WINDING OF MAGNETIC TAPE

The most common ways in which tapes are damaged occur during machine operation: When winding or rewinding a tape, the recorder may produce an uneven "pack" or "wind," with protruding layers that are subject to damage in handling. Holding the reel by the flanges will squeeze them against the tape, and protruding layers will be nicked and permanently deformed.

A transport winding mechanism may also wind the tape with insufficient tension. Later handling will then cause the pack to shift from side to side against the flanges, leading to later edge damage. A loose pack is also subject to tangential slippage between layers, called cinching. Cinching is likely to occur in a reel of tape with one or more regions of too-low tension, especially if subjected to a rapid angular acceleration or deceleration that occurs during the starting and stopping of a tape handler. Figure 31.1 shows cinching. During such slippage, the tape may actually fold over on itself so that permanent creases form immediately, or perhaps later when tension is applied and the tape attempts to return to its original position. Creases cause dropouts by introducing a separation between tape and heads.

Prolonged storage may cause other mechanical changes in a tape or disk coating surface, such as its wearability (durability) and depletion of its lubricant content, which is not firmly locked in with the molecular structure of the coat.

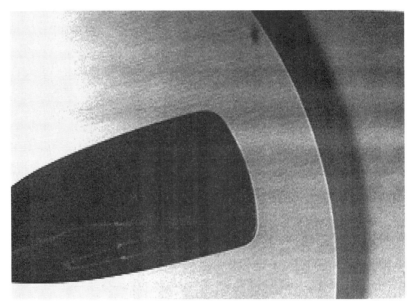

FIGURE 31.1   Cinching in a wound reel of tape.

CARELESS CLAMPING OF DISKETTES

The center hole in a diskette (5¼ or 8 inches) is vulnerable to damage by improper clamping. This occurs when the disk is off-center in the protective jacket, inserted in the disk drive, and quickly clamped. The hole may be severely stressed in one side, causing a permanent wrinkle that makes subsequent centering difficult.

HYDROLYSIS IN TAPE/DISK COATINGS

High temperatures have long been suspected to cause tape degradation, a process that is accelerated in high humidity (Cuddihy's papers). Almost all commercial magnetic tapes employ an oxide binder made from polyester urethanes, which are employed either by themselves or in combination with other polymeric materials intended to impart special properties. Such co-polymers vary in mechanical and physical properties, as well as in processing requirements.

However, all polyester urethanes and all polyester polymers are susceptible to degradation by chemical reaction with water, a chemical reaction called hydrolysis (Cuddihy). The process is complex, but can be essentially expressed as (Bertram):

$$\text{Ester} + \text{Water} \longleftrightarrow \text{Carboxylic acid} + \text{Alcohol}$$

The degradation products (debris, which consists of carboxylic acid) can be extracted from the tape by acetone, and the nature and rates of the chemical reaction can then be monitored. Figure 31.2 shows one set of results covering a three-month span. The debris formation occurs fastest at 100 percent RH, is about zero at 20–24 percent RH—and is reversed at zero relative humidity. The latter observation verifies the reversible process as indicated by the double-arrow in the above expression; it also points to the possible recovery of tapes damaged from hydrolysis during long storage.

PET films used for tape and disk bases are essentially stable against environmental degradation at ordinary use conditions (Cuddihy).

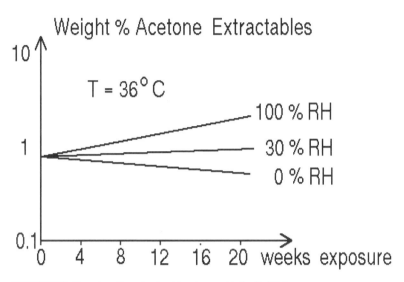

FIGURE 31.2   The aging of tapes in a humid environment (*after Cuddihy*).

HANDLING OF MEDIA

Cinching and nicked tape edges can be avoided if the user follows a simple piece of advice:

Do not rewind a tape after recording or playback, but store it immediately in its container, standing on end.

Numerous tape failures and dropouts are the result of not following this practice.

A tape that is left on a recorder or placed on a shelf outside of its container, as is often the case, will collect dust within a very short time. When the tape is later played back, the dust particles will move onto the coating surface by the airflow that is caused by lifting the tape from the tape pack. This causes dropouts, might permanently damage the tape, and might potentially scratch the magnetic heads. Dust particles can combine with debris from the tape and deposit it on the guides and heads. These protruding particles will scratch the tape surface and further aggravate the dropout situation. There also may be a continuous scratch in the tape surface or backing throughout an entire reel. The slight ridge thus produced can multiply through the layers of tape wound on the reel to produce a much larger ridge in the outer layers of the tape—often large enough to cause permanent deformation in the outer layers.

A word of caution applies in particular to diskettes. Hands and fingers have body oils and salts that attract foreign particles. The oils alone can leave a fingerprint on the exposed surface of a diskette that renders a readout impossible, aggravated of course by any foreign particles attracted to the oil:

Never touch a diskette in the window area!

The same advice applies to the handling of tapes, in particular reel-to-reel videotapes. Each time a tape is threaded on a video transport, the operator may transfer finger oils and salts to the tape or to the tape guides. This, naturally, is a problem whenever a tape is spliced. During a normal splicing or editing session, the tape must necessarily be handled a great deal, and it is a good idea to **wear white cotton gloves**.

In order to make a perfect video splice during an editing session, a solution such as "MagView" or "EdiView" is used. When this is sprayed onto the tape a gray powder remains, revealing the tape tracks and edit pulses, which greatly assists the editor. After editing, the material must be removed completely from the tape with a tissue, or dropouts are inevitable.

GENERAL STORAGE OF MEDIA

For storage of tape, the following rules apply:

- A tape should always be stored in its container with the reel on edge rather than in a flat position. This will tend to eliminate the sideways shifting of the pack against the flanges.

- A tape should be stored under controlled environmental conditions; see Fig. 30.3. Large or sudden changes in environment should be avoided.

- A tape that has been stored under less than ideal environmental conditions should be conditioned by allowing it to remain in a suitable environment for at least 24 hours prior to use.

- When large changes in temperature cannot be avoided, the probability of damage to the tape can be minimized if the reel hub has a thermal coefficient of expansion similar to that of the base film. Most plastic reels have thermal coefficient about twice that of the polyester base film, while the thermal coefficient of aluminum is nearly equal to that of polyester.

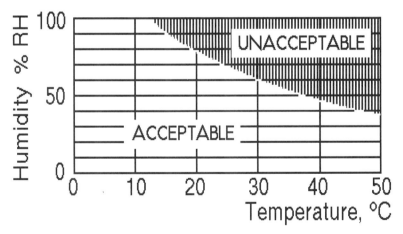

FIGURE 31.3 Tapes should be stored under conditions corresponding to "acceptable," as indicated (*after Cuddihy*).

During shipments of a tape, mechanical agitation will tend to shift the tape pack, especially if wound under improper tension. Any abrupt temperature change during shipment should be avoided, and this is best done by placing reels in special fiberboard shipping containers. This will also protect the reel flanges from being bent to a point where the edges of the tape run on the flange.

Stray magnetic fields may cause some degree of erasure of the information recorded on the tape. There are a few cases on record where tape has been completely erased during shipment, and if such fields are known to exist, special shielding containers are available.

Neither tapes nor disks are affected by the security check equipment in airports. Example 3.2 from Chapter 3 shows the large field required to affect recorded magnetization. Reports about data disks that supposedly have been damaged during travel do nevertheless persist. Here again: Has care been taken in sealing the disks from dust, lint, etc.? Were they under any mechanical stress, or exposed to excessive temperature?

A word of caution: Get rid of all magnetic gadgets on your desk. Never place one of those magnetic holders for notes, etc. on a disk; your data will be erased!

ARCHIVAL STORAGE OF MEDIA

Precautions in addition to the ones just listed are needed when tapes or disks are to be stored for extended periods (many years). A recent study (Bertram) has produced recommendations for the storage conditions for γFe_2O_3 particles in a polyester-urethane binder system: temperature = 65° F, relative humidity = 40 percent (both ±5%).

These binder systems degrade due to hydrolysis. The normal degradation for a new, conditioned tape is typically at a 7 percent hydrolysis state. Taking that value as a "zero" reference, the chart in Fig. 31.3 was prepared. Any temperature/humidity condition that lies on the "zero" hydrolysis contour will preserve the binder at the 7 percent hydrolysis state; this appears to be a satisfactory condition for most applications.

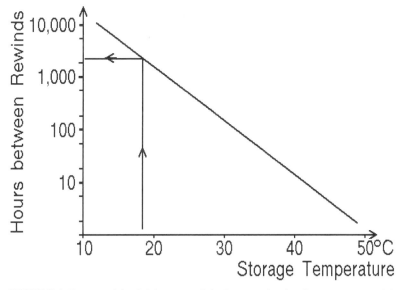

FIGURE 31.4 Recommended periods between rewinds of tapes as a function of storage temperature (*after Bertram*).

An increase due to high temperatures and/or high humidity accelerate the hydrolysis, and when a value of 14 percent has been reached, a tape is damaged; i.e., it produces excessive debris and sticky gunk that deposits onto heads and guides, with resulting signal spacing loss and potential tape/disk damage.

The quality of the stored tape also depends on the internal stresses that build up during winding. Excessive tension during winding may result in deformation of the tape, while a loose wind entraps air and causes cinching. Internal stresses will eventually decrease (Cuddihy, van Keuren, Bertram), provided the temperature is constant. This is rarely the case, and the study makes one further recommendation:

Rewind stored tapes at regular intervals.

Recommended intervals between rewindings are shown in Fig. 31.4 (solid line). The rewinding is best when done on a transport having constant reel torque, not too fast speed, and a pack-follower wheel to prevent air entrapment. Such transports are called rewinders. It may be equipped with a cleaning device (razor blade) that scrapes away excess products from the hydrolysis, whereafter the tape passes over/through a lint-free tissue wiper to remove all dirt and dust. It was found that this process even acted as a conditioner for a damaged tape and improved its *BER* (bit-error-rate).

All the above recommendations and discussions apply only to tapes with γFe_2O_3. For other tapes, detailed information must be obtained from the products' manufacturer. Tapes with Fe particles are known to be sensitive to humidity and subsequent corrosion (Olsen), and little is known about long-time storage of the other particle tapes.

The National Media Laboratory has recently issued a chart with general rules for the behaviour of some fundamental tape types in various environments. The chart is reproduced in Fig. 31.5. See also a new guide from the Commission on Preservation and Access (and NML.)

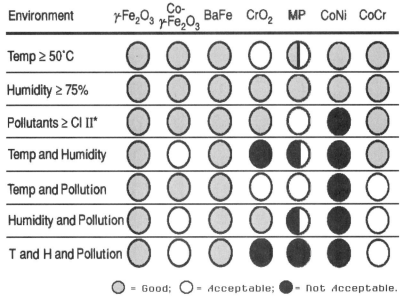

FIGURE 31.5 An overview of archival stability of recording media (*courtesy NML, National Media Laboratory*).

MAINTENANCE OF RECORDING EQUIPMENT

Cleanliness is fundamental to the proper operation of any magnetic recorder or disk drive. Dust particles will not only cause dropouts, but also shorten the life of the recording equipment, especially the heads and bearings. Good cleaning rules apply to home sound recording equipment, but the demands are even greater for instrumentation, video, and computer recording equipment. These machines should ideally be operated in a clean-room environment to avoid the accelerated generation of dropouts. Humidity and temperature conditions should be closely controlled, and the floor area should be cleaned daily using a damp mop. Many professional facilities also have restrictions on the use of food and tobacco; tobacco ashes can easily accumulate on tapes and disks.

Any part touching the magnetic tape on its pass through the transport should be cleaned at regular intervals—tape guides, heads, capstans, and rubber pucks. If this practice is not followed, dirt can accumulate on the tape guides and heads, and act as an abrasive agent and scrape the oxide coating on the tape. Soon the oxide buildup from the scratching may break away and be redeposited elsewhere on the tape. As the tape is wound tightly onto the take-up reel, any loose oxide may be firmly embedded into the tape surface and cause dropouts the next time the tape is used.

Care should be exercised in the selection of a cleaning solution, because some agents do more harm than good. About the only cleaning solution recommended for videotape equipment is Freon TF. The reason is that Freon TF will flush off oxide particles and debris without softening the oxide or the backing. Also, Freon TF will not damage the rubber capstan idler. Table 31.1 states the effects of various cleaning solutions. Cleaning agents and tools are available from companies such as TEXWIPE, Hillsdale, N.J. 07642.

Modern polyurethane coatings are known to deposit debris products, probably formed by hydrolysis (see earlier this chapter). They will collectively form a clear film, invisible to the eye but noticeable in producing dropout and aggravating stick-slip phenomena. This debris film is often referred to as brown stain or varnish, and cannot be removed with any permissible head cleaner short of a mildly abrasive tape.

This film also forms on diskette drive heads and will in some cases cause severe stick-slip-actuated vibrations of the head/mounting arm/flexible disk system. The disk drive will "sing"; when this occurs, beware—either the head or the disk, or both, have a short life ahead. Stop operations and perform a head cleaning with a cleaner disk that has enough abrasive powder embedded to remove the debris on the head.

Another trouble source, often neglected, is head magnetization. A magnetized record head will, in instrumentation and audio recordings, increase second-order harmonic distortion and the overall noise level. In computer tape and disk drives, asymmetrical waveforms will be written. Therefore, a small head degausser should be used at regular intervals (for example, every eight running hours).

Even then, the danger of magnetized heads persists: Any nearby dc fields from permanent magnet motors (rotating in tape drives, linear motion in disk drive head actuators) may link through the head core and act as a "data" signal being recorded onto the head core while the alternating degaussing field acts as a decaying ac-bias field! Even the earth field may prevent complete erasure of a head core, and the outcome is high even-harmonic distortion in audio and instrumentation recorders, and asymmetry in digital equipment. The only remedy is shielding, or when the earth field is the issue, repeated degaussing with a different orientation of the equipment.

With regard to lubrication and head alignment, refer to your recorder manual. Many recorders have bearings that are lubricated for life. But, if lubrication is required in certain parts of the recorder, care should be exercised not to spill any oil on the capstans, rubber pucks, or other sensitive areas.

Preventive maintenance of the tape transport should be made on a regular basis. This should not just be cleaning of guides, rollers, etc., but also a check of the tape tension and tracking plus flutter and TDE.

Magnetic heads may change sensitivities as wear progresses; this mandates a periodic check of the electronics and associated controls. The reproduce level and response should be checked with a standard test tape, and the SNR verified. Bias level in the record chain may need readjustment (see equipment manual for details), and the standard record level may need recalibration, in addition to resetting of the pre-equalization.

Additional details on care and handling are found in a paper by Ford and in two NASA publications, listed in the references under Waites, Heard, Davis, Townsend, and THIC's Head Advisory Committee.

TROUBLESHOOTING TAPE RECORDING EQUIPMENT

The following troubleshooting guide is applicable to most recorders. However, it is necessary that the operating manual and/or the service manual be consulted for a particular recorder prior to repair. For troubleshooting the recorder electronics, instruments similar to those used for amplifiers are normally required (tone generator, voltmeters, and an oscilloscope). It is also useful to have a flutter meter and test or alignment tapes that have signals recorded for alignment of the reproduce head, setting of the record level indicator, frequency-response test, and a tone for wow and flutter tests. To clean various parts in the recorder, it is useful to have on hand a proper cleaning fluid and cotton swabs (see Table 31.1).

TABLE 31.1 Table of Cleaning Agents

Cleaning solvent	Health hazard	Effect on video magnetic tape	Flammable	Effect on rubber
Freon TF	Very slight	None or negligible	No	None or negligible
Acetone	Very slight	Soluble	Yes	None or slight
Carbon tetrachloride	Great	Negligible	No	Slight
Ethyl alcohol	Very slight	Negligible	Yes	None or slight
Heptane	Slight	Softens	Yes	Swells
Methyl alcohol	Some	Negligible	Yes	None or slight
Naphtha	Slight	Softens	Yes	Swells
Mek	Some	Soluble	Yes	None or slight
Trichloroethylene	Some	Soluble	No	Slight
Xylene	Some	Soluble	Yes	Swells

Listed below in tabular form are typical troubles, their possible causes, and steps for their correction.

Noise. dc magnetized heads—Degauss the magnetic heads with a suitable head degausser, and do it very slowly.

Input amplifier—A faulty resistor will cause noise in the input stage and should be replaced. Faulty electrolytic capacitors may also cause noise and should be replaced. Also check for noisy transistors or chips.

Distorted ac bias—Check the oscillator and bias amplifiers for proper operating voltages and freedom from even-order distorted waveforms.

Hum. Faulty shields or ground connections—Ground and shield connections may become corroded and should be scraped clean.

Open-circuited reproduce head—Replace the reproduce head.

Poor power-supply decoupling—Replace faulty electrolytic capacitors.

Distortion. Tape overload—Carefully monitor the record level. If the distortion persists, check the setting of the level indicator on the recorder, which is most easily done with a test tape. For details, see Chapter 23.

No or too little ac bias—Check the oscillator and amplifier. Also check that the tape for which the bias level has been set is used on the recorder.

No or poor erasure. Faulty bias oscillator—Check the oscillator for proper operation. If the oscillator has been malfunctioning, this would also have resulted in a high distortion level (as above).

Debris on erase head—Debris will lift the magnetic tape away from the erase gap. The erase head should be cleaned with a proper solvent and a cotton swab.

Poor frequency response. Debris on heads—Clean all magnetic heads with a suitable solvent and cotton swabs.

Misaligned heads—First check the alignment of the reproduce head using a test tape. Next, check the alignment of the record head by recording a high-frequency note, and then adjust the record-head azimuth screw for maximum output level. The azimuth adjustment should be made only by a person familiar with the recorder.

Faulty equalizer—Poor contacts in the equalizer switching circuit may cause the equalizer to function improperly. Faulty components may likewise cause this problem.

Skew—This condition will normally manifest itself as a variation in the high-frequency output level and is most likely caused by a worn or misaligned capstan rubber idler. A new idler should be installed and/or realigned.

Smeared head gaps—Foreign particles (for example, dust) may cause excessive scratches on the head surfaces, which will cause the material to cold-flow across the gap and thereby destroy otherwise parallel gap edges. This phenomenon also may appear if the mu-metal cores wear away faster than the head shell; a thicker tape will then no longer conform to the head contour and make proper contact with the gap.

Wrong tape—Tapes from different manufacturers require slightly different bias settings for optimum performance. Once selected, the same type tape should be used in future recordings.

No recording. Faulty amplifier—Follow normal amplifier troubleshooting procedures. The record head current is normally referred to in the service manual, and can usually be measured across a 10-ohm resistor in the ground leg of the record head.

Faulty record head—An open record head will result in no voltage across the 10-ohm resistor. (A short-circuited record head or cable connection will allow the current to flow through the 10-ohm resistor, and should be checked very carefully.)

No playback. Faulty amplifier—Troubleshoot the reproduce amplifier.

Faulty reproduce head—This is easy to verify; run a prerecorded test tape through the recorder and check for proper output levels. A shorted reproduce head will result in no output voltage, while an open reproduce head will introduce excessive hum.

Flutter. Debris on capstan—Clean the capstan and rubber puck with a suitable solvent.

Damaged rubber puck—If a power failure occurs while the recorder is in the record or play mode, the rubber puck will remain engaged against the capstan and, if allowed to remain engaged, can indent the rubber puck. Such an indentation can sometimes be removed by letting the recorder run in the play mode several hours without tape. Otherwise, the rubber puck must be replaced.

Worn belts, pucks—Replace worn-out parts.

Tape scraping on reel flanges—Rewind the tape onto a new reel that has no bent or damaged flanges. Also make sure that the inside edges of the reel flanges are free from any nicks or scratches. Reels must also be properly centered and perfectly running.

Heavy oil and dirt in bearings—Clean all bearings as outlined in the service manual and relubricate.

Too slow speed (drift). Reel tension too high—Adjust reel tension in accordance with the service manual instructions.

Debris on rubber puck—Debris may cause a rubber puck to become excessively smooth, in which case it should be cleaned with a suitable solvent, possibly plain water (some organic solvents can damage rubber).

High bearing friction—Clean and lubricate all bearings as described in the service manual.

Speed control error—Many portable recorders and most instrumentation recorders maintain their speed accuracy by an electronic servo system. There may be several causes for malfunction in a servo system, and the reader is referred to the service manual for the particular recorder.

Squeal. Debris on heads and guides—Any disturbance in the tape path through the recorder may cause excessive scrape flutter. Heads and guides should be cleaned with a suitable solvent.

Excessive tape tension—Adjust the tape tension devices in accordance with the service manual.

Worn felt pads; excessive felt pad pressure—Replace pads. Excessive felt pad pressure, either against a guide or against a head, will cause excessive scrape flutter. Adjust the felt pad pressure in accordance with the service manual or by successive experiments.

No tape motion. Broken belts or damaged mechanical parts—Replace damaged parts.

Blown fuse—Check all the fuses and replace ones that are burned out. If a fuse repeatedly burns out, the recorder should be overhauled.

Tape breakage. Worn brake pads—Replace worn brake parts and adjust them in accordance with the service manual.

Maladjusted brakes—Readjust the brakes in accordance with the service manual.

Tape throws a loop. Maladjusted brakes—Adjust the brake system in accordance with the manual instructions.

BIBLIOGRAPHY

Bertram, N.H., and A. Eshel. April 1980. Recording Media Archival Attributes, Rome Air Devel. Center, *Final Tech. Report RADC-TR-80-123*, 112 pages.

Bertram, N.H., and E.F. Cuddihy. Sept. 1982. Kinetics of the Humid Aging of Magnetic Recording Tape. *IEEE Trans. Magn.* MAG-18 (5): 993–999.

Cuddihy, E.F. July 1980. Aging of Magnetic Recording Tape. *IEEE Trans. Magn.* MAG-16 (4): 558–568.

Cuddihy, E.F. March 1976. Hygroscopic Properties of Magnetic Recording Tape. *IEEE Trans. Magn.* MAG-12 (2): 126–135.

Cuddihy, E.F., and W. Keuren. March 1974. Mathematical Description of Heat Transfer in Packs of Magnetic Recording Tapes. *IFT Journal*, pp. 5–7.

Davis, R. April 1982. Cleaning, Packing and Winding of Magnetic Tape. *NASA Ref. Publ. 1075*, pp. 61–75.

Ford, H. Dec. 1984. Handling and Storage of Tape. *Studio Sound and Bcast. Engr.* 26 (2): 64–72.

Head Wear Advisory Committee, THIC. Sept. 1985. Care and Handling of Magnetic Tape Heads. *NASA Ref. Publ. 1111*, pp. 293–296.

Heard, F. April 1982. A Care and Handling Manual for Magnetic Tape Recording. *NASA Ref. Publ. 1075*, pp. 127–147.

Olsen, K.H. Sept. 1974. Microstructure Analysis of Corrosion in Iron Based Recording Tapes. *IEEE Trans. Magn.* MAG-10 (3): 660–662.

Townsend, K. April 1982. Tape Reels, Bands and Packaging. *NASA Ref. Publ. 1075*, pp. 77–84.

van Bogart, J.W.C. June 1995. Magnetic Tape Storage and Handling, A Guide for Libraries and Archives. *Commission on Preservation and Access*, 1400 16th St., NW, Ste 740, Washington, DC 20036-2217. Prepared by National Media Laboratory.

Waites, J.B. April 1982. Care, Handling, and Management of Magnetic Tape. *NASA Ref. Publ. 1075*, pp. 45–59.

LIST OF SYMBOLS

A	area (m^2)
a	arctangent transition parameter
	length (=.5 × Δx) (m)
A_c	cross-sectional area of head core (m^2)
A_{fg}	cross-sectional area of front gap in magnetic head (m^2)
A_g	cross-sectional area of gap in magnetic circuit (m^2)
A_t	irregularity gap loss (dB)
B	magnetic induction (flux density) (Wb/m^2)
	signal bandwidth (Hz)
BL	bit length (μm)
B_g	air-gap flux density (Wb/m^2)
B_r	remanent induction (Wb/m^2)
B_r'	remanent induction after demagnetization (Wb/m^2)
B_r''	remanent induction after recoil (Wb/m^2)
B_{rs}	remanent induction after saturation (Wb/m^2)
B_{rsat}	remanent induction after saturation (Wb/m^2)
B_{rs}'	remanent induction after saturation and demagnetization (Wb/m^2)
B_s	saturation inductance (Wb/m^2)
B_{sat}	saturation inductance (Wb/m^2)
BW	signal bandwidth (Hz)
C	capacitance (Farads)
	capacity, channel (bytes)
c	magnetic coating thickness (μm)
C_e	self-capacitance, inductor (Farads)
d	lamination thickness
d	head-to-medium spacing (μm)
d	code constraint, i.e., minimum number of zeros between transitions
D_g	depth of front gap in magnetic head (μm)
D_{gap}	same as D_g (μm)
E	voltage (volts)
E	Young's modulus of elasticity (Nt/m^2)
e	head output voltage (volts)
E_d	Young's modulus of elasticity, dynamic value
F	force (Nt)
f	frequency (Hz)
f_k	frequency in kHz
f_s	spin resonance frequency in magnetic material (MHz)
f_u	upper signal frequency (Hz)
f_w	wall resonance frequency in magnetic material (MHz)
g	front gap length in magnetic head (μm)
H	magnetic field strength (A/m)
H_a	applied magnetic field (A/m)
H_c	coercivity or coercive force, BH- or MH-loop (A/m)
h_{ci}	coercivity of single particle or grain (A/m)
H_{core}	net field inside a magnetic core material (A/m)
H_d	demagnetizing field (A/m)
H_g	deep gap field in magnetic head (A/m)
H_r	remament coercivity, BH- or MH-loop (A/m)
H_s	sensing function during readout (m^{-1}) (with components H_{sx} and H_{sy})

H_x, H_y, H_z	vector components of a field (A/m)
Δh_r	switching field distribution, ref. to H_r (fraction)
Δh_c	switching field distribution, ref. to H_c (fraction)
I	moment of inertia
i	current (A)
i_{write}	write (record) current through write head winding (A)
i_{100}	100% level write current
J	magnetization (Wb/m^2)
j	magnetic dipole moment of loop (iAn Am^2)
K	magnetic anisotropy constant
K	$2\pi d_{cm}\sqrt{(\mu_{rdc}f_k/\sigma)}$ (from eddy current calculations)
k	Boltzmann's constant = 1.38×10^{-23} (J/Kelvin)
k	code restraint, i.e., maximum number of zeros between transitions
k	coupling coefficient between two inductors
k	number of bits in datawords fed into EDAC encoder
L	length (m)
L	inductance (Henry)
l	length (m)
L'	true inductance (Henry)
L_c	length of magnetic core circuit element (m)
L_g	length of magnetic gap element (μm)
L_{fg}	length of front magnetic gap element (μm)
M	magnetization (A/m)
M	mutual coupling = $k\sqrt{(L_1L_2)}$ (Henry)
m	Bohr's magneton = 9.3×10^{-24} (Am^2)
m	permanent magnet dipole moment = pl (Am^2)
m	number of bits in datawords fed into a modulation encoder
m	amplitude modulation of a recorded signal
M_r	remanent magnetization (A/m)
M_r'	remanent magnetization after demagnetization (A/m)
M_r''	remanent magnetization after demagnetization and recoil (A/m)
M_s	saturation magnetization (A/m)
Δm	small magnetized element in recorded coating. Also $\Delta m = \Delta m_x + \Delta m_y$.
N	magnetic North pole (+) (Am)
N	demagnetization factor (max = 1 in SI, 4π in cgs)
N_x, N_y, N_z	demagnetization factors in x, y, and z directions
n	number of turns in a winding, or coil
n	number of bits in a codeword
P	permeance = magnetic conductance (= 1/Rm) (Hy)
p	magnetic pole strength (Am)
p	length of pole in TFH (in media travel direction)
p	packing fraction of magnetic particles in a coating
p'	magnetic pole strength per unit length in a line pole (A)
P_{eddy}	eddy current power loss in a magnetic material (W)
P_{hyst}	hysteresis power loss in a magnetic material (W)
Q	coil quality = $R_p/\omega L = \omega L/R_s$.
q	electron charge = 1.6022×10^{-19} Coulomb
R	electric resistance (ohm)
R	code rate = k/n (EDAC)
R	code rate = m/n (modulation coding)
r	distance (m)
R'	true electrical resistance (ohm)
R_{bg}	magnetic resistance of back gap in magnetic head (Hy^{-1})
R_c	magnetic resistance of core section in magnetic head (Hy^{-1})

R_{fg}	magnetic resistance of front gap in magnetic head (Hy^{-1})
R_{fgs}	R_{fg} including stray fields in magnetic head (Hy^{-1})
R_m	magnetic resistance (Hy^{-1})
ΣR_m	sum of all magnetic resistance in a magnetic circuit, going "once around" the circuit (Hy^{-1})
ΣR_{ms}	ΣR_m, including stray fields (Hy^{-1})
R_p	loss resistance of coil, parallel equivalent (ohm)
R_s	loss resistance of coil, series equivalent (ohm)
R_{stray}	magnetic resistance of stray flux path (Hy^{-1})
R_{wdg}	winding resistance of a coil (ohm)
S	magnetic South pole (–) (Am)
S	squareness of magnetic MH-loop
S^*	coercivity squareness factor ($SFD = 1 - S^*$)
T	tension (dynes)
T	torque (Nm)
t	time (sec)
t	pole thickness (in MR or perpendicular head)
T_c	Curie temperature (Celsius)
v	velocity (m/s)
v_c	group velocity (m/s)
V_{noise}	noise voltage (V)
w	track width (m)
x	distance (m)
Δx	length of transition zone (=2a) (m)
y	distance (m)
Z	electric impedance (ohm)
z	distance (m)
Z_m	magnetic impedance (Hy^{-1})
Z_{mreal}	magnetic resistance, real component (Hy^{-1})
Z_{mimag}	magnetic resistance, imaginary component (Hy^{-1})

GREEK SYMBOLS

α	angle
α_B	slope of demagnetization line in BH-loop
α_M	slope of demagnetization line in MH-loop
α_x	slope of demagnetization line, longitudinal magnetization
α_y	slope of demagnetization line, perpendicular magnetization
β	angle between sensing function and magnetization
β	skin depth (eddy current)
β	modulation index (FM)
δ	skin depth (cm)
δ	coating thickness (μm)
δ_{eff}	effective coating thickness (that contribute to flux) (μm)
δ_{rec}	recorded thickness (μm)
$\in$	mass in grams per cm, or per cm^2
$\in$	coefficient of expansion /μm/m/
η_r	efficiency, read head
η_w	efficiency. write head
η_{coupl}	coupling efficiency (flux linkage)
λ	wave length (m)

λ_{xyz}	magnetostriction constant in crystal direction xyz
ρ	conductivity (Siemens)
ρ	density (kg/cm^3)
μ	absolute permeability $= \mu_o\mu_r$ (Hy/m)
μ	coefficient of friction
μ_o	permeability of vacuum $= 4\pi10^{-7}$ (Hy/m)
μ_r	relative permeability
μ_t	relative permeability of tape or disk coating $\int\mu_x\mu_y$
μ_x	relative permeability of coating, longitudinal
μ_y	relative permeability of coating, perpendicular
μ_d	dynamic coefficient of friction
μ_s	static coefficient of friction
μ'	real component of permeability (Hy/m)
μ''	imaginary component of permeability
μ_{init}	initial permeability, at low field strength (Hy/m)
μ_{max}	maximum permeability (Hy/m)
μ_{rdc}	relative permeability at dc
ν	Poisson's ratio
ν	field strength value, relative to deep gap field H_g
σ	resistivity ($\mu\Omega$cm)
σ	standard deviation
σ^2	variance
σ_s	specific magnetization (Wb/m^2, or emu/cc)
θ	angle in rads
ϕ	flux, field lines (Wb = Weber = Vs)
ϕ_g	flux in write gap (Wb)
ϕ_m	maximum flux level
ϕ_{rsat}	maximum flux level after saturation (Wb)
ϕ	scalar magnetic potential
χ	magnetic susceptibility
ω	$= 2\pi f\,(\text{s}^{-1})$

INDEX

About the Author

Finn Jorgensen holds an MSEE in telecommunications and electroacoustics and an MSEE in electronics, specifically in RF and VHF solid-state device fabrication and application of thin-film technology. As a lieutenant in the Royal Danish Navy, he served as a telecommunications and radar specialist. Jorgensen has written numerous technical articles and holds several patents.